Lecture Notes in Computer Science   11346

Commenced Publication in 1973
Founding and Former Series Editors:
Gerhard Goos, Juris Hartmanis, and Jan van Leeuwen

More information about this series at http://www.springer.com/series/7407

Donghyun Kim · R. N. Uma
Alexander Zelikovsky (Eds.)

Combinatorial Optimization and Applications

12th International Conference, COCOA 2018
Atlanta, GA, USA, December 15–17, 2018
Proceedings

 Springer

Editors
Donghyun Kim
Department of Computer Science
Kennesaw State University
Marietta, GA, USA

Alexander Zelikovsky
Department of Computer Science
Georgia State University
Atlanta, GA, USA

R. N. Uma
Department of Mathematics
and Computer Science
North Carolina Central University
Durham, NC, USA

ISSN 0302-9743   ISSN 1611-3349   (electronic)
Lecture Notes in Computer Science
ISBN 978-3-030-04650-7   ISBN 978-3-030-04651-4   (eBook)
https://doi.org/10.1007/978-3-030-04651-4

Library of Congress Control Number: 2018962146

LNCS Sublibrary: SL1 – Theoretical Computer Science and General Issues

This Springer imprint is published by the registered company Springer Nature Switzerland AG
The registered company address is: Gewerbestrasse 11, 6330 Cham, Switzerland

Preface

The 12th Annual International Conference on Combinatorial Optimization and Applications (COCOA 2018) was held during December 15–17, 2018, in Atlanta, Georgia, USA. COCOA 2018 provided a forum for researchers working in the area of theoretical computer science, combinatorics, and corresponding applications. The technical program of the conference included 50 regular papers selected by the Program Committee from 106 full submissions received in response to the call for papers. Each submission received at least three reviews from Program Committee members and/or external reviewers. The topics cover most aspects of theoretical computer science and combinatorics related to computing, including combinatorial optimization, geometric optimization, complexity and data structures, graph theory, etc. Some of the papers were selected for publication in special issues of *Theoretical Computer Science* and *Journal of Combinatorial Optimization*. It is expected that the journal version of the selected papers will appear in a more complete form.

We thank everyone who made this meeting possible: the authors for submitting papers, the Program Committee members, and external reviewers for volunteering their time to review the conference papers. We also appreciate the financial sponsorship from Springer. We would also like to extend a special thanks to the general chairs and Organizing Committee for their work in making COCOA 2018 a successful event.

October 2018

Donghyun Kim
R. N. Uma
Alexander Zelikovsky
Zhipeng Cai
Jon A. Preston

Organization

General Co-chairs

Zhipeng Cai   Georgia State University, USA
Jon A. Preston   Kennesaw State University, USA

Technical Program Committee Co-chairs

Donghyun (David) Kim   Kennesaw State University, USA
R. N. Uma   North Carolina Central University, USA
Alex Zelikovsky   Georgia State University, USA

Local Arrangements Chair

Meng Han   Kennesaw State University, USA

Web Chair

Seokjun Lee   Kennesaw State University, USA

Publication Chair

Daehee Seo   Kennesaw State University, USA

Finance Chair

Brian Ellis   Kennesaw State University, USA

Program Committee

Annie Chateau   University of Montpellier 2, France
Yong Chen   Hangzhou University of Electronic Science
     and Technology, China
Ovidiu Daescu   University of Texas at Dallas, USA
Zhenhua Duan   Xidian University, China
Thomas Erlebach   University of Leicester, UK
Neng Fan   University of Arizona, USA
Stanley Fung   University of Leicester, UK
Satish Govindarajan   Indian Institute of Science, India
Meng Han   Kennesaw State University, USA
Micheal Khachay   Ural Federal University, Russia
Milos Kudelka   Technical University of Ostrava, Czech Republic

Joong-Lyul Lee	University of North Carolina at Pembroke, USA
Deying Li	Renmin University of China, China
Ji Li	Kennesaw State University, USA
Xiang Li	Santa Clara University, USA
Xianyue Li	Lanzhou University, China
Xianmin Liu	Harbin Institute of Technology, China
Yingshu Li	Georgia State University, USA
Meghana Nasre	Indian Institute of Technology, India
Viet Hung Nguyen	Sorbonne Université, France
Erfang Shan	Shanghai University, China
Pavel Skums	Georgia State University, USA
Xiang Song	Massachusetts Institute of Technology, USA
Weitian Tong	Georgia Southern University, USA
Wei Wang	Xi'an Jiaotong University, China
Gerhard Woeginger	Eindhoven University of Technology, The Netherlands
Boting Yang	University of Regina, Canada
Yong Zhang	Shenzhen Institutes of Advanced Technology, China
Zhao Zhang	Zhejiang Normal University, China
Martin Ziegler	KAIST, South Korea

Additional Reviewers

Agrawal, Shweta
Armaselu, Bogdan
Ashok, Pradeesha
Augustine, John
Boudet, Vincent
Bourreau, Eric
Chau, Vincent
Chun, Chen
Davot, Tom
Dinesh, Krishnamoorthy
Epstein, Leah
Fox, Kyle
Giroudeau, Rodolphe
Golovach, Petr
Huang, Honhyao
Inkulu, R.
Katzgraber, Helmut G.
Kobylkin, Konstantin

Kochetov, Yury
Kononov, Alexander
Koswara, Ivan
Kotrbcik, Michal
Krishnaswamy,
    Ravishankar
Krithika, R.
Kuzmin, Kirill
Li, Yuchao
Liu, Pengcheng
Liu, Shiping
Lukovszki, Tamas
Mandrà, Salvatore
Mohebbi, Fatemeh
Neznakhina, Katherine
Nimbhorkar, Prajakta
Ogorodnikov, Yuri
Park, Sewon

Pollet, Valentin
Roy, Aniket Basu
Roy, Sasanka
Satti, Srinivasa Rao
Segal-Halevi, Erel
Shannigrahi, Saswata
Shi, Majun
Talmon, Nimrod
Teo, Ka Yaw
Tsyvina, Viachaslau
Wang, Yinling
Wang, Yishui
Wu, Chenchen
Xu, Yicheng
Yang, Zishen
Zhang, Yubai
Zhou, Rong

Contents

Computational Geometry and Combinatorial Optimization

Combinatorial Optimization and Data Structure

Clustering

Miscellaneous

Graph Theory

Fast Approximation of Centrality and Distances in Hyperbolic Graphs

V. Chepoi[1], F. F. Dragan[2]([✉]), M. Habib[3], Y. Vaxès[1], and H. Alrasheed[4]

[1] Laboratoire d'Informatique et Systèmes, Aix-Marseille Univ, CNRS, and Univ. de Toulon Faculté des Sciences de Luminy, Marseille Cedex 9, 13288 Marseille, France
{victor.chepoi,yann.vaxes}@lif.univ-mrs.fr
[2] Algorithmic Research Laboratory, Department of Computer Science, Kent State University, Kent, Ohio, USA
dragan@cs.kent.edu
[3] Institut de Recherche en Informatique Fondamentale, University Paris Diderot - Paris7, Paris Cedex 13, 75205 Paris, France
habib@liafa.univ-paris-diderot.fr
[4] Information Technology Department, King Saud University, Riyadh, Saudi Arabia
halrasheed@ksu.edu.sa

Abstract. We show that the eccentricities (and thus the centrality indices) of all vertices of a δ-hyperbolic graph $G = (V, E)$ can be computed in linear time with an additive one-sided error of at most $c\delta$, i.e., after a linear time preprocessing, for every vertex v of G one can compute in $O(1)$ time an estimate $\hat{e}(v)$ of its eccentricity $ecc_G(v)$ such that $ecc_G(v) \leq \hat{e}(v) \leq ecc_G(v)+c\delta$ for a small constant c. We prove that every δ-hyperbolic graph G has a shortest path tree, constructible in linear time, such that for every vertex v of G, $ecc_G(v) \leq ecc_T(v) \leq ecc_G(v)+c\delta$. We also show that the distance matrix of G with an additive one-sided error of at most $c'\delta$ can be computed in $O(|V|^2 \log^2 |V|)$ time, where $c' < c$ is a small constant. Recent empirical studies show that many real-world graphs (including Internet application networks, web networks, collaboration networks, social networks, biological networks, and others) have small hyperbolicity.

1 Introduction

The *diameter* $diam(G)$ and the *radius* $rad(G)$ of a graph $G = (V, E)$ are two fundamental metric parameters that have many important practical applications in real world networks. The problem of finding the *center* $C(G)$ of a graph G is often studied as a facility location problem for networks where one needs to select a single vertex to place a facility so that the maximum distance from any demand vertex in the network is minimized. In the analysis of social networks (e.g., citation networks or recommendation networks), biological systems (e.g., protein interaction networks), computer networks (e.g., the Internet or peer-to-peer networks), transportation networks (e.g., public transportation or road networks), etc., the *eccentricity* $ecc(v)$ of a vertex v is used to measure the importance of v in the network: the *centrality index* of v is defined as $\frac{1}{ecc(v)}$.

© Springer Nature Switzerland AG 2018
D. Kim et al. (Eds.): COCOA 2018, LNCS 11346, pp. 3–18, 2018.
https://doi.org/10.1007/978-3-030-04651-4_1

Being able to compute efficiently the diameter, center, radius, and vertex centralities of a given graph has become an increasingly important problem in the analysis of large networks. The algorithmic complexity of the diameter and radius problems is very well-studied. For some special classes of graphs there are efficient algorithms [1,7,12,16,29]. However, for general graphs, the only known algorithms computing the diameter and the radius exactly compute the distance between every pair of vertices in the graph, thus solving the all-pairs shortest paths problem (APSP) and hence computing all eccentricities. In view of recent negative results [1,6,36], this seems to be the best what one can do since even for graphs with $m = O(n)$ (where m is the number of edges and n is the number of vertices) the existence of a subquadratic time (that is, $O(n^{2-\epsilon})$ time for some $\epsilon > 0$) algorithm for the diameter or the radius problem will refute the well known Strong Exponential Time Hypothesis (SETH). Furthermore, recent work [2] shows that if the radius of a possibly dense graph ($m = O(n^2)$) can be computed in subcubic time ($O(n^{3-\epsilon})$ for some $\epsilon > 0$), then APSP also admits a subcubic algorithm. Such an algorithm for APSP has long eluded researchers, and it is often conjectured that it does not exist.

Motivated by these negative results, researches started devoting more attention to development of fast approximation algorithms. In the analysis of large-scale networks, for fast estimations of diameter, center, radius, and centrality indices, linear or almost linear time algorithms are desirable. One hopes also for the all-pairs shortest paths problem to have $o(nm)$ time small-constant–factor approximation algorithms. In general graphs, both diameter and radius can be 2-approximated by a simple linear time algorithm which picks any node and reports its eccentricity. A 3/2-approximation algorithm for the diameter and the radius which runs in $\tilde{O}(mn^{2/3})$ time was recently obtained in [10] (see also [4] for an earlier $\tilde{O}(n^2 + m\sqrt{n})$ time algorithm and [36] for a randomized $\tilde{O}(m\sqrt{n})$ time algorithm). For the sparse graphs, this is an $o(n^2)$ time approximation algorithm. Furthermore, under plausible assumptions, no $O(n^{2-\epsilon})$ time algorithm can exist that $(3/2 - \epsilon')$-approximates (for $\epsilon, \epsilon' > 0$) the diameter [36] and the radius [1] in sparse graphs. Similar results are known also for all eccentricities: a 5/3-approximation to the eccentricities of all vertices can be computed in $\tilde{O}(m^{3/2})$ time [10] and, under plausible assumptions, no $O(n^{2-\epsilon})$ time algorithm can exist that $(5/3 - \epsilon')$-approximates (for $\epsilon, \epsilon' > 0$) the eccentricities of all vertices in sparse graphs [1]. Better approximation algorithms are known for some special classes of graphs [13,19,24,25].

Approximability of APSP is also extensively investigated. An additive 2-approximation for APSP in unweighted undirected graphs (the graphs we consider in this paper) was presented in [20]. It runs in $\tilde{O}(\min\{n^{3/2}m^{1/2}, n^{7/3}\})$ time and hence improves the runtime of an earlier algorithm from [4]. In [5], an $\tilde{O}(n^2)$ time algorithm was designed which computes an approximation of all distances with a multiplicative error of 2 and an additive error of 1. Furthermore, [5] gives an $O(n^{2.24+o(1)}\epsilon^{-3}\log(n/\epsilon))$ time algorithm that computes an approximation of all distances with a multiplicative error of $(1 + \epsilon)$ and an additive error of 2.

Better algorithms are known for some special classes of graphs (see [7,13,23] and papers cited therein).

The need for fast approximation algorithms for estimating diameters, radii, centrality indices, or all pairs shortest paths in large-scale complex networks dictates to look for geometric and topological properties of those networks and utilize them algorithmically. The classical relationships between the diameter, radius, and center of trees and folklore linear time algorithms for their computation is one of the departing points of this research. A result from 1869 by C. Jordan [31] asserts that the radius of a tree T is roughly equal to half of its diameter and the center is either the middle vertex or the middle edge of any diametral path. The diameter and a diametral pair of T can be computed (in linear time) by a simple but elegant procedure: pick any vertex x, find any vertex y furthest from x, and find once more a vertex z furthest from y; then return $\{y, z\}$ as a diametral pair. One computation of a furthest vertex is called an *FP scan*; hence the diameter of a tree can be computed via two FP scans. This *two FP scans* procedure can be extended to exact or approximate computation of the diameter and radius in many classes of tree-like graphs. For example, this approach was used to compute the radius and a central vertex of a chordal graph in linear time [12]. In this case, the center of G is still close to the middle of all (y, z)-shortest paths and $d_G(y, z)$ is not the diameter but is still its good approximation: $d(y, z) \geq diam(G) - 2$. Even better, the diameter of any chordal graph can be approximated in linear time with an additive error 1 [25]. But it turns out that the exact computation of diameters of chordal graphs is as difficult as the general diameter problem: it is even difficult to decide if the diameter of a split graph is 2 or 3.

The experience with chordal graphs shows that one have to abandon the hope of having fast exact algorithms, even for very simple (from metric point of view) graph-classes, and to search for fast algorithms approximating $diam(G), rad(G), C(G), ecc_G(v)$ with a small additive constant depending only of the coarse geometry of the graph. *Gromov hyperbolicity* or the *negative curvature* of a graph (and, more generally, of a metric space) is one such constant. A graph $G = (V, E)$ is *δ-hyperbolic* [9,27,28] if for any four vertices w, v, x, y of G, the two largest of the three distance sums $d(w, v) + d(x, y)$, $d(w, x) + d(v, y)$, $d(w, y) + d(v, x)$ differ by at most $2\delta \geq 0$. The *hyperbolicity* $\delta(G)$ of a graph G is the smallest number δ such that G is δ-hyperbolic. The hyperbolicity can be viewed as a local measure of how close a graph is metrically to a tree: the smaller the hyperbolicity is, the closer its metric is to a tree-metric (trees are 0-hyperbolic and chordal graphs are 1-hyperbolic).

Recent empirical studies showed that many real-world graphs (including Internet application networks, web networks, collaboration networks, social networks, biological networks, and others) are tree-like from a metric point of view [3] or have small hyperbolicity [33,37]. It has been suggested in [33], and recently formally proved in [17], that the property, observed in real-world networks, in which traffic between nodes tends to go through a relatively small core of the network, as if the shortest paths between them are curved inwards, is due to the hyperbolicity of the network. Small hyperbolicity in real-world graphs provides also many algorithmic advantages. Efficient approximate solutions are attainable for a number of optimization problems [13,14,17,18,26,38].

In [13] we initiated the investigation of diameter, center, and radius problems for δ-hyperbolic graphs and we showed that the existing approach for trees can be extended to this general framework. Namely, it is shown in [13] that if G is a δ-hyperbolic graph and $\{y,z\}$ is the pair returned after two FP scans, then $d(y,z) \geq diam(G) - 2\delta$, $diam(G) \geq 2rad(G) - 4\delta - 1$, $diam(C(G)) \leq 4\delta + 1$, and $C(G)$ is contained in a small ball centered at a middle vertex of any shortest (y,z)-path. Consequently, we obtained linear time algorithms for the diameter and radius problems with additive errors linearly depending on the input graph's hyperbolicity. In this paper, we advance this line of research and provide a linear time algorithm for approximate computation of the eccentricities (and thus of centrality indices) of all vertices of a δ-hyperbolic graph G, i.e., we compute the approximate values of *all eccentricities* within the same time bounds as one computes the approximation of *the largest* or *the smallest eccentricity* ($diam(G)$ or $rad(G)$). Namely, the algorithm outputs for every vertex v of G an estimate $\hat{e}(v)$ of $ecc_G(v)$ such that $ecc_G(v) \leq \hat{e}(v) \leq ecc_G(v) + c\delta$, where $c > 0$ is a small constant. In fact, we demonstrate that G has a shortest path tree, constructible in linear time, such that for every vertex v of G, $ecc_G(v) \leq ecc_T(v) \leq ecc_G(v) + c\delta$ (a so-called *eccentricity $c\delta$-approximating spanning tree*). This is our first main result of this paper and the main ingredient in proving it is the following interesting dependency between the eccentricities of vertices of G and their distances to the center $C(G)$: up to an additive error linearly depending on δ, $ecc_G(v)$ is equal to $d(v, C(G))$ plus $rad(G)$. To establish this new result, we have to revisit the results of [13] about diameters, radii, and centers, by simplifying their proofs and extending them to all eccentricities.

Eccentricity k-approximating spanning trees were introduced by Prisner in [35]. A spanning tree T of a graph G is called an *eccentricity k-approximating spanning tree* if for every vertex v of G $ecc_T(v) \leq ecc_G(v) + k$ holds [35]. Prisner observed that any graph admitting an additive tree k-spanner (that is, a spanning tree T such that $d_T(v,u) \leq d_G(v,u) + k$ for every pair u,v) admits also an eccentricity k-approximating spanning tree. Therefore, eccentricity k-approximating spanning trees exist in interval graphs for $k = 2$ [32,34], in asteroidal-triple–free graph [32], strongly chordal graphs [8] and dually chordal graphs [8] for $k = 3$. On the other hand, although for every k there is a chordal graph without an additive tree k-spanner [32,34], yet as Prisner demonstrated in [35], every chordal graph has an eccentricity 2-approximating spanning tree. Later this result was extended in [24] to a larger family of graphs which includes all chordal graphs and all plane triangulations with inner vertices of degree at least 7. Both those classes belong to the class of 1-hyperbolic graphs. Thus, our result extends the result of [35] to all δ-hyperbolic graphs.

As our second main result, we show that in every δ-hyperbolic graph G all distances with an additive one-sided error of at most $c'\delta$ can be found in $O(|V|^2 \log^2 |V|)$ time, where $c' < c$ is a small constant. With a recent result in [11], this demonstrates an equivalence between approximating the hyperbolicity and approximating the distances in graphs. Note that every δ-hyperbolic graph G admits a distance approximating tree T [13,14], that is, a tree T (which is not necessarily a spanning tree) such that $d_T(v, u) \leq d_G(v, u) + O(\delta \log n)$ for every pair u, v. Such a tree can be used to compute all distances in G with an additive one-sided error of at most $O(\delta \log n)$ in $O(|V|^2)$ time. Our new result removes the dependency of the additive error from $\log n$ and has a much smaller constant in front of δ. Note also that tree T is not a spanning tree of G and thus cannot serve as an eccentricity $O(\delta \log n)$-approximating spanning tree. Furthermore, as chordal graphs are 1-hyperbolic, for every k there is a 1-hyperbolic graph without an additive tree k-spanner [32,34].

Finally, in the full version of the paper [15], we analyze the performance of our algorithms for approximating eccentricities and distances on a number of real-world networks. Our experimental results show that the estimates on eccentricities and distances obtained are even better than the theoretical bounds proved. Experimental results can be found in the full version of the paper [15].

2   Preliminaries

Center, Diameter, Centrality. All graphs $G = (V, E)$ occurring in this paper are finite, undirected, connected, without loops or multiple edges. We use n and $|V|$ interchangeably to denote the number of vertices and m and $|E|$ to denote the number of edges in G. The *length of a path* from a vertex v to a vertex u is the number of edges in the path. The *distance* $d_G(u, v)$ between vertices u and v is the length of a shortest path connecting u and v in G. The *eccentricity* of a vertex v, denoted by $ecc_G(v)$, is the largest distance from v to any other vertex, i.e., $ecc_G(v) = \max_{u \in V} d_G(v, u)$. The *centrality index* of v is $\frac{1}{ecc_G(v)}$. The *radius* $rad(G)$ of a graph G is the minimum eccentricity of a vertex in G, i.e., $rad(G) = \min_{v \in V} ecc_G(v)$. The *diameter* $diam(G)$ of a graph G is the the maximum eccentricity of a vertex in G, i.e., $diam(G) = \max_{v \in V} ecc_G(v)$. The *center* $C(G) = \{c \in V : ecc_G(c) = rad(G)\}$ of a graph G is the set of vertices with minimum eccentricity.

Gromov Hyperbolicity and Thin Geodesic Triangles. Let (X, d) be a metric space. The *Gromov product* of $y, z \in X$ with respect to w is defined to be $(y|z)_w = \frac{1}{2}(d(y, w) + d(z, w) - d(y, z))$. A metric space (X, d) is said to be δ-*hyperbolic* [28] for $\delta \geq 0$ if $(x|y)_w \geq \min\{(x|z)_w, (y|z)_w\} - \delta$ for all $w, x, y, z \in X$. Equivalently, (X, d) is δ-hyperbolic if for any four points u, v, x, y of X, the two largest of the three distance sums $d(u, v) + d(x, y)$, $d(u, x) + d(v, y)$, $d(u, y) + d(v, x)$ differ by at most $2\delta \geq 0$. A connected graph $G = (V, E)$ is δ-*hyperbolic* (or of *hyperbolicity* δ) if the metric space (V, d_G) is δ-hyperbolic, where d_G is the standard shortest path metric defined on G.

δ-Hyperbolic graphs generalize k-chordal and bounded tree-length graphs: each k-chordal graph has the tree-length at most $\lfloor \frac{k}{2} \rfloor$ [21] and each tree-length λ graph has hyperbolicity at most λ [13]. A graph is k-*chordal* if its induced cycles are of length at most k, and it is of *tree-length* λ if it has a tree-decomposition into bags of diameter at most λ [21].

For geodesic metric spaces and graphs there exist several equivalent definitions of δ-hyperbolicity involving different but comparable values of δ [9,27,28]. *In this paper, we will use the definition via thin geodesic triangles.* Let (X, d) be a metric space. A *geodesic* joining two points x and y from X is a (continuous) map f from the segment $[a, b]$ of $\mathbb{R}^1$ of length $|a - b| = d(x, y)$ to X such that $f(a) = x, f(b) = y$, and $d(f(s), f(t)) = |s - t|$ for all $s, t \in [a, b]$. A metric space (X, d) is *geodesic* if every pair of points in X can be joined by a geodesic. Every graph $G = (V, E)$ can be transformed into a geodesic space (X, d) by replacing every edge $e = uv$ by a segment $[u, v]$ of length 1; the segments may intersect only at common ends. Then (V, d_G) is isometrically embedded in a natural way in (X, d). The restrictions of geodesics of X to the vertices V of G are the shortest paths of G.

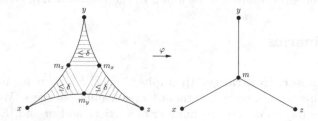

Fig. 1. A geodesic triangle $\Delta(x, y, z)$, the points m_x, m_y, m_z, and the tripod $\Upsilon(x, y, z)$

Let (X, d) be a geodesic metric space. A *geodesic triangle* $\Delta(x, y, z)$ with $x, y, z \in X$ is the union $[x, y] \cup [x, z] \cup [y, z]$ of three geodesic segments connecting these vertices. Let m_x be the point of the geodesic segment $[y, z]$ located at distance $\alpha_y := (x|z)_y = (d(y, x) + d(y, z) - d(x, z))/2$ from y. Then m_x is located at distance $\alpha_z := (y|x)_z = (d(z, y) + d(z, x) - d(y, x))/2$ from z because $\alpha_y + \alpha_z = d(y, z)$. Analogously, define the points $m_y \in [x, z]$ and $m_z \in [x, y]$ both located at distance $\alpha_x := (y|z)_x = (d(x, y) + d(x, z) - d(y, z))/2$ from x; see Fig. 1 for an illustration. There exists a unique isometry φ which maps $\Delta(x, y, z)$ to a tripod $T(x, y, z)$ consisting of three solid segments $[x, m], [y, m]$, and $[z, m]$ of lengths α_x, α_y, and α_z, respectively. This isometry maps the vertices x, y, z of $\Delta(x, y, z)$ to the respective leaves of $T(x, y, z)$ and the points m_x, m_y, and m_z to the center m of this tripod. Any other point of $T(x, y, z)$ is the image of exactly two points of $\Delta(x, y, z)$. A geodesic triangle $\Delta(x, y, z)$ is called δ-*thin* if for all points $u, v \in \Delta(x, y, z)$, $\varphi(u) = \varphi(v)$ implies $d(u, v) \le \delta$. A graph $G = (V, E)$ whose all geodesic triangles $\Delta(u, v, w)$, $u, v, w \in V$, are δ-thin is called a *graph with δ-thin triangles*, and δ is called the *thinness* parameter of G.

The following result shows that hyperbolicity of a geodesic space or a graph is equivalent to having thin geodesic triangles.

Proposition 1 ([9,27,28]). *Geodesic triangles of geodesic δ-hyperbolic spaces or graphs are 4δ-thin. Conversely, geodesic spaces or graphs with δ-thin triangles are δ-hyperbolic.*

In what follows, we will need few more notions and notations. Let $G = (V, E)$ be a graph. By $[x, y]$ we denote a shortest path connecting vertices x and y in G; we call $[x, y]$ a *geodesic* between x and y. A *ball* $B(s, r)$ of G centered at vertex $s \in V$ and with radius r is the set of all vertices with distance no more than r from s (i.e., $B(s, r) := \{v \in V : d_G(v, s) \leq r\}$). The *kth-power* of a graph $G = (V, E)$ is the graph $G^k = (V, E')$ such that $xy \in E'$ if and only if $0 < d_G(x, y) \leq k$. Denote by $F(x) := \{y \in V : d_G(x, y) = ecc_G(x)\}$ the set of all vertices of G that are *most distant* from x. Vertices x and y of G are called *mutually distant* if $x \in F(y)$ and $y \in F(x)$, i.e., $ecc_G(x) = ecc_G(y) = d_G(x, y)$.

3 Fast Approximation of Eccentricities

In this section, we give linear and almost linear time algorithms for sharp estimation of the diameters, the radii, the centers and the eccentricities of all vertices in graphs with δ-thin triangles. Before presenting those algorithms, we establish some conditional lower bounds on complexities of computing the diameters and the radii in those graphs.

3.1 Conditional Lower Bounds on Complexities

Recent work has revealed convincing evidence that solving the diameter problem in subquadratic time might not be possible, even in very special classes of graphs. Roditty and Vassilevska W. [36] showed that an algorithm that can distinguish between diameter 2 and 3 in a sparse graph in subquadratic time refutes the following widely believed conjecture.

The Orthogonal Vectors Conjecture: There is no $\epsilon > 0$ such that for all $c \geq 1$, there is an algorithm that given two lists of n binary vectors $A, B \subseteq \{0, 1\}^d$ where $d = c \log n$ can determine if there is an orthogonal pair $a \in A, b \in B$, in $O(n^{2-\epsilon})$ time.

Williams [39] showed that the Orthogonal Vectors (OV) Conjecture is implied by the well-known Strong Exponential Time Hypothesis (SETH) of Impagliazzo, Paturi, and Zane [30]. Nowadays many papers base the hardness of problems on SETH and the OV conjecture (see, e.g., [1,6] and papers cited therein). Since all geodesic triangles of a graph constructed in the reduction in [36] are 2-thin, we can rephrase the result from [36] as follows.

Statement 1. *If for some $\epsilon > 0$, there is an algorithm that can determine if a given graph with 2-thin triangles, n vertices and $m = O(n)$ edges has diameter 2 or 3 in $O(n^{2-\epsilon})$ time, then the Orthogonal Vector Conjecture is false.*

To prove a similar lower bound result for the radius problem, recently Abboud et al. [1] suggested to use the following natural and plausible variant of the OV conjecture.

The Hitting Set Conjecture: There is no $\epsilon > 0$ such that for all $c \geq 1$, there is an algorithm that given two lists A, B of n subsets of a universe U of size $c \log n$, can decide in $O(n^{2-\epsilon})$ time if there is a set in the first list that intersects every set in the second list.

Abboud et al. [1] showed that an algorithm that can distinguish between radius 2 and 3 in a sparse graph in subquadratic time refutes the Hitting Set Conjecture. Since all geodesic triangles of a graph constructed in [1] are 2-thin, rephrasing that result from [1], we have.

Statement 2. *If for some $\epsilon > 0$, there is an algorithm that can determine if a given graph with 2-thin triangles, n vertices, and $m = O(n)$ edges has radius 2 or 3 in $O(n^{2-\epsilon})$ time, then the Hitting Set Conjecture is false.*

3.2   Fast Additive Approximations

In this subsection, we show that in a graph G with δ-thin triangles the eccentricities of all vertices can be computed in total linear time with an additive error depending on δ. We establish that the eccentricity of a vertex is determined (up-to a small error) by how far the vertex is from the center $C(G)$ of G. Finally, we show how to construct a spanning tree T of G in which the eccentricity of any vertex is its eccentricity in G up to an additive error depending only on δ. For these purposes, we revisit and extend several results from our previous paper [13] about diameters, radii, and centers of δ-hyperbolic graphs.

Define the eccentricity layers of a graph G as follows: for $k = 0, \ldots, diam(G) - rad(G)$ set $C^k(G) := \{v \in V : ecc_G(v) = rad(G) + k\}$. With this notation, the center of a graph is $C(G) = C^0(G)$. In what follows, it will be convenient to define also the eccentricity of the middle point m of any edge xy of G; set $ecc_G(m) = \min\{ecc_G(x), ecc_G(y)\} + 1/2$.

We start with a proposition showing that, in a graph G with δ-thin triangles, a middle vertex of any geodesic between two mutually distant vertices has the eccentricity close to $rad(G)$ and is not too far from the center $C(G)$ of G.

Proposition 2. *Let G be a graph with δ-thin triangles and u, v be a pair of mutually distant vertices of G.*

(a) *If c^* is the middle point of any (u, v)-geodesic, then $ecc_G(c^*) \leq \frac{d_G(u,v)}{2} + \delta \leq rad(G) + \delta$.*

(b) *If c is a middle vertex of any (u, v)-geodesic, then $ecc_G(c) \leq \lceil \frac{d_G(u,v)}{2} \rceil + \delta \leq rad(G) + \delta$.*

(c) *$d_G(u, v) \geq 2rad(G) - 2\delta - 1$. In particular, $diam(G) \geq 2rad(G) - 2\delta - 1$.*

(d) *If c is a middle vertex of any (u, v)-geodesic and $x \in C^k(G)$, then $k - \delta \leq d_G(x, c) \leq k + 2\delta + 1$. In particular, $C(G) \subseteq B(c, 2\delta + 1)$.*

Proof. Let x be any vertex of G and $\Delta(u, v, x) := [u, v] \cup [v, x] \cup [x, u]$ be a geodesic triangle, where $[x, v], [x, u]$ are arbitrary geodesics connecting x with v, u. Let m_x be a point on $[u, v]$ at distance $(x|u)_v = \frac{1}{2}(d(x, v) + d(v, u) - d(x, u))$ from v and at distance $(x|v)_u = \frac{1}{2}(d(x, u) + d(v, u) - d(x, v))$ from u. Since u and

v are mutually distant, we can assume that c^* is located on $[u, v]$ between v and m_x, i.e., $d(v, c^*) \leq d(v, m_x) = (x|u)_v$, and hence $(x|v)_u \leq (x|u)_v$. Since $d_G(v, x) \leq d_G(v, u)$, we also get $(u|v)_x \leq (x|v)_u$.

(a) By the triangle inequality and since $d_G(u, v) \leq diam(G) \leq 2rad(G)$, we get
$d_G(x, c^*) \leq (u|v)_x + \delta + d_G(u, c^*) - (x|v)_u \leq d_G(u, c^*) + \delta = \frac{d_G(u,v)}{2} + \delta \leq rad(G) + \delta$.

(b) Since $c^* = c$ when $d_G(u, v)$ is even and $d_G(c^*, c) = \frac{1}{2}$ when $d_G(u, v)$ is odd, we have $ecc_G(c) \leq ecc_G(c^*) + \frac{1}{2}$. Additionally to the proof of (a), one needs only to consider the case when $d_G(u, v)$ is odd. We know that the middle point c^* sees all vertices of G within distance at most $\frac{d_G(u,v)}{2} + \delta$. Hence, both ends of the edge of (u, v)-geodesic, containing the point c^*, have eccentricities at most $\frac{d_G(u,v)}{2} + \frac{1}{2} + \delta = \lceil \frac{d_G(u,v)}{2} \rceil + \delta \leq \lceil \frac{2rad(G)-1}{2} \rceil + \delta = rad(G) + \delta$.

(c) Since a middle vertex c of any (u, v)-geodesic sees all vertices of G within distance at most $\lceil \frac{d_G(u,v)}{2} \rceil + \delta$, if $d_G(u, v) \leq 2rad(G) - 2\delta - 2$, then $ecc_G(c) \leq \lceil \frac{d_G(u,v)}{2} \rceil + \delta \leq \lceil \frac{2rad(G)-2\delta-2}{2} \rceil + \delta < rad(G)$, which is impossible.

(d) In the proof of (a), instead of any vertex x of G, consider any vertex x from $C^k(G)$. By the triangle inequality and since $d_G(u, v) \geq 2rad(G) - 2\delta - 1$ and both $d_G(u, x), d_G(x, v)$ are at most $rad(G) + k$, we get $d_G(x, c^*) \leq (u|v)_x + \delta + (x|u)_v - d_G(v, c^*) = d_G(v, x) - d_G(v, c^*) + \delta \leq rad(G) + k - \frac{d_G(u,v)}{2} + \delta \leq k + 2\delta + \frac{1}{2}$. Consequently, $d_G(x, c) \leq d_G(x, c^*) + \frac{1}{2} \leq k + 2\delta + 1$. On the other hand, since $ecc_G(x) \leq ecc_G(c) + d_G(x, c)$ and $ecc_G(c) \leq rad(G) + \delta$, by (a) we get $d_G(x, c) \geq ecc_G(x) - ecc_G(c) = k + rad(G) - ecc_G(c) \geq k - \delta$.   □

As an easy consequence of Proposition 2(d), we get that the eccentricity $ecc_G(x)$ of any vertex x is equal, up to an additive one-sided error of at most $4\delta + 2$, to $d_G(x, C(G)) + rad(G)$ (a proof can be found in the full version of this paper [15]).

Corollary 1. *For every vertex x of a graph G with δ-thin triangles, $d_G(x, C(G)) + rad(G) - 4\delta - 2 \leq ecc_G(x) \leq d_G(x, C(G)) + rad(G)$.*

It is interesting to note that the equality $ecc_G(x) = d_G(x, C(G)) + rad(G)$ holds for every vertex of a graph G if and only if the eccentricity function $ecc_G(\cdot)$ on G is unimodal (that is, every local minimum is a global minimum)[22]. A slightly weaker condition holds for all chordal graphs [24]: for every vertex x of a chordal graph G, $ecc_G(x) \geq d_G(x, C(G)) + rad(G) - 1$. Proofs of the following two propositions can be found in the full version of the paper [15].

Proposition 3. *Let G be a graph with δ-thin triangles and u, v be a pair of vertices of G such that $v \in F(u)$.*

(a) *If w is a vertex of a (u, v)-geodesic at distance $rad(G)$ from v, then $ecc_G(w) \leq rad(G) + \delta$.*

(b) *For every pair of vertices $x, y \in V$, $\max\{d_G(v, x), d_G(v, y)\} \geq d_G(x, y) - 2\delta$.*

(c) $ecc_G(v) \geq diam(G) - 2\delta \geq 2rad(G) - 4\delta - 1$.

(d) If $t \in F(v)$, c is a vertex of a (v,t)-geodesic at distance $\lceil \frac{d_G(v,t)}{2} \rceil$ from t and $x \in C^k(G)$, then $ecc_G(c) \leq rad(G) + 3\delta$ and $k - 3\delta \leq d_G(x,c) \leq k + 3\delta + 1$. In particular, $C(G) \subseteq B(c, 3\delta + 1)$.

Proposition 4. *For every graph G with δ-thin triangles, $diam(C^k(G)) \leq 2k + 2\delta + 1$. In particular, $diam(C(G)) \leq 2\delta + 1$.*

Diameter and Radius. For any graph $G = (V, E)$ and any vertex $u \in V$, a most distant from u vertex $v \in F(u)$ can be found in linear $(O(|E|))$ time by a *breadth-first-search* $BFS(u)$ started at u. A pair of mutually distant vertices of a connected graph $G = (V, E)$ with δ-thin triangles can be computed in $O(\delta|E|)$ total time as follows. By Proposition 3(c), if v is a most distant vertex from u and t is a most distant vertex from v, then $d(v,t) \geq diam(G) - 2\delta$. Hence, using at most $O(\delta)$ *breadth-first-searches*, one can generate a sequence of vertices $v := v_1, t := v_2, v_3, \ldots v_k$ with $k \leq 2\delta + 2$ such that each v_i is most distant from v_{i-1} (with, $v_0 = u$) and v_k, v_{k-1} are mutually distant vertices (the initial value $d(v,t) \geq diam(G) - 2\delta$ can be improved at most 2δ times). By Proposition 2 and Proposition 3, we get the following additive approximations for the radius and the diameter of a graph with δ-thin triangles.

Corollary 2. *Let $G = (V, E)$ be a graph with δ-thin triangles.*

1. *There is a linear $(O(|E|))$ time algorithm which finds in G a vertex c with eccentricity at most $rad(G) + 3\delta$ and a vertex v with eccentricity at least $diam(G) - 2\delta$. Furthermore, $C(G) \subseteq B(c, 3\delta + 1)$ holds.*
2. *There is an almost linear $(O(\delta|E|))$ time algorithm which finds in G a vertex c with eccentricity at most $rad(G) + \delta$. Furthermore, $C(G) \subseteq B(c, 2\delta + 1)$ holds.*

All Eccentricities. In what follows, we will show that all vertex eccentricities of a graph with δ-thin triangles can be also additively approximated in (almost) linear time. It will be convenient, for the middle point m of an edge e of G, to define a $BFS(m)$-tree of G; it is nothing else than a $BFS(e)$-tree of G rooted at edge e.

Proposition 5. *Let G be a graph with δ-thin triangles.*

(a) *If v is a most distant vertex from an arbitrary vertex u, t is a most distant vertex from v, c is a vertex of a (v,t)-geodesic at distance $\lceil \frac{d_G(v,t)}{2} \rceil$ from t and T is a $BFS(c)$-tree of G, then $ecc_G(x) \leq ecc_T(x) \leq ecc_G(x) + 6\delta + 1$.*

(b) *If c^* is the middle point of any (u,v)-geodesic between a pair u, v of mutually distant vertices of G and T is a $BFS(c^*)$-tree of G, then, for every vertex x of G, $ecc_G(x) \leq ecc_T(x) \leq ecc_G(x) + 2\delta$.*

Proof. (a) Let x be an arbitrary vertex of G and assume that $ecc_G(x) = rad(G) + k$ for some integer $k \geq 0$. We know from Proposition 3(d) that $ecc_G(c) \leq rad(G) + 3\delta$ and $d_G(c,x) \leq k + 3\delta + 1$. Since T is a $BFS(c)$-tree, $d_G(x,c) = d_T(x,c)$ and $ecc_G(c) = ecc_T(c)$. Consider a vertex y in G such that $d_T(x,y) = ecc_T(x)$. We have $ecc_T(x) = d_T(x,y) \leq d_T(x,c) + d_T(c,y) \leq d_G(x,c) + ecc_T(c) = d_G(x,c) + ecc_G(c) \leq k + 3\delta + 1 + rad(G) + 3\delta = rad(G) + k + 6\delta + 1 = ecc_G(x) + 6\delta + 1$. As T is a spanning tree of G, evidently, also $ecc_G(x) \leq ecc_T(x)$ holds.

(b) Consider an arbitrary vertex x of G and a geodesic triangle $\Delta(x,u,v) := [x,u] \cup [u,v] \cup [v,x]$, where $[u,v]$ is a (u,v)-geodesic containing c^* and $[u,x], [v,x]$ are arbitrary geodesics connecting x with u and v. Let m_x be a point on $[u,v]$ which is at distance $(x|u)_v = \frac{1}{2}(d_G(x,v) + d_G(u,v) - d_G(x,u))$ from v and hence at distance $(x|v)_u = \frac{1}{2}(d_G(x,u) + d_G(v,u) - d_G(x,v))$ from u. Without loss of generality, we can assume that c^* is located on $[u,v]$ between v and m_x. We have, $d_G(x,c^*) \leq (u|v)_x + \delta + d_G(m_x,c^*) = (u|v)_x + \delta + d_G(u,c^*) - (v|x)_u = (u|v)_x + \delta + \frac{d_G(v,u)}{2} - (v|x)_u$, and $ecc_G(x) \geq d_G(x,v) = (u|v)_x + (u|x)_v$. Furthermore, by Proposition 2(a), $ecc_G(c^*) \leq \frac{d_G(v,u)}{2} + \delta$. Hence, $ecc_T(x) - ecc_G(x) \leq d_T(x,c^*) + ecc_T(c^*) - ecc_G(x) = d_G(x,c^*) + ecc_G(c^*) - ecc_G(x) \leq (u|v)_x + \delta + \frac{d_G(v,u)}{2} - (v|x)_u + \frac{d_G(v,u)}{2} + \delta - (u|v)_x - (u|x)_v = 2\delta + d_G(v,u) - ((v|x)_u + (u|x)_v) = 2\delta$. □

Theorem 1. *Every graph $G = (V,E)$ with δ-thin triangles admits an eccentricity (2δ)-approximating spanning tree constructible in $O(\delta|E|)$ time and an eccentricity $(6\delta + 1)$-approximating spanning tree constructible in $O(|E|)$ time.*

Theorem 1 generalizes recent results from [24,35] that chordal graphs and some of their generalizations admit eccentricity 2-approximating spanning trees.

Note that the eccentricities of all vertices in any tree $T = (V,U)$ can be computed in $O(|V|)$ total time. As we noticed already, for trees the following facts are true: (1) $C(T)$ consists of one or two adjacent vertices; (2) $C(T)$ and $rad(T)$ of T can be found in linear time; (3) For any $v \in V$, $ecc_T(v) = d_T(v, C(T)) + rad(T)$. Hence, using $BFS(C(T))$ on T one can compute $d_T(v, C(T))$ for all $v \in V$ in total $O(|V|)$ time. Adding now $rad(T)$ to $d_T(v, C(T))$, one gets $ecc_T(v)$ for all $v \in V$. Consequently, by Theorem 1, we get the following additive approximations for the vertex eccentricities in graphs with δ-thin triangles.

Theorem 2. *Let $G = (V,E)$ be a graph with δ-thin triangles.*

(1) There is an algorithm which in total linear $(O(|E|))$ time outputs for every vertex $v \subset V$ an estimate $\hat{e}(v)$ of its eccentricity $ecc_G(v)$ such that $ecc_G(v) \leq \hat{e}(v) \leq ecc_G(v) + 6\delta + 1$.

(2) There is an algorithm which in total almost linear $(O(\delta|E|))$ time outputs for every vertex $v \in V$ an estimate $\hat{e}(v)$ of its eccentricity $ecc_G(v)$ such that $ecc_G(v) \leq \hat{e}(v) \leq ecc_G(v) + 2\delta$.

4   Fast Additive Approximation of All Distances

Here, we will show that if the δth power G^δ of a graph G with δ-thin triangles is known in advance, then the distances in G can be additively approximated in

$O(|V|^2)$ time. If G^δ is not known, then the distances can be additively approximated in almost quadratic time.

Our method is a generalization of an unified approach used in [23] to estimate (or compute exactly) all pairs shortest paths in such special graph families as k-chordal graphs, chordal graphs, AT-free graphs and many others. For example: all distances in k-chordal graphs with an additive one-sided error of at most $k-1$ can be found in $O(|V|^2)$ time; all distances in chordal graphs with an additive one-sided error of at most 1 can be found in $O(|V|^2)$ time and the all pairs shortest path problem on a chordal graph G can be solved in $O(|V|^2)$ time if G^2 is known. Note that in chordal graph all geodesic triangles are 2-thin.

Let $G = (V, E)$ be a graph with δ-thin triangles. Pick an arbitrary start vertex $s \in V$ and construct a $BFS(s)$-tree T of G rooted at s. Denote by $p_T(x)$ the *parent* and by $h_T(x) = d_T(x, s) = d_G(x, s)$ the *height* of a vertex x in T. Since we will deal only with one tree T, we will often omit the subscript T. Let $P_T(x, s) := (x_q, x_{q-1}, \ldots, x_1, s)$ and $P_T(y, s) := (y_p, y_{p-1}, \ldots, y_1, s)$ be the paths of T connecting vertices x and y with the root s. By $sl_T(x, y; \lambda)$ we denote the largest index k such that $d_G(x_k, y_k) \le \lambda$ (the λ separation level). Our method is based on the following simple fact.

Proposition 6. *For every vertices x and y of a graph G with δ-thin triangles and any BFS-tree T of G, $h_T(x) + h_T(y) - 2k - 1 \le d_G(x, y) \le h_T(x) + h_T(y) - 2k + d_G(x_k, y_k)$, where $k = sl_T(x, y; \delta)$.*

Proof. By the triangle inequality, $d_G(x, y) \le d_G(x, x_k) + d_G(x_k, y_k) + d_G(y_k, y) = h_T(x) + h_T(y) - 2k + d_G(x_k, y_k)$. Consider now an arbitrary (x, y)-geodesic $[x, y]$ in G. Let $\Delta(x, y, s) := [x, y] \cup [x, s] \cup [y, s]$ be a geodesic triangle, where $[x, s] = P_T(x, s)$ and $[y, s] = P_T(y, s)$. Since $\Delta(x, y, s)$ is δ-thin, $sl_T(x, y; \delta) \ge (x|y)_s - \frac{1}{2}$. Hence, $h_T(x) - sl_T(x, y; \delta) \le (s|y)_x + \frac{1}{2}$ and $h_T(y) - sl_T(x, y; \delta) \le (s|x)_y + \frac{1}{2}$. As $d_G(x, y) = (s|y)_x + (s|x)_y$, we get $d_G(x, y) \ge h_T(x) - sl_T(x, y; \delta) + h_T(y) - sl_T(x, y; \delta) - 1$. $\square$

Note that we may regard $BFS(s)$ as having produced a numbering from n to 1 in decreasing order of the vertices in V where vertex s is numbered n. As a vertex is placed in the queue by $BFS(s)$, it is given the next available number. The last vertex visited is given the number 1. Let $\sigma := [v_1, v_2, \ldots, v_n = s]$ be a $BFS(s)$-ordering of the vertices of G and T be a $BFS(s)$-tree of G produced by a $BFS(s)$. Let $\sigma(x)$ be the number assigned to a vertex x in this $BFS(s)$-ordering. For two vertices x and y, we write $x < y$ whenever $\sigma(x) < \sigma(y)$.

First, we show that if G^δ is known in advance (i.e., its adjacency matrix is given) for a graph G with δ-thin triangles, then the distances in G can be additively approximated (with an additive one-sided error of at most $\delta + 1$) in $O(|V|^2)$ time. We consider the vertices of G in the order σ from 1 to n. For each current vertex x we show that the values $\widehat{d}(x, y) := h_T(x) + h_T(y) - 2sl_T(x, y; \delta) + \delta$ for all vertices y with $y > x$ can be computed in $O(|V|)$ total time. By Proposition 6, $d_G(x, y) \le \widehat{d}(x, y) \le d_G(x, y) + \delta + 1$. The values $\widehat{d}(x, y)$ for all y with $y > x$ can be computed using the following simple procedure. We omit the subscripts G and T if no ambiguities arise. Let also $L_i = \{v \in V : d(v, s) = i\}$. In the procedure, S_u represents vertices of a subtree of T rooted at u.

(01) set $q := h(x)$
(02) let $S_u := \{u\}$ for each vertex $u \in L_q$, $u > x$, and denote this family of sets by $\mathcal{F}$
(03) **for** $k = q$ downto 0 **do**
(04)   let x_k be the vertex from $L_k \cap P_T(x, s)$
(05)   **for** each vertex $u \in L_k$ with $u > x$ **do**
(06)   **if** $d_G(u, x_k) \leq \delta$ (i.e., $u = x_k$ or u is adjacent to x_k in G^δ) **then**
(07)   **for** every $v \in S_u$ **do**
(08)   set $\widehat{d}(x, v) := h(x) + h(v) - 2k + \delta$ and remove S_u from $\mathcal{F}$
(09)   /* update $\mathcal{F}$ for the next iteration */
(10)   **if** $k > 0$ **then**
(11)   **for** each vertex $u \in L_{k-1}$ **do**
(12)   combine sets $S_{u_1}, \dots, S_{u_\ell}$ from $\mathcal{F}$ ($\ell \geq 0$) with $p_T(u_1) = \dots = p_T(u_\ell) = u$
(13)   into one new set $S_u := \{u\} \cup S_{u_1} \cup \dots \cup S_{u_\ell}$ /* when $\ell = 0$, $S_u := \{u\}$ */
(14) set also $\widehat{d}(x, s) := h(x)$.

Theorem 3. *Let $G = (V, E)$ be a graph with δ-thin triangles. Given G^δ, all distances in G with an additive one-sided error of at most $\delta + 1$ can be found in $O(|V|^2)$ time.*

To avoid the requirement that G^δ is given in advance, we can use any known fast constant-factor approximation algorithm that in total $T(|V|)$-time computes for every pair of vertices x, y of G a value $\widetilde{d}(x, y)$ such that $d_G(x, y) \leq \widetilde{d}(x, y) \leq \alpha d_G(x, y) + \beta$. We can show that, using such an algorithm as a preprocessing step, the distances in a graph G with δ-thin triangles can be additively approximated with an additive one-sided error of at most $\alpha\delta + \beta + 1$ in $O(T(|V|) + |V|^2)$ time. Although one can use any known fast constant-factor approximation algorithm in the preprocessing step, in what follows, we will demonstrate our idea using a fast approximation algorithm from [5]. It computes in $O(|V|^2 \log^2 |V|)$ total time for every pair x, y a value $\widetilde{d}(x, y)$ such that $d_G(x, y) \leq \widetilde{d}(x, y) \leq 2d_G(x, y) + 1$. Assume that the values $\widetilde{d}(x, y)$, $x, y \in V$, are precomputed. By $\widetilde{sl}_T(x, y; \lambda)$ we denote now the largest index k such that $\widetilde{d}_G(x_k, y_k) \leq \lambda$. We have.

Proposition 7. *For every vertices x and y of a graph G with δ-thin triangles, any integer $\rho \geq \delta$, and any BFS-tree T of G, $h_T(x) + h_T(y) - 2k - 1 \leq d_G(x, y) \leq h_T(x) + h_T(y) - 2k + d_G(x_k, y_k)$, where $k = \widetilde{sl}_T(x, y; 2\rho + 1)$.*

Proof of this propositions can be found in the full version of the paper [15]. Let ρ be any integer greater than or equal to δ. By replacing in our earlier procedure lines (06) and (08) with

(06)′   **if** $\widetilde{d}(u, x_k) \leq 2\rho + 1$ **then**
(08)′   set $\widehat{d}(x, v) := h(x) + h(v) - 2k + 2\rho + 1$ and remove S_u from $\mathcal{F}$

we will compute for each current vertex x all values $\widehat{d}(x,y) := h_T(x) + h_T(y) - 2\widetilde{sl}_T(x,y;2\rho+1)+2\rho+1, y > x$, in $O(|V|)$ total time. By Proposition 7, $d_G(x,y) \le h_T(x) + h_T(y) - 2\widetilde{sl}_T(x,y;2\rho+1)+d_G(x_k,y_k) \le h_T(x)+h_T(y)-2\widetilde{sl}_T(x,y;2\rho+1) + \widetilde{d}(x_k,y_k) \le h_T(x) + h_T(y) - 2\widetilde{sl}_T(x,y;2\rho+1) + 2\rho + 1 = \widehat{d}(x,y)$ and $\widehat{d}(x,y) = h_T(x) + h_T(y) - 2\widetilde{sl}_T(x,y;2\rho+1) + 2\rho + 1 \le d_G(x,y) + 2\rho + 2$. Thus, we have the following result:

Theorem 4. *Let $G = (V,E)$ be a graph with δ-thin triangles.*

(a) *If the value of δ is known, then all distances in G with an additive one-sided error of at most $2\delta + 2$ can be found in $O(|V|^2 \log^2 |V|)$ time.*

(b) *If an approximation ρ of δ such that $\delta \le \rho \le a\delta + b$ is known (where a and b are constants), then all distances in G with an additive one-sided error of at most $2(a\delta + b + 1)$ can be found in $O(|V|^2 \log^2 |V|)$ time.*

The second part of Theorem 4 says that if an approximation of the thinness of a graph G is given, then all distances in G can be additively approximated in $O(|V|^2 \log^2 |V|)$ time. Recently, it was shown in [11] that the converse is also true. From an estimate of all distances in G with an additive one-sided error of at most k, it is possible to compute in $O(|V|^2)$ time an estimation ρ^* of the thinness of G such that $\delta \le \rho^* \le 8\delta + 12k + 4$, proving a $\tilde{O}(|V|^2)$-equivalence between approximating the thinness and approximating the distances in graphs.

Acknowledgements. The research of V.C., M.H., and Y.V. was supported by ANR project DISTANCIA (ANR-17-CE40-0015).

References

1. Abboud, A., Wang, J., Vassilevska Williams, V.: Approximation and fixed parameter subquadratic algorithms for radius and diameter in sparse graphs. In: SODA (2016)
2. Abboud, A., Grandoni, F., Vassilevska Williams, V.: Subcubic equivalences between graph centrality problems, APSP and diameter. In: SODA, pp. 1681–1697 (2015)
3. Abu-Ata, M., Dragan, F.F.: Metric tree-like structures in real-world networks: an empirical study. Networks **67**, 49–68 (2016)
4. Aingworth, D., Chekuri, C., Indyk, P., Motwani, R.: Fast estimation of diameter and shortest paths (w/o matrix multiplication). SIAM J. Comput. **28**, 1167–81 (1999)
5. Berman, P., Kasiviswanathan, S.P.: Faster approximation of distances in graphs. In: Dehne, F., Sack, J.-R., Zeh, N. (eds.) WADS 2007. LNCS, vol. 4619, pp. 541–552. Springer, Heidelberg (2007). https://doi.org/10.1007/978-3-540-73951-7_47
6. Borassi, M., Crescenzi, P., Habib, M.: Into the square - on the complexity of quadratic-time solvable problems. Electron. Notes Theor. Comput. Sci. **322**, 51–67 (2016)
7. Brandstädt, A., Chepoi, V., Dragan, F.F.: The algorithmic use of hypertree structure and maximum neighbourhood orderings. Discrete Appl. Math. **82**, 43–77 (1998)

8. Brandstädt, A., Chepoi, V., Dragan, F.F.: Distance approximating trees for chordal and dually chordal graphs. J. Algorithms **30**, 166–184 (1999)
9. Bridson, M.R., Haefiger, A.: Metric Spaces of Non-Positive Curvature, Grundlehren der Mathematischen Wissenschaften. vol. 319, Springer (1999). https://doi.org/10.1007/978-3-662-12494-9
10. Chechik, S., Larkin, D., Roditty, L., Schoenebeck, G., Tarjan, R.E.: and Williams, V.V.: Better approximation algorithms for the graph diameter. In: SODA (2014)
11. Chalopin, J., Chepoi, V., Dragan, F.F., Ducoffe, G., Mohammed, A., Vaxès, Y.: Fast approximation and exact computation of negative curvature parameters of graphs. In: SoCG (2018)
12. Chepoi, V., Dragan, F.: A linear-time algorithm for finding a central vertex of a chordal graph. In: van Leeuwen, J. (ed.) ESA 1994. LNCS, vol. 855, pp. 159–170. Springer, Heidelberg (1994). https://doi.org/10.1007/BFb0049406
13. Chepoi, V.D., Dragan, F.F., Estellon, B., Habib, M., Vaxès, Y.: Diameters, centers, and approximating trees of δ-hyperbolic geodesic spaces and graphs. In: SoCG (2008)
14. Chepoi, V., Dragan, F.F., Estellon, B., Habib, M., Vaxès, Y., Xiang, Y.: Additive spanners and distance and routing schemes for hyperbolic graphs. Algorithmica **62**, 713–732 (2012)
15. Chepoi, V., Dragan, F.F., Habib, M., Vaxès, Y., Al-Rasheed, H.: Fast approximation of centrality and distances in hyperbolic graphs. arXiv:1805.07232 (2018)
16. Chepoi, V., Dragan, F.F., Vaxès, Y.: Center and diameter problems in plane triangulations and quadrangulations. In: SODA, pp. 346–355 (2002)
17. Chepoi, V., Dragan, F.F., Vaxès, Y.: Core congestion is inherent in hyperbolic networks. In: SODA, pp. 2264–2279 (2017)
18. Chepoi, V., Estellon, B.: Packing and covering δ-hyperbolic spaces by balls. In: Charikar, M., Jansen, K., Reingold, O., Rolim, J.D.P. (eds.) APPROX/RANDOM -2007. LNCS, vol. 4627, pp. 59–73. Springer, Heidelberg (2007). https://doi.org/10.1007/978-3-540-74208-1_5
19. Corneil, D.G., Dragan, F.F., Köhler, E.: On the power of BFS to determine a graph's diameter. Networks **42**, 209–222 (2003)
20. Dor, D., Halperin, S., Zwick, U.: All-pairs almost shortest paths. SIAM J. Comput. **29**, 1740–1759 (2000)
21. Dourisboure, Y., Gavoille, C.: Tree-decompositions with bags of small diameter. Discr. Math. **307**, 208–229 (2007)
22. Dragan, F.F.: Centers of graphs and the Helly property (in Russian), Ph.D. thesis, Moldova State University (1989)
23. Dragan, F.F.: Estimating all pairs shortest paths in restricted graph families: a unified approach. J. Algorithms **57**, 1–21 (2005)
24. Dragan, F.F., Köhler, E., Alrasheed, H.: Eccentricity approximating trees. Discrete Appl. Math. **232**, 142–156 (2017)
25. Dragan, F.F., Nicolai, F., Brandstädt, A.: LexBFS-orderings and powers of graphs. In: d'Amore, F., Franciosa, P.G., Marchetti-Spaccamela, A. (eds.) WG 1996. LNCS, vol. 1197, pp. 166–180. Springer, Heidelberg (1997). https://doi.org/10.1007/3-540-62559-3_15
26. Edwards, K., Kennedy, W.S., Saniee, I.: Fast approximation algorithms for p-centres in large *delta*-hyperbolic graphs, CoRR, vol. abs/1604.07359 (2016)
27. Ghys, E., de la Harpe, P. (eds.) Les groupes hyperboliques d'après Gromov, M. Progress in Mathematics, Vol. 83 Birkhäuser (1990)

28. Gromov, M.: Hyperbolic groups. In: Gersten, S.M. (ed.) Essays in Group Theory. Mathematical Sciences Research Institute Publications, vol. 8. Springer, New York (1987). https://doi.org/10.1007/978-1-4613-9586-7_3

29. Hakimi, S.L.: Optimum location of switching centers and absolute centers and medians of a graph. Oper. Res. **12**, 450–459 (1964)

30. Impagliazzo, R., Paturi, R., Zane, F.: Which problems have strongly exponential complexity? J. Comput. Syst. Sci. **63**, 512–530 (2001)

31. Jordan, C.: Sur les assemblages des lignes. J. Reine Angew. Math. **70**, 185–190 (1869)

32. Kratsch, D., Le, H.-O., Müller, H., Prisner, E., Wagner, D.: Additive tree spanners. SIAM J. Discrete Math. **17**, 332–340 (2003)

33. Narayan, O., Saniee, I.: Large-scale curvature of networks. Phys. Rev. E **84**(6), 066108 (2011)

34. Prisner, E.: Distance approximating spanning trees. In: Reischuk, R., Morvan, M. (eds.) STACS 1997. LNCS, vol. 1200, pp. 499–510. Springer, Heidelberg (1997). https://doi.org/10.1007/BFb0023484

35. Prisner, E.: Eccentricity-approximating trees in chordal graphs. Discr. Math. **220**, 263–269 (2000)

36. Roditty, L., Vassilevska Williams, V.: Fast approximation algorithms for the diameter and radius of sparse graphs. In: STOC, pp. 515–524 (2013)

37. Shavitt, Y., Tankel, T.: Hyperbolic embedding of internet graph for distance estimation and overlay construction. IEEE/ACM Trans. Netw. **16**, 25–36 (2008)

38. Verbeek, K., Suri, S.: Metric embedding, hyperbolic space, and social networks. In: SoCG, pp. 501–510 (2014)

39. Williams, R.: A new algorithm for optimal constraint satisfaction and its implications. In: Díaz, J., Karhumäki, J., Lepistö, A., Sannella, D. (eds.) ICALP 2004. LNCS, vol. 3142, pp. 1227–1237. Springer, Heidelberg (2004). https://doi.org/10.1007/978-3-540-27836-8_101

Rectilinear Shortest Paths Among Transient Obstacles

Anil Maheshwari, Arash Nouri$^{(\boxtimes)}$, and Jörg-Rüdiger Sack

School of Computer Science, Carleton University, Ottawa, Canada
{anil,arash,sack}@scs.carleton.ca

Abstract. This paper presents an optimal $\Theta(n \log n)$ algorithm for determining time-minimal rectilinear paths among n transient rectilinear obstacles. An obstacle is transient if it exists in the scene only for a specific time interval, i.e., it appears and then disappears at specific times. Given a point robot moving with bounded speed among transient rectilinear obstacles and a pair of points s, d, we determine a time-minimal, obstacle-avoiding path from s to d. The main challenge in solving this problem arises as the robot may be required to wait for an obstacle to disappear, before it can continue moving toward the destination. Our algorithm builds on the continuous Dijkstra paradigm, which simulates propagating a wavefront from the source point. We also solve a query version of this problem. For this, we build a planar subdivision with respect to a fixed source point, so that minimum arrival time to any query point can be reported in $O(\log n)$ time, using point location for the query point in this subdivision.

Keywords: Shortest path · Transient obstacles · Time minimal path
Time discretization · Continuous dijkstra

1 Introduction

We study a variant of the classical shortest path problem in which each obstacle exists only during a specific time interval. Such obstacles are called *transient obstacles* (see e.g., [5]). Besides solving an interesting problem in itself, our solutions may find applications in other motion planning problems in time-dependent environments. Transient obstacles can e.g., be used to approximate dynamic obstacles in the plane [6,12]. In such settings, the trajectories of the moving obstacles are divided into a set of small pieces. Each piece is treated as a transient obstacle that exists in the scene only for the time interval in which the moving obstacle and the piece intersect. The approximation quality can be adjusted by varying the sizes of the pieces. This adequately models real world scenarios in which robots are limited by the sampling rate of their sensors acquiring information and executing motion commands.

In general, our model considering transient obstacles can be useful for applications where one can define a discretized representation of time by a set of

© Springer Nature Switzerland AG 2018
D. Kim et al. (Eds.): COCOA 2018, LNCS 11346, pp. 19–34, 2018.
https://doi.org/10.1007/978-3-030-04651-4_2

stages. For instance, in the area of *path planning under uncertainty*, one considers the following problem: Let $\{R_1, ..., R_n\}$ be a set of regions, where each region becomes contaminated at a random time. The probability at which R_i is contaminated at time t, is given by a probability distribution $P_i(t)$. In such a setting, a natural approach is to search for a shortest path which is contamination-free with high probability. This is a class of motion planning referred to as *hazardous region and shelter problems* [11]. A suitable means of planning a low contamination path, is to bound the probability at which the intersecting regions are contaminated. More precisely, for a small value of $\epsilon \in [0, 1]$, the robot cannot enter a region R_i if $P_i(t) > \epsilon$. This can be viewed as a time discretization into a set of "high risk" time intervals for the regions. Using the corresponding probability distribution, we can determine a time interval T_i (or in some cases more than one), which contains the contamination time with a probability of $1 - \epsilon$. This problem is easily transformed into our model where the confidence intervals are mapped into existence intervals for the transient obstacles.

Related Work. The shortest path problem among transient obstacle was first studied by Fujimura [5], who presented an $O(n^3 \log n)$ time algorithm for finding a time-minimal path among transient (non-intersecting) polygonal obstacles. Later [7], he proposed an $O(n^4)$ time algorithm for a variant of this problem in which the obstacles are allowed to occupy the same area of the plane (i.e., intersecting obstacles). A recently introduced model [8] considers another variation of this problem, where the path is allowed to pass through k obstacles. They present an $O(k^2 n \log n)$ time algorithm, where n is the total number of obstacle vertices. A more complex version of this problem has been studied in [2], in which the robot may pass through obstacles at some cost. They proved that this problem is NP-hard even if the obstacles are vertical line segments.

Our Contributions. In this paper, we present an optimal $\Theta(n \log n)$ time algorithm for computing a time-minimal rectilinear path among rectilinear transient obstacles. Although our problem is a special case of the shortest path problem among transient obstacles, the methodology and the results of this work also have the potential to lead to an improvement of the existing $O(n^3 \log n)$ time algorithm for the general case. We first discuss a simple problem instance in which the given obstacles are rectilinear segments. Then, we generalize the algorithm developed for the simpler setting to simple rectilinear polygons. Section 2 describes preliminaries, definitions and introduces some notation. Section 3 presents several techniques that are subsequently employed in this paper. Building on these techniques, Sect. 4 presents an $O(n^2 \log n)$ time algorithm for the problem, which is already an improvement over the existing algorithm applied to our setting. Finally, Sect. 5 details our optimal $\Theta(n \log n)$ time algorithm. **Note:** Due to space limitation some technical details have been moved to the full version of the paper [14].

2 Preliminaries

Let $E = \{E_1, ..., E_n\}$ be a set of rectilinear transient edges, where each edge $E_i \in E$ exists in the scene during a time interval $[T_i^a, T_i^d]$, where $0 \leq T_i^a < T_i^d$. The edges are disjoint, i.e., no two edges are allowed to overlap at any time. We assume that the edges are in general position, which means that, no two edges lie on a common line. Let $\mathcal{R}$ be a point robot having maximum speed $\mathcal{V}_{max}$. For two given points s and d in the plane, our problem is to determine a time-minimal rectilinear path for the point robot from s to d, denoted by $\pi(s, d)$, which is collision free, i.e., the point robot does not pass through the edges during their existence intervals. W.l.o.g., we assume the robot always departs from s at time 0.

Our strategy is to employ the "continuous Dijkstra" paradigm [9,16], which has been applied to solve numerous shortest path problems among permanent (i.e., non-transient) obstacles [9,13,15,16]. We provide here a brief description of this paradigm. The continuous Dijkstra's technique models the effects of sweeping an advancing wavefront from the source point till it reaches the destination. A *wavefront* (in L_1 metric space) is defined as the set of points on the plane at equal L_1 distance from the source. Initially, the wavefront is point located at s. After a short time period, it becomes a rhombus centered at s with diameter ϵ, where ϵ is a small positive constant. The continuous Dijkstra's algorithm proceeds by expanding the rhombus outward from its center point. At any point in time, the wavefront consists of a set of line segments, known as *wavelets*. A wavelet is defined as a maximal set of points on the wavefront, such that each point on the wavelet has a shortest path from s via a common vertex. Each wavelet originates at a vertex, which is called the *source* of the wavelet. Therefore, each wavelets moves in one of the four fixed directions: *north-east, north-west, south-east, south-west*. We abbreviate these four directions as {NE, NW, SE, SW}. More precisely, the wavelets are in four fixed inclinations with respect to the x-axis with angles: $\pi/4$, $3\pi/4$, $5\pi/4$ and $7\pi/4$.

For our setting, we need to modify the continuous Dijkstra's model described above for metric shortest paths, to time-minimal paths among transient obstacles. Note that, on a time-minimal path, the robot's speed alternates between $\mathcal{V}_{max}$ (i.e., the robot is moving) and zero (i.e, the robot is waiting). It is easily seen that, by arriving earlier at some obstacles and then waiting there until the obstacle disappears, the robot can avoid any speed other than $\mathcal{V}_{max}$ and zero. After waiting the robot continues to move towards the next destination. The points on the boundary of obstacles where robots may be waiting are called *wait points*. The behavior of the wavefront changes at portions of obstacles that are "potential" wait point candidates.

Given a point p in the plane, we say $\pi(s, p)$ intersects an edge E_i, if it has a wait point on E_i; and we say it intersects a vertex v if $v \in \pi(s, p)$. Let $S(\pi(s, p))$ be the sequence of edges and vertices that $\pi(s, p)$ intersects. We formally define a wavelet as follows.

Definition 1. *A **wavelet** ω is a maximal set of points, such that for each pair of points $p, q \in \omega$ there exist two paths $\pi(s, p)$ and $\pi(s, q)$ with equal arrival times, for which $S(\pi(s, p)) = S(\pi(s, q))$. Let x be the last element in $S(\pi(s, p))$. We say x is the **origin** of ω (or alternatively, we say that ω is **originating** from x). If x is an edge, we say ω is a **segment wavelet**; otherwise, ω is a **point wavelet**. A **wavefront** is defined as the union of the wavelets at an equal time.*

By the above definition, similar to the original version of continuous Dijkstra, the point wavelets are in four fixed inclinations with respect to the x-axis with angles: $\{\pi/4, 3\pi/4, 5\pi/4, 7\pi/4\}$. Now, observe the following property of the time-minimal paths in our setting.

Observation 1 [7]. *When, after waiting on an edge E_i, the robot departs, at some time T_i^d, it will use a move perpendicular to the orientation of E_i.*

By the above observation, each segment wavelet propagates outwards perpendicularly to its originating segment. Since the edges are axis-parallel, each segment wavelet is oriented in one of four (axis-parallel) directions: $\{0, \pi/2, \pi, 3\pi/4\}$ and moves in one of four directions: *north, south, east or west*; these are abbreviated as: $\{N, S, E, W\}$, respectively.

Our algorithm propagates a wavelet ω, with inclination θ, outwards by using a sweep line through ω (refer to Sect. 4.1 for details). When ω encounters (or "hits") an obstacle, we add new wavelets originating from a vertex or an edge of the obstacle. For each wavelet ω, we designate an area called *search region*, from which ω propagates its interior (refer to Sect. 3.3 for a formal definition). The time at which propagation starts is called *departure time*. A key property that we will be subsequently using is that wavelets are line segments with fixed inclinations. This enables us to efficiently find the next propagation "events" using range searching queries (see Sect. 3) (events are intersections between the wavelets and the obstacles).

We say a path is *monotone* if any axis-parallel line intersects the path in at most one connected set. For any pair of consecutive vertices u and v on a shortest path among non-transient obstacles, in [4], it is proven that the sub-path from u to v is monotone. In the following lemma, whose proof is provided in the full version of the paper [14], we show that the analogous property also holds for time-minimal rectilinear paths among transient obstacles.

Lemma 1 (Monotonicity Property). *In our model, let u and v be two consecutive vertices on $\pi(s, d)$. The sub-path of $\pi(s, d)$ from u to v, denoted by $\pi(u, v)$, is monotone.*

Define a pair (p, t), as *point source*, denoted by $\sigma(p, t)$, where $p = (X_p, Y_p)$ is the x-y location of the robot at time t. We will simply say a path from σ instead of a path leaving p at time t. Suppose the robot departs from σ by moving north at maximum speed. For that motion, we define the *north stop point* for σ, denoted by $\mathcal{U}(\sigma, N)$, as the first point on any obstacle that the robot "hits" during the respective obstacles' existence times. Analogously, we define $\mathcal{U}(\sigma, S)$,

$\mathcal{U}(\sigma, E)$ and $\mathcal{U}(\sigma, W)$ as the south, east and west stop points, respectively. Note that the locations of the stop points may change depending on the departure time t.

A point $q \in E_i$ is called *accessible* from point source $\sigma(p, t)$, if there exists a time-minimal path from σ to q, denoted by $\pi(\sigma, q)$, such that: (1) $\pi(\sigma, q)$ contains no wait points and (2) the robot arrives at q during the existence time interval of E_i. We denote by $T(\sigma, q) = t + \frac{\|pq\|_1}{V_{max}}$ the arrival time of this path, where $\|pq\|_1$ is the L_1 distance between the two points. Observe that, any stop point for σ is an accessible point from σ. Let $\mathcal{U}(\sigma, N) \in E_i$ be the north stop point for σ. We define the *north accessible segment* of σ as a maximal set of accessible points on E_i. Note that, $\mathcal{U}(\sigma, N)$ is a point on the north accessible segment. Analogously, we define the other accessible segments of σ in the three other directions.

Given a sub-segment $e = ((X_1, Y_1), (X_2, Y_2))$ of an edge E_i, we define a *segment source* $\overline{\sigma}(e, T_i^d) = \cup_{p \in e} \{\sigma(p, T_i^d)\}$ as a maximal set of point sources located on e having common departure time $T_i^d)\}$.

A point $q \in E_i$ is a *north accessible point* for $\overline{\sigma}$, if there is exist $\sigma(p, T_i^d) \in \overline{\sigma}(e, T_i^d)$ such that $q = \mathcal{U}(\sigma, N)$. We define the *north stop segment* for $\overline{\sigma}$, denoted by $\mathcal{U}(\overline{\sigma}, N)$, as the maximal set of north accessible points which have minimum distance to e. By the general position assumption, $\mathcal{U}(\overline{\sigma}, N)$ is a connected sub-segment of an edge in E. Intuitively, if we drag the segment e north, the north stop segment is the first intersection between the dragging segment and the obstacles. Analogously, we define $\mathcal{U}(\overline{\sigma}, S)$, $\mathcal{U}(\overline{\sigma}, E)$ and $\mathcal{U}(\overline{\sigma}, W)$.

3   Range Searching Techniques

When propagating a wavelet, we wish to quickly determine the next event, where the wavelet intersects an obstacle. In this section, we present our techniques employed to solve this problem. First, we present a solution to the problem of determining the stop points for a query point source (see Lemma 3). Then, we devise an algorithm to report the stop segments for a query segment source (see Lemma 5). Using these stop points, we define a rectangular range (the search region), which contains potential next points/edges hit by the wavelet. We identify these using a range searching technique, presented in Lemma 6.

3.1   Finding the Stop Points

Given a query point source $\sigma(p = (x, y), t)$, we denote by $\{\mathcal{U}(\sigma, N), \mathcal{U}(\sigma, S), \mathcal{U}(\sigma, E), \mathcal{U}(\sigma, W)\}$ the four stop points for σ. In this section, we present a solution to find $\mathcal{U}(\sigma, N)$ and other stop points can be found analogously. Let $\mathcal{U}(\sigma, N) \in E_i$, where $E_i = ((X_1, Y), (X_2, Y))$. By definition, the robot hits E_i during its existence time interval. So, we must have $T_i^a \leq t + \frac{(Y-y)}{V_{max}} \leq T_i^d$. Hence,

$$T_i^a - \frac{Y}{V_{max}} \leq t - \frac{y}{V_{max}} \leq T_i^d - \frac{Y}{V_{max}}. \tag{1}$$

Also, it is easily seen that,

$$X_1 \leq x \leq X_2. \tag{2}$$

As a result, if the north stop point for σ is located on E_i, Eqs. (1) and (2) must be satisfied. These equations can be viewed as one equation in two dimensions, where the y values are replaced by $(t - \frac{y}{V_{max}})$: let r_i^s be a rectangle where $(X_1, T_i^a - \frac{Y}{V_{max}})$ and $(X_2, T_i^d - \frac{Y}{V_{max}})$ is one of its opposite corner pairs. Given a point $\bar{p} = (x, t - \frac{y}{V_{max}})$, observe that $\bar{p} \in r_i^s$ if and only if σ satisfies the Eqs. (1) and (2). We call r_i^s the *south shadow range* of E_i and $\bar{p}$ the *south shadow point* of σ (see the full version of the paper for an example).

Observation 2. *Let $\sigma(p, t)$ be a query point source and $\mathcal{U}(\sigma, N) \in E_i$ be its north stop point. Then, the south shadow point for σ is located inside the south shadow range of E_i, i.e., $\bar{p} \in r_i^s$.*

Note that the reverse direction of the above observation does not always hold. In other words, there are several edges whose shadow ranges contain $\bar{p}$; however, only one includes the north stop point. Recall that a stop point represents the "first" intersection between the robot and the obstacles. So, $\mathcal{U}(\sigma, N)$ is located on an obstacle whose Y value is minimum among all edges whose south shadow range contain $\bar{p}$. Thus, we assign a weight to a shadow range r_i^s, denoted by $\omega(r_i^s)$, which is the Y value of its corresponding (horizontal) edge E_i. In order to find the north stop point for the query point source σ, we need to find the minimum weight shadow range r that contains $\bar{p}$.

Lemma 2 [1]. *A set H of n axis-parallel rectangles, where each rectangle $h \in H$ has a weight $\omega(h) \in \mathbb{R}$, can be maintained so that the minimum weight rectangle containing a query point can be determined in $O(\log n)$ query time, after $O(n \log n)$ preprocessing time.*

The following is the direct consequence of the above lemma.

Lemma 3. *After $O(n \log n)$ time preprocessing, all stop points of a query point source can be found in $O(\log n)$ time.*

3.2   Finding the Stop Segments

Let $\bar{\sigma}(e, t)$ be a horizontal query segment source and $\mathcal{U}(\bar{\sigma}, N)$ be its north stop segment, where $e = ((x_1, y), (x_2, y))$. Define $\bar{e} = ((x_1, t - \frac{y}{V_{max}}), (x_2, t - \frac{y}{V_{max}}))$ as the *shadow segment* of e. By Eqs. (1) and (2) and Observation 2, the following can be observed.

Observation 3. *Let $\bar{\sigma}(e, t)$ be a horizontal segment source and $\mathcal{U}(\bar{\sigma}, N) \in E_i$ be its north stop segment. Segment $\bar{e}$ intersects the south shadow range of E_i.*

Recall that a stop segment is defined as the first intersection between the dragging segment e and the obstacles. Thus, $\mathcal{U}(\overline{\sigma}, N)$ is located on a segment E_i whose south shadow range's interior intersects $\overline{e}$ and has minimum weight (i.e., Y value). We now consider two cases: firstly, let $\overline{e}$ be located entirely inside the south shadow range of E_i. By definition, it is easily observed that the north stop point for any source point on $\overline{\sigma}(e, t)$, is located on E_i. Thus, by locating a point on $\overline{e}$ we can find the stop segment for $\overline{\sigma}(e, t)$ (see Lemma 2).

For the second case, assume $\overline{e}$ intersects the boundary of r_i^s. Let R be the set of shadow ranges whose boundaries (horizontal and vertical line segments) intersect $\overline{e}$. By the following lemma, we can report the minimum weighted range $r_i \in R$ whose boundary intersects the source segment $\overline{e}$.

Lemma 4 [17]. *Given a family of n rectilinear line segments L and a query rectilinear line segment s, L can be preprocessed in $O(n \log n)$ time, so that a minimum weight segment in L intersecting s, can be reported in $O(\log n)$ query time.*

Thus, the following lemma follows immediately.

Lemma 5. *After $O(n \log n)$ time preprocessing, the stop segment for a query segment source can be found in $O(\log n)$ time.*

3.3 Range Searching for Minimum

Lemma 6 [3]. *Let H be a dynamic set of points in $\mathbb{R}^2$ where insertions and deletions of the points are allowed. In $O(n \log n)$ time, we can preprocess H into a data structure, so that, for a given query axis-parallel rectangle r, we can determine a minimum weight point inside $r \cap H$ in $O(\log n)$ time. H can be updated in $O(\log n)$ time per insertion/deletion.*

Let V be the set of all vertices (end points of the edges in E) union $\{s, d\}$. Let r be a query rectangle and p be one of its corners. We denote by v_m, a vertex of V located inside r with minimum L_1 distance to p. W.l.o.g., assume p is the bottom-left corner of r. Let B be an axis-parallel rectangular bounding box that contains all vertices in V. We assign a weight to each vertex $v \in V$, denoted by $\omega(v)$, which is the L_1 distance between v and the bottom left corner of B. Observe that v_m is the minimum weight vertex in r. By Lemma 6, there is a data structure [3] that allows finding v_m in $O(\log n)$ query time, after $O(n \log n)$ preprocessing time.

Let $\overline{\sigma}(p, t)$ be a query point source and $\mathcal{U}(\sigma, N)$ and $\mathcal{U}(\sigma, E)$ be its north and east stop points, respectively. We define the **north-east search region** for σ as the rectangle where $\mathcal{U}(\sigma, N)$ and $\mathcal{U}(\sigma, E)$ are its two opposite corners (see Fig. 1 (a) as an example). If $\mathcal{U}(\sigma, N)$ or $\mathcal{U}(\sigma, E)$ does not exist, we say that the north-east search region is *undefined*. Thus, we can define, at most, four search regions corresponding to each query point source. For a horizontal (or vertical) segment source $\overline{\sigma}(e, t)$, we define the **segment search region** of $\overline{\sigma}$ as a vertical (or horizontal) strip of width $|e|$ that entirely contains e. In the next section, we use the search regions to locate the events where the wavelets hit the obstacles.

4   Algorithm

We design a simple data structure to represent the wavelets. For each wavelet $\omega(q, t, r)$, the data structure contains the following information:

- The source q, from which the wavelet is propagated. Recall that, wavelets with inclinations $\{0, \pi/2, \pi, 3\pi/2\}$ originate from segment sources. Conversely, wavelets with inclinations $\{\pi/4, 3\pi/4, 5\pi/4, 7\pi/4\}$ originate from the point sources.
- The corresponding departure time of the wavelet, denoted by t.
- A search region r, which contains potential next intersections between the wavelet and the obstacles. In order to propagate ω, we allow the wavelet to sweep the interior of r and report "hits" by the wavelet.

4.1   Propagation

Propagating a wavelet $\omega(q, t, r)$ means to allow the wavelet to sweep in its designated direction, until it hits a vertex v (or alternatively, the body of an edge). We assume that the minimum arrival time at q has been already calculated. Then, we calculate a potential minimum arrival time at v using a shortest L_1 path from q to v. This may involve deleting, updating, and creating wavelets corresponding to the advancing wavefront. Since the source of a wavelet is either a point or a segment, we present two algorithms to propagate these wavelets.

Before discussing the propagation algorithms, we introduce some notation and give some definitions. Let $\sigma(p, t)$ be a point source and r^{NE} be its corresponding north-east search region. We denote by $\omega(\sigma, t, r^{NE})$ a wavelet that originates at σ and is propagating north-east inside r^{NE}. Similarly, we can define (at most) three more wavelets, in the three directions $\{NW, SE, SE\}$, originating from σ. Denote by $\mathcal{W}^a(\sigma)$ the set of wavelets originating from σ in all (at most) four directions. By Lemma 3, $\mathcal{W}^a(\sigma)$ can be found in $O(\log n)$ time, after $O(n \log n)$ preprocessing time. By Lemma 3, for each wavelet $\omega \in \mathcal{W}^a(\sigma)$, we can find the closest vertex to σ inside its corresponding region in $O(\log n)$ time, we denote this vertex by $\Gamma(\omega)$.

We design an algorithm called *PropagatePoint*(ω) (for details, see the full version of the paper [14]) which propagates a point wavelet $\omega(\sigma, t, r)$ inside its corresponding search region r. There are four types of point wavelets depending on their directions (i.e, NE, NW, SE and SW). W.l.o.g., we assume $\omega(\sigma, t, r)$ is propagating north-east; other directions can be treated analogously. Two types of events are discovered by this algorithm, which are explained next.

Firstly, the algorithm finds all accessible segments of σ, on the boundary of r. For an example, see s_1 and s_2 in Fig. 1(a). Since the vertices are located in general position, there are at most four accessible segments on the boundary of r (one for each edge of r). For each discovered accessible segment, the algorithm adds two types of wavelets to the queue: (1) a segment wavelet whose departure time is the disappearance time of its associated edge and (2) a set of point wavelets originating from the end points of the segment (in Fig. 1(a), these points are

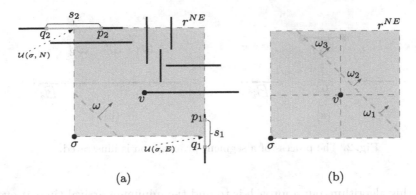

(a)   (b)

Fig. 1. The process of point propagation is illustrated. (a) ω is propagating north-east and v is the first vertex it intersects; (b) when ω intersects v, it is split into three new wavelets ω_1, ω_2 and ω_3.

denoted by q_1, p_1, q_2 and p_2). By Lemmas 5 and 6, we identify these wavelets in $O(\log n)$ time.

Secondly, the algorithm discovers the closest vertex to σ, say v (i.e., $v = \Gamma(\omega)$), as it is the first vertex hit by the wavelet. By Lemma 6, this vertex can be determined in $O(\log n)$ time. When ω hits v, the wavelet is split into three new wavelets. In Fig. 1(b), these wavelets are denoted by ω_1, ω_2 and ω_3. We also create (at most) four new wavelets, corresponding to v and its search regions (i.e., the wavelets in $\mathcal{W}^a(\sigma)$). Since all these operations are executed in $O(\log n)$ time, we can conclude that $PropagatePoint(\omega)$ takes $O(\log n)$ time.

We now describe an algorithm called $PropagateSegment(\omega)$ (see the full version of the paper [14]). This algorithm takes a segment wavelet as input and propagates it inside its corresponding search region. There are four types of segment wavelets depending on their directions (i.e., N, E, W and S). W.l.o.g., assume $\omega(\overline{\sigma}, t, r)$ is propagating north. The algorithm finds the north stop segment for $\overline{\sigma}$, denoted by e'. Note that, by Lemma 5, this can be done in $O(\log n)$ time. When ω hits e', ω is split into smaller wavelets as follows. A new wavelet ω' originating from e', is added to the queue. Then, the algorithm adds smaller wavelets for the parts of ω which do not hit e'. As an example, see ω_1 and ω_2 in Fig. 2(b). Additionally, if a vertex is hit by the segment wavelet, our algorithm adds (at most) four point wavelets originating from the vertex. By Lemma 3, this can be done in $O(\log n)$ time. Thus, $PropagateSegment(\omega)$ runs in $O(\log n)$ time.

4.2   A Naive Algorithm

In this section, we present a "naive" algorithm (see the full version of the paper [14]), which reports the minimum arrival time at the destination in $O(n^2 \log n)$ time. Although the algorithm is not efficient, it illustrates our global approach and serves as the basis for our optimal algorithm describe later in Sect. 5.

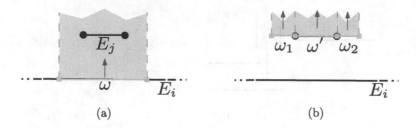

Fig. 2. The process of a segment propagation is illustrated.

In this algorithm, our approach is to find the minimum arrival time at every vertex in V from the source. To achieve this, a set of wavelets is created and maintained in a priority queue, whose keys are their t values (i.e., their corresponding departure times). The queue is initialized with four initial point wavelets originating from the start point s in four directions NE, NW, SE and SW. In each iteration, the algorithm proceeds by extracting a wavelet from the queue with lowest value of t. If the wavelet originates from the destination point, the minimum arrival time at the destination has been found. Otherwise, depending on whether ω is a point wavelet or a segment wavelet, $PropagatePoint(\omega)$ or $PropagateSegment(\omega)$ is executed, respectively. Recall that, these algorithms propagate the given wavelet and add, or update, the wavelets in the queue, if necessary. In the following lemma, we prove that the naive algorithm correctly finds the minimum arrival at the destination.

Lemma 7. *The naive algorithm reports the minimum arrival time at the destination.*

Lemma 8. *The naive algorithm runs in $O(n^2 \log n)$ time.*

Proof. We estimate the total number of calls to the functions $PropagatePoint(\omega)$ and $PropagateSegment(\omega)$. More precisely, we need to bound the total number of wavelets created in the process. Fujimora [7] proved that at any given time, the size of the wavefront (i.e., the number of wavelets in the priority queue) is $O(n)$. Thus, each edge may be hit by $O(n)$ wavelets and consequently generate $O(n)$ new wavelets. This means that the total number of wavelets is bounded by $O(n^2)$. Recall that, each propagation can be executed in optimal $O(\log n)$ time. Therefore, the naive algorithm runs in $O(n^2 \log n)$ time. □

5   An Improved Algorithm

As we proved in Lemma 8, there may be up to $O(n)$ wavelets originating from a single edge. In Sect. 5.1, we will utilize a method called "expanding", to reduce the total number of segment propagations. Note though that the queue may contain wavelets with overlapping search regions. Thus, each vertex may be hit by $O(n)$ wavelets (the maximum size of the queue at any given time). To prevent this from happening, in Sect. 5.2, we propose a procedure called "Narrowing", which shrinks the overlapping search regions, so that they do not sweep the same area.

5.1 Wavelet Expanding

Let $\mathcal{W}_i$ be a maximal set of wavelets, originating from the body of E_i. Recall that, the naive algorithm propagates every wavelet in $\mathcal{W}_i$ individually. In Lemma 8, we proved that the number of these wavelets in the priority queue will be quadratic in the worst case. In this section, we propose an alternative approach in which, we replace all wavelets in $\mathcal{W}_i$ by a single "expanded" wavelet. This wavelet is a segment wavelet whose source is the body of the edge E_i. We will prove that this replacement of wavelets permits avoiding the quadratic number of propagations. The crucial property that we are employing is the following:

Observation 4. *The wavelets in $\mathcal{W}_i$ simulate the robot's motions when: (1) it arrives at the body of the edge E_i in its existence time interval and (2) departs from the edge, at time T_i^d. In other words, the departure time of all wavelets in $\mathcal{W}_i$ is the disappearance time of E_i.*

W.l.o.g, we assume E_i is a horizontal edge. Let $\omega(\sigma{=}(v,t),t,r)$ be a wavelet originating from a vertex v, propagating north-east. Suppose ω hits the interior of edge E_i, i.e., $\mathcal{U}(\sigma,N) \in E_i$. Thus, $PropagatePoint(\omega)$ creates some wavelet(s) originating from the body of E_i (for details of the Algorithm see the full version of the paper). The same process will be repeated for the newly generated wavelets, at a later time. Notice that the sequence at which these wavelets are created is sorted by their departure times. Thus, for each wavelet ω' originating from the body of an edge, we can find a sequence of wavelets, starting with the wavelet ω originating from a vertex, which led to the creation ω'. We say σ is the *root point source* of ω'. Since E_i is horizontal, we observe the following property of the root point sources.

Property 1. Let $\omega(\overline{\sigma},t,r) \in \mathcal{W}_i$ be a segment wavelet where $\overline{\sigma}((X_1,Y),(X_2,Y))$. Then, there is an associated root point source $\sigma{=}(v,t)$ for which: (a) $v{=}(X_v,Y_v)$ is a vertex in V and (b) $X_1 \leq X_v \leq X_2$, i.e., σ is located south of $\overline{\sigma}$.

We store the root point sources of the wavelets of $\mathcal{W}_i$ in a binary search tree (BST) B_i. If E_i is horizontal, B_i is sorted by the x-coordinates of the point sources; and if E_i is vertical, B_i is sorted by their y-coordinates. Next, we use a method, called *Expand*, as follows: remove the wavelets of $\mathcal{W}_i$ from the queue and replace them with an expanded wavelet $\omega_i((E_i,T_i^d),T_i^d,r_i)$, where r_i is the segment search region of E_i. The details of this procedure is presented in the full version of the paper [14]. In this method, we update the binary trees as follows: when ω_i hits its north stop segment on the edge E_j, we construct B_j by applying the appropriate *Split* and *Merge* operations on B_i and B_j. We define these two functions as follows.

- $Split(T, x) : BST \times \mathbb{R} \to BST \times BST$. Given a BST T and a key value x, $split$ divides T into two BSTs T_l and T_r, where T_l consists of all point sources in T with x-coordinates less than x; and T_r includes the rest of the point sources.
- $Merge(T_l, T_r) : BST \times BST \to BST$. Let T_l and T_r be two BSTs, where there exist a value x such that the point sources in T_l have lower (or equal) x-coordinates than x and the point sources in T_r have greater (or equal) x-coordinates than x. Function $Merge$ creates a new BST which is the union of T_l and T_r.

In the expanding algorithm (i.e., $Expand(\omega)$ in the full version of the paper), the input is a wavelet ω originating from the body of an edge E_i. The algorithm proceeds by initializing an empty BST B_i. Next, for any wavelet ω' originating from the body of E_i, it first removes ω' from the queue. Then, it considers two cases: (1) if ω' is a non-expanded wavelet, it inserts the root point source of ω' into B_i and (2) if ω' is an expanded wavelet of edge $E_j = ((X_l, Y), (X_r, Y))$, the algorithm first splits B_j into two sub-trees $T_l, T_r = Split(B_j, X_r)$. Then, it splits T_l again, such that $T_l', T_r' = Split(T_l, X_l)$. At this point, T_r' represents the point sources in B_j with x-coordinates between X_l and X_r. The algorithm merges T_r' with B_i using $B_i = Merge(T_r', B_i)$. The rest of the point sources in T_r and T_l' are maintained in B_j using $B_j = Merge(T_r, T_l')$.

Since updating the binary search trees using the basic operations of $merge$ and $split$, can be done in $O(\log n)$ time, each iteration in the main loop of the expanding algorithm runs in $O(\log n)$ time.

Now, we modify the naive algorithm described in Sect. 4.2 so that it uses the "expanding" method. Let ω be a wavelet extracted from the priority queue. If ω is originating from the body of E_i, we execute the expanding algorithm (see the full version of the paper for details) to replace the wavelets of $\mathcal{W}_i$ with an expanded wavelet. More precisely, if ω is originating from the body of E_i, we execute $Expand(\omega)$ in the naive algorithm. We call this new algorithm $Expanding$ $algorithm$.

W.l.o.g., assume the expanded wavelet ω_i is propagating north and $v=(X_v, Y_v)$ is the first vertex that it hits. Although $PropagateSegment(\omega_i)$ identifies v, it may not report the arrival time at v. This is due to the fact that the wavelets on the body of E_i have been replaced by a single wavelet ω_i. Thus, we need an alternative approach to calculate the minimum arrival time at v. Suppose ω_{min} is the first wavelet in $\mathcal{W}_i$ that hits v (see Fig. 3). In the remaining of this section, we show how to determine the minimum arrival time at v, without explicitly calculating the wavelet ω_{min}. Recall that, the minimum L_1 distance from u to v is denoted by $\|uv\|_1$, and the minimum L_1 distance between E_i and v is denoted by $\|E_i v\|_1$.

Lemma 9. *Let $\sigma(u, t) \in B_i$. If the robot departs from u at time t, the minimum arrival time at v is $T_v(u) = max\left(T_i^d, t + \frac{\|uv\|_1 - \|E_i v\|_1}{v_{max}}\right) + \frac{\|E_i v\|_1}{v_{max}}$.*

Let $\sigma_l = (v_l, t_l)$ be the point source in B_i, where v_l is a vertex with the largest X-coordinate smaller than X_v. For an example, see Fig. 3. Analogously,

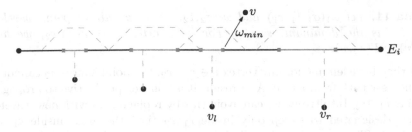

Fig. 3. The segment sources and point sources of an edge E_i are illustrated by red segments and red squares, respectively. Among the wavelets originating from these sources (i.e., the wavelets in $\mathcal{W}_i$), ω_{min} is the first wavelet that hits vertex v.

let $\sigma_r = (v_r, t_r)$, where v_r has the smallest X-coordinate greater than X_v. Using these notations, we state the following lemma whose proof is given in the full version of the paper [14].

Lemma 10. *Among the point sources in B_i, either point source σ_r or σ_l (defined above) has the minimum arrival time at v.*

Finally, by Lemmas 9 and 10, when a vertex v is discovered by an expanded wavelet, we can find the minimum arrival time at v using $t_v = min(T_v(v_r), T_v(v_l))$. Since the calculation of t_v is solely based on E_i, u and v, it is not required to calculate the wavelets in $\mathcal{W}_i$. Therefore, the following corollary is obtained.

Corollary 1. *The Expanding algorithm calculates the minimum arrival time at the destination.*

5.2 Wavelet Narrowing

In the previous section, we described a technique to reduce the number of wavelets originating from the edges by a factor of n. Here, we need to address another challenge: reducing the total number of vertex-originated wavelets. As mentioned before, the search regions may overlap and hence, a vertex may create $O(n)$ point wavelets. To prevent this from happening, in this section, we propose a procedure called "Narrowing" the wavelets.

Let $\omega_1(\sigma_1, t_1, r_1)$ and $\omega_2(\sigma_2, t_2, r_2)$ be a pair of point wavelets whose search regions intersect (i.e., $r_1 \cap r_2 \neq \emptyset$). W.l.o.g., assume that ω_1 and ω_2 are propagating toward the north-east. We denote by p, the bottom left corner of $r_1 \cap r_2$ (see Fig. 4). Recall that $T(\sigma, p)$ is the minimum arrival time at p from σ. Let $T_1 = t_1 + T(\sigma_1, p)$ and $T_2 = t_2 + T(\sigma_2, p)$ be the minimum arrival times at p from σ_1 and σ_2, respectively. The wavelet which arrives at p first is called the *dominant wavelet*. W.l.o.g., assume $T_1 < T_2$ and thus, ω_1 is dominant. Now, let $v \in r_1 \cap r_2$ be a vertex located inside the intersection of the two search regions. Note that, ω_1 hits v at time $T(\sigma_1, v) = T(\sigma_1, p) + \frac{\|pv\|_1}{V_{max}}$; and ω_2 hits v at time $T(\sigma_2, v) = T(\sigma_2, p) + \frac{\|pv\|_1}{V_{max}}$. Since $T(\sigma_1, p) < T(\sigma_2, p)$, we obtain the following:

Lemma 11. *Let $\omega_1(\sigma_1, t_1, r_1)$ and $\omega_2(\sigma_2, t_2, r_2)$ be a pair of point wavelets, where ω_1 is the dominant wavelet. For any vertex $v \in r_1 \cap r_2$, we have $T(\sigma_1, v) < T(\sigma_2, v)$.*

By the above lemma, for any vertex $v \in r_1 \cap r_2$, the point source σ_2 cannot be on a shortest path from s to v. As a result, it is counter-productive to propagate ω_2 inside $r_1 \cap r_2$. Intuitively, we can avoid this by replacing ω_2 with new wavelets that are designated to sweep only inside $r_1 \backslash r_2$ (i.e., the areas inside r_1 and outside r_2). This procedure is called *Narrowing*. Since the underlying search regions are axis-parallel rectangles, we can narrow a wavelet by replacing its search region by at most four smaller rectangles. As an illustration, in Fig. 4, ω_2 is replaced with two new wavelets ω_2' and ω_3' with smaller search regions. The details of this procedure are presented in the full version of the paper [14]. Our approach for reducing the number of point wavelets is based on identifying the dominant wavelets in each iteration. One greedy approach is to compare all pairs of the wavelets and narrow the non-dominant ones. However, this may result in quadratic number of narrowings. Our alternative approach is to execute the narrow procedure for every pair of wavelets originating from two vertices v_1 and v_2, when: (1) there exists two wavelet ω_1 and ω_2 originating from v_1 and v_2, respectively; and (2) either ω_1 hits v_2, or ω_1 and ω_2 hit the same vertex v_3. This modification of the Expanding algorithm results in a new algorithm which we call the *Narrowing algorithm*. The following corollary is a direct consequence of Lemma 11.

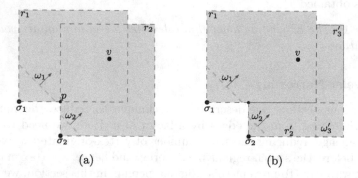

(a)   (b)

Fig. 4. An example of narrowing a wavelet; (a) the minimum arrival time at p is identified from two point sources σ_1 and σ_2. The path from σ_1 is faster, so ω_1 is the dominant wavelet; (b) ω_2 is replaced with two wavelets ω_2' and ω_3' with smaller search regions.

Corollary 2. *The Narrowing algorithm calculates the minimum arrival time at the destination.*

In order to prove that the Narrowing algorithm runs in $O(n \log n)$ time, we first establish a linear bound on the number of point wavelets created in the process.

Lemma 12. *The total number of point wavelets created by the Narrowing algorithm is $O(n)$.*

Proof Sketch. Intuitively, we prove the following: Let ω_1 and ω_2 be two northeast propagating wavelets, originating from u_1 and u_2, respectively. Assume ω_1 and ω_2 both hit a vertex v. If u_2 lies within a $\|u_1 v\|_1$ distance from u_1 (in any direction), then either ω_1 or ω_2 would be narrowed before hitting v, making it impossible to have encountered v. So, there are at most two wavelets hitting v from south-west. Therefore each vertex is hit by a constant number of wavelets in total. See the full version of the paper [14] for the full proof.   □

By the above lemma, there are $O(n)$ point wavelets in the queue. Furthermore, in Sect. 5.1, we proved that the total number of segment wavelets is also $O(n)$. Thus, the running time of the Narrowing algorithm is $O(n \log n)$. Finally, by recording the sequence of the propagations during the process, we can actually construct the time-minimal path among the transient obstacles. Note that, similarly to the optimality argument for the existing $\Theta(n \log n)$ time algorithm [18] for the non-transient obstacles (which is a special case of our problem), our algorithm is also optimal.

Theorem 1. *A time-minimal rectilinear path among transient rectilinear segments can be found in $\Theta(n \log n)$ time.*

As the algorithm proceeds, by recording the trace of the endpoints of the wavelets, we can build a subdivision of the plane. Since the size of this subdivision is proportional to n, by [10], we can construct a data structure to answer point location queries in $O(\log n)$ time. Thus, we can build the shortest path map with respect to a fixed source point in $O(n \log n)$ time. Now, for a given query point q, the minimum arrival time at q from s, can be reported in $O(\log n)$ time.

Theorem 2. *Given a set of n transient edges E, a fixed source point s and a query point q, E can be preprocessed in $O(n \log n)$ time, so that the minimum arrival time at q can be reported in $O(\log n)$ query time.*

References

1. Agarwal, P.K., Arge, L., Yi, K.: An optimal dynamic interval stabbing-max data structure? In: SODA 2005, pp. 803–812 (2005)
2. Agarwal, P.K., Kumar, N., Sintos, S., Suri, S.: Computing shortest paths in the plane with removable obstacles. In SWAT 2018, pp. 5:1–5:15 (2018)
3. Chazelle, B.: Functional approach to data structures and its use in multidimensional searching. SIAM J. Comput. **17**(3), 427–462 (1988)
4. de Rezende, P.J., Lee, D.T., Wu, Y.F.: Rectilinear shortest paths with rectangular barriers. In: SCG 1985, pp. 204–213. ACM, New York (1985)
5. Fujimura, K.: On motion planning amidst transient obstacles. In: ICRA 1992, vol. 2, pp. 1488–1493, May 1992
6. Fujimura, K.: Motion planning using transient pixel representations. In: ICRA 1993, vol. 2, pp. 34–39, May 1993

7. Fujimura, K.: Motion planning amid transient obstacles. Int. J. Robot. Res. **13**(5), 395–407 (1994)
8. Hershberger, J., Kumar, N., Suri, S.: Shortest paths in the plane with obstacle violations. In: ESA 2017, pp. 49:1–49:14 (2017)
9. Hershberger, J., Suri, S.: An optimal algorithm for euclidean shortest paths in the plane. SIAM J. Comput. **28**(6), 2215–2256 (1999)
10. Kirkpatrick, D.: Optimal search in planar subdivisions. SIAM J. Comput. **12**(1), 28–35 (1983)
11. LaValle, S.M.: Planning Algorithms, pp. 495–558. Cambridge University Press, Cambridge (2006)
12. LaValle, S.M., Sharma, R.: Robot motion planning in a changing, partially predictable environment. In: ISIC 1994, pp. 261–266 (1994)
13. Lee, D.T., Yang, C.D., Wong, C.K.: Rectilinear paths among rectilinear obstacles. Discret. Appl. Math. **70**(3), 185–215 (1996)
14. Maheshwari, A., Nouri, A., Sack, J.R.: Rectilinear shortest paths among transient obstacles. CoRR arXiv: abs/1809.08898 (2018)
15. Mitchell, J.S.: L1 shortest paths among polygonal obstacles in the plane. Algorithmica **8**(1–6), 55–88 (1992)
16. Mitchell, J.S.B.: Shortest paths among obstacles in the plane. In: SCG 1993, pp. 308–317. ACM, New York (1993)
17. Vaishnavi, V.K., Wood, D.: Rectilinear line segment intersection, layered segment trees, and dynamization. J. Algorithms **3**(2), 160–176 (1982)
18. Yang, C., Lee, D., Wong, C.: Rectilinear path problems among rectilinear obstacles revisited. SIAM J. Comput. **24**(3), 457–472 (1995)

An Efficient Algorithm for Enumerating Induced Subgraphs with Bounded Degeneracy

Kunihiro Wasa[✉][iD] and Takeaki Uno

National Institute of Informatics, Tokyo, Japan
{wasa,uno}@nii.ac.jp

Abstract. We propose a polynomial delay and polynomial space algorithm for the enumeration of k-degenerate induced subgraphs in a given graph. A graph G is k-degenerate if each of its induced subgraphs has a vertex of degree at most k. The degeneracy is considered as an indicator of the sparseness of the graph. Real-world graphs such as road networks, social networks and internet networks often have small degeneracy. Compared to other kinds of graph classes, bounded degeneracy does not give many structural properties such as induced subgraph free, or minor free. From this, using bounded degeneracy to reduce the time complexity is often not trivial. In this paper, we investigate ways of handling the degeneracy and propose an efficient algorithm for the k-degenerate induced subgraph enumeration. The time complexity is $\mathcal{O}\left(\min\left\{\Delta + kk', (k')^2\right\}\right)$ time per solution with polynomial preprocessing time and the space complexity is linear in the input graph size, where Δ and k' are the maximum degree and the degeneracy of the input graph.

Keywords: Graph algorithms · Enumeration algorithms
Polynomial delay · k-degenerate graphs

1 Introduction

The *subgraph enumeration* is to output all the subgraphs of the given graph satisfying a certain structural condition such as being a tree and the density is no less than a threshold. It is one of the fundamental problems widely studied in the theoretical computer science for more than 40 years ([12] gives the overview of this area). The complexity analysis of subgraph enumeration algorithms has two main streams. One is of a usual style, that is evaluating the time complexity in the input size. Generally speaking, the number of solutions is exponential in the input size, thus the studies are done to reduce the constant factor c of the time complexity of $\mathcal{O}(c^n)$. This has an advantage that at the same time we can often obtain a combinatorial result of bounding the number of subgraphs. On the other hand, considering the practice, enumeration problems have less

---

This work was supported by JST CREST, Grant Number JPMJCR1401, Japan.

D. Kim et al. (Eds.): COCOA 2018, LNCS 11346, pp. 35–45, 2018.
https://doi.org/10.1007/978-3-030-04651-4_3

solutions, say polynomially many, thus evaluation only by the input size is too much overestimating. In such cases, output polynomial time is considered. The time complexity is evaluated by the input size n and the output size N that is the number of solutions. An algorithm is called an *output polynomial time algorithm* if it terminates in a time polynomial in the input size and the output size. Further, we say it runs in $poly(n)$ time for each solution if an algorithm runs in $\mathcal{O}(poly(n)N)$ time plus polynomial time preprocessing. Further, we say the algorithm runs in *polynomial delay* time if the computation time between any two consecutive output solutions is polynomial in the input size.

A graph G is said to be k-*degenerate* [10] if any of its induced subgraphs has at least one vertex of degree at most k. The *degeneracy* of G is the smallest k such that G is k-degenerate. Intuitively speaking, if a graph has a small degeneracy, then the graph is relatively sparse. It is said that real world graphs often have small degeneracies [6,8]. Particularly, some graph classes have constant degeneracies, e.g., trees, grid graphs, outer planar graphs, planar graphs, bounded treewidth graphs, and H-minor-free graphs for fixed H. Bounded-size degenerate graphs have received much attention. However, bounded degeneracy does not give many structural properties such as minor free, compared to other graph classes such as chordal graphs. Thus, despite its importance, there are not so many studies on algorithms that utilize degeneracy. In this paper, we address the problem of enumerating all induced k-degenerate subgraphs. We investigate efficient search strategies that have characterization of irredundant moves, and the way of checking the degeneracy in short time. These yield an efficient enumeration algorithm. There have been several studies on enumeration problems in a bounded degenerate graph [4,6,9], for cliques, dominating sets, induced trees, etc. To the best of our knowledge, there has been little research on the enumeration of k-degenerate subgraphs, or k-degenerate induced subgraphs. The case of $k = 1$ corresponds to the forest enumeration, thus there have been many studies, especially its connected version, that is, trees. Ferreira *et al.* proposed an enumeration algorithm for subtrees of size exactly h in an undirected graph [7]. Wasa *et al.* improved their result to optimal [14], that is, their algorithm runs in $\mathcal{O}(1)$ delay, when the input graphs are trees. For the induced version, the authors proposed an algorithm [13] that enumerates all the vertex subsets that induce a tree. Conte *et al.* proposed an enumeration algorithm for maximal k-degenerate induced subgraphs in a chordal graph that runs in polynomial delay and space [5]. On the other hand, Bauer *et al.* [2] studied the enumeration of all k-degenerate induced graphs having n vertices and m edges.

We propose an efficient algorithm for enumerating k-degenerate induced subgraphs in a given graph $G = (V, E)$. Our algorithm runs in $\mathcal{O}(\min\{\Delta + kk', (k')^2\})$ time per solution and uses $\mathcal{O}(|V| + |E|)$ space, where Δ and k' are the maximum degree and the degeneracy of an input graph. Note that the algorithm outputs also disconnected subgraphs. In the enumeration of k-degenerate subgraphs, the time consuming part is the computation of the degeneracy of a newly generated subgraph. It takes $\mathcal{O}(|V| + |E|)$ time with straightforward ways. Further, if we follow the usual binary partition algorithm, an iteration has to have up to $\mathcal{O}\left(|V|\right)$ trials for finding a vertex whose addition to the current solution yields a k-degenerate induced subgraph. Thus, an iteration has to spend $\mathcal{O}\left(|V|(|V|+|E|)\right)$ time or more in this way. Instead of that, we developed a *reverse search* algorithm [1]. We define a *parent-child relationship* that defines a tree shaped traversal route spanning all k-degenerate induced subgraphs. By generating the child solutions of the current visiting solution recursively, we can traverse the tree in a depth-first manner whilst using polynomial time and polynomial space. We further developed a data structure that enables us to check the degeneracy in short time, that can be updated quickly along the movement on the traversal tree.

The organization of this paper is as follows: In Sect. 2, we give the basic notations and terminologies that are used in this paper. Section 3 shows the reverse search strategy, and Sect. 4 describes the data structure for checking the degeneracy. In Sect. 5, we conclude this paper.

2   Preliminaries

Let $G = (V, E)$ be an undirected graph with vertex set $V = \{1, \ldots, n\}$ and edge set $E \subseteq V \times V$. Let $u, v \in V$ be two distinct vertices in G. Vertices u and v are mutually *adjacent* if $(u, v) \in E$. $N(u)$ denotes the set of vertices that are adjacent to u. We call $d(u) = |N(u)|$ the *degree* of u. Let $\Delta(G) = \max_{u \in V} d(u)$ and $\delta(G) = \min_{u \in V} d(u)$. For a vertex set S, let $d_S(u) = |N(u) \cap S|$ be the number of neighbors of u that are in S. For any edge $e = (u, v) \in E$, we say that u and v are the *endpoints* of e. G is *connected* if any pair of vertices u and v, there is a path between them. For any vertex subset S of V, the subgraph induced by S is the graph whose vertex set is S and edge set is the edges connecting two vertices in S, i.e., $\{(u, v) \in E \mid u, v \in S\}$. For conciseness, for a set S and an element v, we denote $S \cup \{v\}$ by $S \cup v$ and $S \setminus \{v\}$ by $S \setminus v$. In what follows, we assume that G is connected and G has no self loops and multi edges.

2.1 k-degenerate Graphs

A graph is k-*degenerate* [10] if any of its induced subgraphs has a vertex of degree at most k, and the *degeneracy* of G is the smallest k such that G is k-

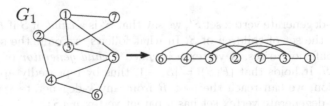

Fig. 1. G_1 is a 2-degenerate graph. A degeneracy ordering of G_1 is shown in the right part of the figure. In the figure, the leftmost vertex 6 is the smallest and the rightmost vertex 1 is the largest. For any vertex v in G_1, the number of larger adjacent vertices of v is at two.

degenerate. We call a vertex set $S \subseteq V$ a *k-degenerate vertex set* if $G[S]$ is an induced subgraph whose degeneracy is k. It is known that G is k-degenerate if and only if it admits a vertex ordering $>^*$, called a *degeneracy ordering*, such that for any vertex v in G, $|\{u \in N(v) \mid u >^* v\}| \leq k$ (See Fig. 1). That is, for each vertex, the number of larger neighbors is at most k. A degeneracy ordering of a graph can be obtained in linear time in the size of a graph [11], by recursively removing the smallest degree vertex from the graph. It is also known that graphs in some graph classes have a constant degeneracy. For example, the degeneracy of trees, grid graphs, outerplanar graphs [3], and planar graphs are respectively at most 1, 2, 2, and 5 [10]. In what follows, we assume the vertices of G are labeled according to some degeneracy ordering of G, and we write $>$ for $>^*$ if no confusion can arise.

We here describe the k-degenerate induced subgraphs enumeration problem.

Problem 1. Given a graph $G = (V, E)$ and positive integer k, enumerate all k-degenerate vertex subsets of G.

Note that those vertex sets may induce disconnected graphs, and are allowed to be solutions.

3   Reverse Search Algorithm

Our algorithm enumerates the solutions by traversing on a tree structure on the solution space, called *a family tree*, in a DFS manner. The basic idea of this strategy is proposed by Avis and Fukuda [1]. In this section, we give an algorithm for k-degenerate induced subgraphs. We first define the tree structure mentioned above. Let $R = (\emptyset, \emptyset)$ be the empty graph, that is a k-degenerate graph. We call R the *root*. The *parent vertex* $pv(S)$ of a k-degenerate vertex set S is defined as the smallest vertex v in S such that v is adjacent to at most k vertices of S. We define the parent of k-degenerate induced subgraphs of G as follows:

Definition 1 (The parent). Let G be a graph and k be a positive integer. We define the *parent* $P(S)$ of a non-empty k-degenerate vertex set $S \subseteq V$ as $S \setminus pv(S)$.

For any k-degenerate vertex set S', we say that S' is a *child* of S if $P(S') = S$. Let $\mathcal{C}_k(S)$ be the set of children of S. In what follows, we omit the subscript k of $\mathcal{C}_k$ if no confusion arises. A vertex u is called a *child generator* of S if $S \cup u$ is a child of S. It holds that $|P(S)| = |S| - 1$, thus by repeatedly applying the parent function, we can reach the root R from any k-degenerate vertex set S since every k-degenerate vertex set has a parent vertex $pv(S)$.

The *family tree* is a tree whose node set is all the k-degenerate vertex sets of G, and an edge connect two nodes if one is the parent of the other. From the above, we can see that the family tree contains no cycle and spans all k-degenerate vertex subsets, thereby is a tree rooted at R. The algorithm traverses

Algorithm 1. Reverse search algorithm

```
1 Procedure Main(G = (V, E), k)
2    R ← (∅, ∅);
3    Rec(G, R, k);
4 Subprocedure Rec(G, S, k)
5    Output S;
6    foreach S' ∈ C(S) do
7        Rec(G, S', k);
```

the tree by recursively moving to the children of the current visiting k-degenerate vertex subset. We can then see the correctness of the algorithm; the algorithm completely outputs the k-degenerate vertex sets without duplication. The algorithm is described in Algorithm 1.

4   Generating Children

The bottle neck part of Algorithm 1 is the computation of the children a current solution in line 6. A naïve way to generate all the children of a k-degenerate vertex set S is as follows: For every vertex $u \notin S$, check whether $S \cup u$ is k-degenerate or not, and $P(S \cup u)$ is S or not; if the answers of the both checking are yes, then add u to the set of child generators. This actually needs $\mathcal{O}\left(|V|(|V| + |E|)\right)$ time. In this section, we propose an efficient method that avoids the above trial-and-error approach. We first consider the sufficient and necessary condition for a vertex u to be a child generator of S. Let k' be the degeneracy of G. In what follows, we assume $k < k'$. Otherwise, our problem can be solved in $\mathcal{O}(1)$ time per solution outputting all subgraphs in G. Let $N_>(u)$ be the set of vertices that are adjacent to u and larger than u. We assume that the graph is stored in the memory by the adjacency lists of its vertices, and the adjacency list is sorted in the degeneracy ordering. $N_>(u)$ is also stored in the memory and sorted in degeneracy ordering.

4.1   Characterization of a Child Generator

We say that v is *black* if $d_S(v) > k$, v is *gray* if $d_S(v) = k$, and v is *white* otherwise. We also say that v is *non-black* if v is not black, that is, v is gray or white. Note that $pv(S)$ is the smallest non-black vertex in S. Let $sw_*(S)$ be the smallest white vertex in S, and $GS(S)$ be the set of the gray vertices in S that are smaller than $sw_*(S)$. Let $h_S(u)$ be the number of vertices in $GS(S)$ that are smaller than u, and $h'_S(u)$ be the number of vertices in $GS(S) \cap N(u)$ that are smaller than u. A vertex u is a child generator of S if and only if $u = pv(S \cup u)$ holds. This condition is characterized as follows.

Lemma 1. *A vertex $u \notin S$ is a child generator of a k-degenerate vertex subset S, i.e., $u = pv(S \cup u)$ if and only if u is non-black in $S \cup \{u\}$ and either one of the followings holds.*

(1) $u < pv(S)$

(2) $pv(S) < u < sw_*(S)$ *and u is connected to all vertices in* $\{v \in GS(S) \mid v < u\}$, *i.e.,* $h_S(u) = h'_S(u)$.

Proof. Suppose that S is a k-degenerate vertex subset and $u \notin S$ is a non-black vertex. Note that $S \cup u$ is k-degenerate. We consider the following three cases.

Case (A): Suppose that (1) holds. We observe that (i) any black vertex in S is also black in $S \cup u$, and (ii) $u < u'$ holds for any non-black vertex $u' \in S$ because $u < pv(S)$. It implies that $u = pv(S \cup \{u\})$.

Case (B): Suppose that (2) holds. Any non-black vertex u' of $S \cup u$ is either gray or white in S. If u' is white in S, then $u < u'$ holds from the condition of (2). If u' is gray in S, then u' is not connected to u. Thus, from the condition of (2), $u < u'$ holds. Therefore, $u < u'$ always holds, and $u = pv(S \cup \{u\})$.

Case (C): Both (1) and (2) do not hold implies that (i) $u > pv(S)$ and (ii) $u > sw_*(S)$ or u is not connected to a gray vertex $v \in GS(S)$ such that $v < u$. If $u > sw_*(S)$ holds, then $pv(S \cup \{u\}) \leq sw_*(S) < u$ holds since $sw_*(S)$ increases its degree by at most one in $S \cup u$, and thereby it is non-black. If u is not connected to a gray vertex $v \in S$ smaller than u, then $pv(S \cup \{u\}) \leq v < u$ holds since v is a non-black vertex of $S \cup \{u\}$. Hence, the statement holds. □

4.2   Data Structure for Finding Child Generators

Lemma 1 shows that we can find child generators by finding the non-black vertices satisfying the condition (1) or (2). Let $\mathcal{L}(S)$ (resp., $\mathcal{L}'(S)$) be the list of all the non-black vertices not in S (resp., in S) that are sorted by the degeneracy ordering. The non-black vertices satisfying (1) are efficiently found by tracing $\mathcal{L}(S)$ from the head. All vertices u in $\mathcal{L}(S)$ that satisfy $u < pv(S)$ are actually child generators of S. Hence, it takes $\mathcal{O}(1)$ time for each child generator. From above discussion, we can immediately obtain the following lema:

Lemma 2. *For any k-degenerate vertex subset S, the child generators smaller than $pv(S)$ are found in $\mathcal{O}(|\mathcal{C}(S)| + 1)$ time by using $\mathcal{L}(S)$.*

For finding those satisfying (2), we construct the list A composed of the first k vertices of $GS(S)$ by tracing the first k elements of $\mathcal{L}'(S)$. By using A, we compute $h_S(u)$ and $h'_S(u)$. We observe that any vertex larger than the $(k+1)$st vertex of $GS(S)$ cannot satisfy (2) since to satisfy (2) the vertex has to be adjacent to at least $k + 1$ vertices of S that means the vertex is not a child generator. We then trace $N_>(v)$ for all vertices v in A. While tracing these larger neighbors, for all such neighbors u, we compute $h'_S(u)$. We first initialize $h'_S(u)$ to zero, and then increase the value by one while tracing. Since the total number of the neighbors is at most kk', this takes $\mathcal{O}(kk')$ time. For any $v \in A$, the computation of $h'_S(u)$ for all vertices u in $N_>(v)$ can be done in $\mathcal{O}(kk')$ time in total by tracing $N_>(v)$ and A, simultaneously. Thus, the computation of $h_S(u)$ and $h'_S(u)$ for all vertices u that are adjacent to at least one vertex of A is done in $\mathcal{O}(kk')$ time. Therefore, child generators satisfying (2) can be found in $\mathcal{O}(kk')$ time.

Lemma 3. *For any k-degenerate vertex subset S, the child generators larger than $pv(S)$ are found in $\mathcal{O}\,(kk')$ time by using $\mathcal{L}'(S)$.*

4.3 Efficiently Updating the Data Structure

The key to efficient computation in a recursive call, called an *iteration*, is the efficiency of the update process of the data structure described above. Suppose that u is a child generator of S and we are going to compute $\mathcal{L}(S \cup u)$ and $\mathcal{L}'(S \cup u)$ from $\mathcal{L}(S)$ and $\mathcal{L}'(S)$.

Lemma 4. *For any k-degenerate vertex subset S, $\mathcal{L}'(S \cup \{u\})$ is obtained from $\mathcal{L}'(S)$ in $\mathcal{O}\,(k)$ time, where $S \cup \{u\}$ is a child of S.*

Proof. Let denote by $S' = S \cup \{u\}$. To obtain $\mathcal{L}'(S')$, we compute $d_{S'}(v)$ from $d_S(v)$ for each v in $\mathcal{L}'(S)$ that is adjacent to u. Since $pv(S') = u$, all vertices in $\mathcal{L}'(S') \setminus \{u\}$ is larger than u. Hence, if a vertex v in $\mathcal{L}'(S)$ is still in $\mathcal{L}'(S')$, $u < v$ and $d_{S'}(v) \le k$. Thus, we first remove all vertices v from $\mathcal{L}'(S)$ that satisfies $d_{S \cup u}(v) > k$. Since the number of such vertices v is at most k, this needs $\mathcal{O}\,(k)$ time. Finally, we insert u to the head of $\mathcal{L}'(S)$, and then we obtain $\mathcal{L}'(S \cup u)$. Hence, the statement holds. □

The computation of $\mathcal{L}(S \cup u)$ is done in the same way as the above, in $\mathcal{O}\,(\Delta)$ time.

Lemma 5. *For any k-degenerate vertex subset S, $\mathcal{L}(S \cup u)$ is obtained from $\mathcal{L}(S)$ in $\mathcal{O}\,(\Delta)$ time.*

From Lemmas 4 and 5, the computation of $\mathcal{L}'(S)$ and $\mathcal{L}(S)$ needs $\mathcal{O}\,(\Delta)$ time since $k < \Delta$. By recording the operations of these update, we can easily restore $\mathcal{L}'(S')$ and $\mathcal{L}(S')$ from $\mathcal{L}'(S)$ and $\mathcal{L}(S)$ in $\mathcal{O}\,(\Delta)$ time.

When Δ is large and k' is small, the following algorithm for updating $\mathcal{L}(S)$ is more efficient. The algorithm deals with the former and latter parts of $\mathcal{L}(S \cup u)$, where the former part is of vertices smaller than u and the latter part is of the other.

Lemma 6. *For any k-degenerate vertex subset S, the part of $\mathcal{L}(S \cup u)$ composed of vertices larger than u is obtained from $\mathcal{L}(S)$ in $\mathcal{O}\,(k')$ time.*

Proof. Removed vertices from the latter part are adjacent to u. Since G is k'-degenerate, the number of removed vertices is at most k'. Hence, by checking k' larger neighbors of u, the latter part is updated in $\mathcal{O}\,(k')$ time. □

For the update of the former part, we prepare another data structure called an *island*. An island of a vertex w is a maximal segment of $\mathcal{L}(S)$ composed only of gray vertices v such that $w \in N_>(v)$ and $v \notin S$. An island is stored in the memory by a doubly linked cyclic list of the segment in that the head and the tail of the list are linked (See Fig. 2).

Lemma 7. *For any k-degenerate vertex subset S, the accumulated size of the islands of all vertices in G is $\mathcal{O}\,(|V| + |E|)$.*

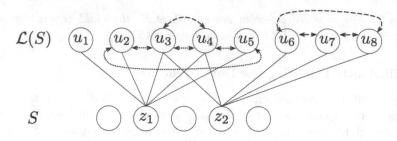

Fig. 2. Example of islands used in the update algorithm. Suppose that $|S \cap N(u_1)| < k$ and $|S \cap N(u_i)| = k$ for $i = 2, \ldots, 8$. Dotted arrows and dashed arrows imply the island of z_1 and z_2, respectively. The island of z_2 consists of two connected doubly linked lists.

Proof. Since each vertex u belongs to at most $|N(u)|$ islands, the sum of the members of islands of all vertices in G is at most $|E|$. Since the doubly cyclic linked list representing an island requires memory linear in the number of its members, the statement holds.   □

We remove the vertices from $\mathcal{L}(S)$ that are not in $\mathcal{L}(S \cup u)$ by tracing $\mathcal{L}(S)$ from its head. When we encounter a vertex to be removed, that is, gray in S and adjacent to u, the vertex is the head of an island of u. We then go to its tail by using the cyclic link, and remove the island from the list by cutting off its head and tail from $\mathcal{L}(S)$. We can find the tail of a head in constant time by an array of at most k' pointers. That is, we can remove the island from the list in constant time. Note that the vertices in the island are all adjacent to u, and thus become black by $S \cup u$. In this way, we can update the former part of $\mathcal{L}(S)$ in time linear in the number of vertices of $\mathcal{L}(S \cup u)$ that are smaller than u. Since $u = pv(S \cup u)$, the computation time is $\mathcal{O}(|\mathcal{C}(S \cup u)| + 1)$. Hence, the following lema holds.

Lemma 8. *For any k-degenerate vertex subset S, the part of $\mathcal{L}(S \cup u)$ composed of vertices smaller than u is obtained from $\mathcal{L}(S)$ in $\mathcal{O}(|\mathcal{C}(S \cup u)| + 1)$ time.*

The computation of $d_{S \cup u}(v)$ for each vertex v of the former part of $\mathcal{L}(S \cup u)$ is done by tracing $N_>(v)$ to check whether u is adjacent to v or not. This is done in $\mathcal{O}(k')$ time per vertex, and thus computation for all the vertices in former part is done in $\mathcal{O}(k'(|\mathcal{C}(S \cup u)| + 1))$ time.

Lemma 9. *For any k-degenerate vertex subset S, $d_{S \cup u}(v)$ for all vertices v in $\mathcal{L}(S \cup u)$ are obtained from $N(v) \cap S$ in $\mathcal{O}(k + k'(|\mathcal{C}(S \cup u)| + 1))$ time.*

Lemma 10. *For any k-degenerate vertex subset S, all the islands in the part of $\mathcal{L}(S \cup u)$ composed of vertices smaller than u are constructed in $\mathcal{O}(k'(|\mathcal{C}(S \cup u)| + 1))$ time by using $\mathcal{L}(S \cup u)$.*

Proof. All the islands in the part of $\mathcal{L}(S \cup u)$ composed of vertices smaller than u are built from the scratch by tracing $\mathcal{L}(S \cup u)$. This is done by tracing all vertices in $N_>(v)$ for all vertices in $\mathcal{L}(S \cup u)$, thus is done in $\mathcal{O}(k'(|\mathcal{C}(S \cup u)| + 1))$ time.   □

Algorithm 2. Island update algorithm

---

1 **Procedure** UpdateIsland(S, u)
2   **foreach** $v \in N(u) \cap \mathcal{L}(S \cup u)$ **do**
3   **if** $d_{S \cup u}(v) = k$ **then**
4   **foreach** *Island I including the smaller neighbor of v on $\mathcal{L}(S \cup u)$* **do**
5   Put together I and v into a new island $I' = I + v$;
6   Link the head of I and v;
7   **foreach** *Island J including the larger neighbor of v on $\mathcal{L}(S \cup u)$* **do**
8   Put together v and J into a new island $J' = v + J$;
9   Link v and the tail of I;
10   **foreach** *Pair of islands I' and J' including v of the same vertex* **do**
11   Put together I' and J' into a new island $I'' = I' + J'$;
12   Link the head of I' and the tail of J';
13   **return** $island(S \cup u)$;

Lemma 11. *For any k-degenerate vertex subset S, all the islands in the part of $\mathcal{L}(S \cup u)$ composed of vertices larger than u are obtained from the islands of $\mathcal{L}(S)$ in $\mathcal{O}\left((k')^2\right)$ time.*

Proof. An island will change by the addition of u to S when a vertex of the island or a vertex neighboring to its head or its tail becomes gray, or becomes black. We observe that at most k' vertices larger than u become gray or black by adding u to S. By the change of a vertex v, at most $3k'$ islands, that include v or a vertex neighboring to v in $\mathcal{L}(S)$, will change. Concatenating two islands, splitting an island and appending a vertex to an island are all done in $\mathcal{O}(1)$ time, thus the update of the islands in the part of $\mathcal{L}(S \cup u)$ composed of vertices larger than u are obtained from the islands of $\mathcal{L}(S)$ in $\mathcal{O}\left((k')^2\right)$ time. □

Algorithm 2 shows the pseudo code for concatenating islands or appending a vertex to an island. We can also implement splitting an island in a similar way. We consider that the update of data structure is done in the iteration with respect to $S \cup u$, as an initialization. Then, an iteration of the algorithm takes $\mathcal{O}\left(\min\left\{\Delta + kk', (k')^2\right\} + k'|\mathcal{C}(S \cup u)|\right)$ time.

Theorem 1. *The vertex subsets of a graph $G = (V, E)$ inducing k-degenerate graphs can be enumerated in $\mathcal{O}\left(\min\left\{\Delta + kk', (k')^2\right\}\right)$ time for each solution with $\mathcal{O}\left(|V| + |E|\right)$ space and $\mathcal{O}\left(|V| + |E|\right)$ preprocessing time, where Δ and k' denote the maximum degree and the degeneracy of G.*

Proof. An iteration of our algorithm takes $\mathcal{O}\left(\min\left\{\Delta + kk', (k')^2\right\} + k'|\mathcal{C}(S)|\right)$ time. By assigning $\mathcal{O}(k')$ to each child, it will be $\mathcal{O}\left(\min\left\{\Delta + kk', (k')^2\right\}\right)$. Our algorithm outputs a solution in each iteration. It implies that the computation time is $\mathcal{O}\left(\min\left\{\Delta + kk', (k')^2\right\}\right)$ time for each solution. In the preprocessing phase, the algorithm needs to sort the vertices in the degeneracy ordering.

This takes $\mathcal{O}\left(|E|+|V|\right)$ time. We also sort the adjacency list of each vertex in $\mathcal{O}\left(|V|+|E|\right)$ time by using bucket sort. The sizes of $\mathcal{L}(S)$, $\mathcal{L}'(S)$ and the memory for remembering $d_S(\cdot)$ are all $\mathcal{O}\left(|V|\right)$. Since the total size of all islands is $\mathcal{O}\left(|E|\right)$, the statement holds. □

Corollary 1. *Let $G = (V, E)$ be a graph with constant degeneracy and k be a positive integer. Then, all k-degenerate induced subgraphs in G can be enumerated in constant time for each solution with $\mathcal{O}\left(|V|+|E|\right)$ space and $\mathcal{O}\left(|V|+|E|\right)$ preprocessing time.*

5   Conclusion

In this paper, we addressed the k-degenerate induced subgraph enumeration problem. As the main result, we proposed an efficient enumeration algorithm that runs in $\mathcal{O}\left(\min\left\{\Delta + kk', (k')^2\right\}\right)$ time per solution with polynomial preprocessing time and linear space. In this paper, we did not consider the connectivity since when we consider it, the parent-child relation in this paper does not work. Investigating other efficient enumeration strategies for connected k-degenerate subgraph is an interesting future research. The variant of the problem, in some graph classes, non-induced version and fixed size are also interesting.

References

1. Avis, D., Fukuda, K.: Reverse search for enumeration. Discret. Appl. Math. **65**(1–3), 21–46 (1996). https://doi.org/10.1016/0166-218X(95)00026-N
2. Bauer, R., Krug, M., Wagner, D.: Enumerating and generating labeled k-degenerate Graphs. In: ANALCO 2010, pp. 90–98. Society for Industrial and Applied Mathematics (2010). https://doi.org/10.1137/1.9781611973006.12
3. Chartrand, G., Harary, F.: Planar permutation graphs. Annales de l'institut Henri Poincaré (B) Probabilités et Statistiques **3**(4), 433–438 (1967)
4. Conte, A., Grossi, R., Marino, A., Versari, L.: Sublinear-space bounded-delay enumeration for massive network analytics: maximal cliques. In: ICALP 2016, pp. 148:1–148:15 (2016). https://doi.org/10.4230/LIPIcs.ICALP.2016.148
5. Conte, A., Kanté, M.M., Otachi, Y., Uno, T., Wasa, K.: Efficient enumeration of maximal k-degenerate subgraphs in a chordal graph. In: Cao, Y., Chen, J. (eds.) COCOON 2017. LNCS, vol. 10392, pp. 150–161. Springer, Cham (2017). https://doi.org/10.1007/978-3-319-62389-4_13
6. Eppstein, D., Strash, D.: Listing all maximal cliques in large sparse real-world graphs. In: Pardalos, P.M., Rebennack, S. (eds.) SEA 2011. LNCS, vol. 6630, pp. 364–375. Springer, Heidelberg (2011). https://doi.org/10.1007/978-3-642-20662-7_31
7. Ferreira, R., Grossi, R., Rizzi, R.: Output-sensitive listing of bounded-size trees in undirected graphs. In: Demetrescu, C., Halldórsson, M.M. (eds.) ESA 2011. LNCS, vol. 6942, pp. 275–286. Springer, Heidelberg (2011). https://doi.org/10.1007/978-3-642-23719-5_24
8. Goel, G., Gustedt, J.: Bounded arboricity to determine the local structure of sparse graphs. In: Fomin, F.V. (ed.) WG 2006. LNCS, vol. 4271, pp. 159–167. Springer, Heidelberg (2006). https://doi.org/10.1007/11917496_15

9. Kanté, M.M., Limouzy, V., Mary, A., Nourine, L.: Enumeration of minimal dominating sets and variants. In: Owe, O., Steffen, M., Telle, J.A. (eds.) FCT 2011. LNCS, vol. 6914, pp. 298–309. Springer, Heidelberg (2011). https://doi.org/10.1007/978-3-642-22953-4_26

10. Lick, D.R., White, A.T.: k-degenerate graphs. Can. J. Math. **22**(5), 1082–1096 (1970). https://doi.org/10.4153/CJM-1970-125-1

11. Matula, D.W., Beck, L.L.: Smallest-last ordering and clustering and graph coloring algorithms. J. ACM **30**(3), 417–427 (1983). https://doi.org/10.1145/2402.322385

12. Wasa, K.: Enumeration of enumeration algorithms. CoRR abs/1605.05102 (2016), http://arxiv.org/abs/1605.05102

13. Wasa, K., Arimura, H., Uno, T.: Efficient enumeration of induced subtrees in a K-degenerate graph. In: Ahn, H.-K., Shin, C.-S. (eds.) ISAAC 2014. LNCS, vol. 8889, pp. 94–102. Springer, Cham (2014). https://doi.org/10.1007/978-3-319-13075-0_8

14. Wasa, K., Kaneta, Y., Uno, T., Arimura, H.: Constant time enumeration of bounded-size subtrees in trees and its application. In: Gudmundsson, J., Mestre, J., Viglas, T. (eds.) COCOON 2012. LNCS, vol. 7434, pp. 347–359. Springer, Heidelberg (2012). https://doi.org/10.1007/978-3-642-32241-9_30

Tree t-Spanners of a Graph: Minimizing Maximum Distances Efficiently

Fernanda Couto[1]($\boxtimes$) and Luís Felipe I. Cunha[2,3]($\boxtimes$)

[1] Universidade Federal Rural do Rio de Janeiro, Nova Iguaçu, Brazil
fernandavdc@ufrrj.br
[2] Universidade Federal Fluminense, Niterói, Brazil
[3] Universidade Federal do Rio de Janeiro, Rio de Janeiro, Brazil
lfignacio@cos.ufrrj.br

Abstract. A tree t-spanner of a graph G is a spanning subtree T in which the distance between any two adjacent vertices of G is at most t. The smallest t for which G has a tree t-spanner is the tree stretch index. The problem of determining the tree stretch index has been studied by: establishing lower and upper bounds, based, for instance, on the girth value and on the minimum diameter spanning tree problem, respectively; and presenting some classes for which t is a tight value. Moreover, in 1995, the computational complexities of determining whether $t = 2$ or $t \geq 4$ were settled to be polynomially time solvable and NP-complete, respectively, while deciding if $t = 3$ still remains an open problem.

With respect to the computational complexity aspect of this problem, we present an inconsistence on the sufficient condition of tree 2-spanner admissible graphs. Moreover, while dealing with operations in graphs, we provide optimum tree t-spanners for 2 cycle-power graphs and for prism graphs, which are obtained from 2 cycle-power graphs after removing a perfect matching. Specifically, the stretch indexes for both classes are far from their girth's natural lower bounds, and surprisingly, the parameter does not change after such a matching removal. We also present efficient strategies to obtain optimum tree t-spanners considering threshold graphs, split graphs, and generalized octahedral graphs. With this last result in addition to vertices addition operations and the tree decomposition of a cograph, we are able to present the stretch index for cographs.

Keywords: Tree t-spanner · Stretch index · Lower bounds
Generalized octahedral graph · Cycle-power graph · Prism graph
Threshold graph · Split graph · Cograph

1 Introduction

The problems of obtaining subgraphs with special restrictions have been considered in several papers, with many motivations and applications in different fields,

This work was partially supported by the Brazilian agencies CAPES, CNPq and FAPERJ.

D. Kim et al. (Eds.): COCOA 2018, LNCS 11346, pp. 46–61, 2018.
https://doi.org/10.1007/978-3-030-04651-4_4

as message routing, computational geometry, and phylogenetic analysis [1–3]. In addition to the inherent difficulty of these problems, another challenge arises when we look for a spanning tree with constraints on the vertices' distances.

A *tree t-spanner* of a graph G is defined as a spanning subtree T of G in which the distance between every pair of vertices is at most t times their distance in G or, equivalently, as the subtree T in which the distance between two adjacent vertices of G is at most t (*cf.* [4]). If a graph has a tree t-spanner, then it is called a *tree t-spanner admissible graph*. The parameter t of a tree t-spanner is called the *tree stretch factor*, and the smallest t for which a graph G is tree t-spanner admissible is called the *tree stretch index* of G, denoted by $\sigma_T(G)$.

Note that the problem of determining the *tree stretch index* of G, called the *minimum stretch spanning tree problem* (MSST), is one of the interesting min-max problems, which are studied not only in graphs, but in several other combinatorial problems, in such a way that bounds, algorithms and computational complexity studies are widely developed [5,6].

An intriguing aspect comes when we want to determine if a graph is tree 3-spanner admissible. In terms of the computational complexity, this task is still the greatest breakthrough we aim to solve, since deciding if $\sigma_T(G) \geq 4$ is NP-complete, whereas for $\sigma_T(G) = 2$ it is polynomially time solvable [4]. There are also some classes for which this problem was settled to be NP-complete, as planar and chordal graphs [7,8], or classes for which the stretch index was proved to be bounded by specific values, as split and cographs (*cf.* [9]). Hence, it is also a great challenge to determine the stretch index even restricted to graph classes. Still in the computational complexity approach, in this work, we can observe that Cai and Corneil's characterization for tree 2-spanner admissible graphs [4], which deals with triconnected components of a graph, is not consistent with the usual definition of k-connected graphs, considering, for instance, complete graphs. In this sense, we present infinite families of split graphs that do not admit tree 2-spanners, but satisfy their sufficient condition, considering either, the convention for K_n graphs connectivity (see [10,11]) or that the connectivity concept does not apply to complete graphs (see [12]).

Studying bounds is an ordinary kind of approach for MSST. A natural lower bound arises when we consider the girth $g(G)$ of a graph G, i.e. the length of its minimum cycle. We have that, if G is a tree t-spanner admissible, then $t \geq g(G) - 1$. Regarding this bound, it is possible to observe some optimum tree t-spanners for some families or classes, for instance complete graphs, cycle graphs, wheel graphs, or complete k-partite graphs, for $k \geq 2$. However, establishing lower bounds is challenging in general, and so it remains when we deal with the MSST problem restricted to graph classes, since the results on it often present tree t-admissible graphs (*cf.* [4]). Another kind of approach considers variant problems, for instance the *minimum diameter spanning tree*. In this problem, the solution tree minimizes the maximum distances between all pairs of vertices, which is polynomially time solvable, and the solution parameter is an upper bound for the MSST problem [13].

We focus on obtaining the stretch index for some graph classes and, although there are already known upper bounds for some of them, in this work we present minimum $t = \sigma_T(G)$ values considering these classes. We also present the stretch index for 2 cycle-power graphs, which is far from the girth's natural lower bound. Furthermore, we are also interested in the stretch index after vertices/edges operations, particularly for generalized octahedral graphs (complete graphs after removing a perfect matching), generalized octahedral graphs after non-universal vertices additions, and for prism graphs (2 cycle-power graphs after removing a perfect matching). Surprisingly, in this last case, the matching removal does not modify the stretch index of 2 cycle-power graphs.

This paper is organized as follows: In Sect. 2, we present basic definitions, considerations about Cai and Corneil's characterization for tree 2-spanner admissible graphs, and previous results. In Sect. 3, we present optimum tree t-spanner for some graph classes, such as 2 cycle-power graphs, prism graphs, generalized octahedral graphs, threshold graphs and their minimal superclasses, split graphs and cographs; In Sect. 4, we present final remarks by considering further investigation on other classes and their properties.

2   Preliminaries

Given a graph $G = (V, E)$, $d_G(x, y)$ denotes the distance between x and y in G and $d_G(v)$, the degree of v in G. We say that a *non-edge* of a spanning tree T is an edge of $G \setminus T$. A p-path is a path of length p.

A *tree t-spanner* of a graph G is a spanning subtree T of G in which the distance between every pair of vertices is at most t times their distance in G. Cai and Corneil proved that this problem is equivalent to the one that considers only adjacent vertices of G [4]. Moreover, they showed what follows.

Theorem 1. *A spanning tree T is a tree t-spanner of G if and only if for every edge $xy \in E(G) \backslash E(T)$ we have $d_T(x, y) \leq t$.*

The *minimum stretch spanning tree* of G (MSST) is an optimization problem of finding a tree t-spanner of G with minimum t. In this case, we say that $\sigma_T(G) = t$, and $\sigma_T(G)$ is called the *stretch index* of G. Upper bounds for $\sigma_T(G)$ can be obtained considering, for instance, the minimum diameter spanning tree, whose smallest parameter is $D_T(G)$, and some other problems [4,14]. In opposite, a natural lower bound can be obtained accordingly to the girth of G, i.e., the length of its minimum induced cycle. Therefore, Theorem 2 states the range of the stretch index of a given graph G.

Theorem 2 [4,13]. *Given $g(G)$ the girth of G, we have that $g(G) - 1 \leq \sigma_T(G) \leq D_T(G)$.*

Consider, for instance, a tree $(n-1)$-spanner of the cycle graph C_n, i.e. a path P_n, and a tree 2-spanner of the complete graph K_n, i.e. a star S_{n-1}. Both spanning trees are optimum, and their associated stretch factors are tight with respect to Theorem 2.

On Cai and Corneil Tree 2-Spanner Characterization. Cai and Corneil [4] proposed a characterization to decide if $\sigma_T(G) = 2$, formulated as follows: a nonseparable graph G has a 2-spanner if and only if G contains a spanning tree T such that for each triconnected component H of G, $T \cap H$ is a spanning star of H.

Indeed, the statement above gives a necessary condition for a graph having a 2-spanner. However, we show in Fig. 1 a nonseparable graph G and a spanning tree T of G such that the intersection of T with the unique triconnected component of G ($H = K_4$) is a spanning star of H, but there is no tree 2-spanner for the split graph in Fig. 1, as a consequence of Proposition 3. Observe that, since the connectivity of a complete graph with n vertices is $n - 1$ [10,11], a K_4 is triconnected and, once this is the unique triconnected component of G, $H = K_4$. Thus, in order that G is tree 2-spanner admissible it must exist a spanning tree T of G such that $T \cap H$ is a star. Figure 1(b) exhibits such a tree. This example can be generalized, for instance, to a graph obtained from a K_{2k} adding k vertices adjacent to two vertices of K_{2k} with no common adjacent vertex. k-sun (see [15]) are also an example of split graphs that satisfy the sufficient condition mentioned above, and thus would be tree 2-spanner admissible graphs, but, accordingly to Proposition 3, they do not admit a tree 2-spanner.

Thus, the Cai and Corneil's sufficient condition for tree 2-spanner admissible graphs is not consistent with the usual definition of the connectivity for complete graphs. Even if we consider that the connectivity concept does not apply for such graphs, the condition does not hold in these families of examples.

Fig. 1. (a) A split graph G with only one triconnected component $H = K_4$. (b) A spanning tree T of G such that $T \cap H$ is a spanning star of H, but G is not a tree 2-spanner admissible graph (see Proposition 3, since there is no vertex of the set $\{1, 2, 3, 4\}$ adjacent to both vertices 5 and 6, then the stretch index of G is equal to 3).

3   Stretch Index for Graph Classes

Next we consider some related graph classes, for which we are able to obtain optimum tree t-spanners even when, in some cases, the lower bound of Theorem 2 is far from the stretch index we obtain. Moreover, seeing whether and how vertices/edges operations affect the stretch index is another goal of this section.

3.1   Cycle-Power Graphs

Any graph is a tree $(n - 1)$-spanner admissible, but, in general, such a bound is far from the stretch index. However, for cycle graphs, $n - 1$ is a tight value, since it reaches the lower bound of Theorem 2. Next, we present classes with extremal bounds, in such a way that $\sigma_T(G)$ is large and far from the lower bound given by Theorem 2.

A *cycle-power* graph [16], C_n^k, is obtained from a C_n by adding edges between two vertices with distance at most k in C_n. We call *external edges* the edges of the *external cycle* C_n, and *internal edges* the added edges. Since $g(C_n^k) = 3$, then $\sigma_T(C_n^k) \geq 2$. We restrict ourselves to $k = 2$ and show an optimum tree $\lfloor \frac{n}{2} \rfloor$-spanner.

Given a graph $G = C_n^2$, we define an *ℓ-come-go* path with respect to a pair of vertices u_i, u_{i+j}, for $j \in \{1, 2\}$, by a path of length ℓ from u_i to u_{i+j}, for $\ell \in \{2, \cdots, n-1\}$, following one of two directions, either: clockwise/counterclockwise direction; or counterclockwise/clockwise direction. When we are not interested in the length, we suppress such a value and refer an *ℓ-come-go* path by a *come-go* path.

A *spot edge* of a come-go path is an external edge that either: changes the way of the path, i.e. from clockwise to counterclockwise or from counterclockwise to clockwise; or immediately precedes an internal edge that changes the way of the path. Figure 2 illustrates the two 7-come-go paths with respect to u_1 and u_2.

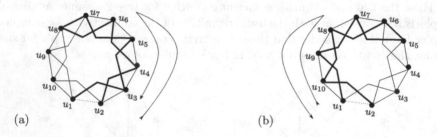

(a)   (b)

Fig. 2. Bold edges belong to 7-come-go paths with respect to u_1 and u_2, such that: (a) path using the counterclockwise/clockwise direction, where $u_7 u_8$ is the spot edge; (b) path using the clockwise/counterclockwise direction, where $u_5 u_6$ is the spot edge.

Lemma 1. *Given a graph $G = C_n^2$ and a come-go path P with respect to $u_i u_{i+j}$, for $j \in \{1, 2\}$, then P contains a unique spot edge.*

Proof. Once P is a come-go path, P must contain a spot edge. Suppose there are more than one of such edges. Following the path P from u_i to u_{i+j}, consider $u_f u_{f+1}$ as the first reachable spot edge of P. After reaching the last spot edge of P, the unique way to achieve u_{i+j} is by crossing again u_f or u_{f+1}, once it is not possible to bypass two consecutive vertices of the external cycle. Since u_f and u_{f+1} already belong to P, then there is a cycle in the come-go path P. □

A path between u_i, u_{i+j}, for $j \in \{1, 2\}$, of length greater than 2 which is not a *come-go* is called a *turn around* path, which is depicted in Fig. 3(a). If $j = 1$, then the length of a turn around path is at least $\lfloor \frac{n}{2} \rfloor$. If $j = 2$, then the length is

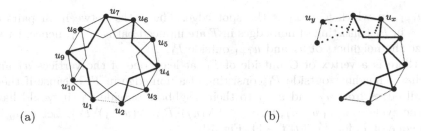

Fig. 3. (a) Bold edges belong to a turn around path with respect to u_1 and u_2. (b) An example of vertices u_x and u_y.

at least $\lfloor \frac{n}{2} \rfloor$ for n odd, and at least $\frac{n}{2} - 1$ for n even. Note that, when $j = 2$, we have between u_i and u_{i+j} either a turn around path or the 2-path $u_i u_{i+1} u_{i+2}$.

For any non-edge of a spanning tree T of a graph G, there is a path which is either: a come-go path, or a turn around path, or the 2-path $u_i u_{i+1} u_{i+2}$, for the non-edge $u_i u_{i+2}$.

Proposition 1. *Given an ℓ-come-go path with respect to u_i and u_{i+j}, for $j \in \{1, 2\}$, if $j = 1$, then there is a unique external edge, otherwise there are exactly two external edge.*

Proof. Considering $j = 1$, since the spot edge is external, let us suppose that there is at least one more external edge $u_f u_{f+1}$, for $i+1 < f < \ell$ in an ℓ-come-go path P. In this case, following the path from u_i to u_{i+1}, at least one of u_f and u_{f+1} will be reached, and after crossing the spot edge, it is necessary to reach u_f or u_{f+1} again, which implies that P is not a path. Similarly, considering $j = 2$, the unique external edges are $u_i u_{i+1}$ and the spot edge. $\square$

Lemma 2. *Given a graph $G = C_n^2$ and an ℓ-come-go path P, with respect to $u_i u_{i+j}$, for $j \in \{1, 2\}$, then P is the unique ℓ-come-go path with respect to $u_i u_{i+j}$ following the same direction of P.*

Proof. Suppose there are at least two ℓ-come-go paths P_1 and P_2 following, w.l.g., the counterclockwise/clockwise direction with respect to $u_i u_{i+j}$, for $j \in \{1, 2\}$. In this case, there is a non-edge in P_1 which is external edge of P_2, and then it is a spot edge of P_2, by Proposition 1. Hence, the length of P_2 is distinct of ℓ. $\square$

Lemma 3. *For any spanning tree T of $G = C_n^2$, there is at least a non-edge $u_i u_{i+j}$, for $j \in \{1, 2\}$, such that the unique path between u_i and u_{i+j} in T is a turn around path.*

Proof. Suppose the path between any pair of vertices of a non-edge of T is not a turn around path. Hence, if the non-edge is external, then the path is a come-go path. If the non-edge is internal, then the path between them is either a 2-path, or it is a come-go path.

Since T must contain an external non-edge, let $u_i u_{i+1}$ be such an external non-edge of T and thus, by hypothesis, there is a come-go path P_1 between u_i

and u_{i+1}, in which $u_f\,u_{f+1}$ is the spot edge. The paths between all pairs of vertices in P_1 consisting of non-edges in T are induced paths of P_1, hence, let us analyze the neighbors of u_f and u_{f+1} outside P_1.

If there is a vertex of G outside of P_1, at least one of the vertices u_f and u_{f+1} has a neighbor outside P_1 consisting of a non-edge in T, because if there were all edges from u_f and u_{f+1} to their neighbors outside P_1, it would have in T the cycle $u_f\,u_{f+1}\,u_y,u_f$, for $u_y \in \{N(u_f)\backslash P_1,\ N(u_{f+1})\backslash P_1\}$. Let $u_x u_y$ be a non-edge of T, for $x \in \{f, f+1\}$, Fig. 3(b).

If $u_x u_y$ is an internal non-edge, then we have two options of a path between u_x and u_y in T:

i. by the 2-path $u_x u_{x+1} u_y$. In this case, go to a non-edge where one of the vertices belongs to the 2-path. This non-edge belongs to: a 2-path, and in this case, we go to a non-edge and the analysis continues; a come-go path with respect to its extremities, and in this case, as it was done with P_1, go to its spot edge and continue the analysis; or $u_{x+1}u_y$ is the spot edge of a come-go path, P_2, following the opposite direction of P_1, with respect to vertices that do not belong to P_1, nor to the 2-path, either. Hence, go to the extremity vertices of P_2 and analyze a non-edge whose vertices are an extremity vertex of P_2 and a vertex that does not belong to P_2 nor to P_1;
ii. by a come-go path with respect to $u_x u_y$, which we call P_3. Note that P_3 must have the same direction of P_1, otherwise, we would visit vertices already in P_1, implying in a cycle. Hence, go to the spot edge of P_3 and consider it similarly as done considering P_1.

If $u_x u_y$ is an external non-edge, then u_x and u_y must be connected by come-go path in T. In this case, proceed as in the previous case ii.

Note that the procedures considered in i and ii. must be finished when we reach either the vertex u_{i-1} (whenever P_1 follows anticlockwise/clockwise direction), or the vertex u_{i+j+1}, for $j \in \{1,2\}$ (whenever P_1 follows clockwise/anticlockwise direction). Let u_w be the last visited vertex, which is neighbor of u_i or u_{i+j}. In T, we have three possible paths between u_w and u_i or u_{i+j}: there is an edge; there is a come and go path; there is a 2-path. For any of such cases, we have created a cycle, because, by P_1, there is a path between u_i and u_{i+j}, which does not include u_w. Therefore, there is path, distinct of P_1, starting from either u_i or u_{i+j} passing through u_w. Thus, there is a turn around path between u_i and u_{i+j}.   □

Lemma 4. *For any cycle-power graph C_n^2, $\sigma_T(C_n^2) \geq \lfloor \frac{n}{2} \rfloor$.*

Proof. Since there is at least a turn around path in any spanning tree T of $G = C_n^2$ (Lemma 3), and if n is odd, then there is a non-edge in T whose corresponding vertices' distance is at least $\lfloor \frac{n}{2} \rfloor$. Therefore, $\sigma_T(C_n^2) \geq \lfloor \frac{n}{2} \rfloor$, for n odd.

Since when n is even, a turn around path has length at least: $\frac{n}{2} - 1$, for an internal non-edge $u_i\,u_{i+2}$; or $\frac{n}{2}$, for an external non-edge. Hence, it remains to analyze the former case. Note that G contains two disjoint internal cycles I_1 and

I_2, each one of length $\frac{n}{2}$. Consider that u_i and u_{i+2} belong to I_1 and the distance between them in T is $\frac{n}{2} - 1$. Since the unique turn around path of length $\frac{n}{2} - 1$ between u_i and u_{i+2} in G includes each edge of the cycle I_1, all internal edges of I_1 must belong to T, except $u_i\ u_{i+2}$. On the other hand, at least one of $u_i\ u_{i+1}$ and $u_{i+1}\ u_{i+2}$ must be non-edge of T. Otherwise, if both edges belong to T, then there would be the 2-path $u_i\ u_{i+1}\ u_{i+2}$ in T, contradicting the assumption of the path between $u_i\ u_{i+2}$ is turn around.

1. If $u_i\ u_{i+1}$ is non-edge of T and $u_{i+1}\ u_{i+2}$ is edge of T (which is similar to the case of $u_i\ u_{i+1}$ being edge of T and $u_{i+1}\ u_{i+2}$ non-edge of T), then the path between u_i and u_{i+1} has length at least $\frac{n}{2}$ considering the path between u_i and u_{i+2}, and the edge $u_{i+2}\ u_{i+1}$. Otherwise, if there is a distinct path P between u_i and u_{i+1}, we would have another path between u_i and u_{i+2}, say $P \cup \{u_{i+1}u_{i+2}\}$.

2. If $u_i\ u_{i+1}$ and $u_{i+1}\ u_{i+2}$ are both non-edges of T, then at least one of the edges $u_{i-1}\ u_{i+1}$ and $u_{i+1}\ u_{i+3}$ must belong to T, otherwise u_{i+1} would be isolated of T. Hence, we analyze the two cases:
   - $u_{i-1}\ u_{i+1}$ is an edge of T and $u_{i+1}\ u_{i+3}$ is a non-edge of T. In this case, note that the path between u_{i+1} and u_{i+2} must be a turn around, because $u_{i+1}\ u_{i+2}$ and $u_{i+1}\ u_{i+3}$ are non-edges of T, Fig. 4(a). Since $u_{i+1}\ u_{i+2}$ is an external non-edge of T, then, the distance between u_{i+1} and u_{i+2} is at least $\frac{n}{2}$.
   - $u_{i-1}\ u_{i+1}$ and $u_{i+1}\ u_{i+3}$ are edges of T. Let us consider the distance between u_{i+1} and u_{i+2} in T. If it is given by a turn around path, then its length is at least $\frac{n}{2}$. Otherwise, it is an come-go path P^1. If P^1 follows the clockwise/anticlockwise direction, then the edge $u_i\ u_{i+2}$ must exist in T, but it contradicts the hypothesis. Hence, P^1 follows the anticlockwise/clockwise direction. Similarly, the path between u_i and u_{i+1} is a turn around path, implying that the distance between u_i and u_{i+1} is at least $\frac{n}{2}$, or it is a come-go path P^2 following the clockwise/anticlockwise direction. In this case, we have that:
     - if there is any path in T between the spot edges of the two come-go paths without passing through u_{i+1}, then we have created a cycle, since u_{i+1} belongs to the two come-go paths just settled;
     - Suppose there is no path in T between the spot edges of the two come-go paths without passing through u_{i+1}, and let P be the path composed by external edges in G that links the P^1 spot edge to the P^2 spot edge following the anticlockwise direction, Fig. 4(b). Note that P has at least one edge, because, otherwise, T would have a cycle. Clearly, there is an edge in P which is a non-edge in T. Thus, there is a path of length at least $\frac{n}{2}$.

Hence, we have that there is a pair of neighbors in G whose distance is at least $\lfloor \frac{n}{2} \rfloor$ in T.   □

Lemma 5. *For any cycle-power graph C_n^2, $\sigma_T(C_n^2) \leq \lfloor \frac{n}{2} \rfloor$.*

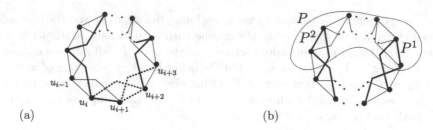

(a)   (b)

Fig. 4. (a) Turn around path between u_{i+1}, u_{i+2} in T. Note that $u_i u_{i+2}$, $u_{i+1} u_{i+2}$ and $u_{i+1} u_{i+3}$ are non-edges of T. (b) Bold edges compose two come-go paths P^1 and P^2, where the bold external edges are their spot edges. The path P is inside the dotted diagram.

Proof. We obtain a spanning tree T of C_n^2 with vertex set $\{u_1, u_2, \ldots, u_n\}$ as follows: add to T the vertex u_1 and its neighbors u_2, u_3, u_n and u_{n-1}. Now, follow the direction in which the next vertex is u_2, set $i = 3$, and: (i) take the vertex u_i; (ii) Add to T the vertices adjacent to u_i which are not in T yet, following the same direction as established initially, i.e., u_4, u_5 in the first step. Increment $i+1$ and return to step (i) until reaching the last vertices not in T yet. It is easy to see that, between two adjacent vertices of C_n^2, the distance between them in T is either 1, 2, 3 or $\frac{n}{2}$. Hence, $\sigma_T(C_n^2) \leq \lfloor \frac{n}{2} \rfloor$.   □

Figure 5 depicts a tree $\lfloor \frac{n}{2} \rfloor$-spanner for C_{10}^2.

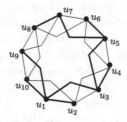

Fig. 5. Bold edges form the tree $\lfloor \frac{n}{2} \rfloor$-spanner T for C_{10}^2. There are: three turn around paths in T, with respect to the internal non-edges $u_7 u_9$ and $u_8 u_{10}$, and the external non-edge $u_8 u_9$; a 2-path between the internal non-edge $u_2 u_{10}$; 3-come-go paths with respect to the internal non-edges $u_2 u_4, u_4 u_6, u_6 u_8$ and $u_8 u_{10}$; and 2-come-go paths with respect to the external non-edges $u_2 u_3, u_4 u_5, u_6 u_7$ and $u_9 u_{10}$.

Theorem 3 follows from Lemmas 4 and 5.

Theorem 3. *For any cycle-power C_n^2 with $n > 5$, $\sigma_T(C_n^2) = \lfloor \frac{n}{2} \rfloor$.*

3.2   Stretch Index After Edges Removal

For several graph classes, we are able to determine the stretch index. But obtaining the stretch index after we consider operations on the vertex/edge sets regarding those classes is a challenge. In this section, we are particularly interested on a perfect matching removal considering 2 cycle-power and complete graphs, which are prism and generalized octahedral graphs, respectively. With this last result, in Sect. 3.3 we obtain the stretch index for cographs.

Removing a Perfect Matching of Cycle-Power Graphs. Considering 2 cycle-power graphs of even order after removing a perfect matching M with respect to external edges, one can note that a $C_{2p}^2 \setminus M$ is the *prism graph* with bases C_p. Lemma 6 presents a lower bound which is far from its girth's lower bound. Moreover, in Lemma 7 we prove that the stretch index is not affected by a perfect matching removal, differently from what happens with the complete graph and the octahedral graph, as proved in Theorem 5.

As in 2 cycle-power graphs, in prism graphs, we also have come-go and turn around paths.

Lemma 6. *Given $G = C_{2p}^2 \setminus M$, a cycle-power graph C_{2p}^2 after removing a perfect matching M with respect to the external edges, we have that $\sigma_T(G) \geq \frac{n}{2}$.*

Proof. Considering any tree t-spanner of G, we analyze two cases: all external edges belong to T; and there is at least an external non-edge in T.

1. All external edges belong to T: In this case, between two consecutive external edges $u_i\ u_{i+1}$ and $u_{i+2}\ u_{i+3}$ it is not possible to exist both internal edges $u_i\ u_{i+2}$ and $u_{i+1}\ u_{i+3}$ in T, otherwise the C_4, u_i, u_{i+1}, u_{i+3}, u_{i+2}, u_i, would belong to T. Next, we analyze two possible subcases: $u_i\ u_{i+2}$ and $u_{i+1}\ u_{i+3}$ are both non-edges of T; only one of such edges belongs to T.

   1.1. $u_i\ u_{i+2}$ and $u_{i+1}\ u_{i+3}$ are both non-edges of T: Considering the edge $u_{i+4}\ u_{i+5}$, it must exist in T either $u_{i+2}\ u_{i+4}$, or $u_{i+3}\ u_{i+5}$, because otherwise, the edge $u_{i+2}\ u_{i+3}$ would be isolated in T.

   Consider, w.l.g., $u_{i+2}\ u_{i+4}$ is in T (and so, $u_{i+3}\ u_{i+5}$ is a non-edge). Although a turn around for an internal non-edge is at least $\frac{n}{2} - 1$, the distance between u_{i+1} and u_{i+3} is at least $\frac{n}{2} + 1$ by a turn around path with respect such an internal non-edge. Note that the length of the turn around path between u_{i+1} and u_{i+3} is equal to $\frac{n}{2} - 1$ only when all edges of $I_1 \setminus \{u_{i+1}u_{i+3}\}$ are in T, where I_1 is the internal cycle of G that contains vertices u_{i+1}, u_{i+3} and u_{i+5}. However, $u_{i+3}\ u_{i+5}$ is a non-edge in T, which implies in a exchange of $u_{i+3}\ u_{i+5}$ by the path $u_{i+5}, u_{i+4}, u_{i+2}, u_{i+3}$ in T. Hence, the length of the turn around path is at least $\frac{n}{2} + 1$ after the edges' exchange.

   1.2. Suppose, w.l.g., $u_i\ u_{i+2}$ is an edge of T and $u_{i+1}\ u_{i+3}$ is a non-edge of T. Now, we prove that in T it must exist a pair of non-edges $u_j\ u_{j+2}$ and $u_{j+1}\ u_{j+3}$ for some j, similarly to Case 1.1. Assume that one of such edges is in T and belongs to the internal cycle I_2 of G. Hence, we create a path starting by the edge $u_i\ u_{i+2}$, and after that we choose one of the two ways of reaching the external edge $u_{i+4}\ u_{i+5}$, by $u_{i+2}\ u_{i+4}$ or $u_{i+3}\ u_{i+5}$. If $u_{i+2}\ u_{i+4}$ is an edge of T, then we are making a path through I_2, otherwise, the path is u_i, u_{i+2}, u_{i+3}, u_{i+5}. Therefore, it is always possible to reach two consecutive external edges by using I_1 or I_2 edges. So, if there is not a pair of non-edges similar to Case 1.1, we can continue this path through external edges of I_1 and I_2, creating then a cycle. Once it is necessary to have non-edges of Case 1.1, then we have the existence of vertices with distance at least $\frac{n}{2} + 1$ in T.

2. There is at least an external non-edge in T: Suppose $u_i \, u_{i+1}$ is a non-edge of T. If the path between u_i and u_{i+1} in T is a turn around, then its length is at least $\frac{n}{2} + 1$, because u_i belongs to I_2 and u_{i+1} belongs to I_1. Hence, assume that the path in T between u_i and u_{i+1} is a come-go.

Note that we have at least one non-edge of I_1 and of I_2 in T, and similarly to Lemma 3 and Case 1.2 above, there is a turn around path between the corresponding vertices of an internal non-edge of I_1 and I_2.

So, each turn around has length at least $\frac{n}{2} - 1$, and such a path with respect to an internal non-edge in T is unique in G. Moreover, in T, a turn around with respect to a non-edge of I_2 (or I_1) has length $\frac{n}{2} - 1$ or any greater value with the same parity, because the path between such vertices does not go through all edges of the corresponding internal cycle, and then we must move to the other internal cycle and return, increasing the path in at least two edges.

Hence, in order to keep in T the distances equal to $\frac{n}{2} - 1$ between the vertices of non-edges e^2 of I_2 and e^1 of I_1, all other edges of both internal cycles of G must belong to T. Let P^2 be the path $I_2 \backslash e^2$ and P^1 be the path $I_1 \backslash e^1$.

Now, P^1 must be linked to P^2. The unique way to do that is by using only one external edge, otherwise, there would be a cycle in T by at least two ways to go through P^2 to P^1, each one using a distinct external edge. Therefore, in T, there is only one external edge of G.

Since there is a come-go path between u_i, u_{i+1}, as well between all other $\frac{n}{2} - 2$ external non-edges, all come-go paths between corresponding vertices of external non-edges must be composed by the same spot edge, say, the external edge we have used to link P^1 and P^2.

Furthermore, the unique way to exist only come-go paths between corresponding vertices of external non-edges in T is by considering $u_{k-1}u_{k+1}$ and $u_{k-2}u_k$ internal non-edges of T. Otherwise, if the non-edges of T were $u_j u_{j+2}$ and $u_{j+2s+1} \, u_{j+2s+3}$, there would be a turn around path with respect to the external edges $u_j \, u_{j+1}$ and $u_{j+2s+1} \, u_{j+2s+2}$.

In this way, we have that the distances between the vertices of the non-edge u_{k-1} and u_k, and between the vertices of the non-edge u_{k+1} and u_{k+2} are $\frac{n}{2} - x$ and $\frac{n}{2} + x$, respectively, according to the place we have chosen the spot edge. Therefore, when $x = 0$, we have that $\sigma_T(G) \geq \frac{n}{2}$.   □

Accordingly to the arguments of Lemma 6, we are able to build a tree $\frac{n}{2}$-spanner as follows.

Lemma 7. *Given $G = C_{2p}^2 \backslash M$, a cycle-power graph C_{2p}^2 after removing a perfect matching with respect to the external edges M, we have that $\sigma_T(G) \leq \frac{n}{2}$.*

Proof. Consider I_1 and I_2 the internal cycles of G, in such a way that $I_1 = u_1, u_3, u_5, \ldots, u_{n-1}, u_1$, $I_2 = u_2, u_4, u_6, \ldots, u_n, u_2$ and $M = \{\{u_2 u_3\}, \{u_4 u_5\}, \{u_6 u_7\}, \ldots, \{u_n u_1\}\}$. We create the spanning tree T by the edge set $\{\{u_3 u_5\}, \{u_5 u_7\}, \ldots, \{u_{n-3} u_{n-1}\} \cup \{u_2 u_4\}, \{u_4 u_6\}, \ldots, \{u_{n-2} u_n\} \cup \{u_{\frac{n}{2}} u_{\frac{n}{2}+1}\}\}$. Note that the unique external edge of G in T is $\{u_{\frac{n}{2}} u_{\frac{n}{2}+1}\}$. The paths between the external non-edges of T have length at most $\frac{n}{2}$, which is equal to this value for the non-edges $u_1 \, u_2$ and $u_{n-1} \, u_n$. Furthermore, there are only two internal non-edges in T, which are $u_n u_2$ and $u_{n-1} u_1$, with distances equal to $\frac{n}{2} - 1$, because all other edges of I_2 and I_2 belong to T. $\qquad\square$

Theorem 4 follows from Lemmas 6 and 7.

Theorem 4. *Given $G = C_{2p}^2 \backslash M$, a cycle-power graph C_{2p}^2 after removing a perfect matching with respect to the external edges M, we have that $\sigma_T(G) = \frac{n}{2}$.*

Generalized Octahedral Graphs. Generalized octahedral graphs figure in several well studied problems [17] because of their regularity and symmetry. A generalized octahedral graph, or simply *octahedral graph* O_k, is the $(2k - 2)$-regular graph, which is exactly a complete graph K_{2k} after removing a perfect matching. This class sounds interesting in here when we deal with cographs in Sect. 3.3, even considering O_k after vertices addition, in Lemma 10.

Theorem 5. *Given an octahedral graph O_k, then $\sigma_T(O_k) = 3$, for $k > 2$.*

Proof. Consider the vertex set $\{u_1, v_1, \ldots, u_k, v_k\}$ in such a way that u_i and v_i are not neighbors, but they are adjacent to all other vertices of O_k. A tree T can be built by first considering two stars, with centers in u_1 and v_2, such that u_1 is adjacent to all u_i's and v_2 is adjacent to all v_i's. Now, we add to T the edge $u_1 v_2$. The distances in T of two vertices of u_i's or of v_i's are equal to 2, and from distinct side are equal to 3, hence $\sigma_T(O_k) \leq 3$. In order to prove that $\sigma_T(O_k) = 3$, suppose we have an optimum tree spanner T for O_k that can be partitioned into two rooted trees, T_1 and T_2, each one with more than two vertices, such that at least one of them is not a star. Suppose, w.l.g., that T_1 is not a star. Let $l \in T_1$ and c be two vertices of T such that $lc \notin E(O_k)$. If $c \in T_1$, then l and c are adjacent to each vertex of T_2. Since T_1 is linked to T_2, there is an edge with one extreme in T_1 and another in T_2. If l is such an extreme, then $d_T(c, v) \geq 3, \forall v \in T_2$. Otherwise, there is a vertex $v \in T_2$ such that $d_T(l, v) \geq 3$. $\qquad\square$

3.3   Threshold Graphs and Their Superclasses

Next, we establish the stretch index for three classes whose graphs are tree 3-spanner admissible (*cf.* [4]).

Threshold Graphs. Threshold graphs [18] can be defined as the intersection of two very well studied classes: split graphs and cographs. Thus, threshold graphs are $\{2K_2,\ P_4,\ C_4\}$-free graphs. Moreover, G is a *threshold graph* if G can constructed from the empty graph by repeatedly adding either an isolated vertex or a universal vertex.

Since to obtain spanning trees we only consider connected graphs, the last vertex of a threshold graph construction must be universal. Hence, a tree can be built as a star whose center is such a universal vertex. Thus we can state the following proposition.

Proposition 2. *If G is a threshold graph, then $\sigma_T(G) = 2$.*

Split Graphs. As just mentioned, split graphs are a superclass of threshold graphs. Formally, a graph $G = (X, Y)$ is a *split graph*, also called a $(1, 1)$-graph, if and only if it can be partitioned into a clique X and a stable set Y. In terms of forbidden subgraphs, they are $\{2K_2,\ C_4, C_5\}$-free graphs.

Lemma 8. *If G is a split graph, then $\sigma_T(G) \leq 3$.*

Proof. We obtain a spanning tree T for a split graph $G = (X, Y)$ as follows. Set any vertex x in X to be the center of a star which includes each other vertex of X. Next, for each vertex $y \in Y$, choose an edge incident to y, arbitrarily, and make y a pendant in T. It remains to show that the distance between two adjacent vertices v, w in G is at most 3 in T. (i) $v, w \in X$: since we have a star in T with respect to X, then $d(v, w) = 2$. (ii) $v \in X$, $w \in Y$: the worst case occurs when $d_G(w) \geq 2$ and v is a leaf of the star in T. In this case, $d(v, w) = 3$ by the path $vxx'w$, where $x'w$ belongs to T. $\square$

Now, we characterize split graphs whose stretch indexes are 2 or 3.

Proposition 3. *Let $G = (X, Y)$ be a split graph which is not a tree. Thus, $\sigma_T(G) = 2$ iff either: (i) $d_G(y) = 1, \forall\ y \in Y$, or (ii) $\exists\ x \in \bigcap_{y \in Y} N_G(y)$, $x \in X$ such that $d_G(y) \geq 2$.*

Proof. If G satisfies (i) or (ii), then G contains a tree 2-spanner which can be constructed following Lemma 8, and, particularly in case (ii), consider any vertex x satisfying conditions required in (ii) to be center of the star. Conversely, by contradiction, since $\sigma_T(G) = 2$, for each pair of vertices in X there is in T either an edge or a P_3 centered in a vertex v of G. If $v \in X$, then the minimum stretch spanning subtree with respect to X is a star. Otherwise, $v \in Y$ and each vertex of the clique would be a leaf of the star centered in v. Once there are two vertices in Y with degree at least 2 without an adjacent vertex in common, in the first case, for any center of the star we have chosen regarding the clique's vertices, there is a vertex of the stable set such that all its neighbors are leaves of the star, which implies $\sigma_T(G) \geq 3$. In the second case, $\sigma_T(G) \geq 3$ anyway, because, by hypothesis, there exist at least two more vertices in Y with degree at least 2, and they will be adjacent only to the leaves of the star centered in v. $\square$

Figure 1 exhibits a split graph G with $\sigma_T(G) = 3$. Another example of split graphs that have stretch index equal to 3 are the k-sun. Such graphs do not satisfy conditions of Proposition 3 either.

Cograph. A *cograph* is a P_4-free graph. A trivial graph is a cograph, and any other can be obtained by disjoint union or join operations of cographs. We can represent the union and join operations of a cograph by a tree decomposition, called *cotree* [19].

Theorem 6. *If G is a cograph, then $\sigma_T(G) \leq 3$.*

Proof. Since G must be connected, its cotree root's label is 1, implying that any vertex of G represented as a leaf node of a root's subtree is adjacent to all vertices of the other root's subtrees. We build a spanning tree T of G as follows. Let f be a leaf node of the leftmost root's subtree, F_1. Since f is adjacent to all vertices of the other root's subtrees, set T as a star with center f and make f adjacent to each vertex of all root's subtrees on F_1's right. Let lf be an edge of the star just obtained. Once the vertex l in G is adjacent to all vertices of F_1, hence we add to T each edge corresponding to a neighbor of l in F_1, except to the edge lf. Therefore, $\sigma_T(G) \leq 3$. $\square$

Lemma 9. *Given a graph G, let k be the number of its cotree root's subtrees. If G does not contain a universal vertex, then G contains an octahedral O_k as an induced subgraph.*

Proof. Since G does not contain a universal vertex, then each root's son of its cotree has label 0. Hence, in each subtree there are at least two leaves corresponding to non-adjacent vertices in G, but these two vertices are adjacent to all vertices of the other cotree root's subtrees. So, the union of each two non-adjacent vertices per subtree induces an O_k in G. $\square$

If a cograph G contains a universal vertex and there exist k' subtrees of the root with more than one leaf each, then there is an octahedral $O_{k'}$ as an induced subgraph of G. Moreover, if there were a universal vertex u with respect to $O_{k'}$, then $\sigma_T(O_{k'} \cup \{u\}) = 2$. However, such a vertex does not exist in a cograph without a universal vertex, because, in this case, all root's subtrees have label 0, and considering two $O_{k'}$ non-adjacent vertices, it does not exist a vertex of a same subtree adjacent to both vertices, otherwise their lowest common ancestor would be 1.

Lemma 10. *Let H be a cograph obtained from O_k by non-isolated vertices addition. If there is not a universal vertex in H with respect to O_k, then $\sigma_T(H) = 3$.*

Proof. Since O_k is an induced subgraph of H, by construction H is a triconnected component. If $\sigma_T(H) = 2$, then the tree 2-spanner of H would be a star. However, it is not possible since H does not have a universal vertex. $\square$

Since a cograph without universal vertex does not contain a universal vertex with respect to some octahedral, then we have that, for cographs, containing a universal vertex is also a necessary condition so that $\sigma_T(G) = 2$.

Theorem 7. *Let G be a cograph. $\sigma_T(G) = 2$ iff G has a universal vertex.*

Proof. If G contains a universal vertex, then $\sigma_T(G) = 2$. Let us prove the converse by contrapositive. If there is no universal vertex in G, then by Lemma 9 we have that G contains an octahedral O_k as induced subgraph, and by Lemma 10 we have that the unique case for decreasing σ_T from 3 to 2 is when there is a universal vertex with respect to O_k, but in a cograph with no universal vertex, there is no universal vertex with respect to an O_k.   □

4   Concluding Remarks and Further Work

In this work, we present an inconsistence on a well known sufficient condition for tree 2-spanner admissible graphs. Moreover, we establish optimum tree t-spanners for some graph classes by considering their characteristics, decompositions and by vertex/edges operations. Following the strategies proposed in this work, we intend to obtain optimum tree t-spanners for generalized split graphs, say (k, ℓ)-graphs, and also for graphs obtained by vertex/edges operations.

References

1. Peleg, D., Ullman, J.D.: An optimal synchronizer for the hypercube. In: Proceedings of the 6th ACM Symposium on Principles of Distributed Computing, Vancouver, pp. 77–85 (1987)
2. Althöfer, I., Das, G., Dobkin, D., Joseph, D., Soares, J.: On sparse spanners of weighted graphs. Discrete Comput. Geom. **9**(1), 81–100 (1993)
3. Bhatt, S., Chung, F., Leighton, T., Rosenberg, A.: Optimal simulations of tree machines. In: 27th Annual Symposium on Foundations of Computer Science, pp. 274–282. IEEE (1986)
4. Cai, L., Corneil, D.G.: Tree spanners. SIAM J. Discrete Math. **8**(3), 359–387 (1995)
5. Bansal, N., et al.: Min-max graph partitioning and small set expansion. SIAM J. Comput. **43**(2), 872–904 (2014)
6. Lanctot, J.K., Li, M., Ma, B., Wang, S., Zhang, L.: Distinguishing string selection problems. Inf. Comput. **185**(1), 41–55 (2003)
7. Fekete, S.P., Kremer, J.: Tree spanners in planar graphs. Discrete Appl. Math. **108**(1–2), 85–103 (2001)
8. Brandstädt, A., Dragan, F.F., Le, H.-O., et al.: Tree spanners on chordal graphs: complexity and algorithms. Theor. Comput. Sci. **310**(1–3), 329–354 (2004)
9. Panda, B., Das, A.: Tree 3-spanners in 2-sep chordal graphs: characterization and algorithms. Discrete Appl. Math. **158**(17), 1913–1935 (2010)
10. West, D.B., et al.: Introduction to Graph Theory, vol. 2. Prentice Hall, Upper Saddle River (2001)
11. Bondy, J., Murty, U.: Graph Theory, vol. 44 (2008)
12. Bondy, J.A., Murty, U.S.R., et al.: Graph Theory with Applications, vol. 290. Citeseer (1976)

13. Hassin, R., Tamir, A.: On the minimum diameter spanning tree problem. Inf. Process. Lett. **53**(2), 109–111 (1995)
14. Bondy, J.: Trigraphs. Discrete Math. **75**(1–3), 69–79 (1989)
15. Farber, M.: Characterizations of strongly chordal graphs. Discrete Math. **43**(2–3), 173–189 (1983)
16. Golumbic, M.C., Hammer, P.L.: Stability in circular arc graphs. J. Algorithm **9**(3), 314–320 (1988)
17. Pizaña, M.A.: The icosahedron is clique divergent. Discrete Math. **262**(1–3), 229–239 (2003)
18. Golumbic, M.C.: Algorithmic Graph Theory and Perfect Graphs, vol. 57. Elsevier (2004)
19. Spinrad, J.P.: Efficient Graph Representations: The Fields Institute for Research in Mathematical Sciences, vol. 19. American Mathematical Society (2003)

On the Approximability of
Time Disjoint Walks

Alexandre Bayen[1], Jesse Goodman[2(✉)], and Eugene Vinitsky[1]

[1] University of California, Berkeley, Berkeley, CA 94720, USA
{bayen,evinitsky}@berkeley.edu
[2] Cornell University, Ithaca, NY 14850, USA
jpmgoodman@cs.cornell.edu

Abstract. We introduce the combinatorial optimization problem Time Disjoint Walks. This problem takes as input a digraph G with positive integer arc lengths, and k pairs of vertices that each represent a *trip demand* from a source to a destination. The goal is to find a path and delay for each demand so that no two trips occupy the same vertex *at the same time*, and so that the sum of trip times is minimized. We show that even for DAGs with max degree $\Delta \leq 3$, Time Disjoint Walks is APX-hard. We also present a natural approximation algorithm, and provide a tight analysis. In particular, we prove that it achieves an approximation ratio of $\Theta(k/\log k)$ on bounded-degree DAGs, and $\Theta(k)$ on DAGs and bounded-degree digraphs.

Keywords: Hardness of approximation
Approximation algorithms · Disjoint Paths problem

1 Introduction

1.1 Related Work

Disjoint Paths is a classic problem in combinatorial optimization that asks: given an undirected graph G, and k pairs of vertices, do there exist vertex-disjoint paths that connect each pair? This problem captures the general notion of *connection without interference*, and has subsequently received much attention due to its applicability in areas like VLSI design [9,12] and communication networks [14,15].

These applications have motivated many variants of this basic problem. For example, one may choose the underlying graph to be undirected or directed, and the disjointness constraint to be over vertices or edges. As an optimization problem, one may consider the *maximum number of pairs* that can be connected with disjoint paths, the *minimum number of rounds* necessary to connect all pairs (where all paths in a round must be disjoint) [7], or the *shortest set of disjoint paths* to connect all pairs (if all pairs can, in fact, be disjointly connected) [8].

A few flavors of Disjoint Paths are tractable: for example, if k is fixed or G has bounded tree-width, then there exists a poly-time algorithm for finding

© Springer Nature Switzerland AG 2018
D. Kim et al. (Eds.): COCOA 2018, LNCS 11346, pp. 62–78, 2018.
https://doi.org/10.1007/978-3-030-04651-4_5

vertex-disjoint paths on undirected graphs [10,11]. Many interesting variants of Disjoint Paths are, however, extremely difficult. Indeed, finding vertex-disjoint paths on undirected graphs is one of Karp's NP-complete problems [6]. Furthermore, nearly-tight hardness results are known for finding the maximum set of edge-disjoint paths in a directed graph with m edges: there exists an $O(\sqrt{m})$-approximation algorithm [7], and it is NP-hard to approximate within a factor of $m^{1/2-\epsilon}$, for any $\epsilon > 0$ [5]. For detailed surveys on the complexity landscape of Disjoint Paths variants, see [7,8].

1.2   Contributions

Despite the great variety of Disjoint Paths problems that have been considered in the literature, it appears that little attention has been given to variants that relax the *disjointness* constraint, even though many natural applications do not always require paths to be completely disjoint. Consider, for example, the application of safely routing a collection of fully autonomous (and obedient) vehicles through an otherwise empty road network. In such a situation, we can certainly prevent collisions by routing all vehicles on disjoint paths. However, it is not difficult to see that if we have full control over the vehicles, using disjoint paths is rarely necessary (and, in fact, can be highly suboptimal).

Applications of this flavor motivate a new variant of Disjoint Paths, which roughly asks: given a graph G and k pairs of vertices that each represent a *trip demand*, how should we assign a delay and a path to each trip so that (1) trips are completed as quickly as possible, and (2) no two trips *collide* (i.e., occupy the same location at the same time). While there are problems in the literature (that do not wield the name "Disjoint Paths") that seemingly come close to capturing this goal, they exhibit some key differences. In particular, multicommodity flows over time [4,13] and job shop scheduling [3] seem, at first glance, very related to our problem. However, the former does not enforce unsplittable flows (as we require), and the latter does not capture the flexibility of scheduling job operations over any appropriate walk in a network.

As such, we are motivated to formalize and study this new variant of Disjoint Paths that relaxes the classical disjointness constraint to a "time disjointness" constraint. In particular, our contributions are threefold:

- We introduce a natural variant of Disjoint Paths, which we call *Time Disjoint Walks* (TDW). To the best of our knowledge, this is the first simple model that captures the notion of collision-free routing of *discrete objects* (i.e., instead of *flows*) over a shared network.
- We prove that Time Disjoint Walks is APX-hard, by providing an L-reduction from a variant of SAT. In fact, our reduction shows that this result holds even for directed acyclic graphs (DAGs) of max degree three ($\Delta \leq 3$).
- We describe an intuitive approximation algorithm for our problem, and provide a tight analysis: we show that it achieves an approximation ratio of $\Theta(k/\log k)$ on bounded-degree DAGs, and $\Theta(k)$ on DAGs and bounded-degree digraphs.

We formally introduce Time Disjoint Walks in Sect. 2. In Sect. 3 we provide some useful definitions regarding approximation. In Sect. 4 we prove our APX-hardness result. In Sect. 5 we describe our approximation algorithm, and provide bounds on its performance for the input classes mentioned above. In Sect. 6 we state our conclusions and present some open problems.

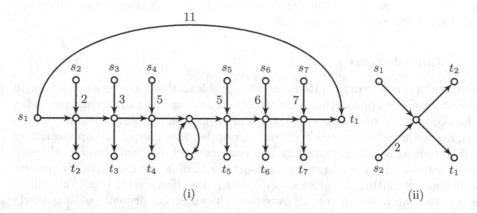

Fig. 1. (Unlabeled arcs have length 1): (i) A TDW instance with an optimal solution that contains cycles and intersecting walks, even though disjoint paths exist. (ii) A TDW instance with an obvious optimal solution, or a Shortest Disjoint Paths instance with no solution.

2   Time Disjoint Walks

We must first mention a few preliminaries: given $a, b \in \mathbb{Z}$, define $[a, b] := \{x \in \mathbb{Z} \mid a \leq x \leq b\}$, and for $b \in \mathbb{Z}$, we write $[b] := [1, b]$. Note that for $b < 1$, $[b] = \emptyset$. Given a directed graph (digraph) $G := (V, E)$, and $u, v \in V$, we define a *walk* W from u to v in G as a tuple $(w_1, w_2, \ldots, w_l)$ of vertices such that $w_1 = u, w_l = v$, and $(w_i, w_{i+1}) \in E$ for each $i \in [l - 1]$. Note that a vertex can be repeated.

Given a digraph G with arc lengths $\lambda : E \to \mathbb{Z}_{\geq 1}$, and a walk $W = (w_1, w_2, \ldots, w_l)$ in G, we let $|W| := l$ denote the *cardinality* of the walk, and we define for every $j \in [l]$ the *length of the walk up to its jth vertex* as

$$\lambda(W, j) := \sum_{i \in [j-1]} \lambda(w_i, w_{i+1}).$$

For convenience, we let $\lambda(W) := \lambda(W, l)$ denote the total length of the walk. Finally, given delays $d_1, d_2 \in \mathbb{Z}_{\geq 0}$ and walks W_1, W_2 in G, we say that (d_1, W_1) and (d_2, W_2) are *time disjoint* if, intuitively, a small object traversing W_1 at constant speed after waiting d_1 units of time does not collide/interfere with a small object traversing W_2 at the same speed after waiting d_2 units of time. We consider walks that have not departed, and walks that have already ended, to no longer exist on the network (and thereby not occupy any vertices). Formally, we have: for every $j_1 \in [|W_1|], j_2 \in [|W_2|]$ such that the j_1^{th} vertex of W_1 is equal to the j_2^{th} vertex of W_2,

$$d_1 + \lambda(W_1, j_1) \neq d_2 + \lambda(W_2, j_2).$$

We are now ready to formally define the problem examined in this paper:

Definition 1 (Time Disjoint Walks). *Let $G := (V, E)$ be a digraph, let $\lambda : E \to \mathbb{Z}_{\geq 1}$ define arc lengths, and let $\mathcal{T} := \{(s_1, t_1), (s_2, t_2), \ldots, (s_k, t_k)\} \subseteq V^2$ define a set of demands across unique vertices. For each $i \in [k]$, find a delay $d_i \in \mathbb{Z}_{\geq 0}$ and walk W_i from s_i to t_i such that the tuples in $\{(d_i, W_i) \mid i \in [k]\}$ are pairwise time disjoint, and $\sum_{i \in [k]} (d_i + \lambda(W_i))$ is minimized.*

We note that one can construct analogous problems by considering undirected graphs as input, edge lengths and delays that are real-valued, or a definition of *time disjoint* that requires large gaps between arrival times at common vertices (whereas the definition above simply requires a nonzero gap). Additionally, one may wish to consider a *min-max* objective instead of our *min-sum* objective.

We leave these variants to future work, noting that our primary goal in this paper is to study a basic flavor of this new combinatorial problem. Furthermore, our selection of this variant is well-motivated by our original application of routing a collection of identical autonomous vehicles over an empty road network (which, for the sake of this futuristic application, we may assume was built specifically for these vehicles). In particular, we may (1) model the road network as a directed graph, (2) assume that all routed vehicles traverse their walk at the same constant velocity, (3) measure road lengths as the time necessary to traverse it at that velocity, and (4) assume that road lengths are integer multiples of the time length of each vehicle. Additionally, we may motivate our *min-sum* objective by the desire to find a socially optimal solution.

Finally, we emphasize the novelty of our *time disjoint* constraint by comparing it to the standard *disjoint* constraint used in classical variants of Disjoint Paths. In particular, observe that if we modify the definition of Time Disjoint Walks to use the latter constraint instead of the former, we arrive at the (Min-Sum) Shortest Disjoint Paths problem [8]. However, this constraint makes all the difference: given an instance of Time Disjoint Walks, it is often the case that a solution under the standard *disjoint* constraint is suboptimal if examined under the *time disjoint* constraint. Indeed, the optimal solution under the latter constraint may even include paths that repeat vertices - hence the name Time Disjoint *Walks*; see (i) in Fig. (1). On the other hand, it is easy to construct an instance of Shortest Disjoint Paths that admits an obvious optimal solution under the time disjoint constraint, but does not yield any solution at all under the classical disjoint constraint; see (ii) in Fig. (1).

These observations strongly suggest that there is no simple reduction, in either direction, between Time Disjoint Walks and Disjoint Paths. Furthermore, using time-expanded networks [13] to reduce Time Disjoint Walks into Disjoint

Paths appears to offer little hope: such reductions will approximately square the size of the original graph, and many variants of Disjoint Paths are hard to approximate within $m^{1/2-\epsilon}$, for any $\epsilon > 0$ [5]. Thus, an approximation algorithm for Disjoint Paths, applied to a transformed Time Disjoint Walks instance, would likely fail to perform better than a trivial approximation algorithm for Time Disjoint Walks. These observations highlight the novelty of our problem and (in)approximability results.

3 Approximation Preliminaries

Given an optimization problem $\mathcal{P}$, we let $I_\mathcal{P}$ denote the instances of $\mathcal{P}$, $SOL_\mathcal{P}$ map each $x \in I_\mathcal{P}$ to a set of feasible solutions, and let $c_\mathcal{P}$ assign a real cost to each pair (x, y) where $x \in I_\mathcal{P}$ and $y \in SOL_\mathcal{P}(x)$. For $x \in I_\mathcal{P}$, we let $OPT_\mathcal{P}(x) := \min_{y^* \in SOL_\mathcal{P}(x)} c_\mathcal{P}(x, y^*)$ if $\mathcal{P}$ is a minimization problem, and $OPT_\mathcal{P}(x) := \max_{y^* \in SOL_\mathcal{P}(x)} c_\mathcal{P}(x, y^*)$ otherwise.

If $\mathcal{A}$ is a polynomial time algorithm with input $x \in I_\mathcal{P}$ and output $y \in SOL_\mathcal{P}(x)$, we say that $\mathcal{A}$ is a ρ-approximation algorithm, or has approximation ratio ρ, if $\mathcal{P}$ is a minimization problem and $c_\mathcal{P}(x, \mathcal{A}(x))/OPT_\mathcal{P}(x) \le \rho$, or $\mathcal{P}$ is a maximization problem and $OPT_\mathcal{P}(x)/c_\mathcal{P}(x, \mathcal{A}(x)) \le \rho$, for all $x \in I_\mathcal{P}$. Note that $\rho \ge 1$.

The class APX contains all optimization problems that admit a ρ-approximation algorithm, for *some* constant $\rho > 1$. An optimization problem is said to be *APX-hard* if every problem in APX can be reduced to it through an approximation-preserving reduction. One reduction of this type is the *L-reduction*:

Definition 2 (L-Reduction). *An L-reduction from an optimization problem $\mathcal{P}$ to an optimization problem $\mathcal{Q}$, denoted $\mathcal{P} \le_L \mathcal{Q}$, is a tuple (f, g, α, β), where:*

- *For each $x \in I_\mathcal{P}$, $f(x) \in I_\mathcal{Q}$ and can be computed in polynomial time.*
- *For each $y \in SOL_\mathcal{Q}(f(x))$, $g(x, y) \in SOL_\mathcal{P}(x)$ and can be computed in polynomial time.*
- *α is a positive real constant such that for each $x \in I_\mathcal{P}$,*

$$OPT_\mathcal{Q}(f(x)) \le \alpha \cdot OPT_\mathcal{P}(x).$$

- *β is a positive real constant such that for each $x \in I_\mathcal{P}, y \in SOL_\mathcal{Q}(f(x))$,*

$$\left| OPT_\mathcal{P}(x) - c_\mathcal{P}(x, g(x, y)) \right| \le \beta \cdot \left| OPT_\mathcal{Q}(f(x)) - c_\mathcal{Q}(f(x), y) \right|.$$

If a problem is APX-hard, it is NP-hard to ρ-approximate for some constant $\rho > 1$; thus, showing APX-hardness is strictly stronger than showing NP-hardness. To show APX-hardness, one can simply L-reduce from a known APX-hard problem. We refer the reader to [1] for a good reference on approximation.

4   Hardness of Approximation

To show the hardness of our problem, we show an L-reduction from MAX-E2SAT(3), which is known to be APX-hard [2]. We remind the reader of the definition, below, and then proceed with our proof.

Definition 3 (MAX-E2SAT(3)). *Let ϕ be a CNF formula in which (i) each clause contains exactly two literals on distinct variables, and (ii) each variable appears in at most three clauses. Find a truth assignment to the variables in ϕ that maximizes the number of satisfied clauses.*

Theorem 1. *Time Disjoint Walks is APX-hard, even for DAGs with $\Delta \leq 3$.*

Proof. We let $\mathcal{P} := $ MAX-E2SAT(3), $\mathcal{Q} := $ TDW with instances restricted to those containing DAGs with $\Delta \leq 3$, and show that $\mathcal{P} \leq_L \mathcal{Q}$. Below, we describe our L-reduction (f, g, α, β).

Description of f: Given an instance $\phi \in I_{\mathcal{P}}$ with n variables and m clauses, we let $X := \{x_1, \ldots, x_n\}$ refer to its variables and $\mathcal{C} := \{C_1, \ldots, C_m\}$ refer to its clauses. We let $L := \{x_1, \ldots, x_n, \overline{x}_1, \ldots, \overline{x}_n\}$ refer to its literals. For convenience, we define $e : L \to X$ that extracts the variable from a given literal; i.e., $e(x_i) = e(\overline{x}_i) = x_i$. We label the literals in clause C_j as l_j^1, l_j^2. For each $l \in L$, we let $S_l := \{l_j^a \mid a \in [2], j \in [m], l_j^a = l\}$ capture all occurrences of literal l in ϕ. Finally, for each $l \in L$, we define an arbitrary bijection $\pi_l : S_l \to [|S_l|]$ to induce an ordering on S_l. We will let π_l^{-1} denote its inverse: i.e., $\pi_l^{-1}(1)$ is the first element in S_l in the order induced by π_l.

We may now describe f, which constructs an instance $(G, \lambda, \mathcal{T}) \in I_{\mathcal{Q}}$ from ϕ. We start with the construction of G (see Fig. (2)), which closely follows the standard proof of NP-hardness for Disjoint Paths: for each clause $C_j = (l_j^1 \vee l_j^2)$ in ϕ, we create a new *clause gadget* and add it to G. That is, for each clause C_j, we add the following vertex and arc set to our construction:

$$V_{C_j} := \{c_j^s, l_j^1, l_j^2, l_j^{1'}, l_j^{2'}, c_j^t\}$$

$$E_{C_j} := \{(c_j^s, l_j^1), (c_j^s, l_j^2), (l_j^1, l_j^{1'}), (l_j^2, l_j^{2'}), (l_j^{1'}, c_j^t), (l_j^{2'}, c_j^t)\}$$

Next, for each $x_i \in X$, we add an *interleaving variable gadget* as follows: first, we add two vertices x_i^s, x_i^t to $V(G)$. Then, we wish to create exactly two directed paths (walks), $W_{\pi_i}^+, W_{x_i}^-$, from x_i^s to x_i^t; we want $W_{x_i}^+$ to travel through all vertices corresponding to positive literals of x_i, and $W_{x_i}^-$ to travel through all vertices corresponding to negative literals of x_i. Formally, for each $l \in \{x_i, \overline{x}_i\}$, we create a path from x_i^s to x_i^t as follows. First, if $|S_l| = 0$, we add arc (x_i^s, x_i^t) to $E(G)$. Otherwise, we add arcs $(x_i^s, \pi_l^{-1}(1))$ and $((\pi_l^{-1}(|S_l|))', x_i^t)$ to $E(G)$, and then for each $j \in [|S_l| - 1]$, we add arc $((\pi_l^{-1}(j))', \pi_l^{-1}(j + 1))$. Note that the prime symbols are merely labels, and are used in our construction to ensure that the max degree of G remains at most three. This completes our construction of G. We now define a set of $n + m$ demands, where each corresponds to a variable or a clause:

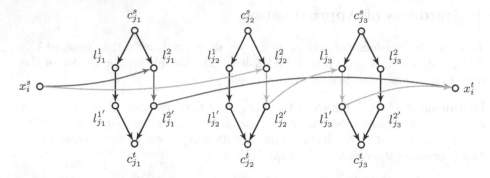

Fig. 2. An interleaving variable gadget (and its affiliated clause gadgets) corresponding to a variable with one negative occurrence (red path) and two positive occurrences (green path). (Color figure online)

$$\mathcal{T} := \{(x_i^s, x_i^t) \mid i \in [n]\} \cup \{(c_j^s, c_j^t) \mid j \in [m]\}$$

Finally, we must define arc lengths $\lambda : E \to \mathbb{Z}_{\geq 1}$. We will do this in a way that for each $j \in [m], a \in [2]$, we have $\lambda(W_{c_j^s}, l_j^a) = \lambda(W_{x_i^s}, l_j^a)$, where $W_{c_j^s}$ is the unique walk in G from c_j^s to l_j^a, and $W_{x_i^s}$ is the unique walk in G from $x_i^s = e(l_j^a)^s$ to l_j^a. Call this property $(*)$. To facilitate our analysis, we will also want every demand-satisfying path in G to have the same length.

Since ϕ is an instance of MAX-E2SAT(3), we know that for each $i \in [n]$, each of the two paths between x_i^s and x_i^t passes through at most 3 clause gadgets. Thus, by our construction, each such path includes at most 7 arcs, and any path from a variable x_i^s to some literal l_j^a with $x_i = e(l_j^a)$ can use at most 5 arcs. Thus, we can successfully force each demand-satisfying path in G to have length 7 while maintaining property $(*)$ by defining $\lambda : E \to \mathbb{Z}_{\geq 1}$ as follows, completing our construction of $(G, \lambda, \mathcal{T}) \in I_{\mathcal{Q}}$:

$$
\lambda(u,v) := \begin{cases}
1, & \text{if } (u,v) = (l_j^a, l_j^{a'}), j \in [m], a \in [2]; \text{ or} \\
& \text{if } (u,v) = (x_i^s, l_j^a), i \in [n], j \in [m], a \in [2]; \text{ or} \\
& \text{if } (u,v) = (l_h^{a'}, l_j^b), h, j \in [m], a, b \in [2]; \\
7, & \text{if } (u,v) = (x_i^s, x_i^t), i \in [n] \\
7 - 2|S_l|, & \text{if } (u,v) = (l_j^{a'}, x_i^t), j \in [m], a \in [2], i \in [n], l_j^a = l \\
2\pi_l(l_j^a) - 1, & \text{if } (u,v) = (c_j^s, l_j^a), j \in [m], a \in [2], l_j^a = l \\
7 - 1 - \lambda(c_j^s, l_j^a), & \text{if } (u,v) = (l_j^{a'}, c_j^t), j \in [m], a \in [2]
\end{cases}
$$

Description of g: Given a solution $y \in SOL_{\mathcal{Q}}(f(\phi))$, we construct a solution $g(\phi, y) \in SOL_{\mathcal{P}}(\phi)$ through two consecutive transformations: z, followed by q. That is, we will define transformations z and q such that g is the composition $g(\phi, y) := q(\phi, z(y))$.

We define z to transform solution y into another solution $y' \in SOL_{\mathcal{Q}}(f(\phi))$ such that $c_{\mathcal{Q}}(f(\phi), y') \leq c_{\mathcal{Q}}(f(\phi), y)$ and such that y' assigns 0 delay to demands

associated with interleaving variable gadgets. To accomplish this, recall that
$y = \{(d_1, W_1), \ldots, (d_{n+m}, W_{n+m})\}$, by definition of SOL_Q. Without loss of generality, we may assume tuples indexed with $[n]$ correspond to demands on interleaving variable gadgets, and tuples indexed with $[n + m] \setminus [n]$ correspond to demands on clause gadgets.

Now, while there exists some $i \in [n]$ such that $d_i > 0$ (and thus $d_i \geq 1$), we perform the following modification on y: first, we reset W_i to be the path traveling through at most one clause gadget - the positive or negative path *must* have this property, because each variable appears in ϕ at most three times, by definition of MAX-E2SAT(3). Now, reset d_i to 0. If W_i shares a vertex with another walk W_j, we know $j \in [n + m] \setminus [n]$, by construction of G. In this case, reset d_j to 1 if and only if d_j is currently 0. By construction of λ, the walks remain time disjoint and the cost of the solution does not increase.

In the second transformation, q, we transform modified solution y' into an assignment $(A : X \rightarrow \{T, F\}) \in SOL_P(\phi)$ as follows: for each $i \in [n]$, set $A(x_i) = T$ if and only if W_{x_i}, the walk from x_i^s to x_i^t, takes the negative literal path.

Valid value for α: We will show that for $\alpha = 29$, $OPT_Q(f(\phi)) \leq \alpha \cdot OPT_P(\phi)$. To see this, we make two observations. First observation: if $A : X \rightarrow \{T, F\}$ is a truth assignment for ϕ, then we can construct a solution to $f(\phi)$ as follows: for each $i \in [n]$, connect demand (x_i^s, x_i^t) using the negative literal path if $A(x_i) = T$, and the positive literal path if $A(x_i) = F$. Either way, assign a delay of 0. Then, for each $j \in [m]$ where clause C_j is satisfied by assignment A, connect demand (c_j^s, c_j^t) using a walk that goes through a literal that evaluates to true under A. Assign a delay of 0 to this demand. For each clause C_j that isn't satisfied by A, select an arbitrary walk to complete the corresponding demand (c_j^s, c_j^t). Assign a delay of 1 to this demand. It is clear that this is a valid solution to $f(\phi)$. Furthermore, the cost of our solution is $7(n + m) + U(A, \phi)$, where $U(A, \phi)$ is the number of clauses in ϕ unsatisfied by A. Second observation: by linearity of expectation, if ϕ is an instance of MAX-E2SAT(3), then there must exist an assignment $A : X(\phi) \rightarrow \{T, F\}$ that satisfies at least $3/4$ of the clauses.

We may now prove the desired inequality for $\alpha = 29$. From our first observation and the fact that $n \leq 2m$ (since each of the m clauses has 2 literals),

$$OPT_Q(f(\phi)) \leq 7(n + m) + (m - OPT_P(\phi)) \leq 22m - OPT_P(\phi). \qquad (1)$$

Now, by our second observation, we know $OPT_P(\phi) \geq 3m/4$. Thus, we have:

$$OPT_Q(f(\phi)) \leq 22 \cdot (4/3) \cdot OPT_P(\phi) - OPT_P(\phi) \leq 29 \cdot OPT_P(\phi).$$

Valid value for β: We will show that for $\beta = 1$ and any $y \in SOL_Q(f(\phi))$, $(OPT_P(\phi) - c_P(\phi, g(\phi, y))) \leq \beta \cdot (c_Q(f(\phi), y) - OPT_Q(f(\phi)))$, as required. As a first step, we recall that transformations z, q define g, and let γ denote the number of clause gadget demands assigned a delay of 0 by solution $z(y)$ to $f(\phi)$. We make the following crucial claim:

$$c_P(\phi, g(\phi, y)) := c_P(\phi, q(\phi, z(y))) \geq \gamma. \qquad (2)$$

To see this, note the following: by construction, $z(y)$ is a valid solution to $f(\phi)$. Thus, if $z(y)$ assigns clause gadget demand (c_j^s, c_j^t) a delay $d_j = 0$ and walk W_j that passes through literal l, then l is a positive literal if and only if the walk selected for the interleaving variable gadget demand (x_i^s, x_i^t) (where $x_i = e(l)$) does not travel through the positive literals of x_i. By definition of q, this occurs if and only if $g(\phi, y)$ assigns $true$ to x_i. Thus, a clause gadget demand given 0 delay by $z(y)$ corresponds to a clause in ϕ satisfied by $g(\phi, y)$, thus proving (2).

Next, by definition of γ and z, we have:

$$7(n+m) + (m - \gamma) \le c_Q(f(\phi), z(y)) \le c_Q(f(\phi), y). \tag{3}$$

Combining inequalities (2) and (3), we get:

$$c_P(\phi, g(\phi, y)) \ge \gamma \ge 7n + 8m - c_Q(f(\phi), y). \tag{4}$$

Finally, using the leftmost inequality in (1) along with inequality (4) gives us:

$$
\begin{aligned}
OPT_P(\phi) - c_P(\phi, g(\phi, y)) &\le \big(7n + 8m - OPT_Q(f(\phi))\big) - \big(7n + 8m - c_Q(f(\phi), y)\big) \\
&= \beta \cdot \big(c_Q(f(\phi), y) - OPT_Q(f(\phi))\big),
\end{aligned}
$$

for $\beta = 1$, as desired. This completes the proof that (f, g, α, β) is a valid L-reduction, and subsequently that TDW on DAGs with $\Delta \le 3$ is APX-hard. □

5   Approximation Algorithm

5.1   Algorithm

We present Algorithm 1, which approximates TDW by finding shortest paths to satisfy each demand, and then greedily assigning delays to each trip (with priority given to shorter trips). To simplify notation, *we assume that the inputted terminal pairs are ordered by nondecreasing shortest path length* (if not, we may simply sort the indices after finding the shortest demand-satisfying paths). The algorithm clearly runs in $\mathrm{poly}(|V|, |E|, k)$ time, and the bad_delay variables ensure its correctness. Next, we briefly note the following easy bound:

Proposition 1. *Algorithm 1 has an approximation ratio of $O(k)$ on general digraphs.*

Proof. Let $x := (G, \lambda, \mathcal{T}) \in I_{TDW}$, and let $\mathcal{A}(x) \in SOL_{TDW}$ be the output of Algorithm 1 on x. First, we show by induction that for each $i \in [k]$,

$$d_i \le 2 \sum_{h \in [i-1]} \lambda(W_h).$$

Algorithm 1. Shortest paths & greedy delays, with priority to shorter paths.

Input: $x := (G := (V, E), \lambda : E \to \mathbb{Z}_{\geq 1}, \mathcal{T} := \{(s_1, t_1), \ldots, (s_k, t_k)\}) \in I_{TDW}$
Output: $y \in SOL_{TDW}(x)$
1: $y \leftarrow \{\}$
2: ▷ Get shortest paths and dummy delays:
3: **for** $i \in [k]$ **do**
4:     $W_i \leftarrow \text{Dijkstra}(G, \lambda, s_i, t_i)$
5:     $d_i \leftarrow 0$
6:     $y \leftarrow y \cup (d_i, W_i)$
7: **end for**
8: ▷ Greedily assign delays, with priority given to shorter paths:
9: **for** $i \in [k]$ **do**
10:     $\text{bad\_delays}_i \leftarrow \{\}$
11:     **for** $h \in [i - 1]$ **do**
12:         $\text{bad\_delays}_{i,h} \leftarrow \{\}$
13:         **for** $v \in W_h \cap W_i$ **do**
14:             $\text{bad\_delay} \leftarrow (d_h + \lambda(W_h, v) - \lambda(W_i, v))$
15:             $\text{bad\_delays}_{i,h} \leftarrow \text{bad\_delays}_{i,h} \cup \{\text{bad\_delay}\}$
16:         **end for**
17:         $\text{bad\_delays}_i \leftarrow \text{bad\_delays}_i \cup \text{bad\_delays}_{i,h}$
18:     **end for**
19:     $d_i \leftarrow \min(\mathbb{Z}_{\geq 0} \setminus \text{bad\_delays}_i)$
20: **end for**
21: **return** y

For the base case $i = 1$, note that $\text{bad\_delays}_1 = \emptyset$ and so $d_1 = 0$. For $i > 1$, first observe that by definition of bad_delay, we have $d_i \leq 1 + \max_{h \in [i-1]}(d_h + \lambda(W_h))$. Thus,

$$d_i \leq 1 + \max_{h \in [i-1]} \left(2 \sum_{h' \in [h-1]} \lambda(W_{h'}) + \lambda(W_h) \right) \qquad \text{(induction hypothesis)}$$

$$\leq 1 + 2 \sum_{h' \in [i-2]} \lambda(W_{h'}) + \lambda(W_{i-1}) \qquad \text{(pick } h = i - 1)$$

$$\leq 2 \sum_{h' \in [i-1]} \lambda(W_{h'}), \qquad \text{(trips have length } \geq 1)$$

completing the induction. Now, recalling that our algorithm uses the shortest paths to satisfy each demand, and that it assigns delays to shorter paths first, we can bound the approximation ratio as follows:

$$\rho \leq \frac{c_{TDW}(x, \mathcal{A}(x))}{OPT_{TDW}(x)} \leq \frac{\sum_{i \in [k]}(d_i + \lambda(W_i))}{\sum_{i \in [k]} \lambda(W_i)} \leq 1 + \frac{2 \sum_{i \in [k]} \sum_{h \in [i-1]} \lambda(W_h)}{\sum_{i \in [k]} \lambda(W_i)}$$

$$\leq 1 + \frac{2k \sum_{i \in [k]} \lambda(W_i)}{\sum_{i \in [k]} \lambda(W_i)} = O(k). \qquad \square$$

5.2 Analysis on Bounded-Degree DAGs

We now show that our algorithm is able to achieve a better approximation ratio on bounded-degree DAGs. In what follows, we call a directed graph a "$(2, l)$-in-tree" if it is a perfect binary tree of depth l, in which every arc points toward the root. Analogously, a "$(2, l)$-out-tree" is a perfect binary tree of depth l, in which every arc points away from the root.

Theorem 2. *Algorithm 1 achieves an approximation ratio of $\Theta(k/\log k)$ on bounded-degree DAGs.*

Proof. <u>Upper bound</u>: Let $x := (G, \lambda, \mathcal{T}) \in I_{TDW}$ such that G is a DAG. Let $\mathcal{A}(x) \in \overline{SOL_{TDW}}$ be the output of Algorithm 1 on x. In what follows, we will justify the following string of inequalities that proves the upper bound:

$$\rho \leq \frac{c_{TDW}(x, \mathcal{A}(x))}{OPT_{TDW}(x)} \leq_{(1)} \frac{\sum_{i \in [k]} \left(d_i + \lambda(W_i) \right)}{\sum_{i \in [k]} \lambda(W_i)}$$

$$\leq_{(2)} 1 + \frac{d_{i^*}}{\lambda(W_{i^*})}, i^* := \max_{i \in [k]} \left(\frac{d_i}{\lambda(W_i)} \right)$$

$$\leq_{(3)} 1 + O(1) \cdot \frac{d_{i^*}}{\log d_{i^*}}$$

$$\leq_{(4)} 1 + O(1) \cdot \frac{k}{\log k} = O(k/\log k).$$

Inequality (1) is clear, because our algorithm takes the shortest path to satisfy each demand. **Inequality (2)** follows (by induction) from the following general observation: given $d_1, d_2 \in \mathbb{Z}_{\geq 0}$ and $\lambda_1, \lambda_2 \in \mathbb{Z}_{\geq 1}$, observe $d_1/\lambda_1 \leq d_2/\lambda_2 \implies (d_1 + d_2)/(\lambda_1 + \lambda_2) \leq d_2/\lambda_2$, and thus $(d_1 + d_2)/(\lambda_1 + \lambda_2) \leq \max(d_1/\lambda_1, d_2/\lambda_2)$.

To show **inequality (3)**, we need two observations. We first observe that for each $i \in [k]$:

$$d_i \leq |\mathsf{bad\_delays}_i| \leq |\{h \in [i - 1] \mid W_h \cap W_i \neq \emptyset\}| =: \mu_i$$

To see this, suppose for contradiction that there exists some $h \in [i - 1]$ with $W_h \cap W_i \neq \emptyset$ and $|\mathsf{bad\_delays}_{i,h}| > 1$. Then, by definition of $\mathsf{bad\_delay}$, there exist vertices $u, v \in W_h \cap W_i$ and delays $\delta_u \neq \delta_v \in \mathbb{Z}_{\geq 0}$ such that:

$$\delta_u + \lambda(W_i, u) = d_h + \lambda(W_h, u),$$
$$\delta_v + \lambda(W_i, v) = d_h + \lambda(W_h, v),$$
$$\lambda(W_h, u) - \lambda(W_h, v) = \lambda(W_i, u) - \lambda(W_i, v) + (\delta_u - \delta_v),$$

where the last equality follows from the first two. But because $\delta_u \neq \delta_v$, this implies that the length of the path that W_h and W_i use to travel between u and v is not the same. Because G is a DAG, W_h and W_i must visit u and v in the same order, implying that one of these walks is not taking the shortest path from u to v, which contradicts the definition of the algorithm. Because $|\mathsf{bad\_delays}_{i,h}| = 0$ if $W_h \cap W_i = \emptyset$, we have $d_i \leq \mu_i$.

Next, we observe that:

$$\mu_i \leq \min(\Delta^{4\lambda(W_i)}, k).$$

Showing $\mu_i \leq k$ is trivial, by definition of μ_i and because $i \in [k]$. To show $\mu_i \leq \Delta^{4\lambda(W_i)}$, first note that in a digraph with max degree Δ, the number of paths that (i) have z arcs, (ii) start at distinct vertices, and (iii) all end at a common vertex, is upper bounded by Δ^z (this is easy to show by induction on z). Thus, the number of paths with $\leq z$ arcs, in addition to properties (ii) and (iii), is upper bounded by $\sum_{l=0}^{z} \Delta^l \leq \Delta^{z+1}$, for $\Delta > 1$. Call this lemma $(*)$.

Now, note that for each $h \in [i-1]$ we may consider each W_h to terminate once it first hits a vertex in W_i (i.e., cut off all vertices that are hit afterwards) without changing the value of μ_i. Now, recall the following facts about our problem and algorithm: (I) each inputted demand has a unique source; (II) each edge in our digraph has length ≥ 1; (III) for all $h \in [i-1]$, $\lambda(W_h) \leq \lambda(W_i)$. Thus, by (III) and lemma $(*)$, each vertex in W_i can be hit by at most $\Delta^{\lambda(W_i)+1}$ walks in $\{W_1, \ldots, W_{i-1}\}$. Furthermore, (II) tells us that the number of vertices in W_i is no more than $\lambda(W_i) + 1$. Thus, recalling that our problem statement ensures $\Delta > 1$, no demands have the same source and destination, and (II), we see that

$$\mu_i \leq (\lambda(W_i) + 1)\Delta^{\lambda(W_i)+1} \leq \Delta^{2(\lambda(W_i)+1)} \leq \Delta^{4\lambda(W_i)},$$

as desired. We now note that we may assume d_i is greater than any constant (otherwise, inequality (2) automatically proves a constant approximation ratio, completing the proof). Thus, from this and the above observations, we have $\log(d_i) \leq 4\lambda(W_i)\log(\Delta)$. This proves inequality (3), because our graph has bounded degree.

Inequality (4) is not difficult: as stated above, we will always have $d_i \leq k$, and we may always assume $d_i \geq 3$. Basic calculus shows the function $x/\log x$ increases over $x \geq 3$.

<u>Lower bound</u>: We show $\forall l \in \mathbb{N}_{\geq 1}, k := 2^l, \exists (G_k, \lambda_k, \mathcal{T}_k) \in I_{TDW}$ such that G_k is a bounded-degree DAG and Algorithm 1 achieves an approximation ratio of $\Omega(k/\log k)$. Construct G_k by taking a $(2, l)$-in-tree A_S and a $(2, l)$-out-tree A_T. Draw an arc from the root of the former to the root of the latter. Then, arbitrarily pair each leaf (source) in A_S with a unique leaf (destination) in A_T. For each such pair, draw an arc from source to destination (called a "bypass arc"), and add a demand to $\mathcal{T}_k$. Finally, define λ_k to assign length $1 + 2l$ to each "bypass" arc, and length 1 to all other arcs. We refer the reader to Fig. (3)(i).

We may assume our algorithm does not satisfy demands using the bypass arcs (as all demand-satisfying paths have length $2l+1$, and no tie-breaking scheme is specified). Thus, each demand-satisfying path uses the root of A_S, which incurs a total delay of $0 + 1 + \ldots + (k-1) = \Omega(k^2)$ and total path length of $k \cdot (1 + 2l)$. Had the bypass arcs been used, no delay would have been required, and the total path length would have still been $k \cdot (1 + 2l)$. Thus, our algorithm achieves an approximation ratio of $(\Omega(k^2) + k \cdot (1 + 2l))/(k \cdot (1 + 2l)) = \Omega(k/\log k)$. □

5.3   Analysis on DAGs

We show that if we no longer require the graph family in Theorem (2) to have bounded degree, our algorithm loses its improved approximation ratio.

Theorem 3. *Algorithm 1 has an approximation ratio of $\Theta(k)$ on DAGs.*

Proof. By Proposition (1), it suffices to construct a family of TDW instances on DAGs, defined over all $k \in \mathbb{N}_{\geq 1}$, for which our algorithm achieves an approximation ratio of $\Omega(k)$. Construct G_k by fixing a "root" vertex and directly attaching $2k$ leaves. Orient half of these arcs towards the root, and half of the arcs away from the root. Call each vertex with out-degree 1 a *source*, and each vertex with in-degree 1 a *destination*. Then, arbitrarily pair each source with a unique destination. For each pair, add an arc from the source to the destination (called a

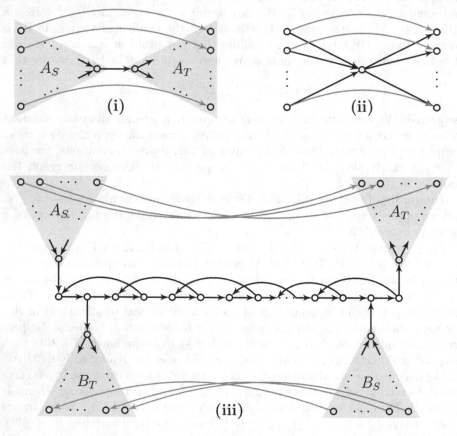

Fig. 3. (i): A bounded-degree DAG G_k upon which Algorithm 1 achieves an approximation ratio of $\Omega(k/\log k)$; (ii): A DAG G_k upon which Algorithm 1 achieves an approximation ratio of $\Omega(k)$; (iii): A bounded-degree digraph G_k upon which Algorithm 1 achieves an approximation ratio of $\Omega(k)$.

"bypass arc"), and add a demand to $\mathcal{T}_k$. Finally, let λ_k assign length 2 to each bypass arc, and length 1 to all other arcs. We refer the reader to Figure (3)(ii).

We may assume our algorithm does not satisfy demands using the bypass arcs (as all demand-satisfying paths have length 2, and no tie-breaking scheme is specified). Thus, each demand-satisfying path uses the root vertex, which incurs a total delay of $0 + 1 + \ldots + (k - 1) = \Omega(k^2)$ and total path length of $2k$. Had the bypass arcs been used, no delay would have been required, and the total path length would have still been $2k$. Thus, our algorithm achieves an approximation ratio of $(\Omega(k^2) + 2k)/(2k) = \Omega(k)$. $\qquad\square$

5.4 Analysis on Bounded-Degree Digraphs

In this section, we show that if we no longer require the graph family in Theorem (2) to be acyclic, our algorithm loses its improved approximation ratio.

Theorem 4. *Algorithm 1 has an approximation ratio of $\Theta(k)$ on bounded-degree digraphs.*

Proof. By Proposition (1), it suffices to construct a family of TDW instances on bounded-degree digraphs, defined over all $l \in \mathbb{N}_{\geq 2}$ with $\hat{k} := 2^l, k := 2\hat{k}$, for which our algorithm achieves an approximation ratio of $\Omega(k)$. Construct G_k by taking two $(2, l)$-in-trees A_S and B_S, and two $(2, l)$-out-trees A_T and B_T. Call their roots $r_{A_S}, r_{B_S}, r_{A_T}$, and r_{B_T}, respectively. Then, add a "central path" C consisting of vertices $\{c_1, c_2, \ldots, c_{\hat{k}}\}$, "forward" arcs $\{(c_i, c_{i+1}) \mid i \in [\hat{k} - 1]\}$, and "backward" arcs $\{(c_j, c_{j-3}) \mid j \in [4, \hat{k}], j \bmod 2 = 0\}$. Attach the directed trees to the central path with arcs $\{(r_{A_S}, c_1), (r_{B_S}, c_{\hat{k}-1}), (c_{\hat{k}}, r_{A_T}), (c_2, r_{B_T})\}$. Next, pair each leaf (source) in A_S with an arbitrary, but unique, leaf (destination) in A_T. Do the same for B_S and B_T. For each such pair, add an arc from the source to destination (called a "bypass arc"), and add a demand to $\mathcal{T}_k$. Finally, let λ_k assign length $2\hat{k} + 2l - 1$ to each bypass arc, length $\hat{k} - 1$ to arcs $(r_{B_S}, c_{\hat{k}-1})$ and $(c_{\hat{k}}, r_{A_T})$, and length 1 to all other arcs. We refer the reader to Figure (3)(iii).

Observe that for each demand, there exist *two* shortest demand-satisfying paths, each of length $2\hat{k} + 2l - 1$. In particular, observe that a demand between leaves of A_S and A_T may be satisfied by a bypass arc, or by a path that travels from the source in A_S, towards the root of A_S, onto the central path vertex c_1, along all forward arcs of C, onto the root of A_T, and towards the destination in A_T. Similarly, a demand between leaves of B_S and B_T may be satisfied by a bypass arc, or by a path that travels from the source in B_S, towards the root of B_S, onto the central vertex $c_{\hat{k}-1}$, across C by alternating between forward and backward arcs (until arriving at c_2), onto the root of B_T, and towards the destination in B_T. We call the paths that do not use the bypass arcs the "meandering paths."

Because our algorithm specifies no tie-breaking scheme, we may assume that it satisfies demands using the meandering paths, and that it alternates between assigning delays to demands from A_S and assigning delays to demands from B_S every *four* iterations. In other words, out of the $2k$ demands created above and fed as input to our algorithm, we may assume that those from A_S to A_T are labeled with indices $I_A := \{i \in [2\hat{k}] \mid \lfloor (i-1)/4 \rfloor \equiv 0 \pmod{2}\}$, while those from B_S to B_T are labeled with $I_B := \{i \in [2\hat{k}] \mid \lfloor (i-1)/4 \rfloor \equiv 1 \pmod{2}\}$.

To understand the suboptimality of this situation, we make several observations that help us determine the values our algorithm assigns to each d_i. First, note that for each $i \in [2\hat{k}]$, $z \in [\hat{k}]$, the length of walk W_i up to vertex c_z on the central path is:

$$\lambda_k(W_i, c_z) = \begin{cases} l + z, & \text{if } i \in I_A \\ l + 2\hat{k} - z - 2 \cdot (z \bmod 2), & \text{if } i \in I_B \end{cases}$$

Using this, we see that the details of Algorithm 1 give us the following relation, which is defined over $i \in [k], h \in [i-1]$:

$$\text{bad\_delays}_{i,h} = \begin{cases} \{d_h\}, & \text{if } i,h \in I_A \text{ or} \\ & \text{if } i,h \in I_B \\ \{d_h - 2\hat{k} + 2z + 2 \cdot (z \bmod 2) \mid z \in [\hat{k}]\}, & \text{if } i \in I_B, h \in I_A \\ \{d_h + 2\hat{k} - 2z - 2 \cdot (z \bmod 2) \mid z \in [\hat{k}]\}, & \text{if } i \in I_A, h \in I_B \end{cases}$$

Because our algorithm defines $\text{bad\_delays}_i := \bigcup_{h \in [i-1]} \text{bad\_delays}_{i,h}$ and $d_i := \min(\mathbb{Z}_{\geq 0} \setminus \text{bad\_delays}_i)$, observe that the above relation is in fact a *recurrence* relation. As such, after noting that $d_1 = 0$, it is straightforward to use the above relation to show by induction that for all $i \in [2\hat{k}]$,

$$d_i = i - 1 + \lfloor \frac{i-1}{8} \rfloor (2\hat{k} - 4).$$

Thus, our algorithm incurs a total delay of $\sum_{i \in [2\hat{k}]} (i - 1 + \lfloor (i-1)/8 \rfloor (2\hat{k} - 4)) = \Omega(\hat{k}^3) = \Omega(k^3)$ and total walk length of $2\hat{k} \cdot (2\hat{k} + 2l - 1)) = \Theta(\hat{k}^2) = \Theta(k^2)$. Had the algorithm opted to use the bypass arcs, no delay would have been required, and the total walk length would have been the same. Thus, our algorithm achieves an approximation ratio of $\Omega(k)$. □

6   Conclusions

In this paper, we introduce Time Disjoint Walks, a new variant of (shortest) Disjoint Paths that also seeks to connect k demands in a network, but relaxes the disjointness constraint by permitting vertices to be shared across multiple walks, as long as no two walks arrive at the same vertex at the same time. We show that Time Disjoint Walks is APX-hard, even for DAGs of max degree three. On the other hand, we provide a natural $\Theta(k/\log k)$-approximation algorithm for directed acyclic graphs of bounded degree. Interestingly, we also show that for general digraphs with just one of these two properties, the approximation ratio of our algorithm is bumped up to $\Theta(k)$.

An interesting future work is to tighten the gap between these inapproximability and approximability results for TDW on bounded-degree DAGs. We conjecture that our approximation algorithm is almost optimal, but that our hardness of approximation result can be strengthened to nearly match our algorithm's approximation ratio of $\Theta(k/\log k)$. This belief is based on the observation that TDW is a complex problem that involves both routing *and* scheduling, and many problems of the latter variety (of size n) are NP-hard to approximate within a factor of $n^{1-\epsilon}$, for any $\epsilon > 0$ [16]. One may also wish to explore similar complexity questions for the many variants of Time Disjoint Walks discussed in Sect. 2.

References

1. Ausiello, G., Crescenzi, P., Gambosi, G., Kann, V., Marchetti-Spaccamela, A., Protasi, M.: Complexity and Approximation: Combinatorial Optimization Problems and Their Approximability Properties. Springer, Heidelberg (2012)
2. Berman, P., Karpinski, M.: On some tighter inapproximability results (extended abstract). In: Wiedermann, J., van Emde Boas, P., Nielsen, M. (eds.) ICALP 1999. LNCS, vol. 1644, pp. 200–209. Springer, Heidelberg (1999). https://doi.org/10.1007/3-540-48523-6_17
3. Graham, R.L.: Bounds for certain multiprocessing anomalies. Bell Syst. Tech. J. **45**(9), 1563–1581 (1966)
4. Groß, M., Skutella, M.: Maximum multicommodity flows over time without intermediate storage. In: Epstein, L., Ferragina, P. (eds.) ESA 2012. LNCS, vol. 7501, pp. 539–550. Springer, Heidelberg (2012). https://doi.org/10.1007/978-3-642-33090-2_47
5. Guruswami, V., Khanna, S., Rajaraman, R., Shepherd, B., Yannakakis, M.: Near-optimal hardness results and approximation algorithms for edge-disjoint paths and related problems. J. Comput. Syst. Sci. **67**(3), 473–496 (2003)
6. Karp, R.M.: On the computational complexity of combinatorial problems. Networks **5**(1), 45–68 (1975)
7. Kleinberg, J.M.: Approximation algorithms for disjoint paths problems. Ph.D. thesis. Massachusetts Institute of Technology (1996)
8. Kobayashi, Y., Sommer, C.: On shortest disjoint paths in planar graphs. Discrete Optim. **7**(4), 234–245 (2010)
9. Lengauer, T.: Combinatorial Algorithms for Integrated Circuit Layout. Springer, Heidelberg (2012). https://doi.org/10.1007/978-3-322-92106-2
10. Robertson, N., Seymour, P.D.: Graph minors. XIII. The disjoint paths problem. J. Combin. Theor. Ser. B **63**(1), 65–110 (1995)
11. Scheffler, P.: A practical linear time algorithm for disjoint paths in graphs with bounded tree-width. TU, Fachbereich 3 (1994)
12. Schrijver, A., Lovasz, L., Korte, B., Promel, H.J., Graham, R.: Paths, Flows, and VLSI-Layout. Springer, New York (1990)
13. Skutella, M.: An introduction to network flows over time. In: Cook, W., Lovász, L., Vygen, J. (eds.) Research Trends in Combinatorial Optimization, pp. 451–482. Springer, Heidelberg (2009). https://doi.org/10.1007/978-3-540-76796-1_21
14. Srinivas, A., Modiano, E.: Minimum energy disjoint path routing in wireless ad-hoc networks. In: Proceedings of the 9th Annual International Conference on Mobile Computing and Networking, pp. 122–133. ACM (2003)

15. Torrieri, D.: Algorithms for finding an optimal set of short disjoint paths in a communication network. IEEE Trans. Commun. **40**(11), 1698–1702 (1992)
16. Zuckerman, D.: Linear degree extractors and the inapproximability of max clique and chromatic number. In: Proceedings of the Thirty-Eighth Annual ACM Symposium on Theory of Computing, pp. 681–690. ACM (2006)

Directed Path-Width of Sequence Digraphs

Frank Gurski[1], Carolin Rehs[1(✉)], and Jochen Rethmann[2]

[1] Institute of Computer Science, Heinrich Heine University Düsseldorf,
40225 Düsseldorf, Germany
carolin.rehs@uni-duesseldorf.de
[2] Faculty of Electrical Engineering and Computer Science,
Niederrhein University of Applied Sciences, 47805 Krefeld, Germany

Abstract. Computing the directed path-width of a digraph is NP-hard even for digraphs of maximum semi-degree 3. In this paper we consider a family of graph classes called sequence digraphs, such that for each of these classes the directed path-width can be computed in polynomial time. For this purpose we define the graph classes $S_{k,\ell}$ as the set of all digraphs $G = (V, A)$ which can be defined by k sequences with at most ℓ entries from V, such that $(u,v) \in A$ if and only if in one of the sequences u occurs before v. We characterize digraphs which can be defined by $k = 1$ sequence by four forbidden subdigraphs and also as a subclass of semicomplete digraphs. Given a decomposition of a digraph G into k sequences, we show an algorithm which computes the directed path-width of G in time $\mathcal{O}(k \cdot (1 + N)^k)$, where N denotes the maximum sequence length. This leads to an XP-algorithm w.r.t. k for the directed path-width problem. As most known parameterized algorithms for directed path-width consider the standard parameter, our algorithm improves significantly the known results for a high amount of digraphs of large directed path-width.

Keywords: Directed path-width · Transitive tournament
Semicomplete digraph · XP-algorithm

1 Introduction

A *sequence digraph* is defined by a set $Q = \{q_1, \ldots, q_k\}$ of k sequences $q_i = (b_{i,1}, \ldots, b_{i,n_i})$, $1 \leq i \leq k$. Further there is a function t which assigns to every item $b_{i,j}$ a type $t(b_{i,j})$. The sequence digraph $g(Q) = (V, A)$ for the set Q has a vertex for every type and an arc $(u,v) \in A$ if and only if there is some sequence q_i in Q where an item of type u is on the left of some item of type v. The set of all sequence digraphs which can be defined by sets Q on at most k sequences that together contain at most ℓ items of each type is denoted by $S_{k,\ell}$.

C. Rehs—The work of the second author was supported by the German Research Association (DFG) grant GU 970/7-1.

We show in Theorem 1 that $S_{1,1}$ is equal to the well known class of transitive tournaments. Since only the first and the last item of each type in every $q_i \in Q$ are important for the arcs in the corresponding digraph all classes $S_{1,\ell}$, $\ell \geq 2$ are equal. We show in Theorem 2 that $S_{1,2}$ is equal to the set of semicomplete $\{\text{co-}(2\overrightarrow{P_2}), \overrightarrow{C_3}, D_4\}$-free digraphs (cf. Fig. 3 for the digraphs). By our Proposition 2 set $S_{k,1}$ can be characterized by only three forbidden subdigraphs. It is also the class of disjoint unions of k transitive tournaments.

Considering the directed path-width problem on sequence digraphs, we get some remarkable results. We show that for digraphs defined by $k = 1$ sequence the directed path-width can be computed in polynomial time. Further we show that for sets Q of sequences of bounded length, of bounded distribution of the items of every type onto the sequences, or bounded number of items of every type computing the directed path-width of $g(Q)$ is NP-hard. We show that for a fixed number k of sequences the directed path-width is computable in polynomial time. Therefore in Theorem 3 we introduce an algorithm which computes the directed path-width of a digraph which is given by a set of k sequences in time $\mathcal{O}(k \cdot (1 + \max_{1 \leq i \leq k} n_i)^k)$. The main idea is to discover an optimal directed path-decomposition by scanning the k sequences left-to-right and keeping in a state the numbers of scanned items of every sequence and a certain number of active types.

From a parameterized point of view our solution leads to an XP-algorithm w.r.t. parameter k. While the existence of FPT-algorithms for computing directed path-width is open up to now, there are further XP-algorithms for the directed path-width problem for some digraph $G = (V, A)$. The directed path-width can be computed in time $\mathcal{O}(|A| \cdot |V|^{2\text{d-pw}(G)}/(\text{d-pw}(G)-1)!)$ by [7] and in time $\mathcal{O}(\text{d-pw}(G) \cdot |A| \cdot |V|^{2\text{d-pw}(G)})$ by [11]. Further in [8] it is shown how to decide whether the directed path-width of an ℓ-semicomplete digraph is at most w in time $(\ell + 2w + 1)^{2w} \cdot |V|^{\mathcal{O}(1)}$. All these algorithms are exponential in the directed path-width of the input digraph while our algorithm is exponential within the number of sequences. Thus our result improves theses algorithms for digraphs of large directed path-width which can be decomposed by a small number of sequences (see Table 1 for examples). Furthermore the directed path-width can be computed in time $3^{\tau(und(G))} \cdot |V|^{\mathcal{O}(1)}$, where $\tau(und(G))$ denotes the vertex cover number of the underlying undirected graph of G, by [9]. Thus our result also improves this algorithm for digraphs of large $\tau(und(G))$ which can be decomposed by a small number of sequences (see Table 1 for examples).

Table 1. Values of parameters within XP-algorithms for directed path-width.

Digraphs $G = (V, A)$, $n = |V|$	d-pw(G)	$\tau(und(G))$	k	ℓ
Transitive tournaments	0	$n - 1$	1	1
Union of k' transitive tournaments	0	$(\sum_{i=1}^{k'} n_i) - k'$	k'	1
Bidirectional complete digraphs $\overleftrightarrow{K_n}$	$n - 1$	$n - 1$	1	2
Semicomplete $\{\overrightarrow{C_3}, D_0, D_4\}$-free	$[0, n-1]$	$n - 1$	1	2
Semicomplete $\{\text{co-}(2\overrightarrow{P_2}), \overrightarrow{C_3}, D_4\}$-free	$[0, n-1]$	$n - 1$	1	2
Union of k' semicomplete $\{\text{co-}(2\overrightarrow{P_2}), \overrightarrow{C_3}, D_4\}$-Free	$[0, n-1]$	$(\sum_{i=1}^{k'} n_i) - k'$	k'	$2k'$

2   Preliminaries

We use the notations of Bang-Jensen and Gutin [1] for graphs and digraphs.

2.1   From Sequences to Digraphs

Let $Q = \{q_1, \ldots, q_k\}$ be a set of k sequences. Every sequence $q_i = (b_{i,1}, \ldots, b_{i,n_i})$ consists of a number n_i of items, such that all $n = \sum_{i=1}^{k} n_i$ items are pairwise distinct. Further there is a function t which assigns to every item $b_{i,j}$ a type $t(b_{i,j})$. The set of all types of the items in some sequence q_i is denoted by $types(q_i) = \{t(b) \mid b \in q_i\}$. For a set of sequences $Q = \{q_1, \ldots, q_k\}$ we denote $types(Q) = types(q_1) \cup \cdots \cup types(q_k)$. For some sequence $q_\ell = (b_{\ell,1}, \ldots, b_{\ell,n_\ell})$ we say item $b_{\ell,i}$ is *on the left of* item $b_{\ell,j}$ in sequence q_ℓ if $i < j$. Item $b_{\ell,i}$ is on the *position i* in sequence q_ℓ, since there are $i - 1$ items on the left of $b_{\ell,i}$ in sequence q_ℓ.

In order to *insert* a new item b on a position j in sequence q_i we first move all items on positions $j' \geq j$ to position $j' + 1$ starting at the rightmost position n_i and then we insert b at position j. In order to *remove* an existing item b at a position j in sequence q_i we move all items from positions $j' \geq j + 1$ to position $j' - 1$ starting at position $j + 1$.

We consider the distribution of the items of a type t onto the sequences by

$$d_Q(t) = |\{q \in Q \mid t \in types(q)\}| \quad \text{and} \quad d_Q = \max_{t \in types(Q)} d_Q(t).$$

For the number of items for type t within the sequences we define

$$c_Q(t) = \sum_{q \in Q} |\{b \in q \mid t(b) = t\}| \quad \text{and} \quad c_Q = \max_{t \in types(Q)} c_Q(t).$$

Obviously it holds $d_Q \leq k$ and $1 \leq d_Q \leq c_Q \leq n$.

The *sequence digraph* $g(Q) = (V, A)$ for a set $Q = \{q_1, \ldots, q_k\}$ has a vertex for every type, i.e. $V = types(Q)$ and an arc $(u, v) \in A$ if and only if there is some sequence q_i in Q where an item of type u is on the left of an item of type v. More formally, there is an arc $(u, v) \in A$ if and only if there is some sequence q_i in Q, such that there are two items $b_{i,j}$ and $b_{i,j'}$ such that (1) $1 \leq j < j' \leq n_i$, (2) $t(b_{i,j}) = u$, (3) $t(b_{i,j'}) = v$, and (4) $u \neq v$.

Sequence digraphs have successfully been applied in order to model the stacking process of bins from conveyor belts onto pallets with respect to customer orders, which is an important task in palletizing systems used in centralized distribution centers [5]. In our examples we will use type identifications instead of item identifications to represent a sequence $q_i \in Q$. For r not necessarily distinct types $t_1, \ldots, t_r$ let $[t_1, \ldots, t_r]$ denote some sequence $q_i = (b_{i,1}, \ldots, b_{i,r})$ of r pairwise distinct items, such that $t(b_{i,j}) = t_j$ for $j = 1, \ldots, r$. We use this notation for sets of sequences as well.

Example 1. Figure 1 shows the sequence digraph $g(Q)$ for $Q = \{q_1, q_2, q_3\}$ with sequences $q_1 = [a, a, d, e, d]$, $q_2 = [c, b, b, d]$, and $q_3 = [c, c, d, e, d]$.

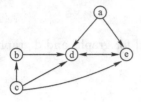
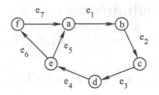

Fig. 1. Sequence digraph $g(Q)$ of Example 1.

Fig. 2. Digraph G of Example 2.

Next we give results in order to compute the sequence digraph $g(Q)$ and also its complement digraph co-$(g(Q))$. Therefore we define the position of the first item in some sequence $q_i \in Q$ of some type $t \in types(Q)$ by $first(q_i, t)$ and the position of the last item of type t in sequence q_i by $last(q_i, t)$. For technical reasons, if there is no item for type t contained in sequence q_i, then we define $first(q_i, t) = n_i + 1$ and $last(q_i, t) = 0$.

Lemma 1 ($\star^1$). *Let $Q = \{q_1\}$ be a set of one sequence, $g(Q) = (V, A)$ the sequence digraph, co-$(g(Q)) = (V, A^c)$ its complement digraph, and $u \neq v$, $u, v \in V$.*

1. *There is an arc $(u, v) \in A$, if and only if $first(q_1, u) < last(q_1, v)$.*
2. *There is an arc $(u, v) \in A^c$, if and only if $last(q_1, v) < first(q_1, u)$.*
3. *If $(u, v) \in A^c$, then $(v, u) \in A$.*
4. *There is an arc $(u, v) \in A$ and an arc $(v, u) \in A^c$, if and only if $last(q_1, u) < first(q_1, v)$.*

Lemma 2 ($\star$). *Let $Q = \{q_1, \ldots, q_k\}$ be a set of k sequences, $g(Q) = (V, A)$ the sequence digraph, co-$(g(Q)) = (V, A^c)$ its complement digraph, and $u \neq v$, $u, v \in V$.*

1. *There is an arc $(u, v) \in A$, if and only if there is some $q_i \in Q$ such that $first(q_i, u) < last(q_i, v)$.*
2. *There is an arc $(u, v) \in A^c$, if and only if for every $q_i \in Q$ we have $last(q_i, v) < first(q_i, u)$.*

By Lemma 2(1) only the first and the last item of each type in every $q_i \in Q$ are important for the arcs in the corresponding digraph. Let $M(q_i)$ be the subsequence of q_i which is obtained from q_i by removing all except the first and last item for each type and $M(Q) = \{M(q_1), \ldots, M(q_k)\}$.

Observation 1. *Let $Q = \{q_1, \ldots, q_k\}$ be a set of k sequences, then $g(Q) = g(M(Q))$.*

---

[1] The proofs of the results marked with a $\star$ are omitted due to space restrictions.

2.2   From Digraphs to Sequences

Let $G = (V, A)$ be some digraph and $A = \{a_1, \ldots, a_\ell\}$ its arc set. The *sequence system* $q(G) = \{q_1, \ldots, q_\ell\}$ for G is defined as follows. (1) There are 2ℓ items $b_{1,1}, b_{1,2}, \ldots, b_{\ell,1}, b_{\ell,2}$. (2) Sequence $q_i = (b_{i,1}, b_{i,2})$ for $1 \leq i \leq \ell$. (3) The type of item $b_{i,1}$ is the first vertex of arc a_i and the type of item $b_{i,2}$ is the second vertex of arc a_i for $1 \leq i \leq \ell$. Thus $types(q(G)) = V$.

Example 2 (Sequence System). For the digraph G of Fig. 2 the corresponding sequence system is $q(G) = \{q_1, q_2, q_3, q_4, q_5, q_6, q_7\}$, where $q_1 = [a, b]$, $q_2 = [b, c]$, $q_3 = [c, d]$, $q_4 = [d, e]$, $q_5 = [e, a]$, $q_6 = [e, f]$, $q_7 = [f, a]$.

By the definition of sequence systems and sequence digraphs we obtain the following result.

Observation 2. *For every digraph G it holds $G = g(q(G))$.*

Lemma 3 (★). *For every digraph $G = (V, A)$ with underlying undirected graph $und(G) = (V, E)$ there is a set Q of at most $|E|$ sequences such that $G = g(Q)$.*

There are digraphs which even can be defined by one sequence (see Theorem 2 for a complete characterization) and there are digraphs for which $|E|$ sequences are really necessary (see Lemma 7). For digraphs of bounded vertex degree the sequence system $Q = q(G)$ leads to sets whose distribution and number of items of each type can be bounded as follows.

Lemma 4. *For every digraph $G = (V, A)$ where $\max(\Delta^-(G), \Delta^+(G)) \leq d$ there is a set Q with $d_Q \leq 2d$ and $c_Q \leq 2d$ such that $G = g(Q)$.*

In case of complete bioriented digraphs, i.e. we have none or both arcs between any pair of vertices, we can improve the latter bounds.

Lemma 5 (★). *For every complete bioriented digraph $G = (V, A)$ such that $\max(\Delta^-(G), \Delta^+(G)) \leq d$ there is a set Q with $d_Q \leq d$ and $c_Q \leq 2d$ (for $d \geq 2$ even $c_Q \leq 2d - 1$) such that $G = g(Q)$.*

3   Properties of Sequence Digraphs

3.1   Graph Classes and Their Relations

We define $S_{k,\ell}$ to be the set of all sequence digraphs defined by sets Q on at most k sequences that contain at most ℓ items of each type in $types(Q)$. By Observation 1 and Lemma 3 we obtain the following bounds.

Corollary 1. *Let Q be a set on k sequences and $g(Q) = (V, A) \in S_{k,\ell}$ the defined graph with $und(g(Q)) = (V, E)$. Then we can assume that $1 \leq \ell \leq 2k$ and $1 \leq k \leq |E|$.*

Lemma 6 (★). *Let $\ell \geq 1$ and $G \in S_{1,\ell}$ be defined by $Q = \{q_1\}$, then $g(Q)$ is semicomplete and graph $und(g(Q))$ is the complete graph on $|types(Q)|$ vertices.*

Next we consider the relations of the defined classes for $k = 1$ sequence. Since $S_{1,1}$ contains only digraphs with exactly one arc between every pair of vertices (cf. Theorem 1) and $S_{1,\ell}$ for $\ell \geq 2$ contains all bidirectional complete digraphs we know that $S_{1,1} \subsetneq S_{1,\ell}$ for $\ell \geq 2$. Further by $S_{k,\ell} \subseteq S_{k,\ell+1}$ and Observation 1 it follows that all classes $S_{1,\ell}$ for $\ell \geq 2$ are equal.

Corollary 2. *For $\ell \geq 2$ the following inclusions hold true.*

$$S_{1,1} \subsetneq S_{1,2} = \ldots = S_{1,\ell}$$

For a set of digraphs $\mathcal{F}$ we denote by $\mathcal{F}$-*free digraphs* the set of all digraphs G such that no induced subdigraph of G is isomorphic to a member of $\mathcal{F}$. If $\mathcal{F} = \{F\}$, we write F-free instead of $\{F\}$-free. For undirected graphs we use this notation as well.

Lemma 7 (★). *Let $G = (V, E)$ be a triangle free graph, i.e. a C_3-free graph, with $|E| \geq 2$, such that $\Delta(G) = \ell$ and let $G' = (V, A)$ be an orientation of G. Then for $k = |E|$ it holds that $G' \in S_{k,\ell}$ but for $k' < k$ or $\ell' < \ell$ it holds that $G' \notin S_{k',\ell'}$.*

Since for every $k \geq 2$ and every $\ell = 2, \ldots, k$ there is a tree T on k edges and $\Delta(T) = \ell$ we know by Lemma 7 that for $k \geq 2$ and $\ell = 2, \ldots, k$ it holds $S_{k,\ell-1} \subsetneq S_{k,\ell}$. Further by Observation 1 we know that for $k \geq 2$ and $\ell \geq 2k$ it holds $S_{k,\ell} = S_{k,\ell+1}$.

Corollary 3. *For $k \geq 2$ the following inclusions hold true.*

$$S_{k,1} \subsetneq S_{k,2} \subsetneq \ldots \subsetneq S_{k,k} \subseteq S_{k,k+1} \subseteq \ldots \subseteq S_{k,2k} = S_{k,2k+1} = \ldots$$

Lemma 8 (★). *Let $G \in S_{k,\ell}$, then for every induced subdigraph H of G it holds $H \in S_{k,\ell}$.*

Graph classes which are closed under taking induced subgraphs are called *hereditary*. Hereditary graph classes are exactly those classes which can be defined by forbidden induced subgraphs.

3.2   Characterizations of Sequence Digraphs for $k = 1$ or $\ell = 1$

In this section we show a finite set of forbidden induced subgraphs for all classes $S_{k,\ell}$ where $\ell = 1$ and for all classes where $k = 1$. These characterizations lead to polynomial time recognition algorithms for the corresponding graph classes. Furthermore we give characterizations in terms of special tournaments and conditions for the complement digraph.

Sequence Digraphs for $k = 1$ and $\ell = 1$. A digraph $G = (V, A)$ is called *transitive* if for every pair $(u, v) \in A$ and $(v, w) \in A$ of arcs with $u \neq w$ the arc (u, w) also belongs to A.

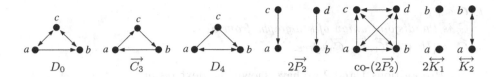

Fig. 3. Special digraphs.

Lemma 9 (★). *Every digraph in $S_{1,1}$ is transitive.*

For a digraph G and an integer d let dG be the disjoint union of d copies of G.

Theorem 1. *For every digraph G the following statements are equivalent.*

1. *$G \in S_{1,1}$*
2. *G is a transitive tournament.*
3. *G is an acyclic tournament.*
4. *G is a $\overrightarrow{C_3}$-free tournament.*
5. *G is a tournament with exactly one Hamiltonian path.*
6. *G is a tournament and every vertex in G has a different outdegree.*
7. *G is $\{2\overleftrightarrow{K_1}, \overleftrightarrow{K_2}, \overrightarrow{C_3}\}$-free.*
8. *$G \in \{(\{v\}, \emptyset)\} \cup \{(\overrightarrow{P_n})^{n-1} \mid n \geq 2\}$, i.e. G is the $(n-1)$-th power of a directed path $\overrightarrow{P_n}$.*

Proof. The equivalence of (2)–(6) is known from [4, Chapter 9]. (1) $\Rightarrow$ (2) By Lemma 9 every digraph $G \in S_{1,1}$ is transitive and by definition of $S_{1,1}$ digraph G is a tournament. (3) $\Rightarrow$ (1) Every acyclic digraph G has a source, i.e. a vertex v_1 of indegree 0, see [1]. Since G is a tournament there is an arc (v_1, v) for every vertex v of G, i.e. v_1 is an out-dominating vertex. By removing v_1 from G, we obtain a transitive tournament G^1 which leads to an out-dominating vertex v_2. By removing v_2 from G^1, we obtain a transitive tournament G^2 which leads to an out-dominating vertex v_3 and so on. The sequence $[v_1, v_2, \ldots, v_n]$ shows that $G \in S_{1,1}$. (4)$\Leftrightarrow$ (7) and (1)$\Leftrightarrow$ (8) can be easily verified. □

By part (3)$\Rightarrow$ (1) of the proof of Theorem 1 we have shown the next result.

Proposition 1. *Let $G = (V, A) \in S_{1,1}$, then a sequence q, such that $G = g(\{q\})$ can be found in time $\mathcal{O}(|V| + |A|)$.*

Sequence Digraphs for $\ell = 1$. The sequence digraph $g(Q) = (V, A)$ for a set $Q = \{q_1, \ldots, q_k\}$ can be obtained by the union of $g(\{q_i\}) = (V_i, A_i)$, $1 \leq i \leq k$ by $V = \cup_{i=1}^k V_i$ and $A = \cup_{i=1}^k A_i$. Since for digraphs in $S_{k,1}$ the vertex sets $V_i = types(q_i)$ are disjoint, all properties of Theorem 1 can be generalized to $k \geq 1$ sequences. Some of them are given next.

Proposition 2. *For every digraph G and every integer $k \geq 1$ the following statements are equivalent.*

1. $G \in S_{k,1}$.
2. G is the disjoint union of k digraphs from $S_{1,1}$.
3. G is $\{(k+1)\overleftrightarrow{K_1}, \overleftrightarrow{K_2}, \overrightarrow{C_3}\}$-free.

By Propositions 1 and 2 we have shown the next result.

Proposition 3. *Let $G = (V, A) \in S_{k,1}$, then a set Q on k sequences, such that $G = g(Q)$ can be found in time $\mathcal{O}(|V| + |A|)$.*

Sequence Digraphs for $k = 1$. The digraph D_0 in Fig. 3 is not transitive, since it has among others the arcs (b, c) and (c, a) but not the arc (b, a). Further D_0 belongs to the set $S_{1,2}$, since it can be defined by set $Q = \{q_1\}$ of one sequence $q_1 = [c, a, b, c]$. Thus for $\ell \geq 2$ items for each type even one sequence can define digraphs which are not transitive. A digraph $G = (V, A)$ is called *quasi transitive* if for every pair $(u, v) \in A$ and $(v, w) \in A$ of arcs with $u \neq w$ there is at least one arc between u and w in A. Since every semicomplete digraph is quasi transitive, by Lemma 6 every digraph in $S_{1,\ell}$, $\ell \geq 1$, is quasi transitive.

To show characterizations for the class $S_{1,2}$ we next give some lemmas.

Lemma 10 (★). *Let $\ell \geq 1$ and $G \in S_{1,\ell}$, then its complement digraph co-G is transitive.*

Lemma 11 (★). *Let $\ell \geq 1$ and $G \in S_{1,\ell}$, then its complement digraph co-G is $2\overrightarrow{P_2}$-free.*

Lemma 12 (★). *Let G be a semicomplete $\{\overrightarrow{C_3}, D_4\}$-free digraph on n vertices, then G has a vertex v such that outdegree$(v) = n - 1$ and a vertex v' such that indegree$(v') = n - 1$.*

Lemma 13 (★). *Let G be a semicomplete $\{\overrightarrow{C_3}, D_4\}$-free digraph, then its complement digraph co-G is transitive.*

Lemma 14 (★). *Every semicomplete $\{\overrightarrow{C_3}, D_4\}$-free digraph has a spanning transitive tournament subdigraph.*

These results allow us to show the following characterizations. Since we use several forbidden induced subdigraphs the semicompleteness is expressed by excluding $2\overleftrightarrow{K_1}$ (see Fig. 3 for the special digraphs).

Theorem 2. *For every digraph G the following statements are equivalent.*

1. $G \in S_{1,2}$
2. $G \in S_{1,\ell}$ for some $\ell \geq 2$
3. *co-G is transitive, co-G is $2\overrightarrow{P_2}$-free, and G has a spanning transitive tournament subdigraph.*
4. G is $\{co\text{-}(2\overrightarrow{P_2}), 2\overleftrightarrow{K_1}, \overrightarrow{C_3}, D_4\}$-free.

Proof. (2) $\Rightarrow$ (1) By Corollary 2. (1) $\Rightarrow$ (4) co-$(2\overrightarrow{P_2})$, $2\overleftrightarrow{K_1}$, $\overrightarrow{C_3}$, $D_4 \notin S_{1,2}$. (4) $\Rightarrow$ (3) By Lemmas 13 and 14. (3) $\Rightarrow$ (2) Let $G' = (V, A')$ be a subdigraph of $G = (V, A)$ which is a transitive tournament. By Theorem 1 we know that $G' \in S_{1,1}$ and thus there is some sequence $q' = [v_1, \ldots, v_n]$ such that $g(\{q'\}) = G'$. If $A' = A$ we know that $G \in S_{1,1} \subseteq S_{1,\ell}$ for every $\ell \geq 2$. So we can assume that $A' \subsetneq A$. Obviously for every arc $(v_i, v_j) \in A - A'$ there are two positions $j < i$ in $q' = [v_1, \ldots, v_j, \ldots, v_i, \ldots, v_n]$. In order to define a subdigraph of G which contains all arcs of G' and arc (v_i, v_j) we can insert (cf. Sect. 2.1 for the definition of inserting an item) an additional item for type v_i on position $k \leq j$, or an additional item for type v_j on position $k > i$, or first an additional item for type v_j and then an additional item for type v_i on a position k, $j < k \leq i$, into q' without creating an arc which is not in A. This is possible if and only if there is some position k, $j \leq k \leq i$, in $q' = [v_1, \ldots, v_j, \ldots, v_{m'}, \ldots, v_k, \ldots, v_{m''}, \ldots, v_i, \ldots, v_n]$ such that for every m', $j < m' \leq k$, it holds $(v_{m'}, v_j) \in A$ and for every m'', $k \leq m'' < i$, it holds $(v_i, v_{m''}) \in A$.

If it is possible to insert all arcs of $A - A'$ by adding a set of additional items into sequence q' resulting in a sequence q such that $G = g(q)$, then it obviously holds $G \in S_{1,\ell}$ for some $\ell \geq 2$. Next we show a condition using the new items of every single arc of $A - A'$ independently from each other.

Claim ($\bigstar$). If for every arc $(v_i, v_j) \in A - A'$ there is a position k, $j < k \leq i$ such that first inserting an additional item for type v_j and then an additional item for type v_i at position k into q' defines a subdigraph of G which contains all arcs of G' and arc (v_i, v_j), then $G \in S_{1,\ell}$ for some $\ell \geq 2$.

Assume that $G \notin S_{1,\ell}$ for every $\ell \geq 2$. By the Claim there is some arc $(v_i, v_j) \in A - A'$ such that for every position k, $j < k \leq i$ inserting an additional item for type v_i and an additional item for type v_j at position k defines an arc which is not in A. That is, for every position k, $j < k \leq i$, in q' there exists some m', $j < m' \leq k$, such that it holds $(v_{m'}, v_j) \notin A$ *or* there exists some m'', $k \leq m'' < i$, such that it holds $(v_i, v_{m''}) \notin A$. By the transitivity of co-G it follows that there is one position k, $j < k \leq i$, in q' such that there exists some m', $j < m' \leq k$, such that it holds $(v_{m'}, v_j) \notin A$ *and* there exists some m'', $k \leq m'' < i$, such that it holds $(v_i, v_{m''}) \notin A$.

If co-$G = (V, A^c)$ is the complement digraph of G we know that

$$(v_{m'}, v_j) \in A^c \text{ and } (v_i, v_{m''}) \in A^c. \tag{1}$$

Since $m' \leq m''$ we know that $(v_{m'}, v_{m''}) \in A$. We also know that $(v_{m''}, v_{m'}) \in A$, since otherwise $(v_{m''}, v_{m'}) \in A^c$, property (1), and the transitivity of co-G would imply that $(v_i, v_j) \in A^c$ which is not possible. Thus we know that

$$(v_{m'}, v_{m''}) \notin A^c \text{ and } (v_{m''}, v_{m'}) \notin A^c. \tag{2}$$

Further the arcs $(v_j, v_{m'})$, $(v_j, v_{m''})$, $(v_{m'}, v_i)$, $(v_{m''}, v_i)$ belong to $A' \subseteq A$ and thus

$$(v_j, v_{m'}) \notin A^c, (v_j, v_{m''}) \notin A^c, (v_{m'}, v_i) \notin A^c \text{ and } (v_{m''}, v_i) \notin A^c. \tag{3}$$

If $(v_i, v_{m'}) \in A^c$ or $(v_{m''}, v_j) \in A^c$ then (1) and the transitivity of co-G would imply that $(v_i, v_j) \in A^c$, thus we know

$$(v_i, v_{m'}) \notin A^c \text{ and } (v_{m''}, v_j) \notin A^c. \tag{4}$$

Properties (1)–(4) imply that $(\{v_i, v_j, v_{m'}, v_{m''}\}, \{(v_i, v_{m'}), (v_{m'}, v_j)\})$ induces a $2\overrightarrow{P_2}$ in co-G, which implies that G contains a co-$(2\overrightarrow{P_2})$. $\qquad\square$

Corollary 4. *Every digraph in $S_{k,\ell}$ can be obtained by the union of at most k many $\{co\text{-}(2\overrightarrow{P_2}), 2\overleftrightarrow{K_1}, \overrightarrow{C_3}, D_4\}$-free digraphs.*

Proposition 4. *Let $G = (V, A) \in S_{1,2}$, then a sequence q, such that $G = g(\{q\})$ can be found in time $\mathcal{O}(|V| + |A|)$.*

Proof. Let $G = (V, A) \in S_{1,2}$ and $q = []$. We perform the following steps until $G = (\emptyset, \emptyset)$.

- Choose $v \in V$ such that $(v, u) \in A$ for all $u \in V - \{v\}$ and append v to q.
- Remove all arcs (v, u) from A.
- If indegree$(v) = $ outdegree$(v) = 0$, remove v from V.
- If there are vertices u such that indegree$(u) = $ outdegree$(u) = 0$, remove u from V and append u to q.

In order to perform the algorithm there has to be an ordering $v_1, \ldots, v_n$ of V such that for $1 \le i < n$ vertex v_i has maximum possible outdegree in subdigraph obtained by removing the outgoing arcs of $v_1, \ldots, v_{i-1}$ and thereby created isolated vertices from G. Since $G \in S_{1,2}$ there is a sequence q' such that $G = g(\{q'\})$. The order in which the types corresponding to the vertices of V appear in subsequence $F(q')$, defined in the proof of Theorem 2, ensures the existence of such an ordering.

Finally it holds $G = g(\{q\})$ by the definition of sequence digraphs and since every vertex which has only outgoing or only incoming arcs will be inserted once into q and every vertex which has outgoing and incoming arcs will be inserted at most twice into q this sequence fulfils the properties stated in the theorem. $\qquad\square$

4   Directed Path-Width of Sequence Digraphs

According to Barát [2], the notion of directed path-width was introduced by Reed, Seymour, and Thomas around 1995 and relates to directed tree-width introduced by Johnson, Robertson, Seymour, and Thomas in [6]. A *directed path-decomposition* of a digraph $G = (V, A)$ is a sequence $(X_1, \ldots, X_r)$ of subsets of V, called *bags*, such that the following three conditions hold true.

(dpw-1) $X_1 \cup \ldots \cup X_r = V$.
(dpw-2) For each $(u, v) \in A$ there is a pair $i \le j$ such that $u \in X_i$ and $v \in X_j$.
(dpw-3) If $u \in X_i$ and $u \in X_j$ for some $u \in V$ and two indices i, j with $i \le j$, then $u \in X_\ell$ for all indices ℓ with $i \le \ell \le j$.

The *width* of a directed path-decomposition $\mathcal{X} = (X_1, \ldots, X_r)$ is

$$\max_{1 \leq i \leq r} |X_i| - 1.$$

The *directed path-width* of G, d-pw(G) for short, is the smallest integer w such that there is a directed path-decomposition for G of width w.

Determining whether the (undirected) path-width of some given (undirected) planar graph with maximum vertex degree 3 is at most some given value w is NP-complete [10]. Since for complete bioriented digraphs the directed path-width (d-pw) is equal to the (undirected) path-width (pw) of the underlying undirected graph it follows that determining whether the directed path-width of some given digraph with maximum semi-degree $\Delta^0(G) = \max\{\Delta^-(D), \Delta^+(D)\} \leq 3$ is at most some given value w is NP-complete, which will be useful to show Proposition 6.

4.1   Hardness of Directed Path-Width on Sequence Digraphs

Next we give some conditions on the sequences in Q such that for the corresponding digraph $g(Q)$ computing its directed path-width is NP-hard.

Proposition 5. *Given a set Q on k sequences such that $n_i = 2$ for $1 \leq i \leq k$ and an integer p, then the problem of deciding whether d-pw($g(Q)$) $\leq p$ is NP-complete.*

Proof. The stated problem is in NP. To show the NP-hardness by a reduction from the directed path-width problem we transform instance (G, p) in linear time into instance $(q(G), p)$ for the stated problem. The correctness follows by Observation 2. □

Proposition 6. *Given a set Q with $d_Q = 3$ or $c_Q = 5$ and an integer p, then the problem of deciding whether d-pw($g(Q)$) $\leq p$ is NP-complete.*

Proof. To show the NP-hardness by a reduction from the directed path-width problem for digraphs G with $\max(\Delta^-(G), \Delta^+(G)) \leq 3$, we transform instance (G, p) in linear time into instance $(q(G), p)$ for the stated problem. The correctness follows by Lemma 5. □

4.2   Polynomial Cases of Directed Path-Width on Sequence Digraphs

We consider the directed path-width of sequence digraphs for $k = 1$ or $\ell = 1$.

Proposition 7. *Let $G \in S_{k,1}$, then d-pw(G) = 0.*

Proof. By Proposition 2 every digraph in $S_{k,1}$ is the disjoint union of k digraphs in $S_{1,1}$. By Theorem 1 every digraph in $S_{1,1}$ is acyclic and thus has directed path-width 0. □

For digraphs in $S_{1,2}$ the directed path-width can be arbitrary large, since this class includes all bidirectional complete digraphs. We can compute this value as follows. Let $Q = \{q\}$. For type $t \in types(q)$ let $I_t = [first(q,t), last(q,t)]$ be the interval representing t, and let $I_q = \{I_t \mid t \in types(q)\}$ be the set of all intervals for sequence q. Let $I(q) = (V, E)$ be the interval graph where $V = types(q)$ and $E = \{\{u, v\} \mid u \neq v, \ I_u \cap I_v \neq \emptyset, I_u, I_v \in I_q\}$.

Proposition 8. *Let $G \in S_{1,2}$ defined by a set $Q = \{q_1\}$ of one sequence, then $d\text{-}pw(G) = \omega(I(q)) - 1 = pw(I(q))$.*

Proof. We obtain $d\text{-}pw(G) \leq \omega(I(q)) - 1$ by an obvious directed path-decomposition along $I(q)$. Further for every integer r the set $I(r) = \{I_t \mid r \in I_t\}$ defines a complete subgraph $K_{|I(r)|}$ in $I(q)$ and also a bidirectional complete subdigraph $\overleftrightarrow{K}_{|I(r)|}$ in G. Thus it holds $d\text{-}pw(G) \geq \omega(I(q)) - 1$. The second equality holds since the (undirected) path-width of an interval graph is equal to the size of a maximum clique [3]. $\square$

Sets Q where $d_Q = 1$ can be handled in polynomial time.

Proposition 9. *Given a set Q with $d_Q = 1$ and an integer p, then the problem of deciding whether $d\text{-}pw(g(Q)) \leq p$ can be solved in time $\mathcal{O}(|types(Q)|^2 + n)$.*

Proof. Let $Q = \{q_1, \ldots, q_k\}$. If $d_Q = 1$ the vertex sets $V_i = types(q_i)$ are disjoint. That is, $g(Q)$ is the disjoint union of digraphs in $S_{1,2}$ for which the directed path-width can be computed in time $\mathcal{O}(\sum_{i=1}^{k} |types(\{q_i\})|^2 + n_i) \subseteq \mathcal{O}(|types(Q)|^2 + n)$ by Proposition 8. $\square$

4.3 An XP-Algorithm for Directed Path-Width

We next give an XP-algorithm for directed path-width w.r.t. the parameter k, which implies that for every constant k for a given set Q on at most k sequences the value $d\text{-}pw(g(Q))$ can be computed in polynomial time. The main idea is to discover an optimal directed path-decomposition by scanning the k sequences left-to-right and keeping in a state the numbers of scanned items of every sequence and a certain number of active types.

Let $Q = \{q_1, \ldots, q_k\}$ be a set of k sequences. Every k-tuple $(i_1, \ldots, i_k)$ where $0 \leq i_j \leq n_j$ for $1 \leq j \leq k$ is a *state* of Q. State $(0, 0, \ldots, 0)$ is the *initial state* and $(n_1, \ldots, n_k)$ is the *final state*. The *state digraph* $s(Q)$ for a set Q has a vertex for each possible state. There is an arc from vertex u labeled by $(u_1, \ldots, u_k)$ to vertex v labeled by $(v_1, \ldots, v_k)$ if and only if $u_i = v_i - 1$ for exactly one element of the vector and for all other elements of the vector $u_j = v_j$. Let $(i_1, \ldots, i_k)$ be a state of Q. We define $L(i_1, \ldots, i_k)$ to be the set of all items on the positions $1, \ldots, i_j$ for $1 \leq j \leq k$ and $R(i_1, \ldots, i_k)$ is the set of all items on the remaining positions $i_j + 1, \ldots, n_j$ for $1 \leq j \leq k$. Further let $M(i_1, \ldots, i_k)$ be the set of all items on the positions i_j for $1 \leq j \leq k$ such that there is exactly one type of these items in Q. Obviously, for every state $(i_1, \ldots, i_k)$ it holds that

$L(i_1, \ldots, i_k) \cup R(i_1, \ldots, i_k)$ leads to a disjoint partition of the items in Q and $M(i_1, \ldots, i_k) \subseteq L(i_1, \ldots, i_k)$.

Further each vertex v of the state digraph is labeled by the value $f(v)$. This value is the number of types t such that either there is at least one item of type t in $L(v)$ and at least one item of type t in $R(v)$ or there is one item of type t in $M(v)$. Formally we define $active(v) = \{t \in types(Q) \mid b \in L(v), t(b) = t, b' \in R(v), t(b') = t\} \cup \{t \in types(Q) \mid b \in M(v), t(b) = t\}$ and $f(v) = |active(v)|$. Obviously for the initial state v it holds $|active(v)| = 0$. Since the state digraph $s(Q)$ is a directed acyclic graph we can compute all values $|active(v)|$ using a topological ordering $topol$ of the vertices. Every arc (u, v) in $s(Q)$ represents one item $b_{i,j}$ if item $b_{i,j-1} \notin M(v)$ and two items $b_{i,j}$ and $b_{i,j-1}$ if item $b_{i,j-1} \in M(v)$ of some types $t(b_{i,j}) = t$ and $t(b_{i,j-1}) = t'$ from some sequence q_j, thus

$$|active((i_1, \ldots, i_{j-1}, i_j + 1, i_{j+1}, \ldots, i_k))|$$
$$= |active((i_1, \ldots, i_{j-1}, i_j, i_{j+1}, \ldots, i_k))| + c_j$$

where

$$c_j = \begin{cases} 1, \text{ if } first(q_j, t) = i_j + 1 \text{ and } first(q_\ell, t) > i_\ell \ \forall \ell \neq j \text{ and} \\ \quad not(first(q_j, t') = last(q_j, t') = i_j \text{ and } last(q_\ell, t') = 0 \ \forall \ell \neq j) \\ 0, \text{ if } first(q_j, t) = i_j + 1 \text{ and } first(q_\ell, t) > i_\ell \ \forall \ell \neq j \text{ and} \\ \quad first(q_j, t') = last(q_j, t') = i_j \text{ and } last(q_\ell, t') = 0 \ \forall \ell \neq j \\ -1, \text{ if } last(q_j, t) = i_j + 1 \text{ and } last(q_\ell, t) \leq i_\ell \ \forall \ell \neq j \text{ and} \\ \quad not(first(q_j, t') = last(q_j, t') = i_j \text{ and } last(q_\ell, t') = 0 \ \forall \ell \neq j) \\ -2, \text{ if } last(q_j, t) = i_j + 1 \text{ and } last(q_\ell, t) \leq i_\ell \ \forall \ell \neq j \text{ and} \\ \quad first(q_j, t') = last(q_j, t') = i_j \text{ and } last(q_\ell, t') = 0 \ \forall \ell \neq j \\ 0, \text{ otherwise.} \end{cases}$$

Thus, the calculation of value $|active(i_1, \ldots, i_k)|$ for the vertex labeled $(i_1, \ldots, i_k)$ depends only on already calculated values, which is necessary in order to use dynamic programming[2].

Let $\mathcal{P}(Q)$ the set of all paths from the initial state to the final state in $s(Q)$. Every $P \in \mathcal{P}(Q)$ has $r = 1 + \sum_{i=1}^{k} n_i$ vertices, i.e. $P = (v_0, \ldots, v_r)$. First we show that every path in $\mathcal{P}(Q)$ leads to a directed path-decomposition for $g(Q)$.

Lemma 15 (★). *Let Q be a set of k sequences and $(v_0, \ldots, v_r) \in \mathcal{P}(Q)$. Then $(active(v_1), \ldots, active(v_{r-1}))$ is a directed path-decomposition for $g(Q)$.*

Lemma 15 leads to an upper bound on the directed path-width of $g(Q)$ using the state graph. The reverse direction is more involved and considered next. The main idea is to use a nice path-decomposition in which the introduce nodes are ordered w.r.t. the existing common order of the types within the sequences.

---

[2] For sets Q such that the number of items for which there is no further item of the same type in Q is small, we suggest to modify Q by inserting a dummy item of the same type at the position after such items. This does not change the sequence digraph and allows to make a case distinct within three instead of five cases when defining c_j. But this modification increases the size of the sequence digraph.

Lemma 16 (★). *Let Q be a set of k sequences. If there is a directed path-decomposition of width $p - 1$ for $g(Q)$, then there is a path $(v_0, \ldots, v_r) \in \mathcal{P}(Q)$ such that for every $1 \leq i \leq r$ it holds $|active(v_i)| \leq p$.*

By Lemmas 15 and 16 we obtain the following result.

Corollary 5. *Given a set Q of k sequences, then*

$$d\text{-}pw(g(Q)) = \min_{(v_0, \ldots, v_r) \in \mathcal{P}(Q)} \max_{1 \leq i \leq r-1} |active(v_i)| - 1.$$

In order to apply Corollary 5 we consider some general digraph problem. Let $G = (V, A, f)$ be a directed acyclic vertex-labeled graph. Function $f : V \to \mathbb{Z}$ assigns to every vertex $v \in V$ a value $f(v)$. Let $s \in V$ and $t \in V$ be two vertices. For some vertex $v \in V$ and some path $P = (v_1, \ldots, v_\ell)$ with $v_1 = s$, $v_\ell = v$ and $(v_i, v_{i+1}) \in A$ we define $val_P(v) := \max_{u \in P}(f(u))$. Let $\mathcal{P}_s(v)$ denote the set of all paths from vertex s to vertex v. We define $val(v) := \min_{P \in \mathcal{P}_s(v)}(val_P(v))$. Then it holds:

$$val(v) = \max\{f(v), \min_{u \in N^-(v)} (val(u))\}.$$

By dynamic programming it is possible to compute all the values of $val(v)$, $v \in V$, in time $\mathcal{O}(|V| + |A|)$. This is possible, since G is acyclic.

Theorem 3. *Given a set Q, such that $g(Q) \in S_{k,\ell}$ for some $\ell \geq 1$, then the directed path-width of $g(Q)$ and also a directed path-decomposition can be computed in time $\mathcal{O}(k \cdot (1 + \max_{1 \leq i \leq k} n_i)^k)$.*

Proof. Let Q be a set, such that $g(Q) \in S_{k,\ell}$. The state digraph $s(Q)$ has at most $(1 + \max_{1 \leq i \leq k} n_i)^k$ vertices and can be found in time $\mathcal{O}(k \cdot (1 + \max_{1 \leq i \leq k} n_i)^k)$ from Q. By Corollary 5 the directed path-width of $g(Q)$ can be computed by considering all paths from the initial state to the final state in $s(Q)$. This can be done by any algorithm for the above general digraph problem on $s(Q) = (V, A)$ using $f(v) = |active(v)|$, $v \in V$, s as the initial state, and t as the final state. Since every vertex of the state digraph has at most k outgoing arcs we have $\mathcal{O}(|V| + |A|) \subseteq \mathcal{O}(k \cdot (1 + \max_{1 \leq i \leq k} n_i)^k)$. Thus we can compute an optimal path in $s(Q)$ in time $\mathcal{O}(k \cdot (1 + \max_{1 \leq i \leq k} n_i)^k)$. $\square$

5   Conclusions

In this paper, we have considered digraphs which can be defined by a set of k sequences. We have shown an XP-algorithm for directed path-width w.r.t. number of sequences k needed to define the input graph. For special digraphs our solution improves known solution w.r.t. the standard parameter as shown in Table 1. This implies that for each constant k, it is decidable in polynomial time whether for a given set Q on at most k sequences the digraph $g(Q)$ has directed path-width at most w. If we know that some digraph can be defined by one sequence, we can find this in linear time (Proposition 4). This implies that for each constant k, it is decidable in polynomial time whether for a digraph G, which is given by the union of at most k many semicomplete $\{\text{co-}(2\overrightarrow{P_2}), \overrightarrow{C_3}, D_4\}$-free digraphs, digraph G has directed path-width at most w.

There are several interesting open questions. **(a)** Is there is an FPT-algorithm for the directed path-width problem w.r.t. parameter k? **(b)** Does the hardness of Proposition 6 also hold for $c_Q \in \{2, 3, 4\}$ and for $d_Q = 2$? **(c)** By Theorem 1, Proposition 2 and Theorem 2 one can decide in polynomial time whether a given digraph belongs to the class $S_{k,\ell}$ for $\ell = 1$, for $k = 1$, or both. It remains to consider this problem for the classes $S_{k,\ell}$ for $k \geq 2$ and $2 \leq \ell \leq 2k$. **(d)** Can we find for a given digraph G a set Q with a smallest number of sequences such that $g(Q) = G$ in polynomial time?

References

1. Bang-Jensen, J., Gutin, G.: Digraphs. Theory, Algorithms and Applications. Springer, Berlin (2009). https://doi.org/10.1007/978-1-84800-998-1
2. Barát, J.: Directed pathwidth and monotonicity in digraph searching. Graphs Combin. **22**, 161–172 (2006)
3. Bodlaender, H.L.: A partial k-arboretum of graphs with bounded treewidth. Theor. Comput. Sci. **209**, 1–45 (1998)
4. Gould, R.: Graph Theory. Dover, Downers Grove (2012)
5. Gurski, F., Rethmann, J., Wanke, E.: On the complexity of the FIFO stack-up problem. Math. Methods Oper. Res. **83**(1), 33–52 (2016)
6. Johnson, T., Robertson, N., Seymour, P.D., Thomas, R.: Directed tree-width. J. Combin. Theory Ser. B **82**, 138–155 (2001)
7. Kitsunai, K., Kobayashi, Y., Komuro, K., Tamaki, H., Tano, T.: Computing directed pathwidth in $O(1.89^n)$ time. Algorithmica **75**, 138–157 (2016)
8. Kitsunai, K., Kobayashi, Y., Tamaki, H.: On the pathwidth of almost semicomplete digraphs. In: Bansal, N., Finocchi, I. (eds.) ESA 2015. LNCS, vol. 9294, pp. 816–827. Springer, Heidelberg (2015). https://doi.org/10.1007/978-3-662-48350-3_68
9. Kobayashi, Y.: Computing the pathwidth of directed graphs with small vertex cover. Inf. Process. Lett. **115**(2), 310–312 (2015)
10. Monien, B., Sudborough, I.H.: Min cut is NP-complete for edge weighted trees. Theor. Comput. Sci. **58**, 209–229 (1988)
11. Nagamochi, H.: Linear layouts in submodular systems. In: Chao, K.-M., Hsu, T., Lee, D.-T. (eds.) ISAAC 2012. LNCS, vol. 7676, pp. 475–484. Springer, Heidelberg (2012). https://doi.org/10.1007/978-3-642-35261-4_50

New Results About the Linearization of Scaffolds Sharing Repeated Contigs

Dorine Tabary[1], Tom Davot[1(✉)], Mathias Weller[2], Annie Chateau[1], and Rodolphe Giroudeau[1]

[1] LIRMM - CNRS UMR 5506, Montpellier, France
{dorine.tabary,tom.davot,annie.chateau,rodolphe.giroudeau}@lirmm.fr
[2] CNRS, LIGM, Paris, France
mathias.weller@u-pem.fr

Abstract. Solutions to genome scaffolding problems can be represented as paths and cycles in a "solution graph". However, when working with repetitions, such solution graphs may contain branchings and, thus, they may not be uniquely convertible into sequences. Having introduced various ways of extracting the unique parts of such solutions, we extend previously known NP-hardness results to the case that the solution graph is planar, bipartite, and subcubic, and show that there is no PTAS in this case.

1 Introduction

Extracting information from genomes has become a very largely spread task, at numerous levels, and most of these need to consider their nucleotidic sequence. Large databases contain genomic sequences of a very large range of organisms, or various individuals of a same species. However, difficulties arise when it comes to extract nucleotidic sequences from the DNA molecule. Technical limitations induce a complex inference process, beginning with the *sequencing* step, where a large amount of overlapping, short sequences are produced, going on with the *assembly* step, which takes those short sequences called *reads*, and exploits overlaps to output longer sequences called *contigs*. Those contigs are usually the final product of most of genomes, called *drafted genomes*. NGS data are going to evolve towards longer and longer sequences, but most of the available sequencing data in public databases are huge collections of billions of *short reads* (*i.e.* words of between fifteen and hundreds of characters) [12]. Those genomes are often sufficient to extract useful information, for instance detect and compare genic content. However, the global structure of the genome may be lacking, depending on how these genomes stay fragmented. Intending to cure this fragmentation, and improve the assembly process, it is possible to perform a *scaffolding* of the contigs, that is the inference of relative order and orientation of contigs, using additional information. Most of scaffolding tools are using information from paired-end sequencing, and using various models and methods (see [7,9] for surveys).

© Springer Nature Switzerland AG 2018
D. Kim et al. (Eds.): COCOA 2018, LNCS 11346, pp. 94–107, 2018.
https://doi.org/10.1007/978-3-030-04651-4_7

Few of them are considering genomic repeats, which are often disturbing both assembly and scaffolding. In numerous organisms, a significant part of the genome *is* repeated. Such repeats may be of various sizes and present variable copy numbers, according to the species and individuals [3]. Due to the conservatism of some assembly methods, a repeat may cover an entire contig which is separated from the other genomic side fragments [13] i.e.

In the Jungle of Problems. We focus in this paper on models, graphs and problems aiming to participate to scaffolding with repeated contigs. To this purpose, we essentially manipulate two kinds of graphs, both modeling contigs and their interactions, the *scaffold graph* and the *solution graph*. We denote by $E(G)$ and $V(G)$ the set of edges and vertices, respectively, of a graph G (or E and V if no ambiguity occurs). A solution graph is a special kind of scaffold graph, the latter being defined the following way:

Definition 1 (scaffold graph). *A graph* (G, M^*, ω, m) *is a* scaffold graph *if* V *corresponds to a set of contig extremities, and* E *is composed of two subsets:* M^* *is the set of edges between extremities belonging to a same contig (thus defining a perfect matching in G), and $E \setminus M^*$ is the set of interactions between contigs. A scaffold graph comes with two functions* $\omega, m : E \to \mathbb{N}$, *defining respectively the confidence level of inter-contigs interactions, and the multiplicity of contigs (their copy number). If m is not provided, then all multiplicities are equal to one.*

An example of scaffold graph, and inference of solutions on this graph, can be found in Fig. 1.

Given a set of contigs, it is possible to infer their multiplicities using various techniques involving for instance the cover depth in a mapping of reads on contigs (using for instance tools like CRAC [11]), or directly infer multiple contigs from kmer counting [8]. Getting links between contigs necessits additional information, for instance mapping of paired-end reads on contigs [4].

Inferring scaffolds, *i.e.* sequences of contigs at the chromosome scale, is modeled by an optimisation problem in the scaffold graph, similar to Traveling Salesman Problem, but taking into account the chromosomal structure (numbers of linear and circular chromosomes). In the simplified case where every contig is supposed to appear just once, this problem is stated as:

SCAFFOLDING (SCA)
Input: a scaffold graph (G, M^*, ω) and integers $\sigma_p, \sigma_c, k \in \mathbb{N}$
Question: Is there some $S \subseteq E \setminus M^*$ such that $S \cup M^*$ is
    a collection of $\leq \sigma_p$ alternating paths and $\leq \sigma_c$ alternating cycles
    and $\sum_{e \in S} \omega(e) \geq k$?

For a vertex v, we define $M^*(v)$ as the unique vertex u with $uv \in M^*$. A path (or a cycle) p is called *alternating* with respect to M^* if, for all vertices u of p, also $M^*(u)$ is a vertex of p.

SCAFFOLDING has been studied in the framework of complexity and approximation [4,14,15]. In this case, the produced solution is a collection of disjoint paths alternating between edges from M^* (contigs) and edges from $E \setminus M^*$ (links between contigs), from which it is easy to infer without any ambiguity a set of nucleotidic sequences by reading contig sequences, and for inter-contig links, either detecting possible overlaps missed by the assembly process, or completing with N's.

To improve the realism of the model, it is convenient to take the multiplicities of contigs into account. The main difference induced by allowing a contig to appear several times in the solution, is that the set of edges which are selected in an optimal solution does not necessarily lead to a unique interpretation as a set of scaffolds. The scaffolding problem with multiplicities thus involves a solution which is a graph, corresponding to the fusion of the right number of walks in the original scaffold graph. For each non-contig edge uv, its multiplicity $m(uv)$ equals the smaller of the multiplicities of the contig edges incident to u and v. A *walk* W is a sequence $(u_1, u_2, \ldots, u_\ell)$ of vertices such that, for each two consecutive vertices u_i and u_{i+1}, we have $u_i u_{i+1} \in E$. Then, W is called *closed* if $u_1 = u_\ell$ and W is called *alternating* with respect to M^* if ℓ is even and, for each odd i, we have $u_i u_{i+1} \in M^*$.

Observation 1. *For each vertex u of a solution graph, the sum of multiplicities of its incident non-matching edges is at most the multiplicity of its incident matching edge.*

The scaffolding problem with multiplicities is thus stated as follows:

SCAFFOLDING WITH MULTIPLICITIES (MSCA)
Input: a scaffold graph (G, M^*, ω, m) and $\sigma_\mathrm{p}, \sigma_\mathrm{c}, k \in \mathbb{N}$
Question: Is there a multiset S of $\leq \sigma_\mathrm{c}$ closed and $\leq \sigma_\mathrm{p}$ non-closed alternating walks in G such that each $e \in M^*$ occurs at most $m(e)$ times in accross all walks of S and $\sum_{e \in E(S) \setminus M^*} \omega(e) \geq k$?

In this setting, a scaffold graph $(G^*, M^*, \omega^*, m^*)$ is called *solution graph* for (G, M^*, ω, m) if (a) G^* is a subgraph of G, (b) ω^* is the restriction of ω to G^*, (c) $m^*(uv) \leq m(uv)$ for all $uv \in E(G)$, (d) G^* can be decomposed into $\leq \sigma_\mathrm{c}$ closed and $\leq \sigma_\mathrm{p}$ non-closed walks. Such a decomposition into walks is called a *linearization* of the solution graph and, in general, it is not necessarily unique (see Fig. 1). Note that decomposability also implies that no vertex can have more incident non-matching multiplicities than the multiplicity of its incident matching edge.

It turns out that, in presence of repeated contigs, a solution graph implies a unique set of sequences if and only if it does not contain so called *ambiguous paths* [16].

Definition 2 (Ambiguous path). *Let p be an alternating u-v-path in a solution graph. If all edges of p have the same multiplicity μ (that is, $m(e) = \mu$ for all $e \in p$), then p is called μ-uniform (or simply uniform if μ is unknown). Further, if p is μ-uniform and each of u and v is incident with a non-matching edge of multiplicity strictly less than μ, then p is called "ambiguous".*

Thus, the task above can be achieved by destroying all ambiguous paths in the solution graph. A brutal way to do this is to cut the non-contig edges incident to both extremities of each ambiguous path. However, this solution may erase potentially important information. Indeed, to destroy an ambiguous path, it is sufficient to remove the non-contig edges incident to one of its extremities. Further, let v be an extremity of an uniform path, we sometimes say "to cut v", by which we mean removing all non-contig edges incident with v, and in that case v is denoted as a *vertex-cut*. The problem of finding a most parsimonious (with respect to some cost function ω') set X of vertex-cuts which destroys all ambiguous paths is called SEMI-BRUTAL CUT. Several cost-functions ω' make sense in this setting.

Definition 3. *A weight-function* $\omega' : 2^V \to \mathbb{N}$ *is called*

1. cut-score, *if* ω' *counts one per vertex-cut (that is, $\omega'(X) = |X|$),*
2. path-score, *if* ω' *counts one per removed edge (that is,*
   $\omega'(X) := \sum\{m(uv) \mid uv \in E \setminus M^* \wedge uv \cap X \neq \varnothing\}$), *and*
3. weight-score, *if* ω' *counts the total weight of the removed edges (that is,*
   $\omega'(X) := \sum\{m(uv) \cdot \omega(uv) \mid uv \in E \setminus M^* \wedge uv \cap X \neq \varnothing\}$).

Note that, from the perspective of computational complexity, the path-score is a special case of the weight score, since we can just set $\omega'(e) = 1$ for all edges e. Thus, when saying "both scores" we refer to cut- and weight-score. Formally, the SEMI-BRUTAL CUT problem on which we focus here, is defined the following way:

SEMI-BRUTAL CUT (SBC)
Input: a solution graph (G, M^*, ω, m) and some $k \in \mathbb{N}$
Question: Is there a set X of vertex-cuts of G which destroys all ambiguous paths and the score of X is at most k?

We consider the functions defined in Definition 3 as scores for X. In context of approximation, SEMI-BRUTAL CUT refers to its optimization variant, minimizing the score of X.

A summary of the different problems involved and their input/output is presented in Table 1.

Table 1. Problems around genome scaffolding.

Problem	SCAFFOLDING	SCAFFOLDING WITH MULTIPLICITIES	SEMI-BRUTAL CUT
Input	Scaffold graph	Scaffold graph	Solution graph
Output	Scaffolds	Solution graph	Scaffolds

Related Works. In previous work [5,16], we proposed the first results concerning the complexity and approximation of SEMI-BRUTAL CUT according to the scoring functions mentioned in Definition 3. Some questions remain open concerning the complexity and (Non)-approximation for the cut and weight-score. In this article, we conclude the study of linearization in the framework of complexity and approximation. We prove that SEMI-BRUTAL CUT according to the cut-score is APX-complete.

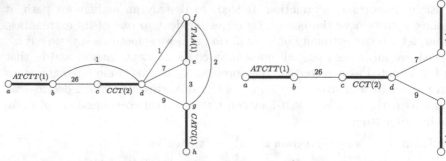

(a) Scaffold graph. This graph illustrates relationship between four contigs, figured by bold edges ab, cd, ef and gh. Labels on these edges show the sequence of the contigs, and their mutliplicity (in parenthesis). Edge cd, whose sequence is CCT, has multiplicity two. Other contigs are of multiplicity one. Links between contigs are labeled by their weight. In the input scaffold graph, the real sequences are both paths (a, b, c, d, e, f) and (c, d, g, h)

(b) Solution graph after solving SCAFFOLDING WITH MULTIPLICITIES. The solution graph is obtained as a solution for the MSCA instance asking for two open walks with total weight ≥ 42. In the solution graph, the contig of multiplicity two labeled CCT constitutes an ambiguous path, yielding two possible sets of sequences {ATCCT..CCT..TAAAA, CCT..CATG} and {ATCCT..CCT..CATG, CCT..TAAAA}.

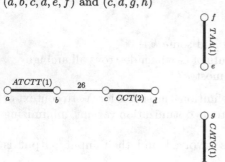

(c) Linearisation using SEMI-BRUTAL CUT. Brutal cut would provide a set of four independent sequences of total weight zero (the initial set of contigs), whereas SEMI-BRUTAL CUT with weight-score provides a unique set of four sequences {ATCCT..CCT, CCT, TAAAA, CATG}, and weight 26 (minimal weight-score 16). After resolving successively MSCA (with $\sigma_p = 2$ and $\sigma_c = 0$) and SBC (dashed edges are cut), the solution is compatible with the initial hypothesis. The only ambiguous path is the matching edge {c, d} and the cut vertex is d.

(d) Direct linearisation from the scaffold graph. Directly searching two maximum weighted alternating paths such that the solution graph does not contain ambiguity yields a chimeric sequence (f, e, g, h).

Fig. 1. Example of scaffold graph (a), a solution graph (b), scaffolds after solving SEMI-BRUTAL CUT (c) and a direct linearization leading to chimeric solution (d).

Table 2. Overview of results for SEMI-BRUTAL CUT.

Topologies	Type of cut	Complexity	Lower and upper bound
General	All	NP-hard [16]	
Trees	All	Linear [16]	
Planar, $\Delta \leq 4$	Cut-score	NP-hard [16]	Approx: 1.37 $(P \neq NP)$ [16], Approx: $2 - \epsilon$ (UGC) [16], Exact: $2^{o(n)}$ (ETH) [16]
Bip. plan., $\Delta \leq 3$	Cut-score	NP-hard [5]	APX-Hard [5] Exact: $2^{o(\sqrt{n+m})}n^{O(1)}$ (ETH) [5] 4-approximation Theorem 3
Bip., planar $\Delta \leq 3$	Weight-score	NP-hard Theorem 1	2-approximation [5] APX-Complete 1
$\Delta \leq 3$			1.108 Theorem 2

Organization of the Article. The Sect. 2 is devoted to the complexity result, we push this hardness result to bipartite, planar, subcubic graphs whereas Sect. 3 propose some lowers bounds according to complexity hypothesis. In the last section, we develop a polynomial-time approximation algorithm which conclude SEMI-BRUTAL CUT. Table 2 summarizes the overall results.

2 Computational Hardness

We consider in this section a very restricted class of graphs, which are planar, bipartite, subcubic graphs. The choice of this class is simultaneously led by biological and theoretical reasons. Biologically, we noticed that solution graphs are really sparse, and once reduced the non-ambiguous paths, are often equivalent to planar graphs with small degrees. However, this is only empirical observation and to our knowledge there are no general properties on real solution graphs that could be directly exploited. The theoretical reason is a wide literature on those classical classes of graphs, and we know hardness and non-approximation results that could be exploited through classical reductions. We mean then to show that, though not capturing the essential of solution graph properties, the results below give a good indication on how hard the problem is to solve, even under structural constraints.

Although it is know that SEMI-BRUTAL CUT is NP-hard under both cut- and weight-score [16], we extend this hardness for the weight-score to planar, bipartite, subcubic graphs. To this end, we reduce the classic NP-complete problem 3-SAT to SEMI-BRUTAL CUT.

MONOTONE 3-SATISFIABILITY (3-SAT)

Input: A boolean formula φ in conjunctive normal form where each clause contains exactly three positive literals or three negative literals.

Question: Is there a satisfying assignment β for φ?

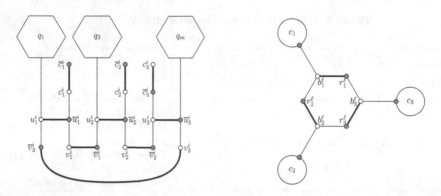

Fig. 2. Matching edges are bold. Left: variable gadget c_{x_i} linked to the clause gadgets q_1, q_3 and q_m, where x_i occurs positively in C_1 and C_3 and negatively in C_m. Right: clause gadget corresponding to the clause $C_\ell = (\overline{x_1} \vee \overline{x_2} \vee \overline{x_3})$.

Construction 1. *Let φ be an instance of 3-SAT with n variables $x_1, x_2, \ldots$ and m clauses $C_1, C_2, \ldots$. For each variable x_i, let ψ_i be the list of indices ℓ such that C_ℓ contains x_i and $|\psi_i|$ is the number of occurrences of x_i in φ. We construct the following solution graph (G^*, M^*, ω, m) along with a 2-coloring of G^* (see Fig. 2).*

- *For each x_i, we construct a cycle c_i on the vertex set $\bigcup_{j \le |\psi_i|} \{u_j^i, \overline{u}_j^i, v_j^i, \overline{v}_j^i\}$ such that, for all $j \le |\psi_i|$,*
  - *$\{u_j^i, \overline{u}_j^i\}, \{v_j^i, \overline{v}_j^i\} \in M^*$, and*
  - *the vertices u_j^i and v_j^i are blue and the vertices $\overline{u}_j^i$ and $\overline{v}_j^i$ are red.*
- *For each clause C_ℓ, we construct an alternating 6-cycle q_ℓ on the vertex set $\bigcup_{j \le 3} \{r_j^\ell, b_j^\ell\}$ such that, for all $j \le 3$, $\{r_j^\ell, b_j^\ell\} \in M^*$, and r_j^ℓ is red and b_j^ℓ is blue.*
- *For each variable x_i and each $j \le |\psi_i|$, let C_ℓ be the jth clause of ψ_i and let t be such that lit_i is the tth literal of C_ℓ. Then,*
  - *create a single matching edge $\{c_j^\ell, \overline{c}_j^\ell\}$, where c_j^ℓ is blue and $\overline{c}_j^\ell$ is red,*
  - *if x_i occurs positively in C_ℓ, introduce the edges $\{r_t^\ell, u_j^i\}$ and $\{c_j^\ell, \overline{u}_j^i\}$, and*
  - *if x_i occurs negatively in C_ℓ, introduce the edges $\{b_t^\ell, \overline{u}_j^i\}$ and $\{\overline{c}_j^\ell, u_j^i\}$.*
- *Each non matching edge has multiplicity 1 and weight 1 and all matching edges have multiplicity 2 (thus, each matching edge except the $\{c_i^\ell, \overline{c}_i^\ell\}$ is an ambiguous path).*

Clearly, Construction 1 can be carried out in polynomial time. Further, the resulting graph G^* is bipartite and the maximum degree $\Delta(G^*) = 3$. In the following, we call a matching edge *clean* if one of its endpoints has degree one. Note that a scaffold graph whose every matching edge is clean does not contain ambiguous paths.

Theorem 1. SEMI-BRUTAL CUT *is NP-complete for the weight-score, even if the graph is planar, bipartite, subcubic.*

In order to prove Theorem 1, we use the following properties of Construction 1, yielding a "canonical" set of cuts.

Lemma 1. *Let $S \subseteq V(G^*)$ be a set of vertex-cuts destroying all ambiguous paths in (G^*, M^*, ω, m), let c_i be a variable gadget and let q_ℓ be a clause gadget. We suppose that we start by cutting the vertices in the variable gadget and then we cut the vertices in the clause gadget. There is a set S' of vertex-cuts with $|S'| \leq |S|$ that also destroys all ambiguous paths and*

(a) $\omega'(S \cap V(c_i)) = \omega'(S' \cap V(c_i)) \geq 2 \times |\psi_i|$ *and* $\omega'(S \cap V(q_\ell)) = \omega'(S' \cap V(q_\ell)) \geq$
    2 (S and S' have the same score in variable gadgets and clause gadgets),
(b) if $\omega'(S' \cap V(c_i)) = 2 \times |\psi_i|$, *then* $S' \cap V(c_i)$ *is either* $\bigcup_{j \leq |\psi_i|} \{u_j^i\}$ *or*
    $\bigcup_{j \leq |\psi_i|} \{\overline{u}_j^i\}$ *(if S' is optimal on a variable gadget, cuts are only on positive sides or only on negative sides),*
(c) $\omega'(S' \cap V(q_\ell)) = 2$ *if and only if S' contains a vertex adjacent to q_ℓ (only two cuts are needed in a clause gadget iff it has been isolated by a cut in an adjacent variable gadget, meaning that the variable satisfies the clause).*

Proof. (a): For each $j \leq |\psi_i|$, we need to remove two edges to linearize the ambiguous paths $\{u_j^i, \overline{u}_j^i\}$. Then we need to remove at least $2 \times |\psi_i|$ edges in c_i. In the clause q_ℓ, we need to remove at least two edges in the inner cycle.

(b): Note that cutting all vertices in either $\bigcup_{j \leq |\psi_i|} \{u_j^i\}$ or $\bigcup_{j \leq |\psi_i|} \{\overline{u}_j^i\}$ suffices to remove all ambiguous path in x_i and in that case $\omega(S \cap V(c_i)) = 2 \times |\psi_i|$. If S contains some $\overline{u}_j^i$ and does not contain $\overline{u}_{j+1}^i$ for some j, then we need a cut to linearize $\{v_j^i, v_j^i\}$ which will increase by one the score of the solution (and analogously for u_j^i). Hence if $\omega'(S \cap V(c_i)) = 2 \times |\psi_i|$, we can suppose that S contains either $\bigcup_{j \leq |\psi_i|} \{u_j^i\}$ or $\bigcup_{j \leq |\psi_i|} \{\overline{u}_j^i\}$. If S contains a cut in some v_j^i or some $\overline{v}_j^i$ then, since the path $\{v_j^i, \overline{v}_j^i\}$ is already linearized by a cut in $\{\overline{u}_j^i, u_{j+1}^i\}$, we can remove the cut in S'.

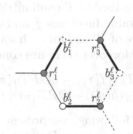

Fig. 3. A cut of size 2 in q_ℓ when one incident edge to q_ℓ is cut. Dashed edges and vertices are part of the cut.

(c): We need to remove at least two edges from the inner cycle of C_ℓ. Suppose that all literals of C_ℓ occur positively. Suppose by symmetry that $\{b_1^\ell, b_2^\ell\} \subseteq S'$. Then if the leaving edge incident to r_3^ℓ is not cut, then we need to remove one more edge from q_ℓ and in that case $\omega'(S' \cap V(q_\ell)) \geq 2$ (see Fig. 3). □

Proof (Proof of Theorem 1). Recall that 3-SAT remains NP-complete if the input formula is planar [1] and, in this case, the graph produced by Construction 1 is also planar. Clearly, SEMI-BRUTAL CUT is in NP. We show that Construction 1 is correct, that is, φ is satisfyable if and only if the scaffold graph (G^*, M^*, ω, m) resulting from Construction 1 can be linearized with a score of $8m$.

"$\Rightarrow$": Let β be a satisfying assignment for φ. Then, for each variable x_i and for all $j \leq |\psi_i|$, we cut the vertices u_j^i if $\beta(x_i) = 1$ and the vertices $\overline{u}_j^i$ otherwise. As β is satisfying, this removes at least one edge adjacent to each clause gadget. Thus, according to Lemma 1(c), we can cut two vertices in each clause gadget q_j to turn every matching edge in q_j clean. Since we also cut either the vertices u_j^i or the vertices $\overline{u}_j^i$ for each vertex gadget, we conclude that all matching edges of the result are clean and we remove exactly $2m + \sum_i 2 \times |\psi_i| = 8m$ edges.

"$\Leftarrow$": Let $S \subseteq V$ be the set of vertices such that cutting each vertex of S destroys all ambiguous paths in (G^*, M^*, ω, m) and $\omega'(S) = 8m$. According to Lemma 1(a), each variable gadget remove $|\psi_i|$ edges and each clause gadget remove two edges. Moreover, by Lemma 1(b), for each variable gadget c_i, we can suppose that $S \cap V(c_i)$ equals $\bigcup_{j \leq |\psi_i|} \{u_j^i\}$ or $\bigcup_{j \leq |\psi_i|} \{\overline{u}_j^i\}$. In the former case, we set $\beta(x_i) = 1$ and, in the latter, we set $\beta(x_i) = 0$. To show that β satisfies φ, assume that there is a clause C_ℓ that is not satisfied by β. Then, none of the edges incident to q_ℓ is cut which, by Lemma 1(c), contradicts the fact that there are two removed edges in q_ℓ. $\qquad\square$

3   Non-approximability

In this section, we prove approximation lower bounds for SEMI-BRUTAL CUT. First recall the definition of *L-reduction* between two hard problems Π and Π', described by Papadimitriou [10]. This reduction consists of polynomial-time computable functions f and g such that, for each instance x of Π, $f(x)$ is an instance of Π' and for each feasible solution y' for $f(x)$, $g(y')$ is a feasible solution for x. Moreover there are constants $\alpha, \beta > 0$ such that:

1. $OPT_{\Pi'}(f(x)) \leq \alpha OPT_\Pi(x)$ and
2. $|val_\Pi(g(y')) - OPT_\Pi(x)| \leq \beta |val_{\Pi'}(y') - OPT_{\Pi'}(f(x))|$.

In the following, we present an *L-reduction* from the classical problem MAX 3-SAT(4) to SEMI-BRUTAL CUT.

MAX 3-SAT(4)
Input: A boolean formula φ in exact 3-CNF where every variable occurs in 4 clauses
Task: Find an assignment that satisfies a maximum number of clauses.

Construction 2. *We reuse Construction 1 and change some variable gadgets and the way we link the variable gadgets to the clause gadgets. First, we change the links between the gadgets: let C_ℓ be a clause and x_i be the jth literal of C_ℓ. Then, attach c_i to r_j^ℓ. The difference with Construction 1 is that we attach the variable gadgets to the red vertices of the clause gadget, no matter if the variable occurs positively or negatively in the clause.*

Now we change some variable gadgets. Let x_i be a variable which occurs positively in the clauses C_p and $C_{p'}$ and negatively in the clauses C_n and $C_{n'}$. We replace the variable gadget associated to x_i by the following gadget r_i:

- *Construct a cycle c_i on the vertex set $\bigcup_{j \leq 2}\{u_j^i, \overline{u}_j^i, v_j^i, \overline{v}_j^i\}$ such that, for all $j \leq 2$, $\{u_j^i, \overline{u}_j^i\}, \{v_j^i, \overline{v}_j^i\} \in M^*$, the vertices u_j^i and v_j^i are blue and $\overline{u}_j^i$ and $\overline{v}_j^i$ are red.*
- *Give multiplicity 1 and weight 1 to all non-matching edges and multiplicity 2 to all matching edges.*
- *Link the clause gadgets $q_p, q_{p'}, q_n$ and $q_{n'}$ to vertices $u_1^i, u_2^i, \overline{u}_1^i$ and $\overline{u}_2^i$ respectively in the same way as previously described.*

Note that all matching edges are ambiguous paths in the variable gadget. The clause gadgets and the other variable gadgets remain unchanged.

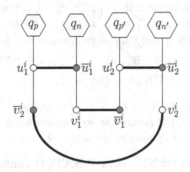

Fig. 4. Matching edges are bold. Example of variable gadget r_{x_i} linked to the clause gadgets q_p, $q_{p'}$, q_n and $q_{n'}$, where x_i occurs positively in C_p and $C_{p'}$ and negatively in C_n and $C_{n'}$.

The resulting graph G^* is bipartite and $\Delta(G^*) = 3$. In the following, when we want to differentiate the variable gadgets, we designate by *rectangle* variable gadget those defined in Construction 2 and by *cycle* variable gadget those defined in Construction 1. An example of a rectangle variable gadget is given in Fig. 4. Notice that the properties (a) and (c) of Lemma 1 hold. We can add the following property:

Lemma 2. *Let $S \subseteq V(G^*)$ be an optimal set of vertex-cuts destroying all ambiguous paths in (G^*, M^*, ω, m), let c_i be a cycle variable gadget and $r_{i'}$ be a rectangle variable gadget. There is a set S' of cuts with $\omega(S') - \omega(S)$ that also destroys all ambiguous paths, and*

(a) $S' \cap V(c_i)$ is either $\bigcup_{j \leq |\psi_i|}\{u_j^i\}$ or $\bigcup_{j \leq |\psi_i|}\{\overline{u}_j^i\}$, and
(b) $S' \cap V(r_{i'})$ is either $\{u_1^{i'}, u_2^{i'}\}$ or $\{\overline{u}_1^{i'}, \overline{u}_2^{i'}\}$.

Proof. Recall that, by Lemma 1(a), $\omega(S \cap V(c_i)) \geq |\psi_i|$.

"(a)": By symmetry, suppose that x_i occurs mostly positively in φ. If x_i occurs four times positively, then replacing $S \cap V(c_i)$ by $\bigcup_{j \leq |\psi_i|}\{u_j^i\}$ in S yields a solution S' as sought. Thus, suppose that x_i occurs three times positively. Let

C_ℓ be the clause where x_i occurs negatively and let z denote the neighbor of $\overline{u}_j^i$ in c_ℓ. If $|S \cap V(c_i)| > |\psi_i|$, then replacing $S \cap c_i$ by $\bigcup_{j \leq |\psi_i|} \{u_j^i\}$ plus z yields a solution S' as sought. Finally, if $|S \cap V(c_i)| = |\psi_i|$, then S already corresponds to (a) as, otherwise, some ambiguous path $\{v_j^i, \overline{v}_j^i\}$ is not destroyed.

"(b)": Note that one cut in $r_{i'}$ is not enough to destroy all ambiguous paths and cutting either the vertices $\{u_1^{i'}, u_2^{i'}\}$ or the vertices $\{\overline{u}_1^{i'}, \overline{u}_2^{i'}\}$ destroys all ambiguous paths in the rectangle variable gadget. By symmetry, suppose that S contains $v_1^{i'}$, if S contains $\overline{u}_1^{i'}$, then we can remove $v_1^{i'}$ from S. Otherwise, since S is optimal, $S \cap V(r_{i'}) \{u_1^{i'}, \overline{u}_2^{i'}, v_1^{i'}\}$. Let $z \notin r_{i'}$ be the vertex adjacent to $u_1^{i'}$. Then, z is clean, since otherwise we can replace $S \cap V(r_{i'})$ by $\{\overline{u}_1^{i'}, \overline{u}_2^{i'}\}$, contradicting the fact that S is optimal. We can then add z in S' and swap $u_1^{i'}$ by $\overline{u}_1^{i'}$. Further if S does not contains any vertices in $\{v_1^{i'}, \overline{v}_1^{i'}, v_2^{i'}, \overline{v}_2^{i'}\}$, then suppose without loss of generality that S contains $\{u_1^{i'}, u_2^{i'}\}$. Let $z_j \notin r_{i'}$ be the vertex incident to $\overline{u}_j^{i'}$. If S contains $\overline{u}_j^{i'}$, then it only serve to remove the leaving edge incident to $\overline{u}_j^{i'}$ and it also removes the edge $\{\overline{u}_j^{i'}, v_{1+(j+1 \mod 2)}^{i'}\}$, which contradicts the fact that S is optimal. Thus, $S \cap V(r_{i'}) = \{u_1^{i'}, u_2^{i'}\}$.

Theorem 2. *There is a constant $\epsilon_4' > 0$ (the value $\epsilon_4' > 0$ is defined in [2]) for which* SEMI-BRUTAL CUT *cannot be approximated to any factor better than $(1 + {}^{7\epsilon_4'}/_{65})$, even on graphs of maximum degree three, unless P=NP.*

Proof. Recall that, unless P=NP, MAX 3-SAT(4) cannot be approximated to a factor better than $\epsilon_4' = 1,00052$ [2] and that, in an optimal solution of MAX 3-SAT(4), at least $^7/_8$ of the clauses are satisfied [6], yielding

$$OPT(\varphi) \geq {}^{7m}/_8. \tag{1}$$

To show that Construction 2 constitutes an L-reduction, let f be a function transforming any instance φ of MAX 3-SAT(4) into an instance I of SEMI-BRUTAL CUT as above, let S be a feasible solution for I corresponding to the properties of Lemma 1(a), Lemma 1(c) and Lemma 2, and let g be the function that transforms S into an assignment β as constructed in the proof of Theorem 1: each variable x_i is set to true if S cuts u_j^i for all j, and false, otherwise. By Lemma 2, for each clause gadget q_ℓ without an adjacent vertex in S, the "extra" cut occurs in q_ℓ. Hence, for each of the at most $^m/_8$ unsatisfied clauses in φ, we have to remove an other edge to linearize I. Thus,

$$OPT(I) \leq 8m + {}^m/_8 \overset{(1)}{\leq} {}^{65}/_7 OPT(\varphi) \tag{2}$$

An important obstacle to overcome (and reason why Construction 1 is not enough for Theorem 2) is that an approximate solution to SBC might spend extra cuts in variable gadgets in order to "change the assignment" of a variable x_i mid-way. However, since each variable occurs at most four times, this only happens for variables that occur two times positively and two times negatively. Now, with our modification to Construction 1, we can observe that each extra cut in any of the variable gadgets allows such a misuse only for a single clause gadget. Thus, the number of satisfied clauses of φ and the clause gadgets in which we have to spend extra cuts adds up to m. Hence,

$$9m = val(g(S)) + val(S) = OPT(I) + OPT(\varphi) \tag{3}$$

Thus, we constructed an L-reduction with $\alpha = {}^{65}/7$, $\beta = 1$ and, since $val(g(S)) < (1 - \epsilon'_4) \cdot OPT(\varphi)$, we conclude

$$val(S) \overset{(3)}{=} OPT(I) + OPT(\varphi) - val(g(S))$$
$$> OPT(I) + \epsilon'_4 OPT(\varphi)$$
$$\overset{(2)}{\geq} (1 + {}^{7\epsilon'_4}/65) \cdot OPT(I)$$

This is conclude the proof.   □

4   Linear-Time Approximation Algorithm

In the following, we present a polynomial-time 4-approximation for SEMI-BRUTAL CUT with cut-score. To this end, we use the following reduction rule introduced by Weller et al. [16].

Rule 1. *Let $\mu \in \mathbb{N}$ and let $uvwx$ be a μ-uniform, alternating path in G. Then, replace $uvwx$ by a matching edge ux with multiplicity μ.*

Rule 1 merges pairs of non-ambiguous contigs into one. Thus, each ambiguous path will consist of a single contig edge. In this sense, we call a contig edge *ambiguous* if it is an ambiguous path and *clean*, otherwise.

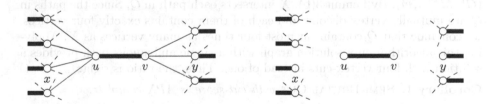

Fig. 5. A forbidden path $xuvy$ (left) and the result of cutting all its vertices (right).

Our approximation algorithm works similarly to the well-known classical 2-approximation for VERTEX COVER that just returns the extremities of any maximal matching. Contrary to VERTEX COVER, our forbidden structures are not edges, but *ambiguous edges*. Thus, we have to consider length-four paths containing an ambiguous edge, and we will cut all four of their vertices. In the following, we call a path $xuvy$ *forbidden* if xu and vy are inter-contig edges and uv is an ambiguous edge such that $m(xu) < m(uv) > m(vy)$ (see Fig. 5).

Lemma 3. *Let Q be a maximal packing of vertex-disjoint forbidden paths in (G^*, M^*, ω, m), let X be any solution for SBC with cut-score on (G^*, M^*, ω, m). Then, (a) cutting all vertices of Q destroys all ambiguous edges in G^* and (b) $X \cap p \neq \varnothing$ for all $p \in Q$.*

Proof. (a): Let H be the result of cutting all vertices of Q in G^*. Towards a contradiction, assume that H contains an ambiguous edge uv. By definition, there are inter-contig edges xu and vy in H such that $m(xu) < m(uv) > m(vy)$. But then, the path $xuvy$ is a forbidden path, contradicting the maximality of Q.

(b): Let H be the result of cutting all vertices of X in G^*. Let $xuvy \in Q$ be a forbidden path in (G^*, M^*, ω, m) and assume towards a contradiction that $X \cap xuvy = \varnothing$. Then, none of the edges of $xuvy$ are removed when cutting the vertices of X, that is, $xuvy$ survives in H. Then, however, uv is an ambiguous path in H, contradicting X being a solution for (G^*, M^*, ω, m). $\qquad\square$

With Lemma 3, we can show that any maximal packing of forbidden paths constitutes a 4-approximation for SEMI-BRUTAL CUT with cut-score.

Theorem 3. *A 4-approximate solution to SEMI-BRUTAL CUT with cut-score can be computed in linear time. This ratio is tight.*

Proof. First, Lemma 1 can be exhaustively applied to (G^*, M^*, ω, m) in linear time since the inner vertices of any μ-uniform alternating path have degree two. Second, a packing of forbidden paths in (G^*, M^*, ω, m) can be computed by scanning all contig edges uv and, if uv is ambiguous, then $xuvy$ is a forbidden path for any inter-contig edges xu and vy. By removing x, u, v, and y from G^*, we make sure that the resulting packing is vertex-disjoint. Thus, such a packing can be produced in linear time.

Let Q be any maximal vertex-disjoint packing of forbidden paths in (G^*, M^*, ω, m). By Lemma 3(a), the vertices of Q form a solution for SBC. To show that this solution is 4-approximate, consider any optimal solution X for (G^*, M^*, ω, m). By Lemma 3(b), X intersects each path in Q. Since the paths in Q are mutually vertex disjoint and each of them contains exactly four vertices, we conclude that Q contains at most four times as many vertices as X. Applying this algorithm on a solution graph with a single ambiguous path provides a solution with four vertex-cuts instead of one. Thus, the ratio is tight. $\qquad\square$

Corollary 1. SEMI-BRUTAL CUT *with cut-score is APX-complete.*

5   Conclusion

We developed results concerning the complexity, lower bounds and approximability of the linearization problem for genome scaffolds sharing repeated contigs with two possible scoring functions. We managed to strengthen previously known NP-hardness to the very restricted class of planar bipartite subcubic graphs with only two multiplicities for the cut-score. We also provided a simple, linear-time 4-approximation of for cut-scores. Natural perspectives of this work are to extend this result to the weight-score, explore the possibility of FPT algorithms and approximations in the difficult cases, and examine the practical performance of the presented approximation algorithm on larger real-world instances.

Acknowledgments. This work was supported by the Institut de Biologie Computationnelle3 (http://www.ibc-montpellier.fr/) (ANR Projet Investissements d'Avenir en bioinformatique IBC) and the "Région Occitanie".

References

1. Berg, M.D., Khosravi, A.: Optimal binary space partitions for segments in the plane. Int. J. Comput. Geom. Appl. **22**(3), 187–206 (2012)
2. Berman, P., Karpinski, M., Scott, A.D.: Approximation hardness and satisfiability of bounded occurrence instances of SAT. Electronic Colloquium on Computational Complexity (ECCC), vol. 10, no. 022 (2003)
3. Biscotti, M.A., Olmo, E., Heslop-Harrison, J.S.: Repetitive DNA in eukaryotic genomes. Chromosome Res. **23**(3), 415–420 (2015)
4. Chateau, A., Giroudeau, R.: A complexity and approximation framework for the maximization scaffolding problem. Theor. Comput. Sci. **595**, 92–106 (2015). https://doi.org/10.1016/j.tcs.2015.06.023
5. Davot, T., Chateau, A., Giroudeau, R., Weller, M.: On the hardness of approximating the linearization of scaffolds sharing repeated contigs, Accepted to RecombCG 2018
6. Håstad, J.: Some optimal inapproximability results. J. ACM **48**(4), 798–859 (2001)
7. Hunt, M., Newbold, C., Berriman, M., Otto, T.: A comprehensive evaluation of assembly scaffolding tools. Genome Biol. **15**(3), R42 (2014)
8. Koch, P., Platzer, M., Downie, B.R.: RepARK-de novo creation of repeat libraries from whole-genome NGS reads. Nucleic Acids Res. **42**(9), e80 (2014)
9. Mandric, I., Lindsay, J., Măndoiu, I.I., Zelikovsky, A.: Scaffolding algorithms. In: Măndoiu, I., Zelikovsky, A. (eds.) Computational Methods for Next Generation Sequencing Data Analysis, Chap. 5, pp. 107–132. Wiley (2016)
10. Papadimitriou, C.H., Yannakakis, M.: Optimization, approximation, and complexity classes. J. Comput. Syst. Sci. **43**(3), 425–440 (1991)
11. Philippe, N., Salson, M., Lecroq, T., Léonard, M., Commes, T., Rivals, E.: Querying large read collections in main memory: a versatile data structure. BMC Bioinform. **12**, 242 (2011). https://doi.org/10.1186/1471-2105-12-242
12. Quail, M., et al.: A tale of three next generation sequencing platforms: comparison of ion torrent, pacific biosciences and illumina miseq sequencers. BMC Genomics **13**(1), 341 (2012)
13. Tang, H.: Genome assembly, rearrangement, and repeats. Chem. Rev. **107**(8), 3391–3406 (2007)
14. Weller, M., Chateau, A., Dallard, C., Giroudeau, R.: Scaffolding problems revisited: complexity, approximation and fixed parameter tractable algorithms, and some special cases. Algorithmica **80**(6), 1771–1803 (2018)
15. Weller, M., Chateau, A., Giroudeau, R.: Exact approaches for scaffolding. BMC Bioinform. **16**(Suppl. 14), S2 (2015)
16. Weller, M., Chateau, A., Giroudeau, R.: On the linearization of scaffolds sharing repeated contigs. In: Gao, X., Du, H., Han, M. (eds.) COCOA 2017. LNCS, vol. 10628, pp. 509–517. Springer, Cham (2017). https://doi.org/10.1007/978-3-319-71147-8_38

Relaxation and Matrix Randomized Rounding for the Maximum Spectral Subgraph Problem

Cristina Bazgan[1], Paul Beaujean[1,2]($\boxtimes$), and Éric Gourdin[2]

[1] Université Paris-Dauphine, Université PSL, CNRS, LAMSADE,
75016 Paris, France
{cristina.bazgan,paul.beaujean}@lamsade.dauphine.fr
[2] Orange Labs, Châtillon, France
{paul.beaujean,eric.gourdin}@orange.com

Abstract. Modifying the topology of a network to mitigate the spread of an epidemic with epidemiological constant λ amounts to the NP-hard problem of finding a partial subgraph with maximum number of edges and spectral radius bounded above by λ. A software-defined network (SDN) capable of real-time topology reconfiguration can then use an algorithm for finding such subgraph to quickly remove spreading malware threats without deploying specific security countermeasures.

In this paper, we propose a novel randomized approximation algorithm based on the relaxation and rounding framework that achieves a $O(\log n)$ approximation in the case of finding a subgraph with spectral radius bounded by $\lambda \in (\log n, \lambda_1(G))$ where $\lambda_1(G)$ is the spectral radius of the input graph and n its number of nodes. We combine this algorithm with a maximum matching algorithm to obtain a $O(\log^2 n)$ approximation algorithm for all values of λ. We also describe how the mathematical programming formulation we give has several advantages over previous approaches which attempted at finding a subgraph with minimum spectral radius given an edge removal budget.

Keywords: Approximation algorithm · Relaxation and rounding Semidefinite programming · Spectral graph theory · Random graphs

1 Introduction

In recent years, a sequence of results [2,4,25,28] have established a relationship between the convergence of Markovian models representing an epidemic spreading over a network and the spectral characteristics of the underlying graph. The generalization of these theorems by Prakash et al. [16] states that in the case of a graph G and an epidemic model with epidemiological characteristic λ, fast convergence of the Markovian model to its absorbing state is guaranteed if the spectral radius of the graph $\lambda_1(G) < \lambda$. This has led the mathematical epidemiology community to look for algorithms that modify the topology of a network to ensure that a given epidemic converges rapidly to extinction.

© Springer Nature Switzerland AG 2018
D. Kim et al. (Eds.): COCOA 2018, LNCS 11346, pp. 108–122, 2018.
https://doi.org/10.1007/978-3-030-04651-4_8

At the same time, the software-defined networking (SDN) paradigm has transformed network administration by allowing real-time statistics [20] and topology reconfiguration [27]. This new paradigm has deep consequences for the management of network security as it is now possible for a SDN controller to automatically detect malware spreading over its network via machine learning [10] and react to such threat by deploying adequate security countermeasures. In this work we are following epidemiological practice and propose to use topology modification as a disease-agnostic countermeasure to the spread of malware in networks.

We are looking to preserve as much as possible the existing network topology by keeping the largest number of edges in the graph while guaranteeing that a given epidemic of epidemiological characteristic λ would rapidly disappear. For this purpose, we introduce the maximum spectral subgraph problem (MSSP) defined formally as follows. Denoting by $\lambda_1(G)$ the spectral radius of G, i.e. the largest eigenvalue of its adjacency matrix A, we have:

Definition 1. MAXIMUM SPECTRAL SUBGRAPH PROBLEM (MSSP)

Input: $G = (V, E)$ an undirected graph and $1 \leq \lambda < \lambda_1(G)$.
Output: $H = (V, E')$ with $E' \subseteq E$ such that $|E'|$ is maximum and $\lambda_1(H) \leq \lambda$.

1.1 Related Work

Spectral graph theory has often been a decisive tool in the design and analysis of algorithms. However, to the best of our knowledge, surprisingly few computational problems have been defined in terms of finding graphs with appropriate spectrum. The mathematical epidemiology community has proposed and analyzed several problems related to the spectrum of the adjacency matrix [19,26] while systems and control researchers have considered optimization problems related to the spectrum of the Laplacian matrix [5]. In a separate effort, the theoretical computer science community has focused on problems related to the design of expander graphs and graphs with high algebraic connectivity i.e. the second smallest eigenvalue of the Laplacian matrix [7,11]. In this line of research, all problems are NP-hard and the algorithms proposed in the literature are often simple to state. We contrast this with the fact that their analysis can be involved and yet, to the best of our knowledge, only amount to conditional approximation guarantees. Throughout this paper we qualify approximation algorithms by their performance guarantee $r > 1$ which corresponds to returning a solution whose value is at least a fraction $1/r$ of the optimal value for maximization problems or at most a factor r of the optimal value for minimization problems.

A minimization version of MSSP has been studied by Saha et al. [19] where the task is to remove the minimum amount of edges from a graph G such that the resulting subgraph H satisfies $\lambda_1(H) \leq \lambda$. They give a $(1 + \varepsilon, \varepsilon^{-1} \log n)$ bi-criteria approximation algorithm which guarantees that if an optimal solution is to remove k edges to achieve a spectral radius less than or equal to λ then the algorithm will remove $O(\varepsilon^{-1} \log n)$ times more edges (with $n = |V|$

the number of nodes in G) and returns a graph with spectral radius less than or equal to $(1 + \varepsilon)\lambda$. Zhang et al. [29] study the problem of maximizing the drop in spectral radius $\lambda_1(G) - \lambda_1(H)$ where H is a subgraph of G obtained by deleting at most k edges. Their randomized algorithm, inspired by the relaxation and rounding framework, has the following conditional guarantees: if the weighted graph obtained from the solution of the relaxed semidefinite programming problem has maximum weighted degree $\Delta^* = \Omega(\log^4 n)$, then the returned subgraph satisfies the constraint on the number of edge deletions in expectation and, with high probability, the remaining graph has a spectral radius within an additive $O(\sqrt{\Delta^*})$ factor of the optimal solution. If the condition on the maximum weighted degree is not satisfied, they do not obtain any performance guarantee.

In this article we introduce the maximum spectral subgraph problem (MSSP) and our main contribution is the design of a $O(\log^2 n)$-approximation algorithm for MSSP obtained by combining a randomized algorithm based on the relaxation and rounding framework with a maximum matching algorithm. We also describe some shortcomings of existing mathematical programming formulation for variants of MSSP that attempt at minimizing the spectral radius of a given graph within a prescribed edge deletion budget.

The rest of this paper is organized as follows. In Sect. 2 we recall some simple facts from spectral graph theory and introduce appropriate notations and known results. In Sect. 3 we describe our relaxation and rounding algorithm and illustrate its usage on star graphs. Then, in Sect. 4, we prove its approximation ratio for the range $\lambda \in (\log n, \lambda_1(G))$ in general graphs. In Sect. 5 we show that a maximum matching is a $O(\lambda^2)$-approximation algorithm for MSSP. Finally, perspectives and concluding remarks are provided in Sect. 6.

2   Preliminaries

We review here useful facts about the spectrum of adjacency matrices of graphs. Unless specified, all graphs are assumed to be undirected. Recall that the adjacency matrix A of a graph $G = (V, E)$ is a symmetric matrix defined as follows:

$$A_{ij} = \begin{cases} 1 & \text{if } ij \in E \\ 0 & \text{otherwise} \end{cases}$$

Property 1. [24] (General bounds) Given a graph $G = (V, E)$, we denote by $\Delta(G)$ its largest degree. The spectral radius of the graph, defined as the largest eigenvalue of its adjacency matrix, lies between the following quantities:

$$\max\left(\sqrt{\Delta(G)}, \frac{2|E|}{|V|}\right) \leq \lambda_1(G) \leq \Delta(G) \tag{1}$$

2.1   Computational Complexity

The problem of deciding whether there exists a subgraph with at least k edges and spectral radius at most λ was studied by van Mieghem et al. [26]. We can see that it is the decision problem associated with both MSSP and the problem of minimum edge removal introduced by Saha et al. [19] that was mentioned in Sect. 1. Van Mieghem et al. proved that the decision problem is NP-complete by reduction from the Hamiltonian path problem. It follows from this result that MSSP is NP-hard.

The reduction uses a fact from extremal spectral graph theory: the path graph on $|V|$ nodes is the graph with minimum spectral radius among all connected graphs with $|V|$ nodes and $|V| - 1$ edges. Setting $\lambda = \lambda_1(P_{|V|}) = 2\cos(\pi/(|V| + 1))$ and $k = |V| - 1$ completes the reduction. Recall that while the spectral radius of a graph might be a real number, verifying a candidate solution amounts to checking whether the eigenvalues of a given adjacency matrix are bounded above by a given value which can be done in polynomial time to any precision [14].

Note that if the bound on the spectral radius $\lambda = 1$, then MSSP becomes the maximum matching problem which can be solved in polynomial time. Indeed, from Property 1, it is easy to see that the problem consists in finding a subgraph of degree at most 1 with maximum number of edges. Furthermore, note that all undirected graphs that are not matchings have a spectral radius larger than or equal to $\sqrt{2}$ which is the spectral radius of the path graph on 3 nodes. From this consideration, we will study the range where the bound on the spectral radius is meaningful, that is $\sqrt{2} \le \lambda < \lambda_1(G)$.

We now present our algorithm based on the relaxation and randomized rounding framework.

3 Relaxation and Matrix Randomized Rounding

The relaxation and randomized rounding framework [17] is a general algorithmic technique composed of two steps: first, solving a continuous relaxation of the original combinatorial programming and then, sampling a discrete solution based on an optimal solution of the relaxed problem. This technique has resulted in the design of a large number of approximation algorithms for a broad range of combinatorial problems and has been the cornerstone of the application of the sum of squares hierarchy developed by Lasserre [8] and Parrilo [15] in combinatorial optimization. There are often two steps in the analysis of a relaxation and randomized rounding algorithm: finding a tight relaxation of the original problem that is solvable in polynomial time and proving that the random discrete solution is feasible with high probability.

Here we propose a mathematical programming formulation of MSSP that uses semidefinite programming (SDP) to model the constraint on the spectral radius. While linear programming allows to define optimization problems with non-negative vector variables written $x \ge 0$, SDP extends to the larger class of problems with positive semidefinite matrix variables written $X \succeq 0$ i.e. all eigenvalues of X are non-negative: $\forall i \in [1, n]$, $\lambda_i(X) \ge 0$. Given an input graph $G = (V, E)$ and a bound on the spectral radius λ, we write the following semidefinite programming problem with binary variables:

$$\max \quad \sum_{ij \in E} y_{ij}$$

$$\text{s.t.} \quad \sum_{ij \in E} y_{ij} A_{ij} \preceq \lambda I$$

$$\sum_{j \in \Gamma(i)} y_{ij} \leq \lambda^2, \ \forall i \in V \qquad \qquad \text{(SDP}_{0,1})$$

$$y_{ij} \in \{0, 1\}, \ \forall \ ij \in E$$

where A_{ij} is the adjacency matrix of the graph $G_{ij} = (V, \{ij\})$ with a single edge ij and I is the identity matrix of size $|V|$. The decision variables y_{ij} represent whether an edge ij belongs to the subgraph when $y_{ij} = 1$ or not when $y_{ij} = 0$. Recall that for a n by n square matrix, $M \preceq tI \iff \forall i \in [1, n], \lambda_i(M) \leq t$. This means that the semidefinite constraint ensures that the adjacency matrix of the subgraph defined by y_{ij} has its spectral radius bounded above by λ. The linear constraint ensures that the degree of each node $i \in V$ in the subgraph is bounded above by λ^2. Note that this constraint is redundant given that the general bounds (1) state that the maximum degree of a graph is bounded above by the square of its spectral radius i.e. $\Delta \leq \lambda_1^2$. However this is in general not the case with weighted graphs, which will be discussed in Sect. 3.3.

The continuous relaxation of Problem (SDP$_{0,1}$) is obtained by relaxing integer constraints into box constraints. We underline that the semidefinite constraint does not originate from the relaxation as is the case for some problems which relax vector variables with quadratic constraints into a SDP problem e.g. the one used in the algorithm given by Goemans and Williamson for the maximum cut problem [6]. Our relaxation is limited to the binary variables.

$$\max \quad \sum_{ij \in E} y_{ij}$$

$$\text{s.t.} \quad \sum_{ij \in E} y_{ij} A_{ij} \preceq \lambda I$$

$$\sum_{j \in \Gamma(i)} y_{ij} \leq \lambda^2, \ \forall i \in V \qquad \qquad \text{(SDP}_{\lambda\Delta})$$

$$y_{ij} \in [0, 1], \ \forall \ ij \in E$$

As semidefinite programming is in P, we can solve Problem (SDP$_{\lambda\Delta}$) in polynomial time. This allows us to state our relaxation and randomized rounding algorithm. In the rest of this article, we denote scalar random variables by lowercase bold letters e.g. $\boldsymbol{x}$ and matrix random variables by uppercase bold letters e.g. $\boldsymbol{X}$. Furthermore, we denote by $\boldsymbol{x} \sim \text{Ber}(\mu)$ the fact that $\boldsymbol{x}$ is a random variable following a Bernoulli distribution of mean μ.

Algorithm 1. RELAXATION & RANDOMIZED ROUNDING

Input: $G = (V, E)$, $\sqrt{2} \leq \lambda < \lambda_1(G)$, and $r > 1$.
Output: $H = (V, E')$ such that $\lambda_1(H) \leq \lambda$ with probability p_r.
$y^* \leftarrow$ arg Problem (SDP$_{\lambda\Delta}$)
Sample $\forall ij \in E$, $x_{ij} \sim$ Ber(y_{ij}^*/r)
return $H = (V, \{ ij \in E : x_{ij} = 1 \})$

We will now turn to a simple application of Algorithm 1 to the case of star graphs and determine the adequate sampling factor r that results in a feasible solution with high probability i.e. $p_r = 1 - 1/n$ where $n = |V|$.

3.1 The Case of Star Graphs

Before giving the complete analysis of our relaxation and randomized rounding algorithm we focus on a specific class of input graphs to illustrate the methodology of relaxation and randomized rounding but also to highlight the importance of the degree constraint in our proposed mathematical formulation.

Recall that a star graph $S_n = K_{1,n}$ is a graph with $V = \{0, \ldots, n\}$ and $E = \{(0, 1), \ldots, (0, n)\}$. It is a well-known fact from spectral graph theory that the spectral radius of a star equals the square root of its number of edges i.e. $\lambda_1(S_n) = \sqrt{n}$. More generally, it is easy to see that a weighted star graph S_w, where each edge ij is associated with weight w_{ij}, has spectral radius $\lambda_1(S_w) = ||w||_2 = \sqrt{\sum_{ij \in E} w_{ij}^2}$. Notice that we recover the non-weighted case by setting every weight to be 1. Using this property, we determine that the number of edges in an optimal solution of Problem (SDP$_{0,1}$) is exactly $\lfloor \lambda^2 \rfloor$ edges. We denote the optimal value of MSSP on a star graph S_n and parameter λ by opt$(S_n, \lambda) = \lfloor \lambda^2 \rfloor$.

To analyze the gap between the combinatorial problem and our relaxation, we now compute the value of an optimal solution of Problem (SDP$_{\lambda\Delta}$). First, we can use the above definition of the spectral radius of a star graph to replace the semidefinite constraint by $||y||_2 \leq \lambda$. Second, we interpret the degree constraint as a constraint on the ℓ_1-norm of y. This means that we can compute an optimal solution of Problem (SDP$_{\lambda\Delta}$) by solving the following second-order cone programming problem:

$$\max_{y \in [0,1]^{|E|}} ||y||_1$$
$$\text{s.t. } ||y||_2 \leq \lambda \qquad\qquad (\text{SOCP}_{\lambda\Delta})$$
$$||y||_1 \leq \lambda^2$$

It is now easy to see that an optimal solution of this problem has value at most λ^2 and that can be achieved by any y such that $||y||_2 \leq \lambda$ e.g. the uniform solution where $\forall ij \in E, y_{ij}^* = \lambda^2/n$ has ℓ_2-norm $||y^*||_2 = \lambda/\sqrt{n}$. We denote by opt$_{\text{rel}}(S_n, \lambda) = \lambda^2$ the optimal value of the relaxation.

We now have a complete description of the integrality gap g_{S_n} of our relaxation for star graphs. The gap is the largest ratio between the optimal value of the relaxation and the optimal value of the original problem:

$$gs_n \stackrel{\text{def}}{=} \max_{S_n, \lambda} \frac{\text{opt}_{\text{rel}}(S_n, \lambda)}{\text{opt}(S_n, \lambda)} = \frac{\lambda^2}{\lfloor \lambda^2 \rfloor} \leq \frac{4}{3} \tag{2}$$

where the last inequality comes from the fact that $\lambda \geq \sqrt{2}$.

3.2 Erdös-Rényi Stars

Now that we have solved our relaxation of MSSP, we will use the computed optimal solution y^* to sample a discrete solution, here a random subgraph S_x of the original star graph S_n. For this purpose we introduce for each edge ij an independent random variable $x_{ij} \sim \text{Ber}(y_{ij}^*/r)$. By definition the random number of edges x of the random subgraph S_x is a sum of independent Bernoulli random variables with mean $\mathbb{E} x = \sum_{ij} y_{ij}^*/r = \lambda^2/r$.

If for some $r > 1$ the random subgraph S_x satisfies the spectral radius constraint with high probability, i.e. $p_r = 1 - O(1/n)$, then we would have a polynomial time randomized r-approximation algorithm. We obtain the following approximation algorithm in the case of star graphs:

Theorem 1. *(Feasible with constant probability) Given a star graph S_n, a bound on the spectral radius $\lambda \geq \sqrt{2}$, and an optimal solution y^* of Problem (SOCP$_{\lambda\Delta}$), the random partial subgraph S_x obtained by keeping edges according to independent random variables $x_{ij} \sim \text{Ber}(y_{ij}^*/r)$ is a feasible solution of MSSP with probability $p_r \geq 2/3$ for $r = 4$.*

Proof. As is common practice in the analysis of randomized algorithms [12], we use the Chernoff bound to get an estimate of the probability that our sampled solution is feasible. Recall that the Chernoff bound gives an upper bound on the probability that a sum of independent random variables exceeds a certain value.

Theorem. [12] *(Chernoff bound) Let $x = \sum_{i=1}^n x_i$ where each x_i is an independent random variable. We have for a given value $a > 0$ the following estimate:*

$$\Pr(x \geq a) \leq \min_{t>0} e^{-ta} \prod_{i=1}^n \mathbb{E} e^{tx_i}.$$

We directly apply the Chernoff bound on the random number of edges $x = \sum_{ij \in E} x_{ij}$ for the value $a = \lambda^2$. After using the fact that $\mathbb{E} e^{tx_{ij}} \leq e^{\frac{y_{ij}^*}{r}(e^t - 1)}$, we obtain the following:

$$\Pr(x \geq \lambda^2) \leq \min_{t>0} e^{-t\lambda^2} \exp\left(\sum_{ij \in E} \frac{y_{ij}^*}{r}(e^t - 1)\right)$$

$$\Pr(x \geq \lambda^2) \leq \min_{t>0} \exp\left(\frac{\lambda^2}{r}(e^t - 1) - t\lambda^2\right).$$

The minimum of the r.h.s. is attained at $t = \log r$ under the condition that $t > 0$ from which we deduce that $r = 1 + h$ for some $h > 0$. The bound then simplifies into:

$$\Pr(\boldsymbol{x} \geq \lambda^2) \leq \exp\left(\lambda^2 \left(\frac{h}{1+h} - \log(1+h)\right)\right).$$

We choose $r = 1 + h = 4$ to produce the following readable bound which remains valid for any $\lambda \geq \sqrt{2}$:

$$\Pr(\boldsymbol{x} \geq \lambda^2) \leq \exp(-0.64\lambda^2) \leq \frac{1}{3}$$

which concludes the proof of Theorem 1. Recall indeed that the event $\lambda_1(S_{\boldsymbol{x}}) \geq \lambda$ is equivalent to the event $\boldsymbol{x} \geq \lambda^2$. $\qquad\square$

Since our success probability for a single sample $p_r \geq 2/3$ we can amplify it by repetition in polynomial time to obtain a solution of expected value $\lambda^2/4$ and such that the solution is feasible with high probability $p_r = 1 - O(1/n)$.

To summarize the case of star graphs, our relaxation and randomized rounding algorithm is a polynomial time algorithm which returns with high probability a feasible star graph of expected size $\lambda^2/4$.

3.3   Without the Degree Constraint

It is important to notice that the degree constraint played a significant role in the tightness of the relaxation in the case of star graphs. Reusing the same analysis as in Sect. 3.1 we can see that Problem (SDP$_{\lambda\Delta}$) without the degree constrained is equivalent to the following problem:

$$\max_{y \in [0,1]^{|E|}} \|y\|_1$$

$$\text{s.t. } \|y\|_2 \leq \lambda \qquad\qquad (\text{SOCP}_\lambda)$$

By a geometrical argument, we notice that the uniform solution $\forall ij \in E, y_{ij} = \lambda/\sqrt{n}$ is the unique optimal solution of Problem (SOCP$_\lambda$). It follows that the associated optimal value $\text{opt}_{\text{rel}'}(S_n, \lambda) = \lambda\sqrt{n}$.

In that case, the integrality gap of the relaxation given by Problem (SOCP$_\lambda$) is

$$g'_{S_n} \overset{\text{def}}{=} \max_{S_n, \lambda} \frac{\text{opt}_{\text{rel}'}(S_n, \lambda)}{\text{opt}(S_n, \lambda)} = \frac{\lambda\sqrt{n}}{\lfloor\lambda^2\rfloor} = O\left(\frac{\sqrt{n}}{\lambda}\right) \qquad (3)$$

which translates into a much higher $r = O(q'_{S_n})$ than the constant obtained in Sect. 3.1. Problem formulations focusing on minimizing the spectral radius given an edge deletion budget cannot a priori bound the maximum degree of the resulting weighted graph. This additional information is a key advantage over problems that optimize the spectral parameter.

We are now ready to describe our matrix randomized rounding whose analysis follows a similar structure to the one for star graphs. However we need to use more powerful concentration inequalities than the Chernoff bound to obtain bounds on the spectral radius of the random matrix we sample. This sampling can be seen as a special case of inhomogeneous Erdős-Rényi random graphs.

4   Spectral Subgraphs in General Graphs

In order to extend the analysis of Algorithm 1 to arbitrary graphs we turn to more advanced concentration inequalities that describe the behavior of random matrices and in particular their spectrum. Fortunately, recent results in the analysis of random matrices (cf. the survey by Tropp [22]) provide tail bounds for the largest eigenvalue of random matrices. These results are directly applicable to the analysis of Algorithm 1 for finding the sampling factor r that guarantees that the returned solution is feasible with high probability $p_r = 1 - 1/n$.

We start by presenting the generic matrix Bernstein bound and its application to adjacency matrices following the work of Radcliffe and Chung [3]. Finally we give the proof that Algorithm 1 is a randomized $O(\log n)$-approximation algorithm with the following property:

Theorem 2. *(Feasible with constant probability) Given a graph $G = (V, E)$ with $|V| = n$, a bound on the spectral radius $\lambda \geq \log n$, and an optimal solution y^* of Problem (SDP$_{\lambda_\Delta}$), the random subgraph H obtained by keeping edges $ij \in E$ according to independent random variables $x_{ij} \sim Ber(y_{ij}^*/r)$ is a feasible solution of MSSP with probability $p_r \geq 2/3$ for $r = O(\log n)$.*

4.1   Following the Matrix Bernstein Bound

The matrix Bernstein bound is a generalization of the classical Bernstein bound to the setting of independent random matrices. The theorem states the following:

Theorem. [3] *(Matrix Bernstein) Let $\boldsymbol{X} = \sum_i \boldsymbol{X}_i$ where each $\boldsymbol{X}_i$ is an independent symmetric random matrix of size n which is centered $\mathbb{E}\,\boldsymbol{X}_i = 0$ and bounded in spectral norm $\lambda_1(\boldsymbol{X}_i) \leq L$. We define the matrix variance of $\boldsymbol{X}$ by $v(\boldsymbol{X}) = \lambda_1(\sum_i \mathbb{E}\,\boldsymbol{X}_i^2)$. The following tail inequality holds:*

$$\Pr(\lambda_1(\boldsymbol{X}) \geq a) \leq n \exp\left(-\frac{a^2}{2v(\boldsymbol{X}) + 2La/3}\right). \tag{4}$$

The output of Algorithm 1 corresponds to a random adjacency matrix $\boldsymbol{A}$ which is the sum of independent random adjacency matrices each corresponding to an edge in the random graph. Let $A_{ij} = (E_{ij} + E_{ji})$ where the E_{ij} form the canonical basis for $M_{n,n}$ and denote by $\boldsymbol{x}_{ij}$ a Bernoulli random variable of mean y_{ij}^*/r. We have the following:

$$\boldsymbol{A} = \sum_{ij \in E} \boldsymbol{x}_{ij} A_{ij} \tag{5}$$

Note that our random adjacency edges have non-zero mean $\mathbb{E}\,\boldsymbol{x}_{ij}A_{ij} = (y_{ij}^*/r)A_{ij}$. Fortunately, applying Weyl's inequalities on $\boldsymbol{A}$ and $\mathbb{E}\,\boldsymbol{A}$ will give us control over the spectral radius of $\boldsymbol{A}$ by proxy.

Theorem. [1] *(Weyl's inequalities) Let X and Y be two symmetric matrices,*

$$\lambda_1(X - Y) \leq \varepsilon \implies |\lambda_1(X) - \lambda_1(Y)| \leq \varepsilon \tag{6}$$

This theorem implies that bounding the spectral radius of our centered random adjacency matrix by $(1 - 1/r)\lambda$ will give us the adequate bound on the spectral radius of $\boldsymbol{A}$. Since we only consider the event where $\boldsymbol{A}$ has greater spectral radius than $\mathbb{E}\,\boldsymbol{A}$, we drop the absolute value:

$$\lambda_1(\boldsymbol{A} - \mathbb{E}\,\boldsymbol{A}) < \left(1 - \frac{1}{r}\right)\lambda \implies \lambda_1(\boldsymbol{A}) - \lambda_1(\mathbb{E}\,\boldsymbol{A}) < \left(1 - \frac{1}{r}\right)\lambda$$

and by feasibility of an optimal solution of the relaxed SDP, i.e. Problem (SDP$_{\lambda\Delta}$), we have $\lambda_1(\mathbb{E}\,\boldsymbol{A}) \leq \lambda/r$ which gives:

$$\lambda_1(\boldsymbol{A} - \mathbb{E}\,\boldsymbol{A}) < \left(1 - \frac{1}{r}\right)\lambda \implies \lambda_1(\boldsymbol{A}) < \lambda.$$

From the general bounds of (1) we know that the spectral radius of the centered adjacency matrix of a random edge ij is either y_{ij}^*/r (no edge) or $1 - y_{ij}^*/r$ (one edge) which lets us bound the spectrum of each summand. In the worst case we have, for each edge ij:

$$\lambda_1\left(\left(\boldsymbol{x}_{ij} - \frac{y_{ij}^*}{r}\right)A_{ij}\right) \leq 1 \tag{7}$$

4.2 Proof of Theorem 2

We start by computing the matrix variance:

$$v(\boldsymbol{A} - \mathbb{E}\,\boldsymbol{A}) = \lambda_1\left(\sum_{ij \in E} \mathrm{Var}(\boldsymbol{x}_{ij}A_{ij})\right).$$

Since $\mathrm{Var}(\boldsymbol{x}_{ij}A_{ij}) = \mathrm{Var}(\boldsymbol{x}_{ij})A_{ij}^2$ and $A_{ij}^2 = D_i + D_j$ where $D_v = E_{vv}$, we obtain a clean expression for the variance of the centered adjacency matrix as the spectral radius of the matrix of degree variances:

$$v(\boldsymbol{A} - \mathbb{E}\,\boldsymbol{A}) = \lambda_1\left(\sum_{ij \in E} \mathrm{Var}(\boldsymbol{x}_{ij})(D_i + D_j)\right)$$

$$= \max_{i \in V} \sum_{j \in \Gamma(i)} \frac{y_{ij}^*}{r}\left(1 - \frac{y_{ij}^*}{r}\right)$$

$$\leq \max_{i \in V} \sum_{j \in \Gamma(i)} \frac{y_{ij}^*}{r}$$

and by feasibility of an optimal solution of the relaxation, the degree constraint holds which means that $\max_{i \in V} \sum_{j \in \Gamma(i)} y_{ij}^* \leq \lambda^2$ and gives:

$$v(\boldsymbol{A} - \mathbb{E}\,\boldsymbol{A}) \leq \frac{\lambda^2}{r}.$$

We now fulfill all the prerequisites to apply the matrix Bernstein bound on $A - \mathbb{E}\,A$ and $L = 1$. To explicitly describe the fact that the approximation ratio $r > 1$ we introduce as earlier $h > 0$ such that $r = 1 + h$. We apply the Bernstein bound for the value $a = (h/(1+h))\lambda$:

$$
\Pr\left(\lambda_1(A - \mathbb{E}\,A) \geq \frac{h}{1+h}\lambda\right) \leq n\exp\left(-\frac{1}{2}\frac{a^2}{v(A - \mathbb{E}\,A) + \frac{a}{3}}\right)
$$

$$
= n\exp\left(-\frac{1}{2}\frac{a^2}{\frac{\lambda^2}{1+h} + \frac{1}{3}\frac{h}{1+h}\lambda}\right)
$$

$$
\leq n\exp\left(-\frac{1}{2}\frac{h^2}{(1+h)^2}\frac{\lambda^2}{\frac{\lambda^2}{1+h} + \frac{1}{3}\frac{h}{1+h}\lambda}\right).
$$

We simplify the above expression to obtain:

$$
\Pr\left(\lambda_1(A - \mathbb{E}\,A) \geq \frac{h}{1+h}\lambda\right) \leq n\exp\left(-\frac{1}{2}\frac{h^2}{(1+h)^2}\frac{\lambda^2}{\frac{\lambda}{1+h}(\lambda + h/3)}\right)
$$

$$
= n\exp\left(-\frac{1}{2}\frac{h^2}{1+h}\frac{\lambda}{\lambda + h/3}\right).
$$

As in the case of star graphs, we will derive possible values for r (resp. for h) such that the probability of our subgraph H being infeasible is less than $1/3$. For this, we attempt to derive an upper bound for the argument of the exponential as $n\exp(-x) \leq \frac{1}{3}$ implies that $x \geq \log 3n$.

We are looking for values of h and λ such that the following inequality holds:

$$
\frac{1}{2}\frac{h^2}{1+h}\frac{\lambda}{\lambda + h/3} \geq \log 3n
$$

We start by deriving a lower bound on λ function of h. In the above inequality, $\lambda/(\lambda + h/3)$ can be arbitrarily small if h is unbounded. To prevent this, we impose that, for a certain constant $c > 0$:

$$
\frac{1}{2}\frac{\lambda}{\lambda + h/3} \geq c
$$

which implies that

$$
\lambda \geq \frac{2c}{3 - 6c}h.
$$

Choosing $c = 1/4$ gives us the condition that $\lambda \geq h/3$.

Now we are left with finding the value of h such that:

$$
\frac{1}{4}\frac{h^2}{1+h} \geq \log 3n.
$$

For all values of n, it is sufficient to take $h = 3 \log n$ which completes the proof. $\square$

Algorithm 1 is a randomized algorithm which returns a feasible solution with probability greater than $2/3$ and of expected value within $1 + 3 \log n$ of the value of an optimal solution whenever $\lambda \geq \log n$. Recall that the success probability of such an algorithm can be amplified to high probability in polynomial time. We now turn to a different algorithm to handle the range $\lambda \in [\sqrt{2}, \log n)$.

5 Maximum Matching

After designing an approximation algorithm for MSSP for the range of the spectral bound $\lambda \in (\log n, \lambda_1(G))$, we turn to the well-studied maximum matching problem: finding a subgraph M consisting of the maximum number of non-adjacent edges in a given graph G. The number of edges in M is often called the matching number $\nu(G)$ of the graph. We use a spectral generalization of a classical lower bound on the matching number due to Stevanović [21] which states the following:

Theorem. [21] *(Spectral lower bound on the matching number) Given a graph $G = (V, E)$ we have the following lower bound:*

$$\nu(G) \geq \frac{|E|}{\lambda_1^2(G) - 1}.$$

This static lower bound can be immediately turned into an approximation algorithm since computing a maximum matching can be done in polynomial time.

Algorithm 2. MAXIMUM MATCHING

Input: $G = (V, E)$, $\sqrt{2} \leq \lambda \leq \lambda_1(G)$
Output: $H = (V, E')$ such that $\lambda_1(H) \leq \lambda$
return $H = \arg \nu(G)$

Theorem 3. *(Approximation by maximum matching) Given $G = (V, E)$ and a spectral bound $\lambda > 0$, a maximum matching of G is a $(\lambda^2 - 1)$-approximation for MSSP.*

Proof. Denoting by H^* an optimal solution of MSSP for a graph G and spectral bound λ, we know that H^* is a partial subgraph of G which implies $\nu(G) \geq \nu(H^*)$. We also know that H^* is feasible i.e. $\lambda_1(H^*) \leq \lambda$. Combining these two statements together with the lower bound of Stevanović, we obtain the following inequality:

$$\nu(G) \geq \nu(H^*) \geq \frac{\text{opt}(G, \lambda)}{\lambda^2 - 1}$$

which shows that the size of a maximum matching is within a factor of $\lambda^2 - 1$ of an optimal solution of MSSP. Furthermore any matching has spectral radius equal to 1 i.e. is trivially feasible. $\square$

Used in the range $\lambda \in [\sqrt{2}, \log n)$ a maximum matching is a $O(\log^2 n)$-approximation algorithm in the worst-case. We then combine Algorithm 1 with Algorithm 2 to obtain a $O(\log^2 n)$-approximation algorithm for all values of λ.

6   Conclusion and Perspectives

We have introduced the maximum spectral subgraph problem and designed a randomized $O(\log^2 n)$-approximation algorithm based on the relaxation and rounding framework to solve it.

In terms of lower bounds, we currently do not have any result regarding hardness of approximation, but we are actively exploring this direction. To the best of our knowledge, no inapproximability results have been established for problems related to the spectrum of a graph. Indeed, NP-hardness results found in the literature [11,26] are based on reductions which relate extremal values in spectral graph theory to classical computational problems. These reductions cannot be directly extended to obtain an approximation gap.

Without a better lower bound than NP-hardness, we are compelled to find new techniques to improve our current upper bound. First, the continuous relaxation used in Algorithm 1 is rather natural aside from the redundant degree constraints. It would be interesting to see if stronger relaxations could be used to obtain more information about the random graph e.g. strong bounds on the variance of the random degrees. For this purpose we would like to consider a sum-of-squares relaxation for the binary semidefinite programming problem. Indeed, Nie [13] has given an extension of the classical sum-of-squares hierarchy to include positivity certificates for matrix variables. This relates to the question of generalizing the results of Raghavendra [18] on maximum constraint satisfaction problems where constraints apply to at most k variables to maximum constraint satisfaction problems with spectral constraints which, by definition, involve all variables at once. Aside from strengthening the relaxation, there is opportunity for improvement in developing more precise tail bounds on the spectrum of random adjacency matrices following recent results by van Handel [23] as well as by Le, Levina, and Vershynin [9]. On a separate note, we are currently working on applying the method of conditional probabilities to derandomize Algorithm 1 in order to obtain a deterministic approximation algorithm. The analysis of Sect. 5 focuses on the maximum matching problem as a way of computing a feasible solution for the range $\lambda \in [\sqrt{2}, \log n)$. It is natural to wonder whether the degree constrained subgraph problem with $\Delta \leq \lambda$ (also known as the simple λ-matching problem) could be proven to return a better solution, and possibly match the $O(\log n)$ ratio obtained by Algorithm 1.

Finally, we are also interested in applying a similar strategy to the problem of adding the smallest number of edges to reach a given algebraic connectivity i.e. a lower bound on the second smallest eigenvalue of the Laplacian matrix of the graph. This problem, proven NP-hard by Mosk-Aoyama [11], is a variant of the problem of finding the maximum algebraic connectivity given an edge addition budget proposed by Ghosh and Boyd [5]. While Kolla et al. have designed an approximation algorithm with conditional guarantees [7] for the original problem, we hope that our methodology could apply to the variant and lead to an unconditional approximation ratio.

References

1. Bhatia, R.: Matrix Analysis, vol. 169. Springer, Heidelberg (2013). https://doi.org/10.1007/978-1-4612-0653-8
2. Chakrabarti, D., Wang, Y., Wang, C., Leskovec, J., Faloutsos, C.: Epidemic thresholds in real networks. ACM Trans. Inf. Syst. Secur. (TISSEC) **10**(4), 1 (2008)
3. Chung, F., Radcliffe, M.: On the spectra of general random graphs. Electron. J. Comb. **18**(1), 215 (2011)
4. Ganesh, A., Massoulié, L., Towsley, D.: The effect of network topology on the spread of epidemics. In: Proceedings of the Annual Joint Conference of the IEEE Computer and Communications Societies (INFOCOM 2005), vol. 2, pp. 1455–1466 (2005)
5. Ghosh, A., Boyd, S.: Growing well-connected graphs. In: Proceedings of the IEEE Conference on Decision and Control (CDC 2006), pp. 6605–6611. IEEE (2006)
6. Goemans, M.X., Williamson, D.P.: Improved approximation algorithms for maximum cut and satisfiability problems using semidefinite programming. J. ACM (JACM) **42**(6), 1115–1145 (1995)
7. Kolla, A., Makarychev, Y., Saberi, A., Teng, S-H.: Subgraph sparsification and nearly optimal ultrasparsifiers. In: Proceedings of the ACM Symposium on Theory of Computing (STOC 2010), pp. 57–66. ACM (2010)
8. Lasserre, J.B.: Global optimization with polynomials and the problem of moments. SIAM J. Optim. **11**(3), 796–817 (2001)
9. Le, C.M., Levina, E., Vershynin, R.: Concentration and regularization of random graphs. Random Struct. Algorithms **51**(3), 538–561 (2017)
10. Mehdi, S.A., Khalid, J., Khayam, S.A.: Revisiting traffic anomaly detection using software defined networking. In: Sommer, R., Balzarotti, D., Maier, G. (eds.) RAID 2011. LNCS, vol. 6961, pp. 161–180. Springer, Heidelberg (2011). https://doi.org/10.1007/978-3-642-23644-0_9
11. Mosk-Aoyama, D.: Maximum algebraic connectivity augmentation is NP-hard. Oper. Res. Lett. **36**(6), 677–679 (2008)
12. Motwani, R., Raghavan, P.: Randomized Algorithms. Chapman & Hall/CRC, Boca Raton (2010)
13. Nie, J.: Polynomial matrix inequality and semidefinite representation. Math. Oper. Res. **36**(3), 398–415 (2011)
14. Pan, V.Y., Chen, Z.Q.: The complexity of the matrix eigenproblem. In: Proceedings of the ACM Symposium on Theory of Computing (STOC 1999), pp. 507–516 (1999)
15. Parrilo, P.A.: Semidefinite programming relaxations for semialgebraic problems. Math. Program. **96**(2), 293–320 (2003)
16. Aditya Prakash, B., Chakrabarti, D., Faloutsos, M., Valler, N., Faloutsos, C.: Threshold conditions for arbitrary cascade models on arbitrary networks. In: Proceedings of the IEEE International Conference on Data Mining (ICDM 2011), pp. 537–546 (2011)
17. Raghavan, P., Tompson, C.D.: Randomized rounding: a technique for provably good algorithms and algorithmic proofs. Combinatorica **7**(4), 365–374 (1987)

18. Raghavendra, P.: Optimal algorithms and inapproximability results for every CSP? In: Proceedings of the ACM Symposium on Theory of Computing (STOC 2008), pp. 245–254. ACM (2008)

19. Saha, S., Adiga, A., Aditya Prakash, B., Vullikanti, A.K.S.: Approximation algorithms for reducing the spectral radius to control epidemic spread. In: Proceedings of the SIAM International Conference on Data Mining (SDM 2015), pp. 568–576 (2015)

20. Shin, S., Gu, G.: CloudWatcher: network security monitoring using OpenFlow in dynamic cloud networks (or: how to provide security monitoring as a service in clouds?). In: Proceedings of the IEEE International Conference on Network Protocols (ICNP 2012), pp. 1–6. IEEE (2012)

21. Stevanović, D.: Resolution of AutoGraphiX conjectures relating the index and matching number of graphs. Linear Algebra Appl. 8(433), 1674–1677 (2010)

22. Tropp, J.A., et al.: An introduction to matrix concentration inequalities. Found. Trends® Mach. Learn. 8(1–2), 1–230 (2015)

23. van Handel, R.: Structured random matrices. In: Carlen, E., Madiman, M., Werner, E.M. (eds.) Convexity and Concentration. TIVMA, vol. 161, pp. 107–156. Springer, New York (2017). https://doi.org/10.1007/978-1-4939-7005-6_4

24. Van Mieghem, P.: Graph Spectra for Complex Networks. Cambridge University Press, Cambridge (2010)

25. Van Mieghem, P., Omic, J., Kooij, R.: Virus spread in networks. IEEE/ACM Trans. Netw. (TON) 17(1), 1–14 (2009)

26. Van Mieghem, P., et al.: Decreasing the spectral radius of a graph by link removals. Phys. Rev. E 84(1), 016101 (2011)

27. Wang, G., Ng, T.S., Shaikh, A.: Programming your network at run-time for big data applications. In: Proceedings of the Workshop on Hot Topics in Software Defined Networks (HotSDN 2012), pp. 103–108. ACM (2012)

28. Wang, Y., Chakrabarti, D., Wang, C., Faloutsos, C.: Epidemic spreading in real networks: an eigenvalue viewpoint. In: Proceedings of the International Symposium on Reliable Distributed Systems (SRDS 2003), pp. 25–34. IEEE (2003)

29. Zhang, Y., Adiga, A., Vullikanti, A., Aditya Prakash, B.: Controlling propagation at group scale on networks. In: Proceedings of the International Conference on Data Mining (ICDM 2015), pp. 619–628 (2015)

Bipartite Communities via Spectral Partitioning

Kelly B. Yancey[1](✉) and Matthew P. Yancey[2](✉)

[1] Department of Mathematics and Institute for Defense Analyses - Center for
Computing Sciences, University of Maryland, Bowie, MD, USA
kbyancey1@gmail.com

[2] Institute for Defense Analyses - Center for Computing Sciences,
University of Maryland, Bowie, MD, USA
mpyancey1@gmail.com

Abstract. In this paper we are interested in finding communities with
bipartite structure. A bipartite community is a pair of disjoint vertex
sets S, S' such that the number of edges with one endpoint in S and the
other endpoint in S' is "significantly more than expected." This additional structure is natural to some applications of community detection.
In fact, using other terminology, they have already been used to study
correlation networks, social networks, and two distinct biological networks.

In 2012 two groups independently ((1) Lee, Oveis Gharan, and Trevisan and (2) Louis, Raghavendra, Tetali, and Vempala) used higher
eigenvalues of the normalized Laplacian to find an approximate solution
to the k-sparse-cuts problem. In 2015 Liu generalized spectral methods
for finding k communities to find k bipartite communities. Our approach
improves the bounds on bipartite conductance (measure of strength of a
bipartite community) found by Liu and also implies improvements to the
original spectral methods by Lee et al. and Louis et al. We also highlight
experimental results found when applying our algorithm to three distinct
real-world networks.

Keywords: Community detection · Spectral graph theory
Network analysis

1 Introduction

For a weighted graph $G = (V, E)$, the problem of finding sets of vertices with
small conductance is a well studied problem. The conductance of a set $S \subset V$ is given by $\phi_G(S) = \frac{w(S, \bar{S})}{w(S, V)}$ where $w(S, T)$ is the sum of weights of edges
between vertex sets S and T. A vertex set with small conductance is called a
community and the sparsest cut problem is to find the optimal community, that
is find $\phi(G) = \min_{2w(S,V) \leq w(V,V)} \phi_G(S)$. In 2012 two groups [11,17] used higher
eigenvalues of the normalized Laplacian to find an approximate solution to the

© Springer Nature Switzerland AG 2018
D. Kim et al. (Eds.): COCOA 2018, LNCS 11346, pp. 123–137, 2018.
https://doi.org/10.1007/978-3-030-04651-4_9

k-sparse-cuts problem, which is to find $\phi_k(G)$ and the optimal communities, where $\phi_k(G) = \min_{S_1,\ldots,S_k} \max_i \phi_G(S_i)$.

This paper is interested in finding vertex sets $T_1,\ldots,T_k$ such that each T_i is a community and induces a graph that is roughly bipartite. Define the bipartite conductance of disjoint vertex sets S, S', with $T = S \cup S'$, to be

$$\tilde{\phi}_G(S, S') = \frac{w(T, \overline{T}) + w(S, S) + w(S', S')}{w(T, V)} = \phi_G(T) + \frac{w(S, S) + w(S', S')}{w(T, V)}.$$

A vertex subset with small bipartite conductance is called a bipartite community. Our problem is then

$$\tilde{\phi}_k(G) = \min_{S_1, S_1', \ldots, S_k, S_k'} \max_i \tilde{\phi}_G(S_i, S_i').$$

This is different than searching for communities when G is bipartite, which is known as biclustering.

We motivate this definition with a hypothetical example. Consider a co-purchasing network: each vertex represents an item for sale and each edge represents a pair of items purchased by the same consumer in a single order. For example, we do not expect a consumer to purchase two televisions or two audio systems in a single purchase, however televisions and audio systems are frequently purchased in tandem. Thus, the set of televisions S and the set of audio systems S' form a bipartite community (S, S'). In this case, the more specific structure of a bipartite community is of greater aid to targeted advertisers than classical communities; when a consumer purchases a television, it is smarter to advertise products that accompany televisions rather than more televisions.

Finding bipartite communities within a graph is an important problem as indicated by several researchers searching for this structure in practice and studying this problem as a dual to classical community detection. Trevisan [21], Bauer and Jost [2], and Liu [15] studied the theory behind bipartite communities, while bipartite communities have been used in several biological networks, such as protein interaction networks [13] and double mutant combination networks [3]. They are also used to describe antagonistic behavior [16] in online social settings. The study of correlation clustering [7] is the special case where an edge may represent similarity or dissimilarity, and a recent approach by Atay and Liu [14] involved bipartite communities. Kleinberg considered a related problem [9] for directed graphs when he developed the famous *Hyperlink Induced Topic Search* (HITS) algorithm to find results for a web search query.

In this paper we present two spectral algorithms for finding many bipartite communities; the first has strong quality guarantees and the second is practical. The quality guarantees are stronger than the analogous results for classical communities in ways that have consequences for k-sparse-partition (see Theorem 4) and small set expansion (see Corollary 1). The practical algorithm replaces randomized steps with respected but non-rigorous methods like k-means clustering. We applied the practical algorithm to real-world data sets, which revealed new scenarios where bipartite community detection is appropriate. Moreover, similar to how the classes of interchangeable products are grouped in our hypothetical example, the bipartite partition of the communities revealed structure in the overall graph that would not be identified by classical community detection. The full version [24] of this extended abstract, which includes complete proofs to all statements (including additional details to the proof of Theorem 7), an expanded introduction, and detailed experiments is available on the arXiv.

1.1 Results

Let L be the normalized Laplacian of graph G, with eigenvalues (with multiplicity) $\lambda_1 \leq \lambda_2 \leq \cdots \leq \lambda_n$. It is well known that if G has k components, k' of which are bipartite, then $0 = \lambda_k < \lambda_{k+1}$ and $\lambda_{n-k'} < \lambda_{n-k'+1} = 2$ (unless $k' = 0$, where we just have $\lambda_n < 2$). In what follows, we assume that G is connected, which implies that the eigenvector associated to λ_1 is constant-valued.

The spectral sweep method accepts any non-constant eigenvector as input and produces a community whose conductance is bounded by some function of the associated eigenvalue. It is a common method to prove Cheeger's Inequality (see [6]). Given multiple eigenvectors, it is then trivial to produce multiple communities in parallel. Producing multiple *disjoint* communities using multiple eigenvectors is difficult. In what follows, all communities are considered to be disjoint. There exist a large number of heuristic approaches to this problem ([18,22]). The three current best rigorous results for eigenvectors with eigenvalues $\lambda_1 \leq \cdots \leq \lambda_k$ are:

(A) k communities, each with conductance at most $O(k^2\sqrt{\lambda_k})$ by [11][1],
(B) $\Theta(k)$ communities, each with conductance at most $O(\sqrt{\log(k)\lambda_k})$ by [17],
(C) $k/2$ communities, each with conductance at most $O(\sqrt{\log(k)\lambda_k})$ by [11].

All three proofs are constructive, each giving a polynomial time and space randomized algorithm.

Independently, Trevisan [21], and Bauer and Jost [2] showed that a modified spectral sweep algorithm can be used to find bipartite communities from eigenvectors, and produces a quality guarantee based on 2 minus the associated eigenvalue. Liu [15] presented an algorithm to turn eigenvectors with eigenvalues $\lambda_n \geq \lambda_{n-1} \geq \cdots \geq \lambda_{n-k+1}$ into k bipartite communities with bipartite conductance at most $O(k^3\sqrt{2 - \lambda_{n-k+1}})$. Our theoretical algorithm efficiently constructs bipartite communities that prove the following bounds:

Theorem 1. *Fix a value for k. For a given graph G*

(1) $\tilde{\phi}_k(G) \leq O(k^2\sqrt{2 - \lambda_{n+1-k}})$
(2) $\tilde{\phi}_{k/4}(G) \leq O(\sqrt{\log(k)(2 - \lambda_{n+1-k})})$.

Theorem 1 is a corollary to the following constructive theorem:

---

[1] See Theorem 4.9 in [11].

Theorem 2. *Fix a value for* k. *There exists disjoint sets* $S_1, S_1', S_2, S_2',$ $\ldots, S_r, S_r'$ *such that for any graph* G *and each* $1 \leq i \leq r$,

(1) $r = k$ *and* $\tilde{\phi}_G(S_i, S_i') \leq \frac{2(8k+1)(4k-1)}{k+1-i} \sqrt{\frac{\sum_{1 \leq i \leq k}(2-\lambda_{n+1-i})}{k}}$

(2) $r \leq k/2$ *and* $\tilde{\phi}_G(S_i, S_i') \leq \frac{10^{1.5}(1280\sqrt{3\ln(200k^2)}+4)k}{9(\frac{k}{2}+1-i)} \sqrt{\frac{\sum_{1 \leq i \leq k}(2-\lambda_{n+1-i})}{k}}.$

Theorem 1 states bounds that match the analogous results for classical communities. A closer inspection of Theorem 2 shows that our results are *stronger*, which we explain in the following two remarks. Section 3 of [15] implies that these strengthenings for the bipartite conductance are also enjoyed by the classical communities.

Remark 1 (No-Separate-But-Equal Principle). Algorithm (A) has a uniform bound for all communities, but Theorem 2 (1) has a bound for community i that depends on i. One effect of this is that for any function $f(k) \to \infty$ there are $k - f(k)$ "super" communities that perform $\Omega(f(k))$ better than the worst-case bound. In particular, half of the communities have bipartite conductance at most $O(k\sqrt{2 - \lambda_{n-k+1}})$.

Remark 2 (Complete Spectrum Principle). Our result uses the average of the eigenvalues rather than the worst in the bound, which to our knowledge is the first evidence to devalue the lauded spectral gap between consecutive eigenvalues [8,10,19,20] in favor of the overall spectral sequence. So while prevailing wisdom is to choose k such that $\lambda_k \ll \lambda_{k+1}$, our bounds are $k^{-1/2}$ better when $2 - \lambda_{n-k+2} \ll 2 - \lambda_{n-k+1}$ (the analogue for classical community detection would be $\lambda_{k-1} \ll \lambda_k$).

It has been communicated to us that the strengthenings of both remarks could be obtained with simple modifications to [11]. While that is true, these remarks are novel to this paper. As evidence, observe that the "No Separate But Equal Principle" was not applied to prior versions of Theorem 4 (see below), and we repeat that the "Complete Spectrum Principle" contradicts the frame of reference provided in [8], which contains some of the same authors as [11]. On a technical level, the "No Separate But Equal Principle" is much harder to establish in bipartite communities, which is a point that we elaborate on in Sect. 1.3.

The bounds on bipartite conductance are strong indicators of the quality of the bipartite communities, as indicated by Liu [15] when he showed that $\tilde{\phi}_k(G) \geq (2 - \lambda_{n+1-k})/2$. It is known for cycles and grids [11] that $\phi_k(G) \geq \Omega(\sqrt{\lambda_k})$. It is also known that when G is bipartite [15], $\phi_k(G) = \tilde{\phi}_k(G)$ and $\lambda_k = 2-\lambda_{n+1-k}$. So there exists an infinite family of graphs such that $\tilde{\phi}_k(G) \geq \Omega(\sqrt{2 - \lambda_{n+1-k}})$. We construct a family of graphs demonstrating that the $\sqrt{\log(k)}$ term is necessary in Theorem 1 (2). We call this example the Bipartite Noisy Hypercube; it is inspired by the Noisy Hypercube construction found in [11].

Theorem 3. *There exists a family of graphs* G_i *such that* $2-\lambda_{n-k} \leq O\left(\frac{1}{\log(k)}\right)$ *and for any set* $T, T' \subset V$ *with* $|T \cup T'| \leq \frac{2}{k}|V|$ *we have that* $\tilde{\phi}(T, T') \geq 1/2$.

Our improvements to classical community detection extend beyond what we can do with bipartite communities. The k-sparse-partition problem is the k-sparse-cuts problem with the additional condition that $V = \cup_i S_i$. See [4] for a survey on applications of k-sparse-partition, including route planning, bioinformatics, and image processing. Formally, define

$$\psi_k(G) = \min_{S_i \cap S_j = \emptyset, V = \cup_i S_i, S_i \neq \emptyset} \max_{1 \leq i \leq k} \phi_G(S_i).$$

Note that $\phi_k(G) \leq \psi_k(G)$. Folklore says that $\psi_k(G) \leq k\phi_k(G)$, which is accomplished by placing $V \setminus \cup_i S_i$ into the largest community in the solution to $\phi_k(G)$ and bounding the conductance of the disturbed community by the sum of the conductance of the other communities. This is the basis for the $O(k^3\sqrt{\lambda_k})$ algorithm in [11] and the $O(k\sqrt{\log(k)\lambda_{O(k)}})$ algorithm in [17]. Remark "No-Separate-But-Equal Principle" improves the former to $O(k^2\log(k)\sqrt{\lambda_k})$ for the k-sparse-partition problem. However, we can accomplish more:

Theorem 4.
$$\psi_k(G) \leq O\left(k^2\sqrt{\lambda_k}\right)$$

Theorem 4 is a corollary to Theorem 5; see Sect. 2.3 for details.

Theorem 5. *Fix a value for k. For any graph G, there exists disjoint sets $S_1, S_1', S_2, S_2', \ldots, S_k, S_k'$ such that*

$$\sum_i \tilde{\phi}_G(S_i, S_i') \leq O\left(k^2\sqrt{\frac{\sum_{1 \leq i \leq k}(2 - \lambda_{n+1-i})}{k}}\right).$$

Louis, Raghavendra, Tetali, and Vempala [17] constructed a family of graphs $\mathcal{J}$ with the property $\psi_k(J) \geq \Omega\left(\sqrt{\lambda_k}\min(k^2 n^{-1/2}, n^{1/12})\right)$ for every $J \in \mathcal{J}$. Using a similar construction, we give a sharper bound that is independent of n.

Theorem 6. *For each $k \geq 2$, there exists a family of unweighted graphs H_ℓ such that $\psi_k(H_\ell) \geq \Omega(\sqrt{k\lambda_k(H_\ell)})$.*

Small set expansion is the problem

$$\phi^{(k)}(G) = \min_{kw(A,V) \leq w(V,V)} \phi_G(A).$$

Note that $\phi(G) = \phi^{(2)}(G)$. Li and Peng [12] prove $\min_{k^{1-\epsilon}w(A,V) \leq w(V,V)} \tilde{\phi}_G(A) \leq \sqrt{(2 - \lambda_{n-k})\log_k(n)}$, which is the bipartite analogue of small set expansion. Small set expansion is intimately tied to complexity theory [1]. It is obvious that $\phi^{(k)}(G) \leq \phi_k(G)$, and so algorithms (B) and (C) give $\phi^{(k)}(G) \leq O(\sqrt{\log(k)\lambda_{O(k)}})$, and algorithm (A) gives $\phi^{(k)}(G) \leq O(k^2\sqrt{\lambda_k})$. Remarks "No-Separate-But-Equal Principle" and "Complete Spectrum Principle" allow us to establish a result between these two bounds.

Corollary 1. *For any function w satisfying $1 \ll w(k) \ll k$ (such as $w(k) = k^d$ for $0 < d < 1$), we have that*

$$\phi^{(k)}(G) \leq O\left(kw(k)\sqrt{\frac{\sum_{i=1}^{k(1+w(k)^{-1})}\lambda_i}{k}}\right) \leq O\left(kw(k)\sqrt{\lambda_{k(1+w(k)^{-1})}}\right).$$

1.2 Related Work

Both Liu's result [15] and ours are constructive in a manner similar to the first three algorithms described below. Moreover, our result is a single algorithm (Theorem 7) that gives both bounds when run with different parameters and an optional projection $\mathbb{R}^k \to \mathbb{R}^{O(\log(k))}$. To better explain the new ideas in our work and contrast them with previous methods, we review previous methods below.

Algorithm (B) [17]. In this algorithm k rays beginning at the origin are chosen at random, and the space $\mathbb{R}^k$ is partitioned into k cones, where cone i is centered around ray i. Each point representing a vertex is projected onto the ray corresponding to the cone containing the point, and the spectral sweep is performed on each ray. A positive proportion of the communities found from the set of spectral sweeps satisfy the desired bound.

Algorithm (A) [11]. The number of ray/cone pairs constructed varies but is always at least k. Every point representing a vertex is in at least one cone. Sets (corresponding to communities) are formed from the union of small cones and partition the set of points representing vertices, where ambiguity of a point contained in multiple cones is resolved through an ordering of the ray/cone pairs. The cones are of variable size and each ray contains some point representing a vertex. The map $F : V \to \mathbb{R}^k$ is transformed into maps $F_1, \ldots, F_k : V \to \mathbb{R}^k$ with disjoint support, each with a Rayleigh quotient bounded by k and the Rayleigh quotient of F. A spectral sweep is applied to each F_i.

Algorithm (C) [11]. This algorithm differs from the previous one (Algorithm A) in the cone size and spectral sweep. Cones of fixed size are used and each ray contains a point near a point representing a vertex. A single spectral sweep is performed to F to form a set T of selected vertices; community i is the intersection of T with set i formed from the union of cones.

Pairs of Cones [15]. To find bipartite communities, the space $\mathbb{R}^k$ is partitioned into pairs of cones C, C', where C' is the reflection of C about the origin. To avoid the problem of C and C' possibly being nondistinct when several pairs of cones are unioned to form one set, k-dimensional projective space is used. In this paradime C and C' are identified. Essentially the same method as (A) is used from there where F_i is projected linearly into different one-dimensional subspaces. While the process of unioning cones into a single set destroys the initial set of central rays, Liu argues that a new central line representing the whole set can be found. The bipartite spectral sweep is preformed on each subspace.

1.3   Theoretical Algorithm

A first look at our approach would summarize it as Liu's method (Pairs of Cones) with Algorithm (C) replacing Algorithm (A), but this is not sufficient to explain the improved bounds or Remarks "No-Separate-But-Equal Principle" and "Complete Spectrum Principle." This simple explanation also masks a deep technical problem about the ordering of the steps: the bipartite spectral sweep is done before the partitioning in Algorithm (C); the projection into a one dimensional space must be done before the bipartite spectral sweep to distinguish S from S'; and the partitioning of the vertices must be done before the projection, as the selected line the points are projected onto depends on the points in the part of the partition. In other words, α must come before β, β must come before γ, and γ must come before α.

One of our contributions is to provide the technical framework for how Algorithm (C) can be altered so that the partitioning of the vertices comes before the spectral sweep, thus eliminating the above contradiction. In Algorithms (A), (B), and (C) the costs incurred from partitioning the vertices and from the spectral sweep are kept distinct and are accumulated at the end; this is not possible with our modification to Algorithm (C). To solve this, we develop a novel method to quantify the cost of the partition as an expression that can be incorporated into the evaluation of the performance of the spectral sweep without modifying the underlying procedure.

Recall, ambiguities that arise from the overlap of small cones are resolved through an ordering of the ray/cone pairs. This may imply that some cones (those earlier in the ordering) perform better than the overall expectation. However, this ordering is lost when the r sets are formed as a union of several cones. A second novel contribution is that some ordering of the sets can be established— with a superior bound for the earlier sets—when Liu re-establishes the central ray for the sets, which results in the improvement mentioned in Remark "No-Separate-But-Equal Principle."

Finally, let us note that despite all of the technicalities posed in this section, our proof is shorter and simpler than the proofs in [11,17].

1.4   Practical Algorithm and Experimental Results

The distinctions between our algorithm and those before it are subtle, but those distinctions are exactly what allow us to achieve better results. In particular, we do the following four things, each of which is different than at least one of (A), (B), and (C). (I) select r lines through the origin that will be the centers of the different communities, (II) each community consists of points in a cone centered around one of the two rays that form the central line, (III) the lines are chosen dependent on the location of the points representing a vertex in the graph, and (IV) each point is projected onto the line corresponding to the cone containing it, and a (bipartite) spectral sweep is performed on that line.

Some of the following notation will be defined in Sect. 2. Our algorithm accepts a map $F : V \to \mathbb{R}^k$ with small signless Rayleigh quotient and returns r bipartite communities. The outline of our algorithm is:

(1) Run weighted r-means using the mirror radial projection distance and weight of a vertex equal to its mass.
(2) For sets $C_1, \ldots, C_r$ from the r-means run for each i: for each $v \in C_i$ with center c_i, calculate $x_v = F(v) \cdot c_i$, and then run (a novel unbalanced version of the) bipartite spectral sweep on the x_v.

We applied this algorithm to three real-world networks (a biological network, political blogs, and a telecommunication network), and we found success two times (in all but the biological network). By examining the eigenvalues, we determined that the issue with the biological network was with the network and not the algorithm. This emphasizes the importance of algorithms with quality guarantees. When successful, the algorithm was competitive with all other attempted methods numerically.

On political blogs, our algorithm found the Authority/Hub framework first described by Kleinberg [9]. On telecommunication networks, our algorithm found a community local to a regional network (Korea) rather than the dense formation at the logical center. Furthermore, the two sets of the community provided information about the peering relationship. This can be used to infer the *level* of a telecommunications company, which approximates how close it is to the logical center of the Internet. Information about levels can be used to efficiently route traffic by idealizing the network as a hyperbolic space. Hence our results do not just score well; they have qualitative significance too.

Moreover, we exclusively worked with networks where the vertices were labeled based on their real-world source, which allowed us to apply an "eyeball test" to the results. In each case the communities found qualitative structure previously unknown to these networks. See the full version [24] of this extended abstract for a thorough description of our algorithm and a detailed analysis of the outcomes.

2   Constructing Bipartite Communities

Let G be a connected graph with adjacency matrix A and degree matrix D. We use $E^< = \{(u,v) : uv \in E, u < v\}$ to simplify equations.

Let $\tilde{L} = 2I - L = I + D^{-1/2}AD^{-1/2}$ be the signless Laplacian. Eigenvalue λ with eigenvector v of L is eigenvalue $\tilde{\lambda} := 2 - \lambda$ with eigenvector v of $\tilde{L}$. Let $H : V \to \mathbb{R}^k$ be a map, and let $\|x - y\|$ be the standard Euclidean distance between points $x, y \in \mathbb{R}^k$. We define the *signless Rayleigh quotient of H* to be
$$\tilde{\mathcal{R}}_G(H) = \frac{\sum_{uv \in E^<} w_{uv}\|H(u)+H(v)\|^2}{\sum_{u \in V} \|H(u)\|^2 d(u)}.$$

Let $e_1, e_2, \ldots, e_k$ be orthonormal eigenvectors of $\tilde{L}$ that correspond to the smallest eigenvalues, and for each i, let $e_i = D^{1/2}f_i$. A direct calculation gives that $\sum_{u \in V} d(u)f_i(u)^2 = 1$ and $\tilde{\mathcal{R}}_G(f_i) = \tilde{\lambda}_i$. We choose $F(u) = (f_1(u), f_2(u), \ldots, f_k(u))$. The *mass* is $\mathcal{M}(S) = \sum_{u \in S} d(u)\|F(u)\|^2$. A vertex at the origin has no mass, and therefore our algorithm will not be disturbed by discarding vertices mapped to the origin.

Define the *mirror radial projection* $d_M(x, y)$ to be

$$\min\left\{\left\|F(x)/\|F(x)\| - F(y)/\|F(y)\|\right\|, \left\|F(x)/\|F(x)\| + F(y)/\|F(y)\|\right\|\right\}.$$

If θ is the angle formed from $F(u)$ and $F(v)$ through the origin and $\theta^* = \min\{\theta, \pi - \theta\}$, then $d_M(x, y) = 2\sin(\theta^*/2)$. For $u \in V$ define the ball $B_t(u) \subseteq V$ to be $B_t(u) = \{w \in V : d_M(u, w) < t\}$ (because of the nature of d_M, this is a pair of cones in $\mathbb{R}^k$). We will use P to denote a partition of the vertex set, and $P(u)$ to denote the part of the partition that contains vertex u.

2.1   Essential Lemmas

We partition our points by the following algorithm.

Lemma 1 ([5]). *There exists a randomized algorithm to generate a partition P such that each part of the partition has diameter at most Δ (with respect to d_M) and*

$$\mathbb{P}[P(u) \neq P(v)] \leq \frac{2\sqrt{k}d_M(u, v)}{\Delta}$$

where k is the dimension of the underlying space.

The output of Lemma 1 is a set of cones. The process in the Pair of Cones Algorithm of unioning small cones into sets is described by the following lemma. It uses the fact that the e_i are orthonormal.

Lemma 2 ([15]). *Let k be the dimension of the underlying space. There exists a randomized algorithm to generate a partition P such that $\mathbb{P}[P(u) \neq P(v)] \leq \frac{2\sqrt{k}d_M(u,v)}{\Delta}$ where*

(1) $\Delta = (2k)^{-0.5}$ implies there are k parts, each satisfying $\mathcal{M}(P(u)) \geq \frac{\mathcal{M}(V)}{2(k-0.25)}$
(2) $\Delta = 0.27$ implies there are $k/2$ parts, each satisfying $\mathcal{M}(P(u)) \geq \mathcal{M}(V)/k$.

The following is a novel method to show that edges uv with $P(u) \neq P(v)$ contribute very little to the term $\sum_{uv \in E^<} w_{uv}\|F(u) + F(v)\|^2$. It follows from the Cauchy-Schwartz formula and a direct calculation showing that $d_M(u, v)\|F(u)\| \leq 2\|F(u) + F(v)\|$.

Lemma 3.

$$\sum_{u \in V} \sum_{v \in N(u)} w_{uv} d_M(u, v)\|F(u)\|^2 \leq \sqrt{8\tilde{\mathcal{R}}(F)^{-1}} \sum_{uv \in E^<} w_{uv}\|F(u) + F(v)\|^2.$$

The following can be proved by expanding all terms and switching the order of summation. In the following lemma, k refers to the number of eigenvectors used to define the function F.

Lemma 4.

$$\tilde{\mathcal{R}}_G(F) = \frac{\sum_i \tilde{\lambda}_i}{k}$$

2.2 General Approach

Theorem 7. *If we have a randomized method to generate a partition P with r parts such that $\mathbb{P}[P(u) \neq P(v)] \leq C_1 d_M(u, v)$ and each part has mass at least $C_2 \mathcal{M}(V(G))$, then there exists vertex sets $S_1, \ldots, S_r, S'_1, S'_2, \ldots, S'_r$ where $\tilde{\phi}(S_i, S'_i) \leq \frac{8C_1 + 4}{C_2(r - i + 1)} \sqrt{\tilde{\mathcal{R}}(F)}$.*

Proof. Let χ denote an indicator variable. Choose a partition P that performs at least as well as the expectation in the sense that

$$\sum_{u \in V} \sum_{v \in N(u)} w_{uv} \chi(P(u) \neq P(v)) \|F(u)\|^2 \leq \sum_{u \in V} \sum_{v \in N(u)} w_{uv} C_1 d_M(u, v) \|F(u)\|^2.$$

(1)

In the following, we assume that P is now fixed.

Fix some i; we will find the communities S_i, S'_i independently. Project F onto one of its coordinates $j^{(i)}$, and use f_j instead of F. When there is no chance for confusion, we will use j as shorthand for $j^{(i)}$. If we choose a j at random then the terms $f_j(u)^2$ and $(f_j(u) + f_j(v))^2$ have expectation $\|F(u)\|^2/k$ and $\|F(u) + F(v)\|^2/k$. We wish to pick a j where the first term shrinks no more than the second term shrinks. We use the coefficient α_i to denote the shrinkage of the first term.

Formally, define α_i for our chosen j such that

$$0 \neq \sum_{u \in P_i} d(u) f_j(u)^2 = \alpha_i \sum_{u \in P_i} d(u) \|F(u)\|^2,$$

and our choice of j then implies

$$\alpha_i^{-1} \sum_{u \in P_i} \sum_{v \in N(u)} w_{uv} \left(C_1 \tilde{\mathcal{R}}(F)^{-1/2} (f_j(u) + f_j(v))^2 + \chi(P(u) \neq P(v)) f_j(u)^2 \right) \quad (2)$$

$$\leq \sum_{u \in P_i} \sum_{v \in N(u)} w_{uv} \left(C_1 \tilde{\mathcal{R}}(F)^{-1/2} \|F(u) + F(v)\|^2 + \chi(P(u) \neq P(v)) \|F(u)\|^2 \right).$$

We have chosen $j^{(i)}$ independently for each fixed i, but (1) is for all i at once. So we bound the sum of the right hand side of (2) for all i using Lemma 3 and (1) with the expression

$$C_1 \tilde{\mathcal{R}}(F)^{-1/2} 2(1 + \sqrt{2}) \sum_{uv \in E^<} w_{uv} \|F(u) + F(v)\|^2.$$

The two terms in the left hand side of (2) are positive so they are independently bounded by the right hand side. The two independent bounds are

$$\sum_i \sum_{u \in P_i} \sum_{v \in N(u)} w_{uv} \chi(P(u) \neq P(v)) f_{j^{(i)}}(u)^2 \alpha_i^{-1} \quad (3)$$

$$\leq 2C_1 \tilde{\mathcal{R}}(F)^{-1/2} (1 + \sqrt{2}) \sum_{uv \in E^<} w_{uv} \|F(u) + F(v)\|^2$$

and

$$\sum_i \sum_{u \in P_i} \sum_{v \in N(u)} (f_{j^{(i)}}(u) + f_{j^{(i)}}(v))^2 \alpha_i^{-1} \leq 2(1 + \sqrt{2}) \sum_{uv \in E} w_{uv} \|F(u) + F(v)\|^2.$$

(4)

Let $\hat{\alpha} = \max_i \alpha_i^{-1} f_{j^{(i)}}^2(u)$. Choose $t \in (0, \hat{\alpha})$ uniformly and randomly and define two sets $S_{i,t} = \{u \in P_i : f_j(u) \geq \sqrt{t\alpha_i}\}$, $S'_{i,t} = \{u \in P_i : f_j(u) < -\sqrt{t\alpha_i}\}$. The expectation for the denominator of $\tilde{\phi}(S_{i,t}, S'_{i,t})$ is the mass of P_i:

$$\mathbb{E}_t[w(S_{i,t} \cup S'_{i,t}, V)] = \hat{\alpha}^{-1} \sum_{u \in P_i} \|F(u)\|^2 d(u) \geq \hat{\alpha}^{-1} C_2 \sum_{u \in V} \|F(u)\|^2 d(u). \quad (5)$$

As shorthand, let $B_{i,t} = w(S_{i,t} \cup S'_{i,t}, \overline{S_{i,t} \cup S'_{i,t}}) + w(S_{i,t}, S_{i,t}) + w(S'_{i,t}, S'_{i,t})$ be the numerator of $\tilde{\phi}(S_{i,t}, S'_{i,t})$. Using (3) and the bipartite spectral sweep method [21] (which involves the Cauchy-Schwartz theorem and (4)),

$$\sum_i \mathbb{E}_t[B_{i,t}] \leq \sum_i \sum_{u \in P_i} \sum_{v \in N(u)} w_{uv} \mathbb{P}[v \notin P_i, u \in S_t \cup S'_t]$$

$$+ \sum_i \mathbb{E}_t[w(S_{i,t}, P_i \setminus S'_{i,t}) + w(S'_{i,t}, P_i \setminus S_{i,t})]$$

$$\leq 2\hat{\alpha}^{-1} C_1(1 + \sqrt{2})\tilde{\mathcal{R}}(F)^{-1/2} \sum_{uv \in E^<} w_{uv} \|F(u) + F(v)\|^2$$

$$+ \hat{\alpha}^{-1} \sum_i \sum_{u \in P_i} \sum_{v \in N(u)} w_{uv} (|f_j(u)| + |f_j(v)|) |f_j(u) + f_j(v)| \alpha_i^{-1}$$

$$\leq 2\hat{\alpha}^{-1} \sqrt{\frac{1 + \sqrt{2}}{\tilde{\mathcal{R}}(F)}} \left(C_1 \sqrt{1 + \sqrt{2}} + 1 \right) \sum_{uv \in E^<} w_{uv} \|F(u) + F(v)\|^2. \quad (6)$$

We have not yet imposed an order on the P_i, and we do so now. Define $\gamma_i \in (0, 1)$ to be such that $\mathbb{E}_t[B_{i,t}] = \gamma_i \mathbb{E}_t[\sum_{i'} B_{i',t}]$. Permute the indices such that $\gamma_1 \leq \gamma_2 \leq \cdots \leq \gamma_r$. Theorem 4 will depend on the fact that $\sum_i \gamma_i = 1$, but for now we use the weaker $\gamma_i \leq (r + 1 - i)^{-1}$. Combining (5) and (6) we have that

$$\mathbb{E}_t \left[\frac{w(S_{i,t} \cup S'_{i,t}, V)}{C_2} \sqrt{\tilde{\mathcal{R}}(F)} - \frac{(r + 1 - i)B_{i,t}}{8C_1 + 4} \right] > 0. \quad (7)$$

If $w(S_{i,t} \cup S'_{i,t}, V) = 0$, then $B_{i,t} = 0$ and so the term inside the expectation of (7) is zero. So we may choose t separately for each i that performs at least as well as the expectation and satisfies $w(S_{i,t} \cup S'_{i,t}, V) \neq 0$.

2.3 Polished Results

The application of Theorem 7 is direct.
Proof of Theorem 2 (1). Apply Theorem 7 with Lemma 2 and $\Delta = (2\sqrt{k})^{-1}$, $C_1 = 4k$, $r = k$, and $C_2 = \frac{1}{2(k-0.25)}$.

Proof of Theorem 2 *(2).* Use the dimension reduction arguments of [11] to project down into $1200(2\ln(k) + \ln(200))$ dimensions. Apply Theorem 7 with Lemma 2 and $\Delta = 0.27$, $C_1 = \frac{2\sqrt{1200\ln(200k^2)}}{0.27}$, $r = \frac{k}{2}$, and $C_2 = \frac{1}{k}$.

Proof of Theorems 4 *and* 5. We repeat the parameter choices from the proof of Theorem 2 (1) so that $C_1 = 4k$ and $C_2 = \frac{1}{2(k-0.25)}$. Let us revisit (7) in slightly modified form. There exists an absolute constant C such that

$$\mathbb{E}\left[Ckw(S_{i,t} \cup S'_{i,t}, V)\sqrt{\tilde{\mathcal{R}}(F)} - \frac{B_{i,t}}{\gamma_i k}\right] > 0,$$

and so $\tilde{\phi}(S_i, S'_i) < C\gamma_i k^2 \sqrt{\tilde{\mathcal{R}}(F)}$. If we use the fact that $\sum_i \gamma_i = 1$ (which was not used in the proof to Theorem 7), we get that

$$\sum_i \tilde{\phi}(S_i, S'_i) \leq O\left(k^2 \sqrt{\frac{\sum_{1\leq i\leq k}(2 - \lambda_{n+1-i})}{k}}\right). \tag{8}$$

It is straightforward to convert Theorem 2 into a statement about classical communities. Simply replace mirror radial projection distance d_M with radial projection distance d_F, convert each signless object into the original form, replace Lemma 3 with the obvious analogue (Lemma 3 relies on the statement $d_M(u,v)\|F(u)\| \leq 2\|F(u) + F(v)\|$; Lemma 3.1 in [11] states that $d_F(u,v)\|F(u)\| \leq 2\|F(u) - F(v)\|$), and use standard spectral sweep instead of bipartite spectral sweep. The two methods strongly parallel each other, and the method for classical communities is the easier of the two. We leave the details of the conversion to the reader.

Let $T = V \setminus \cup_{i<k} S_i$. We use the standard trick of transforming communities $S_1, S_2, \ldots, S_k$ into $S_1, S_2, \ldots, S_{k-1}, T$, where $w(S_k, V) \geq w(S_i, V)$ for all $i < k$. Clearly, $E(T, \overline{T}) \leq \sum_{i<k} E(S_i, \overline{S}_i)$, and therefore $\phi(T) \leq \sum_{i<k} \phi(S_i)$. And by (8), we see that $\phi(T) \leq O\left(k^2 \sqrt{\frac{\sum_{1\leq i\leq k} \lambda_i}{k}}\right)$.

3   Constructing Sharp Examples

3.1   Bipartite Noisy Hypercube: Theorem 3

Let k and c be fixed, with $1 \leq c \leq \frac{10k}{22}$, and let $\epsilon = \frac{1}{\log_{2.2}(k/c)}$. Let $G_{k,c}$ be the weighted complete graph on vertex set $V = \{0,1\}^k$, and the weight of edge xy is $\epsilon^{\|x-y\|_1}$. $G_{k,c}$ is called the *noisy hypercube*. Lee, Oveis Gharan, and Trevisan [11] demonstrated a separation between the eigenvalues of $G_{k,c}$ and the conductance of small sets in the graph. We define $G_{k,c}^{(o)}$ to be a complete bipartite spanning subgraph of $G_{k,c}$ such that $xy \in E(G^{(o)})$ (and keeps the same weight) if and only if $\|x - y\|_1$ is odd. We will show that $G_{k,c}^{(o)}$ satisfies $2 - \lambda_{n-k} \leq 3\epsilon$ and for any set $T, T' \subset V$ with $|T \cup T'| \leq \frac{c}{k}|V|$ we have that $\tilde{\phi}(T, T') \geq 1/2$.

Our proofs will make use of the rich field of study on maps whose domain is $\{0,1\}^k$ with inner product $\langle f, g \rangle = 2^{-k} \sum_{x \in V} f(x)g(x)$ and p-norm $\|f\|_p = \left(2^{-k} \sum_{x \in V} |f(x)|^p\right)^{1/p}$; our notation follows that of [23]. The *Walsh functions* defined by $W_S(x) = (-1)^{\sum_{i \in S} x_i}$ for $S \subseteq [k]$ form an orthonormal basis. Thus, for any f there exist coefficients $\widehat{f}(S) = \langle W_S, f \rangle$ such that $f(x) = \sum_{S \subseteq [k]} \widehat{f}(S)W_S(x)$.

Parseval's Identity states that $\|f\|_2^2 = \sum_{S \subseteq [k]} \widehat{f}(S)^2$. The Bonami-Beckner operator is $N_\eta f(x) = \sum_{S \subseteq [k]} \widehat{f}(S)\eta^{|S|}W_S(x)$, and the Bonami-Beckner inequality is: if $1 \le p \le q$ and $0 \le \eta \le \sqrt{(p-1)/(q-1)}$, then $\|N_\eta f\|_q \le \|f\|_p$. We will not need the full generality of this statement, just that if $0 \le \eta \le 1$, then $\sum_{S \subseteq [k]} \widehat{f}(S)^2 \eta^{2|S|} \le \|f\|_{1+\eta^2}^2$.

Let d_k^o be the degree of a vertex in $G_{k,c}^{(o)}$, so that $d_k^o = \frac{1}{2}\left((1+\epsilon)^k - (1-\epsilon)^k\right) = c_k(1+\epsilon)^k$ when $c_k \in [1/2.2, 1/2]$. For $S \subseteq V$, let $\rho_S = \frac{(1+\epsilon)^{k-|S|}(1-\epsilon)^{|S|} - (1-\epsilon)^{k-|S|}(1+\epsilon)^{|S|}}{2}$. Our first lemma states that W_S is an eigenfunction of the Laplacian with eigenvalue $1 - \rho_S/d_k^o$.

Lemma 5. *If $S \subseteq [k]$ and A is the adjacency matrix for $G_{k,c}^{(o)}$, then $AW_S = \rho_S W_S$.*

By considering the $k+1$ options for S such that $|S| \ge k-1$, we arrive at the conclusion $2 - \lambda_{n-k} \le 3\epsilon$. Lemma 6 will use the bound $\rho_S \le d_k^o \frac{11}{10}\left(\frac{1-\epsilon}{1+\epsilon}\right)^{|S|}$, which is stronger for smaller values of $|S|$ (when $|S| > k/2$, we have $\rho_S < 0$).

We will prove that $\phi(T) \ge \frac{1}{2}$ for any $T \subset V$ with $|T| < \frac{c}{k}|V| = \frac{c}{k}n$. Recall that $\tilde{\phi}(T', T'') = \phi(T' \cup T'') + \frac{w(T',T')+w(T'',T'')}{w(T'\cup T'',V)}$, so this will conclude the details of Example 3. Using the Walsh functions as an orthonormal basis, we bound from above $w(T,T)$ for any vertex set T.

Lemma 6. *Let $T \subseteq V$ and define $\mathbb{1}_T$ to be the characteristic function of T. Under these conditions,*

$$2^{-k}w(T,T) = \langle \mathbb{1}_T, A\mathbb{1}_T \rangle \le d_k^o \frac{11}{10} \sum_{S \subseteq [k]} \left(\frac{1-\epsilon}{1+\epsilon}\right)^{|S|} \left(\widehat{\mathbb{1}_T}(S)\right)^2.$$

It now follows that small vertex sets have large conductance.

Theorem 8. *The conductance $\phi(T) \ge \frac{1}{2}$ for any $T \subset V$ with $|T| < \frac{c}{k}|V| = \frac{c}{k}n$.*

Proof. Let $T \subseteq V$ be such that $|T| \le \frac{c}{k}n$. Note that $\phi(T) = 1 - \frac{w(T,T)}{w(T,V)} = 1 - \frac{w(T,T)}{|T|d_k^o}$. By Lemma 6 and the Bonami-Beckner inequality (with $\eta = \sqrt{\frac{1-\epsilon}{1+\epsilon}}$), we have that

$$\frac{w(T,T)}{|T|d_k^o} \le 1.1 \frac{2^k}{|T|} \|\mathbb{1}_T\|_{1+\eta^2}^2 = 1.1 \left(\frac{|T|}{n}\right)^\epsilon.$$

By choice of ϵ, the theorem follows.

3.2   Expanders and a Universal Vertex: Theorem 6

We assume k is fixed and allow ℓ to grow asymptotically. Let J_ℓ be a k-regular expander graph on ℓ vertices. Let $H_\ell = kJ_\ell \vee K_1$, which is k disjoint copies of J_ℓ plus one universal vertex. Clearly, $\phi_k(H_\ell) \leq 1/(k+1)$, and thus $\lambda_k(H_\ell) \leq 2/(k+1)$. We will show that $\psi_k(H_\ell) \geq \Omega(1)$, which will prove Theorem 6.

Proof of Theorem 6. Let v be the unique vertex of H_ℓ with degree ℓk. Suppose $S_1, \ldots, S_k$ is an optimal partition of H_ℓ and that $v \in S_k$. If S_i for $i < k$ contains vertices in multiple copies of J_ℓ, then it induces a disconnected graph. The conductance of S_i is bounded from below by the smallest conductance of its components. It follows that S_i must have a component with at least $(1 - o(1))\ell$ vertices, or else the expansion of J_ℓ implies $\phi(S_i) \geq \Omega(1)$ and we are done.

So each of $S_1, \ldots, S_{k-1}$ has $(1 - o(1))\ell$ vertices of a copy of J_ℓ. This implies that S_k has at most ℓ vertices in one of the copies of J_ℓ and $o(\ell)$ vertices in each of the others. So $\phi(S_k) \geq \frac{(k-1-o(1))\ell}{(2k+1+o(1))\ell}$.

References

1. Arora, S., Barak, B., Steurer, D.: Subexponential algorithms for unique games and related problems. J. ACM **62**(5), 25 (2015). https://doi.org/10.1145/2775105. Art. 42
2. Bauer, F., Jost, J.: Bipartite and neighborhood graphs and the spectrum of the normalized graph Laplace operator. Commun. Anal. Geom. **21**(4), 787–845 (2013). https://doi.org/10.4310/CAG.2013.v21.n4.a2
3. Bellay, J., et al.: Putting genetic interactions in context through a global modular decomposition. Genome Res. **21**(8), 1375–1387 (2011). https://doi.org/10.1101/gr.117176.110
4. Buluç, A., Meyerhenke, H., Safro, I., Sanders, P., Schulz, C.: Recent advances in graph partitioning. In: Kliemann, L., Sanders, P. (eds.) Algorithm Engineering. LNCS, vol. 9220, pp. 117–158. Springer, Cham (2016). https://doi.org/10.1007/978-3-319-49487-6_4
5. Charikar, M., Chekuri, C., Goel, A., Guha, S., Plotkin, S.: Approximating a finite metric by a small number of tree metrics. In: Proceedings of the 39th Annual Symposium on Foundations of Computer Science. FOCS 1998, p. 379. IEEE Computer Society, Washington, DC (1998). http://dl.acm.org/citation.cfm?id=795664.796406
6. Chung, F.: Four Cheeger-type inequalities for graph partitioning algorithms. In: Proceedings of ICCM, pp. 751–772 (2007)
7. Gallier, J.: Spectral theory of unsigned and signed graphs. Applications to graph clustering: a survey. ArXiv e-prints, January 2016
8. Gharan, S.O., Trevisan, L.: Partitioning into expanders. In: Proceedings of the Twenty-Fifth Annual ACM-SIAM Symposium on Discrete Algorithms, pp. 1256–1266. ACM, New York (2014). https://doi.org/10.1137/1.9781611973402.93
9. Kleinberg, J.M.: Authoritative sources in a hyperlinked environment. J. ACM **46**(5), 604–632 (1999). https://doi.org/10.1145/324133.324140
10. Kolev, P., Mehlhorn, K.: A note on spectral clustering. In: 24th Annual European Symposium on Algorithms. Leibniz International Proceedings in Informatics, LIPIcs, vol. 57, p. 14, Art. No. 57. Schloss Dagstuhl. Leibniz-Zent. Inform., Wadern (2016)

11. Lee, J.R., Oveis Gharan, S., Trevisan, L.: Multi-way spectral partitioning and higher-order Cheeger inequalities. In: Proceedings of the 2012 ACM Symposium on Theory of Computing, STOC 2012, pp. 1117–1130. ACM, New York (2012). https://doi.org/10.1145/2213977.2214078

12. Li, A., Peng, P.: Detecting and characterizing small dense bipartite-like subgraphs by the bipartiteness ratio measure. In: Cai, L., Cheng, S.-W., Lam, T.-W. (eds.) ISAAC 2013. LNCS, vol. 8283, pp. 655–665. Springer, Heidelberg (2013). https://doi.org/10.1007/978-3-642-45030-3_61

13. Li, J., Liu, G., Li, H., Wong, L.: Maximal biclique subgraphs and closed pattern pairs of the adjacency matrix: a one-to-one correspondence and mining algorithms. IEEE Trans. Knowl. Data Eng. 19(12), 1625–1637 (2007). https://doi.org/10.1109/TKDE.2007.190660

14. Liu, F.M.A.S.: Cheeger constants, structural balance, and spectral clustering analysis for signed graphs. Max Planck Institute for Mathematics in the Sciences (2014, Preprint). http://www.mis.mpg.de/de/publications/preprints/2014/prepr2014-111.html

15. Liu, S.: Multi-way dual Cheeger constants and spectral bounds of graphs. Adv. Math. 268, 306–338 (2015). https://doi.org/10.1016/j.aim.2014.09.023

16. Lo, D., Surian, D., Prasetyo, P.K., Zhang, K., Lim, E.P.: Mining direct antagonistic communities in signed social networks. Inf. Process. Manag. 49(4), 773–791 (2013). https://doi.org/10.1016/j.ipm.2012.12.009

17. Louis, A., Raghavendra, P., Tetali, P., Vempala, S.: Many sparse cuts via higher eigenvalues. In: Proceedings of the 2012 ACM Symposium on Theory of Computing, STOC 2012, pp. 1131–1140. ACM, New York (2012). https://doi.org/10.1145/2213977.2214079

18. Nascimento, M.C.V., de Carvalho, A.C.P.L.F.: Spectral methods for graph clustering – a survey. Eur. J. Oper. Res. 211(2), 221–231 (2011). https://doi.org/10.1016/j.ejor.2010.08.012

19. Peng, R., Sun, H., Zanetti, L.: Partitioning well-clustered graphs: spectral clustering works!. SIAM J. Comput. 46(2), 710–743 (2017). https://doi.org/10.1137/15M1047209

20. Rohe, K., Chatterjee, S., Yu, B.: Spectral clustering and the high-dimensional stochastic blockmodel. Ann. Stat. 39(4), 1878–1915 (2011). https://doi.org/10.1214/11-AOS887

21. Trevisan, L.: Max cut and the smallest eigenvalue. SIAM J. Comput. 41(6), 1769–1786 (2012). https://doi.org/10.1137/090773714

22. Verma, D., Meila, M.: A comparison of spectral clustering algorithms. Technical report, University of Washington CSE (2003)

23. Wolf, R.D.: A brief introduction to Fourier analysis on the Boolean cube. Theory of Computing Library Graduate Surveys (2008)

24. Yancey, K., Yancey, M.: Bipartite Communities. ArXiv e-prints, December 2014

Generating Algebraic Expressions for Labeled Grid Graphs

Mark Korenblit[⊠]

Holon Institute of Technology, Holon, Israel
korenblit@hit.ac.il

Abstract. The paper investigates relationship between algebraic expressions and labeled graphs. We consider directed grid graphs having m rows and n columns. Our intent is to simplify the expressions of these graphs. With that end in view, we describe two methods which generate expressions for directed grid graphs. For both methods, lengths of the expressions grow polynomially with n while m is determined as a constant parameter. Besides, we apply these methods to a square grid graph in which the number of rows is equal to the number of columns. We prove that the lengths of the expressions derived by the methods depend exponentially and quasi-polynomially, respectively, on the size of the graph.

1 Introduction

A two-terminal directed acyclic graph (*st-dag* in [4]) has only one source and only one sink. We consider a *labeled graph* in which each edge has a unique label. Each path between the source and the sink (a *spanning path*) in an st-dag can be represented by a product of all edge labels of the path. We define the sum of edge-label products corresponding to all possible spanning paths of an st-dag G as the *canonical expression* of G. The order of labels in every product (from the left to right) is identical to the order of corresponding edges in the path (from the source to the sink). An algebraic expression is called an *st-dag expression* (a *factoring of an st-dag* in [4]) if it is algebraically equivalent to the canonical expression of an st-dag. An st-dag expression consists of literals (edge labels), and the operators + (disjoint union) and · (concatenation, also denoted by juxtaposition). We denote an expression of an st-dag G by $Ex(G)$.

We define the total number of literals in an algebraic expression as its *complexity*. An *optimal representation of the algebraic expression F* is an expression of minimum complexity algebraically equivalent to F.

A *series-parallel graph* is defined recursively so that a single edge is a series-parallel graph and a graph obtained by a *parallel* or a *series composition* of series-parallel graphs is series-parallel [4]. A series-parallel graph expression has a representation in which each literal appears only once [4,17]. This representation is optimal for a series-parallel graph expression. For example, the canonical expression of the series-parallel graph presented in Fig. 1 is $abd + abe + acd + ace + fe + fd$ and it can be reduced to $(a(b + c) + f)(d + e)$.

© Springer Nature Switzerland AG 2018
D. Kim et al. (Eds.): COCOA 2018, LNCS 11346, pp. 138–153, 2018.
https://doi.org/10.1007/978-3-030-04651-4_10

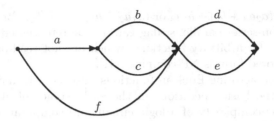

Fig. 1. A series-parallel graph.

A *Fibonacci graph* [9] has vertices $\{1, 2, 3, \ldots, n\}$ and edges $\{(v, v + 1) \mid v = 1, 2, \ldots, n - 1\} \cup \{(v, v + 2) \mid v = 1, 2, \ldots, n - 2\}$. As shown in [5], an st-dag is series-parallel if and only if it does not contain a subgraph which is a homeomorph of the *forbidden subgraph* positioned between vertices 1 and 4 of the Fibonacci graph illustrated in Fig. 2. Thus a Fibonacci graph gives a generic example of non-series-parallel graphs.

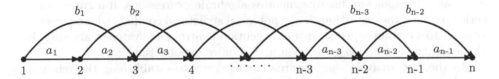

Fig. 2. A Fibonacci graph.

Interrelations between graphs and expressions are discussed in a number of works. In particular, [22, 26] consider the correspondence between series-parallel graphs and read-once functions. A Boolean function is defined as *read-once* if it may be computed by some formula in which no variable occurs more than once (*read-once formula*). On the other hand, a series-parallel graph expression can be reduced to the representation in which each literal appears only once. Hence, such a representation of a series-parallel graph expression can be considered as variety of a read-once formula.

Problems related to computations on graphs have applications in various areas. Specifically, many network problems, which are either intractable or have complicated solutions in the general case are solvable for series-parallel graphs. For example, some efficient algorithms for flow problems on series-parallel networks are presented in [3, 29]. Papers [6, 21] consider sequencing and scheduling in relation to precedence series-parallel constraints. Linear algorithms for reliability problems on series-parallel networks are presented in [25, 30].

An expression of a homeomorph of the forbidden subgraph belonging to any non-series-parallel st-dag has no representation in which each literal appears once. For example, consider the subgraph positioned between vertices 1 and 4 of the Fibonacci graph shown in Fig. 2. Possible optimal representations of its expression are $a_1 (a_2 a_3 + b_2) + b_1 a_3$ or $(a_1 a_2 + b_1) a_3 + a_1 b_2$. For this reason, an expression of a non-series-parallel st-dag can not be represented as a read-once formula. However, for arbitrary functions, which are not read-once, generating the optimum factored form is NP-complete [31].

The problem of factoring Boolean functions into shorter, more compact formulae is one of the basic operations in the early stages of algorithmic logic synthesis since the complexity of a logic circuit and computation time depend on the number of literals. Some algorithms developed in order to obtain good factored forms are described in [8,20].

A symbolic approach to scheduling of a robotic line is considered in [19]. The method uses the max-algebra tools and allows the shortest-path problem to be interpreted as the computation of the st-dag expression. The complexity of this problem is determined by the complexity of the st-dag expression. For a robotic line simulated by a Fibonacci graph, the proposed algorithm generates the processing sequence in polynomial time.

A method for automated composition of algebraic expressions in complex business process modeling based on acyclic directed graph reductions is introduced in [24]. The method transforms business step dependencies described by users into digraphs and finally generates algebraic expressions. If a graph is not series-parallel, the algorithm checks potential structural conflicts whose presence complicates certain aspects, such as execution control and system scalability. In this case, the expression generation may require exponential time.

In the last analysis, our research comes down to establishing the relationship between distributed systems and algebraic expressions. Expressions with a minimum (or, at least, a polynomial) complexity may be considered as a key to generating efficient algorithms on the systems.

In [17] we presented an algorithm, which generates the expression of $O\left(n^2\right)$ complexity for an n-vertex Fibonacci graph. More complicated, *rhomboidal* graphs are considered in [16]. The total numbers of literals in expressions derived for these n-size graphs are $O\left(n^{\log_2 6}\right)$.

In this paper we investigate a *directed grid graph* $G_{m,n}$ [32] having $m \times n$ vertices and $m(n-1) + n(m-1)$ edges. Each vertex in this graph corresponds to a unique pair of integers (i, j) $(1 \leq i \leq m, 1 \leq j \leq n)$ which are vertex coordinates. It has edges $\{((i,j),(i+1,j)) \mid 1 \leq i < m, 1 \leq j \leq n\} \cup \{((i,j),(i,j+1)) \mid 1 \leq i \leq m, 1 \leq j < n\}$. Each edge $((i,j),(i,j+1))$ is labeled by a_{ij}. Each edge $((i,j),(i+1,j))$ is labeled by b_{ij}. In Sects. 2 and 3 we consider m as a constant which determines the *depth* of a grid graph (specifically, a grid graph of depth 1 is a *path graph* and a grid graph of depth 2 is called a *ladder graph* [1,11,23]), while n characterizes the *size* of the graph. Section 4 analyses a *square grid graph*, $G_{n,n}$. For example, a graph $G_{5,6}$ is presented in Fig. 3.

Grid graphs are widely studied in the literature and have many applications. Specifically, routing problems on grid graphs are considered in [10,13,14]. A coloring problem for grids and some other families of grid-like graphs is discussed in [2]. In [27], directed grid graphs are used to model the string edit problem.

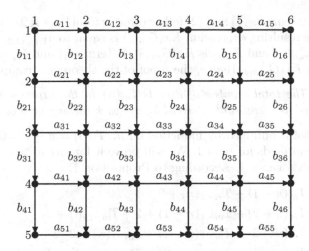

Fig. 3. A directed grid graph.

Our intent is to generate and to simplify the expressions of directed grid graphs. With that end in view, we present two methods one of which we call a *backtracking method* and the second one is named a *decomposition method.*

2 Generating Expressions for Directed Grid Graphs by a Backtracking Method

The method is universal and is appropriate for generating expressions of any st-dag. An expression is derived by using intermediate subexpressions which are accumulated in graph's vertices. Specifically, in our case a subexpression which is accumulated in vertex (i, j) of $G_{m,n}$ corresponds to its subgraph which is positioned between vertices (i, j) and (m, n) and is denoted by $F_{(i,j)}$. The following recursive procedure is used:

1. $F_{(i,n)} \leftarrow b_{i,n} b_{i+1,n} \ldots b_{m-1,n}$
2. $F_{(m,j)} \leftarrow a_{m,j} a_{m,j+1} \ldots a_{m,n-1}$
3. $F_{(i,j)} \leftarrow a_{i,j} F_{(i,j+1)} + b_{i,j} F_{(i+1,j)} \ (i < m, j < n)$

The subexpression accumulated in vertex $(1, 1)$ is the resulting expression.

Proposition 1. *The total number of literals $T_m(n)$ in the expression $Ex(G_{m,n})$ derived by the backtracking method is defined recursively as follows:*

$$T_m(1) = m - 1 \tag{1}$$
$$T_1(n) = n - 1 \tag{2}$$
$$T_m(n) = T_m(n-1) + T_{m-1}(n) + 2 \quad (m > 1, n > 1).$$

Proof. Initial statements (1) and (2) follow directly from lines 1, 2 of the recursive procedure. The resulting expression $Ex(G_{m,n})$ is equal to $a_{1,1}F_{(1,2)} + b_{1,1}F_{(2,1)}$. $F_{(1,2)}$ is $Ex(G_{m,n-1})$ and $F_{(2,1)}$ is $Ex(G_{m-1,n})$. Terms a_{11} and b_{11} are two additional terms in $Ex(G_{m,n})$. Hence, the proof of the theorem is complete.   □

Theorem 1. *The total number of literals $T_m(n)$ in the expression $Ex(G_{m,n})$ (m is considered as a constant) derived by the backtracking method is $O(n^m)$.*

Proof. The proof is obtained by induction on m. $T_1(n) = n - 1 = O(n)$. Therefore, the theorem holds for $m = 1$. We will prove it for any $m > 1$ on condition that it is correct for $m - 1$. According to Proposition 1,

$$
\begin{aligned}
T_m(n) &= T_m(n-1) + T_{m-1}(n) + 2 \\
&= T_m(n-2) + T_{m-1}(n-1) + 2 + T_{m-1}(n) + 2 \\
&= T_m(n-2) + T_{m-1}(n-1) + T_{m-1}(n) + 2 \cdot 2 \\
&= T_m(n-3) + T_{m-1}(n-2) + T_{m-1}(n-1) + T_{m-1}(n) + 2 \cdot 3 \\
&= \ldots = T_m(1) + \sum_{k=2}^{n} T_{m-1}(k) + 2(n-1) \\
&= \sum_{k=2}^{n} T_{m-1}(k) + 2(n-1) + m - 1.
\end{aligned}
$$

Hence, by the induction hypothesis, there exists a positive constant c such that

$$
T_m(n) \leq c \sum_{k=2}^{n} k^{m-1} + 2(n-1) + m - 1.
$$

In accordance to *Faulhaber's formula* for the sum of the p-th powers of the first n positive integers

$$
\sum_{k=1}^{n} k^p = \frac{1}{p+1} \sum_{k=0}^{p} (-1)^k \binom{p+1}{k} B_k n^{p+1-k}, \tag{3}
$$

where B_k are *Bernoulli numbers* $\left(B_1 = -\frac{1}{2}\right)$. The most significant term in the right part of (3) is $\frac{1}{p+1} n^{p+1}$. That is, $\sum_{k=1}^{n} k^p = O\left(n^{p+1}\right)$ and, therefore, $T_m(n) = O\left(n^m\right) + 2(n-1) + m - 1 = O\left(n^m\right)$.   □

Specifically,

$$
T_2(n) = \frac{n^2}{2} + \frac{3}{2}n - 1, \ T_3(n) = \frac{1}{6}n^3 + n^2 + \frac{11}{6}n - 1.
$$

For example, $Ex(G_{3,4})$ derived by the backtracking method is

$$
\begin{aligned}
&a_{11}(a_{12}(a_{13}b_{14}b_{24} + b_{13}(a_{23}b_{24} + b_{23}a_{33})) + \\
&b_{12}(a_{22}(a_{23}b_{24} + b_{23}a_{33}) + b_{22}a_{32}a_{33})) + \\
&b_{11}(a_{21}(a_{22}(a_{23}b_{24} + b_{23}a_{33}) + b_{22}a_{32}a_{33}) + b_{21}a_{31}a_{32}a_{33}).
\end{aligned}
$$

It contains 33 literals.

Hence, the proposed algorithm optimizes prefix parts of all subexpressions. In principle, the backtracking method can be applied by traversing the st-dag in the opposite direction. In this case suffix parts of subexpressions are optimized. The complexity of the derived expression will be the same.

The total numbers of literals $T_m(n)$ in the expressions $Ex(G_{m,n})$ $(m = 1, \ldots, 10;\ n = 1, \ldots, 10)$ derived by the backtracking method are presented in Table 1.

Table 1. Complexities of $Ex(G_{m,n})$ derived by the backtracking method.

$m \backslash n$	1	2	3	4	5	6	7	8	9	10
1	0	1	2	3	4	5	6	7	8	9
2	1	4	8	13	19	26	34	43	53	64
3	2	8	18	33	54	82	118	163	218	284
4	3	13	33	68	124	208	328	493	713	999
5	4	19	54	124	250	460	790	1285	2000	3001
6	5	26	82	208	460	922	1714	3001	5003	8006
7	6	34	118	328	790	1714	3430	6433	11438	19446
8	7	43	163	493	1285	3001	6433	12868	24308	43756
9	8	53	218	713	2000	5003	11438	24308	48618	92376
10	9	64	284	999	3001	8006	19446	43756	92376	184754

One can see that $T_m(n) = T_n(m)$. That is, we have the following claim.

Claim. The total numbers of literals in the expressions $Ex(G_{m,n})$ and $Ex(G_{n,m})$ derived by the backtracking method are equal.

Proof. The proof is obtained by induction on m and n. The claim holds for any $n = m$, including $n = 1$. We will prove it for any $m > 1$ or $n > 1$ on condition that it is correct for $m-1$ and $n-1$. According to Proposition 1 and the induction hypothesis, $T_m(n) = T_m(n-1)+T_{m-1}(n)+2 = T_{n-1}(m)+T_n(m-1)+2 = T_n(m)$. □

Thus the length of $Ex(G_{m,n})$ derived by the backtracking method grows polynomially with n. However, for sufficiently large constant m the problems based on $Ex(G_{m,n})$ can turn out to be intractable.

3 Generating Expressions for Directed Grid Graphs by a Decomposition Method

The method is known as very efficient and is based on recursive revealing subgraphs in the graph of a regular structure. The resulting expression is produced by a special composition of subexpressions describing these subgraphs. The existence of a decomposition method for a graph G is a sufficient condition for the existence of an expression with polynomial complexity for G if the graph is split into the constant number of subgraphs in each recursive step.

The graph $G_{m,n}$ is conditionally split into left and right parts connected by edges $a_{i,j}$ $(i = 1, 2, \ldots, m)$ so that j is chosen in the middle of a row (edges $a_{13}, a_{23}, a_{33}, a_{43}, a_{53}$ in Fig. 3). Each of these edges leaves a sink of a subgraph belonging to the left part (its source is a source of the graph) and enters a source of a subgraph of the right part (its sink is a sink of the graph). Any path from the source to the sink of the graph passes through one of the connecting edges. This decomposition is repeated recursively for every revealed subgraph.

We denote by $F((p_1, p_2), (q_1, q_2))$ a subexpression related to a subgraph with a source (p_1, p_2) and a sink (q_1, q_2) (it has depth $q_1 - p_1 + 1$ and size $q_2 - p_2 + 1$). Therefore,

1. $F((p_1, p_2), (q_1, q_2)) \leftarrow b_{p_1, p_2} b_{p_1+1, p_2} \ldots b_{q_1-1, p_2}$ $(q_2 = p_2)$
2. $F((p_1, p_2), (q_1, q_2)) \leftarrow F((p_1, p_2), (p_1, j)) a_{p_1, j} F((p_1, j + 1), (q_1, q_2)) +$
   $F((p_1, p_2), (p_1 + 1, j)) a_{p_1+1, j} F((p_1 + 1, j + 1), (q_1, q_2)) + \ldots +$
   $F((p_1, p_2), (q_1, j)) a_{q_1, j} F((q_1, j + 1), (q_1, q_2))$ $(q_2 > p_2)$
   $j = \left\lfloor \frac{q_2 + p_2}{2} \right\rfloor$ for even size,
   $j = \left\lceil \frac{q_2 + p_2 - 1}{2} \right\rceil$ or $\left\lfloor \frac{q_2 + p_2 - 1}{2} \right\rfloor$ for odd size.

In the general case $(q_2 > p_2)$ a current subgraph is decomposed into $2(q_1 - p_1 + 1)$ new subgraphs

Proposition 2. *The total number of literals $T_m(n)$ in the expression $Ex(G_{m,n})$ derived by the decomposition method is defined recursively as follows:*

$$T_m(1) = m - 1 \tag{4}$$

$$T_m(n) = T_1 \left(\left\lceil \frac{n}{2} \right\rceil \right) + T_1 \left(\left\lfloor \frac{n}{2} \right\rfloor \right) + T_2 \left(\left\lceil \frac{n}{2} \right\rceil \right) + T_2 \left(\left\lfloor \frac{n}{2} \right\rfloor \right) + \tag{5}$$

$$\ldots + T_m \left(\left\lceil \frac{n}{2} \right\rceil \right) + T_m \left(\left\lfloor \frac{n}{2} \right\rfloor \right) + m \quad (n > 1).$$

Proof. Initial statement (4) follows directly from line 1 of the decomposition procedure. General formula (5) is based on the structure of expression (2) of the same procedure. The graph of size n is decomposed into $2m$ subgraphs (m from the left and m from the right). Every subgraph of depth $m_1 = 1, \ldots, m$ from the left has a complementary subgraph of depth $m_2 = m - m_1 + 1$ from the right. Since location of the split is in the middle of a row, sizes of all revealed subgraphs are $\left\lceil \frac{n}{2} \right\rceil$ or $\left\lfloor \frac{n}{2} \right\rfloor$. Additionally, we have m literals corresponding to m connecting edges. $\qquad \square$

Theorem 2. *The total number of literals $T_m(n)$ in the expression $Ex(G_{m,n})$ (m is considered as a constant) derived by the decomposition method is $O\left(n \log^{m-1} n\right)$.*

Proof. The proof is obtained by induction on m. As follows from Proposition 2, $T_1(n) = T_1\left(\lceil\frac{n}{2}\rceil\right) + T_1\left(\lfloor\frac{n}{2}\rfloor\right) + 1 = O(n)$. Therefore, the theorem holds for $m = 1$. We will prove it for any $m > 1$ on condition that it is correct for $1, 2, \ldots, m-1$. According to Proposition 2,

$$T_m(n) \leq 2T_m\left(\left\lceil\frac{n}{2}\right\rceil\right) + 2T_{m-1}\left(\left\lceil\frac{n}{2}\right\rceil\right) + \ldots + 2T_2\left(\left\lceil\frac{n}{2}\right\rceil\right) + 2T_1\left(\left\lceil\frac{n}{2}\right\rceil\right) + m.$$

By the induction hypothesis, for $r = 1, \ldots, m-1$ there exists a positive constant c_r such that $2T_r\left(\lceil\frac{n}{2}\rceil\right) \leq 2c_r\lceil\frac{n}{2}\rceil\log_2^{r-1}\lceil\frac{n}{2}\rceil \leq c_r(n+1)\log_2^{r-1}n = O\left(n\log^{r-1}n\right)$. Therefore,

$$T_m(n) \leq 2T_m\left(\left\lceil\frac{n}{2}\right\rceil\right) + O\left(n\log^{m-2}n\right) + O\left(n\log^{m-3}n\right) + \ldots$$
$$+ O\left(n\log n\right) + O(n) + O(1)$$
$$= 2T_m\left(\left\lceil\frac{n}{2}\right\rceil\right) + O\left(n\log^{m-2}n\right).$$

In accordance with the *master theorem*, given constants $\alpha \geq 1$, $\beta > 1$, $k \geq 0$ and the recurrence $\Phi(n) = \alpha\Phi\left(\frac{n}{\beta}\right) + f(n)$ ($\frac{n}{\beta}$ is interpreted as $\lfloor\frac{n}{\beta}\rfloor$ or $\lceil\frac{n}{\beta}\rceil$) if $f(n) = O\left(n^{\log_\beta \alpha}\log^k n\right)$ then $\Phi(n) = O\left(n^{\log_\beta \alpha}\log^{k+1} n\right)$. For this reason,

$$T_m(n) \leq 2T_m\left(\left\lceil\frac{n}{2}\right\rceil\right) + O\left(n\log^{m-2}n\right) = O\left(n\log^{m-1}n\right). \qquad \square$$

In particular, for the ladder graph we have the following finding that follows directly from Theorem 2.

Corollary 1. *The total number of literals $T_2(n)$ in the expression $Ex(G_{2,n})$ derived by the decomposition method is $O\left(n\log n\right)$.*

Specifically, for n that is a power of two ($n = 2^p$ for some positive integer $p \geq 1$) we obtained the following explicit formulae:

$$T_2(n) = n\log_2 n + n, \ T_3(n) = \frac{1}{2}n\log_2^2 n + \frac{3}{2}n\log_2 n + 3n - 1.$$

For example, $Ex(G_{3,4})$ derived by the decomposition method is

$$a_{11}a_{12}(a_{13}b_{14}b_{24} + b_{13}a_{23}b_{24} + b_{13}b_{23}a_{33}) +$$
$$(a_{11}b_{12} + b_{11}a_{21})a_{22}(a_{23}b_{24} + b_{23}a_{33}) +$$
$$(a_{11}b_{12}b_{22} + b_{11}a_{21}b_{22} + b_{11}b_{21}a_{31})a_{32}a_{33}.$$

It contains 31 literals.

Thus for a graph $G_{m,n}$ of any depth (m) the total number of literals in $Ex(G_{m,n})$ derived by the decomposition method grows quasi-linearly with the size of the graph (n).

The total numbers of literals $T_m(n)$ in the expressions $Ex(G_{m,n})$ ($m = 1, \ldots, 10; n = 1, \ldots, 10$) derived by the decomposition method are presented in Table 2.

Table 2. Complexities of $Ex(G_{m,n})$ derived by the decomposition method.

$m\backslash n$	1	2	3	4	5	6	7	8	9	10
1	0	1	2	3	4	5	6	7	8	9
2	1	4	8	12	17	22	27	32	38	44
3	2	9	20	31	47	63	79	95	117	139
4	3	16	40	64	104	144	184	224	286	348
5	4	25	70	115	200	285	370	455	602	749
6	5	36	112	188	349	510	671	832	1140	1448
7	6	49	168	287	567	847	1127	1407	1995	2583
8	7	64	240	416	872	1328	1784	2240	3284	4328
9	8	81	330	579	1284	1989	2694	3399	5148	6897
10	9	100	440	780	1825	2870	3915	4960	7754	10548

One can see that asymptotically complexities in Table 2 are significantly less than corresponding complexities in Table 1. However, for small values of n ($n = 2, 3$), the backtracking method gives shorter representations than for the decomposition method.

In addition, Table 2 shows that $T_m(n) < T_n(m)$ for $m < n$ ($m > 1, n > 1$). It is clear that graphs $G_{m,n}$ and $G_{n,m}$ are isomorphic and if $m > n$ in a graph $G_{m,n}$, the graph may be decomposed horizontally into upper and lower parts. That is, the decomposition method may be modified so that to split any subgraph through the longer side and not always vertically.

For example, $Ex(G_{4,3})$ derived by the decomposition method can look like this

$$a_{11}(a_{12}b_{13}b_{23}b_{33} + b_{12}a_{22}b_{23}b_{33} + b_{12}b_{22}a_{32}b_{33} + b_{12}b_{22}b_{32}a_{42}) +$$
$$b_{11}a_{21}(a_{22}b_{23}b_{33} + b_{22}a_{32}b_{33} + b_{22}b_{32}a_{42}) +$$
$$b_{11}b_{21}a_{31}(a_{32}b_{33} + b_{32}a_{42}) + b_{11}b_{21}b_{31}a_{41}a_{42}.$$

It contains 40 literals.

Using the modified decomposition method for $G_{4,3}$ we obtain the following expression which contains only 29 literals:

$$b_{11}b_{21}(a_{31}(a_{32}b_{33} + b_{32}a_{42}) + b_{31}a_{41}a_{42}) +$$
$$(a_{11}b_{12} + b_{11}a_{21})b_{22}(a_{32}b_{33} + b_{32}a_{42}) +$$
$$(a_{11}(a_{12}b_{13} + b_{12}a_{22}) + b_{11}a_{21}a_{22})b_{23}b_{33}.$$

The complexity of the last expression is even less than for $Ex(G_{3,4})$ derived by the regular decomposition method because subgraphs $G_{3,2}$ ($m > n$) are revealed in the course of dividing $G_{3,4}$.

The total numbers of literals $T_m(n)$ in the expressions $Ex(G_{m,n})$ ($m = 1, \ldots, 10$; $n = 1, \ldots, 10$) derived by the modified decomposition method are presented in Table 3.

Table 3. Complexities of $Ex(G_{m,n})$ derived by the modified decomposition method.

$m\backslash n$	1	2	3	4	5	6	7	8	9	10
1	0	1	2	3	4	5	6	7	8	9
2	1	4	8	12	17	22	27	32	38	44
3	2	8	19	29	45	61	76	91	113	135
4	3	12	29	54	87	120	160	200	255	310
5	4	17	45	87	150	211	293	375	493	611
6	5	22	61	120	211	334	475	616	825	1034
7	6	27	76	160	293	475	712	937	1279	1621
8	7	32	91	200	375	616	937	1338	1855	2372
9	8	38	113	255	493	825	1279	1855	2604	3359
10	9	44	135	310	611	1034	1621	2372	3359	4582

As expected, now $T_m(n) = T_n(m)$. One can see that this way provides more compact expressions than the decomposition method.

4   Generating Expressions for Square Grid Graphs

Consider the case when both m and n characterize the size of $G_{m,n}$ (m is not considered as a constant). Specifically, if m grows together with n ($m = n$) we have a *square grid graph* of size n, $G_{n,n}$.

4.1   Generating Expressions for Square Grid Graphs by the Backtracking Method

Firstly, we apply the backtracking method to a graph $G_{m,n}$ in which n and m are not equal in the general case. The following proposition determines the lower bound for complexity of $Ex(G_{m,n})$.

Proposition 3. *The total number of literals $T(m,n)$ in the expression $Ex(G_{m,n})$ derived by the backtracking method is $\Omega\left(\frac{n^m}{m!}\right)$.*

Proof. The proof is obtained by induction on m. $T(1,n) = n - 1 = \Omega\left(\frac{n^1}{1!}\right)$. Therefore, the proposition holds for $m = 1$. We will prove it for any $m > 1$ on condition that it is correct for $m - 1$. As shown in the course of the proof of Proposition 1,

$$T(m,n) = \sum_{k=2}^{n} T(m-1,k) + 2(n-1) + m - 1.$$

Hence, by the induction hypothesis, there exists a positive constant c such that

$$T(m,n) \geq \sum_{k=2}^{n} \frac{ck^{m-1}}{(m-1)!} + 2(n-1) + m - 1 = \frac{c}{(m-1)!} \sum_{k=2}^{n} k^{m-1} + 2(n-1) + m - 1.$$

That is, by formula (3),

$$T(m,n) \geq \frac{c}{(m-1)!m} \sum_{k=0}^{m-1} (-1)^k \binom{m}{k} B_k n^{m-k} + 2(n-1) + m - 1 - \frac{c}{(m-1)!}$$

$$= \frac{c}{m!} n^m + \frac{c}{m!} \sum_{k=1}^{m-1} (-1)^k \binom{m}{k} B_k n^{m-k} + 2(n-1) + m - 1 - \frac{c}{(m-1)!}.$$

Although there are negative summands in the sum obtained, ultimately, $T(m,n)$ is positive and thus there exists a positive constant $c_1 \leq 1$ such that $T(m,n) \geq \frac{cc_1}{m!} n^m = \Omega\left(\frac{n^m}{m!}\right)$.   □

The following theorem for a square grid graph of size n follows directly from Proposition 3.

Theorem 3. *The total number of literals $T(n)$ in the expression $Ex(G_{n,n})$ derived by the backtracking method is $\Omega\left(\frac{n^n}{n!}\right)$.*

In accordance to *Stirling's approximation*, as $n \to \infty$,

$$n! \~ \left(\frac{n}{e}\right)^n \sqrt{2\pi n}.$$

Hence, the lower bound of $T(n)$ may be estimated as

$$\frac{n^n}{n!} \~ \frac{n^n}{\left(\frac{n}{e}\right)^n \sqrt{2\pi n}} = \frac{e^n}{\sqrt{2\pi n}}.$$

Therefore, $T(n) = \Omega\left(\frac{e^n}{\sqrt{n}}\right)$.

Thus the total number of literals in $Ex(G_{n,n})$ derived by the backtracking method grows exponentially with n.

4.2   Generating Expressions for Square Grid Graphs by the Decomposition Method

We denote by $T_m(n)$ the total number of literals in the expression $Ex(G_{m,n})$ (both m and n are not constants) derived by the decomposition method.

Theorem 4. *The total number of literals in the expression $Ex(G_{n,n})$ derived by the decomposition method is $O\left(n^{\lceil \log_2 n \rceil + 2}\right)$.*

Proof. Interpret $\frac{n}{2}$ as $\lceil \frac{n}{2} \rceil$ and denote $z = \lceil \log_2 n \rceil$. In accordance to Proposition 2,

$$
\begin{aligned}
T_n(n) &\leq 2T_n\left(\frac{n}{2}\right) + 2T_{n-1}\left(\frac{n}{2}\right) + 2T_{n-2}\left(\frac{n}{2}\right) + \ldots + 2T_1\left(\frac{n}{2}\right) + n \\
&\leq 2\left[2T_n\left(\frac{n}{4}\right) + 2T_{n-1}\left(\frac{n}{4}\right) + 2T_{n-2}\left(\frac{n}{4}\right) + \ldots + 2T_1\left(\frac{n}{4}\right) + n \right. \\
&\qquad\qquad + 2T_{n-1}\left(\frac{n}{4}\right) + 2T_{n-2}\left(\frac{n}{4}\right) + \ldots + 2T_1\left(\frac{n}{4}\right) + n - 1 \\
&\qquad\qquad\qquad + 2T_{n-2}\left(\frac{n}{4}\right) + \ldots + 2T_1\left(\frac{n}{4}\right) + n - 2 \\
&\qquad\qquad\qquad\qquad \cdot \quad \cdot \quad \cdot \quad \cdot \quad \cdot \quad \cdot \quad \cdot \quad \cdot \\
&\qquad\qquad\qquad\qquad\qquad\qquad\qquad \left. + 2T_1\left(\frac{n}{4}\right) + 1 \right] + n \\
&< 4\left[T_n\left(\frac{n}{4}\right) + 2T_{n-1}\left(\frac{n}{4}\right) + 3T_{n-2}\left(\frac{n}{4}\right) + \ldots + nT_1\left(\frac{n}{4}\right)\right] \\
&\quad + 2n^2 + n \\
&\leq 4\left[2T_n\left(\frac{n}{8}\right) + 2T_{n-1}\left(\frac{n}{8}\right) + 2T_{n-2}\left(\frac{n}{8}\right) + \ldots + 2T_1\left(\frac{n}{8}\right) + n \right. \\
&\qquad + 2\left(2T_{n-1}\left(\frac{n}{8}\right) + 2T_{n-2}\left(\frac{n}{8}\right) + \ldots + 2T_1\left(\frac{n}{8}\right) + n - 1\right) \\
&\qquad\quad + 3\left(2T_{n-2}\left(\frac{n}{8}\right) + \ldots + 2T_1\left(\frac{n}{8}\right) + n - 2\right) \\
&\qquad\qquad\qquad \cdot \quad \cdot \quad \cdot \quad \cdot \quad \cdot \quad \cdot \quad \cdot \quad \cdot \\
&\qquad\qquad\qquad\qquad\qquad \left. + n\left(2T_1\left(\frac{n}{8}\right) + 1\right)\right] \\
&\quad + 2n^2 + n \\
&= 8\left[T_n\left(\frac{n}{8}\right) + (1+2)T_{n-1}\left(\frac{n}{8}\right) + (1+2+3)T_{n-2}\left(\frac{n}{8}\right) + \right. \\
&\qquad\qquad \left. \ldots + \sum_{i=1}^{n} i \cdot T_1\left(\frac{n}{8}\right)\right] \\
&\quad + 4\left(n + 2(n-1) + 3(n-2) + \ldots + n\right) + 2n^2 + n \\
&< 8\left[T_n\left(\frac{n}{8}\right) + 2^2 T_{n-1}\left(\frac{n}{8}\right) + 3^2 T_{n-2}\left(\frac{n}{8}\right) + \ldots + n^2 T_1\left(\frac{n}{8}\right)\right]
\end{aligned}
$$

$$
\begin{aligned}
&\quad + 4n^3 + 2n^2 + n \\
&< 8\left[2T_n\left(\frac{n}{16}\right) + 2T_{n-1}\left(\frac{n}{16}\right) + 2T_{n-2}\left(\frac{n}{16}\right) + \ldots + 2T_1\left(\frac{n}{16}\right) + n \right. \\
&\qquad + 2^2\left(2T_{n-1}\left(\frac{n}{16}\right) + 2T_{n-2}\left(\frac{n}{16}\right) + \ldots + 2T_1\left(\frac{n}{16}\right) + n - 1\right) \\
&\qquad\quad + 3^2\left(2T_{n-2}\left(\frac{n}{16}\right) + \ldots + 2T_1\left(\frac{n}{16}\right) + n - 2\right) \\
&\qquad\qquad\qquad \cdot \quad \cdot \quad \cdot \quad \cdot \quad \cdot \quad \cdot \quad \cdot \quad \cdot \\
&\qquad\qquad\qquad\qquad\qquad \left. + n^2\left(2T_1\left(\frac{n}{16}\right) + 1\right)\right]
\end{aligned}
$$

$$+4n^3 + 2n^2 + n$$

$$= 16 \left[T_n \left(\frac{n}{16} \right) + \sum_{i=1}^{2} i^2 T_{n-1} \left(\frac{n}{16} \right) + \sum_{i=1}^{3} i^2 T_{n-2} \left(\frac{n}{16} \right) + \right.$$

$$\left. \ldots + \sum_{i=1}^{n} i^2 \cdot T_1 \left(\frac{n}{16} \right) \right]$$

$$+ 8 \left(n + 2^2(n-1) + 3^2(n-2) + \ldots + n^2 \right) + 4n^3 + 2n^2 + n$$

$$< 16 \left[T_n \left(\frac{n}{16} \right) + 2^3 T_{n-1} \left(\frac{n}{16} \right) + 3^3 T_{n-2} \left(\frac{n}{16} \right) + \ldots + n^3 T_1 \left(\frac{n}{16} \right) \right]$$

$$+ 8n^4 + 4n^3 + 2n^2 + n$$

$$\cdots \cdots \cdots \cdots \cdots$$

$$\leq 2^z \left[T_n(1) + 2^{z-1} T_{n-1}(1) + 3^{z-1} T_{n-2}(1) + \ldots + (n-1)^{z-1} T_2(1) \right]$$

$$+ \sum_{i=1}^{z} 2^{i-1} n^i$$

$$= 2^z \left[n - 1 + 2^{z-1}(n-2) + 3^{z-1}(n-3) + \ldots + (n-1)^{z-1} \right] + \sum_{i=1}^{z} 2^{i-1} n^i$$

$$< 2^z (n-1)(n-1)^{z-1}(n-1) + O\left(2^z n^z \right)$$

$$= 2^{\lceil \log_2 n \rceil} (n-1)(n-1)^{\lceil \log_2 n \rceil - 1}(n-1) + O\left(2^{\lceil \log_2 n \rceil} n^{\lceil \log_2 n \rceil} \right)$$

$$= O\left(n n^{\lceil \log_2 n \rceil + 1} \right) + O\left(n n^{\lceil \log_2 n \rceil} \right)$$

$$= O\left(n^{\lceil \log_2 n \rceil + 2} + n^{\lceil \log_2 n \rceil + 1} \right) = O\left(n^{\lceil \log_2 n \rceil + 2} \right).$$

$$\square$$

Thus for a square grid graph the decomposition method provides expressions with quasi-polynomial complexity with respect to the size of the graph.

Sometimes it is convenient to present a complexity of an st-dag expression as a function of the number of vertices in the graph. This unified way allows to compare expressions' complexities of different graphs, specifically, when their vertices are linearly ordered by means of a topological sort. Denote the number of vertices in a square grid graph by N. It is clear that $n = \sqrt{N}$. Hence, by Theorem 4, the total number of literals in the expression of an N-vertex square grid graph derived by the decomposition method is $O\left(\sqrt{N}^{\lceil \log_2 \sqrt{N} \rceil + 2} \right) = O\left(N^{\frac{1}{2} \lceil \frac{1}{2} \log_2 N \rceil + 2} \right) = O\left(N^{\frac{1}{2} \left(\frac{1}{2} \log_2 N + 1 \right) + 2} \right) = O\left(N^{0.25 \log_2 N + 2.5} \right)$. The obtained upper bound is not always a tight bound.

5   Conclusion and Future Work

Two methods for generating algebraic expressions of directed grid graphs (having m rows and n columns) have been presented. Both methods give expressions whose complexities polynomially depend on n which characterizes the size of the graph (m is considered as a constant). The first, the backtracking method, derives expressions with complexity $O(n^m)$, while the second, the decomposition method provides expressions of quasi-linear, $O(n \log^{m-1} n)$ complexity. Thus the decomposition method is significantly more efficient than the backtracking one. Moreover, since every edge label of a graph appears in its expression at least once, the total number of literals in the expression complexity for any st-dag of size n is $\Omega(n)$. Therefore, even if expressions generated by the decomposition method for directed grid graphs are not optimal (this is an open problem) their complexities are close to the lower bound.

Additionally, we have shown that the decomposition method being applied to a square grid graph of size $n \times n$, is able to generate expressions of quasi-polynomial complexity, namely, $O(n^{\lceil \log_2 n \rceil + 2})$, whereas the lengths of the same expressions produced by the backtracking method increase exponentially.

We are going to develop the improved version of the decomposition method which is supposed to yield more compact expressions (specifically, by reducing the number of connecting edges and revealing grid subgraphs which adjoin one another at the corners). Eventually we plan to find in this way the expression of a minimum length.

The next object of prospective research is a directed grid graph with diagonal edges. The question is how these additional edges complicate the resulting expression.

It seems interesting to extend the decomposition method to more general classes of graphs of a regular structure (*recursively constructible graphs* [23]).

An undirected graph in which every subgraph has a vertex of degree at most k is called k-*inductive* [12]. The *linkage of a graph* is the smallest value of k for which it is k-inductive [15]. For instance, trees are 1-inductive graphs, underlying graphs of Fibonacci graphs and undirected grid graphs are k-inductive graphs with linkage 2, undirected grid graphs with diagonal edges are k-inductive graphs with linkage 3. Our intent is to extend the decomposition technique to a class of st-dags whose underlying graphs are k-inductive.

The class of *bounded treewidth graphs* [18,28], for which many NP-complete problems can be solved efficiently is based on the concept of a *tree decomposition*. Specifically, series-parallel graphs and grids belong to this class. Extending the decomposition method to bounded treewidth graphs is also well worth for further investigation.

References

1. Ball, W.W.R., Coxeter, H.S.M.: Mathematical Recreations and Essays, 13th edn. The Macmillan Company, Dover, New York (1987)
2. Bar-Noy, A., Cheilaris, P., Lampis, M., Mitsou, V., Zachos, S.: Ordered coloring grids and related graphs. In: Kutten, S., Žerovnik, J. (eds.) SIROCCO 2009. LNCS, vol. 5869, pp. 30–43. Springer, Heidelberg (2010). https://doi.org/10.1007/978-3-642-11476-2_4

3. Bein, W.W., Brucker, P.: Greedy concepts for network flow problems. Discret. Appl. Math. **15**(2–3), 135–144 (1986)
4. Bein, W.W., Kamburowski, J., Stallmann, M.F.M.: Optimal reduction of two-terminal directed acyclic graphs. SIAM J. Comput. **21**(6), 1112–1129 (1992)
5. Duffin, R.J.: Topology of series-parallel networks. J. Math. Anal. Appl. **10**, 303–318 (1965)
6. Finta, L., Liu, Z., Milis, I., Bampis, E.: Scheduling UET-UCT series-parallel graphs on two processors. Theor. Comput. Sci. **162**(2), 323–340 (1996)
7. Golumbic, M.C., Gurvich, V.: Read-once functions. In: Crama, Y., Hammer, P.L. (eds.) Boolean Functions: Theory, Algorithms and Applications, pp. 519–560. Cambridge University Press, New York (2011)
8. Golumbic, M.C., Mintz, A., Rotics, U.: Factoring and recognition of read-once functions using cographs and normality and the readability of functions associated with partial k-trees. Discret. Appl. Math. **154**(10), 1465–1477 (2006)
9. Golumbic, M.C., Perl, Y.: Generalized Fibonacci maximum path graphs. Discret. Math. **28**, 237–245 (1979)
10. Grötschel, M., Martin, A., Weismantel, R.: Routing in grid graphs by cutting planes. ZOR - Math. Methods Oper. Res. **41**(3), 255–275 (1995)
11. Hosoya, H., Harary, F.: On the matching properties of three fence graphs. J. Math. Chem. **12**, 211–218 (1993)
12. Irani, S.: Coloring inductive graphs on-line. Algorithmica **11**(1), 53–72 (1994)
13. Kaufmann, M., Gao, S., Thulasiraman, K.: On Steiner minimal trees in grid graphs and its application to VLSI routing. In: Du, D.-Z., Zhang, X.-S. (eds.) ISAAC 1994. LNCS, vol. 834, pp. 351–359. Springer, Heidelberg (1994). https://doi.org/10.1007/3-540-58325-4_199
14. Kaufmann, M., Mehlhorn, K.: A linear-time algorithm for the homotopic routing problem in grid graphs. SIAM J. Comput. **23**(2), 227–246 (1998)
15. Kirousis, L.M., Thilikos, D.M.: The linkage of a graph. SIAM J. Comput. **25**(3), 626–647 (1996)
16. Korenblit, M.: Decomposition methods for generating algebraic expressions of full square rhomboids and other graphs. Discret. Appl. Math. **228**, 60–72 (2017)
17. Korenblit, M., Levit, V.E.: On algebraic expressions of series-parallel and Fibonacci graphs. In: Calude, C.S., Dinneen, M.J., Vajnovszki, V. (eds.) DMTCS 2003. LNCS, vol. 2731, pp. 215–224. Springer, Heidelberg (2003). https://doi.org/10.1007/3-540-45066-1_17
18. Krause, A.: Bounded Treewidth Graphs - A Survey German Russian Winter School, St. Petersburg, Russia (2003). http://wwwmayr.in.tum.de/konferenzen/Jass03/presentations/krause.pdf
19. Levit, V.E., Korenblit, M.: A symbolic approach to scheduling of robotic lines. In: Intelligent Scheduling of Robots and Flexible Manufacturing Systems. The Center for Technological Education Holon, Israel, pp. 113–125 (1996)
20. Mintz, A., Golumbic, M.C.: Factoring Boolean functions using graph partitioning. Discret. Appl. Math. **149**(1–3), 131–153 (2005)
21. Monma, C.L.: Sequencing with series-parallel precedence constraints. Math. Oper. Res. **4**(3), 215–224 (1979)
22. Mundici, D.: Solution of Rota's problem on the order of series-parallel networks. Adv. Appl. Math. **12**, 455–463 (1991)
23. Noy, M., Ribó, A.: Recursively constructible families of graphs. Adv. Appl. Math. **32**, 350–363 (2004)

24. Oikawa, M.K., Ferreira, J.E., Malkowski, S., Pu, C.: Towards algorithmic generation of business processes: from business step dependencies to process algebra expressions. In: Dayal, U., Eder, J., Koehler, J., Reijers, H.A. (eds.) BPM 2009. LNCS, vol. 5701, pp. 80–96. Springer, Heidelberg (2009). https://doi.org/10.1007/978-3-642-03848-8_7
25. Satyanarayana, A., Wood, R.K.: A linear time algorithm for computing K-terminal reliability in series-parallel networks. SIAM J. Comput. **14**(4), 818–832 (1985)
26. Savicky, P., Woods, A.R.: The number of Boolean functions computed by formulas of a given size. Rand. Struct. Algorithms **13**, 349–382 (1998)
27. Schmidt, J.P.: All highest scoring paths in weighted grid graphs and their application to finding all approximate repeats in strings. SIAM J. Comput. **27**(4), 972–992 (1998)
28. Sesh Kumar, K.S.: Convex relaxations for learning bounded-treewidth decomposable graphs. In: Proceedings of 30th International Conference on Machine Learning (ICML2013), JMLR: W&CP, vol. 28 (2013)
29. Tamir, A.: A strongly polynomial algorithm for minimum convex separable quadratic cost flow problems on two-terminal series-parallel networks. Math. Program. **59**, 117–132 (1993)
30. Wald, J.A., Colbourn, C.J.: Steiner trees in probabilistic networks. Microelectron. Reliabil. **23**(5), 837–840 (1983)
31. Wang, A.R.R.: Algorithms for multilevel logic optimization. Ph.D. thesis, University of California, Berkeley (1989)
32. Weisstein, E.W.: Grid Graph From MathWorld - A Wolfram Web Resource. http://mathworld.wolfram.com/GridGraph.html

Editing Graphs to Satisfy Diversity Requirements

Huda Chuangpishit[1], Manuel Lafond[2(✉)], and Lata Narayanan[3]

[1] Department of Mathematics, Ryerson University, Toronto, Canada
hoda.chuang@gmail.com
[2] Department of Computer Science, Université de Sherbrooke,
Sherbrooke, QC, Canada
manuel.lafond@USherbrooke.ca
[3] Department of Computer Science and Software Engineering, Concordia University,
Montréal, Canada
lata@cs.concordia.ca

Abstract. Let G be a graph where every vertex has a colour and has specified *diversity constraints*, that is, a minimum number of neighbours of every colour. Every vertex also has a *max-degree constraint*: an upper bound on the total number of neighbours. In the Min-Edit-Cost problem, we wish to transform G using edge additions and/or deletions into a graph G' where every vertex satisfies all diversity as well as max-degree constraints. We show an $O(n^5 \log n)$ algorithm for the Min-Edit-Cost problem, and an $O(n^3 \log n \log \log n)$ algorithm for the bipartite case. Given a specified number of edge operations, the Max-Satisfied-Nodes problem is to find the maximum number of vertices whose diversity constraints can be satisfied while ensuring that all max-degree constraints are satisfied. We show that the Max-Satisfied-Nodes problem is $W[1]$-hard, in parameter $r + \ell$, where r is the number of edge operations and ℓ is the number of vertices to be satisfied. We also show that it is inapproximable to within a factor of $n^{1/2-\epsilon}$. For certain relaxations of the max-degree constraints, we are able to show constant-factor approximation algorithms for the problem.

1 Introduction

The ruler of a certain planet is alarmed by the increasing polarization between the red and blue inhabitants of the planet. The two groups disagree on every policy issue, and ascribe the worst motivations to each other. Convinced that increasing the diversity of friendships will help reduce the bitter divisiveness in political debate, the ruler has decreed that everyone needs to have a *minimum* number of friends from the other group. Each person from the red group has been assigned a certain minimum threshold for the number of blue friends and vice versa. However, everyone also has a natural limit on the maximum number of friendships they can have. To satisfy the royal decree, it may be necessary to break off some friendships and initiate new ones. Is it possible to meet the

© Springer Nature Switzerland AG 2018
D. Kim et al. (Eds.): COCOA 2018, LNCS 11346, pp. 154–168, 2018.
https://doi.org/10.1007/978-3-030-04651-4_11

new royal requirements, and if so, what is the minimum number of disruptive changes (i.e. making and breaking friendships) needed?

The problem above can be modelled as a graph editing problem. Let V be a set of entities, and c a colour function $c : V \to \{1, \ldots, k\}$, mapping entities to colours. Let $G = (V, E, c)$ be a graph defined on the set V. Note that there may be edges between vertices of the same colour. The degree of a vertex v in G is denoted $d_G(v)$ and the number of its neighbors of colour i is denoted $d_G^i(v)$. For every vertex v, a certain *minimum* number $\delta_i(v)$ of *desired neighbors of colour i* is specified. At the same time, there is a *maximum allowable degree* $\delta(v)$ specified for every vertex v.

We say a vertex v is *satisfied* in G if and only if it has the desired number of neighbours of every colour, and its degree in G is at most the maximum allowable degree. That is, v is satisfied if and only if it meets the following constraints:

1. **Diversity constraints:** $d_G^i(v) \geq \delta_i(v)$ for all i with $1 \leq i \leq k$, and
2. **Max-degree constraint:** $d_G(v) \leq \delta(v)$.

Given a graph G, we wish to transform G into a new graph G' in which *every* vertex v is satisfied. The *edit distance* between graphs G and G' is the number of graph edit operations needed to transform the graph G to the graph G'. In this paper, we consider only the graph edit operations of *edge deletion*, and *edge addition*. In particular, vertices may not be inserted, deleted, or relabelled.

The problem of transforming a graph to a different graph that has a certain desired property has a long history as described in Sect. 1.2, and has many applications such as in pattern matching, and degree anonymization for privacy. In particular, several variants of the problem of editing a graph to satisfy constraints on vertex degrees have been studied in [2,5,11–13,20,25]. However, to the best of our knowledge, there is no known work on editing a graph to satisfy bounds on number of neighbors of particular colours or types.

In this paper, we study the following two problems:

Problem 1 (Min-Edit-Cost (MEC)). Given a graph $G = (V, E, c)$ and a set S of allowable edit operations, find a graph $G' = (V, E', c)$ with minimum edit distance (with respect to S) from G, such that every vertex in G' is satisfied.

Next we consider a fixed edge distance, and maximize the number of vertices whose diversity constraints can be satisfied while respecting the max-degree constraints of all vertices.

Problem 2 (Max-Satisfied-Nodes (MSN)). Given a graph $G = (V, E, c)$, find a graph $G' = (V, E', c)$ with edit distance at most k from G, that maximizes the number of satisfied vertices in G', while ensuring that all max-degree constraints are respected.

We denote the set of allowable operations by the set S, a non-empty subset of $\{a, d\}$, where a denotes edge addition, and d denotes edge deletion, the corresponding Min-Edit-Cost problem by MEC^S, and the corresponding Max-Satisfied-Nodes problem by MSN^S.

1.1 Our Results

We prove that for any $S \subseteq \{a, d\}$, the MEC^S problem can be solved in polynomial time. In particular, we give an $O(n^5 \log n)$ algorithm that, given a graph G, finds a graph G' which satisfies all diversity and max-degree constraints, and is at minimum possible edit distance from G. In particular when $k = 1$, our result implies a polynomial-time algorithm for adding/deleting a minimum number of edges so that each vertex has its degree within a specified range. This generalizes a result of Mathieson and Szeider [20, Sect. 5] that it is possible in polynomial time, to minimize the number of edge additions/deletions needed so that every vertex v has a specified degree $d(v)$. If G is bipartite, we present an $O(n^3 \log n \log \log n)$ algorithm for MEC^S. We then show that for any $S \subseteq \{a, d\}$, the MSN^S problem is $W[1]$-hard in parameter $r + \ell$ where r is the maximum edit distance and ℓ is the number of vertices to be satisfied. This hardness result holds even if there are two colour classes and no upper bound on the degrees. In fact, the problem is inapproximable to within a factor of $n^{1/2-\epsilon}$. If no upper bounds are specified, MSN^S admits an $O(n^2)$ time algorithm that satisfies at least $\lfloor OPT/2 \rfloor$ vertices, where OPT is the maximum number of satisfiable vertices. Finally, we show that a variant of the MSN problem, in which unsatisfied nodes do not have to obey their max-degree constraints, admits a $\frac{1}{9k}$-approximation where k is the number of colour classes in the input graph G.

1.2 Related Work

Graph editing problems have a long history; the goal is to edit a given graph to achieve a certain graph property. The editing operations that have been studied are vertex and edge deletions and additions, edge contractions, and edge flipping. Problems such as max-clique, vertex cover, independent set, etc. can be seen as graph editing problems. Lewis and Yannakakkis [15] showed that the problem of deleting a given number of vertices to achieve any non-trivial and hereditary graph property is NP-complete. Many edge deletion problems are also NP-hard, for example, deleting edges to achieve a bipartite graph, an outerplanar graph or a cograph [6,18,26]. One particularly well-studied graph editing problem consists in computing the *edit distance* between two given graphs G_1 and G_2, where the objective is to find the minimum number of vertex and/or edge modifications to perform on G_1 to obtain a graph isomorphic to G_2 (see e.g. [22–24] and [9] for a survey). Computing the similarity between two graphs is especially relevant for pattern recognition in machine learning, where a given graph needs to be matched with its closest graph in a set of known samples. Concrete applications include handwriting recognition [8], fingerprint recognition [21] and comparison of biological protein networks [14]. While computing the edit distance between two graphs is known to be APX-hard [27], the parameterized complexity of the problem does not appear to have been studied.

The problem of editing a graph to satisfy given constraints on the *degrees* of vertices has also attracted significant attention. Clearly, the NP-complete problem of finding an r-regular subgraph (see [3,4]) can be seen as a graph editing problem with degree constraints. In [19], Lovász introduced the *general factor problem*, where each vertex v is given a list of possible degrees. Cornuéjols [5] showed that if only edge deletions are allowed, then *general factor problem* is solvable in polynomial time, provided there no given degree set has a *gap* of length > 1. Otherwise, the problem is shown to be NP-complete. Mathieson and Szeider [20] generalized this problem by allowing the graph editing operations of edge addition, edge deletion, and vertex deletion. They also consider weighted nodes and edges, where the weight of a node/edge specifies the cost of an operation involving the node or edge. Their main result is to show that for any subset of allowable edit operations, the problem of graph editing is fixed-parameter tractable for parameter $k + r$ where k is the maximum edit cost, and r is an upper bound on the degree of any vertex, and W[1]-hard for parameter k. If vertex deletion is allowed, then the problem remains W[1]-hard for parameter k even for unweighted graphs where all vertices are to achieve degree exactly r. If only edge addition and deletion are permitted, the problem can be solved in polynomial time if all vertices are to achieve degree exactly $d(v)$, where $d(v)$ is given for each vertex v. Golovach [11] showed that if each vertex is given one desired degree, the problem of inserting/deleting (unweighted) edges to obtain a *connected* graph of given degrees is FPT when parameterized by $k + r$ in general. They also showed that if a connected r-regular graph is desired, the problem is FPT in parameter k. Recently, Subramanya [25] showed that the problem of graph editing to a given neighborhood degree list is also FPT in $k + r$.

The *degree anonymization problem*, introduced in [17], is a variant of graph editing with degree constraints motivated by the protection of privacy in social networks. In this problem, an integer h is given, and we ask if the graph can be modified in k operations so that the graph becomes *h-anonymous*, i.e. for each vertex v, there are at least h other vertices with the same degree as v. The degree anonymization problem was recently shown to be W[1]-hard for parameter k, the number of necessary edge insertion/deletions, but is FPT in Δ, the maximum degree of the input graph [13]. If k is the number of vertex-deletions required to attain h-anonymity, the problem is W[1]-hard for parameter $h + k$ but FPT in $\Delta + \min\{h, k\}$ [2]. A similar problem is considered in [12], where the authors propose the problem of editing a graph to a given degree sequence. The problem is shown to be W[1]-hard for any combination of allowable edit operations, but is FPT in parameter $k + D$, where here D is the maximum desired degree.

2   The Min-Edit-Cost Problem

We give a polynomial time algorithm for MEC when the allowed operations are edge additions and deletions (i.e. $MEC^{\{a,d\}}$). We will assume that $\sum_{1 \leq i \leq k} \delta_i(v) \leq \delta(v)$ for all $v \in V(G)$, as otherwise there is no way of satisfying every vertex.

We reduce MEC to the minimum weight perfect matching problem. Given an instance $G = (V, E, c)$ of the MEC problem, we construct a new graph H from G such that ℓ edit operations are enough in G to satisfy every vertex if and only if H has a perfect matching of cost at most ℓ. The reduction is somewhat technical, but builds upon a relatively intuitive idea initiated in [20].

For each $u \in V(G)$, we will have in H a set $W_1(u)$ of vertices containing $\delta(u)$ "copies" of u, each one representing a potential neighbor of u in a solution G' for G. Suppose for simplicity that in G, there is no "slack" in the degree requirements, meaning that for each $u \in V(G)$, we have $\sum_{i \in [k]} \delta_i(u) = \delta(u)$. Then for each $i \in [k]$, we make it so that exactly $\delta_i(u)$ of the $W_1(u)$ copies represent a desired neighbor of u of colour i. We build H so that in a perfect matching, each copy u' of u will be "paired" with a vertex v', which is a copy of some $v \in V(G)$ such that $c(v)$ is the colour that u' wants (note the symmetry: v' must also want colour $c(u)$). This pairing of u' and v' represents the presence of edge uv in the solution G'. The precise way that u' and v' are "paired" depends on whether $uv \in E(G)$ or not. If $uv \notin E(G)$, then u' and v' must use an *addition gadget* (represented in Fig. 1), which enforces taking an edge of cost 1 in a perfect matching. If $uv \in E(G)$, then u' and v' can be paired using only 0 cost edges of H — see the deletion gadget of Fig. 1. This *deletion gadget* also ensures that if $uv \in E(G)$ but u and v have no paired copies, then we must add a cost 1 edge in a perfect matching of H.

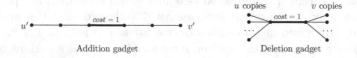

Addition gadget   Deletion gadget

Fig. 1. Addition and deletion gadgets. Heavy edges have cost 1, the others have cost 0. On the left is a gadget used when $uv \notin E(G)$. We say that u' and v' are "paired" in a perfect matching M if the edge incident to u' or the edge incident to v' is in M. This enforces having the heavy edge in M. On the right, a gadget used when $uv \in E(G)$. This allows "pairing" a copy u' and v' at 0 cost. If no such copies are paired, we will be forced to add the heavy edge of cost 1 to M.

This would be sufficient in the case that no slack in the requirements is present. Otherwise, we need to introduce a *partner* to each vertex of H to deal with this technicality. That is, we have in H another set $W_2(u)$ of vertices in 1-to-1 correspondence with $W_1(u)$. To see why this is useful, when there is some slack, H has $\delta(u)$ copies of u, but only $\sum_{i \in [k]} \delta_i(u) < \delta(u)$ that are used to enforce the lower bounds. The extra slack copies of u represent possible neighbors of u, which may or may not be present in G'. In terms of H, these copies desire no particular colour and may or may not be "paired" as described above. To ensure that a slack copy u' always has a neighbor available for a perfect matching, we have its partner $\hat{u}'$ that shares an edge of cost 0 with u'. If u' is not paired, it can use the $u'\hat{u}'$ edge in a perfect matching. If u' is paired however, $\hat{u}'$ must find a neighbor in a perfect matching. For this reason, every vertex of H will have a partner, and all partners of copies from distinct vertices share an edge. This makes it possible to have a perfect matching in H that represents a solution G' for G. We now proceed with the technical details.

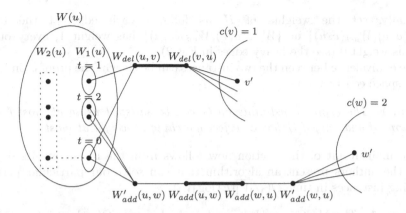

Fig. 2. An illustration of the main elements used in the construction of H. Here, u would have $\delta_1(u) = 1, \delta_2(u) = 2$ and $\delta(u) = 4$. This shows $uv \in E(G)$ and $uw \notin E(G)$. The vertices of $W(u)$ are displayed in two columns: the first column is the set of partners $W_2(u)$, the second is $W_1(u)$.

Construct the edge-weighted graph H from $G = (V, E, c)$ as follows.

1. For each $v \in V(G)$, do the following:
   (a) Add to H a set of vertices $W(v) = W_1(v) \cup W_2(v)$, where $W_1(v)$ and $W_2(v)$ are two disjoint sets each of cardinality $\delta(v)$.
   (b) Assign a colour $t(v) \in \{0, 1, \ldots, k\}$ to each vertex of $W_1(v)$ so that there are exactly $\delta_i(v)$ vertices coloured by i for each $i \in \{1, \ldots, k\}$. Notice that we introduced a new colour 0, and there are $\delta(v) - \sum_{i \in [k]} \delta_i(v)$ vertices coloured by 0 in $W_1(v)$. For $v' \in W_1(v)$, we may call $t(v')$ the *target colour* of v', and 0 is interpreted as "target any".
   (c) Assign each $v' \in W_1(v)$ to a distinct $\hat{v}' \in W_2(v)$, and call $\hat{v}'$ the *partner* of v'. Then for each $v' \in W_1(v)$ such that $t(v') = 0$, add an edge between v' and its partner $\hat{v}'$. This partner will serve in the event that a vertex with target colour 0 cannot be matched.
2. For each distinct $u, v \in V(G)$, add to H the edges $\{\hat{u}'\hat{v}' : \hat{u}' \in W_2(u), \hat{v}' \in W_2(v)\}$. Thus the set of all partners induce a complete n-partite graph.
3. For each edge $uv \in E(G)$, add two new vertices $W_{del}(u, v)$ and $W_{del}(v, u)$, and add an edge between them. Then for each $u' \in W_1(u)$ such that $t(u') = 0$ or $t(u') = c(v)$, add the edge $\{u', W_{del}(u, v)\}$ to H. Do the same for v and $W_{del}(v, u)$: for each $v' \in W_1(v)$ such that $t(v') = 0$ or $t(v') = c(u)$, add the edge $\{v', W_{del}(v, u)\}$ to H.
4. For each *non-edge* $uv \notin E(G)$, add the four vertices $W'_{add}(u, v), W_{add}(u, v),$ $W_{add}(v, u), W'_{add}(v, u)$ that form a path of length 3 - the vertices appear in this order on this path. Then for each $u' \in W_1(u)$ such that $t(u') = 0$ or $t(u') = c(v)$, add the edge $\{u', W'_{add}(u, v)\}$. Then for each $v' \in W_1(v)$ such that $t(v') = 0$ or $t(v') = c(u)$, add the edge $\{v', W'_{add}(v, u)\}$.

Finally, set the weights of H as follows: each edge of the form $\{W_{del}(u,v), W_{del}(v,u)\}$ or $\{W_{add}(u,v), W_{add}(v,u)\}$ has weight 1, every other edge has weight 0 (see the heavy edges in Fig. 2).

The equivalence between the two instances can be shown (the proof is omitted due to space constraints).

Lemma 1. *The degree constraints on G can be satisfied using at most ℓ edit operations if and only if H has a perfect matching of weight at most ℓ.*

The main result of this section now follows from the recent results of [7], where the authors present an algorithm that can solve our particular perfect matching instances in time $O(\sqrt{|V|}|E|\log|V|)$.

Theorem 1. *The MEC^S problem can be solved in time $O(n^5 \log n)$ for all $S \subseteq \{a, d\}$, where n is the number of vertices of the input graph G.*

2.1 A Faster Algorithm for Bipartite Graphs

We now give a more efficient algorithm for bipartite graphs, by reducing the problem to a min-cost network flow problem. Consider a bipartite graph $G = (V, E, c)$ such that $c : V \rightarrow \{1, 2\}$. Let V_j denote the set of vertices v with $c(v) = j$. For $v \in V_j$ we assume that $\delta_i(v) = 0$ when $i = j$. This means that there is no lower bound on the number of the same-coloured neighbours. In other words, we consider the problem of editing a bipartite graph G to another bipartite graph G' that satisfies degree requirements. More precisely a vertex $v \in V_i, i \in \{1, 2\}$ is satisfied whenever $d_G^j(v) \geq \delta_j(v)$ for $j \neq i$, and $d_G(v) \leq \delta(v)$.

We now give the reduction to the min-cost flow problem defined below. Assume $S = \{a, d\}$.

Minimum-Cost Flow Problem. Let $D = (V, E)$ be a directed graph with a cost c_e, a capacity μ_e, and a lower bound $\ell(e)$ associated with every edge $e = (u, v) \in E$. Let $f : E \rightarrow \mathbb{R}$ define a real-valued flow on the graph G satisfying:

$$\forall v \in V \quad f^+(v) = f^-(v) \tag{1}$$

$$\forall e \in E \quad \ell_e \leq f(e) \leq \mu_e \tag{2}$$

where $f^+(v)$ and $f^-(v)$ are the outgoing and incoming flows respectively from vertex v. The cost of a flow f is $\sum_{e \in E} c_e f(e)$. The minimum-cost flow problem is to find a flow f that minimizes $\sum_{e \in E} c_e f(e)$.

Given a bipartite instance of the Min-Edit-Cost problem $G = (V, E, c)$, we build a corresponding (directed) flow network D_G, as described below (see Fig. 3): we have $V(D_G) = V \cup \{s, t\}$ and $E(D_G) = E_1 \cup E_2 \cup E_3 \cup E_4 \cup E_5$ where

$E_1 = \{(u,v)|u \in V_1, v \in V_2, \{u,v\} \in E\}$, and $\forall e \in E_1 \ c_e = 0, \mu_e = 1, \ell_e = 1$

$E_2 = \{(s,u)|u \in V_1\}$, and $\forall e \in E_2 \ c_e = 0, \mu_e = \delta(u), \ell_e = \delta_2(u)$

$E_3 = \{(v,t)|v \in V_2\}$, and $\forall e \in E_3 \ c_e = 0, \mu_e = \delta(v), \ell_e = \delta_1(v)$

$E_4 = \{(u,v)|u \in V_1, v \in V_2, \{u,v\} \notin E\}$, and $\forall e \in E_4 \ c_e = 1, \mu_e = 1, \ell_e = 0$

$E_5 = \{(v,u)|u \in V_1, v \in V_2, \{u,v\} \in E\}$, and $\forall e \in E_5 \ c_e = 1, \mu_e = 1, \ell_e = 0$

Note that $E_1 \cup E_4 = V_1 \times V_2$. However the edges in E_1, which are directed from V_1 to V_2, and correspond to the original undirected edges in E, all have zero cost, but must all have flow 1, since $\ell_e = \mu_e$ for all $e \in E_1$. Meanwhile, the edges in E_4 have cost 1, and can have flow between 0 and 1. These correspond to the edge additions. Also, for each edge $(u,v) \in E_1$, there is a corresponding directed edge $(v,u) \in E_5$, which has cost 1, and can have flow between 0 and 1. The edges in E_5 correspond to edge deletions. Finally, the edges in E_2 and E_3 have zero cost, but their capacity constraints take care of the degree constraints on the vertices in V_1 and V_2 respectively.

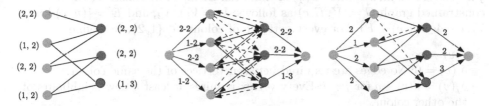

Fig. 3. (left) Graph G, each vertex v s labelled with the pair $(\delta_j(v), \delta(v))$; (middle) the corresponding flow network. Solid edges have zero cost, while dashed edges have cost 1. The arcs incident to the source and sink are annotated with their ℓ_e and μ_e values; (right) A flow that satisfies all constraints (only arcs that have a flow of ≥ 1 are drawn). This solution of cost 2 corresponds to adding one edge (left-to-right dashed arc) and deleting one edge (right-to-left dashed arc).

Lemma 2 establishes the relationship between the Min-Edit-Cost problem and network flow.

Lemma 2. *There exists a bipartite graph G' at edit distance ℓ from G that satisfies all degree constraints if and only if D_G admits a valid flow of cost ℓ.*

Theorem 2. *For any $S \subseteq \{a,d\}$, the MEC^S problem can be solved in $O(n^3 \log n \log \log n)$ time for bipartite graphs.*

Proof. Fix $S = \{a,b\}$. Given a graph G, we construct the corresponding flow network D_G, and solve the min-cost network flow problem. The complexity then follows from that of min cost flow since our maximum cost is 1, and the capacities are $O(n)$ [1]. For $S = \{a\}$, and $S = \{d\}$, we build a similar flow network, but omitting edges in E_4 and E_5 respectively. □

3   The Max-Satisfied-Nodes Problem

We consider the decision version of the MSN problem, which asks the following: given a graph $G = (V, E, c)$, and integers r, ℓ, is there a graph $G' = (V, E', c)$ with edit distance at most r from G such that at least ℓ vertices are satisfied in G', while ensuring that all max-degree constraints are satisfied? We first show that the Max-Satisfied-Nodes problem is not only NP-hard, but also W[1]-hard with respect to $r + \ell$. This holds even for a restricted set of instances.

Theorem 3. *The $MSN^{\{a,d\}}$ and $MSN^{\{a\}}$ problems are W[1]-hard and NP-hard with respect to parameter $r + \ell$, even if all vertices belong to one of two colour classes and $\delta(v) = \infty$ for all vertices v.*

Proof. We give a reduction from the *Balanced biclique* problem for bipartite graphs: Given a bipartite graph $G = (V_1 \cup V_2, E)$, and an integer q, are there subsets $A \subseteq V_1$ and $B \subseteq V_2$, such that $|A| = |B| = q$ and the subgraph of G induced by $A \cup B$ is a complete bipartite graph? The balanced biclique problem is a classic NP-hard problem [10] and was shown to be W[1]-hard with respect to q in [16]. Given an instance (G, q) of Balanced Biclique, we construct the degree-constrained graph $\hat{G} = (\hat{V}, \hat{E}, c)$ as follows: $\hat{V} = V_1 \cup V_2$, and $\hat{E} = \{(u, v) \mid u \in V_1, v \in V_2, (u, v) \notin E\}$. For every vertex v of colour $i \in \{1, 2\}$, we define

- $\delta(v) = \infty$.
- $\delta_i(v) = 0$: No edges are required between vertices of the same colour.
- $\delta_j(v) = d(v) + q$ for $j \neq i$: Every vertex requires at least q new neighbours of the other colour.

Finally, we set $\ell = 2q$ and the maximum edit distance $r = q^2$. We show that G has a balanced biclique (A, B) of size $2q$ if and only if $2q$ vertices of $\hat{G}$ can be satisfied using q^2 edge editions. Since $\ell + r = 2q + q^2$ if a function of q, our W[1]-hardness result follows. Observe that since all max-degree constraints can be met trivially, there is no reason to delete an edge and therefore, only additions are necessary.

If there is a balanced biclique (A, B) of size $2q$ in G, then there are no edges between vertices in A and B in $\hat{G}$, and therefore, by adding all q^2 edges in $A \times B$, all $2q$ vertices in $A \cup B$ can be satisfied. Conversely, suppose that $2q$ vertices in $\hat{G}$ can be satisfied with q^2 edge additions. Since every vertex needs exactly q new edges to vertices of the other colour to be satisfied, there cannot be more than q satisfied vertices of colour 1, as this would require at least $(q+1)q$ edge additions. The same holds for the satisfied vertices of colour 2. It follows that the satisfied vertices constitute a pair of sets (A, B) with $A \subseteq V_1$, $B \subseteq V_2$, $|A| = |B| = q$, such that $(u, v) \notin \hat{E}$ for any pair $(u, v) \in A \times B$. This implies that (A, B) is a balanced biclique of size $2q$ in G. $\qquad\square$

3.1   Inapproximability of Max-Satisfied-Nodes

We establish an approximation hardness result for all versions of MSN^S.

Theorem 4. *For all $S \subseteq \{a, d\}$, MSN^S, it is NP-hard to approximate within a factor $n^{1/2-\epsilon}$ for any $\epsilon > 0$, where n is the number of vertices.*

Proof. We give one reduction for the case $S = \{a\}$, and another for the cases $S \in \{\{d\}, \{a, d\}\}$. Both reductions are from the Independent Set problem, which consists in finding a set of vertices of maximum size that share no edge in a graph G. The problem is NP-hard to approximate within a factor $n^{1-\epsilon}$ for all $\epsilon > 0$ [28]. The first reduction is presented here, the other is similar and omitted. Let $G = (V, E)$ be an instance of Independent Set, letting $n = |V|$ and $m = |E|$. Let $H = (V \cup E, F, c)$ with $c(v) = 1$ for each $v \in V$ and $c(e) = 2$ for each $e \in E$. The edge set F of H has an edge for each $v \in V$ and each $e \in E$ such that v is not an endpoint of e. That is, $F = \{\{v, e\} : v \in V, e \in E \text{ and } v \notin e\}$. Observe that in H, each element of E has exactly $n - 2$ neighbors. For each $v \in V$, put $\delta_2(v) = m$ and $\delta(v) = \infty$. Then for each $e \in E$, put $\delta_1(e) = n$ and $\delta(e) = n - 1$. Thus any $e \in E$ cannot possibly be satisfied, and can receive at most one more neighbor. Allow any number of edge additions by putting $r = \infty$.

We show that for any $t \geq 1$, any independent set of size t in G corresponds to a set of t nodes that can be satisfied in H, and vice versa. Suppose that G has an independent set $I = \{v_1, \dots, v_t\}$ of size t. Note that there are no two vertices of I that are incident to a common edge. In H, we can satisfy every vertex of I by adding, for each $v_i \in I$, every edge from v_i to E that is not already present. Such an edge $\{v_i, e\}$ is added only if $v_i \in e$. Therefore, each $e \in E$ receives at most one new edge. It follows that no vertex of H has a degree above its upper bound, showing that we can satisfy t vertices.

Conversely, suppose that we can satisfy t vertices $T = \{v_1, \dots, v_t\}$ of H. Each $v_i \in T$ must be in V, as all other vertices cannot be satisfied. In H, we must also have inserted each possible edge from $v_i \in T$ to E that was not already present. Since each $e \in H$ can receive at most one new neighbor, no two elements of T can share an edge. Therefore, T is an independent set in G.

To argue the inapproximability, note that the graph H has $p = |V| + |E| \leq n^2$ vertices. For any $\epsilon > 0$, it is NP-hard to decide if $\alpha(G) \leq t$, or if $\alpha(G) \geq n^{1-\epsilon}t$, where $\alpha(G)$ is the maximum size of an independent set of G and t is a given integer. Suppose there is a $p^{1/2-\epsilon}$ approximation algorithm for the Max-Satisfied-Nodes problem, and let $APP(H)$ be the value returned by this algorithm on H. By the above, if $\alpha(G) \leq t$, then $APP(H) \leq t$. Moreover, if $\alpha(G) \geq n^{1-\epsilon}t$, then $APP(H) \geq 1/p^{1/2-\epsilon}n^{1-\epsilon}t \geq n^{\epsilon}t > t$. Hence, the approximation algorithm can distinguish whether G has $\alpha(G) \leq t$ or $\alpha(G) \geq n^{1-\epsilon}t$. $\square$

3.2 Approximation Algorithms with Loose Max-Degree Constraints

In this section, we relax the max-degree constraints in two ways. First, we consider the case when there are no max-degree constraints, that is, $\delta(v) = \infty$ for all vertices v. In the second case, the modified graph is not required to satisfy all the max-degree constraints, but all satisfied nodes in the modified graph must obey both diversity and max-degree constraints.

For a node $v \in V(G)$, let $r_G(v)$ be the minimum number of edges to modify so that v is satisfied (regardless of the other nodes). Define $r_G(v) = 0$ if v is already satisfied in G, and $r_G(v) = \infty$ if v cannot be satisfied in any way. We call $r_G(v)$ the *requirement* of v, and may write $r(v)$ if G is clear from the context. Observe that $r(v)$ can easily be computed in linear time. For the rest of this section, we suppose that we ordered $V(G)$ so that the vertices appear in non-decreasing order of requirement. That is, we have $V(G) = \{v_1, \ldots, v_n\}$ so that $r(v_i) \le r(v_{i+1})$ for all $i \in [n-1]$. We assume that $r(v_1) \le r$, as otherwise no one can be satisfied.

Our approximation algorithms rely on the following idea. One can derive an integer p from r such that at best, we can satisfy the first p vertices $v_1, \ldots, v_p$ of the above ordering. We show that can we always satisfy at least $\lfloor p/2 \rfloor$ vertices of this list if there are no upper bounds, and at least $p/9k$ vertices in the general case. For the remainder of this section, denote by OPT the maximum number of vertices that can be satisfied using r edit operations.

Let $A = (e_1, e_2, \ldots, e_t)$ be a sequence of pairs of vertices of G, i.e. $e_i \in \binom{V}{2}$ for each $i \in [t]$. We call A an *edit sequence*. Denote by $G(A, i)$ the graph obtained after modifying the first i edges of A (adding or deleting, depending on whether the edge is present or not). Define $G(A, 0) = G$. We start with a useful observation.

Proposition 1. *Let $X \subseteq V(G)$ and $h \in \mathbb{N}$ such that $\sum_{x \in X} r_G(x) \le h$. Moreover let $A = (e_1, \ldots, e_t)$ be an edit sequence such that $\sum_{x \in X} r_{G(A,i)}(x) < \sum_{x \in X} r_{G(A,i-1)}(x)$, for each $1 < i \le t$. Then $t \le h$.*

Proof. Each time we apply a modification in A, we lower the sum of requirements by at least 1. This sum is always non-negative and starts at most h, and so A must contain at most h elements. $\qquad\square$

We now obtain our upper bound p on OPT, which is defined as:

$$p = \max\{i \in [n] : \sum_{j=1}^{i} r(v_j) \le 2r\}$$

Lemma 3. *At most p vertices can be satisfied by using r edit operations.*

Proof. If $p = n$, then this is obvious, so assume $p < n$. Inserting or removing an edge can lower the requirement of at most two vertices. Therefore, r edit operations can lower the total requirement by at most $2r$. Since the v_i's are ordered by requirement, for any subset $W \subseteq V(G)$ of $p + 1$ vertices or more, we have $\sum_{v_i \in W} r(v_i) \ge \sum_{i=1}^{p+1} r(v_i) > 2r$. Therefore, it is not possible to satisfy every vertex of W with r edit operations. $\qquad\square$

We can already show that if $\delta(v) \ge n-1$ for every vertex v, there is a simple approximation algorithm.

Theorem 5. *Suppose that $\delta(v) \ge n - 1$ for every $v \in V(G)$. Then there is an $O(n^2)$ time algorithm that satisfies at least $\lfloor OPT/2 \rfloor$ vertices.*

Proof. The algorithm does the following: first, compute $r_G(v)$ for every vertex $v \in V(G)$, and then sort $V(G)$ so that $\{v_1, \ldots v_n\}$ is sorted in nondecreasing order of requirement. We then identify the upper bound p as above. Let $X = \{v_1, \ldots, v_{\lfloor p/2 \rfloor}\}$, observing that $\sum_{v_i \in X} r(v_i) \leq r$. Note that if $x \in X$ is not satisfied, only edge additions are needed since $\delta(x) \geq n-1$. Perform the following sequence of edge additions: while there is $x \in X$ and $y \in V(G)$ such that adding xy reduces $r(x)$ by 1, we add xy. Each edge addition xy decreases the requirement of $x \in X$, and cannot increase the requirement of y (since $\delta(y) \geq n - 1$). After the above loop is finished, each vertex $x \in X$ is satisfied, and as each operation reduces the sum of the X requirements, we know by Proposition 1 that we have used at most r edge operations to do so. We have satisfied $\lfloor p/2 \rfloor \geq \lfloor OPT/2 \rfloor$ vertices. The $O(n^2)$ time complexity of the algorithm is easy to see - it is dominated by the computation of $r_G(v)$.   □

Observe that the algorithm of Theorem 5 yields a $\frac{1}{2}$-approximation for even values of p and a $\frac{1}{3}$-approximation for odd values.

We now show how to extend the above idea to the case where nodes have a max-degree constraint that must be met for satisfied vertices. We present the $S = \{a, d\}$ case, the other cases being similar. The main problem that arises when applying the above algorithm is that lowering the requirement of a vertex might increase the requirement of another. Hence, we need to find a set of vertices that do not interfere with each other. We do this by restricting our attention to only a subset of vertices of the same colour.

Lemma 4. *Suppose that all vertices of $V(G)$ have the same colour c and that only edge additions, or only edge deletions, are allowed. Let $X \subseteq V(G)$ and $h \in \mathbb{N}$ such that $\sum_{x \in X} r_G(x) \leq h$. Then at least $\lceil |X|/2 \rceil$ of the elements of X can be satisfied using at most h edit operations.*

Moreover, if both additions and deletions are allowed, then at least $\lceil |X|/4 \rceil$ of the elements of X can be satisfied using at most h edit operations.

Proof. First observe that each vertex of X must be satisfiable. We give the proof for additions-only first (deletions are similar). In this case, when $r_G(v_i) > 0$ and $r_G(v_i) \neq \infty$, it must be that v_i requires more neighbors. Start by applying the following edit sequence on G: While there exist $x, y \in X$ such that $xy \notin E(G)$ and $r(x), r(y) > 0$, add an edge between x and y. Note that each edge modification reduces the sum of requirements of the X vertices (by 2). Also note that no vertex x of X can have a degree above $\delta(x)$ after these additions (otherwise, it would not be satisfiable). Call G' the graph resulting from these operations.

Now, partition X into two sets X_1 and X_2 as follows: $X_1 = \{x \in X : r_{G'}(x) = 0\}$ and $X_2 = X \setminus X$. If $|X_1| \geq \lceil |X|/2 \rceil$, then we are done as every member of X_1 is satisfied, so assume otherwise. Then $|X_2| \geq \lceil |X|/2 \rceil$. It is not hard to see that in G', X_2 must be a clique: if $x, y \in X_2$ do not share an edge, then the xy edge would have been added in the above loop.

Now apply the following edit sequence on G': for each $x \in X_2$, add a neighbor $y \in V(G') \setminus X_2$ to x until x is satisfied. Call the resulting graph G''. Observe that since each $x \in X_2$ is adjacent to every vertex in X_2, then x must be satisfied in G'' (as x can be made neighbor with every vertex if needed). Moreover, each edge modification from G' to G'' reduces the requirement of one vertex of X_2, and does not increase the requirement of any other vertex of X_2. Therefore, each modification from G to G'' reduces the sum of requirements of the X_2 vertices. By Proposition 1, we have made at most h modifications.

For edge deletions only, the proof is essentially the same. In G', we remove edges between unsatisfied members of X instead of adding them. Then X_2 must be an independent set and its vertices can be satisfied by removing edges with an endpoint outside of X_2. We omit the details.

Finally, let us prove the second statement for the additions and deletions case. Partition X into the sets $X_1 = \{x \in X : d_c(x) \leq \delta_c(x)\}$ and $X_2 = \{x \in X : d(x) > \delta(x)\}$. Note that an x cannot be in both sets, as otherwise it cannot be satisfied. Let $X' = \arg\max_{X_1,X_2}\{|X_1|, |X_2|\}$. Then $|X'| \geq \lceil |X|/2 \rceil$. Moreover, if we only want to satisfy vertices of X', we either need additions only (if $X' = X_1$), or deletions only (if $X' = X_2$). Clearly, $\sum_{x \in X'} r_G(x) \leq h$, and as we have just shown, we can satisfy at least $\lceil |X'|/2 \rceil$ vertices of X', which is at least $\lceil \lceil |X|/2 \rceil /2 \rceil \geq \lceil |X|/4 \rceil$.   □

We can now extend the above ideas to any number k of colours.

Theorem 6. *The problem of finding a graph G' within edit distance r from a given graph G that maximizes the number of vertices for whom both diversity and max-degree constraints are satisfied admits a factor $\frac{1}{9k}$ approximation algorithm, where the set of allowable edit operations is $S \subseteq \{a, d\}$.*

Proof. Let $X = \{v_1, \ldots, v_{\lfloor p/2 \rfloor}\}$, observing that $\sum_{v_i \in X} r(v_i) \leq r$. For each colour $j \in [k]$, let $X_j = \{v_i \in X : c(v_i) = j\}$. Let j^* be the colour that is the most present in X, i.e. $j^* = \arg\max_{j \in [k]}\{|X_j|\}$. We must have $|X_{j^*}| \geq \lceil \frac{\lfloor p/2 \rfloor}{k} \rceil$. Our approximation algorithm ensures that a constant fraction of X_{j^*} is satisfied.

Apply the following edit sequence: while there exists an unsatisfied $x \in X_{j^*}$ and a $y \in V(G) \setminus X_{j^*}$ such that inserting (deleting) xy reduces $r(x)$, then we insert (delete) xy. Let G' be the graph obtained after the above modifications. Suppose that l edges get edited in this manner, and let $r' = r - l$. Clearly, we have $\sum_{x \in X_{j^*}} r_{G'}(x) \leq r'$. Moreover, if we want to reduce the requirement of a vertex of X_{j^*}, only the edges with two endpoints in X_{j^*} can be used to do so (as all other edges that can reduce the requirements of the X_{j^*} vertices have been edited from G to G'). By Lemma 4, at least $\lceil |X_{j^*}|/4 \rceil$ vertices of X_{j^*} can be satisfied using at most r' edit operations, resulting in at least $\lceil \lceil \frac{\lfloor p/2 \rfloor}{k} \rceil /4 \rceil$ satisfied vertices. If $p < 9k$, this does provide a $\frac{1}{9k}$ approximation algorithm. If $p \geq 9k$, then this is at least $\frac{p-1}{8k} \geq \frac{p}{9k}$.   □

Conclusion. In this work, we imposed a single max-degree for every node, but it would be interesting to consider the variant of MEC and MSN were each vertex can have a different upper bound for the number of neighbours of every colour. We believe that the MEC algorithm described above can be adapted to this case. Another variant would have both an overall max-degree, and an upper bound on each colour. As for the MSN problem, we have a W[1]-hardness result for parameter $r + \ell$, but the existence of an intuitively interesting parameter for which the problem is FPT is open. One possible parameter that is worth investigating is $n - k$, the number of *unsatisfied* people in the resulting graph. Finally, it remains to check whether the "loose" MSN variant admits a constant factor approximation that does not depend on k. We also have not considered vertex addition and deletion operations. Although NP-hardness results do not seem difficult to obtain, the approximation and parameterized complexity remains unexplored.

Acknowledgement. We thank Jaroslav Opatrny for useful discussions.

References

1. Ahuja, R.K., Goldberg, A.V., Orlin, J.B., Tarjan, R.E.: Finding minimum-cost flows by double scaling. Math. Program. **53**(1–3), 243–266 (1992)
2. Bredereck, R., Hartung, S., Nichterlein, A., Woeginger, G.J.: The complexity of finding a large subgraph under anonymity constraints. In: Cai, L., Cheng, S.-W., Lam, T.-W. (eds.) ISAAC 2013. LNCS, vol. 8283, pp. 152–162. Springer, Heidelberg (2013). https://doi.org/10.1007/978-3-642-45030-3_15
3. Cheah, F., Corneil, D.G.: The complexity of regular subgraph recognition. Discrete Appl. Math. **27**(1–2), 59–68 (1990)
4. Chvátal, V., Fleischner, H., Sheehan, J., Thomassen, C.: Three-regular subgraphs of four-regular graphs. J. Graph Theory **3**(4), 371–386 (1979)
5. Cornuéjols, G.: General factors of graphs. J. Comb. Theory Ser. B **45**(2), 185–198 (1988)
6. Dondi, R., Lafond, M., El-Mabrouk, N.: Approximating the correction of weighted and unweighted orthology and paralogy relations. Algorithms Mol. Biol. **12**(1), 4 (2017)
7. Duan, R., Pettie, S., Su, H.-H.: Scaling algorithms for weighted matching in general graphs. In: Proceedings of the Twenty-Eighth Annual ACM-SIAM Symposium on Discrete Algorithms, pp. 781–800. Society for Industrial and Applied Mathematics (2017)
8. Fischer, A., Suen, C.Y., Frinken, V., Riesen, K., Bunke, H.: A fast matching algorithm for graph-based handwriting recognition. In: Kropatsch, W.G., Artner, N.M., Haxhimusa, Y., Jiang, X. (eds.) GbRPR 2013. LNCS, vol. 7877, pp. 194–203. Springer, Heidelberg (2013). https://doi.org/10.1007/978-3-642-38221-5_21
9. Gao, X., Xiao, B., Tao, D., Li, X.: A survey of graph edit distance. Pattern Anal. Appl. **13**(1), 113–129 (2010)
10. Garey, M.R., Johnson, D.S.: Computers and Intractability, vol. 29. W. H. freeman, New York (2002)
11. Golovach, P.A.: Editing to a connected graph of given degrees. Inf. Comput. **256**, 131–147 (2017)
12. Golovach, P.A., Mertzios, G.B.: Graph editing to a given degree sequence. Theoret. Comput. Sci. **665**, 1–12 (2017)
13. Hartung, S., Nichterlein, A., Niedermeier, R., Suchý, O.: A refined complexity analysis of degree anonymization in graphs. Inf. Comput. **243**, 249–262 (2015)
14. Hu, H., Yan, X., Huang, Y., Han, J., Zhou, X.J.: Mining coherent dense subgraphs across massive biological networks for functional discovery. Bioinformatics **21**, i213–i221 (2005)

15. Lewis, J.M., Yannakakis, M.: The node-deletion problem for hereditary properties is NP-complete. J. Comput. Syst. Sci. **20**(2), 219–230 (1980)

16. Lin, B.: The parameterized complexity of k-Biclique. In: Proceedings of the Twenty-Sixth Annual ACM-SIAM Symposium on Discrete Algorithms, pp. 605–615. Society for Industrial and Applied Mathematics (2015)

17. Liu, K., Terzi, E.: Towards identity anonymization on graphs. In: Proceedings of the 2008 ACM SIGMOD International Conference on Management of Data, pp. 93–106. ACM (2008)

18. Liu, Y., Wang, J., Guo, J., Chen, J.: Cograph editing: complexity and parameterized algorithms. In: Fu, B., Du, D.-Z. (eds.) COCOON 2011. LNCS, vol. 6842, pp. 110–121. Springer, Heidelberg (2011). https://doi.org/10.1007/978-3-642-22685-4_10

19. Lovász, L.: The factorization of graphs. ii. Acta Mathematica Academiae Scientiarum Hungarica, **23**(1–2), 223–246 (1972)

20. Mathieson, L., Szeider, S.: Editing graphs to satisfy degree constraints: a parameterized approach. J. Comput. Syst. Sci. **78**(1), 179–191 (2012)

21. Neuhaus, M., Bunke, H.: A graph matching based approach to fingerprint classification using directional variance. In: Kanade, T., Jain, A., Ratha, N.K. (eds.) AVBPA 2005. LNCS, vol. 3546, pp. 191–200. Springer, Heidelberg (2005). https://doi.org/10.1007/11527923_20

22. Neuhaus, M., Bunke, H.: Bridging the Gap Between Graph Edit Distance and Kernel Machines, vol. 68. World Scientific, Singapore (2007)

23. Riesen, K., Bunke, H.: Approximate graph edit distance computation by means of bipartite graph matching. Image Vis. Comput. **27**(7), 950–959 (2009)

24. Riesen, K., Neuhaus, M., Bunke, H.: Bipartite graph matching for computing the edit distance of graphs. In: Escolano, F., Vento, M. (eds.) GbRPR 2007. LNCS, vol. 4538, pp. 1–12. Springer, Heidelberg (2007). https://doi.org/10.1007/978-3-540-72903-7_1

25. Subramanya, V.: Graph editing to a given neighbourhood degree list is fixed-parameter tractable. Master's thesis, University of Waterloo (2016)

26. Yannakakis, M.: Edge-deletion problems. SIAM J. Comput. **10**(2), 297–309 (1981)

27. Zeng, Z., Tung, A.K.H., Wang, J., Feng, J., Zhou, L.: Comparing stars: on approximating graph edit distance. Proc. VLDB Endow. **2**(1), 25–36 (2009)

28. Zuckerman, D.: Linear degree extractors and the inapproximability of max clique and chromatic number. In: Proceedings of the Thirty-Eighth Annual ACM Symposium on Theory of Computing, pp. 681–690. ACM (2006)

Computing a Rectilinear Shortest Path amid Splinegons in Plane

Tameem Choudhury and R. Inkulu$^{(\boxtimes)}$

Department of Computer Science and Engineering, IIT Guwahati, Guwahati, India
{tameem.choudhury,rinkulu}@iitg.ac.in

Abstract. We reduce the problem of computing a rectilinear shortest path between two given points s and t in the given splinegonal domain $\mathcal{S}$ to the problem of computing a rectilinear shortest path between two points in the polygonal domain. Our reduction algorithm defines a polygonal domain $\mathcal{P}$ from $\mathcal{S}$ by identifying a coreset of points on the boundaries of splinegons in $\mathcal{S}$. Further, it transforms a shortest path between s and t amid polygonal obstacles in $\mathcal{P}$ to a shortest path between s and t amid splinegonal obstacles in $\mathcal{S}$. When $\mathcal{S}$ is comprised of h pairwise disjoint splinegons defined with a total of n vertices, excluding the time to compute a rectilinear shortest path amid polygons in $\mathcal{P}$, our reduction algorithm takes $O(n + h \lg n + (\lg h)^{1+\epsilon})$ time. Here, ϵ is a small positive constant (resulting from the triangulation of the free space using [2]). For the special case of $\mathcal{S}$ comprising concave-in splinegons, we have devised another reduction algorithm which does not rely on the structures used in the algorithm (Inkulu and Kapoor [14]) to compute a rectilinear shortest path in the polygonal domain. Further, we have characterized few of the properties of rectilinear shortest paths amid splinegons which could be of independent interest.

1 Introduction

Computing obstacle avoiding shortest path between two points is both fundamental and well-known in computational geometry. The case of polygonal obstacles has been well studied (Ghosh and Mount [11], Kapoor and Maheshwari [18], Hershberger and Suri [12], Kapoor and Maheshwari [17], Kapoor et al. [19], Mitchell [21], Rohnert [22], Storer and Reif [23], and Inkulu et al. [16]). Algorithms for visibility on which many Euclidean shortest path algorithms rely are detailed in Ghosh [10]. Clarkson et al. [7], Inkulu and Kapoor [14], and Chen et al. [3] devised algorithms to compute a rectilinear shortest path amid polygonal obstacles. In this paper, we devise an algorithm to compute a rectilinear shortest path amid planar curved obstacles. In specific, as in Dobkin and Souvaine [8], Dobkin et al. [9], and Melissaratos and Souvaine [20], we use splinegons to model planar curved objects. Chen and Wang [5], and Hershberger et al. [13]

R. Inkulu—This research is supported in part by NBHM grant 248(17)2014-R&D-II/1049.

D. Kim et al. (Eds.): COCOA 2018, LNCS 11346, pp. 169–182, 2018.
https://doi.org/10.1007/978-3-030-04651-4_12

devised algorithms to compute a Euclidean shortest path amid curved obstacles in the plane. To our knowledge, this paper is the first work to compute an optimal rectilinear shortest path amid splinegons in the plane, especially, the reduction algorithms. Since splinegons model the real-world obstacles more closely than the simple polygons, all the applications of computing shortest paths amid simple polygonal obstacles extend to computing shortest paths amid splinegons as well.

We first introduce terminology from [8,9,20]. A (simple) *splinegon* S is a simple region formed by replacing each edge e_i of a simple polygon P by a curved edge s_i joining the endpoints of e_i such that the region S-seg_i bounded by the curve s_i and the line segment e_i is convex [9]. The new edge need not be smooth; a sufficient condition is that there exists a left-hand and a right-hand derivative at each point on the splinegon. The vertices of S are the vertices of P. The polygon P is called the *carrier polygon* of the splinegon S. If S-$seg_i \subseteq P$, then we say that the edge e_i is *concave-in*. Otherwise, e_i is *concave-out*. We call a splinegon *concave-in* whenever each of its edges is concave-in.

We assume that the combinatorial complexity of each splinegon edge is $O(1)$. We also assume that each of the primitive operations on a splinegon edge can be performed in $O(1)$ time. These operations include the following: computing the points of intersections of a splinegon edge with a line, computing the tangents (if any) between two given splinegon edges, computing the tangents between a point and a splinegon edge, computing the distance between two points along a splinegon edge, and finding a point on an edge that has a horizontal or vertical tangent to that edge at that point. We assume no two carrier polygons of splinegons in $\mathcal{S}$ intersect, and the carrier polygon of any splinegon $S \in \mathcal{S}$ does not intersect with any other splinegon $S' \neq S$ and $S' \in \mathcal{S}$.

The input *splinegonal domain* $\mathcal{S}$ is comprised of h pairwise disjoint splinegons, together defined with n vertices in $\mathbb{R}^2$. The *free space* $\mathcal{F}(\mathcal{S})$ of a splinegonal domain $\mathcal{S}$ is defined as the closure of $\mathbb{R}^2$ excluding the union of the interior of splinegon obstacles in $\mathcal{S}$. Given a splinegonal domain $\mathcal{S}$ and two given points $s, t \in \mathcal{F}(\mathcal{S})$, our algorithm computes a shortest path in rectilinear metric, termed *rectilinear shortest path*, between s and t that lie in $\mathcal{F}(\mathcal{S})$. We reduce the problem of computing a rectilinear shortest path between s and t amid splinegon obstacles in $\mathcal{S}$ to the problem of computing a rectilinear shortest path between two points amid polygonal obstacles in a polygonal domain $\mathcal{P}$. Here, $\mathcal{P}$ is computed from $\mathcal{S}$. We assume points s and t are exterior to the carrier polygons of splinegons in $\mathcal{S}$. In specific, we prove that this path is a shortest one with respect to the rectilinear metric in $\mathcal{F}(\mathcal{S})$ between s and t. Analogous to $\mathcal{F}(\mathcal{S})$, the *free space* $\mathcal{F}(\mathcal{P})$ of a polygonal domain $\mathcal{P}$ is defined as the closure of $\mathbb{R}^2$ excluding the union of the interior of polygonal obstacles in $\mathcal{P}$.

Chen and Wang [6] extended the corridor structures defined for polygonal domains in [17,19] to splinegonal domains. In computing corridors, this result used bounded degree decomposition of $\mathcal{F}(\mathcal{S})$ which is analogous to the triangulation of the free space of the polygonal domain [2]. Similar to [14], using a corridor decomposition of $\mathcal{F}(\mathcal{S})$, we characterize rectilinear shortest paths in splinegonal domains. These properties facilitate in computing a connected undirected graph G_S that contains a rectilinear shortest path between s and t. We compute a set $\mathcal{P}$ of pairwise disjoint simple polygons in $\mathbb{R}^2$ such that each polygon $P \in \mathcal{P}$ corresponds to a unique splinegon $S \in \mathcal{S}$ and the vertices of P lie on the boundary of S while ensuring $s, t \in \mathcal{F}(\mathcal{P})$. The vertices of $\mathcal{P}$ are essentially the *coreset* of points on the boundaries of splinegons in $\mathcal{S}$. Further, we introduce s and t points in $\mathcal{F}(\mathcal{P})$ at the same respective coordinate locations as they are in $\mathcal{F}(\mathcal{S})$. [7,14] gave a constructive proof showing that there exists a connected undirected graph $G_{\mathcal{P}}$ that contains a rectilinear shortest path between s and t amid polygons in $\mathcal{P}$. We extended this proof to splinegons and shown that there exists a connected undirected graph G_S that contains a rectilinear shortest path between s and t amid splinegons in $\mathcal{S}$. The coreset of each $S \in \mathcal{S}$ is determined so that it ensures the graph G_S is same as $G_{\mathcal{P}}$. In the context of computing shortest paths, coresets for polygonal obstacles were defined in [1,4,15]. However, the purpose and the definitions of coresets in our context are different. We use the algorithm given in [14] to compute a rectilinear shortest path R between s and t in $\mathcal{F}(\mathcal{P})$. Further, we modify R to a rectilinear shortest path between s and t in $\mathcal{F}(\mathcal{S})$. Hence, the reduction of the problem of computing a rectilinear shortest path between s and t amid splinegon obstacles in $\mathcal{S}$ to the problem of computing a rectilinear shortest path between two points located in $\mathcal{F}(\mathcal{P})$ amid polygonal obstacles in $\mathcal{P}$. Excluding the time to triangulate $\mathcal{F}(\mathcal{S})$, our reduction algorithm takes $O(n + h \lg n)$ time.

Since the above reduction algorithm assumes that a rectilinear shortest path in $\mathcal{F}(\mathcal{P})$ is computed using the algorithm from [14], we devise another algorithm which works independent of the structures used in the algorithm to compute a rectilinear shortest path in the polygonal domain. This algorithm is applicable only when $\mathcal{S}$ is comprised of concave-in splinegon obstacles. For every concave-in simple splinegon S, we partition the boundary of S into xy-monotone pieces first. The endpoints of these monotone chains together with the points on the boundary of S that have vertical or horizontal tangents are defined as the coreset of S. These vertices are used in defining a simple polygon P corresponding to S. We show the rectilinear shortest path between s and t in the polygonal domain $\mathcal{P}$ comprising of all such polygons can be efficiently transformed to a rectilinear shortest path between s and t amid splinegons in $\mathcal{S}$. In specific, this reduction algorithm works whenever the shortest path computed in the polygonal domain is polygonal. Our reduction algorithm takes $O(n + (h + k) \lg n + (h + k + k') \lg (h + k))$ time (while excluding the time to compute a rectilinear shortest path amid polygonal obstacles in $\mathcal{P}$). Let R be the polygonal rectilinear shortest path between s and t amid polygonal obstacles in $\mathcal{P}$ that was output by the algorithm used. Then k is the number of line segments in R and k' is the number of points of intersections of that path with splinegons in $\mathcal{S}$.

We call the vertices of a graph as nodes and the vertices of a polygonal/splinegonal domain as vertices. For any splinegon S (resp. polygon P), the boundary of S (resp. P) is denoted with $bd(S)$ (resp. $bd(P)$). A shortest path between s and t amid splinegons (resp. polygons) in $\mathcal{S}$ (resp. $\mathcal{P}$) is denoted with

$SP_S(s,t)$ (resp. $SP_P(s,t)$). The rectilinear distance between s and t amid spline-gons in S (resp. P) is denoted with $dist_S(s,t)$ (resp. $dist_P(s,t)$). Unless specified otherwise, a shortest path is shortest with respect to rectilinear metric and the distance is the shortest distance in rectilinear metric.

Section 2 extends the staircase structures for the polygonal domain from [14] to splinegonal domain. The reduction algorithm for the case of concave-in spline-gon obstacles is given in Sect. 3. For the case of arbitrary splinegon obstacles, the reduction algorithm is given in Sect. 4. Conclusions are in Sect. 5.

2   Staircase Structures for Splinegonal Domain

In the context of polygonal domains, corridor and hourglass structures were first described by Kapoor et al., in [17,19]. Chen and Wang in [6] extended them to the splinegonal domain. First, we detail these structures. Later, analogous to [14], we define staircase structures for the splinegonal domain and use them in computing a visibility graph that contains a rectilinear shortest path between s and t.

For convenience, we assume that no splinegon in S has an edge that is parallel to either of the coordinate axes. In decomposing $\mathcal{F}(S)$ into corridors, points s and t are considered as two special (degenerate) splinegons in S. [6] first decomposes $\mathcal{F}(S)$ into $O(n)$ bounded degree regions by introducing $O(n)$ non-intersecting diagonals. This decomposition of $\mathcal{F}(S)$ is termed *bounded degree decomposition*, denoted with $BDD(\mathcal{F}(S))$. In $BDD(\mathcal{F}(S))$, two regions are neighboring if they share a diagonal on their boundaries. Each such region has at most four sides, and each side is either a diagonal or (part of) a splinegon edge and has at most three neighboring regions. In addition to the splinegon vertices, the endpoints of the diagonals of $BDD(\mathcal{F}(S))$ are also treated as the vertices of $BDD(\mathcal{F}(S))$. Let $G(\mathcal{F}(S))$ denote the planar dual graph of $BDD(\mathcal{F}(S))$. Since each region of $BDD(\mathcal{F}(S))$ has at most three neighbors, $G(\mathcal{F}(S))$ is a planar graph whose vertex degrees are at most three.

Based on $G(\mathcal{F}(S))$, [6] computes a planar 3-regular graph, denoted by G^3 (the degree of each node in it is three), possibly with loops and multi-edges, as follows. First, it removes every degree-one node from $G(\mathcal{F}(S))$ along with its incident edge; repeats this process until no degree-one node exists. Second, the algorithm removes every degree-two node from $G(\mathcal{F}(S))$ and replaces its two incident edges by a single edge; it repeats this process until no degree-two node exists. The number of faces, nodes, and edges in the resulting graph G^3 is proved to be $O(h)$. Each node of G^3 corresponds to a region

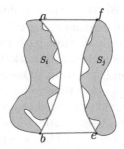

Fig. 1. Illustrating an open hourglass (blue). (Color figure online)

of $BDD(\mathcal{F}(S))$, which is called a *junction*. Removal of all junctions from G^3 results in $O(h)$ corridors, each of which corresponds to one edge of G^3.

The boundary of each *corridor* C consists of four parts (refer Figs. 1, 2): (1) A boundary portion of a splinegon obstacle $S_i \in \mathcal{S}$, from a point a to a point b, (2) a diagonal of a junction triangle from b to a point e on an obstacle $S_j \in \mathcal{S}$ ($S_i = S_j$ is possible), (3) a boundary portion of the obstacle S_j from e to a point f, and (4) a diagonal of a junction triangle from f to a.

Let $\pi(a,b)$ (resp., $\pi(e,f)$) be the Euclidean shortest path from a to b (resp., e to f) in C. The region H_C bounded by $\pi(a,b), \pi(e,f), \overline{be}$, and $\overline{fa}$ is called an *hourglass*, which is *open* if $\pi(a,b) \cap \pi(e,f) = \emptyset$ and *closed* otherwise. If H_C is open (refer Fig. 1.), then both $\pi(a,b)$ and $\pi(e,f)$ are convex chains and are called the *sides* of H_C. Otherwise, H_C consists of two *funnels* and a path $\pi_C = \pi(a,b) \cap \pi(e,f)$ joining the two apices of the two funnels, and π_C is called the *corridor path* of C (refer Fig. 2). The paths $\pi(b,x), \pi(e,x), \pi(a,y)$, and $\pi(f,y)$ are termed the *sides of funnels* of hourglass H_C. The sides of the funnels are convex chains.

Following [14], we define the staircase structures for the given splinegonal domain $\mathcal{S}$ next. The set of vertices V_{ortho} is defined such that $v \in V_{\text{ortho}}$ if and only if either of the following is true: (i) v is an endpoint of a corridor convex chain; (ii) v is a point on some corridor convex chain CC, with the property that there exists either a horizontal or a vertical tangent to CC at v.

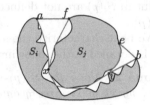

Fig. 2. Illustrating a closed hourglass (blue). (Color figure online)

Let $\mathcal{O}(p)$ be the orthogonal coordinate system defined with $p \in V_{\text{ortho}}$ as the origin, and horizontal x-axis and vertical y-axis passing through p. We next adopt and redefine the staircase structures from [7,14]. For $i \in \{1,2,3,4\}$, we define a set of points $\pi_i(p)$ as follows: a point $r \in \pi_i(p)$ if and only if $r \in V_{\text{ortho}}$ and r is located in the ith quadrant of $\mathcal{O}(p)$.

Further, we define a set of points $S_i(p)$ as follows: a point q is in the set $S_i(p)$ (refer Fig. 3) if and only if (i) $q \in \pi_i(p)$; (ii) there is no p' (distinct from p) such that p' is in $\pi_i(p)$ and q is in $\pi_i(p')$; and (iii) q is visible from p. We assume that the points in $S_i(p)$ are sorted in increasing x-order.

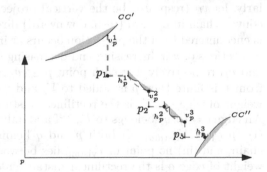

The proofs of the lemmas and theorems stated in this section are similar to the ones provided for the polygonal domain in [14].

Fig. 3. Illustrating staircase structure with $S_1(p) = p_1, p_2, p_3$.

Lemma 1. *Sorting the set of points in $S_1(p)$ in increasing x-order results in the same set of points being sorted in decreasing y-order (or, vice versa).*

We term two points $\{p_u, p_v\} \subseteq S_i(p)$ as *adjacent* in $S_i(p)$ if no point $p_l \in S_i(p)$ occurs between p_u and p_v when the points in $S_i(p)$ are ordered by either the x- or y-coordinates. Let $p_1, p_2, \ldots, p_k$ be the points in $S_1(p)$ in increasing x-order. Let h_j be the rightward horizontal ray from p_j. And, let v_j be the upward vertical ray from p_j. The ray h_j intersects either a corridor convex chain or v_{j+1}. Let this point of intersection be h_j^p. The ray v_j first intersects either a corridor convex chain or h_{j-1}. Let this point of intersection be v_j^p. If the ray does not intersect any other line or line segment then the point h_j^p or v_j^p could be at infinity. Let R_j ($j \in \{1, \ldots, k\}$) denote the unique sequence of sections of corridor convex chains/bounding edges that join h_j^p and v_{j+1}^p. As will be proved, R_j is continuous. Note that for the case in which $h_j^p = v_{j+1}^p$, R_j is empty. The elements in the set $\bigcup_{\forall j \in \{1,2,\ldots,k\}} (v_j \cup h_j \cup R_j)$ form a contiguous sequence, termed as the $S_1(p)$-*staircase* (refer Fig. 3). Analogously, $S_i(p)$ for $i \in \{2, 3, 4\}$ are defined. Note that the convex chains which may intersect the coordinate axes and do not contain a point in $S_i(p)$ are not defined to be part of the staircases in the i-th quadrant of $\mathcal{O}(p)$.

We next characterize the structure of a staircase in the splinegonal domain. This is detailed in the following theorem (whose proof is detailed in full version).

Lemma 2. *Along the $S_1(p)$-staircase, any two adjacent points in $S_1(p)$ are joined by at most three geometric entities. These entities ordered by increasing x-coordinates are: (i) a horizontal line segment, (ii) a section of a convex chain where the tangent to each point in that section of splinegon has a negative slope, and (iii) a vertical line segment.*

We now define the weighted restricted visibility graph $G_{\text{vistmp}}(V_{\text{vistmp}} = V_{\text{ortho}} \cup V_1, E_{\text{vistmp}} = E_{\text{occ}} \cup E_1 \cup E_{\text{tmp}})$:

- For each $v \in V_{\text{ortho}}$, let v_L (resp. v_R) be the horizontal projection of v onto the first corridor convex chain in leftward (resp. rightward) direction. Similarly, let v_U (resp. v_D) be the vertical projection of v onto the first corridor convex chain in upward (resp. downward) direction. If no such corridor chain is encountered then the projection occurs at infinity. Similarly, let the vertical projections of v in increasing and decreasing direction of y-coordinates be v_U and v_D respectively. For each point $p \in \{v_L, v_R, v_D, v_U\}$, if the distance of p from v is finite then p is added to V_1 and the edge pv is added to E_1. The weight of edge $e \in E_1$ is the rectilinear distance between its two endpoints.
- An edge $e = (p, q)$ belongs to E_{occ} if and only if the following conditions hold (i) $\{p, q\} \subseteq V_{\text{vistmp}}$, (ii) both p and q belong to the same corridor convex chain, and (iii) no point in V_{vistmp} lies between p and q along the chain. The weight of edge e is the rectilinear distance along the section of convex chain between p and q.
- An edge $e' = (p', q')$ with $p' \in V_{\text{ortho}}$ belongs to E_{tmp} if and only if $q' \in S_i(p')$. The weight of e' is the rectilinear distance along e'.

Theorem 1. *Let $\{p, q\} \subseteq V_{\text{vistmp}}$. Then a shortest path from p to q in G_{vistmp} defines a shortest path in L_1 metric from p to q that does not intersect any of the splinegon obstacles in S.*

3   Reduction for Concave-In Splinegons

In this section, we devise an algorithm to find a shortest path in rectilinear metric when $\mathcal{S}$ is comprised of concave-in splinegons. Our algorithm reduces this problem to the problem of computing a shortest path in rectilinear metric amid simple polygonal obstacles. This is accomplished by computing a polygonal domain $\mathcal{P}$ comprising of h simple polygonal obstacles from h simple concave-in splinegonal obstacles in $\mathcal{S}$.

For each splinegon $S \in \mathcal{S}$ that has n' vertices, we compute its corresponding simple polygonal obstacle $P \in \mathcal{P}$ with $O(n')$ vertices. Further, we introduce points s and t in $\mathcal{F}(\mathcal{P})$ at the same respective coordinate locations as they are in $\mathcal{F}(\mathcal{S})$.

Fig. 4. Illustrating carrier polygon (dashed) and the polygon constructed (brown). (Color figure online)

As mentioned, for computing shortest paths, coresets for polygonal obstacles were defined in [1,4,15]. Here for every splinegon $S \in \mathcal{S}$, we define the *coreset* of points on the $bd(S)$, suiting to our reduction. These points define the vertices of a simple polygon $P \in \mathcal{P}$ that correspond to $S \in \mathcal{S}$. First, we define the coreset for every splinegon S as described herewith: every vertex of S is a vertex of P; for every point $p \in bd(S)$, if tangent to S at p is either horizontal or vertical then p is a vertex of P. Apart from these two sets of points, no additional point is a vertex of P. Let V_P be the coreset of vertices of P. For every two successive vertices $v', v'' \in V_P$ that occur successively while traversing $bd(S)$, we add an edge between v' and v'' to obtain polygon P (refer Fig. 4). Since every $S \in \mathcal{S}$ is a simple concave-in splinegon and from the way the coreset of every $S \in \mathcal{S}$ is defined, it is immediate that every $P \in \mathcal{P}$ is a simple polygon.

Lemma 3. *If $Q \in \mathcal{F}(\mathcal{S})$ is a shortest path between s and t in $\mathcal{F}(\mathcal{P})$, then Q is a shortest path between s and t in $\mathcal{F}(\mathcal{S})$.*

Proof: Suppose there exists a path $Q' \in \mathcal{F}(\mathcal{S})$ between s and t whose length is less than the distance along Q. Since Q' belongs to $\mathcal{F}(\mathcal{S})$ and since $\mathcal{F}(\mathcal{P}) \subseteq \mathcal{F}(\mathcal{S})$, Q' belongs to $\mathcal{F}(\mathcal{P})$ as well. □

Lemma 4. *There exists a rectilinear shortest path Q between s and t amid polygonal obstacles in $\mathcal{P}$ such that no point of Q belongs to any of the open S-seg regions of splinegons in $\mathcal{S}$.*

Proof: Let Q be a shortest path that enters $S\text{-}seg_i$ region at a point p_1 and exits it at p_2. (Refer to Fig. 5). Let s_i be the section of spline to which p_1 and p_2 belong. Also, let s_i bounds a side of $S\text{-}seg_i$. Since we have introduced vertices of P at every point on the boundary of splinegon where there is a horizontal and/or vertical tangent to splinegon, s_i is guaranteed to be xy-monotone. Hence, replacing the simple path Q from p_1 to p_2 with the section of s_i from p_1 to p_2 does not increase the length of Q. □

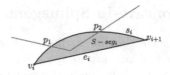

Fig. 5. Illustrating the case of a path entering S-seg_i region.

Lemma 5. *The $dist_\mathcal{P}(s,t)$ is equal to the $dist_\mathcal{S}(s,t)$.*

Proof: There are two cases to consider. Suppose $SP_\mathcal{P}(s,t)$ does not intersect with any of the open S-seg regions. In this case, since $\mathcal{F}(\mathcal{S}) \subseteq \mathcal{F}(\mathcal{P})$, $SP_\mathcal{P}(s,t)$ is a shortest path amid splinegons in $\mathcal{S}$. On the other hand, suppose that $SP_\mathcal{P}(s,t)$ does intersect with an open S-seg_i region. From the proof of Lemma 4, we know that there exists a path between s and t that avoids the open S-seg_i region. Let Q be the path between s and t resultant of all such sub-path replacements in $SP_\mathcal{P}(s,t)$. The rectilinear distance along Q is same as the rectilinear distance along $SP_\mathcal{P}(s,t)$. Further, Q belongs to $\mathcal{F}(\mathcal{S})$. Hence, due to Lemma 3, Q is a shortest path between s and t amid splinegons in $\mathcal{S}$.   □

Lemma 6. *Computing $\mathcal{P}$ from $\mathcal{S}$ takes $O(n)$ time.*

Proof: To find a set T comprising of points on the boundary of a spline S such that every point $p \in T$ has either a horizontal or a vertical tangent to the spline on which p resides, considering our model of computation, it takes $O(|T|)$ time. Since there are n edges in $\mathcal{S}$ and each edge has $O(1)$ points that belong to T, there are $O(n)$ vertices that define $\mathcal{P}$. Including the cost of traversal of each spline to compute polygons in $\mathcal{P}$, reduction algorithm takes $O(n)$ time to compute $\mathcal{P}$.   □

To transform $SP_\mathcal{P}(s,t)$ to $SP_\mathcal{S}(s,t)$, a plane sweep algorithm is used to find the points of intersection of the $SP_\mathcal{P}(s,t)$ with the splinegons in $\mathcal{S}$. We sort the endpoints of the line segments in $SP_\mathcal{P}(s,t)$ with respect to their y-coordinates. For every splinegon $S_i \in \mathcal{S}$, let S_i^{max} (resp. S_i^{min}) be a point on the $bd(S_i)$ that has the largest (resp. smallest) y-coordinate among all the points of S_i. We sort all the points in the set T comprising of $\bigcup_i (S_i^{max} \cup S_i^{min})$. We use balanced binary search trees to respectively implement the event queue and the status structure needed for the plane sweep. The left to right order of the segments along the sweep line corresponds to the left to right order of the leaves in the balanced binary search tree (status structure). We sweep the plane with a horizontal line from the point that has the maximum coordinate in T to the point that has the minimum coordinate in T. Let p be an endpoint of the line segment $e \in SP_\mathcal{P}(s,t)$. When p is encountered by the sweep line, we check if there is a splinegon, say S, immediately to the right or left of the edge e in the status structure; if S exists, we find the points of intersection of e with the S using the algorithm given in [9].

Lemma 7. *The plane sweep algorithm involved in transforming a polygonal path in $\mathcal{F}(\mathcal{P})$ to $\mathcal{F}(\mathcal{S})$ takes $O((h+k)\lg n + (h+k+k')\lg(h+k))$ time. Here, h is the number of obstacles, k is the number of line segments in $SP_\mathcal{P}(s,t)$, and k' is the number of intersection points of $SP_\mathcal{P}(s,t)$ with the boundaries of splinegons in $\mathcal{S}$.*

Proof: If there is a splinegon S immediately to the left or right of a line segment l of $SP_\mathcal{P}(s,t)$, then we can find the intersection of l with S in $O(\lg n')$ time using the algorithm given in [9], where n' is the number of vertices of S. Computing and sorting the event points take $O((h+k)\lg(h+k))$ time. We check whether a line segment of $SP_\mathcal{P}(s,t)$ intersects a splinegon when the sweep line reaches endpoints of segments of $SP_\mathcal{P}(s,t)$ or when it encounters points that belong to T; and the number of these event points is $O(h+k)$. Since to check the points of intersection at each event point requires $O(\lg n)$ time, the total time needed to find the points of intersection at event points take $O((h+k)\lg n)$ time. We update the status structure whenever the sweep line encounters either of these points: intersection points of line segments of $SP_\mathcal{P}(s,t)$ with splinegons in $\mathcal{S}$; points belonging to T; endpoints of line segments of $SP_\mathcal{P}(s,t)$. Considering that updating the status structure per one such event takes $O(\lg(h+k))$ time, the time required for all updates together is $O((h+k)\lg n + (h+k+k')\lg(h+k))$. $\square$

Theorem 2. *Given a splinegonal domain $\mathcal{S}$ comprising of h pairwise disjoint simple concave-in splinegons together defined with n vertices and two points $s,t \in \mathcal{F}(\mathcal{S})$, the reduction procedure to compute a rectilinear shortest path between s and t amid splinegons in $\mathcal{S}$, excluding the time to compute a rectilinear shortest path amid polygons in $\mathcal{P}$, takes $O(n + (h+k)\lg n + (h+k+k')\lg(h+k))$ time. Here, $\mathcal{P}$ is the computed polygonal domain from $\mathcal{S}$, k is the number of line segments in the polygonal shortest path $SP_\mathcal{P}(s,t)$ between s and t amid polygonal obstacles in $\mathcal{P}$, and k' is the number of points of intersections of $SP_\mathcal{P}(s,t)$ with the boundaries of splinegons in $\mathcal{S}$.*

Proof: Immediate from Lemmas 5, 6 and 7. $\square$

4   Reduction for Simple Splinegons

We first describe the algorithm when the decomposition of $\mathcal{F}(\mathcal{S})$ has only open corridors. Later we extend this algorithm to handle closed corridors.

In the following, similar to Sect. 3, we define a coreset $V_\mathcal{P}$ of points on the boundaries of splinegons in $\mathcal{S}$; these are used in defining the polygonal domain $\mathcal{P}$. For every edge s of a splinegon $S \in \mathcal{S}$, the endpoints of s belong to $V_\mathcal{P}$; further if a tangent to s at a point $p \in S$ is either horizontal or vertical and p does not lie inside the convex hull of the carrier polygon of S, then p is included into $V_\mathcal{P}$. Also, for every side s' of every open hourglass H_C in the splinegonal domain $\mathcal{S}$, the endpoints of s belong to $V_\mathcal{P}$. Further, for every vertex $v \in V_{ortho}$ in $\mathcal{S}$ (see Sect. 2 for the definition of V_{ortho}), the horizontal and vertical projections of v onto sides of hourglasses are added to $\mathcal{P}$. For every splinegon $S \in \mathcal{S}$, for any

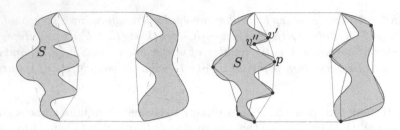

Fig. 6. Illustrating two splinegons whose sections of boundary belong to an open corridor: splinegons are in black, carrier polygon is red (left), hourglass in blue, and polygons belonging to $\mathcal{P}$ are in brown (right). (Color figure online)

two vertices $v', v'' \in V_{\mathcal{P}}$ that occur successively while traversing the $bd(S)$, we add an edge between v' and v'' to obtain $P \in \mathcal{P}$ corresponding to S. We also introduce points s and t in $\mathcal{F}(\mathcal{P})$ at the same coordinate locations as they are in $\mathcal{F}(\mathcal{S})$.

Let $\mathcal{S}'$ be the set comprising of convex hulls corresponding to each of the carrier polygons of splinegons in $\mathcal{S}$. Since carrier polygons are simple polygons and since no point in $V_{\mathcal{P}}$ belongs to the interior of any of the convex hulls in $\mathcal{S}'$, every polygon in $\mathcal{P}$ is guaranteed to be a simple polygon. (Refer to Fig. 6.)

Lemma 8. *If $Q \in \mathcal{F}(\mathcal{P})$ is a rectilinear shortest path between s and t amid polygons in $\mathcal{P}$ computed using the algorithm given in [14], then Q is a shortest path between s and t amid splinegons in $\mathcal{S}$.*

Proof: Consider the graph $G_{\mathcal{P}}$ from which a rectilinear shortest path amid simple polygons in $\mathcal{P}$ between s and t is computed in [14]. Let $G_{\mathcal{S}}$ be the graph corresponding to $\mathcal{S}$, as defined in Sect. 2. We prove that $G_{\mathcal{P}}$ is same as $G_{\mathcal{S}}$. Analogous to V_{ortho} and V_1 defined for the splinegonal domain (Sect. 2), as in [14], we define V'_{ortho} and V'_1 for the polygonal domain $\mathcal{P}$. Let $V = V_{ortho} \cup V_1$ be the vertex set of $G_{\mathcal{S}}$ and $V' = V'_{ortho} \cup V'_1$ be the vertex set of $G_{\mathcal{P}}$. We prove that a vertex belongs to V if and only if it belongs to V'. Suppose $v \in V'_{ortho}$ but v does not belong to V_{ortho}. Then it must be the case that v is hidden by a convex chain ab in the splinegonal domain. Since a and b are endpoints of an hourglass side in the decomposition of $\mathcal{F}(\mathcal{S})$, these two are vertices of polygons in $\mathcal{P}$. Suppose v is an endpoint of an hourglass side in $\mathcal{P}$. Then this would lead to a contradiction as we could extend the convex chain to a or b in $\mathcal{P}$. Suppose v is residing on an hourglass side ab in $\mathcal{P}$ but not an endpoint of the hourglass. Since the hourglass side is the shortest path between a and b in $\mathcal{P}$ and since every vertex of the chain lies on the boundary of a splinegon, v being hidden by the convex chain ab would contradict the fact that the chain from a to b is the shortest path between a and b in $\mathcal{P}$. Thus v lies on a convex chain in $\mathcal{S}$, and v does not lie inside the convex hull of the carrier polygon as v is part of the shortest path between a and b. Since there is a horizontal (resp. vertical) tangent to v in $\mathcal{P}$, there exists a horizontal (resp. vertical) tangent at v to a splinegon in $\mathcal{S}$. This contradicts our assumption that v does not belong to V_{ortho}, therefore if

$v \in V'_{ortho}$ then $v \in V_{ortho}$. Analogously we can prove the converse. By the way we defined V'_1, it is immediate to note that a vertex $v \in V_1$ if and only if $v \in V'_1$.

Now we show that for every edge $e' \in G_{\mathcal{P}}$ we introduce a corresponding edge $e \in G_{\mathcal{S}}$ such that the weights of the corresponding edges are same. Let e be an edge in G_s of length l. Also let p and q be the endpoints of e. We prove that there is a path of length l between p and q amid polygonal obstacles in $\mathcal{P}$ as well. The definitions of E_1, E_{occ} and E_{tmp} are given in Sect. 2.

- Suppose $e \in E_1$. Here l is the rectilinear distance between $p \in V_{ortho}$ and $q \in V_1$. Since we had proven that if a vertex belongs to V_{ortho} (resp. V_1) in $\mathcal{S}$ then the vertex also belongs to V'_{ortho} (resp. V'_1) in $\mathcal{P}$. Thus the rectilinear distance between p and q will be same in both $\mathcal{F}(\mathcal{P})$ as well as in $\mathcal{F}(\mathcal{S})$.
- Suppose $e \in E_{occ}$. Here l is the rectilinear distance along the (splinegonal) convex chain between p and q, where p and q are the consecutive points on the side of an hourglass obtained due to the decomposition of $\mathcal{F}(\mathcal{S})$. Since every section of convex chain in the splinegon domain is xy monotone, the rectilinear distance between p and q in $\mathcal{F}(\mathcal{P})$ equals to the rectilinear distance between p and q in $\mathcal{F}(\mathcal{S})$.
- Suppose $e \in E_{tmp}$. Here l is the rectilinear distance between $p \in V_{ortho}$ and let $q \in S_1(p)$. We prove that if $q \in S_1(p)$ in the splinegonal domain then $q \in S'_1(p)$ in the polygonal domain as well. Suppose q does not belong to $S'_1(p)$. Then q is not visible from p amid polygons in $\mathcal{P}$. This means that a convex chain of an open hourglass of the decomposition of $\mathcal{F}(\mathcal{S})$ intersects the line segment pq. However, since the convex chain in the polygonal domain is always bounded by a convex chain in $\mathcal{S}$, this would imply pq is intersected by a convex chain in the splinegonal domain as well.

Therefore, if Q is a shortest s-t path obtained from $G_{\mathcal{P}}$ then it is also the shortest path from s to t in $G_{\mathcal{S}}$. This together with Theorem 1 leads to conclude that Q is also a shortest path amid splinegons in $\mathcal{S}$.   □

To find the horizontal and vertical projections of points in V_{ortho}, we extend the plane sweep algorithm from [14] to splinegons. Essentially, we sweep a vertical line from left to right to find the horizontal rightward projections of every $v \in V_{ortho}$. The status of the vertical sweep line is maintained as a set of points in V_{ortho} that lie on the sweep line, sorted by their y-coordinates. Let p be the first point of a convex chain CC struck by sweep line and let r be a point in the sweep-line status structure at the time the sweep line encounters p. If p projects onto CC at p' then p' is a projection of p. After finding p', we remove p from the sweep-line status structure. Analogously, projections of points in V_{ortho} are determined.

Lemma 9. *Given the open corridor decomposition of $\mathcal{F}(\mathcal{S})$, computing a polygonal domain $\mathcal{P}$ from $\mathcal{S}$, so that the $dist_{\mathcal{P}}(s,t)$ is equal to the $dist_{\mathcal{S}}(s,t)$ for two given points $s, t \in \mathcal{F}(\mathcal{S})$, takes $O(n + h \lg n)$ time.*

Proof: Since there are n edges in $\mathcal{S}$ and we are adding a constant number of points to each edge, computing $\mathcal{P}$ from $\mathcal{S}$ takes $O(n)$ time. Each vertex in V_{ortho} is inserted into (resp. deleted from) sweep line data structures' only once, together taking $O(h \lg h)$ time. With binary search, the intersection of a horizontal (resp. vertical) line from a point with a convex chain can be found in $O(\lg n)$ time. Hence all the points of projections can be computed as stated.   □

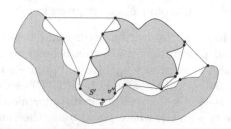

Fig. 7. Illustrating sections of boundaries of two splinegons participating in a closed corridor and their carrier polygon (red) and sections of polygons computed (brown). The point v' is introduced into the coreset as it belongs to V_{ortho}. The point v'' is introduced into the coreset as it is an endpoint of a contiguous maximal section S' of a spline that belongs to the shortest path between apieces of the closed corridor shown. (Color figure online)

Now we extend this algorithm to handle closed corridors. For each side of every funnel, very similar to sides of open hourglasses, we introduce points into $V_\mathcal{P}$; these include projections of points V_{ortho} onto sides of funnels. Let Q be the rectilinear shortest path between apices of a closed corridor in $\mathcal{S}$. For every contiguous maximal section S' of every spline that belongs to Q, we add the endpoints of S' to $V_\mathcal{P}$. Further, any point in V_{ortho} that belongs to S' is also included into $V_\mathcal{P}$. For every two vertices $v', v'' \in V_\mathcal{P}$ that occur successively along the boundary of a splinegon, $v'v''$ is introduced as an edge of a polygon in $\mathcal{P}$. (Refer to Fig. 7). Note that for any two splinegons that participate in a closed corridor, their corresponding polygons in $\mathcal{P}$ are guaranteed to be disjoint.

Lemma 10. *If a and b are the apices of a closed corridor and the shortest distance between them is d in $\mathcal{F}(\mathcal{S})$ then the shortest distance between a and b in $\mathcal{F}(\mathcal{P})$ is d.*

Proof: Suppose there is a path Q in $\mathcal{F}(\mathcal{P})$ whose length is less than d. Now Q must intersect with a splinegon S. Let v' and v'' be the successive points of intersection of Q with S. But Q can only intersect with an edge of splinegon which is outside the convex hull of its corresponding carrier polygon. Since each of these edges is partitioned into xy-monotone pieces, we can replace the section of path Q between v' and v'' with the section of spline edge between v' and v''. The modified path has length less than d as well.   □

Lemma 11. *If a path Q in $\mathcal{F}(\mathcal{S})$ is a shortest path between s and t amid polygons in $\mathcal{F}(\mathcal{P})$ then Q is a shortest path between s and t amid splinegons in $\mathcal{S}$.*

Proof: The edges that occur in G_S due to any closed corridor C are those that join two apices corresponding to C. Let R be the sequence of edges of G_S that connect the apieces of C. A The length of R equals the rectilinear distance between apices of C in C. This together with the Lemma 10 and the proof of Lemma 8 proves that the graph G_S corresponding to S is same as the graph G_P corresponding to P. □

Lemma 12. *Computing a polygonal domain P from the given splinegonal domain S takes $O(n + h \lg n + (\lg h)^{1+\epsilon})$ time. Here, ϵ is a small positive constant.*

Proof: Triangulating $\mathcal{F}(S)$ and partitioning the same into hourglasses takes $O(n + h((\lg n) + (\lg h)^{1+\epsilon})$ time. For the triangulation of $\mathcal{F}(S)$, we use the algorithm from [2]. Since funnels are processed analogous to open hourglasses and because of Lemma 9, computing polygonal chains corresponding to sides of open corridors and funnels together take $O(n + h \lg n)$ time. Computing a shortest path between apices of any closed corridor C in $\mathcal{F}(S)$ takes $O(k)$ time [8,20], where k is the number of vertices that belong to that closed corridor. Further, traversing along a shortest path Q between two apices of C and introducing vertices of P along Q takes $O(k)$ time. □

Theorem 3. *Given a splinegonal domain S comprising of h pairwise disjoint simple splinegons together defined with n vertices and two points $s, t \in \mathcal{F}(S)$, the reduction procedure to compute a rectilinear shortest path between s and t amid splinegons in S, excluding the time to compute a rectilinear shortest path amid polygons in P, takes $O(n + h \lg n + (\lg h)^{1+\epsilon})$ time. Here, P is the polygonal domain computed from S and ϵ is a small positive constant (resulting from the triangulation of $\mathcal{F}(S)$ using [2]).*

Proof: Immediate from Lemmas 8, 11 and 12. □

5   Conclusions

We have devised an algorithm to reduce the problem of computing a rectilinear shortest path between two points in the splinegonal domain to the problem of computing a rectilinear shortest path between two points in the polygonal domain. The reduction algorithm given for the case of concave-in simple splinegon obstacles does not rely on the details of the algorithm to compute a rectilinear shortest path between two points amid polygonal obstacles. Further, as part of this, we have generalized few of the properties given for rectilinear shortest paths in the polygonal domain to the case of rectilinear shortest paths amid splinegons. It would be interesting to devise rectilinear shortest path algorithms when the obstacles in the plane are more generic.

References

1. Agarwal, P.K., Sharathkumar, R., Yu, H.: Approximate Euclidean shortest paths amid convex obstacles. In: Proceedings of Symposium on Discrete Algorithms, pp. 283–292 (2009)

182   T. Choudhury and R. Inkulu

2. Bar-Yehuda, R., Chazelle, B.: Triangulating disjoint Jordan chains. Int. J. Comput. Geom. Appl. **4**(4), 475–481 (1994)
3. Chen, D.Z., Inkulu, R., Wang, H.: Two-point L_1 shortest path queries in the plane. J. Comput. Geom. **7**(1), 473–519 (2016)
4. Chen, D.Z., Wang, H.: A nearly optimal algorithm for finding L_1 shortest paths among polygonal obstacles in the plane. In: Demetrescu, C., Halldórsson, M.M. (eds.) ESA 2011. LNCS, vol. 6942, pp. 481–492. Springer, Heidelberg (2011). https://doi.org/10.1007/978-3-642-23719-5_41
5. Chen, D.Z., Wang, H.L.: Computing shortest paths among curved obstacles in the plane. In: Proceedings of Symposuim on Computational Geometry, pp. 369–378 (2013)
6. Chen, D.Z., Wang, H.: Computing shortest paths among curved obstacles in the plane. ACM Trans. Algorithms **11**(4), 26:1–26:46 (2015)
7. Clarkson, K.L., Kapoor, S., Vaidya, P.M.: Rectilinear shortest paths through polygonal obstacles in $O(n(\lg n)^2)$ time. In: Proceedings of Symposium on Computational Geometry, pp. 251–257 (1987)
8. Dobkin, D.P., Souvaine, D.L.: Computational geometry in a curved world. Algorithmica **5**(3), 421–457 (1990)
9. Dobkin, D.P., Souvaine, D.L., Van Wyk, C.J.: Decomposition and intersection of simple splinegons. Algorithmica **3**, 473–485 (1988)
10. Ghosh, S.K.: Visibility Algorithms in the Plane. Cambridge University Press, New York (2007)
11. Ghosh, S.K., Mount, D.M.: An output-sensitive algorithm for computing visibility graphs. SIAM J. Comput. **20**(5), 888–910 (1991)
12. Hershberger, J., Suri, S.: An optimal algorithm for Euclidean shortest paths in the plane. SIAM J. Comput. **28**(6), 2215–2256 (1999)
13. Hershberger, J., Suri, S., Yildiz, H.: A near-optimal algorithm for shortest paths among curved obstacles in the plane. In: Proceedings of Symposuim on Computational Geometry, pp. 359–368 (2013)
14. Inkulu, R., Kapoor, S.: Planar rectilinear shortest path computation using corridors. Comput. Geom. **42**(9), 873–884 (2009)
15. Inkulu, R., Kapoor, S.: Approximate Euclidean shortest paths amid polygonal obstacles. CoRR, abs/1506.01769 (2015)
16. Inkulu, R., Kapoor, S., Maheshwari, S.N.: A near optimal algorithm for finding Euclidean shortest path in polygonal domain. CoRR, abs/1011.6481 (2010)
17. Kapoor, S., Maheshwari, S.N.: Efficent algorithms for Euclidean shortest path and visibility problems with polygonal obstacles. In: Proceedings of Symposium on Computational Geometry, pp. 172–182 (1988)
18. Kapoor, S., Maheshwari, S.N.: Efficiently constructing the visibility graph of a simple polygon with obstacles. SIAM J. Comput. **30**(3), 847–871 (2000)
19. Kapoor, S., Maheshwari, S.N., Mitchell, J.S.B.: An efficient algorithm for Euclidean shortest paths among polygonal obstacles in the plane. Discrete Comput. Geom. **18**(4), 377–383 (1997)
20. Melissaratos, E.A., Souvaine, D.L.: Shortest paths help solve geometric optimization problems in planar regions. SIAM J. Comput. **21**(4), 601–638 (1992)
21. Mitchell, J.S.B.: Shortest paths among obstacles in the plane. Int. J. Comput. Geom. Appl. **6**(3), 309–332 (1996)
22. Rohnert, H.: Shortest paths in the plane with convex polygonal obstacles. Inf. Process. Lett. **23**(2), 71–76 (1986)
23. Storer, J.A., Reif, J.H.: Shortest paths in the plane with polygonal obstacles. J. ACM **41**(5), 982–1012 (1994)

Graph Problems with Obligations

Alexis Cornet and Christian Laforest[✉]

LIMOS (UMR CNRS 6158), Université Clermont-Auvergne,
Clermont-Ferrand, France
cornet.alexis.j@gmail.com, christian.laforest@isima.fr

Abstract. In this paper we study variants of well-known graph problems: *vertex cover, connected vertex cover, dominating set, total dominating set, independent dominating set, spanning tree, connected minimum weighted spanning graph, matching* and *hamiltonian path*. Given a graph $G = (V, E)$, we add a partition Π_V (resp. Π_E) of its vertices (resp. of its edges). Now, any solution S containing an element (vertex or edge) of a part of this partition must also contain all the others ones. In other words, elements can only be added set by set, instead of one by one as in the classical situation (corresponding to obligations that are singletons). A motivation is to give a general framework and to study the complexity of combinatorial problems coming from systems where elements are interdependent. We propose hardness and approximation results.

Keywords: Graph problems · Approximation algorithms · Hardness

1 Obligations

Systems (production, distribution, network,...) are composed of elements (factories, vehicles, softwares, nodes, links, people...) and must supply outputs (services or goods). These elements are linked (to communicate, to exchange materials,...) and these links form a network modeled as a graph $G = (V, E)$. For the production of outputs or to manage the network, elements must work to complete a task and must be organized. For example, a spanning tree can be useful (to broadcast pieces of information), or a vertex cover (to monitor the links of G) or a dominating set (to monitor the elements). But, in some situations, some sets of these elements must be simultaneously active. This is the case for example when the treatment of a task involves a tool that is distributed on several nodes and to use one of these nodes, all the other ones must also be active. Another case is when nodes are people that are member of teams: if one member of a given team is mobilized for the task then all the other members are also mobilized.

We can model this interdependence between two elements a and b as follows: element a is active (or selected for the task) if and only if element b is active (or selected). We write this dependence $<a,b>$, or equivalently $<b,a>$. However, by its nature, this relation $<.,.>$ is transitive (if $<a,b>$ and $<b,c>$ then we necessarily get $<a,c>$) and reflexive (we have $<a,a>$ for any element a). $<.,.>$

© Springer Nature Switzerland AG 2018
D. Kim et al. (Eds.): COCOA 2018, LNCS 11346, pp. 183–197, 2018.
https://doi.org/10.1007/978-3-030-04651-4_13

is then an equivalence relation and it creates a *partition* of the elements, where all the parts are called *obligations* in this article. This means that when an element x is involved, *all* the elements in relation with x in the transitive closure of $<.,.>$ are also involved. Note that if an element y is involved in *no* $<.,.>$ relation (except with itself), then it is alone in its obligation (singleton $\{y\}$).

We do not address here any specific practical problem but we give a general framework and we treat the underlying combinatorial optimisation problems. Hence, in this paper we deal with classical graph problems with additional constraints. Let $G = (V, E)$ be any undirected graph. We call *system of obligations on vertices of G* a partition $\Pi_V = V_1, \ldots, V_k$ of V and a *system of obligations on edges of G* a partition $\Pi_E = E_1, \ldots, E_k$ of E. Each element V_i (resp. E_i) is called a *part* (or *obligation*) of Π_V (resp. Π_E). Now, given G and an associated system Π_V (resp. Π_E) of obligations on vertices (resp. edges), any solution S to a problem on G must *respect (or satisfy) the (constraints on) obligations*, that is must have the following property: if $u \in S$ (resp. $e \in S$) and $u \in V_i$ (resp. $e \in E_i$) then V_i (resp. E_i) must be entirely included in S, that is $V_i \subseteq S$ (resp. $E_i \subseteq S$). In other words, once an "object" x (vertex or edge), element of a part X, is in a solution, *all* the others elements of X *must* also be included in the solution. As mentioned at the beginning, obligations can be useful to model situations in which some set of elements (captors, computers, softwares, people, etc.) are interdependent and the presence of one element induces the presence of *all* the other ones. From an algorithmic point of view, it is clear that introducing obligations constraints in a classical graph problem $\mathcal{PROB}$ leads to a direct generalization of $\mathcal{PROB}$ (where obligations are all singletons). But we will see that in most cases the problems with obligations become much harder than the original ones.

In addition to the motivations mentioned above, this study comes to complete many recent works on a sort of opposite problem, implying what is called *conflit* which is a pair $\{x, y\}$ of edges or vertices of a graph that *cannot* be both in a solution (x and y are incompatible). Here an instance is then a graph G and a set of conflicts. Obtaining a solution without conflict is hard in general for many graph problems, as it is shown in these papers [4–6, 9–15, 17].

In what follows we give useful notations for the rest of the paper (undefined terms can be found in [7] for example). Let $G = (V, E)$ be any non directed graph, with V its set of vertices and E its set of edges. Two vertices u and v are *neighbors* if G contains the edge uv. The *degree* of a vertex u is its number of neighbors. We call *graph induced* by a set of edges $E_i \subseteq E$, the graph whose set of edges is E_i and whose vertices are the ones that are at the extremity of at least an edge of E_i. The graph induced by a set S of vertices of G, noted $G[S]$, is the graph whose set of vertices is S and whose edges are the ones of G connecting two vertices of S. A *stable* (or *independent*) S of G is a subset of its vertices having the property that $G[S]$ contains no edge. In our paper we reduce some of our problems to well-known NP-complete problems like *set cover, X3C (exact cover by 3 sets), minimum size stable,...* whose strict description can be found for example in the classical textbook [8].

2   Vertex Cover with Obligations on Vertices

Let $G = (V, E)$ be any graph and $\Pi_V = V_1, \ldots, V_k$ a partition of V, a system of obligations on vertices of G. A *vertex cover with obligations* $(\mathcal{VCO})$ S, of (G, Π_V) is: a vertex cover of G (each edge $e = uv \in E$ is covered by S ($u \in S$ or $v \in S$ (both can be in S))) and $\forall u \in S$, if $u \in V_i$, then $V_i \subseteq S$ (i.e. S respects the constraints on obligations).

It is easy to see that any instance $(G = (V, E), \Pi_V)$ always contains at least a $\mathcal{VCO}$, namely $S = V$. A $\mathcal{VCO}$ S^* of the instance (G, Π_V) is said *optimal*, and noted $\mathcal{VCO}_{OPT}$, if it is of minimum size. Constructing a $\mathcal{VCO}_{OPT}$ is hard since even in the very particular case where each part of Π_V is a singleton, this is the classical NP-complete vertex cover problem [8]. In what follows we propose an approximation algorithm for the $\mathcal{VCO}_{OPT}$ problem. But first we can easily simplify the instance in some cases. Indeed, we can remark that if $e = uv \in E$ and u and v are in the same part V_i of the partition Π_V ($u \in V_i$ and $v \in V_i$) then any $\mathcal{VCO}$ (thus any $\mathcal{VCO}_{OPT}$) must contain V_i since the edge $e = uv$ must be covered and u or v must be in any solution and thus also V_i. Before running any algorithm, we can include in any solution, all the parts V_i of Π_V such that G contains an edge e with both extremities in V_i. This can be done in polynomial time. We suppose now that this pre-treatment has been done and that G does not contain these vertices anymore and Π_V does not contain these parts anymore.

We describe now a 2-approximation algorithm for the $\mathcal{VCO}_{OPT}$ problem. At this point we can suppose that an instance is now $(G = (V, E), \Pi_V = V_1, \ldots, V_k)$ where each V_i is a stable of G.

1. Construct as follows a new weighted graph $G_c = (V_c, E_c)$ called *contracted graph*: each stable V_i of Π_V is associated to a vertex v_i of G_c. The *weight* of v_i is the number of vertices of V_i ($|V_i|$). Add an edge between v_i and v_j in G_c iff G contains (at least) an edge having an extremity in V_i and the other in V_j.
2. Construct a 2-approximated *weight vertex cover* S_c in G_c (i.e. a vertex cover of G_c whose total weight is at most twice the minimum one. This approximation can be done in polynomial time, see [1]).
3. Return $S = \bigcup_{i : v_i \in S_c} V_i$ (for each v_i of S_c, put the corresponding V_i in S).

Theorem 1. *The algorithm described above is a 2-approximation algorithm for the $\mathcal{VCO}_{OPT}$ problem.*

Proof. This algorithm is polynomial. It constructs a vertex cover of G that satisfies the constraints on obligations.

Note that to respect the conditions on obligations, any $\mathcal{VCO}$ of (G, Π_V) is a union of some parts of Π_V. We construct now a one-to-one correspondance respecting the weights and the sizes between the $\mathcal{VCO}$ of (G, Π_V) and the weighted vertex covers of G_c.

Let S be any $\mathcal{VCO}$ of (G, Π_V). The set $S_c = \{v_i : i : V_i \subseteq S\}$ associated to S is a vertex cover of G_c, of weight $|S|$.

Conversely, let $S_c = \{v_1, \ldots, v_l\}$ be any weighted vertex cover of G_c. In this case, $S = \{V_i : i : v_i \in S_c\}$ is a $\mathcal{VCO}$ of (G, Π_V) whose size is equal to the weight of S_c.

A 2-approximation of an optimal weighted vertex cover of G_c corresponds to a 2-approximated $\mathcal{VCO}_{\mathcal{OPT}}$ of (G, Π_V). Hence the proposed algorithm is a 2-approximation algorithm for the $\mathcal{VCO}_{\mathcal{OPT}}$ problem.   □

Connected Vertex Cover with Obligations on Vertices
In this part, $G = (V, E)$ is a *connected* graph. As previously, the *obligations* are given by a partition $\Pi_V = V_1, \ldots, V_k$ of V. A $\mathcal{CVCO}$, *connected vertex cover with obligations*, S of the instance (G, Π_V) is: a vertex cover of G (for any edge $uv \in E$, $u \in S$ or $v \in S$ (both can be in S)), a connected set of vertices: $G[S]$ (the induced graph of S in G) is connected, and S respects the constraints of obligations of Π_V. It is easy to see that any instance $(G = (V, E), \Pi_V)$ always contains at least a $\mathcal{VCO}$, namely $S = V$ since G is connected. A $\mathcal{CVCO}_{\mathcal{OPT}}$ is a $\mathcal{CVCO}$ of minimum size. Constructing a $\mathcal{CVCO}_{\mathcal{OPT}}$ is a hard problem, even if Π_V is a partition of singletons (in this case this is the classical NP-complete connected vertex cover problem [8]).

Theorem 2. *Any α-approximation algorithm for the $\mathcal{CVCO}_{\mathcal{OPT}}$ problem can be transformed into a 2α-approximation algorithm for the minimum size set cover problem.*

Proof. Let (A, X) be any instance of the set cover problem: $A = \{a_1, \ldots, a_n\}$ is a set of n elements and $X = X_1, \ldots, X_k$ is a family of subsets of A ($X_i \subseteq A$) covering A: $A = \cup_{i=1}^{k} X_i$. An optimal set cover is a sub-family of X, of minimum size, covering A. We note t^* the size of such an optimal solution of (A, X).

From (A, X) let us construct an instance of our problem. Each element a_i is associated to a vertex, also noted a_i. Each set X_i of X is associated to a set noted V_i of $n+1$ new vertices, forming a stable. Each of the $n+1$ vertices of the set V_i is connected to a vertex a_j iff the set X_i contains the element a_j. Create now a new vertex r and connect it to all the vertices of the k sets V_i. The degree of r is then $k(n+1)$. We note $G = (V, E)$ the final graph that is bipartite.

The obligations are the following. Each V_i is an obligation containing exactly $n+1$ independent vertices. Add the obligation V_0 containing r and the n vertices of A. V_0 is then also a stable of G composed of $n+1$ vertices. $\Pi_V = V_0, V_1 \ldots, V_k$ is a partition of the set V of vertices of G and is the system of obligations that we consider here; each V_i is a stable of $n+1$ vertices of G. The instance (G, Π_V) can be constructed in polynomial time from the instance (A, X). Consider now the following one-to-one mapping between the $\mathcal{CVCO}$ of (G, Π_V) and the set covers of (A, X).

Let $S_X = X_{i_1}, \ldots, X_{i_t}$ be any set cover of size t of (A, X). Consider now the following set S of vertices of G: $S = V_0 \cup \bigcup_{j=1}^{t} V_{i_j}$. S is a vertex cover of G (all the edges of G are covered by the vertices of V_0), $G[S]$ is connected (because the vertices of V_{i_j} are interconnected *via* r and each a_i is connected to at least all the vertices of a set V_{i_j} because S_X is a covering) and satisfies the obligations of Π_V (S is composed of a union of obligations of Π_V). The size of S is: $|S| = n + 1 + t(n+1) = (n+1)(t+1)$.

Consider now any $\mathcal{CVCO}$ S of (G, Π_V). As S satisfies the constraints on obligations, it is composed of a union of obligations. As $G[S]$ is connected and G is bipartite, it must contain some of the obligations $V_i, i \geq 1$. But as S must contain r or a vertex a_i to ensure the connectivity it must contain the obligation V_0. Note $V_0, V_{i_1}, \ldots, V_{i_t}$ the obligations composing S: $S = V_0 \cup V_{i_1} \cup \cdots \cup V_{i_t}$. Let $S_X = X_{i_1}, \ldots, X_{i_t}$ be the sub-family associated to this $\mathcal{CVCO}$ S. As $V_0 \subseteq S$, each vertex a_i is connected to the other vertices of S via the vertices of at least a V_{i_j}. Thus S_X is a set cover of (A, X). We get: $|S_X| = t$ and $|S| = (t+1)(n+1)$.

This one-to-one mapping associates to each set cover of size t a $\mathcal{CVCO}$ of size $(t+1)(n+1)$ and reciprocally. The transformations in one direction or the other can be done in polynomial time.

Suppose that a $\mathcal{CVCO}_{\mathcal{OPT}}$ can be approximated with a ratio α in polynomial time. Then, for any instance (A, X) one can: construct the associated instance (G, Π_V), then use this approximation algorithm to construct a α-approximated $\mathcal{CVCO}$ S: $(t+1)(n+1) = |S| \leq \alpha |S^*|$. Then with the one-to-one transformation, one can construct the associated set cover S_X, of size t. This chain of constructions is polynomial. Let S_X^* be an optimal set cover, of size t^*. By the one-to-one transformation, this corresponds to a $\mathcal{CVCO}$ of size $(t^*+1)(n+1)$. This $\mathcal{CVCO}$ is optimal (otherwise it would be possible to construct a smaller one with the one-to-one transformation). Hence, $|S| = (t+1)(n+1) \leq \alpha(t^*+1)(n+1)$, then, $t + 1 \leq \alpha(t^*+1)$ and $t \leq \alpha t^* + (\alpha - 1) \leq \alpha(t^*+1) \leq 2\alpha t^*$ (because $1 \leq t^*$). The algorithm described above is then a 2α-approximation algorithm for the set cover problem. $\square$

Corollary 1. *The* $\mathcal{CVCO}_{\mathcal{OPT}}$ *problem cannot be approximated by a ratio better than* $c \log(n)/2$ *unless* $P = NP$.

Proof. Theorem 2 shows that the $\mathcal{CVCO}_{\mathcal{OPT}}$ problem cannot be approximated by a ratio better than $c \log(n)/2$ since the optimal set cover problem cannot be approximated within $c \log(n)$ for some $c > 0$, unless $P = NP$, see [1]. $\square$

3 Dominating Set with Obligations on Vertices

In this section, we study the complexity of 3 domination problems with obligations: Dominating set with obligations, Total Dominating set with obligations and Independant dominating set with obligations.

Dominating Set with Obligations on Vertices
In this part, an instance is $(G = (V, E), \Pi_V = V_1, \ldots, V_k)$ where G is a graph and Π_V is a partition of V. A *dominating set with obligations* S ($\mathcal{DO}$) of (G, Π_V) satisfies: S dominates G (for any $u \in V - S$, u has at least a neighbor in S), and S respects the constraints of obligations of Π_V.

We can remark that there is always a $\mathcal{DO}$: V, the set of vertices of G. The minimization problem is NP-complete and cannot be approximated with a better ratio than $c \cdot \log |V|$ for some $c > 0$ (unless $P = NP$): indeed when the obligations are all singletons, we get the classical dominating set problem having this bound on approximation ratio, see [16].

Let us show now that it is possible to construct a $\mathcal{O}(\log(|V|))$-approximation for our problem of dominating set with obligations. For that purpose we reduce it to the *weighted set cover* for which there is such an approximation ratio $\mathcal{O}(\log(|V|))$, see [3].

Theorem 3. *Given* (G, Π_V), *it is possible to approximate an optimal* $\mathcal{DO}$ *with ratio* $\mathcal{O}(\log(|V|))$.

Proof. From instance $(G = (V, E), \Pi_V)$, we construct (U, S, w) an instance of the weighted set cover. Let $U = V$. For any obligation $V_i \in \Pi_V$, we construct a set S_i composed of the union of the closed neighborhoods of vertices of V_i (the closed neighborhood of x is the set of neighbors of x plus x itself). We can remark that a set S_i contains exactly the vertices dominated by V_i. The weight of this set is the size of the obligation (which is, in general, different from the size of S_i) i.e. $w(S_i) = |V_i|$. The family S of sets of the instance (U, S, w) is composed of all these S_i. Figure 1 shows an example of construction of S_1 from V_1. Here, the set constructed has weight 3 (the size of V_1) and dominates V_1 and its neighbors. We construct now a one-to-one mapping between the dominating sets with obligations of (G, Π_V) and the set covers of (U, S, w).

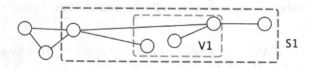

Fig. 1. Construction of S_1 from V_1.

Let D be any dominating set with obligations of (G, Π_V). As D respects the obligations, D is a union of obligations $V_{i_1}, \ldots, V_{i_t}$. Construct $C = \bigcup_{j=1}^{t} S_{i_j}$. As D is a dominating set of G, each vertex u of V is dominated by a vertex in a set V_{i_j} and, hence, each element u of $U = V$ is covered by S_{i_j}, i.e. by C. We also have $|D| = \sum_{j=1}^{t} |V_{i_j}| = \sum_{j=1}^{t} w(S_{i_j}) = w(C)$.

Reciprocally, let $C = S_{i_1}, \ldots, S_{i_l}$ be a set cover of (U, S, w). Construct $D = \bigcup_{i=1}^{l} V_{i_l}$. As C is a set cover, each element v is covered by at least a S_{i_j}, and then each corresponding vertex v is dominated by itself if it is in V_{i_j}, or by one of its neighbors in V_{i_j}, D is then a dominating set of G. Moreover, by construction, D respects the obligations. As previously, $w(C) = \sum_{j=1}^{l} w(S_{i_j}) = \sum_{j=1}^{l} |V_{i_j}| = |D|$.

The final result follows from this polynomial transformation and one-to-one mapping, preserving size/weight and the result of [3]. $\square$

Total Dominating Set with Obligations on Vertices
In this part, an instance is $(G = (V, E), \Pi_V = V_1, \ldots, V_k)$ where G is a graph and Π_V is a partition of V. A *total dominating set with obligations* S ($\mathcal{TDO}$) of (G, Π_V) satisfies: S totally dominates G (for any $u \in V$, u has at least a neighbor in S) and S respects the constraints of obligations of Π_V.

We can remark that $(G = (V, E), \Pi_V)$ contains a $\mathcal{TDO}$ (the set V) iff G has no isolated vertices. An *optimal* $\mathcal{TDO}$ is a $\mathcal{TDO}$ of minimum size.

The minimization problem is NP-complete and cannot be approximated with a better ratio than $c \cdot \log |V|$ for some $c > 0$: indeed when the obligations are all singletons, we get the classical total dominating set problem having this bound on approximation ratio, see [2].

Theorem 4. *Given* (G, Π_V), *it is possible to approximate an optimal* $\mathcal{TDO}$ *with ratio* $\mathcal{O}(\log(|V|))$.

Proof. The proof of Theorem 4 is very similar to the one of Theorem 3, hence we only give a sketch of proof. The reduction and mapping are the same as in Theorem 3, except that sets S_i are composed of the union of the *open* neighborhoods of vertices of V_i (the open neighborhood of x is the set of neighbors of x, x excluded), to ensure that each vertex of S_i is dominated by an other vertex of V_i. □

Independent Dominating Set with Obligations on Vertices
In this part, an instance is $(G = (V, E), \Pi_V = V_1, \ldots, V_k)$ where G is a graph and Π_V is a partition of V. An *independent dominating set with obligations* S ($\mathcal{IDO}$) of (G, Π_V) satisfies: S dominates G (for any $u \subset V - S$, u has at least a neighbor in S), S is a stable of G (no edges between vertices of S) and S respects the constraints of obligations of Π_V. In this particular variant of domination, a solution is not always guaranteed.

Theorem 5. *Determining if* (G, Π_V) *contains an* $\mathcal{IDO}$ *is NP-complete.*

Proof. The problem is clearly in NP. Let (X, Z) be a X3C instance (exact cover by 3 sets) where X is a set of $3q$ elements and each Z_i is a subset of 3 elements of X ($Z_i \subseteq X$ and $|Z_i| = 3$) with the property: $X = \bigcup_{i=1}^{k} Z_i$ (the sets Z_i cover X). The X3C problem consists in deciding if this instance contains an *exact cover* of X (each element of X is in exactly one subset of the solution). This problem is NP-complete, see [8].

Let us construct an instance of our problem from (X, Z). For each element x of X, a P_3 (a path with 3 vertices) is created and one extremity is called the *vertex representing the element*. For each subset z of Z, a path P_2 is created and one extremity is called the *vertex representing the subset*. Additional edges are added between: each vertex representing a subset and each vertex representing an element inside this subset; each pair of vertices representing subsets whose associated subsets have non-empty intersection.

For each element x, an obligation containing the vertex representing x and its neighbor in its P_3 is created. They are called *obligations of elements*. All the other obligations are singletons. An exemple of result of this (polynomial) construction is given in Fig. 2.

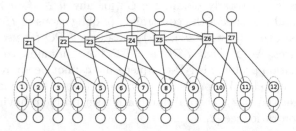

Fig. 2. Construction of (G, Π_V) from (X, Z).

Let D be an independent dominating set respecting the obligations of (G, Π_V). D contains no obligation of elements because these obligations are between two vertices linked by an edge. Hence, each vertex representing an element can only be dominated by vertices representing subsets. Let S be the family of subsets corresponding to the vertices representing subsets of D. Then as each vertex representing an element is dominated by D, each element is covered by S. Moreover, as only subsets with non-empty intersection are neighbors, D is an independent set, and the subsets of S are pairwise disjoint: S is then an exact cover of (X, Z).

Now, let S be an exact cover of (X, Z). Let us construct D. For each Z_i, the corresponding vertex is added to D iff Z_i is in S. Otherwise, the neighbor of Z_i in the P_2 is added to D. Also add to D all the vertices that are the opposite extremities of the vertices representing an element in each P_3. It is easy to see that D is an independent set. Moreover, D respects the obligations (since each vertex of D is in a singleton obligation). Finally, each element is covered by S: each vertex representing an element is then dominated by a vertex representing a subset. The paths P_2 are dominated either by the vertex representing the subset or by the other extremity. For each path P_3 the vertex, opposite extremity of the vertex representing the element, is in D and covers itself and its unique neighbor. D is then an independent dominating set, respecting the obligations of (G, Π_V). □

4   Spanning Graph with Obligations on Edges

Spanning Tree with Obligations on Edges
In this part, an instance is $(G = (V, E), \Pi_E = E_1, \ldots, E_k)$ where G is any connected graph and Π_E, the obligations, is a partition of E. The objective is, given an instance, $(G = (V, E), \Pi_E = E_1, \ldots, E_k)$, to decide if there is a *tree spanning G with obligations* $(\mathcal{TSO})$ $T = (V, E_T)$ which is a tree spanning G and such that for any $e \in E_T$, if $e \in E_i$ then all the edges of E_i must also be in T.

Theorem 6. *Deciding if* $(G = (V, E), \Pi_E = E_1, \ldots, E_k)$ *contains a $\mathcal{TSO}$ is NP-complete, even if: G is bipartite, of maximum degree 4 and each E_i induces a star (that is a tree with a vertex directly connected to all the others) with exactly 3 edges ($|E_i| = 3$).*

Proof. The problem is in NP. Let $(X = \{x_1, \ldots, x_{3q}\}, Z_1, \ldots, Z_k)$ be any instance of the X3C problem (exact cover by 3 sets) where X is a set of $3q$ elements and each Z_i is a subset of 3 elements of X ($Z_i \subseteq X$ and $|Z_i| = 3$) with the property: $X = \bigcup_{i=1}^{k} Z_i$ (the sets Z_i cover X). The X3C problem consists in deciding if this instance contains an *exact cover* of X, i.e., if there exist $Z_{i_1}, \ldots, Z_{i_q}$ pairwise disjoint sets such that $X = \bigcup_{j=1}^{q} Z_{i_j}$. This is a well-known NP-complete problem, even if each element x_i is in at most 3 sets, see [8]. It is this restricted formulation that we consider here.

From this instance, let us construct a graph G. For each element x_i of X create a new vertex, also noted x_i. For each set Z_i create a new vertex, also noted Z_i. Add an edge between each vertex Z_i and the 3 vertices that are in the set Z_i. Now, create a tree T_r to connect the k vertices Z_i that will become leaves of T_r. The Z_i are connected two-by-two by new vertices. Then these $\lceil k/2 \rceil$ new vertices are connected two-by-two by new vertices, and so on until there is only one final new vertex that we call r (as "root" of T_r). Each vertex u, except the leaves Z_i, have one or two children. For each such u we add a new vertex l_u (or two if necessary) that is only connected to u (l_u is a leave). These 3 vertices are called the 3 children of u. We get now the final tree. All of these vertices and edges form the final graph $G = (V, E)$ that is bipartite and, thanks to the restriction on X3C instances, the maximum degree of G is 4. An illustration of the construction is given in Fig. 3: the bottom vertices are elements of X, squared vertices are the Z_i, black ones are the additional children and colored vertices are the internal ones of tree T_r. The dashed ellipses represent the obligations that are described now.

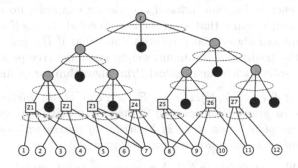

Fig. 3. Construction of G from a X3C instance.

For each vertex Z_i we group in a same obligation noted E_i the 3 edges connecting Z_i to the 3 vertices representing the 3 elements that are in set Z_i.

We group in a same obligation the 3 edges connecting any internal vertex u to its 3 children. All these obligations are called *tree obligations*. Each edge of G is now in exactly one obligation (tree one or in a E_i) and the set of all these obligations is Π_E, composed of stars of exactly 3 edges. This construction is polynomial.

Suppose that the X3C instance has a solution $Z_{i_1}, \ldots, Z_{i_q}$. In this case, we can select the following obligations: all the tree obligations and all the edges in the obligations $E_{i_1}, \ldots, E_{i_q}$. This gives a tree spanning G (each vertex x_i is a leaf because it is a neighbor of exactly one vertex Z_{i_j} and each vertex Z_l is connected to the other vertices *via* the tree T_r). This tree respects the obligations of Π_E and is then a $\mathcal{TSO}$ of (G, Π_E).

Conversely, suppose that the instance (G, Π_E) has a $\mathcal{TSO}$ noted T. As T respects the obligations, it necessarily contains *all* the tree obligations, this is mandatory to include the leaves of the form l_u. It also contains other obligations. But each vertex x_i is a leaf of T. Otherwise, if it is neighbor of 2 vertices, Z_a and Z_b then we would have a cycle with some edges of the tree obligations, that is not allowed because T is a tree. As T covers all the $3q$ vertices/leaves x_i it must contain exactly q vertices of type Z_i, noted $Z_{i_1}, \ldots, Z_{i_q}$, and their 3 edges incident from the associated obligations $E_{i_1}, \ldots, E_{i_q}$. The sets $Z_{i_1}, \ldots, Z_{i_q}$ cover X and are pairwise disjoint and is then a solution for the X3C instance. □

Connected Spanning Graph of Minimum Weight with Obligations on Edges
In this part $G = (V, E)$ is a weighted connected graph: each edge $e \in E$ has a weight $w(e) > 0$. The obligations form a partition $\Pi_E = E_1, \ldots, E_k$ of E. The objective is to extract from G a subset S of edges, inducing a connected graph spanning all the vertices of V, having a minimum weight and respecting the obligations. Such an object is called a $\mathcal{CSGO}_{\mathcal{OPT}}$ (*Minimum Weight Connected Spanning Graph with obligations*). We call $\mathcal{CSGO}$ a *Connected spanning Graph with obligations* (a $\mathcal{CSGO}_{\mathcal{OPT}}$ is a minimum weight $\mathcal{CSGO}$).

We can note that, because of the obligations, a $\mathcal{CSGO}$ is not necessarily a tree. Indeed, if each obligation induced a cycle for example, no spanning tree is possible. We can remark that since G is connected, G itself is a $\mathcal{CSGO}$ of (G, Π_E) (the problem always has a solution) and that if Π_E only contains singletons, this is the traditional minimum weight spanning tree problem that can be polynomially solved with the classical Prim algorithm for example.

Theorem 7. *Let $(G = (V, E), \Pi_E = E_1, \ldots, E_k)$ be an instance with G a weighted connected graph. Determining if there is a $\mathcal{CSGO}$ of weight at most $|V| - 1$ is NP-complete, even if: G is bipartite, of maximum degree 4, all the weights are 1 and each obligation induces a star with 3 edges.*

Proof. This problem is clearly in NP. Any spanning graph contains at least $n-1$ edges, with $n = |V|$. Hence, in the case where each edge has weight 1, there is no $\mathcal{CSGO}$ with weight strictly less than $n-1$. Deciding if there exists a $\mathcal{CSGO}$ of weight at most $n-1$ is then strictly equivalent to decide if there exists a $\mathcal{TSO}$ in this instance, which is NP-complete, even if G verifies the hypotheses, thanks to Theorem 6. □

Theorem 7 shows that deciding whether an instance contains a $\mathcal{CSGO}$ is NP-complete, even if all the weights are equal. The next result shows that there is no constant approximation algorithm for the weighted case (unless P $\neq$ NP).

Theorem 8. *Any α-approximation algorithm for the $\mathcal{CSGO}$ problem in bipartite graphs where obligations induce stars can be transformed into a α-approximation algorithm for the* minimum size set cover *problem.*

Proof. Let $(X = \{x_1, ...x_n\}, F = \{F_1, ...F_k\})$ be any instance of the set cover problem. Let us construct an instance of $\mathcal{CSGO}$. Put in V the n vertices corresponding to $x_1, ...x_n$, k vertices corresponding to sets $F_1, ...F_k$ and an additional new vertex r. Link r to each vertex F_i and link each vertex F_i to all the x_j such that $x_j \in F_i$. Clearly the graph obtained is bipartite. Put in a same obligation, noted O_0, all the incident edges of r and assign a weight ϵ/k on each edge of O_0 (where ϵ can be as small as desired). For each F_i, put all incident edges to F_i, except the one between F_i and r, in an obligation noted O_i and assign to each such edge of O_i a weight $1/|F_i|$. Hence, the total weight of each obligation is 1, except O_0 with weight ϵ. Each obligation induces a star.

We show now that each solution of the set cover problem can polynomialy be transformed in a solution of equivalent weight for the $\mathcal{CSGO}$ problem, and reciprocally.

Let S be a solution of the set cover, of size t. We construct C the set of edges as follows: put O_0 in C, and for each $0 < i \leq k$, put O_i in C iff F_i is in S. The vertices r and F_i are connected in C (via O_0). Each element x_j is covered by a set F_i of the set cover: the corresponding vertex x_j is connected to the vertex F_i via the obligation O_i, hence C is a $\mathcal{CSGO}$. C contains O_0 and t other obligations, its weight is then $t + \epsilon$.

Let now C be a $\mathcal{CSGO}$. C contains O_0 (to connect r) and t other obligations. Its weight is then $t + \epsilon$. Let us construct a solution S of the set cover. For each $0 < i \leq k$, put F_i in S iff $O_i \in C$. Let x_j be an element of X. The corresponding vertex is connected by an edge which is an element of an obligation O_i. Hence, the corresponding set F_i belongs to S and the element is covered, S is then a set cover. Moreover, the size of S is t.

As ϵ can be arbitrary small, using these transformations, one can use a α-approximation algorithm for our $\mathcal{CSGO}$ to create a α-approximation algorithm for the set cover problem. $\square$

Corollary 2. *The minimum weight $\mathcal{CSGO}$ cannot be approximated with a constant approximation ratio (unless $P = NP$), even if G is a bipartite graph and if each obligation induces a star.*

Proof. Theorem 8 shows that it is not possible to approximate the $\mathcal{CSGO}$ problem with a better ratio than the one of the minimum size set cover, even in bipartite graphs where each obligation induces a star. But this last problem cannot be approximated within $c \log(c)$ for some c, unless $P = NP$, see [1]. $\square$

5   Matchings with Obligations on Edges

In this section, an instance is (G, Π_E) where $G = (V, E)$ is any graph and $\Pi_E = E_1, \ldots, E_k$ is a partition of E, the set of edges of G.

A *matching with obligations* (MO) M of the instance (G, Π_E) is a matching of G (set of pairwise non incident edges of G) respecting the obligations.

It is polynomial to determine if (G, Π_E) contains a MO. Indeed, there is a non empty MO iff at least an obligation E_i induces a matching. From this we can simplify an instance (G, Π_E): if a part E_i of Π_E induces a graph in which a vertex has more than one neighbor then E_i can be deleted from Π_E and the edges of E_i can be deleted from G. This pretreatment can be done in polynomial time. From now we suppose that $(G = (V, E), \Pi_E = E_1, \ldots, E_k)$ is an instance where each E_i induces a matching of G and thus contains a MO (possibly empty). A MO *of maximum size* is noted MO_{OPT}.

Theorem 9. *Let $(G = (V, E), \Pi_E = E_1, \ldots, E_k)$ be an instance where each E_i induced a matching of G. Any α-approximation algorithm for the MO_{OPT} problem can be transformed into a α-approximation algorithm for the maximum size stable problem.*

Proof. Let $H = (V_H, E_H)$ be any graph, instance of the maximum size stable problem. Note $V_H = \{h_1, \ldots, h_n\}$ the n vertices of H. We construct an instance of our problem from H.

For each edge $h_i h_j$ of H, we create a new P_3 (path with 3 vertices) associated to this edge. The union of these $|E_H|$ pairwise disjoint paths form a graph noted Q (not yet the final graph G). Now, for each i, $1 \leq i \leq n$, we create D_i a subset of edges of Q as follows. For each edge $h_i h_j$ of H, put an edge of the associated P_3 path in D_i and the other one in D_j. These n sets $D_1, \ldots, D_n$ form a partition of the edges of Q and each D_i is a matching. Figure 4 gives an example of this construction.

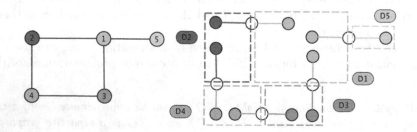

Fig. 4. Construction of Q (right) from H (left). Sets D_i are in dashed boxes.

The sets D_i can have different sizes. Let D_a be the one of maximum size (this corresponds to the maximum degree of H). The next steps consist in adding new independent edges, between new vertices, to each D_i such that the n sets all have the same size $|D_a|$.

Note $G = (V, E)$ the graph obtained from Q by the addition of these new vertices and edges. We note $\Pi_E = E_1, \ldots, E_n$ obtained by the previous operation of homogenization of size. We now have the following properties: all the sets E_i have the same size noted t, $E_1, \ldots, E_n$ is a partition of the set E of edges of G, each E_i induces a matching in G, $E_i \cup E_j$ is a matching of G iff $h_i h_j \notin E$. This instance (G, Π_E) can be constructed in polynomial time from the instance H of the maximum size stable problem.

Let $S = \{h_{i_1}, \ldots, h_{i_q}\}$ be any stable, of size q, in H. Let us consider the associated obligations to S: $E_{i_1}, \ldots, E_{i_q}$. As S is a stable, $M_S = \bigcup_{j=1}^{q} E_{i_j}$ is a matching of G of size qt.

Conversely, let M be any matching of G, composed of the $E_{i_1}, \ldots, E_{i_q}$. As M is a matching of G of size qt, $S = \{h_{i_1}, \ldots, h_{i_q}\}$ is a stable of size q in H.

There is a one-to-one mapping between the $\mathcal{MO}$ of (G, Π_E) and the stables of H. The sizes are all the same, up to a factor t. Hence, if an approximation algorithm of ratio α exists for the $\mathcal{MO}_{\mathcal{OPT}}$ problem then it would be possible to approximate the maximum size stable problem with a ratio α via the previously described transformations: transform H into the instance (G, Π_E), then apply the approximation algorithm on this instance and then transform its result into a stable of H. The conservation of the sizes (up to a factor t) by these transformations insures the approximation ratio. $\qquad\square$

With this result and the fact that the maximum size stable problem cannot be approximated within $|V|^{1/2-\epsilon}$ for any $\epsilon > 0$, unless $P = NP$, see [1] we get:

Corollary 3. *The $\mathcal{MO}_{\mathcal{OPT}}$ problem cannot be approximated with a ratio better than $|V|^{1/2-\epsilon}$ unless $P = NP$.*

6   Hamiltonian Path in Complete Graphs with Obligations on Edges

Here an instance is $(G = (V, E), \Pi_E = E_1, \ldots, E_k)$ where G is any connected graph and Π_E a partition of E. A *hamiltonian path with obligation* $(\mathcal{HPO})$ of (G, Π_E) is a hamiltonian path of G (a path of $|V| - 1$ edges, spanning V) satisfying all the constraints of obligations of Π_E (if an edge e is in the path then all the edges belonging to the same obligation must also be in the path).

Theorem 10. *Deciding if (G, Π_E) contains a $\mathcal{HPO}$ is NP-complete, even if G is a complete graph.*

Proof. The problem is in NP. Let $H = (V, E)$ be any connected graph, instance of the hamiltonian path problem, which is an NP-complete problem, see [8]. Let $n = |V|$. We suppose here that $n \geq 4$ (if n is smaller then the problem can easily be solved in constant time). The graph for our problem is K_n, the complete graph on the n vertices V of H. The obligations are the following. For each edge uv of H, the edge uv of K_n is the only element (singleton) of this part. All the edges uv outside H ($uv \notin E$) are grouped in a single obligation E_0. This instance

(K_n, Π_E) can be constructed in polynomial time. We divide our study in two cases. Case 1: E_0 induces a graph of maximum degree greater than or equal to 3. In this case, the edges of E_0 cannot be in a $\mathcal{HPO}$ of K_n. Hence, H contains a hamiltonian path iff (K_n, Π_E) contains a $\mathcal{HPO}$.

Case 2: E_0 induces a graph of maximum degree at most 2. In this case, each vertex u has degree at least $n - 2$ in H. But, by hypothesis $n \geq 4$, this implies that the degree in H of each vertex is at least $n/2$. This is the well-known (see [7] for example) Dirac sufficient condition for H to have a hamiltonian cycle, i.e. also a hamiltonian path. Hence, H has a hamiltonian path and (K_n, Π_E) has a $\mathcal{HPO}$. In all cases, H has a hamiltonian path iff (K_n, Π_E) has a $\mathcal{HPO}$.   □

7   Conclusion

In this paper we shown that adding obligations drastically increases the approximation ratio of classical graph problems. This is the case for the connected vertex cover with obligations that has no constant approximation ratio algorithm (while there is a 2-approximation algorithm for the original problem), the minimum connected weighted spanning graph and the maximum size matching. For the last two problems, the classical versions are polynomial but the version with obligations are as hard as set cover or maximum stable problems. For some other problems, the situation is even worst: it becomes NP-complete to know whether there is a solution, regardless of its size (while it is trivial or polynomial in the original problem). This is the case for the following problems: independent dominating set, spanning tree and hamiltonian path in complete graphs. For the dominating and total dominating set problem, the approximation ratios are almost the same with or without obligations constraints (but these ratios are not constant). Only the vertex cover problem keeps the same constant approximation ratio 2.

One might imagine that a perspective could be to refine our results by studying more specific/restricted instances. Unfortunately in some cases, the problem is "equivalent" to another hard problem (minimum set cover, maximum size stable problem) that already received a lot of attention and improving them is known as a hard challenge in itself since a long time. In other cases, the instances for which our problem is hard are basic in a sense: bipartite graphs of maximum degree 4 and very small obligation sizes for the spanning tree problem, complete graph for the hamiltonian path problem.

Other combinatorial problems can be studied with our framework. But our results show that dealing with obligations can lead to very complex problems that could be unsolvable. Organizing practical systems with obligations should be done with a lot of attention.

References

1. Ausiello, G., et al.: Complexity and Approximation: Combinatorial Optimization Problems and Their Approximability Properties. Springer, Heidelberg (2012). https://doi.org/10.1007/978-3-642-58412-1

2. Chlebík, M., Chlebíková, J.: Approximation hardness of dominating set problems. In: Albers, S., Radzik, T. (eds.) ESA 2004. LNCS, vol. 3221, pp. 192–203. Springer, Heidelberg (2004). https://doi.org/10.1007/978-3-540-30140-0_19

3. Chvatal, V.: A greedy heuristic for the set-covering problem. Math. Oper. Res. **4**(3), 233–235 (1979)

4. Cornet, A., Laforest, C.: Total domination, connected vertex cover and Steiner tree with conflicts. Discret. Math. Theor. Comput. Sci. **19**(3), Article 17 (2017)

5. Cornet, A., Laforest, C.: Domination problems with no conflicts. Discret. Appl. Math. **244**, 327–338 (2018)

6. Delbot, F., Laforest, C., Phan, R.: Hardness results and approximation algorithms for discrete optimization problems with conditional and unconditional forbidden vertices. Research report hal-01257820, January 2016

7. Diestel, R.: Graph Theory. GTM, vol. 173, 4th edn. Springer, Heidelberg (2017). https://doi.org/10.1007/978-3-662-53622-3

8. Garey, M.R., Johnson, D.S.: Computers and Intractability. W.H. Freeman, New York (1979)

9. Kanté, M.M., Laforest, C., Momège, B.: An exact algorithm to check the existence of (elementary) paths and a generalisation of the cut problem in graphs with forbidden transitions. In: van Emde Boas, P., Groen, F.C.A., Italiano, G.F., Nawrocki, J., Sack, H. (eds.) SOFSEM 2013. LNCS, vol. 7741, pp. 257–267. Springer, Heidelberg (2013). https://doi.org/10.1007/978-3-642-35843-2_23

10. Kanté, M.M., Laforest, C., Momège, B.: Trees in graphs with conflict edges or forbidden transitions. In: Chan, T.-H.H., Lau, L.C., Trevisan, L. (eds.) TAMC 2013. LNCS, vol. 7876, pp. 343–354. Springer, Heidelberg (2013). https://doi.org/10.1007/978-3-642-38236-9_31

11. Kanté, M.M., Moataz, F.Z., Momège, B., Nisse, N.: Finding paths in grids with forbidden transitions. In: Mayr, E.W. (ed.) WG 2015. LNCS, vol. 9224, pp. 154–168. Springer, Heidelberg (2016). https://doi.org/10.1007/978-3-662-53174-7_12

12. Kolman, P., Pangrác, O.: On the complexity of paths avoiding forbidden pairs. Discret. Appl. Math. **157**(13), 2871–2876 (2009)

13. Kováč, J.: Complexity of the path avoiding forbidden pairs problem revisited. Discret. Appl. Math. **161**(10), 1506–1512 (2013)

14. Laforest, C., Momège, B.: Some hamiltonian properties of one-conflict graphs. In: Kratochvíl, J., Miller, M., Froncek, D. (eds.) IWOCA 2014. LNCS, vol. 8986, pp. 262–273. Springer, Cham (2015). https://doi.org/10.1007/978-3-319-19315-1_23

15. Laforest, C., Momège, B.: Nash-Williams-type and Chvátal-type conditions in one-conflict graphs. In: Italiano, G.F., Margaria-Steffen, T., Pokorný, J., Quisquater, J.-J., Wattenhofer, R. (eds.) SOFSEM 2015. LNCS, vol. 8939, pp. 327–338. Springer, Heidelberg (2015). https://doi.org/10.1007/978-3-662-46078-8_27

16. Raz, R., Safra, S.: A sub-constant error-probability low-degree test, and a sub-constant error-probability PCP characterization of NP. In: STOC, pp. 475–484. ACM (1997)

17. Yinnone, H.: On paths avoiding forbidden pairs of vertices in a graph. Discret. Appl. Math. **74**(1), 85–92 (1997)

Bipartizing with a Matching

Carlos V. G. C. Lima[1(✉)] [iD], Dieter Rautenbach[2] [iD], Uéverton S. Souza[3] [iD],
and Jayme L. Szwarcfiter[4]

[1] Computer Science Department, Federal University of Minas Gerais, Belo Horizonte, Brazil
carloslima@dcc.ufmg.br
[2] Institute of Optimization and Operations Research, Ulm University, Ulm, Germany
dieter.rautenbach@uni-ulm.de
[3] Institute of Computing, Fluminense Federal University, Niterói, Brazil
ueverton@ic.uff.br
[4] PESC, COPPE, Federal University of Rio de Janeiro, Rio de Janeiro, Brazil
jayme@nce.ufrj.br

Abstract. We study the problem of determining whether a given graph $G = (V, E)$ admits a matching M whose removal destroys all odd cycles of G (or equivalently whether $G - M$ is bipartite). This problem is equivalent to determine whether G admits a $(2, 1)$-coloring, which is a 2-coloring of $V(G)$ in which each color class induces a graph of maximum degree at most 1. We determine a dichotomy related to the NP-completeness of such a decision problem, where it is NP-complete even for 3-colorable planar graphs of maximum degree 4, while it is linear-time solvable for graphs of maximum degree at most 3. In addition, we present polynomial-time algorithms for many graph classes including those in which every odd-cycle subgraph is a triangle, graphs having bounded dominating sets, and P_5-free graphs. Additionally, we show that this problem is fixed-parameter tractable when parameterized by the clique-width, which implies that it is polynomial-time solvable for many interesting graph classes, such as distance-hereditary, outerplanar, and chordal graphs.

Keywords: Graph modification problems · Edge bipartization ·
Defective coloring · Planar graphs

1 Introduction

Given a graph $G = (V, E)$, an integer $k \geq 0$, and a graph property Π, the Π *edge-deletion problem* asks for a set $F \subseteq E(G)$ with $|F| \leq k$, such that the graph obtained by the removal of F satisfies Π. This problem and its optimization version have received wide attention on the study of their complexity, where we can cite [11,27]. When the obtained graph is required to be bipartite, the corresponding edge- (vertex-) deletion problem is called *edge* (*vertex*) *bipartization* [14,21].

This study was financed in part by the Coordenação de Aperfeiçoamento de Pessoal de Nível Superior - Brasil (CAPES) - Finance Code 001, by the Conselho Nacional de Desenvolvimento Científico e Tecnológico - Brasil (CNPq) - CNPq/DAAD2015SWE/290021/2015-4, and FAPERJ.

© Springer Nature Switzerland AG 2018
D. Kim et al. (Eds.): COCOA 2018, LNCS 11346, pp. 198–213, 2018.
https://doi.org/10.1007/978-3-030-04651-4_14

Choi, Nakajima, and Rim [14] showed that edge bipartization is NP-complete even for cubic graphs. Furmańczyk, Kubale, and Radziszowski [21] considered vertex bipartization of cubic graphs by removing an independent set. Vertex bipartization by removing an independent set has also been considered from a computational perspective [5], where it is called INDEPENDENT ODD CYCLE TRANSVERSAL. Considering a distributed system, such a problem addresses situations in which we need to solve a distributed computation that is in deadlock, so that we need to remove elements of the network that do not conflict with each other, in order to release the computation [13].

In this paper we study the analogous edge-deletion decision problem, that is, the problem of determining whether a finite, simple, and undirected graph G admits a removal of a matching in order to obtain a bipartite graph. Formally, for a set $S \subseteq E(G)$, let $G - S$ be the graph with vertex set $V(G)$ and edge set $E(G) \setminus S$. We say that a matching $M \subseteq E(G)$ is a *bipartizing matching* of G if $G - M$ is bipartite. Denoting by $\mathcal{BM}$ the set of all graphs admitting a bipartizing matching, we deal with the complexity of the following decision problem.

BIPARTIZING MATCHING (BM)
Input: A finite, simple, and undirected graph G.
Question: Is $G \in \mathcal{BM}$?

A more restricted version was considered by Schaefer [32], where he asked whether a given graph G admits a 2-coloring of the vertices such that each vertex has *exactly* one neighbor with same color as itself. We can see that the removal of the set of edges whose endvertices have same color, which is a perfect matching of G, generates a bipartite graph. He proved that such a problem is NP-complete even for planar cubic graphs.

BM can also be seen as a *defective coloring*. A (k,d)-coloring of a graph G is a k-coloring of $V(G)$ such that each vertex has at most d neighbors with same color as itself. Such colorings were introduced independently by Andrews and Jacobson [1], Harary and Jones [23], and Cowen, Cowen and Woodall [17] and have received wide attention in the literature [2, 3, 7, 16, 19]. We can see that a graph belongs to $\mathcal{BM}$ if and only if it admits a $(2, 1)$-coloring.

Lovász [28] proved that if a graph G satisfies $(d_1 + 1) + (d_2 + 1) + \cdots + (d_k + 1) \geq \Delta(G) + 1$, then $V(G)$ can be partitioned into $V_1, \ldots, V_k$, such that each *induced subgraph* $G[V_i]$ has maximum degree at most d_i, $1 \leq i \leq k$, where $\Delta(G)$ is the *maximum degree of G*. Hence *subcubic graphs* G, those where $\Delta(G) \leq 3$, are $(2, 1)$-colorable.

Angelini et al. [2] present a linear-time algorithm which determines that partial 2-trees, a subclass of planar graphs, are $(2, 1)$-colorable. We emphasize that k-tree graphs have treewidth at most k, for any $k \geq 1$. Eaton and Hull [19] proved that all triangle-free outerplanar graphs are $(2, 1)$-colorable. Borodin, Kostochka, and Yancey [7] studied $(2, 1)$ colorable graphs with respect to the *maximum average degree*, $mad(G) = \max\left\{ \frac{2|E(H)|}{|V(H)|}, \text{ for all } H \subseteq G\right\}$. They proved that every graph G of $mad(G) \leq \frac{14}{5}$ is $(2, 1)$-colorable, where this bound is sharp. By Euler's formula, a planar graph G with *girth* g, the size of a smallest cycle, has $mad(G) < \frac{2g}{g-2}$. Hence if G has girth at least 7, then it is $(2, 1)$-colorable. Differently to the previous results, we consider the study of bipartizing matchings instead of trying to improperly color the graph, which can inspire future works as those in [26, 30] and improvements on defective coloring studies.

Our Results. Cowen, Goddard, and Jesurum [16] proved that it is *NP*-complete to determine whether a given graph is $(2, 1)$-colorable, even for graphs of maximum degree 4 and for planar graphs of maximum degree 5, however they were unable to tell if this holds for planar graphs of maximum degree 4. Here we solve this.

Theorem 1. BM *is NP-complete for* 3-*colorable planar graphs of maximum degree* 4.

On the positive side, we present polynomial-time algorithms for several graph classes.

Theorem 2. *Every graph with maximum degree at most* 3 *is in* $\mathscr{BM}$. *In addition, a bipartizing matching for such graphs can be found in linear time.*

As previously observed, the result of Lovász [28] implies that all subcubic graphs are in BM. However no algorithm to find a bipartizing matching has been given. The next result comprises some other graph classes.

Theorem 3. BM *can be solved in polynomial time for the following graph classes:*

(a) graphs having bounded dominating set;
(b) P_5-free graphs.
(c) graphs in which every odd-cycle subgraph is a triangle;

We also study parameterized complexity aspects. In particular, we consider the complexity of BM parameterized by the clique-width.

Theorem 4. BM *is FPT when parameterized by the clique-width.*

From Theorem 4 we can solve several interesting graph classes in polynomial time, as for example *distance-hereditary*, *series-parallel*, *control-flow*, and some subclasses of planar graphs such as *outerplanar*, *Halin*, and *Apollonian networks* [4,8,22,33]. The same follows for (P_6, claw)-free and (claw, co-claw)-free graphs [9,10]. Moreover, since clique-width generalizes several graph parameters [25], it follows that BM is in FPT when parameterized by the following parameters: neighborhood diversity; treewidth; pathwidth; feedback vertex set; and vertex cover. In addition, it also follows that

Corollary 5. BM *is polynomial-time solvable for chordal graphs.*

Organization of the Paper. In Sect. 2 we present some definitions and notation used throughout the paper and some initial properties of graphs in $\mathscr{BM}$. In Sect. 3 we prove that BM is *NP*-complete. In Sect. 4 we present the positive results, where we give a linear-time algorithm for subcubic graphs and show that graphs with only triangles as odd-cycles admits a polynomial-time algorithm, as well as graphs of bounded dominating set. In Sect. 5 we show that BM is FPT when parameterized by clique-width, presenting a *Monadic Second Order Logic (MSOL$_1$)* formulation.

2 Preliminaries and Additional Concepts

We use standard notation and definitions of graph theory, where we consider only simple undirected graphs. For any undefined terminology and notation, see [18].

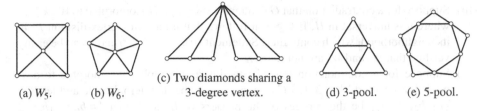

(a) W_5.   (b) W_6.   (c) Two diamonds sharing a 3-degree vertex.   (d) 3-pool.   (e) 5-pool.

Fig. 1. Some examples of forbidden subgraphs.

Given a graph $G = (V, E)$, we denote by $n(G)$ and $m(G)$ the number of vertices and edges of G, respectively. For a vertex $v \in V(G)$, let $N_G(v)$ be the *neighborhood* of v in G and $N_G[v] = \{v\} \cup N_G(v)$ its *closed neighborhood* in G. The *degree* of $v \in V(G)$, $|N_G(v)|$, is denoted as $d_G(v)$. The subscripts can change for a subgraph H of G when necessary.

Given a set $S \in V(G)$, let $H = G[S]$ be the *induced subgraph* of G by S, such that $V(H) = S$ and $uv \in E(H)$ if and only if $uv \in E(G)$ and $u, v \in S$. We also say that S *induces* H and that H is the graph induced by S.

Let $P_n = v_1 v_2 \ldots v_n$ and $C_n = v_1 v_2 \ldots v_n v_1$ be the induced *path* and induced *cycle* of order n, respectively. Furthermore, we denote by K_n and $K_{n,m}$ the *complete graph* of order n and the *complete bipartite graph* with parts of order n and m, respectively. A *diamond* is the graph obtained by removing one edge from the K_4.

Let W_k be the *wheel graph* of order $k + 1$, that is, the graph obtained by connecting a universal vertex v, called *central*, to all the vertices of an induced cycle C_k.

We say that a graph G is a *k-pool* if it is formed by k edge-disjoint triangles whose the union of all the vertices of their bases induce a C_k. Formally, a k-pool is obtained from a cycle $C_{2k} = v_1 v_2 \ldots v_{2k} v_1$, $k \geq 3$, such that the odd-indexed vertices induce the *internal cycle* $p_1 p_2 \ldots p_k p_1$ of the k-pool, where $p_i = v_{2i-1}$, $1 \leq i \leq k$. The even-indexed vertex $b_i = v_{2i}$, $1 \leq i \leq k$, is called the i-th-*border* of the k-pool.

Figure 1 depicts some examples of forbidden subgraphs, that is, graphs that are not in $\mathscr{BM}$, while Lemma 6 collects some properties of graphs in $\mathscr{BM}$.

Lemma 6. *For a graph $G \in \mathscr{BM}$ and a bipartizing matching M of G,*

(i) For every diamond D of G, M matches both vertices of degree 3 of D.

(ii) For every $v \in V(G)$, $G[N_G(v)]$ cannot contain two vertex-disjoint P_3.

(iii) G cannot contain a W_k as a subgraph, for all $k \geq 5$.

(iv) G cannot contain a k-pool as a subgraph, for all odd $k \geq 3$.

Proof. (i) Let D a subgraph of G that is a diamond and such that $V(D) = \{u, v_1, v_2, v_3\}$ and $d_D(u) = d_D(v_2) = 3$. We can see that $M \cap E(D)$ must be one of the following sets: $\{uv_1, v_2v_3\}$, $\{v_1v_2, uv_3\}$, $\{uv_2\}$. In each one, u and v_2 are matched by M.

(ii) Suppose for a contradiction that $v \in V(G)$ is such that $G[N_G(v)]$ contains two vertex-disjoint P_3: P and P'. This implies that $G[\{v\} \cup P]$ and $G[\{v\} \cup P']$ are diamonds sharing a vertex of degree three in each one. Then, by Statement (i) it follows that M contains two incident edges, one for each diamond, a contradiction.

(iii) Suppose for a contradiction that G contains a subgraph H isomorphic to W_k, $k \geq 5$, where u is universal in H. If $k \geq 7$, it follows that u has two vertex-disjoint P_3 in its neighborhood, which contradicts Statement (ii). Then $k \leq 6$ and it can be easily verified that W_5 and W_6 are not in $\mathscr{BM}$.

(iv) Suppose for a contradiction that G contains a subgraph H isomorphic to a k-pool, for some odd $k \geq 3$. Let $C = p_1 p_2 \ldots p_k p_1$ be its internal cycle and let $B = \{b_1, b_2, \ldots, b_k\}$ be the vertices of the borders of H, such that $\{p_i b_i, p_{i+1} b_i\} \subset E(H)$, for all $1 \leq i \leq k$ modulo k. Clearly M must contain some edge of C and one edge of every triangle $p_i b_i p_{i+1} p_i$. W.l.o.g., consider $p_1 p_2 \in M \cap E(C)$. This implies that M contains no edge in $\{p_2 b_2, p_2 p_3, p_1 b_k, p_1 p_k\}$. Therefore, $p_k b_k$ and $p_3 b_2$ must be in M, which forbids two more edges from the triangles $p_{k-1} b_{k-1} p_k p_{k-1}$ and $p_3 b_3 p_4 p_3$. Continuing this process, it follows that $c_{\lfloor \frac{k+3}{2} \rfloor}$, which is at the same distance of p_1 and p_2 in C, must contain two incident edges in M, a contradiction. $\qquad\square$

Clearly every graph in $\mathscr{BM}$ admits a proper 4-coloring. Hence every graph in $\mathscr{BM}$ is K_5-free. More precisely, every graph in $\mathscr{BM}$ is W_5-free, as in Lemma 6, which implies that even some properly 3-colorable graphs do not have a decycling matching.

3   NP-Completeness for BM

In order to prove Theorem 1, we first show a polynomial-time reduction from the well known problem POSITIVE PLANAR 1-IN-3-SAT [29] to PLANAR 1-IN-3-SAT$_3$ – a variant of PLANAR 1-IN-3-SAT where each clause has either 2 or 3 literals and each variable occurs at most 3 times. Moreover, each positive literal occurs at most twice, while every negative literal occurs at most once in the given instance.

Let F be a Boolean formula in 3-CNF such that $\mathbf{X} = \{X_1, X_2, \ldots, X_n\}$ is the variable set and $\mathbf{C} = \{C_1, C_2, \ldots, C_m\}$ is the clause set of F. The *associated graph* $G_F = (V, E)$ of F is the bipartite graph such that there exists a vertex for every variable and clause of F, where $(\mathbf{X}, \mathbf{C})$ is the bipartition of $V(G_F)$, and there exists an edge $X_i C_j \in E(G_F)$ if and only if C_j contains either x_i or $\overline{x_i}$. We say that F is a *planar formula* if G_F is planar.

Theorem 7. PLANAR 1-IN-3-SAT$_3$ *is NP-complete.*

Proof. Let F be a Boolean planar formula in 3-CNF such that $\mathbf{X} = \{X_1, X_2, \ldots, X_n\}$ is the variable set and $\mathbf{C} = \{C_1, C_2, \ldots, C_m\}$ is the clause set of F. Since verifying whether a graph is planar can be done in linear time [24], as well as whether a formula in 3-CNF has a truth assignment, the problem is in NP.

We construct a formula F' from F as follows. If X_i is such that $d_{G_F}(X_i) = k \geq 3$, then we create k new variables X_i^z replacing the jth ($1 \leq j \leq k$) occurrence of X_i by a variable X_i^j, where a literal x_i (resp. $\overline{x_i}$) is replaced by a literal x_i^j (resp. $\overline{x_i^j}$). In addition, we create k new clauses $C_i^j = \left(x_i^j, \overline{x_i^{j+1}} \right)$, for $j \in \{1, \ldots, k-1\}$, and $C_i^k = \left(x_i^k, \overline{x_i^1} \right)$.

Let S be the set of all vertices $X_i \in V(G_F[\mathbf{X}])$ with $d_{G_F}(X_i) = k \geq 3$. For such a vertex $X_i \in S$, let $X_i' = \{X_i^1, \ldots, X_i^k\}$ and $C_i' = \{C_i^1, \ldots, C_i^k\}$.

(a) The head graph H.

(b) The bipartizing matching of H.

Fig. 2. The head graph and its unique bipartizing matching.

Note that, the associated graph $G_{F'}$ can be obtained from G_F by replacing the corresponding vertex of $X_i \in S$ by a cycle of length $2d_{G_F}(X_i)$ induced by the corresponding vertices of the new clauses in C'_i and the new variables in X'_i. In addition, for each $X_i \in S$ and $C_j \in N_{G_F}(X_i)$, an edge $X^l_i C_j$ is added in $E(G_{F'})$, such that every corresponding vertex $X^l_i \in X'_i$ has exactly one neighbor $C_j \notin C'_i$.

As we can see, every variable X occurs at most 3 times in the clauses of F', since every variable X_i with $d_{G_F}(X_i) \geq 3$ is replaced by $d_{G_F}(X_i)$ new variables that are in exactly 3 clauses of F'. By the construction, each literal occurs at most twice. Moreover, if F has no negative literals, then only the new variables have a negated literal and each one occurs exactly once in F'.

Consider a planar embedding Ψ of G_F. We construct $G_{F'}$ replacing each corresponding vertex $X_i \in S$ by a cycle of length $2d_{G_F}(X_i)$, as described above. After that, in order to preserve the planarity, we can follow the planar embedding Ψ to add a matching between vertices corresponding to variables in such a cycle and vertices corresponding to clauses $C_j \notin C'_i$ and that $X_i \in C_j$. This matching indicates in which clause of C'_i a given new variable will replace X_i in F'. Thus, without loss of generality, if G_F is planar, then we can assume that F' is planar as well.

As we can observe, for any truth assignment of F', all $X^l_i \in X_i$ (for a given variable X_i of F) have the same value. Hence, any clause of F' with exactly two literals has true and false values. At this point, it is easy to see that F has an 1-in-3 truth assignment if and only if F' has an 1-in-3 truth assignment. □

Now we show the NP-completeness of BM. Let us call the graph depicted in Fig. 2a by *head*. We also call vertex v as the *neck* of the head. The next lemmas are used in the correctness of our reduction.

Lemma 8. *An induced head H with neck v admits exactly one bipartizing matching M. Moreover v is matched by M.*

Proof. We can see that the thicker edges in Fig. 2b composes a bipartizing matching M of H, where the color of the vertices indicates the part of each one in $G - M$. Moreover M satisfies the lemma.

Now let us suppose that H admits another bipartizing matching M' which does not match v. In this case, we get that vh_1 and vh_4 does not belong to M', which implies that $h_1 h_4 \in M'$. By the triangle $h_1 h_2 h_5$, it follows that $h_2 h_5$ must be in M'. Hence the cycle $vh_1 h_2 h_3 h_4 v$ remains in $G - M'$, a contradiction.

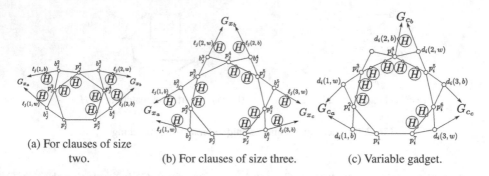

(a) For clauses of size two.

(b) For clauses of size three.

(c) Variable gadget.

Fig. 3. a and b are the clause gadgets and c is the variable gadget in Theorem 10. Each pair of arrow edges connects G_{X_i} to one clause gadget G_C such that $X_i \in C$.

Let M'' be a bipartizing matching that contains vh_4. In this case, the edge h_1h_2 cannot be in M'', otherwise the cycle $h_1h_4h_3h_2h_5h_1$ survives in $G - M''$. In the same way, the edge $h_1h_5 \notin M''$, otherwise the cycle $h_1h_2h_5h_3h_4h_1$ is not destroyed by M''. Therefore we get that h_2h_5 must be in M'', which implies that the cycle $h_1h_4h_3h_2h_5h_1$ belongs to $G - M''$, a contradiction. □

Lemma 9. *Let G be an odd k-pool with internal cycle $C = p_1p_2 \ldots p_kp_1$. Let b be a border of G, where $N_G(b) = \{p_1, p_k\}$. Then every bipartizing matching M of $G - b$ contains exactly one edge of C. Moreover, $c_1c_k \notin M$ for such bipartizing matchings.*

Proof. Let b_i be the i-th-border of G, such that $N_G(b_i) = \{p_i, p_{i+1}\}$, $1 \le i \le k - 1$. Since k is odd, every bipartizing matching of $G - b$ must contain at least one edge of C.

First, suppose that $G - b$ has a bipartizing matching M containing p_1p_k. In this case, we get that the edges in $\{p_1p_2, p_1b_1, p_kp_{k-1}, p_kb_{k-1}\}$ cannot be in M. Thus M must contain edges b_1p_2 and $b_{k-1}p_{k-1}$. In the same way, we can see that M cannot contain the edges $\{p_2p_3, p_2b_2, p_{k-1}p_{k-2}, p_{k-1}b_{k-2}\}$. Hence, it can be seen that all edges incident to $p_{\frac{k+1}{2}}$ are forbidden to be in M, which implies that the triangles containing $p_{\frac{k+1}{2}}$ have no edge in M, a contradiction by the choice of M.

Let p_ip_{i+1} be an edge of C in a bipartizing matching M of $G - b$, with $i \ne k$. In a same fashion, the edges in $\{p_ip_{i-1}, p_ib_{i-1}, p_{i+1}p_{i+2}, p_{i+1}b_{i+1}\}$ cannot be in M. Following this pattern, we can see that every edge b_jp_{j+1} must be in M, for every $j \in \{1, \ldots, k\} \setminus \{i\}$. Since M contains only one edge of C and one edge of every triangle of $G - b$, it follows that M is unique, for each edge p_ip_{i+1}, $i \ne k$. □

NP-Completeness for Planar Graphs of Maximum Degree 4.

We prove the *NP*-completeness by a reduction from PLANAR 1-IN-3-SAT₃. We first prove it for 3-colorable planar graphs of maximum degree 5. The circles with an H in the figures are induced head graphs, whose neck is the vertex touching the circle.

Theorem 10. BM *is NP-complete for 3-colorable planar graphs G of $\Delta(G) = 5$.*

Proof. Let F be an instance of PLANAR 1-IN-3-SAT₃, with $X = \{X_1, X_2, \ldots, X_n\}$ and $C = \{C_1, C_2, \ldots, C_m\}$ be the sets of variables and clauses of F, respectively. We construct a planar graph $G = (V, E)$ of maximum degree 5 as follows:

– For each clause $C_j \in C$, we construct a gadget G_{C_j} as depicted in Fig. 3a and b. Such gadgets are just a 5-pool and a 7-pool that we remove a border vertex, for clauses of size 2 and 3, respectively. Moreover, for the alternate edges of the internal cycle we subdivide them twice and append a head to each such a new vertex. Finally, we add two vertices $\ell_j(k, w)$ and $\ell_j(k, b)$, such that $b_j^{2k-1}\ell_j(k, w) \in E(G)$ and $b_j^{2k}\ell_j(k, b) \in E(G)$, for $k \in \{1, 2, 3\}$. We append a head to all such new vertices.

– For each variable $X_i \in X$, we construct a gadget G_{X_i} as depicted in Fig. 3c. This gadget is a 7-pool that we remove a border vertex. Moreover, we subdivide twice the edges $p_i^2 p_i^3$, $p_i^3 p_i^4$, $p_i^4 p_i^5$, and $p_i^6 p_i^7$, appending a head to each new vertex. We rename each border vertex b_i^{2k-1} as $d_i(k, b)$, $k \in \{1, 3\}$, and b_i^{2k} as $d_i(k, w)$, $k \in \{1, 2, 3\}$. Finally, a vertex $d_i(2, b)$ with a pendant head and adjacent to p_i^4 is added.

– The connection between clause and variable gadgets is as in Fig. 3. Each pair of arrow edges in a variable gadget G_{X_i} corresponds to the same pair in a clause gadget G_{C_j}, where $X_i \in C_j$. Precisely, if $x_i \in C_j$, then we add the edges $\ell_j(k, b)d_i(k', b)$ and $\ell_j(k, w)d_i(k', w)$, for some $k \in \{1, 2, 3\}$ and some $k' \in \{1, 2\}$. Note that each $\ell_j(k, b)$ (and $\ell_j(k, w)$) represents precisely one variable and it is connected to only one of $d_i(k', b)$ (and $d_i(k', w)$) in G_{X_i}. However, if $\overline{x}_i \in C_j$, then we add the edges $\ell_j(k, b)d_i(3, b)$ and $\ell_j(k, w)d_i(3, w)$, for some $k \in \{1, 2, 3\}$. Note that $d_i(3, b)$ and $d_i(3, w)$ represent $\overline{x}_i$.

– For a variable occurring exactly twice in F, just consider those connections corresponding to literals of X_i in the clauses of F, i.e., the pair $d_i(3, b)$, $d_i(3, w)$ (resp. $d_i(2, b)$, $d_i(2, w)$) will be used to connect to a clause gadget only if $\overline{x}_i$ occurs (resp. does not occur) in F.

Let G be the graph obtained from F by the above construction. We can see that G has maximum degree 5, where the only vertices with degree 5 are those p_i^4, for each variable gadget. Furthermore, it is clear that G is 3-colorable.

It remains to show that if G_F is planar, then G is planar. Consider a planar embedding ψ of G_F. We replace each vertex v_{X_i} of G_F by a variable gadget G_{X_i}, as well as every vertex v_{C_j} of G_F by a clause gadget G_{C_j}. The clause gadgets correspond to clauses of length two or three, which depends on the degree of v_{C_j} in G_F. Since the clause and variable gadgets are planar, we just need to show that the connections among them keep the planarity. Given an edge $v_{X_i}v_{C_j} \in E(G_F)$, we connect G_{X_i} and G_{C_j} by duplicating such an edge as parallel edges $\ell_j(k, w)d_i(k', w)$ and $\ell_j(k, b)d_i(k', b)$, for some $k \in \{1, 2, 3\}$ and some $k' \in \{1, 2\}$, or $\ell_j(k, b)d_i(3, b)$ and $\ell_j(k, w)d_i(3, w)$, for some $k \in \{1, 2, 3\}$, as previously discussed. Hence G is also planar.

In order to prove that F is satisfiable if and only if $G \in \mathscr{BM}$, we discuss some considerations related to bipartizing matchings of the clause and variable gadgets. By Lemma 9, we know that the graph obtained by removing a border from an odd k-pool admits a unique bipartizing matching for each edge of the internal cycle, except that whose endvertices are adjacent to the removed border. Furthermore, Lemma 8 implies that each external edge incident to the neck of an induced head cannot be in any bipartizing matching of G. Figure 4 shows the possible bipartizing matchings M, stressed edges, for the clause gadget G_{C_j} of clauses of size three. The black and white vertex assignment represents the bipartition of $G_{C_j} - M$. Exactly one pair of vertices $\ell_j(k, w)$ and $\ell_j(k, b)$ ($k \in \{1, 2, 3\}$) is such that they have the same color, while the others pairs have

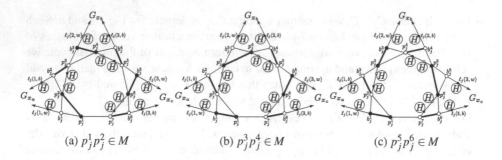

(a) $p_j^1 p_j^2 \in M$ (b) $p_j^3 p_j^4 \in M$ (c) $p_j^5 p_j^6 \in M$

Fig. 4. Configurations by removing a bipartizing matching clause gadget of size three.

opposite colors. Precisely, we can see that $\ell_j(k,w)$ has the same color for each pair with opposite color vertices as well as $\ell_j(k,b)$, for each bipartizing matching of G_{C_j}. Hence, we can associate one literal x_j^1, x_j^2, and x_j^3 to each pair of vertices $\ell_j(k,w)$ and $\ell_j(k,b)$, $k \in \{1,2,3\}$. A similar analysis can be done for clause gadgets of clauses of size two.

In the same way, each variable gadget G_{X_i} admits two possible bipartizing matchings M, as depicted in Fig. 5. We can see that the pair $d_i(3,b)$ and $d_i(3,w)$ has a different assignment for the other two pairs $d_i(k,b)$ and $d_i(k,w)$, $k \in \{1,2\}$. Moreover, the last two pairs have the same assignment, as depicted in Fig. 5a and b. One more detail is that the unique possibilities for such pairs is that $d_i(3,b)$ and $d_i(3,w)$ have opposite assignments if and only if the vertices $d_i(k,b)$ and $d_i(k,w)$ have the same assignment, $k \in \{1,2\}$. Therefore we can associate the positive literal x_i to the pairs $d_i(k,b)$ and $d_i(k,w)$, $k \in \{1,2\}$, while $\overline{x_i}$ can be represented by $d_i(3,b)$ and $d_i(3,w)$.

As observed above for clause gadgets, we can associate true value to the pair of vertices $\ell_j(k,w)$ and $\ell_j(k,b)$ with same color, $k \in \{1,2,3\}$. Hence exactly one of them is true, that is, exactly one literal of C_j is true. Moreover, each variable gadget has two positive literals and one negative. Hence, if $G \in \mathcal{BM}$, then every clause gadget has exactly one true literal and every variable has a correct truth assignment, which implies that F is satisfiable.

Conversely, if F is satisfiable, then each clause has exactly one true literal. Thus, for each clause gadget G_{C_j} we associate a same color to the pair of vertices corresponding to its true literal. By Fig. 4, there is an appropriate choice of a bipartizing matching for each true literal of C_j. The same holds for each variable gadget.   □

Proof of Theorem 1. Let G be the graph obtained by the construction in Theorem 10. Since the only vertices of degree 5 are those p_i^4 in the variable gadgets, we slightly modify the variable gadget as in Fig. 6a. In Fig. 2 we can see that vertex h_6 has degree 3, which allows us to use it to connect the variable gadget to the clause one. Figure 6b and c show the possible bipartizing matchings of the new variable gadget. Since such configurations are analogous to those of the original variable gadget, with respect to the vertex that connect to clause gadgets, the theorem follows.   □

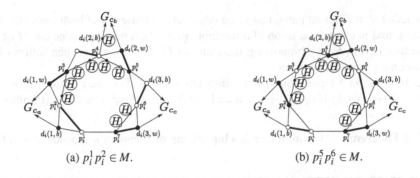

(a) $p_i^1 p_i^2 \in M$.   (b) $p_i^5 p_i^6 \in M$.

Fig. 5. All configurations by removing a bipartizing matching M of a variable gadget.

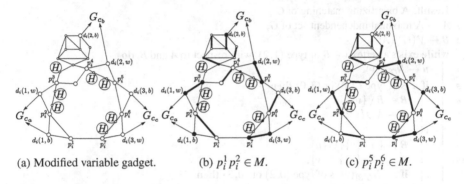

(a) Modified variable gadget.   (b) $p_i^1 p_i^2 \in M$.   (c) $p_i^5 p_i^6 \in M$.

Fig. 6. The modified variable gadget and its all configurations given by removing a bipartizing matching M.

4   Polynomial Time Results

Subcubic Graphs. Bondy and Locke [6] presented the following lemma, which was also obtained by Erdős [20] by induction on $n(G)$.

Lemma 11 (Bondy and Locke [6]). *Let G be a graph and let B be a largest bipartite subgraph of G. Then $d_B(v) \geq \frac{1}{2}d_G(v)$, for every $v \in V(G)$.*

Lemma 11 shows that every subcubic graph G admits a bipartizing matching, since every vertex has at most one incident edge not in a largest bipartite subgraph of G. This result was also obtained by Lovász [28] with respect to 1-improper 2-coloring of graphs with maximum degree at most 3. Although we know that every subcubic graph admits a bipartizing matching, to the best of our knowledge, there is no algorithm able to give such a matching in polynomial time. Therefore, one of our contributions is to provide a linear-time algorithm that returns such a matching for a subcubic graph G.

Given an initial bipartition of $V(G)$ into sets A and B, Algorithm 1 consists on swapping a vertex from one part $X \in \{A, B\}$ to the other whenever it has at least two neighbors in X. This procedure is based on the observation that the number of edges whose endvertices are in different parts, that is, an edge cut, increases by at least one after the swapping, and in every bipartition of a subcubic graph G, a maximal edge cut M of G is such that $E(G) \setminus M$ is a bipartizing matching of G. The algorithm then follows by constructing a maximal edge cut.

Let us consider a bipartition of $V(G)$ into sets A and B. For every vertex v, we say that v is of type (a,b) if $d_{V(G) \setminus X}(v) = a$ and $d_X(v) = b$, where X is the part (either A or B) which contains v.

Proof of Theorem 2. Algorithm 1 finds a bipartizing matching of a subcubic graph G.

Algorithm 1. A linear-time algorithm for BM on subcubic graphs.

Data: A subcubic graph $G = (V,E)$.
Result: A bipartizing matching of G.
1 $A \leftarrow$ A maximal independent set of G;
2 $B \leftarrow V(G) \setminus A$;
3 **while** exists a vertex $v \in B$ of type $(1,2)$, with respect to A and B, **do**
4 $\quad$ $u \leftarrow N_{G[A]}(v)$;
5 $\quad$ **if** u is of type $(2,0)$ or $(3,0)$ **then**
6 $\quad\quad$ $B \leftarrow B \setminus \{v\}$;
7 $\quad\quad$ $A \leftarrow A \cup \{v\}$;
8 $\quad$ **else**
9 $\quad\quad$ $B \leftarrow \{B \setminus \{v\}\} \cup \{u\}$;
10 $\quad\quad$ $A \leftarrow \{A \setminus \{u\}\} \cup \{v\}$;
11 $\quad\quad$ **if** $z \in N_{G[B]}(v)$ is of type $(0,2)$ or $(0,3)$ **then**
12 $\quad\quad\quad$ $B \leftarrow B \setminus \{z\}$;
13 $\quad\quad\quad$ $A \leftarrow A \cup \{z\}$;

14 **return** $E(G[A] \cup G[B])$;

Correctness of the Algorithm. Lines 1–2 initialize set A as a maximal independent set and $B = V(G) \setminus A$. Then every vertex of A is of type $(k,0)$ and there is no vertex in B of type $(0,k)$, $k \in \{1,2,3\}$. Therefore, if there exists a vertex v of type (a,b) with $a < b$, then it must be in B and it is of type $(1,2)$. In order to prove the correctness, it suffices to show that the operations within the loop never add a vertex of type (a,b) with $a < b$ in part A and whenever increase the number of edges from A towards B, which means that such edges are a maximal edge cut at some time step.

For a vertex $v \in B$ of type $(1,2)$, let $N_{G[A]}(v) = \{u\}$. If u is of type $(2,0)$ or $(3,0)$, then v is moved from B to A by lines 6–7. When u is of type $(2,0)$ it is of type $(1,1)$ after swapping, otherwise it is of type $(2,1)$ after swapping. Moreover, in both cases v modifies to type $(2,1)$ and the number of edges from A towards B increases by exactly one. Otherwise, lines 9–10 modify the type of v to $(3,0)$ and the type of u as follows.

- If u is of type $(1,0)$, then u continues of type $(1,0)$;
- If u is of type $(1,1)$, then u is modified to type $(2,0)$;
- If u is of type $(2,1)$, then u is modified to type $(2,1)$.

Since each swapping modifies only the types of u and v and of their neighbors, all the remaining vertices do not change their types. More specifically, if u is of type $(2,0)$ or $(3,0)$, then no other vertex in A changes its type besides u, and the two neighbors of v in B lose one neighbor in B. This implies that each vertex of A has at most one neighbor in A and the number of vertices of type $(1,2)$ in B decreases after the swapping. On the other hand, if u has a neighbor $z \neq v$ in A (which must be unique), then the swapping turns u into a vertex with one neighbor in B. Moreover, the number of edges from A towards B increases by at least one in each case and no vertex of A has two neighbors in A after the swapping. The unique problem occurs when z is either of type $(1,1)$ or $(1,2)$ before the swapping, when it turns into a vertex of type $(0,2)$ and $(0,3)$, respectively. These cases are considered in lines 12–13, where z is moved from B to A, modifying its type for $(2,0)$ or $(3,0)$.

We can easily observe that after the swapping, in any case, part A induces a subgraph of maximum degree at most one and the number of edges from A towards B whenever increases. This implies that the only vertex with two neighbors in itself part must be in B and they are of type $(1,2)$. Hence the algorithm finishes when it reaches a maximal edge cut with edges not in $G[A]$ or $G[B]$ and no vertex of type $(1,2)$ exists in B. Finally the algorithm returns the set of edges in $G[A]$ and $G[B]$ in line 14. $\Box$

Graphs Having Bounded Dominating Sets. Now, consider that the domination number of the input graph G is bounded by a constant k.

Proof (Theorem 3(a)). A dominating set of order at most k can be found in time $O(n^{k+2})$, by enumerating each vertex subset of size k and checking in time $O(n(G) + m(G))$ whether it is a dominating set or not. Let D be such a dominating set of G of order at most k. Let $\mathscr{P}_{\mathscr{D}}$ be the set of all bipartitions P_D of D into sets A_D and B_D, such that $G[A_D]$ and $G[B_D]$ do not have any vertex of degree 2. Note that $|\mathscr{P}_{\mathscr{D}}| = O(2^k)$.

Let $P_D \in \mathscr{P}_{\mathscr{D}}$ be a bipartition of D. We partition all the other vertices of $V(G) \setminus V(D)$ in such a way that P_D defines a bipartition of $G - M_D$, if one exists, where M_D is a matching that will be removed, given the choice of D. We do the following tests and operations for each vertex $v \in V(G) \setminus V(D)$:

- If $d_{G[A_D \cup \{v\}]}(v) \geq 2$ and $d_{G[B_D]}(v) \geq 2$, then P_D is not a valid partition;
- If $d_{G[A_D \cup \{v\}]}(v) \geq 2$, then $B_D \leftarrow B_D \cup \{v\}$;
- If $d_{G[B_D \cup \{v\}]}(v) \geq 2$, then $A_D \leftarrow A_D \cup \{v\}$.

Iteratively we allocate the vertices of $V(G) \setminus V(D)$ as described above into the respective sets A_D e B_D, or we stop if it is not possible to acquire a valid bipartition. After these operations, set $V' = V(G) \setminus \{A_D \cup B_D\}$ can be partitioned into three sets:

- $X = \{u \in V' : d_{G[A_D \cup \{u\}]}(u) = 1 \text{ and } d_{G[B_D \cup \{u\}]}(u) = 0\}$;
- $Y = \{u \in V' : d_{G[A_D \cup \{u\}]}(u) = 0 \text{ and } d_{G[B_D \cup \{u\}]}(u) = 1\}$;
- $Z = \{u \in V' : d_{G[A_D \cup \{u\}]}(u) = 1 \text{ and } d_{G[B_D \cup \{u\}]}(u) = 1\}$;

Since every vertex in $V(G) \setminus V(D)$ has a neighbor in D, it follows that the neighborhood of all the vertices of $V'' = X \cup Y \cup Z$ in $A_D \cup B_D$ is in D. In this way, we can make a choice of a matching M_D to be removed, where the vertices of V'' are allocated either in A_D or in B_D, and $G - M_D$ is bipartite. Since each vertex of D can be matched to at most one vertex of V'', there are $O\left((n-k)^k\right)$ possibilities of choices for M_D.

Hence we obtain the following complexity:

$$O\left(\sum_{i=1}^{k} n^{i+2} \cdot 2^i \cdot (n-i)^i\right) = O\left(k \cdot 2^k \cdot n^{k+2} \cdot (n-k)^k\right) = O\left(n^{2k+2}\right).$$

$\square$

Theorem 3(a) allows us to prove Theorem 3(b).

Proof (Theorem 3(b)). Every connected P_5-free graph has a dominating clique or a dominating P_3 [12], and graphs in $\mathscr{BM}$ do not admit K_5 as a subgraph. Thus, P_5-free graphs in $\mathscr{BM}$ have domination number at most four. $\square$

Graphs with Only Triangles as Odd Cycles. Consider now a slightly general version of BM, where some edges are forbidden to be in any bipartizing matching.

ALLOWED BIPARTIZING MATCHING (ABM)
Instance: A graph G and a set F of edges of G.
Task: Decide whether G has a bipartizing matching M that does not intersect F, and determine such a matching if it exists.

A matching M as in ABM is called an *allowed bipartizing matching* of (G, F).

We may clearly assume G as connected and bridge-free. Moreover, note that if (G, F) has an allowed bipartizing matching, then $G \in \mathscr{BM}$.

Proof (Theorem 3(c)). Let G be a graph having no C_{2k+1}, for $k > 1$, and let $F \subseteq E(G)$.

First, consider G a non-bipartite graph with no cut vertex, and let $v_1 v_2 v_3 v_1$ be an triangle of G. Without loss of generality, we can assume that there is a vertex, say v_1, such that $\{v_1 v_2, v_1 v_3\}$ is not an edge cut, otherwise G would be a triangle. Then $G - \{v_1 v_2, v_1 v_3\}$ has a path P from v_1 to $\{v_2, v_3\}$. Consider P as a longest one of length at least 2, and let v_2 be the first vertex reached by P between v_2 and v_3. Thus P must be of the form $v_1 u v_2$, otherwise either $G[V(P) \cup \{v_3\}]$ or $G[V(P)]$ contains an odd cycle of length at least 5, when P has either an even or odd number of vertices, respectively.

If $w \in V(G)$ has exactly one neighbor $z \in \{v_1, v_2, v_3, u\}$, then w is in a path P' of length at least two between z and $z' \in \{v_1, v_2, v_3, u\}$, $z \neq z'$. Hence $P' \cup \{v_1, v_2, v_3, u\}$ contains an odd cycle of length at least 5. Hence consider that w has at least two neighbors in $\{v_1, v_2, v_3, u\}$. If $uv_3 \in E(G)$, then $G[\{v_1, v_2, v_3, u\}]$ is a K_4, which implies that $V(G) = \{v_1, v_2, v_3, u\}$, since $\{w, v_1, v_2, v_3, u\}$ induces an odd cycle. Thus G has an allowed bipartizing matching if and only if a maximal matching of G no intersecting F.

Otherwise if $uv_3 \notin E(G)$, then w must be adjacent to either u and v_3 or to v_1 and v_2, and no other vertex in $\{v_1, v_2, v_3, u\}$. Moreover, the vertices adjacent to both u and v_3 induce an independent set of G, as well as the vertices adjacent to both v_1 and v_2. In this case we can see that G has an allowed bipartizing matching if and only if $v_1 v_2 \notin F$.

Now, we consider a block decomposition of G with block-cut tree T. Let B be a block containing exactly one cut-vertex v, that is, B is a leaf in T. If (B, F) has an allowed bipartizing matching which is not incident to v, then (G, F) has an allowed bipartizing matching if and only if (G',F) admits an allowed bipartizing matching, where $G' = ((V(G) \setminus V(B)) \cup \{v\}, E(G) \setminus E(B))$. Otherwise, (G,F) has an allowed bipartizing matching if and only if (B,F) and (G'',F'') admit allowed bipartizing matchings, where $G'' = ((V(G) \setminus V(B)) \cup \{v\}, E(G) \setminus E(B))$ and $F'' = F \cup \{uv \mid u \in N_{G''}(v)\}$.

As in a block the desired matchings can be found, if any exists, in polynomial time, it is easy to see that we can solve ABM in polynomial time.   □

5 Taking the Clique-Width as Parameter

Definition 12. *The clique-width of a graph G, denoted by $cwd(G)$, is the minimum number of labels needed to construct G, using the following operations [8]:*

1. *Create a single vertex v with an integer label ℓ (denoted by $\ell(v)$);*
2. *Disjoint union of two graphs (i.e. co-join) (denoted by $\oplus$);*
3. *Join by an edge every vertex labeled i to every vertex labeled j for $i \neq j$ (denoted by $\eta(i,j)$);*
4. *Relabeling all vertices with label i by label j (denoted by $\rho(i,j)$).*

Some graph classes with bounded clique-width include cographs [8], graphs with bounded tree-width, and distance-hereditary graphs [22].

For any graph G with clique-width bounded by a constant k and for each graph property Π that can be formulated in a *monadic second order logic* ($MSOL_1$), we can use the result of Courcelle, Makowsky, and Rotics [15] which establishes that there exits an algorithm running in time $f(cwd(G)) \cdot n(G)$ that decides whether G satisfies Π. In $MSOL_1$, we obtain a graph that is described by a set of vertices V and a binary adjacency relation $edge(.,.)$, and the graph property in question may be defined in terms of vertex sets, but not in terms of edge sets.

Proof (Theorem 4). Recall that BM is equivalent to determining whether G admits a $(2,1)$-coloring. Thus, using Courcelle, Makowsky and Rotics's meta-theorem based on monadic second order logic for graphs G with bounded clique-width [15], it is enough to observe that the property "G has a $(2,1)$-coloring" is $MSOL_1$-expressible. We construct a formula $\varphi(G)$ such that $G \in \mathcal{BM} \Leftrightarrow \varphi(G)$ as follows:

$$\exists S_1, S_2 \subseteq V(G) : (S_1 \cap S_2 = \emptyset) \wedge (S_1 \cup S_2 = V(G)) \wedge$$
$$(\forall v_1 \in S_1 [\nexists u_1, w_1 \in S_1 : (u_1 \neq w_1) \wedge edge(u_1, v_1) \wedge edge(w_1, v_1)]) \wedge$$
$$(\forall v_2 \in S_2 [\nexists u_2, w_2 \in S_2 : (u_2 \neq w_2) \wedge edge(u_2, v_2) \wedge edge(w_2, v_2)]).$$

□

Since K_5 is a forbidden subgraph, chordal graphs in $\mathcal{BM}$ have bounded treewidth [31], and thus, BM is polynomial-time solvable for such a class, proving Corollary 5.

References

1. Andrews, J., Jacobson, M.: On a generalization of chromatic number. In: Proceedings of Sixteenth Southeastern International Conference on Combinatorics, Graph Theory and Computing (SEICCGTC 1985), vol. 47, pp. 18–33 (1985)
2. Angelini, P., et al.: Vertex-coloring with defects. J. Graph Algor. Appl. **21**(3), 313–340 (2017). https://doi.org/10.7155/jgaa.00418
3. Axenovich, M., Ueckerdt, T., Weiner, P.: Splitting planar graphs of girth 6 into two linear forests with short paths. J. Graph Theory **85**(3), 601–618 (2017). https://doi.org/10.1002/jgt.22093
4. Bodlaender, H.L.: A partial k-arboretum of graphs with bounded treewidth. Theor. Comput. Sci. **209**(1–2), 1–45 (1998). https://doi.org/10.1016/S0304-3975(97)00228-4
5. Bonamy, M., Dabrowski, K.K., Feghali, C., Johnson, M., Paulusma, D.: Independent feedback vertex set for P_5-free graphs. Algorithmica (2018). https://doi.org/10.1007/s00453-018-0474-x
6. Bondy, J.A., Locke, S.C.: Largest bipartite subgraphs in triangle-free graphs with maximum degree three. J. Graph Theory **10**(4), 477–504 (1986)
7. Borodin, O., Kostochka, A., Yancey, M.: On 1-improper 2-coloring of sparse graphs. Discrete Math. **313**(22), 2638–2649 (2013). https://doi.org/10.1016/j.disc.2013.07.014
8. Brandstädt, A., Dragan, F.F., Le, H., Mosca, R.: New graph classes of bounded clique-width. Theory Comput. Syst. **38**(5), 623–645 (2005). https://doi.org/10.1007/s00224-004-1154-6
9. Brandstädt, A., Engelfriet, J., Le, H., Lozin, V.V.: Clique-width for 4-vertex forbidden subgraphs. Theory Comput. Syst. **39**(4), 561–590 (2006)
10. Brandstädt, A., Klembt, T., Mahfud, S.: P_6- and triangle-free graphs revisited: structure and bounded clique-width. Discrete Math. Theor. Comput. Sci. **8**, 173–188 (2006)
11. Burzyn, P., Bonomo, F., Durán, G.: NP-completeness results for edge modification problems. Discrete Appl. Math. **154**(13), 1824–1844 (2006). https://doi.org/10.1016/j.dam.2006.03.031
12. Camby, E., Schaudt, O.: A new characterization of P_k-free graphs. Algorithmica **75**(1), 205–217 (2016). https://doi.org/10.1007/s00453-015-9989-6
13. Carneiro, A.D.A., Protti, F., Souza, U.S.: Deletion graph problems based on deadlock resolution. In: Cao, Y., Chen, J. (eds.) COCOON 2017. LNCS, vol. 10392, pp. 75–86. Springer, Cham (2017). https://doi.org/10.1007/978-3-319-62389-4_7
14. Choi, H.A., Nakajima, K., Rim, C.S.: Graph bipartization and via minimization. SIAM J. Discrete Math. **2**(1), 38–47 (1989). https://doi.org/10.1137/0402004
15. Courcelle, B., Makowsky, J.A., Rotics, U.: Linear time solvable optimization problems on graphs of bounded clique-width. Theory Comput. Syst. **33**(2), 125–150 (2000)
16. Cowen, L., Goddard, W., Jesurum, C.E.: Defective coloring revisited. J. Graph Theory **24**(3), 205–219 (1997)
17. Cowen, L.J., Cowen, R., Woodall, D.: Defective colorings of graphs in surfaces: partitions into subgraphs of bounded valency. J. Graph Theory **10**(2), 187–195 (1986)
18. Diestel, R.: Graph Theory. Springer, Heidelberg (2017)
19. Eaton, N., Hull, T.: Defective list colorings of planar graphs. Bull. Inst. Combin. Appl **25**, 79–87 (1999)
20. Erdős, P.: On some extremal problems in graph theory. Israel J. Math. **3**(2), 113–116 (1965)
21. Furmańczyk, H., Kubale, M., Radziszowski, S.: On bipartization of cubic graphs by removal of an independent set. Discrete Appl. Math. **209**, 115–121 (2016)
22. Golumbic, M.C., Rotics, U.: On the clique-width of some perfect graph classes. Int. J. Found. Comput. Sci. **11**(03), 423–443 (2000). https://doi.org/10.1142/S0129054100000260

23. Harary, F., Jones, K.: Conditional colorability ii: bipartite variations. Congr. Numer. **50**, 205–218 (1985)
24. Hopcroft, J., Tarjan, R.: Efficient planarity testing. J. ACM **21**(4), 549–568 (1974)
25. Lampis, M.: Algorithmic meta-theorems for restrictions of treewidth. Algorithmica **64**(1), 19–37 (2012). https://doi.org/10.1007/s00453-011-9554-x
26. Lima, C.V.G.C., Rautenbach, D., Souza, U.S., Szwarcfiter, J.L.: Decycling with a matching. Inf. Proc. Lett. **124**, 26–29 (2017). https://doi.org/10.1016/j.ipl.2017.04.003
27. Liu, Y., Wang, J., You, J., Chen, J., Cao, Y.: Edge deletion problems: branching facilitated by modular decomposition. Theor. Comput. Sci. **573**, 63–70 (2015)
28. Lovász, L.: On decomposition of graphs. Studia Sci. Math. Hungar. **1**, 237–238 (1966)
29. Mulzer, W., Rote, G.: Minimum-weight triangulation is NP-hard. J. ACM **55**(2), 1–29 (2008)
30. Protti, F., Souza, U.S.: Decycling a graph by the removal of a matching: characterizations for special classes. CoRR abs/1707.02473 (2017). http://arxiv.org/abs/1707.02473
31. Robertson, N., Seymour, P.: Graph minors. ii. Algorithmic aspects of tree-width. J. Algorith. **7**(3), 309–322 (1986). https://doi.org/10.1016/0196-6774(86)90023-4
32. Schaefer, T.J.: The complexity of satisfiability problems. In: Proceedings 10th Symposium on Theory of Computing (STOC 1978), pp. 216–226. ACM Press, New York (1978). https://doi.org/10.1145/800133.804350
33. Thorup, M.: All structured programs have small tree width and good register allocation. Inf. Comput. **142**(2), 159–181 (1998). https://doi.org/10.1006/inco.1997.2697

Network Flow and Security

Removing Undesirable Flows by Edge Deletion

Gleb Polevoy[1]([✉]), Stojan Trajanovski[1,2], Paola Grosso[1], and Cees de Laat[1]

[1] University of Amsterdam, Amsterdam, The Netherlands
G.Polevoy@uva.nl
[2] Philips Research, Eindhoven, The Netherlands

Abstract. Consider mitigating the effects of denial of service or of malicious traffic in networks by deleting edges. Edge deletion reduces the DoS or the number of the malicious flows, but it also inadvertently removes some of the desired flows. To model this important problem, we formulate two problems: (1) remove all the undesirable flows while minimizing the damage to the desirable ones and (2) balance removing the undesirable flows and not removing too many of the desirable flows. We prove these problems are equivalent to important theoretical problems, thereby being important not only practically but also theoretically, and very hard to approximate in a general network. We employ reductions to nonetheless approximate the problem and also provide a greedy approximation. When the network is a tree, the problems are still MAX SNP-hard, but we provide a greedy-based $2l$-approximation algorithm, where l is the longest desirable flow. We also provide an algorithm, approximating the first and the second problem within $2\sqrt{2|E|}$ and $2\sqrt{2(|E| + |\text{undesirable flows}|)}$, respectively, where E is the set of the edges of the network. We also provide a fixed-parameter tractable (FPT) algorithm. Finally, if the tree has a root such that every flow in the tree flows on the path from the root to a leaf, we solve the problem exactly using dynamic programming.

1 Introduction

Attacks such as the (distributed) Denial of Service (DDoS) [14] are widespread [19] and heavily impede the functionality, especially when the system is required to be quick (soft real time, for example) [13]. One of the options to fight the problem is deleting network edges or disabling them (anyway, deleting from the network graph) [11]. Another practically important problem is having malicious connections, like Trojans. The danger is not only the bandwidth these connections take but primarily the information they transfer.

We define a flow as a path from the source to the sink and we model DoS or merely malicious communication as a set of undesirable (name them *bad*) flows. The system has also desirable (call them *good*) flows. We aim to remove the undesirable flows by deleting some edges on their paths, while minimizing the resulting damage to the desirable flows. The model assumes no rerouting of

© Springer Nature Switzerland AG 2018
D. Kim et al. (Eds.): COCOA 2018, LNCS 11346, pp. 217–232, 2018.
https://doi.org/10.1007/978-3-030-04651-4_15

flows. If we delete an edge on the path of a bad flow, we consider that flow to be removed, but we also inadvertently remove all the good flows that pass through the deleted edge.

We can identify the DoS flows by frequent access trials from the same IP group, and the malicious flows can be identified by information leaks. Therefore, we assume we know which flows are good and which are bad.

Judicious deletion is crucial, as Example 1 shows. We need to autonomously decide which edges to delete as suggested in [11]. This calls for an algorithm to find which edges to delete. In order to cope with large instances in real time, the algorithm has to be polynomial.

Example 1. In Fig. 1, removing all the bad flows by deleting their only common edge e would remove no good flows. This is infinitely better than removing each bad flow b_i by deleting any of its edges other than e, because that would also remove g_i.

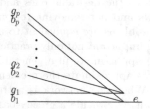

Fig. 1. The bad flows are denoted by b with an index, while the good ones are denoted by g with an index. The paths of flow b_i and g_i coincide, besides edge e, belonging only to the bad flows.

Consider this example of using an algorithm that finds which edges to delete.

Example 2. In a (for example, software-defined) network, assume that the intrusion detection system discovers a DDoS attack, and determines which flows are attacking. We need to respond quickly and efficiently by removing certain links from the network. Assuming we know which flows are desirable and which are attacking, we first estimate how important it is not to remove each desirable flow by deleting a link. Now, we run our algorithm to obtain an (approximately) easiest set of the edges to delete, such that all the attacking flows are removed (disconnected) while the minimum damage is inflicted on the desirable flows.

We present the necessary background in Sect. 1.1. To solve the problem, in Sect. 2 we model the situation as two possible problems. First, we show in Sect. 3 that our problems are equivalent w.r.t. approximation to notorious hard problems. Those hard and important problems admit polynomial approximations, but those are extremely loose. We then reduce the problems to submodular set cover to use approximation known for that problem and also suggest a greedy approximation in Sect. 4. We approximate the important particular case when the network is a tree in Sect. 5, which is still MAX SNP-hard. We assume in Sect. 5.3 that the tree has a root such that every flow is on the path from the root to a leaf, as often happens when communicating through service providers to the backbones, and solve this case exactly using dynamic programming. Finally, we provide a fixed parameter tractable (FPT) algorithm for trees in Sect. 5.4. These results suggest determining the kind of network we have at hand and subsequently applying the best applicable algorithm. We summarize our approach and suggest further research directions in Sect. 6.

We approximate the problem in general and provide additional approximation and exact algorithms for special cases such as trees.

1.1 Related Work

We study edge deletion that removes the bad flows and does not remove too many good ones, which is a new problem in the realm of edge deletion problems. The simpler problem of deleting the minimum number of edges that can disconnect all the flows from a source to a sink is a famous problem, and Menger's theorem [2, Chap. 3.2] characterizes the minimum number of edges one has to remove in order to disconnect the source from the sink. Finding a minimum cut in a graph [4, Chap. 26] and disconnecting it is optimal for this problem.

There are many other problems of edge deletion, such as deleting the minimum number of edges to obtain certain properties like no cycles of various lengths [18], or removing forbidden graphs [17]. Similar network design studies include other problems, such as edge addition [10]. In practice, edge deletion can be especially easily implemented in a software-defined network. See [11] for a practical example.

There are many important vertex deletion problems as well, such as the famous vertex cover problem [7]. Other vertex deletion problems include the feedback vertex set, where we aim to break all the cycles in a graph [1], and the problem of breaking all the cycles of a given length or at most a given length [18].

2   Model

Let the flow network be a directed graph $G = (N, E)$. A flow f from node a to node z in this network is a path from source a to sink z, each of which edges carries the flow. Formally, $f = (P(f))$, where $P(f)$ is the set of the edges of the path that the flow takes from a to z. Flow in this paper are not splittable, meaning that a flow takes a single path. This can also model a splitting flow as separate flows with partially overlapping paths. We do not model the capacities of the edges and the values of the flows, because these notions are irrelevant to the problem. Of course, real flows have values and real edges (say, wires or roads) have capacities, and the total flow value on an edge cannot exceed the capacity of the edge.

We model removal of undesirable flow as two problems. The first one follows:

Definition 1. *The* Bad Flow Removing (BFR) *problem receives the input* $(G = (N, E), F, GF, BF, w \colon GF \to \mathbb{R}_+)$. *Here,* $G = (N, E)$ *is a network with flows* $F = \{f_i\}$, *where some flows, denoted* $GF = \{g_i\} \subseteq F$, *are marked as good (desirable), and the rest, denoted* $BF = \{b_i\} \overset{\triangle}{=} F \backslash GF$, *are bad (undesirable). Every good flow* f *is endowed with a weight* $w(f)$, *designating how important the good flow is.*

A solution $S \subseteq E$ *is a subset of edges to delete.*

Flow f *is removed by* S *if* $S \cap P(f) \neq \emptyset$; *otherwise, it is remaining. A feasible solution is a solution such that all the bad flows are removed.*

We aim to find a feasible solution with the minimum total weight of the removed good flows. Intuitively, we aim to remove all the bad flows while minimizing the weight of the removed good flows.

BFR assumes we have to get rid of all the bad flows. For instance, when flows can spread malicious content or steal information, leaving even few flows can harm the network. Since sometimes bad flows are mostly taking resources and are not so dangerous, we may merely want to remove most of bad flows while still not losing too many good ones, so we relax the requirement to remove all the bad flows and allow leaving them for a cost:

Definition 2. *The* Balanced Bad Flow Removing (BBFR) *problem receives* $(G = (N, E), F, GF, BF, w \colon F \to \mathbb{R}_+)$; *the difference from the BFR is that all the flows are weighted.*

Here, a solution is the same as in BFR, but any solution is feasible here.

We aim to find a feasible solution such that the total weight of the remaining bad flows plus the total weight of the removed good flows is minimized. Intuitively, we aim to balance removing the bad flows and not removing too many good ones.

Denote by $D(S)$ the flows removed from the set of flows D by deleting the edges S. We define the weight of a set of flows $D \subset F$ as $\sum_{f \in D} w(f)$, which we denote as $w(D)$, abusing the notation w. Therefore, BFR removes all the bad flows $(BF(S) = BF)$ while aiming to minimize $w(GF(S))$, while BBFR targets to minimize the $w(BF \backslash BF(S)) + w(GF(S))$. We do not add a balancing parameter such as in $w(BF \backslash BF(S)) + \alpha w(GF(S))$, because the weights can be defined already with balancing in mind, modeling such a balancing.

This formalism allows to prune all the edges that do not belong to good flows, so we can assume that all edges belong to at least one good flow. On the other hand, if one wants to avoid deleting edges lightly, one can model this by introducing dummy good flows of length 1 through all the edges.

3   Equivalence Resulting in Hardness and Approximation

We prove that Problems BFR and BBFR, are equivalent w.r.t. approximation to the classical Red-Blue Set Cover [3] and to the Partial Set Cover [12] problems, respectively. This immediately implies strong inapproximability and provides some approximation algorithms.

3.1 BFR

We first formally define the RBSC problem.

Definition 3. *The* Red-Blue Set Cover (RBSC) *problem* [3] *receives the input* $(R, B, S, w \colon R \to \mathbb{R}_+)$, *where* R *and* B *are two disjoint sets of red and blue elements and* $S \subseteq 2^{R \cup B}$, *i.e. every set* S *in the collection* S *is a subset of* $R \cup B$. *These subsets can cover all the blue elements, i.e.* $B \subseteq \cup_{S \in S} S$. *Finally,* w *denotes the weight of the red elements.*

A solution is a subcollection C *of* S.

A feasible solution is a solution C *that covers all the blue elements, i.e.* $B \subseteq \cup_{S \in C} S$. *The aim is to find a feasible solution with the minimum total weight of the covered red elements,* $w(C) \triangleq \sum_{r \in R \cap \{\cup_{S \in C} S\}} w(r)$.

Theorem 1. *BFR is equivalent w.r.t. approximation to RBSC. Reducing from RBSC to BFR, we can ensure that* $|E| = 2 |S|$, $|GF| = |R| + 1$ *and* $|BF| = |B|$. *Reducing from BFR to RBSC,* $|S| = |E|$, $|R| = |GF|$ *and* $|B| = |BF|$.

Proof. We first reduce RBSC to BFR. For each set $S \in S$, define edge e_S and let all the edges have a common node. Define good flow g_i for each red element $r_i \in R$, with the same weight, and bad flow b_j for each blue element $b_j \in B$, such that the path of a flow contains edge s_S if and only if the element from which the flow has been created is included in S. To be able to route flows through the required edges from $E(S) \triangleq \{e_S | S \in S\}$, we add edges, called E', that connect the non-common nodes of the edges from $E(S)$, as illustrated in Fig. 2. To prevent choosing the edges from E', we define an additional good flow that contains every edge from E', and we give this good flow a prohibitively high weight, say $2 \sum_{g \in GF} w(s)$. The solutions to the constructed BFR are directly transformed to solutions for RBSC, besides the case when a solution for BFR contains an edge from E'. Such solutions are transformed to the trivial solution for RBSC that contains all the sets. Unless edges from E' are selected for the solution for BFR, the weight of the covered red elements is equal to the weights of the removed good flows. If edges from E' are selected, then the cost of that solution is at least twice higher than the cost of the corresponding solution for RBSC. Therefore, the reduction preserves approximation.

We now reduce BFR to RBSC. Make a red element from a good flow, a blue element from a bad flow and a set from an edge, such that the set contains exactly the flows that have the edge on their paths. The weights are transferred as they are. The solution to the obtained RBSC instance is mapped to a solution for the original BFR in the reverse manner. This mapping preserves being a solution, feasibility and the weights.   ■

This theorem immediately implies that all the hardness and all the positive results for RBSC transfer to BFR. In particular, Theorems 3.1 and 3.2 from Sect. 3 of [3] imply the following.

RBSC   BFR

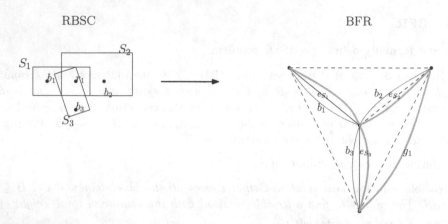

Fig. 2. On the right, the bad flows are denoted by b with an index, while the good ones are denoted by g with an index. The dashed lines are the E' edges.

Corollary 1. *1. Unless $NP \subseteq DTIME(n^{\text{polylog}(n)})$, it is impossible to approximate BFR within $O(2^{\log^{1-\delta}|E|})$, for any $\delta > 0$.*

2. Unless $P = NP$, it is impossible to approximate BFR within $O(2^{\log^{1-\delta}|E|})$, where $\delta = 1/\log\log^c|E|$, for any constant $c < 0.5$.

Proof. The existing hardness results state the impossibility to approximate within $O(2^{\log^{1-\delta}|\mathcal{S}|})$, for certain values of δ. If BFR could be approximated within $O(2^{\log^{1-\delta}|E|})$, reducing RBSC to BFR would allow approximating RBSC within $O(2^{\log^{1-\delta}2|\mathcal{S}|})$, contrary to the impossibility.   ∎

On the approximation side, Theorems 3.5 and 3.6 from [15] imply that.

Corollary 2. *BFR can be approximated within $2\sqrt{|E|\log|BF|}$.*

Proof. The approximation follows by reducing the given instance of BFR to an RBSC, employing Algorithm Low_Deg2 from Sect. 3 of [15] on the obtained instance of RBSC, and then translating the obtained cover to a set of the corresponding edges. The transformations from BFR to RBSC and back preserve the approximation ratio, and the approximation of the algorithm, $2\sqrt{|\mathcal{S}|\log|B|}$, is equal to $2\sqrt{|E|\log|BF|}$, since each edge defines a set in the reduction of BFR to RBSC.   ∎

3.2 BBFR

For the sake of handling the BBFR problem, we define a problem we will reduce to and from, namely the Positive-Negative Partial Set Cover problem, which generalizes the unweighted definition from Sect. 1.1 from [12].

Definition 4. *The Positive-Negative Partial Set Cover (±PSC) problem receives the input $(R, B, \mathcal{S}, w : R \cup B \to \mathbb{R}_+)$, where R and B are two disjoint* *sets of red and blue elements and every set S in the collection $\mathcal{S}$ is a subset of $R \cup B$. Finally, we denote by w the weight of the elements.*

A solution *is a subcollection $\mathcal{C}$ of $\mathcal{S}$.*

Any solution is feasible *here. The aim is to find a feasible solution with the minimum total weight of the uncovered blue elements plus the covered red elements,*
$w(\mathcal{C}) \triangleq \sum_{b \in B \setminus \cup_{S \in \mathcal{C}} S} w(b) + \sum_{r \in R \cap \{\cup_{S \in \mathcal{C}} S\}} w(r).$

We now reduce BBFR to and from $\pm$PSC, omitting the proofs to save space.

Theorem 2. *BBFR is approximation-equivalent to $\pm$PSC. Reducing from $\pm$PSC to BBFR, we can ensure that $|E| = 2|\mathcal{S}|$, $|GF| = |R|+1$ and $|BF| = |B|$. Reducing from BBFR to $\pm$PSC, $|\mathcal{S}| = |E|$, $|R| = |GF|$ and $|B| = |BF|$.*

This theorem implies that the hardness results and the approximation results transfer from $\pm$PSC to BBFR. In particular, the following holds:

Corollary 3. *1. Unless $NP \subseteq DTIME(n^{\mathrm{polylog}(n)})$, it is impossible to approximate BBFR within $O(2^{\log^{1-\delta}|E|})$, for any $\delta > 0$.*
2. Unless $P = NP$, it is impossible to approximate BBFR within $O(2^{\log^{1-\delta}|E|})$, where $\delta = 1/\log\log^c|E|$, for any constant $c < 0.5$.
3. Unless $P = NP$, there is no approximation to BBFR within $O(2^{\log^{1-\delta}|BF|})$, for any $\delta > 0$.

As for approximation, Corollary 3 from [12] implies.

Corollary 4. *BBFR is approximable within $2\sqrt{(|E| + |BF|)\log(|BF|)}$.*

4   Approximation

First, consider the approximation for submodular cost set cover within the maximal frequency of an element from [9, Sect. 4.4]. Since the total weight of the removed good flows is a submodular function of the chosen edge subset, that approximation applies to BFR. The maximal frequency of an element becomes the maximal number of edges a bad flow contains, i.e. l', which is the approximation ratio for BFR. We can also approximate BBFR using the following theorem.

Theorem 3. *BBFR can be reduced to BFR while preserving approximation by extending the path of each bad flow at its end by an edge with a new second node and defining a new good flow with the path that consists solely of this new edge and with the weight $w(b_i)$.*

Proof. Miettinen [12, Sect. 2.2] reduces $\pm$PSC to RBSC, inspiring us to the following reduction of BBFR to BFR. Given a BBFR instance $(G = (N, E), F, GF, BF, w: F \to \mathbb{R})$, construct the following BFR instance $(G = (N', E'), F', GF', BF', w': GF \to \mathbb{R})$. We extend the path of each bad flow

b_i at its end by an edge with a new second node and we also define a new good flow with the path that consists solely of this new edge. The weight of the new good flow is defined to be $w(b_i)$, and the function w is restricted to GF, to obtain a BFR. We transform a solution S to this BFR to the solution $S \cap E$ for the original BBFR.

Similarly to [12, Sect. 2.2], the approximation $S \cap E$ provides for the original BBFR is at least as good as the approximation S provides for BFR.   ∎

This reduction increases the maximum length of a bad flow by 1, and thus, allows approximating BBFR withing $(l' + 1)$.

We also propose the following greedy algorithm, inspired by the famous greedy algorithm for set cover [16, Algorithm 2.2] and the greedy algorithm from [15, Sect. 3.1].

ALGORITHM 1. GreedyBFR$(G = (N, E), F, GF, BF, w)$

---

1. Given a BFR instance, define the following weighted set cover instance.
   (a) the elements are the bad flows with all edges intersecting good flows;
   (b) the sets are the good flows, a good flow covering all the bad flows it intersects.
2. Approximately solve this set cover instance, obtaining the output $S \subseteq GF$.
3. Return the edge set of S, i.e. $\cup_{g \in S} P(g)$, augmented with edges of bad flows intersecting no good flows.

Proposition 1. *Algorithm 1 approximates BFR within factor* $k \cdot (\ln(|BF|) + 1)$.

Proof. Algorithm 1 returns a feasible solution, since all the bad flows intersecting at least one good flow can be covered by the edges of the good flows.

As for the approximation ratio, call the original problem I and let I' denote the problem we construct in line 1. Problem I' models directly removing the bad flows by removing the good ones that intersect them, ignoring the fact that removing occurs through deleting edges, which can remove several intersecting good flows simultaneously. Therefore, $\text{OPT}(I') \leq \text{OPT}(I)$, and so the $\ln(|BF|)+1$-approximation to I' costs at most $\ln(|BF|) + 1$ times $\text{OPT}(I)$. We do have to add the intersecting good flows to ensure feasibility. This action can require the k factor, implying the proposition.   ∎

Theorem 3 implies we can approximate BBFR within the same factor. Theorems 1 and 2 imply that this algorithm allows approximating the Red-Blue Set Cover and the Positive-Negative Partial Set Cover within $k \cdot (\ln(|B|) + 1)$, k being the maximum number of the red elements that a red elements can have a common set with.

Remark 1. Using the 2-approximation for set cover on trees [8, Sect. 7] instead of the general approximation for set cover, we can adopt Algorithm 1 to approximate BFR on trees within $2k$. As before, Theorem 3 implies the same approximation factor for BBFR as well.

Similarly to Peleg [15, Sect. 3.2], we could continue with dividing the good flows to those intersecting many other good flows and not, approximating the problem with the good flows intersecting not too many other good flows and bounding the number of good flows that intersect many others. We omit the details, because this approach would yield the same approximation ratios as Peleg receives for Red-Blue Set Cover.

5 Trees

We showed that the problems BFR and BBFR are extremely hard to approximate, though we still provided approximations. Given the proven hardness of the general case, we now concentrate on trees, often found in communication networks. Trees subsume star networks, for instance. For trees, we can do more.

5.1 Hardness

We will show that BFR and BBFR problems are MAX SNP-hard even for trees. First, we prove an important result, connecting the two problems.

Theorem 4. *BFR is reducible to BBFR with an approximation-preserving reduction that preserves the problem instance, besides assigning weights to the bad flows.*

Proof. Consider the following reduction. Given an instance x of BFR, construct an instance x' of BBFR by defining the weight of each bad flow to be $2 \sum_{g \in GF} w(y)$. Then, a solution y' of x' is directly transformed to a solution y of x, unless y' does not cover all the bad flows. In the latter case, y' is transformed to the trivial solution for x that simply contains all the edges. Now, the weights of the solutions y and y' are equal, unless y' does not cover all the bad flows. In the latter case, however, the weight of y' is at least twice larger than that of y, by the definition of the weights of the bad flows.   ∎

We are now set to prove the following hardness result.

Proposition 2. *Even on trees of height 1 and good flows of unit weights, the problems BFR and BBFR are MAX SNP-hard and not approximable within 1.166.*

The proof reduces BFR to the tree set cover and is omitted for lack of space.

5.2 Approximation

Having shown the hardness, we design two approximation algorithms for BFR. Later, we show that similar results hold for BBFR as well.

We now suggest other algorithms for BFR on trees that approximate it within $2l$ and $2\sqrt{2|E|}$. These algorithms refine the **Greedy_RB** and **Low_Deg2** algorithms, respectively, for RBSC from [15]. Recall that l is defined as

$\max\{|P(g)| : g \in GF|\}$, the maximum length of a good flow. We first assume a non-weighted BFR; the extension for the weighted case is straight-forward.

First, in a tree Set Cover can be approximated within the factor of 2 [8, Sect. 7]. This allows, extending [15, Sect. 3.1], to reduce BFR to Set Cover on trees and obtain the approximation ratio of $2l$. Call this algorithm **SCTreeGreedy**. This is a useful algorithm on its own right, and we continue now to obtain a $2\sqrt{2|E|}$- approximation.

In order to present the final approximation, we first present Algorithm 2, inspired by **Low_Deg** from [15].

ALGORITHM 2. Low_Deg_TreeBFR$(G = (N, E), F, GF, BF, w, x)$

---

1. Remove the edges in E that belong to more than x good flows, creating E' and defining a new problem instance $(G' = G[E'], F', GF', BF', w, x)$, where $G[E']$ is the subgraph induced by E', and so are the flows F', GF' and BF'.
2. If the new instance is infeasible, **return** E (the trivially feasible solution).
3. Let the long good flows be $GF'_l \triangleq \left\{g \in GF' : |P(g)| > \sqrt{|E|/2}\right\}$.
4. Leave only the not long good flows, i.e. $GF'_s \triangleq GF' \setminus GF'_l$.
5. **Return SCTreeGreedy$(G' = G[E'], F', GF'_s, BF', w, x)$.**

Similarly to Lemma 3.3 from [15], we prove Lemma 1.

Lemma 1. *Let x be the last input to Algorithm 2 and let the long good flows of GF' be $GF'_l \triangleq \left\{g \in GF' : |P(g)| > \sqrt{|E|/2}\right\}$. Then, $|GF'_l| < \sqrt{2|E|}x$.*

Proof. Since every edge in E' belongs to at most x good flows, we have

$$|GF'_l|\sqrt{|E|/2} < \sum_{g \in GF'_l} |P(g)| \leq \sum_{g \in GF'} |P(g)| = \sum_{e \in E'} |\{g \in G | e \in P(g)\}| \leq |E'|x \leq |E|x,$$

the first inequality stemming from the definition of GF'_l, and the equality being a reversal of the summation order. Therefore, $|GF'_l| < |E|x/\sqrt{|E|/2} = \sqrt{2|E|}x$. ∎

And we can now prove.

Lemma 2. *Let S^* be an optimal solution for the BFR instance at hand (if we knew it). If we activate Algorithm 2 with $\hat{x} \triangleq \max|\{g : GF | e \in P(g) \cap S^*\}|$, then it returns a $2\sqrt{2|E|}$-approximation.*

Proof. By definition of $\hat{x}$, the algorithm returns a feasible solution in line 5.

Since **SCTreeGreedy** yields an $2l$-approximation and its input has $l \leq \sqrt{|E|/2}$, then its solution, say S, fulfills $|GF'_s(S)| \leq \sqrt{2|E|}|GF'_s(S^*)|$. Lemma 1 applies that $|GF'_l| < \sqrt{2|E|}\hat{x}$. Therefore, the total number of removed good flows is $|GF(S)| < \sqrt{2|E|}|GF'_s(S^*)| + \sqrt{2|E|}\hat{x}$. Since $\hat{x} \leq |GF(S^*)|$, we can bound $|GF(S)|$ by $2\sqrt{2|E|}|GF(S^*)|$. ∎

We are now ready to present the full algorithm for approximating the actual problem, where we do not know the $\hat{x}$ in advance.

ALGORITHM 3. Low_Deg_TreeBFR2$(G = (N, E), F, GF, BF, w)$

---

1. min_sol $\leftarrow E$
2. **for** $x = 1 \ldots |GF|$ **do:**
   (a) $S \leftarrow$ **Low_Deg_TreeBFR** $(G = (N, E), F, GF, BF, w, x)$
   (b) **if** $w(GF(S)) < w(GF(\text{min\_sol}))$:
          min_sol $\leftarrow S$
3. **return** min_sol

Lemma 2 implies the following.

Theorem 5. *Algorithm 3 approximates the solution to BFR within* $2\sqrt{2|E|}$.

This has been proven for the non-weighted case, but is straight-forward to extend the result to the weighted case.

Having found approximations for BFR within $2l$ and $2\sqrt{2|E|}$, we approximate BBFR:

Theorem 6. *BBFR can be approximated within $2l$ and within* $2\sqrt{2(|E| + |BF|)}$.

Proof. We employ the approximation preserving reduction from Theorem 3. The new graph is still a tree, because we have added connected edges and created no cycles. We can, therefore, solve it using an algorithm for BFR on trees.

If we employ the $2l$ approximation, we obtain the same approximation ratio for BBFR, since only the length of the bad flows has increases when reducing from BBFR to BFR. Alternatively, if we employ the $2\sqrt{2|E|}$-approximation, we obtain a $2\sqrt{2(|E| + |BF|)}$-approximation, since the reduction introduces $|BF|$ new edges. ∎

5.3   Trees with Root-to-Leaf Flows

Since communication often goes from the clients to the Internet Service Providers (ISPs) and then to the backbone, we assume that the network is a *tree*, and there exists a fixed node r called root, such that every flow is on a path from the root to a leaf.

This assumption allows us to solve both BFR and BBFR exactly using Dynamic Programming (DP). We define a subproblem of our DP to be a subtree and the flows that strictly flow through its root after possible edge deletions outside this subtree. Let $v \in N$ be a node and let $P(v)$ be the set of the edges on the (only) path from v to the root of the tree. Let $F(v) \triangleq \{f \in F : v \text{ is on } P(f)\}$ be the set of the originally given flows that pass through v. Denote by $T(v)$ the subtree rooted on v. The possible subsets of flows that enter $T(v)$ after deleting some edges outside of $T(v)$ are $\mathcal{F}(v) = \{F(v) \backslash F(v)(\{e\}) : e \in P(v)\}$[1]. We do not consider deleting a subset of edges on $P(v)$, because for $\mathcal{F}(v)$ it would be equivalent to deleting the edge of this subset that is closest to v.

The DP Algorithm 4 receives a root r such that every flow is on a path from r to a leaf and solves BFR and BBFR exactly.

ALGORITHM 4. DPAlg_TreeRoottoLeaf($G = (N, E), F, GF, BF, w, r$)

---

1. The algorithm maintains the DP-table indexed by $\{v, \mathcal{F}(v) : v \in N \setminus \{r\}\}$.
2. **For each** node $v \in N \setminus \{r\}$ in a **post-order** traversal (i.e. its subtree has been handled):
   (a) **For each** $S \in \mathcal{F}(v)$:
       i. Delete the edge from v to its parent $\iff$ it maximizes the total objective function in $T(v)$. This uses the optimal solutions that we have memoized for the children of v.
       ii. Memoize the resulting edge deletion and the resulting objective function for the current entry $(v, S) \in \{v, \mathcal{F}(v) : v \in N \setminus \{r\}\}$.
3. The completed DP-table contains an optimal set of edge deletions.

Theorem 7. *Algorithm 4 optimally solves the problem on trees when all the flows are from the root to leaves in $O(|N|^3 |F|)$.*

Proof. The algorithm is correct, since all the flows go from the root to a leaf, thereby making such a traversal consider all the relevant edge deletions.

For each entry in the DP-table, the algorithm looks at all the children of the current node. There are at most $|N|^2$ entries, because each $\mathcal{F}(v)$ contains $|P(v)|$ elements, as defined and explained above. Each entry requires looking at all the flows that pass through the node, providing the factor of $|F|$. In addition, a node has at most $|N| - 1$ children, implying the theorem. ∎

In the rest of the section, when a flow from a to z does not flow from the root r to leaves, we may split it to two parts that do by looking at the first node on the (only) path from a to r that is also on the (only) path from r to z.

Approximation when All the Bad Flows are Root-to-Leaf. If we find a root such that the bad flows can be guaranteed to be from that root to leaves, we can 2-approximate the problem as follows. Given such an instance I and a root r we define another instance I' by splitting each good flow g that does not flow from r to a leaf to two good flows that do. Define the weight of each one of the obtained good flows to be $w(g)$. Denote the weight of the optimal solution to an instance by OPT(instance). Then, OPT(I') $\leq$ 2OPT(I). Therefore, we can solve I' using Algorithm 4 and this will constitute a 2-approximation for I.

---

[1] We remind that for a set of flows D and a set of edges S, we denote by $D(S)$ the flows from D removed by deleting S.

FPT when All the Good Flows are Root-to-Leaf. If we actually find a root such that the good flows are from the root to leaves, we can split each bad flow that does not go from the given root to a leaf by splitting each bad flow b that does not flow from r to a leaf to two bad flows that do. Remove one of the parts and leave the other part. If we are given a BBFR instance, assign the remaining part of the bad flow the weight of the original bad flow. We solve each of the obtained problems for each such a split of the bad flows and output the best solution. This constitutes an optimal algorithm that runs in $O(2^{|BF|})$ times a polynomial time.

5.4 FPT for Trees

Section 5.3 assumes some of the flows are from the root to leaves. We now advance and prove that when the flows follow any simple paths, meaning that a flow does not intersect itself, then the problem is fixed-parameter tractable, parametrized by the number of the bad and the number of the good flows.

We first present the definition of a parametrized optimization problem and fixed-parameter tractability with several parameters, adapted from [5,6].

Definition 5. *A parametrized optimization problem with $t \in \mathbb{N}$ parameters is a set of instances $\Sigma^* \times \mathbb{N}^t$, where Σ is a finite alphabet, encoding the object at hand, and the t natural numbers are called the parameters.*

For example, set cover parametrized by the total number of the elements is a parametrized optimization problem. Tractability is defined as follows.

Definition 6. *A parametrized optimization problem with t parameters, consisting of the instances $\Sigma^* \times \mathbb{N}^t$, is called fixed-parameter tractable (FPT) if there exists an algorithm A, a computable function $f : \mathbb{N}^t \to \mathbb{N}$, and a constant c such that if A receives instance $(x, k_1, \ldots, k_t) \in \Sigma^* \times \mathbb{N}^t$, it computes an optimal solution to it within at most $f(k_1, \ldots, k_t)(|x| + k_1 + \ldots + k_t)^c$ time.*

Intuitively speaking, only the parameters may contribute more than a polynomial to the run time. We are now ready to prove that BFR and BBFR, when the network is a tree and the flows have simple paths, is FPT. Consider Algorithm 5.

Notice that step 2 follows the approach of Sect. 5.3, only that now there is no assumption about flowing from the root to the leaves.

We summarize Algorithm 5 in the following theorem.

Theorem 8. *Algorithm 5 runs in time $O(2^{|BF|}4^{|GF|}|BF||GF||N|^3|F|)$.*

Proof. The algorithm is correct, because it goes over all the possibilities to remove bad flows and good flows. A bad flow can be removed be deleting either edge on its path, so we just choose the best option. A good flow has to be paid for only once, if it is removed at all, and so we nullify the weight of a good flow if the other split part has been removed.

As for the time complexity, maintaining the splits of the bad and of the good flows takes the factor of $|BF||GF|$. The algorithm first goes over $O(2^{|BF|})$

ALGORITHM 5. DPAlg_Tree$(G = (N, E), F, GF, BF, w)$

---

1. Arbitrarily pick a node to be the root. Call it r.
2. Split each bad flow that does not flow from the root to a leaf to two parts that do.
3. Delete one part, and if this is a BBFR instance, define the weight of the remaining part to be the weight of the original flows. For each bad flow that does
   not flow from r to a leaf, there are 2 options as to which part to delete. For each set of options, do:
   (a) For each good flow that has a path not from r to a leaf, split it to two parts that do, and assign each part the weight of the original good flow.
   (b) Solve the obtained instance using Algorithm 4 with the following adjustment. If the dynamic programming decides to delete an edge from a split part of a good flow,
       then it has to assign the second part of that flow zero weight (in its subtree). Accounting for 2 options per a split good flow each time requires 2 attempts per each split good flow that enters the subtree.
4. Return the best solution from all the solutions in the above tried combinations.

---

splitting options for bad flows (i.e. which part of each split bad flow to delete). For each such an option, it splits the good flows and runs Algorithm 4, while trying all the options for the weights of each split good flow, i.e. $O(2^{2|GF|})$. Employing Theorem 7, the total resulting time is

$$O(|BF||GF| \cdot 2^{|BF|} \cdot 2^{2|GF|} |N|^3 |F|) = O(2^{|BF|} 4^{|GF|} |BF||GF| |N|^3 |F|).$$

∎

This immediately implies the following.

Corollary 5. *BFR and BBFR parametrized by $|BF|$ and $|GF|$ are FPT.*

6 Conclusion

We study two problems that model fighting DoS and malicious communication: BFR and BBFR. We need to delete edges so that the bad (undesirable) flows are disconnected. Unlike the usual network design problems [18], we do not merely minimize the number of the deleted edges, but rather the resulting number of the disconnected good (desirable) flows. We prove that in the general setting, these problems are extremely hard to approximate, being approximation equivalent to hard problems. We reduce our problems to submodular set cover to provide a approximation and provide a greedy approximation as well. In the important case when the network is a tree, the problems are still MAX SNP-hard, and we provide an approximation algorithm. Furthermore, if the tree can be rooted such that every flow is on the path from the root to a leaf, we solve the problems exactly using dynamic programming (DP). This also inspires us to 2-approximate the case where just the bad flows are known to be from the root to leaves and to provide fixed parameter tractable algorithms for the case of just the good flows being from the root to leaves and for the general case of a tree network.

These results suggest removing all the edges that do not pass through a good flow, being free, and then checking for every connected component of the resulting graph whether it is a tree. If yes, we can employ the designed algorithm for trees. Furthermore, if a root can be chosen such that every flow in such a tree flows on the path from that root to a leaf, then the suggested DP solves the problem exactly. In case the number of the bad and the good flows are small, we can also employ the suggested fixed parameter algorithms. We can also postprocess and delete only the edges that uniquely remove a bad flow.

We have a continuous ranking of the bad flows by weight, but the distinction between the bad and the good is binary. In the future, exploring other rankings would allow modeling other domains of congestion problems. Another possibility one can model is rerouting the disconnected flows, when the tree contains cycles. We would then need to consider the edge capacities, which were not needed so far. Another challenge is also avoiding disconnecting the network or at least minimizing the number of the connected components in the resulting network.

To conclude, we have modeled and approximated two important NP-complete problems at various topology-dependent complexity levels, providing the basis for future research.

Acknowledgments. This research is funded by the Dutch Science Foundation project SARNET (grant no: CYBSEC.14.003/618.001.016)

References

1. Bafna, V., Berman, P., Fujito, T.: A 2-approximation algorithm for the undirected feedback vertex set problem. SIAM J. Discret. Math. **12**(3), 289–297 (1999)
2. Bondy, J., Murty, U.: Graph Theory with Applications. Elsevier Science Publishing Co., Inc., New York (1976)
3. Carr, R.D., Doddi, S., Konjevod, G., Marathe, M.: On the red-blue set cover problem. In: Proceedings of the Eleventh Annual ACM-SIAM Symposium on Discrete Algorithms, SODA 2000, pp. 345–353. Society for Industrial and Applied Mathematics, Philadelphia (2000)
4. Cormen, T., Leiserson, C., Rivest, R., Stein, C.: Introduction to Algorithms, 3rd edn. MIT Press, Cambridge (2009)
5. Cygan, M., et al.: Parameterized Algorithms, 1st edn. Springer Publishing Company, Incorporated, Cham (2015). https://doi.org/10.1007/978-3-319-21275-3
6. Downey, R.G., Fellows, M.R.: Fixed parameter tractability and completeness I: basic results. SIAM J. Comput. **24**(4), 873–921 (1995). https://doi.org/10.1137/S0097539792228228
7. Garey, M.R., Johnson, D.S.: Computers and Intractability: A Guide to the Theory of NP-Completeness. W. H. Freeman & Co., New York (1979)
8. Garg, N., Vazirani, V.V., Yannakakis, M.: Primal-dual approximation algorithms for integral flow and multicut in trees. Algorithmica **18**(1), 3–20 (1997)
9. Iwata, S., Nagano, K.: Submodular function minimization under covering constraints. In: 2009 50th Annual IEEE Symposium on Foundations of Computer Science, pp. 671–680, October 2009

10. Khuller, S., Thurimella, R.: Approximation algorithms for graph augmentation. J. Algorithms **14**(2), 214–225 (1993)
11. Koning, R., de Graaff, B., de Laat, C., Meijer, R., Grosso, P.: Interactive analysis of SDN-driven defence against distributed denial of service attacks. In: 2016 IEEE NetSoft Conference and Workshops (NetSoft), pp. 483–488, June 2016
12. Miettinen, P.: On the positive-negative partial set cover problem. Inf. Process. Lett. **108**(4), 219–221 (2008)
13. Mirkovic, J., Dietrich, S., Dittrich, D., Reiher, P.: Internet Denial of Service: Attack and Defense Mechanisms (Radia Perlman Computer Networking and Security). Prentice Hall PTR, Upper Saddle River (2004)
14. Mirkovic, J., Reiher, P.: A taxonomy of DDoS attack and DDoS defense mechanisms. SIGCOMM Comput. Commun. Rev. **34**(2), 39–53 (2004)
15. Peleg, D.: Approximation algorithms for the label-covermax and red-blue set cover problems. J. Discret. Algorithms **5**(1), 55–64 (2007)
16. Vazirani, V.: Approximation Algorithms. Springer, Heidelberg (2001). https://doi.org/10.1007/978-3-662-04565-7
17. Watanabe, T., Ae, T., Nakamura, A.: On the removal of forbidden graphs by edge-deletion or by edge-contraction. Discret. Appl. Math. **3**(2), 151–153 (1981)
18. Yannakakis, M.: Node-and edge-deletion NP-complete problems. In: Proceedings of the Tenth Annual ACM Symposium on Theory of Computing, STOC 1978, pp. 253–264. ACM, New York (1978)
19. Zargar, S.T., Joshi, J., Tipper, D.: A survey of defense mechanisms against distributed denial of service (DDoS) flooding attacks. IEEE Commun. Surv. Tutor. **15**(4), 2046–2069 (2013, Fourth)

Min-Max-Flow Based Algorithm
for Evacuation Network Planning
in Restricted Spaces

Yi Hong[1(✉)], Jiandong Liu[1], Chuanwen Luo[2], and Deying Li[2]

[1] Information Engineering College, Beijing Institute of Petrochemical Technology,
Beijing 102617, People's Republic of China
hongyi@bipt.edu.cn
[2] School of Information, Renmin University of China, Beijing 100872,
People's Republic of China

Abstract. Recently, emergency evacuation management, which is a social work around the world, has been getting lots of attentions due to its importance and necessity. The primary task of emergency evacuation management is evacuation route planning. Considering the particularity of restrict space scenarios, it is more important to guarantee the security and promptness of evacuation routes than that in open space scenarios. In this paper, we introduce a new evacuation route planning problem in restricted spaces, namely Congestion-Avoidable Evacuation Route Network Planning (CA-ERNP) problem. Based on the minimum cost maximum flow (Min-Max Flow) problem, we propose a batch scheduling algorithm based on node-slitting transformation. In addition, we evaluate the average performance of the algorithms via simulation and the results indicate the proposed algorithm outperforms the existing alternatives in terms of efficiency and time cost.

Keywords: Evacuation route planning · Restrict space
Min-max flow · Batch scheduling

1 Introduction

The primary mission of emergency evacuation is the planning of evacuation route/path, which can be modeled as a multi-source to multi-destination route network. In restricted space scenarios, the emergency evacuation planning has more difficulties and complexity because of the properties of structure and environment. For example, there are strict access limitations of the entrance and exit, and the emergency will cause the trapped subspaces and obstacles easily, e.g., urban underground tunnels, factory workshop. Thus we focus on the evacuation network planning problem in restricted spaces. The existing research mostly considered one-to-one or many-to-one route planning problem. And in these research, handling of congestion in evacuation adopted the way of queueing or buffering, which cannot be applied to restricted spaces. Therefore, it is

D. Kim et al. (Eds.): COCOA 2018, LNCS 11346, pp. 233–245, 2018.
https://doi.org/10.1007/978-3-030-04651-4_16

necessary and crucial to design a route network planning strategy with avoiding congestion for restricted space evacuation.

Motivated by our observations, we study a congestion-avoidable route network planning problem in restricted spaces in this paper. Our motivation is to construct the maximum population route network and schedule the evacuees, which provides a guidance of emergency management. According to the guidance, the evacuees can be scheduled in batches based on the maximum population route network. Thus to avoid the evacuation congestion happened in the corners in only one batch, the goal of the problem is to plan a maximum evacuation in each batch and minimize the total number of batches. In this paper, we introduce a new router network planning problem in restricted space evacuation. The contribution of this paper is as follows.

(i) We introduce the congestion-avoidable evacuation route network planning problem. The goal is to compute a maximum population route network from multi-source to multi-destination and schedule the evacuee to evacuate based on the route network in batches.
(ii) We propose an algorithms based on the minimum cost maximum flow (Min-Max Flow) problem. The algorithm consists of three phases: The first one is computing an auxiliary graph based on the properties of space structure and environment; the second one is scheduling in batches to construct a minimum cost maximum flow in the auxiliary graph; and the last one is restoring the evacuation route network of the original space in each round.
(iii) We finally conduct simulations to evaluate the average performance of the proposed algorithm in terms of the scheduling cost and the evacuation time consumption.

The rest of this paper is organized as follows. Section 2 introduces the related work. The restricted space model and the problem definition are given in Sect. 3. The description of the min-max-flow based algorithm for the problem is presented in Sect. 4. Our simulation results and corresponding analysis are in Sect. 5. Section 6 concludes this paper and presents future works.

2   Related Work

Due to the importance of evacuation planning, a large number of researches focused on the evacuation path planning problems. Most existing researches solved the problems based on the classical algorithms like Dijkstra algorithm [1,2], Floyd-Warshall algorithm [3], K shortest paths algorithm [4–6], Dynamic programming [7], and Maximum flow algorithm. And in recent years, some research works solved the planning problems based on the Minimum Weighted Set Cover problem [8,9]. Furthermore, some research works focused on the evacuation model in evacuation management: [10] studied a large-scale optimization problem and proposed a clustering technique via divide-and-conquer method; [11] and [12] considered the evacuation betweenness centrality and evacuation centrality and proposed algorithms to make the evacuation routes sufficiently safe.

The evacuation paths from multi-source to multi-destination can be generally modeled as a path network and evacuees' traffic capacity can be reviewed as the flow quantity in the network. Thus a certain amount of the exiting solutions for evacuation planning problems were based on the classical algorithms for the network flow problem, which has been largely and deeply studied for several year [13]. Among the algorithms for the maximum flow problem, the cycle canceling is a general primal method [14] and the minimum mean cycle canceling has strongly polynomial running time [15]. And based on the Ford-Fulkerson algorithm, [16] proposed two dual algorithm, the successive shortest path and the capacity scaling; based on the linear programming simplex method, [17] proposed the network simplex algorithm. In the evacuation planning studies based on the network flow problem, [18] summarized a systematic collection of network flow models applied to emergency evacuation and their applications, e.g., max flows and min cost flows, lexicographic flows, quickest flows, and earliest arrival flows and so on; [19] proposed greedy algorithms for building evacuation problems which were modeled as the maximum flow, minimum cost flow, and minimax flow problems; [20] proposed a corridor-based emergency evacuation system based on contraflow design.

Most exiting research paid attentions on the evacuation path planning for open space scenarios. And although the flow algorithms treated the evacuees' traffic situation via the flow quantity, the congestion handling cannot be considered or directly implemented. Thus we consider the evacuation path planning in restricted space and design the scheduling without congestion and with minimum evacuation time consumption.

3 Restricted Space Model and Problem Definition

3.1 Hierarchical Space Model

In the evacuation planning problem in a restricted space, the restricted space can be modeled as a three-dimensional (3D) undirected graph $G = (V, E)$. The node set $V = \{v_1, \cdots, v_n\}$ is composed of the observation points, corners, and other critical positions, and the edge set $E = \{(v_i, v_j) | 1 \leq i < j \leq n\}$ includes all the available path segments connecting the nodes.

In most application scenarios, the restricted space has a system structure with several relatively independent components, which are relevant to each neighboring one via stair edges. Such a component of the space is deployed based on the division requirement of structure function, e.g., in certain range of height, $(-300\text{m}, -200\text{m})$, $(-200\text{m}, -100\text{m})$, $(-100\text{m}, 0\text{m})$. Thus a 3D undirected graph G can be regarded as an equivalent multi-layer 3D model via the transformation, which can be reformulated into $G = \{G_1, G_2, ..., G_{L-1}, G_L\}$. Note that the value of L is predetermined and decided by the structure of the restricted space.

3.2   Node Capacity and Edge Weight

Since the restricted space cannot allow longstanding stay on edges, the evacuees can be assumed to maintain moving state until they reach the exits. Thus the potential congestions are prone to occur on the nodes rather than the edges, which is the reason for the consideration of the capacities of nodes. Here we assign each node a capacity as $capacity(v_i)$ $(1 \leq i \leq n)$ and denote the capacity set as $C = \{capacity(v_i)|\forall v_i \in V\}$.

Considering the influence of the emergency environment and the disaster spreading on evacuation, it is necessary to assign an equivalent length to each edge in G, and the assignment measure for each edge $e_k \in E$ $(1 \leq k \leq |E|)$ is according to the following influence factors. *a. Euclidean distance: $length(e_k)$*. $length(e_k)$ is the geometric length of the edge; *b. Environmental factor: $EF(e_k)$*. When the disaster or accident happens, the spread characteristic of disaster will effect evacuation. Note that the spreading patten can be modeled as a statistical model depending on the disaster itself, which is denoted as $EF(e_k)$; *c. Exit priority factor: $EP(e_k)$*. In the restricted space, each edge has the nearest exit based on the layered structure. We use the layer difference between the edge and its nearest exit to represent the escaping priority of the evacuees through the edge, $EP(e_k) = |layer(e_k) - layer(nearest(e_k, D))|$, where $layer(e_k) = l$ if $e_k \in E(G_l)$, $layer(v_i) = l$ if $v_i \in V(G_l)$ and $nearest(e_k, D) = v_{d^*}$ if $dist(e_k, v_{d^*}) = \min\{dist(e_k, v_d)|\forall v_d \in D\}$.

Based on these influence factors, the equivalent length of edge can be expressed in terms of $weight(e_k) = length(e_k) \cdot w_1 + EF(e_k) \cdot w_2 + EP(e_k) \cdot w_3, 1 \leq k \leq |E|$ and we denote the weight set as $W = \{weight(e_k)|\forall e_k \in E\}$. Thus G can be further modeled as an 3D edge-weighted graph $G = (V, E, C, W)$.

3.3   Problem Formulation

Based on the space model, we consider the evacuation route network planning problem in restricted space scenarios. Given the following conditions: (i) A restricted space modeled as a 3D undirected graph $G = (V, E, C, W)$, where locates *sou* evacuation beginning positions (denoted as source set S) and *des* evacuation exits (denoted as destinations set D); (ii) The relevant measures are known or predefined: the number of evacuees on each source e_s $(1 \leq s \leq sou)$ and the flow quantity on each destination per unit timeslot c_d $(1 \leq d \leq des)$, which can be regarded as their capacities. We aim to construct the evacuation route network from S to D and the scheduling strategy for the objective that the total time consumption of entire evacuation can be minimized. We give the formal definiation of the problem as follows.

Definition 1 (Congestion-Avoidable Evacuation Route Network Planning (CA-ERNP) problem). *Given a 3D undirected graph $G = (V, E, C, W)$, source set S and destination set D in V, with the constraints of V's capacities, the CA-ERNP problem is to find a evacuation planning scheduling from S to D with the goal of evacuating all the evacuees and the minimum time consumption.*

4 Min-Max-Flow-Based Algorithm for CA-ERNP Problem

To solve the CA-ERNP problem, we adopt the idea of batch scheduling under iteration form and the objectives of the scheduling are as follows: (i) The time consumption can be minimized and the population flow can be maximized for each evacuation batch; (ii) The total number of batches in evacuation scheduling can be minimized. Here we assume the maximum allowable time consumption T_{max} of a security evacuation can be separated into unit timeslot 1, which is for the convenience of measuring the flow quantity of nodes in G. Based on the model transformation, the main idea of the scheduling algorithm is as follows.

- Phase 1: We construct an auxiliary graph G^* via node slitting. Based on the node weights of three sets S, D and $V \setminus \{S \bigcup D\}$, we transform the 3D undirected graph $G = (V, E, C, W)$ into a directed graph $G^* = (V^*, E^*, C^*, W^*)$ via adding virtual source, virtual destination, and node slitting. The procedure is denoted as $Transform(G)$.
- Phase 2: We perform a batch scheduling on G^* via iterations. In each iteration i, based on the current remaining evacuation requirements S_{rem} of the source set S, we compute the minimum cost maximum flow in G^*. In the auxiliary directed graph G^*, the flow quantity and original node capacity, C^*, can be viewed as the edge capacity and the edge weights, W^*, are regarded as the edge cost. Thus constructing the maximum population evacuation route network can be transformed to finding a minimum cost maximum flow $Flow_i$ in G^*, which is denoted as $Min\_Max\_Flow(G^*, S_{rem})$.
- Phase 3: We restore the batch scheduling on G. The flow in G^* in each iteration i should be restored into the actual feasible route network in G, which is denoted as $Restore(G^*, Flow_i)$.

Before presenting the whole scheduling algorithm, we introduce the pivotal procedures in the following subsections.

4.1 $Transform(G)$: Construct an Auxiliary Graph G^* via Node Slitting

We construct G^* based on the assumption that the 3D undirected graph G is connected. It is because that a restricted space has a connected structure or it is composed of several independent and connected components in practice, and we consider the connected space or subspace for the evacuation plan. By relying on this assumption, the details of auxiliary graph constructing are as follows and V^*, E^*, C^*, W^* are all set to be $\emptyset$ originally.

(i) *Adding virtual source and destination:* We logically set two virtual nodes s_0 and d_0, i.e., they may not exist in the actual restricted space. And the virtual source s_0 has sou directed edge pointing to all the actual sources in S, respectively. Similarly, each actual destination in D has a directed edge pointing to the virtual destination d_0. The updates are as follows and shown in Fig. 1(i):

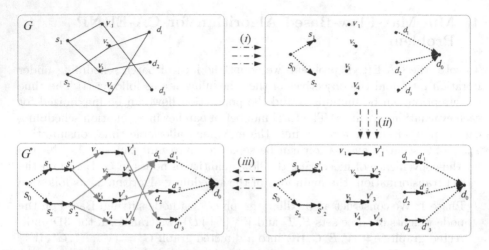

Fig. 1. The illustration of $Transform(G)$

(a) $V^* \leftarrow V \bigcup \{s_0, t_0\}$,
(b) $E^* \leftarrow \{< s_0, v_s > | \forall v_s \in S\} \bigcup \{< v_d, d_0 > | \forall v_d \in D\}$,
(c) $C^* = \{capacity(e_k) = \infty | \forall e_k \in E^*\}$,
(d) $W^* = \{weight(e_k) = 0 | \forall e_k \in E^*\}$.

(ii) *Slitting nodes:* For each node in V, we slit it into an directed edge with the same capacity of the original node, which is also an equivalent transformation. And G^* is updated as follows and shown in Fig. 1(ii):

(a) $V^* \leftarrow V^* \bigcup \{v_i' | \forall v_i \in V\}$,
(b) $E^* \leftarrow E^* \bigcup \{< v_i, v_i' > | \forall v_i \in V\}\}$. Since each slit node v_i corresponds two nodes v_i and v_i' in the new V^*, for v_i's associated edges in the old V^*, we appoint v_i associated with all the incident edges and v_i' associated with all the reflected edges in the new V^*, e.g., edge $< s_0, v_s >$ is changed to $< s_0, v_s >$ and $< v_s, v_s' >$ and edge $< v_d, d_0 >$ is replaced by $< v_d, v_d' >$ and $< v_d', d_0 >$,
(c) $C^* = \{capacity(< v_i, v_i' >) = capacity(v_i) | \forall v_i \in V\}$. Note that for each actual source in S, $capacity(< v_s, v_s' >) = e_s$ $(1 \leq s \leq sou)$; and for each actual destination in D, $capacity(< v_d, v_d' >) = c_d$ $(1 \leq d \leq des)$,
(d) $W^* = \{weight(< v_i, v_i' >) = 0 | \forall v_i \in V\}$.

(iii) *Adding edges' directions:* For each undirected edge in E, it can be converted into a pair of directed edges with the original weight without damaging its property. And this transformation is as follows and shown in Fig. 1(iii):

(a) $E^* \leftarrow E^* \bigcup \{< v_i', v_j >, < v_j', v_i > | \forall e_k = (v_i, v_j) \in E\}$,
(b) $C^* \leftarrow C^* \bigcup \{capacity(< v_i', v_j >) = capacity(< v_j', v_i >) = \infty | \forall e_k = (v_i, v_j) \in E\}$,
(c) $W^* \leftarrow W^* \bigcup \{weight(< v_i', v_j >) = weight(< v_j', v_i >) = weight(e_k) | \forall e_k = (v_i, v_j) \in E\}$.

Based on above rules, we obtain a 3D directed graph $G^* = (V^*, E^*, C^*, W^*)$ which is a structure-equivalent auxiliary graph of G for three reasons: **i.** The virtual source and destination are both logical nodes without weight cost ($= 0$) and capacity limitation ($= \infty$); **ii.** The node slitting is for converting node capacity to edge capacity without breaking the relevance relationship with the edges in original graph; **iii.** The edge directing is bidirectional, which maintains the original weight and capacity.

4.2 $Min\_Max\_Flow(G^*, S_{rem})$: Compute a Min-Max Flow in G^*

To efficiently schedule the whole evacuation scheduling, we adopt evacuation in batches as the scheduling way and construct a maximum population route network in each batch to reduce the whole time consumption, i.e., the number of batches. The construction basis of each batch is the remaining evacuation requirement S_{rem}: If there are e'_s unscheduled evacuees on v_s, the capacity of edge $< v_s, v'_s >$ is updated as e'_s in G^* ($0 \le s \le sou$). Here we regard maximizing the evacuation population in each iteration as an important subproblem and it can be modeled as: Given a 3D directed graph $G^* = (V^*, E^*, C^*, W^*)$, source s_0 and destination d_0 in V^*, the subproblem is to find a flow $Flow_i = \{f(e')|\forall e' \in E^*\}$ from s_0 to d_0 such that

Minimize: $\sum_{\forall e' \in E^*} (weight(e') \cdot f(e'))$;
Maximize: $\sum_{\forall v_s \in S} f(< s_0, v_s >)$ ($/ \sum_{\forall v_d \in D} f(< v_d, d_0 >)$);
Subject to:

(i) $\sum_{1 \le j \le |V^*|} f(< v'_i, v_j >) - \sum_{1 \le j \le |V^*|} f(< v'_j, v_i >) = 0$ ($1 \le i \le |V^*|$),
(ii) $0 \le f(e') \le capacity(e')$ ($\forall e' \in E^*$).

The first objectives is to minimize the edge weight of the flow $Flow_i$ and the second is to maximize the flow quantity of $Flow_i$. Note that the total outflow quantity of source s_0 is equivalent to the total inflow quantity of destination d_0. The constraints are to conserve the equality between inflow and outflow on each intermediate node and to satisfy the edges' capacities, i.e., the flow quantity cannot exceed the capacity on each edge. And note that $f(e')$ stands for the number of the evacuees on edge e' thus it is an integer.

To solve the subproblem, we design an algorithm based on the classical problem, the minimum cost maximum flow (Min-Max Flow). Referring the minimum cost flow problem, it can be solved by Linear Programming, Ford-Fulkerson algorithm based on augmenting path or Shortest Path Faster Algorithm.

4.3 $Restore(G^*, Flow_i)$: Restore the Evacuation Flow in G

In the batch scheduling on G^*, we obtain a set of flows in iterations and they are needed to restore the actual path set in G. The restoring procedure is to map the paths in each $Flow_i$ on G^* to the paths on G for each scheduling Sch_i. As shown in Fig. 2, the details are as follows.

(i) *Deleting virtual source and destination:* We delete the virtual nodes s_0, d_0 and their associated edges as follows:

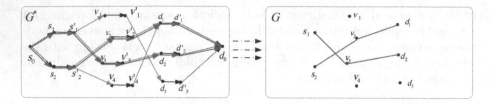

Fig. 2. The illustration of $Restore(G^*, Flow_i)$

(a) $V[Flow_i] \leftarrow V[Flow_i] \setminus \{s_0, t_0\}$,
(b) $E[Flow_i] \leftarrow E[Flow_i] \setminus (\{< s_0, v_s > |\forall v_s \in S\} \bigcup \{< v_d, d_0 > |\forall v_d \in D\})$,
    and the deleted edges' capacities and weights are removed correspondingly.
(ii) *Shrinking edge* $< v_i, v_i' >s$: We return all the directed edges in node slitting to these slitted nodes in V as follows:
(a) $V[Flow_i] \leftarrow V[Flow_i] \setminus \{v_i' | \forall v_i \in V\}$,
(b) $f[Flow_i] = \{f(v_i) = f(< v_i, v_i' >) | \forall v_i \in V\}$,
(c) $E[Flow_i] \leftarrow E[Flow_i] \setminus \{< v_i, v_i' > | \forall v_i \in V\}$. And the edges from all the v_i's change the endpoint v_i' as v_i, and the deleted edges' weights are removed correspondingly.
    Note that for each pair of directed edges $< v_i', v_j >$ and $< v_j', v_i >$ in $Flow_i$, they are changed to $< v_i, v_j >$ and $< v_j, v_i >$.
(iii) *Removing edges' direction:* For each pair of directed edges $< v_i, v_j >$ and $< v_j, v_i >$ in $Flow_i$, we convert them into an undirected edge (v_i, v_j) with the same weight of their as follows:
(a) $E[Flow_i] \leftarrow E[Flow_i] \setminus \{< v_i, v_j >, < v_j, v_i > | \forall v_i, v_j \in V[Flow_i]\}$, and the deleted edges' capacities and weights are removed correspondingly,
(b) $E[Flow_i] \leftarrow E[Flow_i] \bigcup \{(v_i, v_j) | \forall v_i, v_j \in V[Flow_i]\}$,
(c) $W[Flow_i] = \{weight((v_i, v_j)) | \forall (v_i, v_j) \in E[Flow_i]\}$.

The variant of $Flow_i$ based on the above rules is the scheduling scheme Sch_i in the i-th iteration. $f[Flow_i]$ is composed of each node's flow quantity/evacuee number and $W[Flow_i]$ stands for the edge weights in the current scheduling. And we can calculate the evacuation time of Sch_i based on $W[Flow_i]$.

4.4 Min-Max-Flow Based Algorithm Description

Based on the three pivotal procedures, we proposed the min-max-flow based algorithm for evacuation network planning in restricted spaces in Algorithm 1.

Initialization: We set the global measures and that of the local measures. The former kind (step 1) is composed of the whole scheduling (SCH), time consumption (T), and the round number of the flow planning ($round$). The later kind (steps 2–5) includes the temporal set of the unscheduled/remaining evacuees (S_{rem}), the constructed flow ($Flow_i$), the time consumption (t_i) and

Algorithm 1. Min-Max-Flow based Evacuation Network Planning Algorithm $(G = (V, E, C, W), S, D)$

1: $round = 0$, $T = 0$, $SCH \leftarrow \emptyset$.//Initialization 1 - Global measures: step 1
2: $S_{rem} \leftarrow S$
3: **for** $i = 1$ to T_{max} **do**
4:    $Flow_i \leftarrow \emptyset$, $t_i = 0$, $Sch_i \leftarrow \emptyset$
5: **end for**//Initialization 2 - Local measures: steps 2-5
6: $G^* = (V^*, E^*, C^*, W^*) \leftarrow Transform(G)$.//Phase 1: step 6
7: $i = 0$
8: **while** $S_{rem} \neq \emptyset$ **do**
9:    $Flow_i \leftarrow Min\_Max\_Flow(G^*)$, t_i is calculated by $W[Flow_i]$
10:    $S_i \leftarrow \{f(< v_s, v'_s >) | \forall < v_s, v'_s > \in Flow_i$ and $1 \leq s \leq sou\}$
11:    **for** $s = 1$ to sou **do**
12:       $capacity(< v_s, v'_s >) = capacity(< v_s, v'_s >) - f(< v_s, v'_s >)$
13:       **if** $capacity(< v_s, v'_s >) \leq 0$ **then**
14:          $S_{rem} \leftarrow S_{rem} \setminus \{v_s\}$
15:       **end if**
16:    **end for**
17:    $T = t_i + i$
18:    $i + +$
19: **end while**//Phase 2: steps 7-19
20: $round = i$.
21: **for** $i = 1$ to $round$ **do**
22:    $Sch_i \leftarrow Restore(G^*, Flow_i)$, $SCH \leftarrow SCH \bigcup \{Sch_i\}$
23: **end for**//Phase 3: steps 20-23
24: **return** SCH and T.

the scheduling (Sch_i) in each i-th iteration, which is the preparation of flow planning with iteration form.

Phase 1 (Step 6): We obtain an auxiliary directed graph G^* by $Transform(G)$.

Phase 2 (Steps 7–19): We perform the batch scheduling on G^* through a WHILE loop until all the evacuees on S have been scheduled (i.e., $S_{rem} = \emptyset$). In each iteration i: **i.** We apply Ford-Fulkerson algorithm based on augmenting path to compute a min-max flow in the current G^* to maximum the evacuee number in the current schedule; **ii.** Based on the constructed flow $Flow_i$, we obtain the flow's time consumption t_i and update the capacities of edge $< v_s, v'_s >$ (steps 10–16) and the remaining evacuation requirement S_{rem} (step 14); **iii.** To avoid the congestion, the scheduling is implemented timeslot by timeslot, i.e., $Flow_{i+1}$ begins in the second timeslot of $Flow_i$. Thus the current whole time consumption is the sum of the previous round number and the current flow's time (step 17).

Phase 3 (Steps 20–23): In each round, we restore the flow scheduling $Flow_i$ on G^* to the network planning Sch_i on G based on $Restore(G^*, Flow_i)$.

Finally, the algorithm outputs the evacuation network planning SCH and the evacuation time consumption T.

5   Simulation Results and Analysis

In this section, we conduct a series of simulations to compare the average performance of our algorithm (MMF Algorithm) for the CA-ERNP problem with the classical algorithm, the shortest path algorithm (SP Algorithm). We apply SP Algorithm for each source nodes for the CA-ERNP problem. And when multi-evacuee choose the same shortest path, they should queue in the evacuation with the constraint of the traffic capacity, which is calculated as the waiting time consumption.

Based on the main idea of batch scheduling, it can be found that the scheduling duration mostly depends on the round number, which can be regarded as an average performance of the algorithms. Furthermore, the whole evacuation time consumption can be measured as the evacuation length of the longest path, which is a global performance indicator. Thus we adopt the scheduling round number and the evacuation length as the **performance evaluation criteria**. We evaluate these performance of the two algorithms under the changes of three **influence parameters**, the number of space nodes n, the number of source nodes sou and the number of destination nodes des.

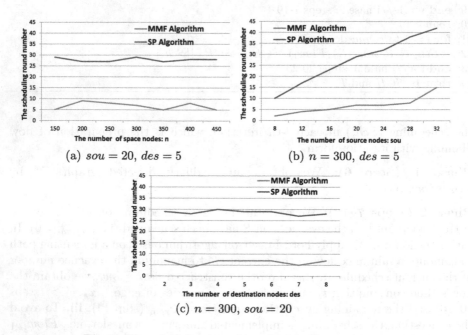

(a) $sou = 20$, $des = 5$ (b) $n = 300$, $des = 5$

(c) $n = 300$, $sou = 20$

Fig. 3. The scheduling round number vs. the parameters

To perform simulation experiments, we randomly generate a connected 3D graph $G = (V, E)$, source set S and destination set D in a $500 * 500 * 500$ space, in which each node has a capacity in range of $[5, 10]$ and each edge has a weight decided by its length. And each node in S has an evacuation requirement in range of $[5, 15]$. The parameter settings are as follows:

(a) n varies from 150 to 450 by the step of 50 and $sou = 20$, $des = 5$;
(b) sou varies from 8 to 32 by the step of 4 and $n = 300$, $des = 5$;
(c) des varies from 2 to 8 by the step of 1 and $n = 300$, $sou = 20$.

For each parameter setting, we run 100 instances and compute their average.

We firstly observe the scheduling round numbers of the algorithms with the changes of n, sou and des. As shown in Fig. 3, MMF Algorithm has significant advantage on the round number and the round number differences between MMF Algorithm and SP Algorithm present less fluctuate when n and des change in Fig. 3(a) and (c). But with the increasing of sou, such difference grows in Fig. 3(b). It is because that the increasing of evacuee number will increase the possibility of congestion, and it is necessary to scheduling more round for avoiding congestion. In this case, the advantage of MMF Algorithm is more prominent, i.e. it schedules more evacuees' evacuation in each round and the round number is less than that of SP Algorithm.

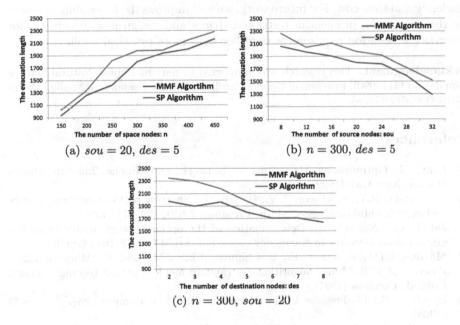

(a) $sou = 20$, $des = 5$ (b) $n = 300$, $des = 5$

(c) $n = 300$, $sou = 20$

Fig. 4. The evacuation length vs. the parameters

Secondly, we investigate the evacuation length of the algorithms under the parameters variation. Since SP Algorithm gives priority to minimizing the path length for a certain evacuee in one round and ignores avoiding congestion, the global evacuation lengths obtained by it are larger than that of MMF Algorithm as shown in each subfigures in Fig. 4. With the growth of the space scale, n, the evacuation lengths in the algorithms are increased in Fig. 4(a) and the opposing results happen when *sou* and *des* increases in Fig. 4(b) and (c). It can be explained that with the enlarge of the space scale, the number of the candidate paths becomes large, which is followed by the increasing of capacity limitations on the intermediate nodes on these paths. To meet the capacity limitations and avoid congestion, the whole evacuation lengths are increased.

Based on the above two groups of simulation results, we can find that MMF Algorithm outperforms SP Algorithm in terms of the scheduling round number and the evacuation length.

6   Conclusion

In this paper, we investigate the evacuation path planning problem with the consideration of avoiding congestion. To solve the problem with the goal of minimizing the evacuation time consumption, we propose a 3-phase algorithm based on the Min-Max Flow problem. In the performance evaluation, the simulation results indicate our algorithm outperforms the existing alternatives in terms of efficiency and time cost. For future work, we will improve the scheduling strategy for the evacuation problem in restricted spaces and design new algorithms for difference evacuation requirement together with more practical applications.

Acknowledgment. This research was supported in part by Beijing Natural Science Foundation (4174090), Program of Beijing Excellent Talents Training for Young Scholar (2016000020124G056).

References

1. Evans, J.: Optimization Algorithms for Networks and Graphs, 2nd edn. Marcel Dekker, New York (1992)
2. Cherkassky, B.V., Goldberg, A.V., Radzik, T.: Shortest paths algorithms: theory and experimental evaluation. Math. Program. **73**(2), 129–174 (1996)
3. Jalali, S.E., Noroozi, M.: Determination of the optimal escape routes of underground mine networks in emergency cases. Saf. Sci. **47**, 1077–1082 (2009)
4. Martins, E.D.Q.V., Queir, E., Dos Santos, J.L.E., Martins, V., Margarida, M., Pascoal, M.M.B.: A new algorithm for ranking loopless paths. Technical report, Univ. de Coimbra (1997)
5. Eppstein, D.: Finding the k shortest paths. SIAM J. Comput. **28**(2), 652–673 (1998)
6. Jin, W., Chen, S., Jiang, H.: Finding the K shortest paths in a time-schedule network with constraints on arcs. Comput. Oper. Res. **40**, 2975–2982 (2013)
7. Sever, D., Dellaert, N., Van Woensel, T., De Kok, T.: Dynamic shortest path problems: hybrid routing policies considering network disruptions. Comput. Oper. Res. **40**, 2852–2863 (2013)
8. Hong, Y., Li, D., Wu, Q., Xu, H. Dynamic route network planning problem for emergency evacuation in restricted space scenarios. J. Adv. Transp. Article ID 4295419, 13 p. (2018)

9. Hong, Y., Li, D., Wu, Q., Xu, H.: Priority-oriented route network planning for evacuation in constrained space scenarios. J. Optim. Theory Appl. (2018). https://doi.org/10.1007/s10957-018-1386-2

10. Vogiatzis, C., Walteros, J.L., Pardalos, P.M.: Evacuation through clustering techniques. In: Goldengorin, B., Kalyagin, V., Pardalos, P. (eds.) Models, Algorithms, and Technologies for Network Analysis. Springer Proceedings in Mathematics & Statistics, vol. 32, pp. 185–198. Springer, New York (2013). https://doi.org/10.1007/978-1-4614-5574-5_10

11. Lujak, M., Giordani, S.: Centrality measures for evacuation: finding agile evacuation routes. Futur. Gener. Comput. Syst. **83**, 401–412 (2018)

12. Vogiatzis, C., Pardalos, P.M.: Evacuation modeling and betweenness centrality. In: Kotsireas, I., Nagurney, A., Pardalos, P. (eds.) DOD 2015 2016. Springer Proceedings in Mathematics & Statistics, vol. 185, pp. 345–359. Springer, Cham (2016). https://doi.org/10.1007/978-3-319-43709-5_17

13. Ahuja, R.K., Magnanti, T.L., Orlin, J.B.: Network Flows: Theory, Algorithms, and Applications. Prentice-Hall, Inc., Upper Saddle River (1993). ISBN 0-13-617549-X

14. Klein, M.: A primal method for minimal cost flows with applications to the assignment and transportation problems. Manag. Sci. **14**, 205–220 (1967)

15. Goldberg, A.V., Tarjan, R.E.: Finding minimum-cost circulations by canceling negative cycles. J. ACM **36**(4), 873–886 (1989)

16. Edmonds, J., Karp, R.M.: Theoretical improvements in algorithmic efficiency for network flow problems. J. ACM **19**(2), 248–264 (1972)

17. Orlin, J.B.: A polynomial time primal network simplex algorithm for minimum cost flows. Math. Program. **78**, 109–129 (1997)

18. Tanka Nath Dhamala: A survey on models and algorithms for discrete evacuation planning network problems. J. Ind. Manag. Optim. **11**(1), 265–289 (2015)

19. Choi, W., Hamacher, H.W., Tufekci, S.: Modeling of building evacuation problems by network flows with side constraints. Eur. J. Oper. Res. **35**(1), 98–110 (1988)

20. Liu, Y., Chang, G.L., Liu, Y., Lai, X.: Corridor-based emergency evacuation system for Washington, DC: system development and case study. Transp. Res. Rec.: J. Transp. Res. Board **2041**, 58–67 (2008)

Practical and Easy-to-Understand Card-Based Implementation of Yao's Millionaire Protocol

Daiki Miyahara[1,4](✉), Yu-ichi Hayashi[2], Takaaki Mizuki[3], and Hideaki Sone[3]

[1] Graduate School of Information Sciences, Tohoku University,
6–3–09 Aramaki-Aza-Aoba, Aoba-ku, Sendai 980-8578, Japan
daiki.miyahara.q4@dc.tohoku.ac.jp
[2] Graduate School of Information Sciences,
Nara Institute of Science and Technology, 8916–5 Takayama, Ikoma,
Nara 630-0192, Japan
[3] Cyberscience Center, Tohoku University, 6–3 Aramaki-Aza-Aoba,
Aoba-ku, Sendai 980-8578, Japan
[4] National Institute of Advanced Industrial Science and Technology,
2–3–26, Oume, Koto-ku, Tokyo 135-0064, Japan

Abstract. Yao's millionaire protocol enables Alice and Bob to know whether or not Bob is richer than Alice by using a public-key cryptosystem without revealing the actual amounts of their properties. In this paper, we present a simple and practical implementation of Yao's millionaire protocol using a deck of playing cards; we straightforwardly implement the idea behind Yao's millionaire protocol so that even non-experts can easily understand its correctness and secrecy. Our implementation is based partially on the previous card-based scheme proposed by Nakai, Tokushige, Misawa, Iwamoto, and Ohta; their scheme admits players' private actions on a sequence of cards called Private Permutation (PP), implying that a malicious player could make an active attack (for example, he/she could exchange some of the cards stealthily when doing such a private action). In contrast, our implementation relies on a familiar shuffling operation called a random cut, and hence, it can be conducted completely publicly so as to avoid any active attack.

Keywords: Card-based protocols · Real-life hands-on cryptography
Secure multi-party computations · Yao's millionaire protocol
Deck of cards

1 Introduction

Assume that Alice and Bob have a and b dollars, respectively, such that $a, b \in \{1, 2, \ldots, m\}$ for some natural number m. They want to know who is richer without revealing any information about their values (more than that is necessary), i.e., they want to determine only whether $a < b$ or not. This is the

© Springer Nature Switzerland AG 2018
D. Kim et al. (Eds.): COCOA 2018, LNCS 11346, pp. 246–261, 2018.
https://doi.org/10.1007/978-3-030-04651-4_17

famous *millionaires' problem* proposed by Yao [21] in 1982, and he designed a protocol, which we call *Yao's millionaire protocol*, to solve the problem based on a public-key cryptosystem. The fundamental principle behind Yao's millionaire protocol could be interpreted as follows. If Alice arranges m symbols consisting of a number a of ♠s and a number $(m-a)$ of ◇s as

$$\underset{1}{♠}\underset{2}{♠} \cdots \underset{a}{♠} \underset{a+1}{◇} \underset{a+2}{◇} \cdots \underset{m}{◇},$$

and Bob points at the $b-$th symbol, then the $b-$th symbol being ◇ implies $a < b$, and the $b-$th symbol being ♠ implies $a \geq b$:

$$\underset{1}{♠}\underset{2}{♠} \cdots \underset{a}{♠} \underset{a+1}{◇} \underset{a+2}{◇} \cdots \underset{\underset{b\text{-th}}{\uparrow}}{◇} \cdots \underset{m}{◇} \iff a < b,$$

$$\underset{1}{♠}\underset{2}{♠} \cdots \underset{\underset{b\text{-th}}{\uparrow}}{♠} \cdots \underset{a}{♠} \underset{a+1}{◇} \underset{a+2}{◇} \cdots \underset{m}{◇} \iff a \geq b.$$

While Yao's millionaire protocol relies on the public-key cryptosystem to implement the above principle without leaking actual values a and b, Nakai, Tokushige, Misawa, Iwamoto, and Ohta [15] considered the use of a deck of physical cards in 2016. That is, following the fundamental principle above, they constructed a card-based scheme using cards of two types such as

$$\boxed{♣}\boxed{♣} \cdots \boxed{♣}\boxed{♡}\boxed{♡} \cdots \boxed{♡}$$

whose backs are all identical $\boxed{?}$. Roughly speaking, in their scheme, Alice first encodes her secret value a with a sequence of face-down cards, and then Bob "privately" changes the positions of cards according to his secret value b. We will describe the details in Sect. 2. Since many people on earth are familiar with playing cards, their card-based scheme is human-friendly and useful. Its only drawback is that it requires a player's "private" action, called *Private Permutation* (PP) [15], which permits Bob to rearrange the sequence of cards privately (for example, he is allowed to manipulate the cards behind his back). Private Permutation is considered to be such a strong assumption that a malicious player may do an active attack. Hereinafter, we refer to their scheme as the *NTMIO protocol with PP*.

Thus, it is preferable to construct a card-based easy-to-understand scheme which does not rely on Private Permutation, in order to avoid possible malicious actions. To this end, in this paper, we present a "PP-free" scheme, which implements the fundamental principle behind Yao's millionaire protocol; instead of using Private Permutation, we use a familiar shuffling operation called the *random cut* (RC). A random cut is a cyclic shuffle, which can be easily implemented by humans as in the case of usual card games (e.g. [2,6,20]). Therefore, our scheme, named the *PP-free protocol with RC*, can be conducted completely publicly, and hence, any malicious action can be detected. As will be seen in Sect. 3, we straightforwardly implement the above principle. Therefore, we believe that even non-experts can easily understand the correctness and secrecy of our scheme, and can practically use it in everyday life.

It should be noted that Nakai et al. [15] proposed a PP-free protocol as well; they presented a card-based scheme, which follows not the above-mentioned fundamental principle but a logical circuit representing the comparison $a < b$. This PP-free circuit-based protocol relies on a shuffling operation called the random bisection cut [13] (instead of Private Permutation). In this paper, we also improve upon this existing protocol; we will reduce the number of required random bisection cuts to around $1/3$. We will explain the details in Sects. 4 and 5.

Table 1 summarizes the performance of the PP-free protocols.

Table 1. The PP-free millionaire protocols.

	#Cards	#Shuffles	Section
Our implementation with RC	$3m+1$	1	Section 3
The previous circuit-based [15]	$4\lceil \log m \rceil + 4$	$6\lceil \log m \rceil - 5$	Section 4
Our improved circuit-based	$4\lceil \log m \rceil + 2$	$2\lceil \log m \rceil - 1$	Section 5

The remainder of this paper is organized as follows. In Sect. 2, we introduce the NTMIO protocol with PP [15]. In Sect. 3, we present our implementation, the PP-free protocol with RC. As for circuit-based protocols, we introduce the previous protocol in Sect. 4, and give an improved protocol in Sect. 5. We conclude this paper in Sect. 6.

2   The Previous Scheme: The NTMIO Protocol with PP

In this section, we introduce the NTMIO protocol with PP [15].

Recall the fundamental principle behind Yao's millionaire protocol; Alice arranges m symbols:

$$\overset{1}{\spadesuit}\,\overset{2}{\spadesuit}\,\cdots\,\overset{a}{\spadesuit}\,\overset{a+1}{\diamondsuit}\,\overset{a+2}{\diamondsuit}\,\cdots\,\overset{m}{\diamondsuit}.$$

Using a pair of physical cards ♣ and ♡, let us encode each symbol as follows:

$$\boxed{♣}\,\boxed{♡} = \spadesuit, \quad \boxed{♡}\,\boxed{♣} = \diamondsuit.$$

Thus, Alice can encode her private value a using m pairs of $\boxed{♣}\boxed{♡}$, and put the cards with their faces down such that Bob does not see the order of the cards. For such a sequence of m pairs encoding Alice's secret value a, Bob needs to point at the b-th pair without leaking any information about his secret value b; to this end, Bob is permitted to use Private Permutation. Specifically, the NTMIO protocol with PP proceeds as follows.

1. Alice holding m ♣ s and m ♡ s places a number a of ♣ ♡ s on a table with their faces down, and then puts $(m-a)$ ♡ ♣ s next to them:

$$
\overset{1}{\boxed{?}\,\boxed{?}}\;\overset{2}{\boxed{?}\,\boxed{?}}\cdots\overset{a}{\boxed{?}\,\boxed{?}}\;\overset{a+1}{\boxed{?}\,\boxed{?}}\;\overset{a+2}{\boxed{?}\,\boxed{?}}\cdots\overset{m}{\boxed{?}\,\boxed{?}},
$$

while Bob does not see the order of each pair.

2. Bob uses Private Permutation; he takes the sequence of cards and move them behind his back. Then, he moves the b-th pair to the first without Alice seeing which pair comes first:

$$
\overset{1}{\boxed{?}\,\boxed{?}}\cdots\overset{b-1}{\boxed{?}\,\boxed{?}}\;\overset{b}{\boxed{?}\,\boxed{?}}\;\overset{b+1}{\boxed{?}\,\boxed{?}}\cdots\overset{m}{\boxed{?}\,\boxed{?}}
$$

$$
\rightarrow\;\overset{b}{\boxed{?}\,\boxed{?}}\;\overset{1}{\boxed{?}\,\boxed{?}}\cdots\overset{b-1}{\boxed{?}\,\boxed{?}}\;\overset{b+1}{\boxed{?}\,\boxed{?}}\cdots\overset{m}{\boxed{?}\,\boxed{?}}.
$$

3. The first pair of cards is revealed.
   - If the revealed cards are ♡ ♣ , $a < b$.
   - If the revealed cards are ♣ ♡ , $a \geq b$.

This is the existing card-based solution to the millionaires' problem using Private Permutation[1]. Let us stress that Bob needs to use Private Permutation in Step 2.

The use of Private Permutation is so powerful as to contribute to improving the efficiency of card-based protocols [14,15], and also it is used in other physical secure protocols [1,9]; however, it might lead to some issues. To implement Step 2 of this protocol, the following issues are considered. (1) If Bob were malicious, he could make an active attack; for instance, he could replace the sequence of cards with another prepared one so that he would be able to peep the exact value of a later. (2) Alice and/or audience watching the execution of the protocol could learn Bob's secret value b by observing his tiny shoulder movement. (3) Permuting some cards behind one's back might be challenging because one only has to rely on the sense of hands; the case of $b = 1$ or $b = m$ might be no problem, but if $b = m/2$, Bob might have difficulty in searching the desired pair of cards.

In the next section, we design a simple PP-free protocol.

3 Our Implementation Using a Random Cut

In this section, we present our card-based implementation of Yao's millionaire protocol; instead of relying on Private Permutation, we use

- a random cut (RC), which is a well-known and easy-to-perform shuffle, and
- cards whose backs are # , which is a different pattern from ? .

---

[1] It should be noted that Fagin, Naor, and Winkler proposed a similar idea to solve the socialist millionaires' problem [5] where Alice and Bob want to know whether they think the same person in mind or not (see Solution 11 in [3]). In addition, Nakai et al. [15] presented another card-based scheme with Private Permutation, which compares a and b bit by bit with the help of "storage" cards.

3.1 How to Proceed

Our PP-free protocol with RC proceeds as follows.

1. Alice holds m ♣s and $(m-1)$ ♡s. Depending on her secret value a, she places a number a of ♣s on a table with their faces down, and then puts a number $(m-a)$ of ♡s next to them. The resulting sequence is Alice's input:

$$
\overset{1}{\boxed{?}}\;\overset{2}{\boxed{?}}\cdots\overset{a}{\boxed{?}}\;\overset{a+1}{\boxed{?}}\;\overset{a+2}{\boxed{?}}\cdots\overset{m}{\boxed{?}}\,.
$$
$$
\clubsuit\;\clubsuit\;\;\clubsuit\;\heartsuit\;\heartsuit\;\;\heartsuit
$$

On the other hand, Bob holds $(m-1)$ cards of ♣ whose backs are $\boxed{\#}$ and a card of ♡ whose back is also $\boxed{\#}$. Then, he places these m cards with their faces down on the table such that only the b-th card is ♡. The resulting sequence is Bob's input:

$$
\overset{1}{\boxed{\#}}\;\overset{2}{\boxed{\#}}\cdots\overset{b-1}{\boxed{\#}}\;\overset{b}{\boxed{\#}}\;\overset{b+1}{\boxed{\#}}\cdots\overset{m}{\boxed{\#}}\,.
$$
$$
\clubsuit\;\clubsuit\;\;\clubsuit\;\heartsuit\;\clubsuit\;\;\clubsuit
$$

2. Take every card from Alice's input sequence and Bob's input sequence from the left alternately one by one, and put it to the right of the previous card:

$$
\overset{1}{\boxed{?}}\,\boxed{\#}\;\overset{2}{\boxed{?}}\,\boxed{\#}\cdots\overset{m}{\boxed{?}}\,\boxed{\#}\,.
$$

We further add two cards to the sequence:

$$
\overset{1}{\boxed{?}}\,\boxed{\#}\;\overset{2}{\boxed{?}}\,\boxed{\#}\cdots\overset{m}{\boxed{?}}\,\boxed{\#}\;\overset{m+1}{\boxed{?}}\,\boxed{\#}\,;
$$
$$
\heartsuit\quad\clubsuit
$$

these two cards are put for handling the case of $a = b = m$. Note that recalling the fundamental principle behind Yao's millionaire protocol, the left card of Bob's ♡-card determines whether $a < b$ or not:

$$
\overset{1}{\boxed{?}}\,\boxed{\#}\cdots\overset{a}{\boxed{?}}\,\boxed{\#}\;\overset{a+1}{\boxed{?}}\,\boxed{\#}\cdots\overset{b}{\boxed{?}}\,\boxed{\#}\cdots\overset{m+1}{\boxed{?}}\,\boxed{\#}\iff a < b,
$$
$$
\clubsuit\;\clubsuit\;\;\clubsuit\;\clubsuit\;\heartsuit\;\clubsuit\;\;\heartsuit\;\heartsuit\;\;\heartsuit\;\clubsuit
$$

$$
\overset{1}{\boxed{?}}\,\boxed{\#}\cdots\overset{b}{\boxed{?}}\,\boxed{\#}\cdots\overset{a}{\boxed{?}}\,\boxed{\#}\;\overset{a+1}{\boxed{?}}\,\boxed{\#}\cdots\overset{m+1}{\boxed{?}}\,\boxed{\#}\iff a \geq b.
$$
$$
\clubsuit\;\clubsuit\;\;\clubsuit\;\heartsuit\;\;\clubsuit\;\clubsuit\;\heartsuit\;\clubsuit\;\;\heartsuit\;\clubsuit
$$

Note, furthermore, that when $a \geq b$, the $(b+1)$-st pair determines whether $a = b$ or $a > b$: if the $(b+1)$-st pair is $\boxed{♡}\,\boxed{♣}$ then $a = b$; if it is $\boxed{♣}\,\boxed{♣}$ then $a > b$. Of course, we cannot open Bob's cards $\boxed{\#}$ now; hence, we add a randomization in the next step.

3. Apply a random cut to the sequence of $(2m+2)$ cards, which means shuffling the card sequence cyclically (we denote this operation by $\langle\,\cdot\,\rangle$):

$$
\Big\langle\,\boxed{?}\,\boxed{\#}\;\boxed{?}\,\boxed{\#}\cdots\boxed{?}\,\boxed{\#}\,\Big\rangle\,.
$$

The random cut can be securely implemented by the shuffle operation called the "Hindu cut" [20]; the shuffle may be repeated by Alice and Bob, or even other people until they are all satisfied with the result. Note that the random cut can be done completely publicly [20], and hence, each player can notice any illegal action if any.

4. Reveal all the cards whose backs are $\boxed{\#}$ (namely, the m cards placed by Bob and the additional card); then, one card of $\heartsuit$ appears. Reveal the card on its left.
   - If the revealed card is $\boxed{\heartsuit}$, $a < b$.
   - If the revealed card is $\boxed{\clubsuit}$, we have $a \geq b$. To see whether equality holds or not, open the card to the right of Bob's $\heartsuit$-card (apart from cyclic rotation). If the opened card is $\boxed{\heartsuit}$, $a = b$. If it is $\boxed{\clubsuit}$, $a > b$.

This is our PP-free protocol with RC. It uses $(3m+1)$ cards in total and uses one shuffle. In Step 1, Alice places a $\boxed{\clubsuit}$s; if Alice has only a $\boxed{\clubsuit}$s at first, the value a might be leaked from the number of cards that Alice holds. Therefore, Alice needs to have m $\boxed{\clubsuit}$s at first (the number of $\boxed{\heartsuit}$ is similar). Since we apply a random cut in Step 3, revealing Bob's cards in Step 4 does not expose where Bob placed the $\heartsuit$-card. If $a = b$, Alice and Bob will learn the exact value; note that their values are not leaked to any other people watching the execution of the protocol.

As for the use of a different back $\boxed{\#}$, we were inspired by the technique called the "Chosen Cut" that Koch and Walzer proposed [6][2]. If the back-side symbol of the cards is vertically asymmetric, we do not need cards of different backs like $\boxed{\#}$: It suffices that Bob puts his cards upside down as follows:

$$\boxed{?}\,\boxed{\text{\textit{ʓ}}}\,\boxed{?}\,\boxed{\text{\textit{ʓ}}}\cdots\boxed{?}\,\boxed{\text{\textit{ʓ}}}.$$

Our protocol can be executed completely publicly. Any malicious action will be noticed. Moreover, we can automatically confirm that Bob put his input in a correct format when we reveal all Bob's cards in Step 4. We can even confirm that Alice put her input in a correct format by applying the idea in [11] with some additional cards.

3.2 A Pseudocode

In this subsection, we present a more formal description of our protocol, that is, we show a pseudocode that follows the computational model of card-based protocols, which was formalized in [7,10,12].

First, let us describe an input card sequence. Remember that, for example, if $a = b = 1$, then Alice and Bob will arrange their inputs with two additional cards as:

$$\Gamma^{(1,1)} = (\overbrace{\frac{?}{\clubsuit}, \frac{?}{\heartsuit}, \ldots, \frac{?}{\heartsuit}}^{m \text{ cards}}, \overbrace{\frac{\#}{\heartsuit}, \frac{\#}{\clubsuit}, \ldots, \frac{\#}{\clubsuit}}^{m \text{ cards}}, \frac{?}{\heartsuit}, \frac{\#}{\clubsuit}).$$

---

[2] Koch and Walzer [6] showed that one can securely "choose" a permutation from a specific set using helping cards with a different color.

Generally, for $a, b \in \{1, 2, \ldots, m\}$, we define

$$\Gamma^{(a,b)} = (\overset{1}{\underset{\clubsuit}{?}}, \ldots, \overset{a-1}{\underset{\clubsuit}{?}}, \overset{a}{\underset{\clubsuit}{?}}, \overset{a+1}{\underset{\heartsuit}{?}}, \ldots, \overset{m}{\underset{\heartsuit}{?}}, \overset{m+1}{\underset{\clubsuit}{\#}}, \ldots, \overset{m+b-1}{\underset{\clubsuit}{\#}}, \overset{m+b}{\underset{\heartsuit}{\#}}, \overset{m+b+1}{\underset{\clubsuit}{\#}}, \ldots, \overset{2m}{\underset{\clubsuit}{\#}}, \overset{2m+1}{\underset{\heartsuit}{?}}, \overset{2m+2}{\underset{\clubsuit}{\#}}).$$

Next, we need to define the following operations applied to a card sequence $\Gamma = (\alpha_1, \alpha_2, \ldots, \alpha_d)$:

- (turn, T) for $T \subseteq \{1, 2, \ldots, d\}$, i.e., turning over the cards is denoted by a set T, which specifies the turned positions of cards;
- (perm, π) for $\pi \in S_d$, where S_i denotes the symmetric group of degree i, i.e., a rearranging operation is denoted by permutation π;
- (shuf, $\Pi, \mathcal{F}$) for $\Pi \subseteq S_d$ and a probability distribution $\mathcal{F}$ on Π, i.e., a shuffling operation is denoted by a permutation set Π and a probability distribution $\mathcal{F}$. If $\mathcal{F}$ is uniform, we simply write it as (shuf, Π);
- (result, e) for some expression e. This indicates that the protocol terminates with the output e.

Based on the above formalization, a pseudocode of our PP-free protocol with RC is shown as follows, where "visible seq." denotes what we can look at for a card sequence on the table, and we define

$$\pi \overset{\text{def}}{=} \begin{pmatrix} 1 & 2 & 3 & \cdots & m & m+1 & m+2 & \cdots & 2m & 2m+1 & 2m+2 \\ 1 & 3 & 5 & \cdots & 2m-1 & 2 & 4 & \cdots & 2m & 2m+1 & 2m+2 \end{pmatrix},$$

and

$$\mathsf{RC}_{2m+2} \overset{\text{def}}{=} \{(1\ 2\ 3\ \cdots\ 2m+2)^j \mid 1 \le j \le 2m+2\}.$$

The PP-free protocol with RC
input set: $\left\{ \Gamma^{(a,b)} \mid 1 \le a, b \le m \right\}$
(perm, π)
(shuf, RC_{2m+2})
if visible seq. $= (\#, ?, \#, ?, \ldots, \#, ?)$ **then**
  (perm, $(2m+2\ \ 2m+1 \cdots 1)$)
(turn, $\{2, 4, \ldots, 2m+2\}$)
let r **s.t.** visible seq. $= (\overset{\text{1st}}{\overbrace{?, \clubsuit}}, \ldots, \overset{(r-1)\text{-st}}{\overbrace{?, \clubsuit}}, \overset{r\text{-th}}{\overbrace{?, \heartsuit}}, \overset{(r+1)\text{-st}}{\overbrace{?, \clubsuit}}, \ldots, \overset{(m+1)\text{-st}}{\overbrace{?, \clubsuit}})$
(turn, $\{2r-1\}$)
if visible seq. $= (?, \clubsuit, \ldots, \overset{r\text{-th}}{\overbrace{\heartsuit}}, \heartsuit, \ldots, ?, \clubsuit)$ **then** (result, "$a < b$")
else if visible seq. $= (?, \clubsuit, \ldots, \overset{r\text{-th}}{\overbrace{\clubsuit}}, \heartsuit, \ldots, ?, \clubsuit)$ **then**
  (turn, $\{2r+1 \pmod{2m+2}\}$)

$$\mathbf{if} \text{ visible seq.} = (?, \clubsuit, \ldots, \overbrace{\clubsuit}^{r\text{-th}}, \heartsuit, \overbrace{\heartsuit, \clubsuit}^{(r+1)\text{-st}}, \ldots, ?, \clubsuit) \mathbf{\ then\ } (\text{result}, \text{"} a = b \text{"})$$

$$\mathbf{else\ if} \text{ visible seq.} = (?, \clubsuit, \ldots, \overbrace{\clubsuit}^{r\text{-th}}, \heartsuit, \overbrace{\clubsuit, \clubsuit}^{(r+1)\text{-st}}, \ldots, ?, \clubsuit) \mathbf{\ then\ } (\text{result}, \text{"} a > b \text{"})$$

3.3 Example of Real Execution

Our protocol is quite simple and easy-to-implement. For example, two colleagues, Alice and Bob, in a company are easily able to compare their bonuses by using our protocol, where Alice's bonus is 10^a dollars and Bob's bonus is 10^b dollars. The protocol falls into real world cryptography; Fig. 1 shows a real execution of our protocol for $m = 4$, i.e., $10^a, 10^b \in \{\$10, \$100, \$1000, \$10000\}$. Card-based protocols are far more practical than might be imagined.

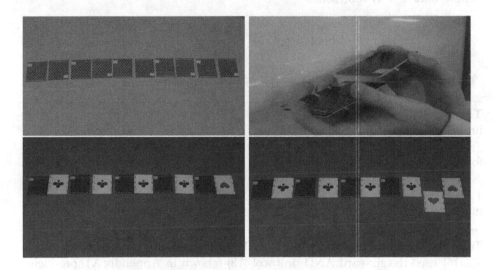

Fig. 1. An implementation of our PP-free protocol with RC when $m = 4$.

4 The Existing Circuit-Based Protocol

In this section, we introduce the existing circuit-based protocol [15] for solving the millionaires' problem.

Consider the following encoding:

$$\boxed{\clubsuit}\,\boxed{\heartsuit} = 0, \quad \boxed{\heartsuit}\,\boxed{\clubsuit} = 1. \tag{1}$$

Then, Alice and Bob can place sequences of cards corresponding to the binary representations of $a = (a_n, \ldots, a_1)_2$ and $b = (b_n, \ldots, b_1)_2$, respectively, where

$n = \lceil \log_2 m \rceil$:

$$\boxed{?}\boxed{?} \cdots \boxed{?}\boxed{?} \quad \boxed{?}\boxed{?} \cdots \boxed{?}\boxed{?} .$$
$$\underbrace{\quad}_{a_n} \quad \underbrace{\quad}_{a_1} \quad \underbrace{\quad}_{b_n} \quad \underbrace{\quad}_{b_1}$$

Such a pair of face-down cards

$$\underbrace{\boxed{?}\boxed{?}}_{x}$$

corresponding to a bit $x \in \{0,1\}$ is called a *commitment* to x. Given the above card sequence along with some additional cards, the existing circuit-based protocol given by Nakai et al. [15] determines whether $a < b$ or not:

$$\underbrace{\boxed{?}\boxed{?}}_{a_n} \cdots \underbrace{\boxed{?}\boxed{?}}_{a_1} \underbrace{\boxed{?}\boxed{?}}_{b_n} \cdots \underbrace{\boxed{?}\boxed{?}}_{b_1}\boxed{\clubsuit}\boxed{\heartsuit}\boxed{\clubsuit}\boxed{\heartsuit} \rightarrow \cdots \rightarrow \underbrace{\boxed{?}\boxed{?}}_{\text{bool}(a<b)} ,$$

where $\text{bool}(a < b)$ represents

$$\text{bool}(a < b) \overset{\text{def}}{=} \begin{cases} 0 \text{ if } a \geq b, \\ 1 \text{ if } a < b. \end{cases}$$

Their protocol proceeds based on the following logical circuit.

The circuit-based protocol [15]
input : $a = (a_n, \ldots, a_1)_2, b = (b_n, \ldots, b_1)_2$;
$f_1 = \bar{a}_1 \wedge b_1$;
for $(i : 2 \text{ to } n)$ {
 $f_i = (\bar{a}_i \wedge b_i) \vee ((\bar{a}_i \vee b_i) \wedge f_{i-1})$;
}
output : $f_n (= \text{bool}(a < b))$.

To implement this circuit, one needs AND (OR) and COPY protocols; Nakai et al. [15] used the six-card AND protocol [13] (shown in Appendix A), producing a commitment to $x \wedge y$ from the input commitments to x and y:

$$\underbrace{\boxed{?}\boxed{?}}_{x}\boxed{\clubsuit}\boxed{\heartsuit}\underbrace{\boxed{?}\boxed{?}}_{y} \rightarrow \cdots \rightarrow \underbrace{\boxed{?}\boxed{?}}_{x \wedge y} ,$$

and the six-card COPY protocol [13] (shown in Appendix B), producing two commitments to x from an input commitment to x:

$$\underbrace{\boxed{?}\boxed{?}}_{x}\boxed{\clubsuit}\boxed{\heartsuit}\boxed{\clubsuit}\boxed{\heartsuit} \rightarrow \cdots \rightarrow \underbrace{\boxed{?}\boxed{?}}_{x}\underbrace{\boxed{?}\boxed{?}}_{x} .$$

Therefore, their circuit-based protocol uses the AND (OR) protocol [13] $(4\lceil \log m \rceil - 3)$ times and the COPY protocol [13] $(2\lceil \log m \rceil - 2)$ times for duplicating the commitments to $\bar{a}_i$ and b_i. Thus, the number of required shuffles is

$(6\lceil \log m \rceil - 5)$ in total (because each of the AND and COPY protocols [13] uses one shuffle; see Appendices A and B for the details). Regarding the number of cards, six additional cards are required to duplicate the commitments to a_2 and b_2 in order to compute $f_2 = (\bar{a}_2 \wedge b_2) \vee ((\bar{a}_2 \vee b_2) \wedge f_1)$. Thus, six cards need to be left directly after the AND computation of $f_1 = \bar{a}_1 \wedge b_1$, and hence, four additional cards are required before the computation of f_1. Consequently, the total number of required cards is $(4\lceil \log m \rceil + 4)$.

5   Our Improved Circuit-Based Protocol

In this section, we improve upon the circuit-based protocol introduced in Sect. 4, i.e., we present an improved circuit-based protocol that uses a less number of shuffles and cards. We first show the idea behind our improved circuit-based protocol in Sect. 5.1, and then show the procedure of our improved circuit-based protocol in Sect. 5.2.

5.1   The Idea

We borrow the idea behind the storage protocol [15]; it uses Private Permutation and regards f_i shown in Sect. 4 as:

$$f_i = \begin{cases} f_{i-1} & \text{if } a_i = b_i, \\ b_i & \text{if } a_i \neq b_i. \end{cases} \tag{2}$$

That is, the storage protocol is supposed to choose f_{i-1} or b_i depending on whether $a_i = b_i$ or not. More specifically,

- f_{i-1} is equal to $\text{bool}((a_{i-1}, \ldots, a_1) < (b_{i-1}, \ldots, b_1))$;
- $a_i = b_i$ implies $f_i = f_{i-1}$;
- $a_i = 0$ and $b_i = 1$ imply $(a_i, \ldots, a_1) < (b_i, \ldots, b_1)$, and hence $f_i = 1 = b_i$ while $a_i = 1$ and $b_i = 0$ imply $(a_i, \ldots, a_1) > (b_i, \ldots, b_1)$, and hence $f_i = 0 = b_i$.

Such a choice can be made without Private Permutation; if we can let a six-card sequence be either

then, we can obtain a commitment to f_i by revealing the first two cards as follows:

The above flow can be accomplished by using the procedure of the six-card AND protocol [13] shown in Appendix A.

Moreover, we can easily obtain commitments to $a_i \oplus b_i$ and b_i by using the six-card COPY protocol [13] shown in Appendix B. That is, from the following sequence where r is a uniform random bit:

$$\boxed{?}\,\boxed{?}\;\boxed{?}\,\boxed{?}\;\boxed{?}\,\boxed{?},$$
$$\underbrace{}_{b_i \oplus r}\;\underbrace{}_{a_i \oplus r}\;\underbrace{}_{r}$$

it is determined whether $r = b_i$ or $r = \bar{b}_i$ by revealing the first two cards, and then we obtain commitments to $a_i \oplus b_i$ and b_i as:

$$\boxed{\clubsuit}\,\boxed{\heartsuit}\;\boxed{?}\,\boxed{?}\;\boxed{?}\,\boxed{?} \;\text{ or }\; \boxed{\heartsuit}\,\boxed{\clubsuit}\;\boxed{?}\,\boxed{?}\;\boxed{?}\,\boxed{?}.$$
$$\underbrace{}_{a_i \oplus b_i}\;\underbrace{}_{b_i}\qquad\underbrace{}_{\overline{a_i \oplus b_i}}\;\underbrace{}_{b_i}$$

Note that revealing the first two cards leaks no information about b_i because r is a random bit.

As described above, by using the procedure of the COPY and AND protocols [13], we can obtain a commitment to f_i according to Eq. 2 without revealing the values of a_i, b_i, and f_{i-1}. It should be noted that f_i in Eq. 2 is the three-input majority function of a_i, b_i, and f_{i-1}. An efficient card-based protocol for the three-input majority function was proposed by Nishida et al. [16] in 2013, which was based on the same idea mentioned above.

5.2 The Description of Our Protocol

Based on the idea presented in Sect. 5.1, we construct an improved circuit-based protocol. Given the input card sequence

$$\boxed{?}\,\boxed{?}\;\cdots\;\boxed{?}\,\boxed{?}\quad\boxed{?}\,\boxed{?}\;\cdots\;\boxed{?}\,\boxed{?}$$
$$\underbrace{}_{a_n}\qquad\underbrace{}_{a_1}\quad\underbrace{}_{b_n}\qquad\underbrace{}_{b_1}$$

and two additional cards, our protocol proceeds as follows.

1. Compute $f_1 = \bar{a}_1 \wedge b_1$ by using the six-card AND protocol [13]:

$$\underbrace{\boxed{?}\,\boxed{?}}_{\bar{a}_1}\;\underbrace{\boxed{?}\,\boxed{?}}_{b_1}\;\boxed{\clubsuit}\,\boxed{\heartsuit} \to \cdots \to \underbrace{\boxed{?}\,\boxed{?}}_{f_1}.$$

Now, four reusable cards remain.

2. Repeat the following computation from $i = 2$ to $i = n$.
   (a) Obtain commitments to $a_i \oplus b_i$ and b_i from the commitments to a_i and b_i by using the six-card COPY protocol [13] (and the NOT computation):

$$\underbrace{\boxed{?}\,\boxed{?}}_{b_i}\;\underbrace{\boxed{?}\,\boxed{?}}_{a_i}\;\underbrace{\boxed{?}\,\boxed{?}}_{0} \to \underbrace{\boxed{?}\,\boxed{?}}_{a_i \oplus b_i}\;\underbrace{\boxed{?}\,\boxed{?}}_{b_i}.$$

(b) Obtain a commitment to f_i from the commitments to $a_i \oplus b_i$, b_i, and f_{i-1} by using the six-card AND protocol:

$$\underbrace{\boxed{?}\boxed{?}}_{a_i \oplus b_i}\underbrace{\boxed{?}\boxed{?}}_{f_{i-1}}\underbrace{\boxed{?}\boxed{?}}_{b_i} \rightarrow \underbrace{\boxed{?}\boxed{?}}_{f_i} \, .$$

3. Then, a commitment to $f_n = \mathrm{bool}\,(a < b)$ can be obtained:

$$\underbrace{\boxed{?}\boxed{?}}_{\mathrm{bool}(a<b)} \, .$$

Due to the use of two additional cards, this protocol uses $(4\lceil \log m \rceil + 2)$ cards in total. The number of required shuffles is $(2\lceil \log m \rceil - 1)$ in total, because this protocol repeats each procedure of the AND protocol and the COPY protocol from $i = 2$ to $i = n$ after AND computation for f_1, as shown in Table 1.

This improved circuit-based protocol is a combination of the existing information-theoretically secure card-based protocols, and hence, it is guaranteed to be secure.

6   Conclusion

In this paper, we proposed two card-based protocols to solve the millionaires' problem without using Private Permutation. See Table 1 again for the performance of the PP-free millionaire protocols. In particular, the PP-free protocol with RC proposed in Sect. 3 uses only one random cut, and its correctness and secrecy are clear. Therefore, we believe that even non-experts such as high school students can easily understand and use it practically. Note that a random cut can be easily and securely implemented by using the Hindu cut [20].

Moreover, we can use our protocols in didactic contexts in order to invite young people and students to cryptography; they would be an ideal tool to exhibit the concept of secure multiparty computations, as often pointed out, e.g., [4,8].

Regarding the number of required cards in the improved circuit-based protocol, it will be best to use the four-card AND protocol [7], because four cards are necessary (and hence minimal) to represent two inputs based on the encoding rule (1). Moreover, there is the five-card COPY protocol [18], and hence we can theoretically reduce the number of required cards in the improved circuit-based protocol. However, the four-card AND protocol requires an average of eight shuffling operations, and the five-card COPY protocol requires ideal cases [17,19] for execution. Therefore, for practicality, we considered only the six-card AND protocol [13] and the six-card COPY protocol [13].

Acknowledgments. We thank the anonymous referees, whose comments have helped us to improve the presentation of the paper. This work was supported by JSPS KAKENHI Grant Number JP17K00001.

A   The Six-Card and Protocol

In 2009, Mizuki and Sone [13] invented the following six-card AND protocol, which securely outputs a commitment to $x \wedge y$ from the commitments to x and y (and two additional cards).

1. Place input commitments and additional cards of black and red, and then turn over the two cards in the center:

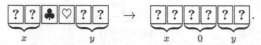

2. Rearrange the sequence:

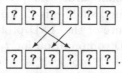

3. Apply a *random bisection cut*, which means bisecting the sequence and switching the two halves randomly:

$$[\boxed{?}\boxed{?}\boxed{?}|\boxed{?}\boxed{?}\boxed{?}] \quad \rightarrow \quad \boxed{?}\boxed{?}\boxed{?}\boxed{?}\boxed{?}\boxed{?}.$$

After applying this shuffling operation, the six-card sequence results in either the same sequence as the original one or a sequence whose left and right halves are switched; each case occurs with a probability of $1/2$.

4. Rearrange the sequence:

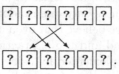

After this rearranging operation, the six-card sequence will be as follows:

$$\underbrace{\boxed{?}\boxed{?}}_{x}\underbrace{\boxed{?}\boxed{?}}_{0}\underbrace{\boxed{?}\boxed{?}}_{y} \quad \text{or} \quad \underbrace{\boxed{?}\boxed{?}}_{\bar{x}}\underbrace{\boxed{?}\boxed{?}}_{y}\underbrace{\boxed{?}\boxed{?}}_{0}.$$

5. Reveal the first two cards. Then, a commitment to $x \wedge y$ can be obtained as:

$$\boxed{\clubsuit}\boxed{\heartsuit}\underbrace{\boxed{?}\boxed{?}}_{x \wedge y}\boxed{?}\boxed{?} \quad \text{or} \quad \boxed{\heartsuit}\boxed{\clubsuit}\underbrace{\boxed{?}\boxed{?}}_{x \wedge y}\boxed{?}\boxed{?}.$$

Note that we can reuse the two revealed two cards, and moreover, the other two cards not being a commitment to $x \wedge y$ can be reused by revealing them after shuffling.

As mentioned above, we can obtain a commitment to $x \wedge y$ (keeping its value secret). It is well known that in the literature [20], a random bisection cut can be implemented by humans securely so that nobody knows the resulting card sequence.

An OR protocol can be obtained in a similar way.

B The Six-Card COPY Protocol

The following six-card COPY protocol proposed by Mizuki and Sone [13] produces two commitments to x from a commitment to x and four additional cards.

1. Place an input commitment and black and red additional cards, and then turn over the additional cards:

$$\underbrace{\boxed{?}\ \boxed{?}}_{x}\ \boxed{\clubsuit}\boxed{\heartsuit}\boxed{\clubsuit}\boxed{\heartsuit} \rightarrow \underbrace{\boxed{?}\ \boxed{?}}_{x}\ \underbrace{\boxed{?}\ \boxed{?}}_{0}\ \underbrace{\boxed{?}\ \boxed{?}}_{0}.$$

2. Rearrange the sequence:

$$\overset{1\ \ 2\ \ 3\ \ 4\ \ 5\ \ 6}{\boxed{?}\boxed{?}\boxed{?}\boxed{?}\boxed{?}\boxed{?}} \rightarrow \overset{1\ \ 3\ \ 5\ \ 2\ \ 4\ \ 6}{\boxed{?}\boxed{?}\boxed{?}\boxed{?}\boxed{?}\boxed{?}}.$$

3. Apply a random bisection cut:

$$\left[\boxed{?}\boxed{?}\boxed{?}\ \middle|\ \boxed{?}\boxed{?}\boxed{?}\right] \rightarrow \boxed{?}\boxed{?}\boxed{?}\boxed{?}\boxed{?}\boxed{?}.$$

4. Rearrange the sequence:

$$\overset{1\ \ 2\ \ 3\ \ 4\ \ 5\ \ 6}{\boxed{?}\boxed{?}\boxed{?}\boxed{?}\boxed{?}\boxed{?}} \rightarrow \overset{1\ \ 4\ \ 2\ \ 5\ \ 3\ \ 6}{\boxed{?}\boxed{?}\boxed{?}\boxed{?}\boxed{?}\boxed{?}}.$$

5. Reveal the first two cards. Then, two commitments to x can be obtained as follows (two revealed cards can be reused in the next protocol):

$$\boxed{\clubsuit}\boxed{\heartsuit}\underbrace{\boxed{?}\ \boxed{?}}_{x}\underbrace{\boxed{?}\ \boxed{?}}_{x}\ \text{ or }\ \boxed{\heartsuit}\boxed{\clubsuit}\underbrace{\boxed{?}\ \boxed{?}}_{\bar{x}}\underbrace{\boxed{?}\ \boxed{?}}_{\bar{x}}.$$

In the latter case, we can easily get two commitments to x (from commitments to $\bar{x}$) by using the NOT protocol (swapping the two cards of each commitment).

References

1. Balogh, J., Csirik, J.A., Ishai, Y., Kushilevitz, E.: Private computation using a PEZ dispenser. Theor. Comput. Sci. **306**(1), 69–84 (2003). http://www.sciencedirect.com/science/article/pii/S030439750300210X
2. Boer, B.: More efficient match-making and satisfiability *the five card trick*. In: Quisquater, J.-J., Vandewalle, J. (eds.) EUROCRYPT 1989. LNCS, vol. 434, pp. 208–217. Springer, Heidelberg (1990). https://doi.org/10.1007/3-540-46885-4_23
3. Fagin, R., Naor, M., Winkler, P.: Comparing information without leaking it. Commun. ACM **39**(5), 77–85 (1996). https://doi.org/10.1145/229459.229469
4. Hanaoka, G.: Towards user-friendly cryptography. In: Phan, R.C.-W., Yung, M. (eds.) Mycrypt 2016. LNCS, vol. 10311, pp. 481–484. Springer, Cham (2017). https://doi.org/10.1007/978-3-319-61273-7_24

5. Jakobsson, M., Yung, M.: Proving without knowing: on oblivious, agnostic and blindfolded provers. In: Koblitz, N. (ed.) CRYPTO 1996. LNCS, vol. 1109, pp. 186–200. Springer, Heidelberg (1996). https://doi.org/10.1007/3-540-68697-5_15

6. Koch, A., Walzer, S.: Foundations for actively secure card-based cryptography. Cryptology ePrint Archive, Report 2017/423 (2017). https://eprint.iacr.org/2017/423

7. Koch, A., Walzer, S., Härtel, K.: Card-based cryptographic protocols using a minimal number of cards. In: Iwata, T., Cheon, J.H. (eds.) ASIACRYPT 2015. LNCS, vol. 9452, pp. 783–807. Springer, Heidelberg (2015). https://doi.org/10.1007/978-3-662-48797-6_32

8. Marcedone, A., Wen, Z., Shi, E.: Secure dating with four or fewer cards. Cryptology ePrint Archive, Report 2015/1031 (2015). https://eprint.iacr.org/2015/1031

9. Mizuki, T., Kugimoto, Y., Sone, H.: Secure multiparty computations using the 15 puzzle. In: Dress, A., Xu, Y., Zhu, B. (eds.) COCOA 2007. LNCS, vol. 4616, pp. 255–266. Springer, Heidelberg (2007). https://doi.org/10.1007/978-3-540-73556-4_28

10. Mizuki, T., Shizuya, H.: A formalization of card-based cryptographic protocols via abstract machine. Int. J. Inf. Secur. **13**(1), 15–23 (2014)

11. Mizuki, T., Shizuya, H.: Practical card-based cryptography. In: Ferro, A., Luccio, F., Widmayer, P. (eds.) Fun with Algorithms. Lecture Notes in Computer Science, vol. 8496, pp. 313–324. Springer, Cham (2014). https://doi.org/10.1007/978-3-319-07890-8_27

12. Mizuki, T., Shizuya, H.: Computational model of card-based cryptographic protocols and its applications. IEICE Trans. Fundam. Electron. Commun. Comput. Sci. **E100.A**(1), 3–11 (2017)

13. Mizuki, T., Sone, H.: Six-card secure AND and four-card secure XOR. In: Deng, X., Hopcroft, J.E., Xue, J. (eds.) FAW 2009. LNCS, vol. 5598, pp. 358–369. Springer, Heidelberg (2009). https://doi.org/10.1007/978-3-642-02270-8_36

14. Nakai, T., Shirouchi, S., Iwamoto, M., Ohta, K.: Four cards are sufficient for a card-based three-input voting protocol utilizing private permutations. In: Shikata, J. (ed.) ICITS 2017. LNCS, vol. 10681, pp. 153–165. Springer, Cham (2017). https://doi.org/10.1007/978-3-319-72089-0_9

15. Nakai, T., Tokushige, Y., Misawa, Y., Iwamoto, M., Ohta, K.: Efficient card-based cryptographic protocols for millionaires' problem utilizing private permutations. In: Foresti, S., Persiano, G. (eds.) CANS 2016. LNCS, vol. 10052, pp. 500–517. Springer, Cham (2016). https://doi.org/10.1007/978-3-319-48965-0_30

16. Nishida, T., Mizuki, T., Sone, H.: Securely computing the three-input majority function with eight cards. In: Dediu, A.-H., Martín-Vide, C., Truthe, B., Vega-Rodríguez, M.A. (eds.) TPNC 2013. LNCS, vol. 8273, pp. 193–204. Springer, Heidelberg (2013). https://doi.org/10.1007/978-3-642-45008-2_16

17. Nishimura, A., Hayashi, Y., Mizuki, T., Sone, H.: Pile-shifting scramble for card-based protocols. IEICE Trans. Fundam. Electron. Commun. Comput. Sci. **E101.A**(9), 1494–1502 (2018)

18. Nishimura, A., Nishida, T., Hayashi, Y., Mizuki, T., Sone, H.: Five-card secure computations using unequal division shuffle. In: Dediu, A.-H., Magdalena, L., Martín-Vide, C. (eds.) TPNC 2015. LNCS, vol. 9477, pp. 109–120. Springer, Cham (2015). https://doi.org/10.1007/978-3-319-26841-5_9

19. Nishimura, A., Nishida, T., Hayashi, Y., Mizuki, T., Sone, H.: Card-based protocols using unequal division shuffles. Soft Comput. **22**, 361–371 (2017). https://doi.org/10.1007/s00500-017-2858-2

20. Ueda, I., Nishimura, A., Hayashi, Y., Mizuki, T., Sone, H.: How to implement a random bisection cut. In: Martín-Vide, C., Mizuki, T., Vega-Rodríguez, M.A. (eds.) TPNC 2016. LNCS, vol. 10071, pp. 58–69. Springer, Cham (2016). https://doi.org/10.1007/978-3-319-49001-4_5

21. Yao, A.C.: Protocols for secure computations. In: Proceedings of the 23rd Annual Symposium on Foundations of Computer Science, pp. 160–164. SFCS 1982. IEEE Computer Society, Washington, DC, USA (1982). https://doi.org/10.1109/SFCS.1982.88

Defend the Clique-based Attack
for Data Privacy

Meng Han[1]([⊠]), Dongjing Miao[2,3], Jinbao Wang[3], and Liyuan Liu[1]

[1] Data-driven Intelligence Research (DIR) Laboratory, Kennesaw State University,
Kennesaw, GA 30060, USA
mhan9@kennesaw.edu
[2] Georgia State University, Atlanta, GA 30303, USA
[3] Harbin Institute of Technology, Harbin, Heilongjiang, China

Abstract. Clique, as the most compact cohesive component in a graph,
has been employed to identify cohesive subgroups of entities and explore
the sensitive information in the online social network, crowdsourcing network, and cyber physical network, *etc.* In this study, we focus on the
defense of clique-based attack and target at reducing the risk of entities
security/privacy issues in clique structure. Since the ultimate resolution
for defending the clique-based attack and risk is wrecking the clique
with minimum cost, we establish the problem of clique-destroying (CD)
in the network from a fundamental algorithm aspect. Interestingly, we
notice that the clique-destroying problem in the directed graph is still an
unsolved problem, and complexity analysis also does not exist. Therefore, we propose an innovative formal clique-destroying problem and
proof the *NP-complete* problem complexity with solid theoretical analysis, then present effective and efficient algorithms for both undirected
and directed graph. Furthermore, we show how to extend our algorithm
to data privacy protection applications with controllable parameter k,
which could adjust the size of a clique we wish to destroy. By comparing our algorithm with the up-to-date anonymization approaches, the
real data experiment demonstrates that our resolution could efficaciously
defend the clique-based security and privacy attacks.

1 Introduction

The capabilities of smartphone, smart home [1], and smart earth [2], coupled
with the almost ubiquitous availability of internet connectivity promoted the
ever increasing number of exponential data generation. The report [3] shows,
by 2020, about 50–100 billion devices are going to be connected to internet and
generating a huge number of data for analysis and knowledge discovery. The volume, variety, and velocity of big data are further amplified by the popularity of
social networks, especially by mobile social networks, crowdsourcing networks,
and cyber physical networks, *etc.* When the data above is released for justified
mining and analytical purposes, truly beneficial consequences will be available
in sociology, economics, and advertisers, *etc.* However, benefits of these data also

© Springer Nature Switzerland AG 2018
D. Kim et al. (Eds.): COCOA 2018, LNCS 11346, pp. 262–280, 2018.
https://doi.org/10.1007/978-3-030-04651-4_18

give rise to potentially significant privacy issues in the sense that, for deleterious purposes, the malicious entities may violate the privacy of the data usage by analyzing and deliberately conducting these privacy violations [4–6]. The sensitive information such as the user's physical identity, location history, individual behavior patterns, and personal social connections are elementary to be used for illegal purposes.

Clique, in networks represents a complete subgraph $\mathcal{C}$ such that every two vertices are the two endpoints of an edge in $\mathcal{C}$. The interior connectivity makes the vertex in clique be strongly connected to other vertices. In social networks, undirected graph like Facebook[1], LinkedIn[2], or directed graph like Twitter[3], Instagram[4], the users with similar properties, interests, and background are commonly linked together and construct cliques; many topic groups also cluster the crowdsourcing network with strong connectivity; the devices within one WIFI system or under one uniform platform are commonly linked together as well. All correlations mentioned above within one clique will potentially exhibit the sensitive information or expose the anonymization of identifiable data. Recently, the clique-based techniques were used to attack the data privacy in multiple scenarios [7–9]. Although many privacy preserving approaches such as differential privacy [10], k-anonymity [11], and deep learning [12], *etc.* have been developed in the past couple of years, most of the privacy-preserving techniques are still only focusing on the data content itself instead of the structure or the unit correlation of data.

Take the famous and popular differential privacy mechanism as an example; we now give a concrete example to illustrate the violation of individual's privacy in a social network setting with clique structure. Differential privacy addresses the question that given the total number of particular information, whether or not an adversary can learn if an individual has the specific information. Differential privacy requires that when one particular user's data is added or removed from the database, the output of the database only changes slightly with a specific notion of closeness ϵ. Suppose the same political bias is commonly based on social class, then similar political directives are always within the same community. Consider a group of people in the social network construct a clique community with the average size of 200. In each clique, either at most 20% of the people are Republicans, or at most 20% are Democrats. We assume that on average the population of Republicans and Democrats are roughly same. To protect the privacy of political preference, we add the Laplacian noise to make the data satisfy the differential privacy for parameter $\epsilon = 0.1$. Let function $Lap(\lambda)$ be the Laplace distribution that follows the probability density function $f_\lambda(x) = \frac{1}{2\lambda e^{|x|/\lambda}}$. Any individual who changes her or his political preference's proportion would be $1/200$ in this situation. To achieve ϵ-differential privacy, the $Lab(1/200\epsilon)$ noise needs to be added to the clique in the social network

---
[1] https://www.facebook.com/.
[2] https://www.linkedin.com/.
[3] https://twitter.com/.
[4] https://www.instagram.com.

independently. Although the setting above satisfies the ϵ-differential privacy for the small parameter ϵ, releasing such information will still violate the privacy of the particular person.

According to the privacy definition of Dalenius [13] and the demonstration of [14], even though we cannot achieve the most desirable notion of privacy that "anything about an individual that can be learned from the data can also be learned without access to the data", the privacy protection in our example should ensure that the released data will not allow the adversary to guess correctly with the probability greater than 1/2 apparently, no matter what a particular individual is a Republican or a Democrat. However, under differential privacy mechanism, with $\epsilon = 0.1$, the $Lap(1/200\epsilon)$ is a small amount of noise. If releasing the data by the protection above, according to the main political preference of the corresponding clique, an adversary could easily guess the political preference for particular individuals with probability approximate 80% in any clique.

However, adding the noise by differential privacy, or use $k-1$ anonymities to the related client essentially are still changing the data itself. Because the privacy is still under the deduction of the group or implicit structural information, a straightforward idea would be collecting the affiliation between the individual and the group, then inferring the sensitive information from the affiliation and the collective attribution.

Besides the differential privacy, many other privacy protection mechanisms are also interfered by the correlations in the relational data and the connections in graph data. Many techniques are developed to preserve the privacy, but surprisingly ignore the important clique structure altogether. One main reason is that the privacy preserving is still mainly developed for services, and the loss of privacy is the price of services somehow. In fact, in most situations, the services quality is mainly based on the individual's data but not the structural information. Therefore, from the structure aspect, to take care the correlation between individual and the group would be an exceptional choice to protect the privacy while preserving the services quality.

Therefore, the key point is how to hide the information between an individual and their closely related group. A naïve way would be hiding the information of a specific attribute, height in the above example. But the utility cost is too high to accept in most of the applications, then what would be the resolution to hide the subordinated relationship and preserving the cost of the utility? We proposed to build an adjacent graph to organize the information and break the relationship between the individual and the implicit group. Since the relationship between each could still hold as a chain, the redundancy can help the data publishing.

Based on the above observations, we believe destroying the implicit compact structure, clique, in a graph could help to protect the data privacy to a great extent. However, cutting off the connection between different units in a graph or separating the correlation in the structure of a network is not a trivial job; therefore, in this study, we concentrate on destroying the clique in a graph. Even working for the simplified graph model, wrecking the clique in a graph is still a very challenge task.

Since the public networks include two types: directed and undirected, both undirected networks like Facebook, LinkedIn as well as directed networks like Twitter, Instagram are all typical target of our attention. Surprisingly, when we look at the clique-destroying problems, we find that although the edge-deletion problems in the undirected graph have been investigated previously [15], the clique destroy problems in the directed graph unexpectedly have never been addressed yet. We neither know the problem hardness nor the resolution at all. We further show the finding that the problem of clique destroying in directed graph dominate the problem of the undirected graph version. Therefore, we theoretical analyze the unresolved problem hardness first; then propose the algorithm for both directed and undirected clique destroying problems. We summarize our contributions in this work as follows:

- We proposed the problem of defending the implicit clique-based attacks and modelled the challenges of the clique destroying problem.
- Surprisingly, we noticed an unsolved fundamental problem of destroying clique in the directed graph, which has not been investigated in graph theory yet.
- We first modeled the clique-destroying problem in a directed graph, and proved the *NPcomplete* hardness of the problem by reducing it to the *SATthree* problem.
- Also, we showed the relationship between the clique-destroying problem in both undirected and directed graph, and illustrated that our proof dominates the conclusion in undirected graph previously.
- We proposed integer linear programming algorithm and the relax version algorithm to linear programming, then rounded the solutions to solve our proposed problem very deftly.
- The approximate ratio of our proposed algorithm has been proved, and we extended our algorithm to a generalized version with a controllable parameter k.
- By conducting the experiments on six real-life datasets and the up-to-date comparable algorithms, we demonstrated that our approach could be a very efficient and effective algorithm to improve the data privacy protection techniques further.

In this study, we worked on the defense of clique-based attack and proposed a couple of fundamental results in theoretical complexity analysis, then introduced the efficient and effective algorithms to solve the proposed problem. Firstly, we illustrated the motivation and the background with examples of the typical problems as an introduction in Sect. 1. Secondly, we summarized the up-to-date related works in Sect. 2. Thirdly, some preliminary knowledge regarding the novel problem and the detailed algorithm complexity analysis are presented in Sect. 3. In Sect. 4, the detailed algorithm complexity analysis is introduced. Then, we pointed out some new challenges and opportunities based on our proposed problems. We also showed the comprehensive evaluation of our proposed resolution for the novel problem in Sect. 5. Finally, Sect. 6 concludes this paper.

2   Related Works

A large amount of online social networks and diverse information from many cyber physical aspects pose an unparalleled data privacy threat to users. Regarding data privacy, along with Dalenius's desirable privacy goal that database doesn't provide any extra knowledge to reveal the sensitive information [13], Johannes, Edward, et al. also developed a zero-knowledge definition of privacy. Although the general impossibility of the ideal privacy goal has been demonstrated in [14], the challenges and the goals of data privacy protection still attracted a lot of efforts from both academia [16–18] and industry [19–21].

2.1   Privacy Attacks and Defences

A common way to handle the unwelcome intrusion on the data privacy is somehow anonymizing the data by removing most potentially identifying attributes, or adding noise to hide the sensitive information. Early works on data security privacy learning were done in the secure function evaluation (SFE) framework [22] and secure multi-party computations (MPC) [23], which tried to split the input between two or more parties, and target at minimizing the information leaked during the joint computation. Our study is working for the situation that data is stored centrally, and we are concerned with the leakage of data knowledge. Besides, with the social networks come to our life, privacy concerns with social networking services is an essential subset of data privacy problems [24, 25]. There are many researchers focus on finding an efficient mechanism to defenses the cyber system attack [26–28]. And some are paying more attentions to specific threats, such as location privacy [4, 29, 30], IoT security [31–33], transportation cyber systems [34], blockchain privacy [35], and other society problems [36, 37].

In recent years, k-anonymity and differential privacy have become two of the most popular tools with strong theoretical and empirical limitations [38]. By generalizing and suppressing identifying attributes, k-anonymity seeks to offer a degree of protection to underlying data [16].

But hiding information to $k - 1$ dummy instances cannot apply to deanonymization of high-dimensional, diverse input datasets. In the situation of clique-based attack, k-anonymity may not provide any satisfying privacy even if all k isomorphic neighborhoods have the same value of some sensitive attributes. Differential privacy constitutes a reliable standard for privacy guarantees for an aggregate algorithm on databases [17, 39]. Applications of differential privacy had been extended to boosting for learning [40], linear and logistic regression [41], principal component analysis [42], continuous data processing [43], and many location-based privacy protections [44–46], etc. However, as we demonstrated earlier in Sect. 1, differential privacy doesn't work well on the clique structure.

On another hand, the clique is a very densely-connected group of nodes in a network. It applies to social, biological, crowdsourcing, and cyber physical network. Yildiz particularly investigates the social cliques and Kruegel [47] to control the privacy. Many clique-based attack methods had been proposed [7, 8, 48] also. The authors of [8] used the 3-clique to find the most vulnerable member of a targeted community. The attack algorithm in [7] is one of the most typical passive attack work, which executed in seed identification and propagation two phases: (1) the attacker tries to find in the anonymized graph the counterpart of a unique 4-clique presented in the source graph. Within a 4-clique in the source graph, the degree of each vertex and the number of common neighbors for each pair of nodes are computed by the attacker; the attacker will look for similar 4-cliques with similar values in the target graph. (2) the algorithm iteratively adds nodes to the mapping until there are unmapped vertices with proper mappings. The second phase will never run if the attacker fails in the first phase. Further, this attack does not allow any structural modifications within the cliques; and the identification will fail if one or more edges erase from the clique. Later, Narayanan et al. introduced a similar attack with a less rigid, non-pattern-based seed identification phase [49]. But they still followed the same propagation phase and with same unchangeable structural limitation. Srivatsa and Hicks [50] proposed a re-identified approach by matching the location traces to a social network on small datasets. Ji et al. [51] also confirmed the result on the same datasets. Then, [9] argued that the similarity function used to re-identify nodes is a key component of the de-anonymization attack, and they designed a measure tailored for the social network to process their attack. However, the main novelty of this work is the similarity measurement but not change the situation that structure can still explore the privacy of the data. Gulyás and Imre [52] analyzed the protective strength of identity separation again the clique-based de-anonymization attack by introducing a statistical user behavior model. However, this work is only based on the statistical result and can adapt to the specific distribution which is not practical in real application. The clique structure in findings above actually provided the vulnerability of data and the attackers' opportunities to complete their attack. This is what our work addresses to fundamentally update the structure to protect the data from the attack.

2.2 Destroying Clique in Graph

Since we figured that destroying clique was a fundamental way to protect the data privacy from the structure aspect, we researched the existing technologies of destroying cliques. Apparently, to destroy clique, we can directly go to the problem of destroying 3-clique, the smallest clique, since there is no k-clique exists in a graph if no $k - 1$-clique exist. In the undirected graph, the *NPcomplete* hardness of deleting the 3-clique was investigated decades ago [15]. Even restricting the graph to planar of maximum degree seven, the problem is still *NPcomplete* [53]. The latest progress of this problem was proved by [54], which provided polynomial-time data reduction rules for 3-clique and obtained problem kernels consisting of $6k$ vertices for general graphs and $11k/$ three vertices for planar graphs.

Surprisingly, as far as we know, there was still no result about the theoretical analysis of clique destroying in the directed graph. The very first reason might be that it was not very common to use a clique in directed graph since clique asked any nodes in a subgraph could reach out to any other nodes directly.

Another possible reason was the directed graph actually could be a superset of the undirected graph if added the two side direction to the undirected graph. Then the reduction of clique-destroying in the directed graph would be much harder than the problem in the undirected graph.

In this work, we employed a full step to prove the hardness of clique-destroying problem in the directed graph and showed that our problem actually could dominate the problem in undirected graph version. Besides, we proposed the corresponding algorithms to solve the problem in both types graph and demonstrated the effectiveness and efficiency of our resolutions in the data privacy protection application.

3   Preliminaries and Complexity Analysis

3.1   Preliminaries

We provide a generalized definition to model the clique destroying problem in the directed graph so that we can give a complete study on the computational complexity of this problem. Below, we introduce some necessary notations followed by the problem definition.

Definition 1 (clique). *Given a graph $G(V, E)$ where G is a graph, V and E are the vertices and edges set. A clique $\mathbb{C}$ in G is a complete subgraph of G. That is, it is a subset $\mathbb{C}$ of the vertices such that every two vertices in $\mathbb{C}$ are the two endpoints of an edge in G.*

In the above definition, clique $\mathbb{C}$ can be adapted to both undirected and directed graph. With specific size of $\mathbb{C}$, we have the definition of k-clique.

Definition 2 (k-clique). *Given a graph G and a specific number k, which represents the size of clique $\mathbb{C}$. k-clique is a complete subgraph in G with size k.*

To destroy all cliques with a size larger than or equal to k in a graph G, means destroy all k-cliques in G, because it is impossible for a graph G has $(k + i)$-clique without the desistance of k-clique, for $i > 0$. The atomic clique in a graph G in this study, is required to be 3-clique.

Let $\triangle_{uvw}$ be a 3-clique consisting of edges (u, v), (v, w), (w, u) where u, v, w are different vertices.

Definition 3 (k-Clique Destroying Problem). *Given a graph $G(V, E)$, a fixed constant k, find a minimum edge set D such that there is no k-clique existing in $G(V, E \setminus D)$*

The decision problem of k-clique destroying problem is stated as follow,

Definition 4 (Decision problem of k-Clique Destroying). *Given a graph $G(V, E)$, a fixed k, and an input number h, decide if there a edge set D with a size $|D| < h$ such that there is no k-clique existing in $G(V, E \setminus D)$*

If $k = 3$, k-clique destroying problem is the triangle removing problem, and it's one of 21 well-known *NPcomplete* problems [55] if the given graph is an undirected graph.

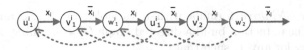

Fig. 1. A 6 vertices section of the component G_i in the hardness proof for Δq: A minimum edge set chooses either all the solid lines marked e_i, or all the solid lines marked $\overline{e_i}$. All dotted lines are crossed because each of them is only part of one single 3-clique, say rgb clique, thus they are never chosen.

3.2 Computational Complexity Analysis

We've already known that the k-clique destroying problem is *NPcomplete* for a fixed $k \geq 3$, since triangle removing is *NPcomplete*. However, we surprisingly found that 3-clique destroying problem in the directed graph is also *NPhard*. This is non-trivial because the solution space is different so that the hardness result of k-clique destroying problem in the undirected graph could not imply its counterpart in the directed graph.

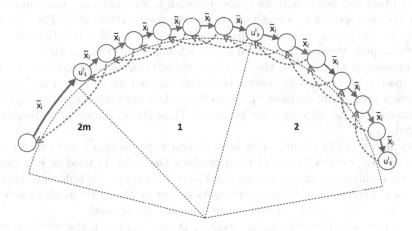

Fig. 2. Each component G_i in the proof is a 2 m 6-vertices section and a total of 12 m rgb cliques. By removing the 6 m edges marked x_i or 6 m edges marked $\overline{x}_i$ can be eliminated all of them. The crossed sign on all even numbered sections is because they are never used for connecting different components; they only separate the odd ones, then preventing the spurious 3-cliques.

Theorem 1. *k-clique destroying problem in directed graph is* NPcomplete.

Proof. Observe that for a fixed $k > 0$, k-clique checking could be carried out polynomially, even in *textit*directed graph, therefore this problem is in *NP*.

Now, we reduce every instance f of *SATthree* to corresponding instance (G, h) of 3-clique destroying problem in directed graph.

Let f be a *CNFthree* formula with n variables $x_1, \ldots, x_n$ and m clauses $c_0, \ldots, c_{m-1}$. The reduction build a directed graph G_f containing 3-cliques, and set a number h_f for any f, such that

$$f \text{ is satisfiable} \Leftrightarrow \exists D \text{ such that } |D| < h_f \text{ and any } \triangle_{uvw} \in G_f, \text{ it follows}$$
$$\triangle_{uvw} \notin G'(V_f, E_f \setminus D)$$

In our construction, if f is satisfiable, then there is an edge set of size $h_f = 6mn$ that destroying all 3-cliques in G_f, whereas if f is unsatisfiable, then every edge set destroying all 3-cliques in G_f is of a size larger than h_f.

As shown in Figs. 1, 2, and 3, we build each 3-clique in G_f by a red edge, a green edge and a blue edge, one can check that there is no any other combinations could form a clique in G_f. The directed edges $u \rightarrow v$, $v \rightarrow w$, $w \rightarrow u$ are shown in the figures. The idea of an edge set is to remove all rgb cliques.

G_f contains a ring-like component $\mathbf{r}_i$ for each variable x_i. The ring consists of $12m$ solid edges, as shown in Figs. 1 and 2, half of them marked x_i and the other half marked $\overline{x}_i$. Note that there are $12m$ rgb cliques and they can be minimally broken by choosing the $6m$ x_i edges or the $6m$ $\overline{x}_i$ edges.

Any other way requires more edges removed. Intuitively, each minimum edge set for G_f corresponds to a truth assignment to the variables of f. There will be an edge set of size $h_f = 6mn + 1$ destroying all 3 cliques iff f is satisfiable.

We complete the construction of G_f by adding one rgb clique for each clause c_j. Suppose c_j is $x_1 + \overline{x}_2 + x_3$, the rgb clique we add consists of a red edge marked x_1, a green edge marked $\overline{x}_2$, and a blue edge marked x_3 as shown in Fig. 3. If the chosen assignment satisfies c_j, then all x_1 edges are removed, or all $\overline{x}_2$ edges are removed, or all x_3 edges are removed. Thus the c_j clique is subsequently removed.

To create c_j's rgb clique, as we have chosen $\mathbf{r}_i$ to contain 2 sections for each clause, we use section $2j + 1$ of $\mathbf{r}_i$ to produce the x_i or $\overline{x}_i$ used in c_j's clique. The even numbered sections are not used since they served as buffers to prevent spurious rgb clique from being created. As shown in Fig. 2, we mark these even segments with the "cross" mark because they are never used.

More precisely, the red edge x_1 from $\mathbf{r}_1$ is (u_{4j+1}^1, v_{4j+1}^1), the green edge $\overline{x}_2$ from $\mathbf{r}_2$ is (v_{4j+1}^2, w_{4j+1}^2), and the blue edge x_3 from $\mathbf{r}_3$ is (w_{4j+1}^3, u_{4j+2}^3) as shown in Fig. 3.

Observe that, in order to make this no rgb clique in G_f, we identify the two u-vertices, the two v-vertices, and the two w-vertices. That is, $\mathbf{r}_1$'s u-vertex u_{4j+1}^1 is equal to $\mathbf{r}_3$'s u-vertex u_{4j+2}^3. Then, we construct c_j's rgb clique.

Because there is no other way to get back to $\mathbf{r}_1$ from $\mathbf{r}_2$ in two steps, the key idea is that these identifications can only create this single new rgb clique. All other identifications will involve different sections and are at least 6 steps away.

One can verify the correctness of the reduction, *i.e.*, f is satisfiable *iff* there is a D, $|D| < h_f$, any $\triangle_{uvw}$ of G_f, is no longer in $G'(V_f, E_f \setminus D)$.

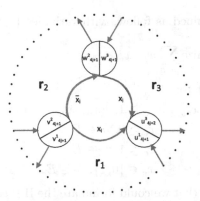

Fig. 3. For an example clause $c_j = (x + \overline{y} + z)$ in f, we identify vertices $v_{4j+1}^1 \in r_1$ with $y_{4j+1}^2 \in r_2$; $z_{4j+1}^2 \in r_2$ with $z_{4j+1}^3 \in r_3$; and $x_{4j+2}^3 \in r_3$ with $x_{4j+1}^1 \in r_1$. The 3-clique will be deleted iff the chosen variable assignment satisfies C_j.

3.3 k-Clique Destroying in Directed Graph

The solutions of the directed graph case and undirected graph case are very different. It is interesting that we show the directed graph will also make this problem harder. Surprisingly, we find that the clique destroying in the directed graph is *NPcomplete*, while all previous results only tell the *NPhardness* of the undirected graph and planar graph.

3.4   Brief Summary

In the above proof, we have demonstrated the interesting clique destroying problem in the directed graph. Practically, the clique destroying will not go directly to destroy all 3-clique in a graph, we usually only want to protect the sensitive information from the data and preserve the privacy from the referring of individual's affiliation. Our analysis proof the hardness of 3-clique, which could be easily extended to k-clique, since 3-clique is the hardest case of the clique destroying. Later in the next section, we will show the approximate algorithm for both 3-clique and k-clique.

4   LP-Based Approximation Algorithm

Due to the *NPhardness*, algorithm is too expensive, this section conducts the algorithm development for optimal clique destroying, and results in the experiment section will show the effectiveness. We extend the clique destroying problem in directed graph one step by adding weight w one each edge e, and solve the problem via an integer linear programming (ILP) relaxation. Then, we relax it to get linear programming (LP), and then we round the solution.

First, the ILP is defined as follow, with variables $\{x_e | e \in E\}$, one for each edge:

$$\text{minimize} \sum_{e \in E} w_e \cdot x_e$$

subject to:

$$x_{e_u} + x_{e_v} + x_{e_w} \geq 1, \forall (x_{e_u}, x_{e_v}, x_{e_w}) \; is \; clique \qquad (1)$$

$$x_e \geq 0, \forall e \in E$$

$$x_e \leq 1, \forall e \in E$$

$$x_e \in \{0, 1\}, \forall e \in E$$

It worth to mention that we could formulate the ILP equations polynomially since if there are m edges, we can enumerate all triples of edges in $O(m^3)$ time and check if they form a 3-clique. Thus this is a polynomial number of variable and constraints in ILP. It is obvious that any solution to this ILP results in a 3-clique free graph, and conversely that the optimal edge set of S is a feasible solution to this ILP. Simply relax the ILP to an LP by allowing the variable x_e to be an arbitrary real number between 0 and 1. Therefore, we can solve the LP using any polynomial-time LP solver. Finally, we round the solution as follow: if $x_e \geq \frac{1}{3}$, then we add edge e to the set S.

Theorem 2 (Correctness of rounding). *Any rounded Solution of LP algorithm guarantees no 3-clique remaining.*

Proof. First, we argue that the resulting graph is 3-clique-free. Notice that in any 3-clique (e_u, e_v, e_w), the three variables $x_{e_u} + x_{e_v} + x_{e_w}$ must sum to at least 1, according to the LP's constraints, and hence at least one of the three variables must be at least $\frac{1}{3}$. Thus, no 3-clique remains after removing edge set S.

Practically, it is not heavily necessary to destroy all atomic cliques in a graph; this constraint is too harsh, it's not necessary to hide structural information completely just to strengthen protection to some degree. Therefore, we extend the LP algorithm above to destroy k-clique, prove a general case of the approximation ratio analysis, so that we can adjust the k to trade off the publication and protection.

The idea of algorithm is straightforward, that we extend the set of edge (e_u, e_v, e_w) to set $(e_1, \ldots, e_k)$ by enumerating k-cliques within a $O(m^k)$ time, then the number of variable extend from 3 to k correspondingly, which is shown as follows,

$$\text{minimize} \sum_{e \in E} w_e \cdot x_e$$

subject to:

$$\sum_{1 \leq i \leq k} x_{e_i} \geq 1, \forall (x_{e_1}, x_{e_2}, \ldots, x_{e_k}) \text{ is a clique} \tag{2}$$

$$x_e \geq 0, \forall e \in E$$
$$x_e \leq 1, \forall e \in E$$
$$x_e \in \{0, 1\}, \forall e \in E$$

Theorem 3. *LP algorithm based on Eq. 3 has an approximation ratio of k for destroying k-clique.*

Proof. We argue that the result is a k-approximation of the optimal solution. Let OPT be the cost of the optimal edge set to produce a k-clique-free graph. *i.e.*, OPT is the cost of the solution found by the ILP. Let L be the cost of the solution found by the LP. Because LP is a relaxation of the ILP, then we know that $L \leq OPT$. Because we have effectively increased some variables x_e from the value of at least $\frac{1}{k}$ to 1, by rounding the solution of LP, we have only increased the cost by a factor of k. That is, if x_e is the variable from the LP, and if y_e is the rounded variable, we know that $y_e \leq k \cdot x_e$. We can calculate the total cost of our solution as shown in Eq. 3:

$$cost(S) = \sum_{e \in E} y_e \cdot w(e)$$
$$\leq \sum_{e \in E} k \cdot x_e w(e) \tag{3}$$
$$\leq k \cdot L$$
$$\leq k \cdot OPT$$

Our solution is a k-approximation.

Above all, the lightweight method we applied has been detailed. With k approximation ratio, it is easy to implement, and any existing fast LP-solver can be utilized to solve it.

5   Experiment

In this section, we evaluate the clique destroying algorithm as a framework upon other algorithm and as an individual algorithm to defend the clique-based attack also. All the experiments are performed on a PC running Windows 10 with Intel(R) Core(TM) i3-2120 CPU 3.30 GHz and 12GB memory. The six real life datasets are used to evaluate our algorithm and approach. The basic statistic information of the large scale dataset is listed as shown in Table 1.

Table 1. Statistic of real life data sets

Name	Type	# of vertices	# of edges	Average degree $\bar{d}$	# of 3-clique	Fraction of 3-clique
com-Amazon	Undirected	334,863	925,872	21.85	1612,010	0.2647
email-Enron	Undirected	36,692	183,831	5.01	727,044	0.0302
social-Facebook	Undirected	4,039	88,234	21.84	1612,010	0.2647
Epinions	Directed	75,879	508,837	6.71	1624,481	0.0229
wiki-Vote	Directed	7,115	103,689	14.57	608,389	0.0456
wiki-Talk	Directed	1234,385	5021,410	2.09	9203,519	0.0011

- **com-Amazon(AM)** is a dataset collected by crawling Amazon website[5]. The links among vertices are based on the "Customers Who Bought This Item Also Bought" feature that an undirected edge will be established if a product is frequently co-purchased with another product [56].
- **email-Enron(EN)** is a dataset of email communication network covers about half million emails. The data was originally posted to the web publicly by the Federal Energy Regulatory Commission[6] during its investigation. Each vertex is an email address, and the connection is based on their communication record without direction [57].
- **Facebook(FA)** is dataset collected from survey participants Facebook app[7]. The original data include the node profiles, circles, and ego networks. We only use the combined ego-network main structure in our evaluation. [58].
- **Epinions(EP)** is a who-trust-who online social network of a general consumer review site from Epinions.com[8]. The members of Epinions could decide whether to "trust" or "not trust" others. The trust relationships interact and form the network of trust which is combined with review rating to decide which reviews are shown up [59].
- **wiki-Vote(WV)** is a network extracted by Wikipedia[9] page edit history. Wikipedia community via a public discussion or a vote decides who to promote to adminship. In this data set, above half of the votes are raised by existing admins, while the remaining comes from ordinary Wikipedia users [60].
- **wiki-Talk(WT)** is a network contains all the users and discussion from the inception of Wikipedia. The nodes in the network represent Wikipedia users and a directed edge from node i to node j represents that i edited a talk page of user j at least once [61].

We conduct our experiment based on the open source software SecGraph[10]. SecGraph [62] provides a recent selection and comparison of structural social network de-anonymization attacks. With the sampling based perturbation method

---

[5] https://www.amazon.com/.

[6] https://www.ferc.gov/.

[7] https://www.facebook.com/apps/application.php?id=201704403232744.

[8] http://www.epinions.com/.

[9] https://www.wikipedia.org/.

[10] http://cap.ece.gatech.edu/.

implemented in SecGraph, we recreated the datasets source and target graph G_s and G_t as well as [62] by randomly sampling edges with probability s. We add our clique-destroying (CD) algorithm upon the previous anonymization algorithms and take CD as an own algorithm to anonymize the graph also. For comparison, we use *Switch*[63], k-DA [64], and DP [65] as baselines.

For the de-anonymization algorithm, we select Narayanan-Shmatikov (Nar) attack [7] and Yartseva-Grossglauser attack (Yar) [66] to test the effectiveness of our CD algorithm (Table 2).

Table 2. Robustness of attacks against different anonymization schemes

s	Switch(k)				k-DA(k)				DP(ϵ)				CD(k)					
	5		10		50		300		50		300		3		5		10	
	Re	Pr	Re	Pr	Re	Pr	Re	Pr	Re	Pr	Re	Pr	Re	Pr	Re	Pr	Re	Pr
Nar 0.85	31.2	94.1	15.8	89.3	39.6	97.8	11.2	93.2	38.1	97.9	0.45	96.5	25.2	45.5	33.1	56.3	27.8	83.2
0.9	33.5	95.5	16.3	91.7	40.5	98.2	35.5	94.6	38.9	98.6	0.54	97.2	27.8	51.2	36.2	58.4	29.4	86.5
0.95	37.6	96	30.1	93.9	42.2	99	37.1	96.5	39.4	99.2	0.66	98.3	28.6	58.6	37.9	62.9	30.9	90.1
Yar 0.85	13.2	90.2	12.2	85.6	15.6	93.7	1.54	23.2	15.1	90.9	9.6	76.2	16.8	38.9	22.5	46	28.7	79.2
0.9	15.1	91.6	12.9	87.9	16.6	95.2	2.65	35.5	16.8	92.6	10.1	77.0	18.3	41.1	24.8	57.2	28.9	85.1
0.95	16.8	94.3	14.1	92.4	17.1	95.4	4.37	48.2	18.2	94.2	9.5	74.5	19.1	46.8	29.6	59.7	32.1	88.9

Due to space limitation, we do not show the evaluation results of all the network, particularly, we take the network Enron as a typical example to demonstrate the performance of our algorithm. As shown in Table 3, compare to other anonymization algorithms, our algorithm significantly improve the protection of data. We reduce the recall and precision more than 30% for both Nar and Yar two algorithms.

It worth to mention that the average degree of the email network Enron is 5.01, which means not a lot of nodes could construct a very large clique. That is the reason why the performance of our anodization approach doesn't work that well compare to the situation k equals to 5, or 3 when we increase the destroying clique size k to 10. But if we destroy the clique with size 3, means if we destroy all cliques in the graph, then only the circle or chain remain. It will be a very exacting term which may not very applicable to some situation.

Besides, we take our CD algorithm as a framework adding to previous anonymization schemes to test the improvement of our protection. Adding CD to other approaches could significantly improve their performance since the CD algorithm is also a good anonymization strategy independently. The idea we'd like to show is the clique structure is a very important key which inherently affect the performance and the result of anonymization, then the light algorithm CD with controllable k could reduce the risk of group reference easily.

By attacking Nar algorithm, we evaluate the precision of the six real life datasets as shown in Table 4. For different datasets, the performance is entirely different. Especially for the data like FA and WV, the results under the CD column are quite effective for 92.1 and 91.6. The main reason for that difference

Table 3. Improved robustness of attacks against different anonymization schemes add the CD as a framework

s		Switch(k)+CD($k=15$)				k-DA(k)+CD($k=15$)				DP(ϵ)+CD($k=15$)			
		5		10		50		300		50		300	
		Re	Pr	Re	Pr	Re	Pr	Re	Pr	Re	Pr	Re	Pr
Nar	0.85	28.6	90.0	13.4	65.2	36.3	89.6	10.1	89.3	35.7	95.1	0.38	92.2
	0.90	31.2	93.3	14.1	88.5	38.4	90.0	31.9	92.3	35.7	94.2	0.49	92.7
	0.95	30.9	91.0	28.7	87.4	35.2	91.3	32.5	92.1	33.6	92.7	0.51	94.2
Yar	0.85	11.5	86.1	9.8	81.2	13.2	90.5	1.19	19.8	13.2	88.1	8.9	70.5
	0.90	12.6	87.4	11.5	84.2	13.5	90.9	2.31	31.2	14.2	88.4	9.2	77.0
	0.95	10.3	94.3	12.4	88.1	15.7	89.6	3.89	42.9	16.1	86.8	8.1	69.7

is both graphs are very locally compacted, then the number of the clique is relatively high. However, the performance of AM is not that well even with a very high degree and a large number of the clique. We believe the behind reason is that Amazon is a co-purchased network, while the clique existing in small size is pretty high, but the large size cliques are relatively small. Since we set our DC algorithm with k equals to 10, which is not a quite fit to destroy the main part of cliques in that network.

Table 4. Precision of the attack algorithm nar with different schemes.

	Switch(5)	k-DA(25)	DP(200)	CD(10)
AM	90.5	96.8	98.1	75.5
EN	94.5	94.6	96.1	90.2
FA	97.2	92.2	97.3	52.1
EP	92.6	90.5	94.6	92.1
WV	94.1	86.3	92.5	65.5
WT	93.7	87.7	91.4	91.6

6   Conclusion

In this work, we tackled the problem of destroying the clique in the graph for cutting off the connection between individual to their affiliation. Surprisingly, we noticed a new fundamental algorithm problem, clique destroying in a directed graph, had not been investigated yet. Then we demonstrated the problem hardness by reducing the instance to *SATthree* problem. We showed that the clique destroying in directed graph is *NPcomplete*, while all previous results only displayed the *NPhardness* of the undirected graph and planar graph. We provided solid theoretical problem complexity analysis and showed the complete proof of the problem hardness in multiple situations. Furthermore, we demonstrated the effectiveness and efficiency by testing our algorithms on six real life networks.

Acknowledgement. This work is partly supported by the Foundation of Guizhou Provincial Key Laboratory of Public Big Data (No. 2018BDKFJJ002) and the National Science Foundation (NSF) under grant NOs. 1252292, 1741277, 1704287, and 1829674.

References

1. Zhang, J., Li, Q., Schooler, E.M.: iHEMS: an information-centric approach to secure home energy management. In: 2012 IEEE Third International Conference on Smart Grid Communications (SmartGridComm), pp. 217–222. IEEE (2012)
2. Aberer, K., Alonso, G., Kossmann, D.: Data management for a smart earth: the swiss NCCR-MICS initiative. ACM SIGMOD Rec. **35**(4), 40–45 (2006)
3. Perera, C., Zaslavsky, A., Christen, P., Georgakopoulos, D.: Context aware computing for the internet of things: a survey. IEEE Commun. Surv. Tutor. **16**(1), 414–454 (2014)
4. Liang, Y., Cai, Z., Han, Q., Li, Y.: Location privacy leakage through sensory data. Secur. Commun. Netw. **2017**, 1–12 (2017)
5. Zheng, X., Cai, Z., Li, J., Gao, H.: Location-privacy-aware review publication mechanism for local business service systems. In: INFOCOM 2017-IEEE Conference on Computer Communications, pp. 1–9. IEEE (2017)
6. He, Z., Cai, Z., Yu, J., Wang, X., Sun, Y., Li, Y.: Cost-efficient strategies for restraining rumor spreading in mobile social networks. IEEE Trans. Veh. Technol. **66**(3), 2789–2800 (2017)
7. Narayanan, A., Shmatikov, V.: De-anonymizing social networks. In: 2009 30th IEEE Symposium on Security and Privacy, pp. 173–187. IEEE (2009)
8. Potharaju, R., Carbunar, B., Nita-Rotaru, C.: iFriendU: leveraging 3-cliques to enhance infiltration attacks in online social networks. In: Proceedings of the 17th ACM Conference on Computer and Communications Security. ACM (2010) 723–725
9. Gulyás, G.G., Simon, B., Imre, S.: An efficient and robust social network de-anonymization attack. In: Proceedings of the 2016 ACM on Workshop on Privacy in the Electronic Society, pp. 1–11. ACM (2016)
10. Dwork, C.: Differential privacy: a survey of results. In: Agrawal, M., Du, D., Duan, Z., Li, A. (eds.) TAMC 2008. LNCS, vol. 4978, pp. 1–19. Springer, Heidelberg (2008). https://doi.org/10.1007/978-3-540-79228-4_1
11. Niu, B., Li, Q., Zhu, X., Cao, G., Li, H.: Achieving k-anonymity in privacy-aware location-based services. In: 2014 Proceedings IEEE INFOCOM, pp. 754–762. IEEE (2014)
12. Shokri, R., Shmatikov, V.: Privacy-preserving deep learning. In: Proceedings of the 22nd ACM SIGSAC Conference on Computer and Communications Security, pp 1310–1321. ACM (2015)
13. Dalenius, T.: Towards a methodology for statistical disclosure control. Statistik Tidskrift **15**, 429–444 (1977)
14. Dwork, C., Naor, M.: On the difficulties of disclosure prevention in statistical databases or the case for differential privacy. J. Priv. Confid. **2**(1), 8 (2008)
15. Yannakakis, M.: Edge-deletion problems. SIAM J. Comput. **10**(2), 297–309 (1981)
16. Sweeney, L.: k-anonymity: a model for protecting privacy. Int. J. Uncertain. Fuzziness Knowl. Based Syst. **10**(05), 557–570 (2002)
17. Dwork, C., Roth, A., et al.: The algorithmic foundations of differential privacy. Found. Trends® Theor. Comput. Sci. **9**(3–4), 211–407 (2014)

18. Acquisti, A., Brandimarte, L., Loewenstein, G.: Privacy and human behavior in the age of information. Science **347**(6221), 509–514 (2015)
19. Young, A.L., Quan-Haase, A.: Privacy protection strategies on facebook: the internet privacy paradox revisited. Inf. Commun. Soc. **16**(4), 479–500 (2013)
20. Bettini, C., Riboni, D.: Privacy protection in pervasive systems: state of the art and technical challenges. Pervasive Mob. Comput. **17**, 159–174 (2015)
21. Zhao, J., Liu, J., Qin, Z., Ren, K.: Privacy protection scheme based on remote anonymous attestation for trusted smart meters. IEEE Trans. Smart Grid **9**, 3313–3320 (2016)
22. Naor, M., Nissim, K.: Communication preserving protocols for secure function evaluation. In: Proceedings of the Thirty-Third Annual ACM Symposium on Theory of Computing, pp. 590–599. ACM (2001)
23. Goldwasser, S.: Multi party computations: past and present. In: Proceedings of the Sixteenth Annual ACM Symposium on Principles of Distributed Computing, pp. 1–6. ACM (1997)
24. Han, M., Li, J., Cai, Z., Han, Q.: Privacy reserved influence maximization in GPS-enabled cyber-physical and online social networks. In: 2016 IEEE International Conferences on Big Data and Cloud Computing (BDCloud), Social Computing and Networking (SocialCom), Sustainable Computing and Communications (SustainCom) (BDCloud-SocialCom-SustainCom), pp. 284–292. IEEE (2016)
25. Albinali, H., Han, M., Wang, J., Gao, H., Li, Y.: The roles of social network mavens. In: 2016 12th International Conference on Mobile Ad-Hoc and Sensor Networks (MSN), pp. 1–8. IEEE (2016)
26. Cai, Z., Zheng, X.: A private and efficient mechanism for data uploading in smart cyber-physical systems. IEEE Trans. Netw. Sci. Eng. (2018)
27. Zheng, X., Luo, G., Cai, Z.: A fair mechanism for private data publication in online social networks. IEEE Trans. Netw. Sci. Eng. (2018)
28. Li, J., Cai, Z., Wang, J., Han, M., Li, Y.: Truthful incentive mechanisms for geographical position conflicting mobile crowdsensing systems. IEEE Trans. Comput. Soc. Syst. **5**(2), 324–334 (2018)
29. Han, M., Wang, J., Yan, M., Ai, C., Duan, Z., Hong, Z.: Near-complete privacy protection: cognitive optimal strategy in location-based services. Procedia Comput. Sci. **129**, 298–304 (2018)
30. Ling, X., Wu, C., Ji, S., Han, M.: H_2DoS: an application-layer DoS attack towards HTTP/2 protocol. In: Lin, X., Ghorbani, A., Ren, K., Zhu, S., Zhang, A. (eds.) SecureComm 2017. LNICST, vol. 238, pp. 550–570. Springer, Cham (2018). https://doi.org/10.1007/978-3-319-78813-5_28
31. Han, M., Li, L., Peng, X., Hong, Z., Li, M.: Information privacy of cyber transportation system: opportunities and challenges. In: Proceedings of the 6th Annual Conference on Research in Information Technology, pp. 23-28. ACM (2017). https://dl.acm.org/citation.cfm?id=31256
32. Zheng, X., Cai, Z., Yu, J., Wang, C., Li, Y.: Follow but no track: privacy preserved profile publishing in cyber-physical social systems. IEEE Internet Things J. **4**(6), 1868–1878 (2017)
33. Zhou, Y., Han, M., Liu, L., He, J.S., Wang, Y.: Deep learning approach for cyber-attack detection. In: IEEE INFOCOM 2018-IEEE Conference on Computer Communications Workshops (INFOCOM WKSHPS), pp. 262–267. IEEE (2018)
34. Han, M., Duan, Z., Li, Y.: Privacy issues for transportation cyber physical systems. In: Sun, Y., Song, H. (eds.) Secure and Trustworthy Transportation Cyber-Physical Systems. SCS, pp. 67–86. Springer, Singapore (2017). https://doi.org/10.1007/978-981-10-3892-1_4

35. Joshi, A.P., Han, M., Wang, Y.: A survey on security and privacy issues of blockchain technology. Math. Found. Comput. **1**(2), 121–147 (2018)
36. Liu, L., Han, M., Wang, Y., Zhou, Y.: Understanding data breach: a visualization aspect. In: Chellappan, S., Cheng, W., Li, W. (eds.) WASA 2018. LNCS, vol. 10874, pp. 883–892. Springer, Cham (2018). https://doi.org/10.1007/978-3-319-94268-1_81
37. Liang, Y., Cai, Z., Yu, J., Han, Q., Li, Y.: Deep learning based inference of private information using embedded sensors in smart devices. IEEE Netw. **32**(4), 8–14 (2018)
38. Wang, J., Cai, Z., Li, Y., Yang, D., Li, J., Gao, H.: Protecting query privacy with differentially private k-anonymity in location-based services. Pers. Ubiquit. Comput. **22**, 1–17 (2018)
39. Dwork, C.: A firm foundation for private data analysis. Commun. ACM **54**(1), 86–95 (2011)
40. Dwork, C., Rothblum, G.N., Vadhan, S.: Boosting and differential privacy. In: 2010 51st Annual IEEE Symposium on Foundations of Computer Science (FOCS), pp. 51–60. IEEE (2010)
41. Chaudhuri, K., Monteleoni, C.: Privacy-preserving logistic regression. In: Advances in Neural Information Processing Systems pp. 289–296 (2009)
42. Chaudhuri, K., Sarwate, A., Sinha, K.: Near-optimal differentially private principal components. In: Advances in Neural Information Processing Systems, pp. 989–997 (2012)
43. Sarwate, A.D., Chaudhuri, K.: Signal processing and machine learning with differential privacy: algorithms and challenges for continuous data. IEEE Sig. Process. Mag. **30**(5), 86–94 (2013)
44. Ho, S.S., Ruan, S.: Differential privacy for location pattern mining. In: Proceedings of the 4th ACM SIGSPATIAL International Workshop on Security and Privacy in GIS and LBS, pp. 17–24. ACM (2011)
45. Dewri, R.: Local differential perturbations: location privacy under approximate knowledge attackers. IEEE Trans. Mob. Comput. **12**(12), 2360–2372 (2013)
46. Xiao, Y., Xiong, L.: Protecting locations with differential privacy under temporal correlations. In: Proceedings of the 22nd ACM SIGSAC Conference on Computer and Communications Security, pp. 1298–1309. ACM (2015)
47. Yildiz, H., Kruegel, C.: Detecting social cliques for automated privacy control in online social networks. In: 2012 IEEE International Conference on Pervasive Computing and Communications Workshops (PERCOM Workshops), pp. 353–359. IEEE (2012)
48. Pan, X., Xu, J., Meng, X.: Protecting location privacy against location-dependent attacks in mobile services. IEEE Trans. Knowl. Data Eng. **24**(8), 1506–1519 (2012)
49. Narayanan, A., Shi, E., Rubinstein, B.I.: Link prediction by de-anonymization: How we won the Kaggle social network challenge. In: The 2011 International Joint Conference on Neural Networks (IJCNN), pp. 1825–1834. IEEE (2011)
50. Srivatsa, M., Hicks, M.: Deanonymizing mobility traces: using social network as a side-channel. In: Proceedings of the 2012 ACM Conference on Computer and Communications Security, pp. 628–637. ACM (2012)
51. Ji, S., Li, W., Srivatsa, M., He, J.S., Beyah, R.: Structure based data de-anonymization of social networks and mobility traces. In: Chow, S.S.M., Camenisch, J., Hui, L.C.K., Yiu, S.M. (eds.) ISC 2014. LNCS, vol. 8783, pp. 237–254. Springer, Cham (2014). https://doi.org/10.1007/978-3-319-13257-0_14
52. Gulyás, G.G., Imre, S.: Analysis of identity separation against a passive clique-based de-anonymization attack. Infocommunications J. **4**(3), 11–20 (2011)

53. Niedermeier, R.: Invitation to Fixed-parameter Algorithms (2006)
54. Brügmann, D., Komusiewicz, C., Moser, H.: On generating triangle-free graphs. Electron. Notes Discret. Math. **32**, 51–58 (2009)
55. Karp, R.M.: Reducibility among combinatorial problems. In: Miller, R.E., Thatcher, J.W., Bohlinger, J.D. (eds.) Complexity of Computer Computations, pp. 85–103. Springer, Heidelberg (1972). https://doi.org/10.1007/978-1-4684-2001-2_9
56. Yang, J., Leskovec, J.: Defining and evaluating network communities based on ground-truth. Knowl. Inf. Syst. **42**(1), 181–213 (2015)
57. Leskovec, J., Lang, K.J., Dasgupta, A., Mahoney, M.W.: Community structure in large networks: natural cluster sizes and the absence of large well-defined clusters. Internet Math. **6**(1), 29–123 (2009)
58. Leskovec, J., Mcauley, J.J.: Learning to discover social circles in ego networks. In: Advances in Neural Information Processing Systems, pp. 539–547 (2012)
59. Richardson, M., Agrawal, R., Domingos, P.: Trust management for the semantic web. In: Fensel, D., Sycara, K., Mylopoulos, J. (eds.) ISWC 2003. LNCS, vol. 2870, pp. 351–368. Springer, Heidelberg (2003). https://doi.org/10.1007/978-3-540-39718-2_23
60. Leskovec, J., Huttenlocher, D., Kleinberg, J.: Predicting positive and negative links in online social networks. In: Proceedings of the 19th International Conference on World Wide Web, pp. 641–650. ACM (2010)
61. Leskovec, J., Huttenlocher, D., Kleinberg, J.: Signed networks in social media. In: Proceedings of the SIGCHI Conference on Human Factors in Computing Systems, pp. 1361–1370. ACM (2010)
62. Ji, S., Li, W., Mittal, P., Hu, X., Beyah, R.A.: SecGraph: a uniform and open-source evaluation system for graph data anonymization and de-anonymization. In: USENIX Security Symposium, pp. 303–318 (2015)
63. Ying, X., Wu, X.: Randomizing social networks: a spectrum preserving approach. In: Proceedings of the 2008 SIAM International Conference on Data Mining, pp. 739–750. SIAM (2008)
64. Liu, K., Terzi, E.: Towards identity anonymization on graphs. In: Proceedings of the 2008 ACM SIGMOD International Conference on Management of Data, pp. 93–106. ACM (2008)
65. Sala, A., Zhao, X., Wilson, C., Zheng, H., Zhao, B.Y.: Sharing graphs using differentially private graph models. In: Proceedings of the 2011 ACM SIGCOMM Conference on Internet Measurement Conference, pp. 81–98. ACM (2011)
66. Yartseva, L., Grossglauser, M.: On the performance of percolation graph matching. In: Proceedings of the First ACM Conference on Online Social Networks, pp. 119–130. ACM (2013)

Exact Computation of Strongly Connected Reliability by Binary Decision Diagrams

Hirofumi Suzuki[1]($\boxtimes$), Masakazu Ishihata[2]($\boxtimes$), and Shin-ichi Minato[3]($\boxtimes$)

[1] Graduate School of Information Science and Technology, Hokkaido University,
Sapporo, Japan
h-suzuki@ist.hokudai.ac.jp
[2] NTT Communication Science Laboratories, Kyoto, Japan
ishihata.masakazu@lab.ntt.co.jp
[3] Graduate School of Informatics, Kyoto University, Kyoto, Japan
minato@i.kyoto-u.ac.jp

Abstract. *Network reliability* is the probability that a network system can perform a desired operation, such as communication between facilities, against stochastic equipment failures. On analyzing network systems that are represented by undirected graphs, the *all-terminal reliability* (ATR) is commonly used as one of the network reliability. As a natural extension of the ATR for the directed version, the *strongly connected reliability* (SCR) is known. The SCR should be computed on various network systems, such as ad-hoc network, that demand the property called *strongly connected*. Unfortunately, computing the SCR is known to be #P-complete, and little studies challenge the computation of the exact or an approximate SCR on limited graph classes. In this study, we propose the first practically efficient algorithm to compute the exact SCR in general. The algorithm constructs a *binary decision diagram* (BDD) representing all the *strongly connected spanning subgraphs* (SCSSs) in a given directed graph. Subsequently, the algorithm computes the exact SCR by a dynamic programming on the BDD. To efficiently construct BDDs, we designed a new variant of the *frontier based search* (FBS). We conducted computational experiments to evaluate the proposed algorithm. The results demonstrated that the proposed algorithm succeeded in computing the SCR in real-world networks with a few hundred edges within a reasonable time, which was previously impossible.

Keywords: Network reliability · All-terminal reliability · Strongly connected reliability · Binary decision diagram · Dynamic programming

1 Introduction

Network reliability is the probability that a network system can perform a desired operation, such as communication between facilities, against stochastic equipment failures. A network system is often represented by a graph; the vertices

© Springer Nature Switzerland AG 2018
D. Kim et al. (Eds.): COCOA 2018, LNCS 11346, pp. 281–295, 2018.
https://doi.org/10.1007/978-3-030-04651-4_19

represent facilities such as servers, and the edges represent connections between facilities such as cables. On undirected graphs, connectivity-based network reliability called the *all-terminal reliability* (ATR) is commonly used [1,2]. The ATR is the probability that all the vertices are connected (i.e., they can communicate with each other) after some of the edges stochastically dropped from the graph. As a natural extension of the ATR, the directed version can be considered. A directed graph is said to be *strongly connected* if any pair of vertices has bidirectional paths. The directed version of the ATR is the probability that the graph maintains strongly connected, and called the *strongly connected reliability* (SCR) [3]. Because strongly connected is demanded in various network systems such as ad-hoc network, computing the SCR is important to manage network systems.

Unfortunately, the exact computations of the ATR and the SCR are #P-complete [4]. For the ATR, there exist an efficient approximation algorithm [5] and a mildly exponential exact algorithm [6], and both algorithms have been applied to various practical network systems. On the other hand, to the best of our knowledge, no research paper computes the exact SCRs of practical networks whereas only a few studies compute the SCRs of limited graphs: Polynomial time exact algorithms are known for double-loop directed graphs [7] and complete directed graphs [3]. An approximation algorithm is known for Eulerian directed graphs [5].

A mildly exponential exact algorithm for the ATR [6] uses a property of the ATR: the ATR is the sum of the probabilities of all the spanning trees in a given graph, i.e, the ATR can be computed by the spanning tree enumeration. Because the number of the spanning trees is exponential in the size of the graph, the explicit enumeration is unrealistic. The algorithm conducts an implicit enumeration by constructing the *binary decision diagram* (BDD) [8] representing all the spanning trees compactly. Subsequently, the algorithm computes the ATR by a dynamic programming on the BDD. Similarly, the SCR can be computed by the enumeration of all the *strongly connected spanning subgraphs* (SCSSs) in a given directed graph [3]. However, it is NOT trivial generalizing the above algorithm to the SCR because it utilizes the property of the undirected graph.

In this paper, we propose the first nontrivial algorithm to compute the exact SCR in general. The proposed algorithm is to construct the BDD representing all the SCSSs in a given directed graph. The algorithm is based on a new variant of the *frontier-based search* (FBS), which is known to be an efficient way to construct several types of BDDs. Once a BDD for the SCSSs is obtained, we can efficiently compute the exact SCR by a dynamic programming on the BDD. As a secondary application of our algorithm, we can also easily obtain the *minimum* SCSS. Obtaining the minimum SCSS is NP-hard [9], and has applications for the visualization of network structures such as social networks [10], robotic networks [11], and biological networks [12,13]. Particularly, because the BDD has several functions to search solutions under various conditions, it can be flexibly utilized. We conducted computational experiments to evaluate our algorithm. We used several real-world and synthetic networks with a few hundred edges, and showed that our algorithm can construct the BDDs for the SCSSs in a few minutes. We also computed the SCR on each instance using the constructed BDDs.

2 Preliminaries

In this section, we describe the formal definition of the SCR. We also introduce the BDD that is a data structure used in our algorithm.

2.1 Strongly Connected Reliability

Let $G = (V, E)$ be a directed graph with vertices V and directed edges $E \subseteq V^2$, where $E = \{e_1, \ldots, e_m\}$. For any edge subset $X \subseteq E$, $V[X]$ denotes the induced vertices that is the set of endpoints of each edge in X, i.e., $V[X] := \bigcup_{(u,v) \in X} \{u, v\}$. Similarly, $G[X]$ denotes the edge induced subgraph such that $G[X] := (V[X], X)$.

For any $u, v \in V$, v is *reachable* from u on G, which is denoted by $u \rightsquigarrow_G v$, if G contains a directed path from u to v. Similarly, $u \not\rightsquigarrow_G v$ denotes that v is not reachable from u on G. A directed graph G is said to be *strongly connected* if $u \rightsquigarrow_G v$ for all $(u, v) \in V^2$. Given a strongly connected graph G, $G[X]$ ($X \subseteq E$) is an SCSS if $V[X] = V$ and $G[X]$ is strongly connected. If $G[X]$ is an SCSS, X is called a *strongly connected edge subset* (SCES).

The SCR of G is the probability that G remains strongly connected after stochastic edge dropping. Let $\sigma(G)$ be the SCR of G, and let $\mathcal{S}_G \subseteq 2^E$ be the set of all the SCESs on G. In this paper, we assume that each edge e_i independently drops with probability $p(e_i)$. Then $\sigma(G)$ is defined as:

$$\sigma(G) := \sum_{X \in \mathcal{S}_G} p(X) \tag{1}$$

where

$$p(X) := \prod_{e_i \in X} (1 - p(e_i)) \prod_{e_j \in E \setminus X} p(e_j). \tag{2}$$

The most naive method for the exact computation of the SCR is to explicitly enumerate $\mathcal{S}_G$; however, this requires an exponential amount of time in m because the number of possible edge subsets is also exponential in m. Therefore, we avoid the explicit enumeration of $\mathcal{S}_G$.

2.2 Binary Decision Diagrams

For enumerating $\mathcal{S}_G$ implicitly, we use the BDD [8] that is a compact graph representation of Boolean functions based on the Shannon decomposition. Note that a Boolean function can be used to represent a set family as an indicator function: We use each edge in E as a variable such that, for an edge subset $X \subseteq E$, $e_i \notin X$ ($e_j \in X$) is assigned to *False* (*True*).

A BDD is a rooted directed acyclic graph $\mathcal{B} = (N, A)$ with a node set N, and an arc set A.[1] It has exactly one root node ρ and exactly two terminal nodes $\bot$ and $\top$. Each non-terminal node $\alpha \in N$ has a label $\ell(\alpha) \in \{1, \ldots, m\}$ (i.e., α is associated with a variable $e_{\ell(\alpha)} \in E$), and has exactly two outgoing arcs called the 0-arc and the 1-arc. The node pointed by the x-arc of α is called the x-child for each $x \in \{0, 1\}$, and denoted by α_x where $\ell(\alpha) < \ell(\alpha_x)$ if α_x is not a terminal.

A BDD represents a Boolean function as follows: A directed path from ρ to $\top$ represents a (possibly partial) variable assignment for which the represented Boolean function is *True*. If the path descends a 0-arc (1-arc) of a node α, the variable $e_{\ell(\alpha)}$ is assigned to *False* (*True*).

Any BDD has the unique *reduced* form. A BDD is reduced if the following two rules are applied as long as possible:

– Delete α if $\alpha_0 = \alpha_1$.
– Share any two nodes β, β' where if $\ell(\beta) = \ell(\beta')$, $\beta_0 = \beta'_0$ and $\beta_1 = \beta'_1$.

These rules eliminate the redundant nodes in the BDD.

Figure 1 shows an example of a reduced BDD, which represents the set family $\{\{e_1, e_2, e_3, e_4, e_5, e_6\}, \{e_1, e_2, e_3, e_4, e_5\}, \{e_1, e_3, e_4, e_5, e_6\}, \{e_2, e_3, e_4, e_5, e_6\}, \{e_2, e_3, e_4, e_5\}\}$. Its indicator function is $f(e_1, e_2, e_3, e_4, e_5, e_6) = e_2 e_3 e_4 e_5 \vee e_1 \bar{e}_2 e_3 e_4 e_5 e_6$. The size of a BDD $\mathcal{B}$ is the number of nodes $|N|$. For convenience, in the following sections, we write the size of $\mathcal{B}$ as $|\mathcal{B}|$.

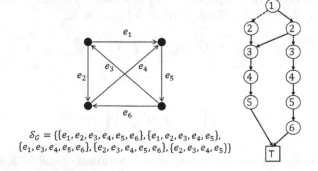

$$S_G = \{\{e_1, e_2, e_3, e_4, e_5, e_6\}, \{e_1, e_2, e_3, e_4, e_5\},$$
$$\{e_1, e_3, e_4, e_5, e_6\}, \{e_2, e_3, e_4, e_5, e_6\}, \{e_2, e_3, e_4, e_5\}\}$$

Fig. 1. Sample graph and the BDD representing SCESs of the graph. The 0-arcs and the 1-arcs are denoted by the dotted lines and the solid lines, respectively. The arcs to $\bot$ are omitted.

---

[1] To avoid the confusion, we use the terms "vertex" and "edge" for a vertex and edge in the graph G, and "node" and "arc" for a vertex and edge in the BDD $\mathcal{B}$. Vertices and nodes are denoted using Roman letters $(u, v, \ldots)$ and Greek letters $(\alpha, \beta, \ldots)$, respectively.

3 Proposed Algorithm

In this section, we present an algorithm to compute the SCR exactly. Firstly, the algorithm constructs a BDD for $\mathcal{S}_G$ using a new variant of the FBS. Secondly, it computes $\sigma(G)$ by a dynamic programming on the BDD.

3.1 SCR Computation

Once a BDD $\mathcal{B}$ for $\mathcal{S}_G$ is obtained, we can compute $\sigma(G)$ by a bottom-up dynamic programming as follows: Each node $\alpha \in N$ stores a value $\psi(\alpha)$ which is the sum of the probability of each edge subset represented by the descendants of α. The value of $\bot$ and $\top$ are initialized to $\psi(\bot) = 0$ and $\psi(\top) = 1$, respectively. For each non-terminal node $\alpha \in N$, its value $\psi(\alpha)$ is computed by

$$\psi(\alpha) = \psi(\alpha_0)p(e_{\ell(\alpha)}) + \psi(\alpha_1)(1 - p(e_{\ell(\alpha)})). \qquad (3)$$

By the definition, $\sigma(G)$ is equal to $\psi(\rho)$. This yields a dynamic programming algorithm (Algorithm 1) that requires the computation time of $O(|\mathcal{B}|)$.

Algorithm 1. Computing SCR

1: Construct BDD $\mathcal{B}$ for $\mathcal{S}_G$
2: $\psi(\bot) \leftarrow 0, \psi(\top) \leftarrow 1$
3: **for** $\alpha \in N \setminus \{\bot, \top\}$ in the reversal topological order **do**
4: $\quad \psi(\alpha) \leftarrow \psi(\alpha_0)p(e_{\ell(\alpha)}) + \psi(\alpha_1)(1 - p(e_{\ell(\alpha)}))$
5: **end for**
6: **return** $\psi(\rho)$

3.2 Existing Framework of Frontier-Based Search

For constructing a BDD for $\mathcal{S}_G$, we design an FBS [14], which is a general procedure for enumerating all constrained edge subsets implicitly. Here, we describe the general framework of the FBS.

Given a function $C\colon 2^E \to \{0,1\}$, if $C(X) = 1$ for an edge subset $X \subseteq E$, then we say that X has the property C. Let $P = \langle G, C \rangle$ be a problem to obtain all the possible edge subsets of E having the property C. The solution of P is a set of edges subsets defined as:

$$\mathcal{E}_P := \{X \in 2^E \mid C(X) = 1\}. \qquad (4)$$

Given a problem P, the algorithm constructs a BDD for $\mathcal{E}_P$ by processing the edges individually as the exhaustive search; the algorithm constructs the node set $N_i := \{\alpha \mid \ell(\alpha) = i\}$ for $i = 1, \ldots, m$, and the x-arc set $A_x := \{(\alpha, \alpha_x) \mid \alpha \in N \setminus \{\bot, \top\}\}$ for each $x \in \{0, 1\}$.

The processed edges at the i-th step is denoted by $E^{\leq i-1} := \{e_1, \ldots, e_{i-1}\}$, and the unprocessed edges at the i-th step is denoted by $E^{\geq i} := \{e_i, \ldots, e_m\}$.

Let $\mathcal{E}(\alpha) \subseteq 2^{E^{\leq i-1}}$ be a set of edge subsets corresponding to the paths from the root to a node $\alpha \in N_i$. Each node $\alpha \in N_i$ is associated with a subproblem of P denoted by $P_\alpha := \langle G[E^{\geq i}], C_\alpha \rangle$ where the property C_α is defined as

$$C_\alpha(X) = 1 \iff \forall Y \in \mathcal{E}(\alpha),\ C(X \cup Y) = 1. \tag{5}$$

For any pair of nodes $\beta, \beta' \in N_i$, β and β' are *equivalent* if $C_\beta(X) = 1 \iff C_{\beta'}(X) = 1$ for any $X \in 2^{E^{\geq i}}$. The algorithm merges some equivalent nodes into one node.

The primary process of the algorithm is as follows. Initially, the algorithm generates the node set $N_1 = \{\rho\}$. At the i-th step, the algorithm constructs N_{i+1} using N_i as follows. For each node $\alpha \in N_i$, the algorithm generates its children; $\mathcal{E}(\alpha_0)$ (resp. $\mathcal{E}(\alpha_1)$) is the set of the edge subsets that e_i is excluded from (resp. included in) the edge subsets of $\mathcal{E}(\alpha)$. On generating a new child, the algorithm conducts the following procedures to reduce the number of nodes:

- $\perp$-*pruning*: Let $\perp$-prune(α, e_i, x) be the functions defined as follows:

$$\perp\text{-prune}(\alpha, e_i, x) := \begin{cases} True & \mathcal{E}_{P_{\alpha_x}} = \emptyset, \\ False & \text{otherwise.} \end{cases} \tag{6}$$

If $\perp$-prune(α, e_i, x) outputs *True*, the x-child of α is $\perp$. Then the algorithm adds the x-arc $(\alpha, \perp)$ to A_x.

- $\top$-*pruning*: Let $\top$-prune(α, e_i, x) be the functions defined as follows:

$$\top\text{-prune}(\alpha, e_i, x) := \begin{cases} True & \mathcal{E}_{P_{\alpha_x}} = 2^{E^{\geq i+1}}, \\ False & \text{otherwise.} \end{cases} \tag{7}$$

If $\top$-prune(α, e_i, x) outputs *True*, the x-child of α is $\top$. Then the algorithm adds the x-arc $(\alpha, \top)$ to A_x.

- *merging*: Let β be a child of α. If β and a node $\beta' \in N_{i+1}$ are equivalent, the algorithm sets β' to β.

To apply these procedures efficiently, each node β maintains an additional information $\phi(\beta)$, referred to as a *configuration* that satisfies the condition that if $\phi(\beta) = \phi(\beta')$, then β and β' are equivalent. Note that the inverse is not required, which causes redundant node expansions.

Essentially, the FBS is a dynamic programming using the configuration as the state. It constructs a BDD as the structure derived from the table of the dynamic programming.

The general framework of the FBS is shown in Algorithm 2. The function generateNode(α, e_i, x) generates the x-child of α. The constructed BDD is not necessarily reduced because redundant node expansions can be caused. We apply the reduction rules until the reduced BDD is obtained if necessary.

Algorithm 2. Frontier-based Search

1: $N_1 \leftarrow \{\rho\}$, $N_i \leftarrow \emptyset$ for $i = 2, \ldots, m$
2: Generate the terminals $\bot$ and $\top$
3: $A_x \leftarrow \emptyset$ for each $x \in \{0, 1\}$
4: **for** $i = 1, \ldots, m$ **do**
5:   **for** $\alpha \in N_i$ **do**
6:     **for** $x \in \{0, 1\}$ **do**
7:       **if** $\bot$-prune(α, e_i, x) **then**
8:         $A_x \leftarrow A_x \cup \{(\alpha, \bot)\}$
9:       **else if** $\top$-prune(α, e_i, x) **then**
10:        $A_x \leftarrow A_x \cup \{(\alpha, \top)\}$
11:      **else**
12:        $\beta \leftarrow$ generateNode(α, e_i, x)
13:        **if** $\exists \beta' \in N_{i+1}$, $\phi(\beta) = \phi(\beta')$ **then**
14:          $\beta \leftarrow \beta'$
15:        **else**
16:          $N_{i+1} \leftarrow N_{i+1} \cup \{\beta\}$
17:        **end if**
18:        $A_x \leftarrow A_x \cup \{(\alpha, \beta)\}$
19:      **end if**
20:    **end for**
21: **end for**
22: **end for**
23: $N \leftarrow (\bigcup_{i=1,\ldots,m} N_i) \cup \{\bot, \top\}$, $A \leftarrow A_0 \cup A_1$
24: **return** $\mathcal{B} = (N, A)$

3.3 Proposed Frontier-Based Search for SCSSs

We adapt the FBS to our problem by designing four primary components: configuration, $\bot$-prune function, $\top$-prune function, and generateNode function. Because our aim is to enumerate $\mathcal{S}_G$, the property C is defined as

$$C(X) = 1 \iff X \text{ is an SCES of } G. \tag{8}$$

Configuration. Here, we design the configuration for enumerating $\mathcal{S}_G$ by the FBS. We first explain the basic idea of our configuration. Subsequently, we show that our configuration satisfies the condition above. In the following, we assume that $e_i = (s_i, t_i)$ for each $i = 1, \ldots, m$ to simplify the notation.

For each $i = 1, \ldots, m$, the i-th *frontier* is the vertex subset defined as

$$F_i := V[E^{\leq i-1}] \cap V[E^{\geq i}] \tag{9}$$

that has both processed and unprocessed edges. We use the reachability on the frontier vertices as the configuration: for any node $\alpha \in N_i$, $\phi(\alpha)$ is defined as a reachability matrix indexed by F_i^2, which represents the reachability on $G[Y]$ ($Y \in \mathcal{E}(\alpha)$):

$$\phi(\alpha)_{u,v} := \begin{cases} 1 & \forall Y \in \mathcal{E}(\alpha), u \rightsquigarrow_{G[Y]} v, \\ 0 & \text{otherwise.} \end{cases} \tag{10}$$

We assume that the proposed algorithm must perform $\perp$-pruning if possible. This assumption deduces that each vertex excluded from the past frontier is reachable from (resp. to) the new frontier on $G[Y]$ ($Y \in \mathcal{E}(\alpha)$). Based on this assumption, we can obtain the equivalent definition of the (5) as follows.

For any node $\alpha \in N_i$, let E_α be a *contracted edge set* of G, which is derived from the configuration $\phi(\alpha)$, defined as

$$E_\alpha := \{(u,v) \in F_i^2 \mid \phi(\alpha)_{u,v} = 1\}. \tag{11}$$

Similarly, let G_α be a *contracted graph* of G defined as

$$G_\alpha := (V[E^{\geq i}], E^{\geq i} \cup E_\alpha). \tag{12}$$

According to the assumption above, each edge subset $X \in \mathcal{E}_{P_\alpha}$ satisfies that $X \cup E_\alpha$ is an SCES of G_α. Therefore, we obtained an equivalent definition of (5) that depends on only the configuration as follows:

$$C_\alpha(X) = 1 \iff X \cup E_\alpha \text{ is an SCES of } G_\alpha. \tag{13}$$

Thus, under the assumption above, our configuration satisfies the condition that: for any pair of nodes $\beta, \beta' \in N_i$, if $\phi(\beta) = \phi(\beta')$, then β and β' are equivalent.

$\perp$-**prune**(α, e_i, x). Here, we design a $\perp$-prune function that does not contradict the assumption above. We use the following properties of C_α:

- $C_\alpha(X) = 1$ implies $C_\alpha(X \cup \{e_i\}) = 1$ for any $X \subseteq E^{\geq i}$, because an edge subset that contains an SCES is also an SCES.
- If $\phi(\alpha)_{s_i,t_i} = 1$, then $C_\alpha(X) = 1$ implies $C_\alpha(X \setminus \{e_i\}) = 1$ for any $X \subseteq E^{\geq i}$, because $e_i = (s_i, t_i) \in E_\alpha$.

Thus, we have a chance to conduct $\perp$-pruning for the case that exclude e_i and $\phi(\alpha)_{s_i,t_i} = 0$.

Let $G - e := (V, E \setminus \{e\})$ be a graph that is obtained by removing an edge e from G. A graph $G_\alpha - e_i$ has no SCES if and only if $\{e_i\}$ forms a cut set from s_i to t_i. Therefore, we design $\perp$-prune(α, e_i, x) as:

$$\perp\text{-prune}(\alpha, e_i, x) = \begin{cases} True & x = 0, \phi(\alpha)_{s_i,t_i} = 0, \text{ and } s_i \not\rightsquigarrow_{G_\alpha - e_i} t_i, \\ False & \text{otherwise.} \end{cases} \tag{14}$$

To evaluate $\perp$-prune(α, e_i, x) efficiently, we precompute the transitive closure of $G[E^{\geq i+1}]$. Using the transitive closure of $G[E^{\geq i+1}]$, the reachability from s_i to t_i on $G_\alpha - e_i$ is verified in $O(|F_i|^2)$ time by the BFS/DFS on the frontier. The precomputation of the transitive closures can be efficiently performed in decreasing order $i = m, \ldots, 1$; We compute the transitive closure of $G[E^{\geq i}]$ as the extension of the transitive closure of $G[E^{\geq i+1}]$ by the BFS/DFS on the $V[E^{\geq i}]$. Although the precomputation requires $O(\sum_{i=1}^{m} |V[E^{\geq i}]|^2)$ time, it is typically much faster than the FBS.

T-prune(α, e_i, x). Similarly, we have a chance to conduct T-pruning for the case that include e_i. The following property is observed: $C_\alpha(\{e_i\}) = 1$ implies $C_\alpha(\{e_i\} \cup X) = 1$ for any $X \subseteq E^{\geq i+1}$. Therefore, we design T-prune(α, e_i, x) as

$$\text{T-prune}(\alpha, e_i, x) = \begin{cases} True & x = 1, \text{ and } E_\alpha \cup \{e_i\} \text{ is an SCES of } G_\alpha, \\ False & \text{otherwise.} \end{cases} \quad (15)$$

To evaluate T-prune(α, e_i, x) efficiently, we precompute an integer

$$r := \min\{i \in \{1, \ldots, m\} \mid V[E^{\leq i}] = V\}. \quad (16)$$

Subsequently, $i < r$ implies T-prune($\alpha, e_i, 1$) $= False$ as $V[X \cup \{e_i\}] \neq V$ for any $X \in \mathcal{E}(\alpha)$. If $i \geq r$, T-prune($\alpha, e_i, 1$) is evaluated by the BFS/DFS on the frontier with computation time $O(|F_i|^2)$.

generateNode(α, e_i, x). The primary role of generateNode(α, e_i, x) is to compute the configuration of the new node. The function first generates a new node β labeled with $i + 1$ and copies configuration $\phi(\alpha)$ to $\phi(\beta)$. Subsequently, the configuration $\phi(\beta)$ is updated as follows:

1. Remove (resp. Insert) the rows and columns corresponding to the vertices excluded from the frontier, i.e, $F_i \setminus F_{i+1}$ (resp. included in the frontier, i.e, $F_{i+1} \setminus F_i$).
2. If $x = 0$, end the update. Otherwise, including a new edge changes the reachability; thus, we update $\phi(\beta)$ as the transitive closure of the frontier.

The number of removed (resp. Insert) rows and columns is constant because $|F_i \setminus F_{i+1}|$ (resp. $|F_{i+1} \setminus F_i|$) is 0, 1, or 2. Updating $\phi(\beta)$ is performed in $O(|F_i|^2)$ time by the BFS/DFS on the frontier. Thus, our generateNode function can be processed in $O(|F_i|^2)$ time.

3.4   Edge Ordering

Finally, we introduce a technique of *edge ordering* to accelerate the FBS. The time complexity of the FBS depends on the size of the frontier, because $|N_i|$ is equal to the number of the types of the configuration ϕ, which is at most $2^{|F_i|^2}$. Since the functions $\perp$-prune, T-prune, and generateNode take $O(|F_i|^2)$ time per node as mentioned above, the time complexity of our FBS is $O(\sum_{i=1}^{m} |F_i|^2 2^{|F_i|^2})$. Thus, it is important to optimize the edge ordering for reducing the frontier size.

It is known that the frontier size is closely related to the graph parameter named *pathwidth* [15]. In this paper, we use the *path-decomposition-based ordering* proposed in [16]. The algorithm first conducts optimization with beam search-based heuristics to compute a path decomposition with a small pathwidth. Subsequently, it computes an edge ordering using the path decomposition information.

4 Other Applications

In the previous section, we present an algorithm that constructs a BDD for $\mathcal{S}_G$ to compute the exact SCR. The constructed BDD also allows us to solve SCSS-related problems efficiently.

4.1 Finding Minimum SCSS

Given a weight function $w \colon E \to \mathbb{N}$, let $w(X) := \sum_{e \in X} w(e)$. An SCSS that has the smallest total weight of edges is called the *minimum* SCSS. The SCES of the minimum SCSS is defined as follows:

$$X^* \in \arg \min_{X \in \mathcal{S}_G} w(X). \tag{17}$$

Once a BDD $\mathcal{B}$ for $\mathcal{S}_G$ is obtained, we can obtain X^* by an algorithm that is similar to the Algorithm 1 as follows: Each node $\alpha \in N$ stores a value $\theta(\alpha)$ that is the minimum weight of the edge subsets represented by the descendants of α. The value of $\bot$ and $\top$ are initialized to $\theta(\bot) = \infty$ and $\theta(\top) = 0$, respectively. For each non-terminal node $\alpha \in N$, its value $\theta(\alpha)$ is computed by

$$\theta(\alpha) = \min\{\theta(\alpha_0), \theta(\alpha_1) + w(e_{\ell(\alpha)})\}. \tag{18}$$

Subsequently, $w(X^*) = \theta(\rho)$. This yields a dynamic programming algorithm that requires computation time $O(|\mathcal{B}|)$. The shortest path from ρ to $\top$ represents X^*: Starting with ρ, we descend x-arc such that $\theta(\alpha) = \theta(\alpha_x) + x \times w(e_{\ell(\alpha)})$ where α is the current node. Then X^* has the edges assigned to *True*.

4.2 Obtaining SCSSs with Constraints

An important function of the BDD is that it allows the efficient manipulation of set families. In particular, when two set families are represented by $\mathcal{B}_1$ and $\mathcal{B}_2$ with the same variable ordering, the union and intersection operations of these BDDs are performed in $O(|\mathcal{B}_1||\mathcal{B}_2|)$ time [17]. Moreover, we can obtain the BDDs for various logical constraints for example:

- Transforming a $\mathcal{B}$ to represent the subgraphs including the specified edges requires $O(|\mathcal{B}|)$ time [8].
- A BDD for the subgraphs with a bounded weight $\hat{w} \in \mathbb{N}$ (i.e., $\{X \subseteq E \mid w(X) \le \hat{w}\}$) can be constructed in $O(\hat{w}|E|)$ time [18].

Using these functions, we can obtain SCSSs with various constraints.

5 Experiments

We conducted computational experiments to evaluate the proposed algorithm. All the codes were implemented in C++ (g++4.8.4 with the -O3 option) using the TdZdd library (https://github.com/kunisura/TdZdd), which is a highly optimized implementation for the FBS framework. All experiments were conducted on a 64-bit Ubuntu 16.04 LTS with an Intel Core i7-3930K 3.2 GHz CPU and 64 GB RAM.

5.1   Scalability on Synthetic Networks

First, to observe the performance of the proposed algorithm, we applied our method to two classes of synthetic networks. The first class was $5 \times w$ gird graphs that had $5w$ vertices, $18w - 10$ directed edges (undirected edges were replaced with two directed edges in both directions), and a pathwidth of 5. The second class was random graphs that had the same number of vertices of $5 \times w$ grid, $9w - 5$ directed edges, and a pathwidth of $\Theta(n)$. We used the algorithm proposed in [19] to generate the strongly connected random graphs. For each $w \in \{5, \ldots, 20\}$, one hundred random graphs were generated, and we evaluated the average performance on the random graphs.

The results of the synthetic networks are shown in Fig. 2. The grid graphs shows that the computation time increased slowly; our algorithm executed in 16 seconds for $n = 100$. However, for the random networks, the computation time increased rapidly for networks with $80 \leq |V|$ vertices. These results show that the large pathwidth affected the computation time of our algorithm.

The size of the constructed BDDs had a similar tendency with the computation times. The BDD size was increased slowly for the grid graphs and rapidly for the random networks with $80 \leq |V|$ vertices. Meanwhile, the sizes of the reduced BDDs were sufficiently small.

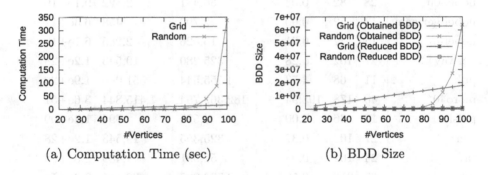

(a) Computation Time (sec)   (b) BDD Size

Fig. 2. Computational results on $5 \times w$ grid graphs and random graphs.

5.2   Scalability on Real-World Networks

Next, to evaluate the practical performance of the proposed algorithm, we applied our method to real-world networks. The real-world graphs were obtained from SNDlib (http://sndlib.zib.de/home.action). All the self-loops and multiple edges are deleted. Because all the graphs were undirected, we replaced each edge with two directed edges in both directions.

The results are shown in Table 1. The algorithm succeeded in constructing the BDDs for $\mathcal{S}_G$ on almost all the networks, however failed on three networks

Table 1. Computational results on real-world networks. Time denotes the time to construct the BDDs, BDD Size 1 denotes the size of the constructed BDDs, BDD Size 2 denotes the size of the reduced BDDs, and Cardinality denotes the number of SCESs. For the last three networks, the algorithm failed due to the memory limit.

Network	$|V|$	$|E|$	Time (sec)	BDD Size 1	BDD Size 2	Cardinality
abilene	12	30	0.00	284	94	1.2e+06
atlanta	15	44	0.01	4,964	630	5.9e+10
brain	161	332	2.31	9,095	2,990	1.6e+07
cost266	37	114	0.46	382,680	22,221	5.3e+28
france	25	90	0.05	45,715	5,420	2.2e+23
geant	22	72	0.07	81,391	5,611	1.3e+18
germany50	50	176	138.56	124,052,168	3,454,355	1.3e+47
giul39	39	172	972.38	781,882,756	15,878,463	2.2e+49
india35	35	160	90.05	83,212,903	2,422,816	4.8e+45
janos-us-ca	39	122	0.37	416,769	29,749	7.4e+30
janos-us	26	84	0.08	59,486	5,320	3.4e+21
newyork	16	98	12.70	12,178,145	607,550	9.8e+28
nobel-eu	28	82	0.03	35,481	2,822	9.1e+19
nobel-germany	17	52	0.00	2,790	458	6.3e+12
nobel-us	14	42	0.01	15,526	2,285	6.1e+10
norway	27	102	0.28	325,280	19,543	1.2e+28
pdh	11	68	4.73	4,555,544	551,065	1.9e+20
pioro40	40	178	198.94	182,362,754	1,415,844	3.6e+51
polska	12	36	0.00	4,525	794	1.4e+09
sun	27	102	0.37	325,355	19,543	1.2e+28
ta1	24	102	0.08	52,991	4,489	2.5e+28
ta2	65	216	2.56	2,573,009	98,115	2.3e+54
zib54	54	160	0.18	158,979	13,571	1.2e+37
dfn-bwin	10	90	-	-	-	-
dfn-gwin	11	94	-	-	-	-
di-yuan	11	84	-	-	-	-

due to the memory limit. Although each succeeded instance might have a few hundred edges, the algorithm executed in a few seconds or a few minutes. By comparing a brain network and the failed ones, we found that the computational cost of the algorithm depends on the network structure.

As shown in the cardinality column, the numbers of SCESs are enormous. This implies that the naive exhaustive search is unrealistic. Particularly, the ta2 network has

2, 320, 225, 475, 355, 945, 207, 334, 621, 674, 664, 990, 580, 848, 170, 679, 757, 701, 120

($\simeq 2.3 \times 10^{54}$) SCESs. This shows the advantage of our approach; the BDD representation of the SCESs is efficient.

5.3   SCR Computation

We also conducted the experiments to compute the exact SCR using the constructed BDDs for analyzing the reliability of each network used in the experiments above. Once the BDDs are obtained, we can easily compute the exact SRC iteratively for various settings of the edge dropping probability. Therefore, we used the probability of the edge dropping that was moved from 1.0 to 0.0 and decreased by 0.01. The results are shown in Fig. 3.

For the synthetic networks, the grid graphs have relatively high reliability until the edge dropping probability is less than 0.1, whereas the reliability of the random graphs was much lower. We consider that it is attributable to the

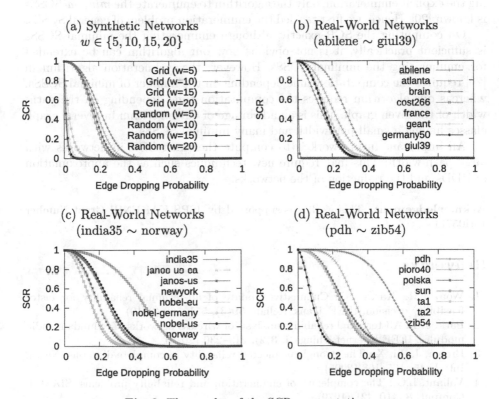

Fig. 3. The results of the SCRs computation.

sparsity of the random graphs; the number of edges is $9w - 5$ in contrast with the number of vertices $5w$.

The real-world networks brain, ta2, and zib54 have low reliability; meanwhile, the newyork and pdh networks have much higher reliability. By comparing these networks, each of the networks with high reliability tends to have relatively many edges in contrast with the number of vertices. Although this may be a natural consequence, its practical verification was impossible previously.

6   Discussion and Conclusion

In this study, we proposed an algorithm to compute the SCR exactly. The proposed algorithm first constructs a BDD representing all SCSSs using a new variant of the FBS. Subsequently, it computes the SCR by a bottom-up dynamic programming over the constructed BDD. The experimental results showed that the proposed algorithm can compute the SCR of the real-world networks with a few hundred edges, even though they have a large number of SCSSs such as 2.3×10^{54}. As a secondary application, our algorithm can also compute the minimum SCSS.

Our algorithm implicitly solved the enumeration problem of SCSSs. Regarding the explicit enumeration, only the algorithm to enumerate the *minimal* SCSSs is known [20]. Hence, we first tackled the enumeration problem of general SCSSs.

On computing the SCR exactly, although enumerating the minimal SCSSs is sufficient practically, it is not obvious how our algorithm can be extended for enumerating the minimal SCSSs. However, the enumeration algorithm in [20] requires the computation time depending on the number of minimal SCSSs, whereas our algorithm requires the computation time depending on the path-width of the given graph. This is an advantage of our algorithm in several graph classes having a small pathwidth and many minimal SCSSs.

An important future work is to compute the exact SCR in networks with large path-width. This may require new techniques such as the approximation of BDDs and the reduction of the networks.

Acknowledgement. This work was supported by JSPS KAKENHI Grant Number 15H05711.

References

1. Won, J.M., Karray, F.: Cumulative update of all-terminal reliability for faster feasibility decision. IEEE Trans. Reliab. **59**(3), 551–562 (2010)
2. Park, J.H.: All-terminal reliability analysis of wireless networks of redundant radio modules. IEEE Internet Things J. **3**(2), 219–230 (2016)
3. Brown, J., Li, X.: The strongly connected reliability of complete digraphs. Netw.: Int. J. **45**, 165–168 (2005)
4. Valiant, L.G.: The complexity of enumeration and reliability problems. SIAM J. Comput. **8**, 410–421 (1979)

5. Karger, D.R.: A randomized fully polynomial time approximation scheme for the all terminal network reliability problem. In: Proceedings of the Twenty-seventh Annual ACM Symposium on Theory of Computing, STOC 1995, pp. 11–17. ACM, New York (1995)

6. Imai, H., Sekine, K., Imai, K.: Computational investigations of all-terminal network reliability via BDDs. IEICE Trans. Fundam. Electron. Commun. Comput. Sci. **82**, 714–721 (1999)

7. Hwang, F.K., Wright, P.E., Hu, X.: Exact reliabilities of most reliable double-loop networks. Networks **30**, 81–90 (1997)

8. Bryant, R.E.: Graph-based algorithms for boolean function manipulation. IEEE Trans. Comput. **35**(8), 677–691 (1986)

9. Frederickson, G.N., JáJá, J.: Approximation algorithms for several graph augmentation problems. SIAM J. Comput. **10**(2), 270–283 (1981)

10. Vincent, D., Cecile, B.: Transitive reduction for social network analysis and visualization. In: Proceedings of the 2005 IEEE/WIC/ACM International Conference on Web Intelligence, WI 2005, pp. 128–131. IEEE Computer Society, Washington, D.C. (2005)

11. Ardito, C.F., Paola, D.D., Gasparri, A.: Decentralized estimation of the minimum strongly connected subdigraph for robotic networks with limited field of view. In: 2012 IEEE 51st IEEE Conference on Decision and Control (CDC), pp. 5304–5309, December 2012

12. Albert, R., DasGupta, B., Dondi, R., Sema Kachalo, E.S., Zelikovsky, A., Westbrooks, K.: A novel method for signal transduction network inference from indirect experimental evidence. J. Comput. Biol. **14**(7), 927–949 (2007)

13. Aditya, S., DasGupta, B., Karpinski, M.: Algorithmic perspectives of network transitive reduction problems and their applications to synthesis and analysis of biological networks. Biology **3**(1), 1–21 (2014)

14. Kawahara, J., Inoue, T., Iwashita, H., Minato, S.: Frontier-based search for enumerating all constrained subgraphs with compressed representation. IEICE Trans. Fundam. **E100-A**(9), 1773–1784 (2017)

15. Kinnersley, N.G.: The vertex separation number of a graph equals its path-width. Inf. Process. Lett. **42**(6), 345–350 (1992)

16. Inoue, Y., Minato, S.: Acceleration of ZDD construction for subgraph enumeration via path-width optimization. TCS-TR-A-16-80. Hokkaido University (2016)

17. Yoshinaka, R., Kawahara, J., Denzumi, S., Arimura, H., Minato, S.: Counterexamples to the long-standing conjecture on the complexity of BDD binary operations. Inf. Process. Lett. **112**, 636–640 (2012)

18. Bergman, D., Ciré, A.A., van Hoeve, W., Hooker, J.N.: Decision Diagrams for Optimization. Artificial Intelligence: Foundations, Theory, and Algorithms. Springer, Cham (2016). https://doi.org/10.1007/978-3-319-42849-9

19. Maurer, P.: Generating strongly connected random graphs. In: Proceedings of the 2017 International Conference on Modeling, Simulation and Visualization Methods, MSV 2017, pp. 3–6. CSCE, Las Vegas (2017)

20. Boros, E., Elbassioni, K., Gurvich, V., Khachiyan, L.: Enumerating minimal dicuts and strongly connected subgraphs and related geometric problems. In: Bienstock, D., Nemhauser, G. (eds.) IPCO 2004. LNCS, vol. 3064, pp. 152–162. Springer, Heidelberg (2004). https://doi.org/10.1007/978-3-540-25960-2_12

Combinatorial Optimization

Upper and Lower Bounds for Different Parameterizations of (n,3)-MAXSAT

Tatiana Belova[✉] and Ivan Bliznets

St. Petersburg Department of Steklov Institute of Mathematics of the Russian
Academy of Sciences, Saint Petersburg, Russia
yukikomodo@gmail.com, iabliznets@gmail.com

Abstract. In this paper, we consider the (n,3)-MAXSAT problem. The
problem is a special case of the Maximum Satisfiability problem with
an additional requirement that in input formula each variable appears
at most three times. Here, we improve previous upper bounds for $(n,3)$-
MAXSAT in terms of n (number of variables) and in terms of k (number
of clauses that we are required to satisfy). Moreover, we prove that satisfying more clauses than the simple all true assignment is an NP-hard
problem.

Keywords: Maximum satisfiability
Exact algorithms · Parameterization above guarantee

1 Introduction

Satisfiability problem (SAT for short) is probably the most known NP-complete
problem with enormous importance in computer science and with lots of applications. Its optimization version Maximum Satisfiability also has a lot of applications [5,18,24]. There is a special conference SAT dedicated to the problem
and there is a special contest for determining the best SAT-solver (SAT-solver
is a program that tries to solve instances of the SAT problem). Many heuristics,
approximation and exact algorithms [4,6,8–10,22] were constructed for Maximum Satisfiability problem.

Here, we consider only a special case of Maximum Satisfiability defined in
the following way:

(n,3)-MAXSAT
Input: A formula F in which each variable appears at most three
               times.
Question: Find an assignment that simultaneously satisfies a maxi-
               mum number of clauses.

It is well-known that even with this restrictions Maximum Satisfiability is
NP-hard [23]. The (n,3)-MAXSAT problem has a chain of improvements and is

Research is supported by Russian Science Foundation (project 18-71-10042).

well-studied. The details can be found in Table 1. We would like to emphasize that such special cases are very important in understanding Maximum Satisfiability and constructing efficient algorithms for it. For example, in 2012 Bliznets and Golovnev [6] improved the best-known algorithm for Maximum Satisfiability in terms of k (number of clauses that one is asked to satisfy) and it was achieved by careful analysis of variables that appear at most three times. The currently best-known algorithm for Maximum Satisfiability in terms of k was developed by Chen, Xu and Wang [9] and is based on analysis of variables which appears at most four times in the formula. We note that even more restricted special cases attract attention. For example, $(n,3)$-MAX-2-SAT (a special case of $(n,3)$-MAXSAT with an additional requirement that all clauses in input formula have length at most 2) was considered in papers [15,16].

We note that lower bounds or intractability results are harder to prove for more restricted input formulas. So such results are more preferable for $(n,3)$-MAXSAT than just Maximum Satisfiability. Moreover, such results have more potential to be suitable for further reductions and proof of intractability. And this is an additional reason to study (n,3)-MAXSAT.

Our Results: All our results are for (n,3)-MAXSAT problem but for different parameterizations. In the work, we improve previous upper bound $O^*(1.194^n)$ in terms of n (number of variables in input formula) down to $O^*(1.191^n)$. In order to obtain the improvement we employ new reduction rule 7 that automatically leads to a situation where we need to consider only two cases of how literal $\bar{x}$ appears in the formula. The fact might be useful in future works of case analysis simplification.

In terms of k we achieve improvement from $O^*(1.175^k)$ to $O^*(1.168^k)$ without any case analysis by a very simple algorithm and establish a useful connection with the problem of maximum satisfiability parameterized above maximum matching in a variable-clause graph. Moreover, we show that any improvement of upper bound for (n,3)-MAXSAT in terms of n leads to an improvement of upper bound in terms of k.

Also, we show that parameterization above assignment that assigns true values to each variable is intractable. To be specific: to check that there is an assignment satisfying more clauses than the trivial all true assignment is already an NP-hard problem.

2   Preliminaries

A *clause* C is a disjunction of literals, for example, $C = x_1 \lor x_2 \lor x_3$. A *CNF formula* F is a conjunction of clauses. Let V be the set of variables in formula F. An assignment is a function $f : V \to \{0,1\}$. A clause C is *satisfied by* f if there is a literal $x \in C$ such that $f(x) = 1$, or a literal $\bar{x} \in C$ such that $f(x) = 0$. *Maximum Satisfiability* is a problem in which one is given a CNF formula and is asked to find assignment that simultaneously satisfies a maximum possible number of clauses. In (n,3)-MAXSAT problem we restrict input only to formulas that contain each variable at most three times.

Table 1. Known results. Here, n denotes the number of variables in input formula and k denotes the number of clauses that one is asked to satisfy.

Bound w.r.t k	Bound w.r.t n	Reference	Year
	$O^*(1.732^n)$	Raman, Ravikumar, Rao [23]	1998
	$O^*(1.3248^n)$	Bansal, Raman [2]	1999
$O^*(1.3247^k)$		Chen, Kanj [8]	2002
	$O^*(1.27203^n)$	Kulikov [16]	2005
$O^*(1.2721^k)$		Bliznets, Golovnev [6]	2012
	$O^*(1.2600^n)$	Bliznets [7]	2013
$O^*(1.194^k)$	$O^*(1.237^n)$	Xu, Chen, Wang [9]	2016
$O^*(1.175^k)$	$O^*(1.194^n)$	Li, Xu, Wang, Yang [17]	2017
$\mathbf{O^*(1.168^k)}$	$\mathbf{O^*(1.191^n)}$	**This paper**	**2018**

We call x and $\bar{x}$ *literals* of the boolean variable x. We denote by $v(a)$ the boolean variable corresponding to literal a, so for example $v(x) = v(\bar{x})$. We call literal x *positive* and we call literal $\bar{x}$ *negative*. Clause is called *unit* clause if it contains only one literal. If a clause contains at least one positive literal we call such a clause *positive*. Without loss of generality we can assume that all negative literals appear at most once. Since otherwise, we can always replace variable x with a variable y such that the positive literal x corresponds to the negative literal $\bar{y}$ and the literal $\bar{x}$ corresponds to y i.e. $x = \bar{y}$. We denote literals by letters x, y, z, t if we know that the literals are positive. If we need to denote some literal and we do not know that it is positive or negative then such literal we denote by letters a, b, c, d. A literal a is an *(i,j)-literal* if a and $\bar{a}$ occur exactly i and j times in F, respectively. $F_{a=1}$ and $F_{a=0}$ are the formulas obtained from F by assigning $a = 1$ and $a = 0$, respectively. By $\tilde{a}$ we denote a literal which corresponds to the same variable as the literal a does, i.e. $\tilde{a}$ may stand for either a or $\bar{a}$.

Algorithm for (n,3)-MAXSAT in this paper is based on branching technique. Such algorithms are described by reduction rules, that are used to simplify a problem instance, and branching rules, that are used to solve an instance by recursively solving smaller instances. Generally, branches are just subproblems in which instead of original formula F we work with formulas $F_{x=0}$ and $F_{x=1}$, i.e. substitute value of x. If a branching rule branches an instance of size n into r instances of size $n - t_1, n - t_2, \ldots, n - t_r$, we call $(t_1, t_2, ..., t_r)$ a branching vector of this branching rule. By a branching factor of a branching rule we understand a constant c that is a solution of a linear reccurence corresponding to some branching vector of this rule; such constants are used to bound the running time of an algorithm by $O^*(c^n)$. By the worst branching factor of branching rules we understand the largest c among such constants. We refer the reader to [12] for the detailed explanation of these aspects.

3 Reduction Rules

In this section, we introduce several reduction rules. A reduction rule is a polynomial algorithm which transforms the original problem into an equivalent and somewhat simpler problem. The following five reduction rules 1–5 are well-known and were used in papers [2,8,17,21]. That is why we omit their proof here.

Reduction rule 1. *Let x be literal. If there is a clause $(x \lor x \lor C)$, then replace it with $(x \lor C)$, i.e. $F = F' \land (x \lor x \lor C) \to F = F' \land (x \lor C)$.*

Reduction rule 2. *Let x be literal. If there is a clause $(x \lor \bar{x} \lor C)$, then remove the clause, since it is always satisfied, i.e., $F = F' \land (x \lor \bar{x} \lor C) \to F = F'$.*

Reduction rule 3. *If there is an $(i,0)$-literal x, then replace F with the formula $F_{x=1}$.*

Reduction rule 4. *If there is a $(1,1)$-literal x, then replace $F = F' \land (x \lor C_1) \land (\bar{x} \lor C_2)$ with the formula $F = F' \land (C_1 \lor C_2)$, i.e. replace the two clauses $x \lor C_1$, $\bar{x} \lor C_2$ with the clause $C_1 \lor C_2$.*

Proposition 1. *If variable x appears in F at most two times then we can construct an equivalent formula without x. Moreover, such construction takes polynomial time (proof follows from Reduction rules 1–4).*

Reduction rule 5. *If there is a $(2,1)$-literal x and there is at least one unit clause x in F, then replace F with $F_{x=1}$.*

If none of Reduction rules 1–5 is applicable, each variable occurs exactly three times. So, every literal a is either a $(1,2)$-literal or a $(2,1)$-literal. In addition, for each $(2,1)$-literal x, there is no unit clause x.

Reduction rule 6 [17]. *If there are three clauses $(x \lor y), (\bar{x}), (\bar{y})$ in the formula F, then replace these three clauses with the clause $(\bar{x} \lor \bar{y})$, i.e. $F = (x \lor y) \land (\bar{x}) \land (\bar{y}) \land F' \to F = (\bar{x} \lor \bar{y}) \land F'$.*

We use the following definition introduced in [25].

Definition 1. *A CNF formula F is called linear if no two variables appear in more than one clause together.*

The following proposition was shown in [25].

Proposition 2. *Given formula F as an instance of (n,3)-MAXSAT, it is possible to construct a simplified equivalent instance with formula F' such that F':*

- *is linear*
- *depends on at most the same number of variables*
- *each variable occurs at most three times.*

Moreover, careful analysis of the simplification shows that if initially formula was non-linear then after such simplification the number of variables in the formula decrease at least by one.

Reduction rule 7. *If* $F = (\bar{x}_1 \vee x_2 \vee A_1) \wedge (\bar{x}_2 \vee x_3 \vee A_2) \wedge \cdots \wedge (\bar{x}_k \vee x_1 \vee A_k) \wedge F'$, *where all* x_i *are* $(2,1)$-*literals, then assignment* $x_1 = x_2 = \cdots = x_k = 1$ *satisfies all clauses containing variables* $x_1, x_2, \ldots, x_k$.

For a given formula F we construct the following graph G_F. For each variable we introduce a vertex. We will refer to a vertex by the name of the corresponding variable. If there is a clause $(\bar{x} \vee y \vee A)$ in F we add an oriented edge going from x to y into G_F. The existence of the pattern considered in the previous rule is equivalent to the existence of a cycle in the graph G_F. So, if the graph G_F contains a cycle then we can reduce our formula by reduction rule 7. Hence, without loss of generality, we can assume that G_F does not contain cycles. It means that one of the following situations take place (as otherwise each vertex have an outgoing edge and this leads to the existence of a cycle):

1. All negative literals appear as singletons, i.e. as clauses containing only one literal.
2. F can be represented as $(\bar{x} \vee \bar{y} \vee C'_x) \wedge F'$, where C'_x contains only negative literals.
3. F can be represented as $(\bar{x} \vee y \vee C'_x) \wedge \bar{y} \wedge F'$ and in graph G_F there is no path of length two starting from vertex corresponding to variable x.

Indeed, if our formula does not fit to the above stated situations 1–2 then we pick an arbitrary vertex $v_1 \in G_F$ with at least one outgoing edge(if there are no edges then we are in situation 1 or 2). Let us consider the longest directed path $v_1, v_2, \ldots v_q$ in G_F starting from vertex v_1. We take variable corresponding to v_{q-1} as x and variable corresponding to v_q as y. Note that from definition of G_F follows that F has the following representation $(\bar{x} \vee y \vee C'_x) \wedge F_1$. As F does not fit in the case 2 and $v_1, v_2, \ldots v_q$ was the longest path starting from v_1 we conclude that $F = (\bar{x} \vee y \vee C'_x) \wedge \bar{y} \wedge F_2$. Moreover, in graph G_F there is no path of length two starting from vertex corresponding to variable x as the path $v_1, v_2, \ldots v_q$ was the longest path starting from v_1.

In [7] it was shown that in the first situation the problem admits a polynomial time solution. Hence, we have the following reduction rule:

Reduction rule 8. *If all negative literals appear as singletons then solve the problem by polynomial time algorithm from [7].*

Note, that the two remaining situations can be rewritten in the following way

$1'$. $F = (\bar{x} \vee \bar{y} \vee C'_x) \wedge F'$
$2'$. $F = (\bar{x} \vee y \vee C'_x) \wedge \bar{y} \wedge F'$ where C'_x contains only positive literals and each $z \in C'_x$ has a unit clause $(\bar{z})$.

Clearly, situation 2 is included in situation $1'$. In situation 3 if there exists negative literal $\bar{z} \in C'_x$ then it fits to situation $1'$. So, we can assume that all literals in C'_x are positive. Moreover, we can require that its negation literals occurs as unit clauses because if there is a clause $(\bar{z} \vee \bar{t})$ then this is situation $1'$. Moreover, $\bar{z}$ cannot be in a clause with positive literals, since otherwise, in graph G_F there is a path of length at least two starting at vertex x which contradict to situation 3.

4   Branching Rules

Lemma 1. *For any formula F there is an optimal assignment that satisfies all positive clauses.*

Proof. Consider an optimal assignment τ that satisfies the maximum number of positive clauses. If the assignment satisfies all positive clauses then the lemma statement is true. Otherwise, assume that there is a clause $x \vee C$, which is not satisfied by an assignment τ. It means that $\tau(x) = 0$. If we flip value of x i.e. put $x = 1$, we will satisfy at least the same number of clauses (since we lose at most one clause containing literal $\bar{x}$ and satisfy clause $x \vee C$) but the number of satisfied positive clauses will increase which contradict choice of τ. Hence, there was no unsatisfied positive clause in the formula.   □

Corollary 1. *In branch $x = 0$ it is enough to consider only assignments that satisfy clauses containing literal x.*

Proof. The result immediately follows from Lemma 1.   □

Let F be a formula and x be a variable such that one of the cases 1' or 2' take place. In this case we may assume that x appears in clauses $x \vee C_1, x \vee C_2, \bar{x} \vee C_3$. Note that $|C_3| > 0$ and if no reduction rules are applicable to F then all variables in C_1, C_2, C_3 are different. We emphasize that in future we treat C_1 and C_2 interchangeably, so there is no reason simultaneously consider case $|C_1| = 1, |C_2| = 2$ and case $|C_1| = 2, |C_2| = 1$.

First of all, we present nine branching rules that consider different cases of sizes of sets C_1, C_2, C_3. First three branching rules without any changes are taken from paper [17] so we do not present its proof here. Other six branching rules are an adaptation of some rules from [17] and its proofs are omitted here due to space constraints. All proofs can be found in full version.

Branching rule 1 [17]. *If $|C_1| = 1, |C_2| = 2, |C_3| = 2$ then there is $(4,4)$-branching on x.*

Branching rule 2 [17]. *If $|C_1| = 2, |C_2| = 2, |C_3| = 1$ then there is $(7,2)$-branching on x.*

Branching rule 3 [17]. *If $|C_1| \geq 3, |C_2| \geq 2, |C_3| = 1$ then there is a $(7,2)$-branching on x.*

Branching rule 4. *If $|C_1| \geq 3, |C_2| \geq 2, |C_3| \geq 2$ there is a $(6,3)$-branching.*

Branching rule 5. *If $|C_1| \geq 3, |C_2| = 1, |C_3| = 1$ there is a $(6,3)$-branching.*

Branching rule 6. *If $|C_1| = 1, |C_2| = 1, |C_3| \geq 3$ then there is a $(3,6)$-branching.*

Branching rule 7. *In case $|C_1| \geq 2, |C_2| = 1, |C_3| \geq 3$ there is a $(4,5)$-branching.*

Branching rule 8. *In case* $|C_1| = 2, |C_2| = 2, |C_3| \geq 3$ *there is a* $(5,4)$-*branching.*

Branching rule 9. *In case* $|C_1| \geq 3, |C_2| = 1, |C_3| = 2$ *there is a* $(5,4)$-*branching.*

From now on we assume that none of the branching rules 1–9 is applicable to our formula. In Table 2 we summarize achieved branching factors and considered subcases.

Table 2. Branching factors of rules

$|C_1|$	$|C_2|$	$|C_3|$	Branching vector	Branching factor	Branching rule
1	2	2	$(4,4)$	1.190	Branching rule 1
2	2	1	$(7,2)$	1.191	Branching rule 2
≥ 3	≥ 2	1	$(7,2)$	1.191	Branching rule 3
≥ 3	≥ 2	≥ 2	$(6,3)$	1.174	Branching rule 4
≥ 3	1	1	$(6,3)$	1.174	Branching rule 5
1	1	≥ 3	$(3,6)$	1.174	Branching rule 6
≥ 2	1	≥ 3	$(4,5)$	1.168	Branching rule 7
2	2	≥ 3	$(5,4)$	1.168	Branching rule 8
≥ 3	1	2	$(5,4)$	1.168	Branching rule 9

Note that only the following values $\{(1,1,1),(1,2,1),(1,1,2),(2,2,2)\}$ of $(|C_1|,|C_2|,|C_3|)$ have not been considered.

As was noted before our formula F has one of the following representations:

1. $F = (\bar{x} \vee \bar{y} \vee C'_x) \wedge F'$
2. $F = (\bar{x} \vee y \vee C'_x) \wedge \bar{y} \wedge F'$ where C'_x contains only positive literals and each $z \in C'_x$ has a unit clause $(\bar{z})$.

Now, we consider these two cases and in each of them show how to branch. We organize these branching rules in a different style as they contain new ideas while branching rules 1–9 more-or-less was taken or inspired by work [17].

Case 1. $F = (\bar{x} \vee \bar{y} \vee C'_x) \wedge F'$

If C'_x consists of more than 1 literal then using Branching rules 6, 7, 8 we can get a good branching. So there are only two remaining subcases: (i) $C'_x = \emptyset$; (ii) $C'_x = d$ (recall that d can be positive or negative literal).

Without loss of generality $F = (\bar{x} \vee \bar{y} \vee C'_x) \wedge (x \vee A_x) \wedge (x \vee B_x) \wedge (y \vee A_y) \wedge (y \vee B_y) \wedge F'$. We will omit F' in further cases writing down only some clauses of F.

Case 1.1. $C'_x = \emptyset$
$(\bar{x} \vee \bar{y}) \wedge (x \vee A_x) \wedge (x \vee B_x) \wedge (y \vee A_y) \wedge (y \vee B_y)$

Due to space constraints we omit the analysis of this case The whole analysis can be found in full version of the paper.

Case 1.2. $C'_x = d$

$(\bar{x} \vee \bar{y} \vee d) \wedge (x \vee A_x) \wedge (x \vee B_x) \wedge (y \vee A_y) \wedge (y \vee B_y)$ Because of the branching rules 1–9 and the fact that $\bar{x}$ share clause with two other literals we have that either $|A_x| = |B_x| = 1$ or $|A_x| = |B_x| = 2$. Similar situation take place with A_y, B_y. We consider two subcases: (i) $|A_x| = |B_x| = |A_y| = |B_y| = 1$; (ii) at least two numbers from $|A_x|, |B_x|, |A_y|, |B_y|$ are greater than 1.

Case 1.2.1. $|A_x| = |B_x| = |A_y| = |B_y| = 1$

$(\bar{x} \vee \bar{y} \vee d) \wedge (x \vee a_1) \wedge (x \vee b_1) \wedge (y \vee a_2) \wedge (y \vee b_2)$

We branch on x. In branch $x = 1$ by Proposition 1 we can remove a_1 and b_1. So we remove at least 3 variables.

In branch $x = 0$ by Corollary 1 $a_1 = b_1 = 1$ and by reduction rule 5 we can assign $y = 1$. Moreover, from Proposition 1 follows that we can remove d, a_2 and b_2. So we remove $x, y, d, a_1, b_1, a_2, b_2$. However, some of the variables may coincide. So we consider several cases depending on these literals.

Case 1.2.1.1. If $\{a_1, b_1, a_2, b_2\}$ contains at least 3 different variables then we have a $(3, 6)$-branching.

Since we assume that the formula is linear we have that variables a_1, b_1 are different as well as variables a_2, b_2. So if $\{a_1, b_1, a_2, b_2\}$ consists only of 2 variables (note that it cannot be only one variable as each variable occur at most three times) then WLOG we can assume that $a_2 = \tilde{a}_1, b_2 = \tilde{b}_1$.

Case 1.2.1.2. $a_2 = \tilde{a}_1, b_2 = \tilde{b}_1$

$(\bar{x} \vee \bar{y} \vee d) \wedge (x \vee a_1) \wedge (x \vee b_1) \wedge (y \vee \tilde{a}_1) \wedge (y \vee \tilde{b}_1)$

There is no reason to assign $x = y = 0$ because in this case we can reassign one of them to 1 and we satisfy at least the same number of clauses.

It means we may consider two branches: $x = y = 1$ and $x \neq y$.

In branch $x = y = 1$ by Proposition 1 we remove a_1 and b_1. So we remove at least 4 variables.

Consider a branch with $x \neq y$, i.e. $y = \bar{x}$. Then F will become $(x \vee a_1) \wedge (x \vee b_1) \wedge (\bar{x} \vee \tilde{a}_1) \wedge (\bar{x} \vee \tilde{b}_1)$ and d will disappear.

There are two subcases here.

Case 1.2.1.2.1. $\tilde{a}_1 = a_1$ or $\tilde{b}_1 = b_1$

WLOG we assume that $\tilde{a}_1 = a_1$, then our formula contains clauses $(x \vee a_1) \wedge (x \vee b_1) \wedge (\bar{x} \vee a_1) \wedge (\bar{x} \vee \tilde{b}_1)$. There is an optimum assignment τ such that $\tau(a_1) = 1$. Assume that $\tau(a_1) = 0$, then consider τ' such that for any variable z except a_1 we have $\tau'(z) = \tau(z)$ and $\tau'(a_1) = 1$. Note that τ satisfy exactly one clause among clauses $(x \vee a_1)$ and $(\bar{x} \vee a_1)$. Hence, if we flip the value of variable a_1 we satisfy at least one additional clause and lose at most one clause since a_1 is positive literal (as literal a_1 appears two times in the formula). So τ' satisfies at least the same number of clauses as τ. It means that we can assume that in an optimal solution $a_1 = 1$ and we can remove clauses $(x \vee a_1)$ and $(\bar{x} \vee a_1)$. Hence, x occurs now only in two clauses $(x \vee b_1)$ and $(\bar{x} \vee \tilde{b}_1)$. So, by reduction rule 4 we replace these two clauses with one clause $(b_1 \vee \tilde{b}_1)$. And now using reduction rules we can remove b_1 as well. It means we remove at least 5 variables: x, y, d, a_1, b_1. So we achieve a $(4, 5)$-branching.

Case 1.2.1.2.2. $\widetilde{a}_1 = \bar{a}_1$ and $\widetilde{b}_1 = \bar{b}_1$

So the part of our formula can be rewritten as $(x \vee a_1) \wedge (x \vee b_1) \wedge (\bar{x} \vee \bar{a}_1) \wedge (\bar{x} \vee \bar{b}_1)$.

Let us show that there is an optimum assignment τ such that $\tau(a_1) = \tau(b_1)$. Any assignment τ' such that $\tau'(a_1) \neq \tau'(b_1)$ satisfies exactly three clauses from $(x \vee a_1) \wedge (x \vee b_1) \wedge (\bar{x} \vee \bar{a}_1) \wedge (\bar{x} \vee \bar{b}_1)$. Note that if we flip value of variable a_1 or b_1 (depending on the situation) such that the fourth clause becomes satisfied we will satisfy one more clause and at most one clause previously satisfied becomes unsatisfied since both a_1 and b_1 have one more occurrence. So we obtain an assignment that satisfies at least the same number of clauses and the values of literals a_1 and b_1 are equal. Hence we may consider only assignments τ such that $\tau(a_1) = \tau(b_1)$. So we can rewrite our formula as $(x \vee a_1) \wedge (x \vee a_1) \wedge (\bar{x} \vee \bar{a}_1) \wedge (\bar{x} \vee \bar{a}_1)$. Note that there are at most two occurrences of the variable corresponding to literal a_1. It is easy to see that it is always beneficial to assign the value of x equal to $\bar{a}_1$ as otherwise variable x is not satisfying any additional clauses. So our four clauses can be rewritten as: $(\bar{a}_1 \vee a_1) \wedge (\bar{a}_1 \vee a_1) \wedge (a_1 \vee \bar{a}_1) \wedge (a_1 \vee \bar{a}_1)$. We can simply omit these clauses as they are always satisfied. So we already removed variables x, d, y, b_1. Moreover, a_1 now appears at most twice, so using Proposition 1 we can remove it too. So we remove at least 5 variables. And we have $(5, 5)$-branching.

Case 1.2.2. This case can be described as $|A_x| = |B_x| = 2, |A_y| = |B_y| \geq 1$
$$(\bar{x} \vee \bar{y} \vee d) \wedge (x \vee a_1 \vee a_2) \wedge (x \vee b_1 \vee b_2) \wedge (y \vee a_3 \vee A'_y) \wedge (y \vee b_3 \vee B'_y)$$

We branch on the value of x. In branch $x = 1$ by Proposition 1 we remove a_1, a_2, b_1, b_2, so we remove at least 5 variables.

In branch $x = 0$ by reduction rule 3 we can assign $y = 1$. After this by Proposition 1 we remove d, a_3, b_3, so we remove at least 5 variables. So we have $(4, 5)$-branching in this situation.

Case 2. $F = (\bar{x} \vee y \vee C'_x) \wedge \bar{y} \wedge F'$ where C'_x contains only positive literals and each $z \in C'_x$ has a unit clause $(\bar{z})$.

Note that if $|C'_x|$ has two or more variables than we can apply one of branching rules 6, 7, 8. So it is enough to consider two cases $C'_x = z$ and $C'_x = \emptyset$.

Case 2.1. $C'_x = z$
$$(\bar{x} \vee y \vee z) \wedge \bar{y} \wedge \bar{z} \wedge (x \vee A_x) \wedge (x \vee B_x) \wedge (y \vee A_y) \wedge (z \vee C)$$

Since none of branching rules 1, 4 is applicable we have that either $|A_x| = |B_x| = 1$ or $|A_x| = |B_x| = 2$. Consider two subcases.

Case 2.1.1. $|A_x| = |B_x| = 1$
$$(\bar{x} \vee y \vee z) \wedge \bar{y} \wedge \bar{z} \wedge (x \vee a) \wedge (x \vee b) \wedge (y \vee A_y) \wedge (z \vee C)$$

Branch on the value of x. In branch $x = 1$ by Proposition 1 we remove a and b. After this by reduction rule 6 we simplify the formula and then apply Proposition 1 which removes variables y and z. So we remove at least 5 variables.

In branch $x = 0$ by Corollary 1 we may assign $a = b = 1$ and using Proposition 1 we will remove y and z. So we will remove at least 5 variables. So, we have $(5, 5)$-branching.

Case 2.1.2. $|A_x| = |B_x| = 2$

$(\bar{x} \vee y \vee z) \wedge \bar{y} \wedge \bar{z} \wedge (x \vee a_1 \vee a_2) \wedge (x \vee b_1 \vee b_2) \wedge (y \vee A_y) \wedge (z \vee C)$

Branch on the value of x. In branch $x = 1$ by Proposition 1 we remove a_1, a_2, b_1, b_2. Moreover, using reduction rule 6 and again Proposition 1 we remove y and z. So we remove at least 7 variables.

In branch $x = 0$ by Proposition 1 we remove y and z. So we remove at least 3 variables. So we have $(7, 3)$-branching.

Case 2.2. $C'_x = \emptyset$

$(\bar{x} \vee y) \wedge \bar{y} \wedge (x \vee A_x) \wedge (x \vee B_x) \wedge (y \vee C)$

Since branching rules 5, 3, 2 are not applicable we can assume that one clause with literal x has exactly two literals. So the part of the formula has the following structure:

$(\bar{x} \vee y) \wedge \bar{y} \wedge (x \vee a) \wedge (x \vee B_x) \wedge (y \vee C)$

Again since branching rules 5, 3, 2 are not applicable $|B_x|$ can take only values 1 or 2. We consider these situations in Case 2.2.1 and Case 2.2.2.

Case 2.2.1. $|B_x| = 1$

$(\bar{x} \vee y) \wedge \bar{y} \wedge (x \vee a) \wedge (x \vee b) \wedge (y \vee C)$

Branch on variable x. In branch $x = 1$ by Proposition 1 we remove a and b. Moreover, using reduction rule 5 we assign $y = 1$. So we remove at least 4 variables in this branch.

In branch $x = 0$ by Proposition 1 we remove y and by Corollary 1 we assign $a = b = 1$. So we remove at least 4 variables which leads to $(4, 4)$-branching.

Case 2.2.2. $|B_x| = 2$

$(\bar{x} \vee y) \wedge \bar{y} \wedge (x \vee a) \wedge (x \vee b_1 \vee b_2) \wedge (y \vee C)$

Branch on variable y. In branch $y = 1$ by reduction rule 3 we assign $x = 1$. Now we can use Proposition 1 and remove a, b_1, b_2, C. So we remove at least 5 variables plus all new variables in C.

In branch $y = 0$ by Corollary 1 we assign $x = 0$ and $a = 1$. It means we remove at least 3 variables. So far we achieve $(5, 3)$-branching. However, we aim for a smaller branching factor, so we consider several subcases depending on the structure of $|C|$.

Case 2.2.2.1. $c \in C$, where c is a new variable, i.e. a variable different from a, b_1, b_2. In such case, we already have $(6, 3)$-branching.

So it is left to consider cases when C consists only of variables $v(a), v(b_1)$ and $v(b_2)$. C cannot be empty since reduction rule 5 is not applicable. Moreover, from Proposition 2 follows that C does not contain $\widetilde{b}_1$ and $\widetilde{b}_2$ simultaneously.

Case 2.2.2.2. $C = \widetilde{b}_1$

In branch $y = 0$ by Corollary 1 we additionally remove b_1 which leads to $(5, 4)$-branching. (Similar result holds for b_2 instead of b_1, i.e. $C = \widetilde{b}_2$.)

Case 2.2.2.3. $C = a \vee \widetilde{b}_1$

In branch $y = 0$ we have $a = 1$ so by Proposition 1 we remove b_1 which leads to $(5, 4)$-branching. (Similar result holds for b_2 instead of b_1, i.e. $C = a \vee \widetilde{b}_2$.)

Case 2.2.2.4. $C = \bar{a} \vee \widetilde{b}_1$

Again in branch $y = 0$ we have $a = 1$ and now by Corollary 1 we can assign $b_1 = 1$. So we have $(5,4)$-branching in this situation. (Similar result holds for b_2 instead of b_1.)

Case 2.2.2.5. $C = \bar{a}$

Our formula contains clauses $(\bar{x} \vee y), (x \vee a), (y \vee \bar{a})$. All these clauses are positive and, as was observed before, there is an optimum assignment that satisfies all positive clauses. And we are looking only for such assignments. Note that if $y = 0$ then we cannot satisfy all three clauses. It means that there is an optimal solution with $y = 1$ and the branching with $y = 0$ is redundant. So we reduce formula without any branching in this case.

Case 2.2.2.6. $C = a$

$(\bar{x} \vee y) \wedge \bar{y} \wedge (x \vee a) \wedge (x \vee b_1 \vee b_2) \wedge (y \vee a) \wedge (\bar{a} \vee D)$

Case 2.2.2.6.1. $c \in D$, where c is a new variable.

In this situation, we do not branch on the variable y instead we branch on the value of a. In branch $a = 1$ using reduction rule 4 two times we remove y and x. So we remove at least 3 variables.

In branch $a = 0$ by Corollary 1 we assign $x = y = 1$. Moreover, by Proposition 1 we remove b_1, b_2, c. So we remove at least 6 variables and we have $(3,6)$-branching.

From now on we can assume that D consists only of $\widetilde{b}_1$ and $\widetilde{b}_2$. Since the formula is linear D does not contain both $\widetilde{b}_1$ and $\widetilde{b}_2$.

Without loss of generality it is enough to consider cases $D = \emptyset$ and $D = \widetilde{b}_1$.

Case 2.2.2.6.2. $D = b_1$

$(\bar{x} \vee y) \wedge \bar{y} \wedge (x \vee a) \wedge (x \vee b_1 \vee b_2) \wedge (y \vee a) \wedge (\bar{a} \vee b_1)$

We branch in the same way as in the previous subcase 2.2.2.6.1. Note that we do not remove any new variables. However, in branch $a = 1$ by reduction rule 5 we assign $b_1 = 1$ which leads to $(4,5)$-branching.

Case 2.2.2.6.3. $D = \bar{b}_1$

$(\bar{x} \vee y) \wedge \bar{y} \wedge (x \vee a) \wedge (x \vee b_1 \vee b_2) \wedge (y \vee a) \wedge (\bar{a} \vee \bar{b}_1)$

Variable b_1 appears with two variables in one clause and with one in another. Since branching rules 1, 2, 3 are not applicable the third occurrence of variable b_1 is in a clause of length 2.

We consider two subcases whether the third appearance of variable b_1 is literal b_1 or literal $\bar{b}_1$. Note that this third appearance shares clause with a literal c which corresponds to a new variable that is different from variables x, y, a, b_1, b_2.

Case 2.2.2.6.3.1. $(\bar{x} \vee y) \wedge \bar{y} \wedge (x \vee a) \wedge (x \vee b_1 \vee b_2) \wedge (y \vee a) \wedge (\bar{a} \vee \bar{b}_1) \wedge (b_1 \vee c)$

Branch on value of b_1. In branch $b_1 = 1$ by Proposition 1 we remove x, b_2, c. So remove at least 4 variables.

In branch $b_1 = 0$ by reduction rule 3 we assign $a = 1$. Now we can use Proposition 1 and remove x, y. Moreover, by Corollary 1 we assign $c = 1$(since literal b_1 occur two times we conclude that clause $b_1 \vee c$ is positive). So we remove at least 5 variables which leads to $(4,5)$-branching.

Case 2.2.2.6.3.2. $(\bar{x} \vee y) \wedge \bar{y} \wedge (x \vee a) \wedge (x \vee b_1 \vee b_2) \wedge (y \vee a) \wedge (\bar{a} \vee \bar{b}_1) \wedge (\bar{b}_1 \vee c)$

Branch on the value of b_1. In branch $b_1 = 1$ by Corollary 1 we assign $c = 1$(since clause $\bar{b}_1 \vee c$ is positive as $\bar{b}_1$ occurs twice in formula). After this using Proposition 1 we remove x and b_2. So we remove at least 4 variables.

In branch $b_1 = 0$ by reduction rule 3 we assign $a = 1$. After this by Proposition 1 we can remove x, y, c. So we remove at least 5 variables which leads to $(4, 5)$-branching.

Case 2.2.2.6.4. $D = \emptyset$

$(\bar{x} \vee y) \wedge \bar{y} \wedge (x \vee a) \wedge (x \vee b_1 \vee b_2) \wedge (y \vee a) \wedge \bar{a}$

It is impossible to satisfy all 6 clauses because there are three clauses $\bar{y}, (y \vee a)$ and $\bar{a}$ that cannot be satisfied together. However we note that variables x, y, a occur only in these clauses and by assignment $a = 0, y = 1, x = 1$ we satisfy 5 clauses. So the assignment $a = 0, y = 1, x = 1$ is the best possible and we do not need any branching here.

We considered all possible cases and the worst branching is $(7, 2)$-branching. Hence, we prove the following theorem:

Theorem 1. *There is a $O^*(1.191^n)$-time algorithm for $(n, 3)$-MAXSAT.*

5   Parameterization by Number of Satisfied Clauses

In this section we consider the following problem:

---

k-(n,3)-MAXSAT

Input: A formula F in which each variable appears at most three times and integer k

Question: Is there an assignment that satisfies at least k clauses in formula F?

We show that the problem can be solved in $O^*(1.168^k)$ time, while the previous best upper bound was $O^*(1.175^k)$ [17]. We note that the result is obtained not by careful and tedious case analysis but by a combination of two algorithms. We note that speed up of any of the used algorithms will lead to an improvement of our upper bound. One of the algorithms is the algorithm from the previous section which running time is bounded in terms of the number of variables. And the second algorithm is an algorithm for maximum satisfiability parameterized beyond the number of variables [3, 10].

Before we proceed with the upper bound we present some reduction rules that simplify the input formula. First of all, we construct a bipartite variable-clause graph H_F corresponding to input formula F in the following way:

- For each variable x of F introduce a new vertex v_x.
- For each clause C of F create a new vertex v_C.
- If a variable x is in clause C then connect vertex v_x with vertex v_C by an edge.

Let us denote the set of all vertices corresponding to the variables by A and the set of all vertices corresponding to the clauses by B. We use the following well-known lemma.

Lemma 2. *For a given bipartite graph (A, B, E) in polynomial time we can either find a matching that saturates left side A or we can find a minimal inclusion-wise subset $A' \subseteq A$ such that $|N(A')| < |A'|$.*

For formula F, we denote by $\nu(F)$ the size of a maximum matching in the clause-variables graph. We consider the following auxiliary problem:

$(\nu(F) + k')$-Maximum Satisfiability
Input: A formula F and integer k'
Question: Is there an assignment that satisfies at least $\nu(F) + k'$
            clauses?

Theorem 2. *If $(\nu(F) + k')$-MAXSAT can be solved in $O^*(c_1^{k'})$ time and $(n, 3)$-MAXSAT can be solved in $O^*(c_2^n)$ time then k-$(n,3)$-MAXSAT can be solved in $O^*(c^k)$ time where $c = c_2^{\frac{\log c_1}{\log c_1 + \log c_2}}$.*

Proof. Applying Lemma 2 to graph H_F we find a set $A' \subseteq A$ such that $|N(A')| < |A'|$ or find a matching that saturates all vertices corresponding to the variables. Let us consider the first case, i.e. there is $A' \subseteq A$ such that $|N(A')| < |A'|$. It means that there is some set of variables of size $|A'|$ such that these variables occur less than in $|A'|$ clauses. Since A' is an inclusion-wise minimal subset with such properties we can delete an arbitrary vertex v from A' and then construct a perfect matching between $A' \setminus v$ and $N(A')$. It means that for each clause corresponding to the vertices from $N(A')$ we can assign a variable such that this variable appears only in clauses corresponding to the vertices from $N(A')$. Hence, we can assign values to the variables from set A' in such a way that all clauses containing them will be satisfied. So we reduce our problem to a smaller one by deleting variables corresponding to the vertices from A' and deleting clauses corresponding to the vertices from $N(A')$.

We perform this reduction until we obtain formula F' such that corresponding graph $H_{F'}$ contains a bipartite matching between the set of all variables and the set of all clauses. In the beginning, we wanted to satisfy at least k clauses in formula F, after reductions we will want to satisfy at least k' clauses in formula F' where $k' \leq k$ and $|F'| \leq |F|$. So, if we solve this problem in $O^*(c^{k'})$ time then we solve the original problem in at most $O^*(c^k)$ time. Hence, from now on without loss of generality, we assume that F is a formula such that in graph H_F there is a matching between variables and clauses that saturates vertices corresponding to variables i.e. $\nu(F) = n$.

Now we pick the fastest algorithm among algorithms solving $(\nu(F) + (k - \nu(F)))$-MAXSAT and the algorithm from Sect. 4. Recall that here $\nu(F) = n$.

Let $k = \alpha n$, then the first algorithm runs in $O^*(c_1^{k'}) = O^*(c_1^{(1-\frac{1}{\alpha})k})$ time and the second runs in $O^*(c_2^n) = O^*(c_2^{\frac{1}{\alpha}k})$ time. Since the running time of one of the algorithms is increasing in terms of α and the other is decreasing, then the maximum running time of the combined algorithm in terms of k is achieved when $c = c_1^{(1-\frac{1}{\alpha})} = c_2^{\frac{1}{\alpha}}$. Hence, $c = c_2^{\frac{\log c_1}{\log c_1 + \log c_2}}$ and the running time of the algorithm is at most $O^*(c^k)$. $\square$

Corollary 2. *k-(n,3)-MAXSAT can be solved in $O^*(1.168^k)$ time.*

Proof. Taking into account that $(\nu(F) + k')$-MAXSAT can be solved in $O^*(4^{k'})$ time [3] and (n,3)-MAXSAT can be solved in $O^*(1.191^n)$ time we have that k-(n,3)-MAXSAT can be solved in $O^*(1.168^k)$ time.   □

6   Parameterization Above All True Assignment

It is easy to see that it is possible to satisfy all clauses containing at least one positive literal. In order to do this, it is enough to assign all variables to true. It is natural to ask whether it is possible to satisfy at least k clauses more than this trivial lower bound. Such parameterization by k is called parameterization above guarantee, as we want to satisfy k clauses more than some guaranteed lower bound. Parameterization above guarantee is a natural question that was studied in [11,13,14,19]. Maximum satisfiability problem from this point of view was studied in papers [1,10,20]. Here, we show that parameterization above all true assignment is intractable. We prove that even to satisfy one more clause above all true assignment is an NP-hard problem. Hence, unless P = NP, we have that (n,3)-MAXSAT problem does not admit FPT or XP algorithm parametrized above considered lower bound.

Theorem 3. *Let F be a CNF formula in which each variable appears at most three times. Let ℓ be the number of positive clauses in F. It is NP-hard to figure out whether it is possible to satisfy simultaneously at least $\ell + 1$ clauses in F.*

Due to space constraints we omit proof here. It can be found in full version of the paper.

Acknowledgments. We would like to thank Sergey Kopeliovich for discussions in early stage of the project and Danil Sagunov as well as the anonymous reviewers for valuable comments that improved the presentation of this paper.

References

1. Alon, N., Gutin, G., Kim, E.J., Szeider, S., Yeo, A.: Solving MAX-r-SAT above a tight lower bound. Algorithmica **61**(3), 638–655 (2011)
2. Bansal, N., Raman, V.: Upper bounds for MaxSat: further improved. ISAAC 1999. LNCS, vol. 1741, pp. 247–258. Springer, Heidelberg (1999). https://doi.org/10. 1007/3-540-46632-0_26
3. Basavaraju, M., Francis, M.C., Ramanujan, M., Saurabh, S.: Partially polynomial kernels for set cover and test cover. SIAM J. Discrete Math. **30**(3), 1401–1423 (2016)
4. Battiti, R., Protasi, M.: Reactive search, a history-sensitive heuristic for MAX-SAT. J. Exp. Algorithmics (JEA) **2**, 2 (1997)
5. Berg, J., Hyttinen, A., Järvisalo, M.: Applications of MaxSAT in data analysis. Pragmatics of SAT (2015)

6. Bliznets, I., Golovnev, A.: A new algorithm for parameterized MAX-SAT. In: Thilikos, D.M., Woeginger, G.J. (eds.) IPEC 2012. LNCS, vol. 7535, pp. 37–48. Springer, Heidelberg (2012). https://doi.org/10.1007/978-3-642-33293-7_6
7. Bliznets, I.A.: A new upper bound for (n, 3)-MAX-SAT. J. Math. Sci. **188**(1), 1–6 (2013)
8. Chen, J., Kanj, I.A.: Improved exact algorithms for MAX-SAT. Discrete Appl. Math. **142**(1–3), 17–27 (2004)
9. Chen, J., Xu, C., Wang, J.: Dealing with 4-variables by resolution: an improved MaxSAT algorithm. In: Dehne, F., Sack, J.-R., Stege, U. (eds.) WADS 2015. LNCS, vol. 9214, pp. 178–188. Springer, Cham (2015). https://doi.org/10.1007/978-3-319-21840-3_15
10. Crowston, R., Gutin, G., Jones, M., Raman, V., Saurabh, S., Yeo, A.: Fixed-parameter tractability of satisfying beyond the number of variables. Algorithmica **68**(3), 739–757 (2014)
11. Crowston, R., Jones, M., Mnich, M.: Max-Cut parameterized above the edwards-erdős bound. Algorithmica **72**(3), 734–757 (2015)
12. Fomin, F.V., Kratsch, D.: Exact Exponential Algorithms. Springer, Heidelberg (2010). https://doi.org/10.1007/978-3-642-16533-7
13. Gutin, G., Kim, E.J., Lampis, M., Mitsou, V.: Vertex cover problem parameterized above and below tight bounds. Theory Comput. Syst. **48**(2), 402–410 (2011)
14. Gutin, G., Rafiey, A., Szeider, S., Yeo, A.: The linear arrangement problem parameterized above guaranteed value. Theory Comput. Syst. **41**(3), 521–538 (2007)
15. Kojevnikov, A., Kulikov, A.S.: A new approach to proving upper bounds for MAX-2-SAT. In: Proceedings of the Seventeenth Annual ACM-SIAM Symposium on Discrete Algorithm, pp. 11–17. Society for Industrial and Applied Mathematics (2006)
16. Kulikov, A.S., Kutskov, K.: New upper bounds for the problem of maximal satisfiability. Discrete Math. Appl. **19**(2), 155–172 (2009)
17. Li, W., Xu, C., Wang, J., Yang, Y.: An improved branching algorithm for $(n, 3)$-MaxSAT based on refined observations. In: Gao, X., Du, H., Han, M. (eds.) COCOA 2017 Part II. LNCS, vol. 10628, pp. 94–108. Springer, Cham (2017). https://doi.org/10.1007/978-3-319-71147-8_7
18. Lin, P.C.K., Khatri, S.P.: Application of MAX-SAT-based atpg to optimal cancer therapy design. BMC Genomics **13**(6), S5 (2012)
19. Madathil, J., Saurabh, S., Zehavi, M.: MAX-CUT ABOVE SPANNING TREE is fixed-parameter tractable. In: Fomin, F.V., Podolskii, V.V. (eds.) CSR 2018. LNCS, vol. 10846, pp. 244–256. Springer, Cham (2018). https://doi.org/10.1007/978-3-319-90530-3_21
20. Mahajan, M., Raman, V.: Parameterizing above guaranteed values: MaxSat and MaxCut. J. Algorithms **31**(2), 335–354 (1999)
21. Niedermeier, R., Rossmanith, P.: New upper bounds for maximum satisfiability. J. Algorithms **36**(1), 63–88 (2000)
22. Poloczek, M., Schnitger, G., Williamson, D.P., Van Zuylen, A.: Greedy algorithms for the maximum satisfiability problem: simple algorithms and inapproximability bounds. SIAM J. Comput. **46**(3), 1029–1061 (2017)
23. Raman, V., Ravikumar, B., Rao, S.S.: A simplified NP-complete MAXSAT problem. Inf. Processing Letters **65**(1), 1–6 (1998)
24. Walter, R., Zengler, C., Küchlin, W.: Applications of MAXSAT in automotive configuration. In: Configuration Workshop, vol. 1, p. 21 (2013)
25. Xu, C., Chen, J., Wang, J.: Resolution and linear CNF formulas: improved (n, 3)-MaxSAT algorithms. Theor. Comput. Sci. (2016)

Related Machine Scheduling
with Machine Speeds Satisfying
Linear Constraints

Siyun Zhang[1], Kameng Nip[2(✉)], and Zhenbo Wang[1]

[1] Department of Mathematical Sciences, Tsinghua University, Beijing, China
[2] School of Mathematics (Zhuhai), Sun Yat-sen University, Zhuhai, China
niejm3@mail.sysu.edu.cn

Abstract. We propose a related machine scheduling problem in which the speeds of machines are variables and must satisfy a system of linear constraints, and the processing times of jobs are given and known. The objective is to decide the speeds of machines and minimize the makespan of the schedule among all the feasible choices. The problem is motivated by some practical application scenarios. This problem is strongly NP-hard in general, and we discuss various cases of it. In particular, we obtain a polynomial time algorithm when there is one linear constraint. If the number of constraints is more than one and the number of machines is a fixed constant, then we give a $(2 + \epsilon)$-approximation algorithm. For the case where the number of machines is an input of the problem instance, we propose several approximation algorithms, and obtain a PTAS when the number of distinct machine speeds is a fixed constant.

Keywords: Related machine scheduling · Linear programming
Approximation algorithm

1 Introduction

The scheduling problem is a fundamental combinatorial optimization problem. The parallel machine scheduling problem is one of the most basic and widely-studied scheduling model. Depending on the relation between the machines and the processing times of jobs, the classic parallel machine scheduling environment is usually classified as the identical parallel machines, the (uniformly) related parallel machines, and the unrelated parallel machines. The related machine scheduling problem is described as follows: given a set of n jobs with processing times p_i for job i and m machines with speeds x_j for machine j, the actual execution time t_{ij} of job i on machine j satisfies that $p_i = x_j t_{ij}$, and the goal is to find a schedule of jobs minimizing a specific objective, e.g., the makespan or the total completion time. The identical parallel machine scheduling is a special case of this problem in which all the machines have the same speeds (and are usually assumed to be 1), and thus the actual execution time of job i is p_i throughout all the machines. The unrelated parallel machine scheduling is a generalization

© Springer Nature Switzerland AG 2018
D. Kim et al. (Eds.): COCOA 2018, LNCS 11346, pp. 314–328, 2018.
https://doi.org/10.1007/978-3-030-04651-4_21

of these problems, where the actual execution time of job i depends arbitrarily on the machine j that it is assigned to, and is usually expressed by its processing time p_{ij}.

In the classic model of scheduling problems, the parameters such as the processing times of jobs or the speeds of machines are usually deterministic and given in advance. However, it is possibly in practice that people do not have exact or enough information on the values of the processing times or machine speeds when they are facing a decision, e.g., in the stochastic scheduling problem [1] or the robust scheduling problem [2]. In some cases, these values could also be a part of the decision, which is studied in this current work.

In this paper, we investigate a related machine scheduling problem under linear constraints (RSLC for short) in which the machine speeds are a part of the decision and must satisfy a system of linear constraints, and the processing times of the jobs are known and given in advance. The goal of the problem is to determine the speed of each machine, and minimize the makespan of the schedule among all the feasible choices. The problem can be seen as a variant of the generalization of the identical parallel machine scheduling problem (SLC) studied in [3]. In the SLC problem, the processing times of jobs are decision variables satisfying several given linear inequalities, and its goal is to determine the processing time for each job, and to schedule the jobs to the machines such that the makespan is minimized. It is shown in [3] that when there is only one machine or there are at most two constraints, or the numbers of machines and constraints are fixed constants, the SLC problem is polynomially solvable. If the number of machines is a fixed constant and the number of constraints are arbitrary, then the SLC problem is NP-hard and has a PTAS. If the numbers of machines and constraints are arbitrary, then the SLC problem is strongly NP-hard and several approximation algorithms are proposed in [3]. There have also been some studies on several combinatorial optimization problems in which certain values are not given in advance and should be determined, such as the bin packing under linear constraints [4] and knapsack under linear constraints [5].

Here we give a real-world application scenario that motivates our research of the RSLC problem. Consider a manufacturer which has m machines and n jobs to be processed. The speeds of machines depend on the resources such as electricity and labor force allocated to each machine. Commonly, the larger amount of resources is allocated to one machine, the faster speed it gains. The quantities of available resources are usually limited. In many cases, these constraints can be formulated as some linear constraints of machine speeds. A concrete example is given in Table 1.

Let x_i be the speed of machine i. In the example shown in Table 1, the total speed of all machines should be at most 100, which leads to the constraint $x_1 + x_2 + \cdots + x_m \leq 100$. It is often that the higher speed of a machine leads to higher costs, for example, machine 1 costs 200 per speed in the example. The decision maker has a total budget no more than 25000, thus we have a constraint $200x_1 + 100x_2 + 300x_3 + \cdots + 400x_m \leq 25000$. The restrictions of

Table 1. An example in industrial production

	Machines					
	1	2	3	$\cdots$	m	
Total speed	1	1	1	$\cdots$	1	≤ 100
Total budget	200	100	300	$\cdots$	400	≤ 25000
$\vdots$	$\vdots$	$\vdots$	$\vdots$	$\vdots$	$\vdots$	$\vdots$
Resource A	1	2	0	$\cdots$	10	≤ 300
Resource B	0	3	4	$\cdots$	0	≤ 100
$\vdots$	$\vdots$	$\vdots$	$\vdots$	$\vdots$	$\vdots$	$\vdots$
Maximum speed for machine 1	1	0	0	$\cdots$	0	≤ 20
Minimum speed for machine 1	1	0	0	$\cdots$	0	≥ 10
$\vdots$	$\vdots$	$\vdots$	$\vdots$	$\vdots$	$\vdots$	$\vdots$

resources A, B and so on can be similarly formulated as some linear constraints, e.g., $x_1 + 2x_2 + \cdots + 10x_m \leq 300$ for resource A. Due to different restrictions of the facilities, the speed of each machine has its lower and upper bounds, e.g. $10 \leq x_1 \leq 20$. The manufacturer has to decide the speeds of all machines, which satisfy the above linear constraints, and to schedule the jobs such that they can be completed as early as possible. The whole decision can be regarded as the RSLC problem.

In this research, we discuss the computational complexity of the RSLC problem, and design polynomial time or approximation algorithms for various cases. We point out that the problem is strongly NP-hard in general. If the number of machines or the number of constraints is one, then we find an optimal solution in polynomial time. Then we give a $(2 + \epsilon)$-approximation algorithm for the case where the number of machines is fixed, which is based on the binary search and the linear programming rounding technique proposed in [6]. For the general case when the number of machines is an input of the problem, we propose an $O(m)$-approximation algorithm. Furthermore, let h be the number of different machine speeds, which is assumed to be known for the problem. For the case where h is a fixed constant, we propose an $O(h)$-approximation algorithm for some special case, and a PTAS which combines the techniques of guessing the optimal machine speeds and the PTAS of the related machine scheduling problem [7].

The rest of the paper is organized as follows: In Sect. 2, we give a formal definition of the problem studied in the paper, and discuss its computational complexity. In Sect. 3, we study some simple cases which can be solved in polynomial time. In Sect. 4, we study the case where the number of machines is fixed. We investigate several cases where the number of machines is an input of the instance in Sect. 5. Finally, we conclude our work in Sect. 6.

2   Problem Description and Complexity

First, we formally define the related machine scheduling problem under linear constraints.

Definition 1. *There are m uniformly related machines and n available jobs. Each job i has a positive processing time p_i, and each machine j has a nonnegative speed x_j. The speeds of machines are determined by k linear constraints. The goal of the related machine scheduling problem under linear constraints (RSLC) is to determine the speeds of the machines such that they satisfy the linear constraints and to assign the jobs to the machines to minimize the makespan.*

In other words, the machine speeds $x = (x_1, ..., x_m)^T$ should satisfy

$$Ax \leq b, \quad x \geq 0, \tag{1}$$

where $A \in \mathbb{R}^{k \times m}$ and $b \in \mathbb{R}^k$. Let the actual execution time of job i on machine j be t_{ij}, and then $p_i = x_j t_{ij}$. The makespan $C_{\max}$ is the completion time of the last job, i.e. let J_j be the set of jobs assigned to machine j in a schedule, the makespan of the schedule is $C_{\max} = \max_{j=1,...,m} \sum_{i \in J_j} t_{ij}$. Note that we would not assign jobs on the machines whose speeds are zero as otherwise the execution time on this machine could be arbitrarily large. Therefore, if a machine always has zero speed among all the feasible solutions, then it is without loss of generality to remove such machine. The problem can be formulated as the following mathematical program:

$$
\begin{aligned}
\min \quad & t \\
s.t. \quad & \sum_{i=1}^{n} p_i y_{ij} \leq x_j t \quad \forall j = 1, \ldots, m \\
& \sum_{j=1}^{m} y_{ij} = 1 \quad \forall i = 1, \ldots, n \\
& Ax \leq b \\
& y_{ij} \in \{0, 1\} \quad \forall i = 1, \ldots, n, \ j = 1, \ldots, m \\
& x \geq 0,
\end{aligned}
$$

where $y_{ij} = 1$ means that job i is assigned to machine j. This is a mixed integer quadratic programming [8,9], and is very hard to solve and approximate in general.

It is known that the related machine scheduling problem ($Q||C_{\max}$) is NP-hard even if there are only two machines and it is strongly NP-hard in general [10]. It can be seen that the $Q||C_{\max}$ problem is a special case of the RSLC problem, since we can set A to be $\left(\begin{smallmatrix} I_m \\ -I_m \end{smallmatrix}\right)$ where I_m is an $m \times m$ identity matrix and b to be $\left(\begin{smallmatrix} c \\ -c \end{smallmatrix}\right)$ where c denotes the vector of predetermined processing times of the jobs in $Q||C_{\max}$. Therefore, the hardness result of the $Q||C_{\max}$ problem also holds for the RSLC problem, which implies that the RSLC problem is strongly NP-hard in general, and is NP-hard even if there are only two machines and four inequality constraints.

The $Q||C_{max}$ problem has been extensively studied in the literature. Since the problem is strongly NP-hard, it is unlikely that there is a polynomial time algorithm or a fully polynomial time approximation scheme (FPTAS) unless $P = NP$. List scheduling algorithm and longest processing time algorithm are classic algorithms for the identical parallel machine scheduling $P||C_{max}$, and their performance ratios for the $Q||C_{max}$ problem have also been analyzed in [11–15]. Moreover, MULTIFIT algorithm has a better approximation ratio than LPT for $Q||C_{max}$ [16]. Based on the PTAS of $P||C_{max}$ [17], a PTAS of $Q||C_{max}$ is proposed in [7]. The special case of the $Q||C_{max}$ problem where there is at most one machine and its speed is different from the others has also been studied [12,13,18], and the algorithms often have better approximation ratios than the general case. For the more general unrelated parallel machine scheduling problem $R||C_{max}$, the authors in [6] proposed a 2-approximation algorithm, which is based on the linear programming rounding technique that will be subsequently used in our research. Readers can refer to [19] for a more detailed review of the parallel machine scheduling and related problems.

3   Several Polynomially Solvable Cases

In this section, we discuss several simple cases of the RSLC problem, where the number of machines or jobs or constraints is one. We show that all of them can be solved in polynomial time, and thus we will assume that $m \geq 2$, $n \geq 2$ and $k \geq 2$ in the subsequent sections for convenience of exposition. Note that the RSLC problem is NP-hard even if there are two machines and four inequality constraints.

3.1   Single Machine or Single Job

If there is only one machine or one job, then it is straightforward that an optimal solution must assign all the jobs to exactly one machine. Consequently, the optimal solution is to first find the largest possible speed by solving the linear programs:

$$\begin{aligned} \max \quad & x_j \\ s.t. \quad & A\boldsymbol{x} \leq \boldsymbol{b} \\ & \boldsymbol{x} \geq 0, \end{aligned}$$

for $j = 1, ..., m$ ($m = 1$ when there is only one machine), and then schedule all the jobs on the machine which has the maximum speed. This can be done in a polynomial time of k and m.

3.2   Single Constraint

In this case, the linear constraint can be written as

$$\sum_{j=1}^{m} a_j x_j \leq b, \quad \boldsymbol{x} \geq 0, \tag{2}$$

where $\boldsymbol{x} = (x_1, ..., x_m)^T$. Note that it is without loss of generality to assume that all $a_j > 0, j = 1, ..., m, b > 0$. To see this, suppose that there is an $a_j \leq 0$ for some $j \in \{1, ..., m\}$. It is clear that (2) must have feasible solution $\boldsymbol{x} = (x_1, ..., x_m)^T$ to be machine speeds. Then we can set the speed of machine j to be a sufficiently large value and the speeds of other machines keep unchanged. The resulted machine speeds still satisfy (2). Assign all the jobs to machine j, and the makespan can be arbitrarily close to zero. Thus we can assume that $a_j > 0$ for all $j = 1, ..., m$. Thus $\sum_{j=1}^{m} a_j x_j \geq 0$, which leads to that $b \geq 0$. If $b = 0$, then $\sum_{j=1}^{m} a_j x_j = 0$. This means that $\boldsymbol{x} = 0$ is the unique feasible solution to (2), which contradicts to the assumption that each machine could have nonzero speed. For the RSLC problem with single constraint, we have the following theorem.

Theorem 1. *The optimal schedule is given by assigning all the jobs in the machine l with $a_l = \min_{1 \leq j \leq m} a_j$, the optimal speeds are $x_l = b/a_l$ and $x_j = 0$ otherwise, and the makespan of the optimal schedule is $a_l \sum_{i=1}^{n} p_i / b$.*

Proof. First we show that there exists a schedule to the RSLC problem, which has makespan exactly $a_l \sum_{i=1}^{n} p_i / b$. It can be seen by simply setting $x_l = b/a_l$ and $x_j = 0$ otherwise, and assigning all the jobs to machine l.

Now, we prove that any feasible schedule must have makespan at least $a_l \sum_{i=1}^{n} p_i / b$. Suppose not, there must be a schedule whose makespan is less than $a_l \sum_{i=1}^{n} p_i / b$. Let J_j be the set of jobs assigned to machine j in the schedule and $P(J_j)$ be the total processing time of jobs assigned to machine j. For each machine j with positive speed $x_j > 0$, its completion time satisfies $P(J_j)/x_j < a_l \sum_{i=1}^{n} p_i / b$, or equivalently, $P(J_j) < x_j a_l \sum_{i=1}^{n} p_i / b$. For each machine j with zero speed $x_j = 0$, since no jobs would be assigned to such machine, it follows that $P(J_j) = 0 = x_j a_l \sum_{i=1}^{n} p_i / b$. Note that the makespan of the schedule is finite, there must be at least one machine whose speed is positive, as otherwise the makespan would be infinite. Summing up all the total processing times of assigned jobs on all the machines, we obtain

$$\sum_{i=1}^{n} p_i = \sum_{j=1}^{m} P(J_j) < \frac{a_l \sum_{i=1}^{n} p_i}{b} \sum_{j=1}^{m} x_j \leq \frac{\sum_{i=1}^{n} p_i}{b} \sum_{j=1}^{m} a_j x_j \leq \sum_{i=1}^{n} p_i,$$

where the second last inequality holds since $a_l = \min_{1 \leq j \leq m} a_j$, and the last inequality holds from the constraint (2). This leads to a contradiction. Therefore, the schedule described above with makespan $a_l \sum_{i=1}^{n} p_i / b$ is best possible. $\square$

Theorem 1 shows that although the RSLC problem is strongly NP-hard in general, it can be solved in polynomial time when the number of constraint is one.

4   Fixed Number of Machines

In this section, we consider the case where the number of machines is a fixed constant, and the number of constraints is at least two and is an input of instance. As mentioned in Sect. 2, this case is NP-hard.

Suppose that the value of makespan T of a schedule is fixed. We denote I_j as the set of jobs that can be processed on machine j with execution time at most T, i.e. $I_j = \{i \in \{1, ..., n\} | p_i \le x_j T\}$ for each machine $j = 1, ..., m$. Consider the following linear program:

$$
\begin{array}{ll}
\sum_{i \in I_j} p_i y_{ij} \le x_j T & \forall j = 1, \dots, m \\
\sum_{j \in K_i} y_{ij} = 1 & \forall i = 1, \dots, n \\
(\text{LP1}) \quad A\boldsymbol{x} \le \boldsymbol{b} & \\
p_i \le x_j T & \forall i \in I_j, j = 1, \dots, m \\
\boldsymbol{x}, y_{ij} \ge 0 & \forall i \in I_j, j = 1, \dots, m
\end{array}
$$

in which $K_i = \{j | i \in I_j\}$ is the set of machines that job i can be processed by with execution time at most T. Note that for any fixed x_j and T, if $i \in I_j$, then all the jobs with processing times no more than p_i must belong to I_j. Therefore, each I_j has $n + 1$ possible job sets. We can determine all the I_js in $O(n^m)$ enumerations, which is a polynomial to n if the number of machines m is a fixed number.

We design an algorithm based on the binary search of T. The algorithm is summarized as Algorithm 1. Before describing our algorithm, we define X_0 as the optimal value of the following linear program:

$$
\begin{array}{ll}
& \max \sum_{i=1}^{m} x_i \\
(\text{LP2}) & \\
& s.t. \quad A\boldsymbol{x} \le \boldsymbol{b} \\
& \qquad \boldsymbol{x} \ge 0.
\end{array}
$$

It can be seen that the range of the minimal makespan $C_{\max}^*$ is that $C_{\max}^* \in \left[\frac{\sum_{i=1}^n p_i}{X_0}, m \frac{\sum_{i=1}^n p_i}{X_0} \right]$.

The following theorem indicates that Algorithm 1 can always return a solution if the RSLC problem is feasible.

Lemma 1. *If (LP1) is feasible, then (LP3) in Step 9 of Algorithm 1 must be feasible.*

Proof. Suppose (LP1) is feasible fixing $T, I_1, \dots, I_m$. One of its solutions is $\boldsymbol{x}, y_{ij}, i \in I_j, j = 1, \dots, m$. Then it can be seen that $\boldsymbol{x}, y_{ij}, i \in I_j, j = 1, \dots, m'$ is also a feasible solution to the following linear program (LP4):

$$
\begin{array}{ll}
& \sum_{i \in I_j} p_i y_{ij} \le x_j T & \forall j = 1, \dots, m' \\
(\text{LP4}) & \sum_{j \in K_i} y_{ij} = 1 & \forall i = 1, \dots, n \\
& y_{ij} \ge 0 & \forall i \in I_j, j = 1, \dots, m'.
\end{array}
$$

Algorithm 1. Approximation algorithm for fixed m

1: Set $L = \frac{\sum_{i=1}^{n} p_i}{X_0}$, $U = m \frac{\sum_{i=1}^{n} p_i}{X_0}$.
2: $q = 0$.
3: **while** $q < \log \frac{2(m-1)}{\epsilon}$ **do**
4:     $T = \frac{1}{2}(L + U)$.
5:     **for** each possible $I_1, \ldots, I_m$ **do**
6:         Find a feasible solution to (LP1).
7:         **if** (LP1) is feasible **then**
8:             Let $\boldsymbol{x}$ be the solution found in Step 6, and remove the machines whose speeds are 0. Let m' be the total number of the remaining machines. Suppose the remaining machines is machine 1 to m' for simplicity.
9:             Find a basic feasible solution to the following (LP3):

$$(\text{LP3}) \quad \begin{array}{ll} \sum_{i \in I'_j} p_i y_{ij} \le x_j T & \forall j = 1, \ldots, m' \\ \sum_{j \in K'_i} y_{ij} = 1 & \forall i = 1, \ldots, n \\ y_{ij} \ge 0 & \forall i \in I'_j, j = 1, \ldots, m' \end{array}$$

in which $I'_j = \{i | p_i \le x_j T\}$ and $K'_i = \{j | p_i \le x_j T\}$.
10:            Use the rounding technique proposed in [6] to find a feasible schedule, and record its makespan.
11:            Set $U = T$, and break from "for" loop.
12:        **if** $U \ne T$ **then**
13:            Set $L = T$.
14:    Set $q = q + 1$, and move to the next iteration.
15: If there is no feasible schedule returned by the previous iterations, then set $T = U$, and run Step 5-11 again to obtain a feasible schedule.
16: **return** the schedule with minimum makespan obtained among all the iterations.

Since $I_j \subset I'_j$ and $K_i \subset K'_i$, we can set

$$y'_{ij} = \begin{cases} y_{ij}, i \in I_j, j = 1, \ldots, m' \\ 0, \text{otherwise}, \end{cases}$$

and it is clear that $y'_{ij}, i \in I'_j, j = 1, \ldots, m'$ is the feasible solution of (LP3). This completes the proof.   □

Although the value of L could be changed in Step 13, the following lemma shows that it always provides a lower bound of the problem.

Lemma 2. *The value L is always the lower bound of $C^*_{\max}$ during the execution of Algorithm 1. Moreover, the value $(m \sum_{i=1}^{n} p_i)/X_0$ is an upper bound of $C^*_{\max}$.*

Proof. Let $\boldsymbol{x}^*$ be the optimal machine speed vector, and $\boldsymbol{x}^* = (x_1^*, \ldots, x_m^*)^T$. Then $C^*_{\max} \ge \frac{\sum_{i=1}^{n} p_i}{\sum_{i=1}^{m} x_i^*} \ge \frac{\sum_{i=1}^{n} p_i}{X_0}$, which means the initial L is a lower bound of $C^*_{\max}$.

If L is changed during the execution of Algorithm 1, then for all possible $I_1, \ldots, I_m$, (LP1) is infeasible. This means that there cannot be any schedule with makespan no more than L. If so, suppose $\tilde{x}$ and $\tilde{y}_{ij}, i = 1, \ldots, n, j = 1, \ldots, m$ is such schedule. Thus the following equations hold:

$$
\begin{aligned}
\sum_{i=1}^{n} p_i \tilde{y}_{ij} &\le \tilde{x}_j L & \forall j = 1, \ldots, m \\
\sum_{j=1}^{m} \tilde{y}_{ij} &= 1 & \forall i = 1, \ldots, n \\
A\tilde{x} &\le b \\
\tilde{x} &\ge 0, \tilde{y}_{ij} \ge 0 & \forall i = 1, \ldots, n, j = 1, \ldots, m.
\end{aligned}
$$

Set $\tilde{I}_j = \{i | p_i \le \tilde{x}_j L\}$ and $\tilde{K}_i = \{j | p_i \le \tilde{x}_j L\}$. It is clear that $\tilde{y}_{ij} = 0$ for $i \notin \tilde{I}_j, j = 1, \ldots, m$. Thus $\sum_{i \in \tilde{I}_j} p_i \tilde{y}_{ij} \le \tilde{x}_j L$ and $\sum_{j \in \tilde{K}_i} \tilde{y}_{ij} = 1$. Furthermore, the following linear program is feasible:

$$
\begin{aligned}
\sum_{i \in \tilde{I}_j} p_i y_{ij} &\le x_j L & \forall j = 1, \ldots, m \\
\sum_{j \in \tilde{K}_i} y_{ij} &= 1 & \forall i = 1, \ldots, n \\
Ax &\le b \\
p_i &\le x_j L & \forall i \in \tilde{I}_j, j = 1, \ldots, m \\
x &\ge 0, y_{ij} \ge 0 & \forall i \in \tilde{I}_j, j = 1, \ldots, m.
\end{aligned}
$$

It contradicts to the fact that for all possible $I_1, \ldots, I_m$, (LP1) is infeasible. Hence, L is always the lower bound of $C^*_{\max}$.

For the upper bound, let the optimal solution $\bar{x} = (\bar{x}_1, \ldots, \bar{x}_m)$ to (LP2) be the machine speeds, and $\bar{C}_{\max}$ be the makespan of the schedule—assigning all jobs to the fastest machine whose speed is denoted by $\bar{x}_{\max}$. Since $\bar{x}_{\max} \ge \frac{X_0}{m}$, we have $C^*_{\max} \le \bar{C}_{\max} = \frac{\sum_{i=1}^{n} p_i}{\bar{x}_{\max}} \le m \frac{\sum_{i=1}^{n} p_i}{X_0}$.   □

Applying the lemmas above, we have the following property:

Theorem 2. *Given $0 < \epsilon < 1$, Algorithm 1 is a $(2 + \epsilon)$-approximation algorithm for the RSLC problem with fixed number of machines.*

Proof. First we discuss the computational complexity of Algorithm 1. Solving the linear program (LP2) requires $O((k + m)^3 N)$ operations, where N is the length of data. Fixing ϵ, Step 3 requires $\lceil \log \frac{2(m-1)}{\epsilon} \rceil$ iterations. Note that there are $n+1$ possible I_js for each machine j, and then Step 5 requires at most $O(n^m)$ enumerations. Determining the feasibility of (LP1), finding its feasible solution and finding a basic feasible solution of (LP3) all require polynomial time. And the rounding process in Step 10 also requires polynomial time [6]. By the fact that m and ϵ are fixed, the running time of Algorithm 1 is a polynomial on the input instance.

Next we prove that the returned schedule has a makespan no more than $2 + \epsilon$ of the optimal makespan. If U is changed during the execution of the algorithm, then the makespan of the returned schedule is no larger than $2\hat{U}$ by [6], where $\hat{U}$ denotes the final value of U. If U is not changed, which means $\hat{U} = m \frac{\sum_{i=1}^{n} p_i}{X_0}$, then there exists a schedule whose makespan is no more than $\hat{U}$ because $m \frac{\sum_{i=1}^{n} p_i}{X_0}$

is an upper bound of $C_{\max}^*$ by Lemma 2. There must exist some $I_1, \ldots, I_m$ such that (LP1) is feasible, and so the makespan of the returned schedule is also no larger than $2\hat{U}$ by [6].

Let $C_{\max}$ be the returned makespan. Since $m \geq 2$ and $0 < \epsilon < 1$, then $\log \frac{2(m-1)}{\epsilon} \geq 1$. Therefore, the while loop in Step 3 must be done at least once. Thus

$$\hat{U} - \hat{L} \leq 2^{-\log \frac{2(m-1)}{\epsilon}} \left(m \frac{\sum_{i=1}^n p_i}{X_0} - \frac{\sum_{i=1}^n p_i}{X_0} \right) = \frac{\epsilon}{2} \frac{\sum_{i=1}^n p_i}{X_0} \leq \frac{\epsilon}{2} C_{\max}^*,$$

the last inequality holds since L is a lower bound of $C_{\max}^*$ by Lemma 2. Therefore, we have

$$C_{\max} \leq 2\hat{U} = 2(\hat{U} - \hat{L} + \hat{L}) \leq 2 \left(\frac{\epsilon}{2} C_{\max}^* + C_{\max}^* \right) = (2 + \epsilon) C_{\max}^*.$$

Hence Theorem 2 holds.   □

5   Arbitrary Number of Machines

In this section, we discuss the case where the number of machines m is an input of the instance. Assume we know that there are at most h different types of machines for the problem, where the speeds of machines within each group are identical, as well as which type each machine belongs to (if $h = m$, then we do not have any additional information and this leads to the normal RSLC problem). Each type consists of $m_i (i = 1, ..., h)$ machines. In other words, the number of distinct values of machine speeds of the problem is at most h. Let s_i be the speed of machines in group i, and $\boldsymbol{s} = (s_1, ..., s_h)$. The linear constraints (1) can be reformulated as $A\boldsymbol{s} \leq \boldsymbol{b}$, where $A \in \mathbb{R}^{k \times h}, \boldsymbol{b} \in \mathbb{R}^k$ (we use the same notations A and $\boldsymbol{b}$ for simplicity). Let $\boldsymbol{s}^* = (s_1^*, ..., s_h^*)$ denote the optimal solution, and $C_{\max}^*$ denote the optimal makespan.

5.1   Approximation Algorithm for Arbitrary h

We first propose an approximation algorithm for the general case of the RSLC problem, i.e., when the number of different machine types h is arbitrary. Let $m_{\min} = \min_{1 \leq i \leq h} \{m_i\}$ be the minimum number of machines in a type. The algorithm can be shown in Algorithm 2.

Theorem 3. *Algorithm 2 is a* $\left(1 + \frac{m}{m_{\min}} - \frac{1}{m_{\min}}\right)$-*approximation algorithm for arbitrary m and h.*

Proof. First we consider the computational complexity. Solving (LP6) requires $O((k + h)^3 N)$. The computational complexity of scheduling is $O(mn)$. Hence, Algorithm 2 is a polynomial time algorithm.

Algorithm 2. Approximation algorithm for arbitrary m and h

1: Solve the linear program:

$$\text{(LP6)} \qquad \begin{aligned} \max\ & \sum_{i=1}^{h} m_i s_i \\ s.t.\ & As \leq b \\ & s \geq 0. \end{aligned}$$

2: Let machine speeds be the solution obtained by the above linear program, and schedule by list scheduling algorithm [12].

Let $\hat{s} = (\hat{s}_1, \ldots, \hat{s}_h)$ be the optimal solution to (LP6), and $C_{\max}$ be the makespan returned by the algorithm. Note that $\sum_{i=1}^{h} m_i \hat{s}_i \geq \sum_{i=1}^{h} m_i s_i^*$. Denote job q as the last finished job in the schedule returned by the algorithm. By the property of list scheduling algorithm, we have

$$
\begin{aligned}
C_{\max} \ &\leq\ \frac{\sum_{i=1}^{n} p_i}{\sum_{i=1}^{h} m_i \hat{s}_i} + \frac{m-1}{\sum_{i=1}^{h} m_i \hat{s}_i} p_q \\
&\leq\ \frac{\sum_{i=1}^{n} p_i}{\sum_{i=1}^{h} m_i s_i^*} + \frac{m-1}{\sum_{i=1}^{h} m_i s_i^*} p_q \\
&\leq\ \frac{\sum_{i=1}^{n} p_i}{\sum_{i=1}^{h} m_i s_i^*} + \frac{m-1}{m_{\min} \sum_{i=1}^{h} s_i^*} p_q.
\end{aligned}
$$

Let $s_{\max}^* = \max_{1 \leq i \leq h}\{s_i^*\}$. It is clear that $\sum_{i=1}^{n} p_i / \sum_{i=1}^{h} m_i s_i^* \leq C_{\max}^*$ and $p_q / \sum_{i=1}^{h} s_i^* \leq p_q / s_{\max}^* \leq C_{\max}^*$. Therefore,

$$
C_{\max} \leq \left(1 + \frac{m}{m_{\min}} - \frac{1}{m_{\min}}\right) C_{\max}^*.
$$

$\square$

Note that $m_{\min} = 1$ in the worst case or $h = m$, thus Algorithm 2 is an m-approximation algorithm for the RSLC problem in general.

5.2 Approximation Algorithms for Fixed h

In this section, we consider the case when h is a fixed constant. If $m_1 = \cdots = m_h$, then we have $m_{\min} = \frac{m}{h}$, and Algorithm 2 is an $\left(h + 1 - \frac{h}{m}\right)$-approximation algorithm. Below we show that we can obtain an approximation algorithm with better approximation ratio $(h + \epsilon)$. The detail of the algorithm is summarized as Algorithm 3.

Theorem 4. *Given any $\epsilon > 0$, Algorithm 3 is an $(h + \epsilon)$-approximation algorithm for arbitrary m with fixed h and $m_1 = \cdots = m_h$.*

Algorithm 3. Approximation algorithm for arbitrary m with fixed h and $m_1 = \cdots = m_h$

1: Solve a series of linear programs for $i = 1, ..., h$:

$$\text{(LP7)} \quad \begin{aligned} \max \quad & s_i \\ \text{s.t.} \quad & A\mathbf{s} \le \mathbf{b} \\ & \mathbf{s} \ge 0. \end{aligned}$$

Let the optimal values are $\tilde{s}_i, i = 1, ..., h$. Set $\tilde{s}_{\max} = \max_{1 \le i \le h} \tilde{s}_i$, and suppose that $\tilde{s}_{\max}$ is obtained by the j_0th linear program, i.e. $\tilde{s}_{\max} = \tilde{s}_{j_0}$.

2: Schedule all the jobs on the j_0th type of machines by the PTAS of identical machine scheduling problem $P||C_{\max}$ [17].

Proof. First we consider the computational complexity of the algorithm. Solving linear programs in Step 1 requires $O(h(k + h)^3 N) = O(k^3 N)$ operations. Obtaining the schedule in Step 2 needs polynomial time. Thus the total running time of the algorithm is polynomial time.

Let $C_{\max}$ be the makespan of the schedule returned by the algorithm, and $C_{\max}^{(1)}$ be the minimum makespan among the schedules where jobs are only scheduled on the j_0th type machines, i.e., the group of machines which has speed $\tilde{s}_{\max}$ in Step 1. Since we apply a PTAS of $P||C_{\max}$ in Step 2, we have $C_{\max} \le (1 + \epsilon/h)C_{\max}^{(1)}$ for any $\epsilon > 0$.

Let $C_{\max}^*$ be the makespan of the optimal schedule. Consider the schedule that we move the jobs from the machines of all other types to the machine of type j_0 with the same order, that is, we move the jobs scheduled in the kth machine of any other types on the optimal solution to the kth machine of type j_0, for each $k = 1, \ldots, m/h$. Moreover, the speed of type j_0 machines in this schedule is set to be $\tilde{s}_{\max}$ obtained in Step 1. It can be seen that the makespan of this schedule is at most $hC_{\max}^*$, since $m_1 = \cdots = m_h = m/h$ and the machines in type j_0 have the maximum speed, i.e., $\tilde{s}_{\max} \ge \tilde{s}_i \ge s_i^*$ for all $i = 1, \ldots, h$. By definition, $C_{\max}^{(1)}$ is no more than the makespan of this schedule, hence $C_{\max} \le (1 + \epsilon/h)C_{\max}^{(1)} \le h(1 + \epsilon/h)C_{\max}^* = (h + \epsilon)C_{\max}^*$. □

Next we propose a PTAS for fixed h. Let M_i denote the set of machines with speed s_i. First we show that if the machines in M_i are not all empty(here empty means no jobs are assigned to that machine) in the optimal schedule, then s_i^* has a positive lower bound.

The optimal values $\tilde{s}_i, i = 1, ..., h$ of the linear programs (LP7) are the largest possible values for $s_i, i = 1, ..., h$. We assume w.l.o.g. that $\tilde{s}_i > 0$ for all $i = 1, ..., h$ as otherwise we can remove such machine.

Lemma 3. *If the machines in M_i are not all empty in the optimal schedule, then $s_i^* \ge \tilde{s}_i P_{\min} / \sum_{j=1}^n p_j$, where $P_{\min} = \min_{1 \le i \le n} p_i$.*

Proof. The makespan of assigning all jobs on the machine having speed $\tilde{s}_i$ is $\sum_{j=1}^n p_j / \tilde{s}_i$. Thus $\sum_{j=1}^n p_j / \tilde{s}_i \ge C_{\max}^*$. Since there is at least one machine in

M_i that is not empty, we have $P_{\min}/s_i^* \leq C_{\max}^*$. It follows that $P_{\min}/s_i^* \leq \sum_{j=1}^{n} p_j/\tilde{s}_i$, and hence $s_i^* \geq \tilde{s}_i P_{\min}/\sum_{j=1}^{n} p_j$.   □

Lemma 3 implies that as long as a machine is used in the optimal schedule, its optimal speed has an explicit lower bound. Based on this fact, we design a PTAS and describe it in Algorithm 4.

Theorem 5. *Algorithm 4 is a PTAS for the RSLC problem with arbitrary m and fixed h.*

Algorithm 4. Approximation algorithm for arbitrary m and fixed h

---

1: Solve the linear programs (LP7) ($i = 1, ..., h$), whose optimal solutions are denoted by $\tilde{s}_i, i = 1, ..., h$.
2: Set $U_i = \tilde{s}_i$, $L_i = \tilde{s}_i P_{\min}/\sum_{j=1}^{n} p_j (i = 1, ..., h)$. Let $r = \lceil \log(\sum_{j=1}^{n} p_j/P_{\min})/\log(1 + \frac{\epsilon}{3}) \rceil$, and divide $[L_i, U_i]$ into $B_r^i = \left[L_i, \frac{U_i}{(1+\frac{\epsilon}{3})^{r-1}} \right]$,
$B_{r-1}^i = \left[\frac{U_i}{(1+\frac{\epsilon}{3})^{r-1}}, \frac{U_i}{(1+\frac{\epsilon}{3})^{r-2}} \right], ..., B_1^i = \left[\frac{U_i}{1+\frac{\epsilon}{3}}, U_i \right]$.
3: **for** each subset S^d of $\{1, ..., h\}$ **do**
4:     **for** each possible combination of $l_i \in \{1, ..., r\}, \forall i \in S^d$ **do**
5:         Find a feasible solution of the following linear program:

$$
\begin{aligned}
& As \leq b \\
(\text{LP8}) \quad & s \geq 0 \\
& s_i \in B_{l_i}^i, \forall i \in S^d.
\end{aligned}
$$

6:     **if** (LP8) is feasible **then**
7:         Let the machine speed vector s be the solution found in (LP8). Use the PTAS for scheduling on uniform processors in [7] to find a schedule.
8: **return** the schedule with the smallest makespan among all generated results and its corresponding machine speeds.

Proof. First we consider the running time of Algorithm 4. Solving the linear programs in Step 1 requires $O(h(h+k)^3 N) = O(k^3 N)$ time. Set $|S^d| = h_d$. Step 3 requires $2^h - 1$ iterations, while Step 4 requires r^{h_d} iterations. Since $h_d \leq h$, there are totally less than $2^h r^h = O(r^h)$ iterations. Determining the feasibility of (LP8), solving (LP8) and scheduling can be done in polynomial time. Since h is a constant and r is polynomial, the total computational complexity of Algorithm 4 is a polynomial of the input instance.

Then we prove that Algorithm 4 returns a solution whose makespan is no more than $1 + \epsilon$ of the minimum makespan $C_{\max}^*$. There must be an iteration in which s^* satisfies (LP8). Among these iterations, we consider the iteration that S^d is exactly the set of nonempty machine types (which have nonzero speeds by Lemma 3) in this optimal solution. Let s be the solution obtained by that iteration, and $C_{\max}$ be the makespan obtained by that iteration.

Note that $n \geq 2$, we must have $r \geq 1$. For each $i \in S^d$, if $l_i = r$, where r is defined in Step 2, then $s_i \geq L_i$. Since

$$r \geq \log \left(\sum_{i=1}^{n} p_i / P_{\min} \right) \Big/ \log \left(1 + \frac{\epsilon}{3} \right) = \log_{1+\frac{\epsilon}{3}} \left(\sum_{i=1}^{n} p_i / P_{\min} \right),$$

we get

$$\left(1 + \frac{\epsilon}{3} \right)^{l_i} \geq \left(1 + \frac{\epsilon}{3} \right)^{\log_{1+\frac{\epsilon}{3}} \left(\sum_{i=1}^{n} p_i / P_{\min} \right)} = \sum_{i=1}^{n} p_i / P_{\min} = U_i / L_i.$$

Thus $L_i \geq U_i / (1 + \frac{\epsilon}{3})^{l_i}$, which means $s_i \geq U_i / (1 + \frac{\epsilon}{3})^{l_i}$. If $l_i < r$, then $s_i \geq U_i / (1 + \frac{\epsilon}{3})^{l_i}$. Therefore,

$$s_i \geq \frac{U_i}{\left(1 + \frac{\epsilon}{3} \right)^{l_i}} = \frac{U_i}{\left(1 + \frac{\epsilon}{3} \right)^{l_i - 1}} \frac{1}{\left(1 + \frac{\epsilon}{3} \right)} \geq \frac{s_i^*}{\left(1 + \frac{\epsilon}{3} \right)}, \forall i \in S^d.$$

Let $\tilde{C}_{\max}$ be the makespan of the schedule where jobs are scheduled exactly the same as the optimal schedule under machine speed vector s. It can be observed that $\tilde{C}_{\max} \leq \left(1 + \frac{\epsilon}{3} \right) C_{\max}^*$, since we assume that S^d is the same as the optimal solution. Let $\tilde{C}_{\max}^*$ be the minimum makespan of the schedule under speed s. Since Step 7 returns a solution by the PTAS algorithm, we have

$$C_{\max} \leq \left(1 + \frac{\epsilon}{3} \right) \tilde{C}_{\max}^* \leq \left(1 + \frac{\epsilon}{3} \right) \tilde{C}_{\max} \leq \left(1 + \frac{\epsilon}{3} \right)^2 C_{\max}^*.$$

Note that the makespan returned by Algorithm 4 is no larger than $C_{\max}$. By adjusting an appropriate value, we can find an $\epsilon' > 0$ such that $C_{\max} \leq (1 + \epsilon'/3)^2 C_{\max}^* < (1 + \epsilon) C_{\max}^*$. Therefore, the algorithm is a PTAS and Theorem 5 holds. $\qquad \square$

6   Conclusion

In this paper, we propose a related machine scheduling problem where machine speeds satisfy a series of linear constraints. We discuss the computational complexity of the problem, and propose different polynomial time algorithms and approximation algorithms for various cases of the problem. A further direction of this work is to improve the approximation algorithms for different cases, e.g. to propose approximation algorithm with performance ratio better than $2 + \epsilon$ for the case where the number of machines is fixed, or to consider whether there are approximation algorithms with constant factor for the general case with arbitrary number of machines. Moreover, other forms of constraints of the problem, such as convex constraints, quadratic constraints, or other combinatorial optimization problems under constraints are also worth considering.

Acknowledgments. This work has been supported by NSFC No. 11801589, No. 11771245 and No. 11371216. We also thank Tianning Shi for helpful discussions on this work.

References

1. Möhring, R.H., Radermacher, F.J., Weiss, G.: Stochastic scheduling problems I-general strategies. Zeitschrift für Oper. Res. **28**(7), 193–260 (1984)
2. Daniels, R.L., Kouvelis, P.: Robust scheduling to hedge against processing time uncertainty in single-stage production. Manag. Sci. **41**(2), 363–376 (1995)
3. Nip, K., Wang, Z., Wang, Z.: Scheduling under linear constraints. Eur. J. Oper. Res. **253**(2), 290–297 (2016)
4. Wang, Z., Nip, K.: Bin packing under linear constraints. J. Comb. Optim. **34**(4), 1198–1209 (2017)
5. Nip, K., Wang, Z., Wang, Z.: Knapsack with variable weights satisfying linear constraints. J. Glob. Optim. **69**(3), 713–725 (2017)
6. Lenstra, J.K., Shmoys, D.B., Tardos, É.: Approximation algorithms for scheduling unrelated parallel machines. Math. Program. **46**(1), 259–271 (1990)
7. Hochbaum, D.S., Shmoys, D.B.: A polynomial approximation scheme for scheduling on uniform processors: using the dual approximation approach. SIAM J. Comput. **17**(3), 539–551 (1988)
8. Burer, S., Letchford, A.N.: Non-convex mixed-integer nonlinear programming: a survey. Surv. Oper. Res. Manag. Sci. **17**(2), 97–106 (2012)
9. Köppe, M.: On the complexity of nonlinear mixed-integer optimization. In: Lee, J., Leyffer, S. (eds.) Mixed Integer Nonlinear Programming. The IMA Volumes in Mathematics and its Applications, vol. 154, pp. 533–557. Springer, New York (2012). https://doi.org/10.1007/978-1-4614-1927-3_19
10. Garey, M.R., Johnson, D.S.: Computers and Intractability: A Guide to the Theory of NP-Completeness. W.H. Freeman, San Francisco (1979)
11. Dobson, G.: Scheduling independent tasks on uniform processors. SIAM J. Comput. **13**(4), 705–716 (1984)
12. Cho, Y., Sahni, S.: Bounds for list schedules on uniform processors. SIAM J. Comput. **9**(1), 91–103 (1980)
13. Gonzalez, T., Ibarra, O.H., Sahni, S.: Bounds for LPT schedules on uniform processors. SIAM J. Comput. **6**(1), 155–166 (1977)
14. Friesen, D.K.: Tighter bounds for LPT scheduling on uniform processors. SIAM J. Comput. **16**(3), 554–560 (1987)
15. Kovács, A.: New approximation bounds for LPT scheduling. Algorithmica **57**(2), 413–433 (2010)
16. Kunde, M.: A multifit algorithm for uniform multiprocessor scheduling. In: Cremers, A.B., Kriegel, H.-P. (eds.) GI-TCS 1983. LNCS, vol. 145, pp. 175–185. Springer, Heidelberg (1982). https://doi.org/10.1007/BFb0036479
17. Hochbaum, D.S., Shmoys, D.B.: Using dual approximation algorithms for scheduling problems: theoretical and practical results. J. ACM **34**(1), 144–162 (1987)
18. Kovács, A.: Tighter approximation bounds for LPT scheduling in two special cases. In: Calamoneri, T., Finocchi, I., Italiano, G.F. (eds.) CIAC 2006. LNCS, vol. 3998, pp. 187–198. Springer, Heidelberg (2006). https://doi.org/10.1007/11758471_20
19. Chen, B., Potts, C.N., Woeginger, G.J.: A review of machine scheduling: complexity, algorithms and approximability. In: Du, D., Pardalos, P.M. (eds.) Handbook of Combinatorial Optimization: Volume 1–3, pp. 1493–1641. Springer, Boston (1998). https://doi.org/10.1007/978-1-4613-0303-9_25

Open-Shop Scheduling for Unit Jobs
Under Precedence Constraints

An Zhang[1,2], Yong Chen[1,2], Randy Goebel[2], and Guohui Lin[2(✉)]

[1] Department of Mathematics, Hangzhou Dianzi University, Hangzhou, China
{anzhang,chenyong}@hdu.edu.cn
[2] Department of Computing Science, University of Alberta, Edmonton, AB, Canada
{rgoebel,guohui}@ualberta.ca

Abstract. We study open-shop scheduling for unit jobs under precedence constraints, where if one job precedes another job then it has to be finished before the other job can start to be processed. For the three-machine open-shop to minimize the makespan, we first present a simple 5/3-approximation based on a partition of the job set into agreeable layers using the natural layered representation of the precedence graph. We then show a greedy algorithm to reduce the number of singleton-job layers, resulting in an improved partition, which leads to a 4/3-approximation. Both approximation algorithms apply to the general m-machine open-shops too.

Keywords: Open-shop scheduling · Precedence constraint
Directed acyclic graph · Approximation algorithm

1 Introduction

Machine scheduling with precedence constraints on the jobs has received much attention in the past few decades, and several algorithmic techniques such as the *critical path method* and the *project evaluation and review technique* [9] have been developed from the line of research. Job precedence constraints are common in construction and manufacturing industries, for example, the bicycle assembly problem is an earliest precedence constrained scheduling application introduced by Graham [7].

Precedence constraints describe the job processing order in a way that one or more jobs have to be finished before another job is allowed to start its processing. Such relationships together are usually represented as a *directed acyclic graph* (DAG) $G = (V, E)$, called the *precedence graph*, where V is the set of jobs and an edge $(v_i, v_j) \in E$ states that the job v_i precedes the job v_j, that is, v_i needs to be finished before v_j can start to be processed.

In this paper, we discuss the open-shop scheduling environment and use Om to denote the m-machine open-shop for some constant m, and O to denote the open-shop in which the number of machines is part of the input. In either Om or O, every job needs to be processed non-preemptively by each machine, in any

© Springer Nature Switzerland AG 2018
D. Kim et al. (Eds.): COCOA 2018, LNCS 11346, pp. 329–340, 2018.
https://doi.org/10.1007/978-3-030-04651-4_22

machine order, and it is *finished* (or said *completed*) when it has been processed by all the machines. Note that the usual scheduling rules apply to a feasible schedule, that is, at any time point, a job can be processed by at most one machine and each machine can be processing at most one job. The makespan of the schedule is the maximum job completion time. The open-shop scheduling to minimize the makespan is denoted as $Om \parallel C_{\max}$ or $O \parallel C_{\max}$, which has received much study [6,9,11,12,15]. In particular, $O2 \parallel C_{\max}$ is solvable in $O(n)$-time, where n denotes the number of jobs [6,9]; $Om \parallel C_{\max}$ becomes weakly NP-hard when $m \geq 3$ [6] but admits a polynomial-time approximation scheme (PTAS) [11,12]; $O \parallel C_{\max}$ is strongly NP-hard and cannot be approximated within 1.25 [15].

Open-shop scheduling with precedence constraints, denoted as $Om \mid prec \mid C_{\max}$ or $O \mid prec \mid C_{\max}$, is more difficult than its classical counterparts, which can be considered as scheduling without precedence constraints. Several special classes of precedence graphs have been investigated in the literature. If every job has at most one predecessor and at most one successor, the precedence graph is referred to as *chains*. If every job has at most one successor (one predecessor, respectively), the precedence graph is referred to as an *intree* (an *outtree*, respectively). The fact that the precedence graph belongs to a particular class may change the computational complexity of the scheduling problem. In general, one can expect that the precedence constraints increase the problem complexity. For example, $O2 \mid chains \mid C_{\max}$ becomes NP-hard [13]. For more complexity results on precedence constrained scheduling, the interested readers can refer to Lenstra and Rinnooy Kan [8], or Prot and Bellenguez-Morinea [10].

Unlike most past results which are on computational complexity, in this paper we aim to develop algorithmic positive results for open-shop scheduling with precedence constraints, from the approximation algorithm perspective. We focus on the problems restricted to unit jobs, that is, the jobs have the same processing times on all the machines (*i.e.*, $p_{ij} = 1$); most of these problems remain NP-hard, or their complexity are still open. To name a few, for an arbitrary precedence graph, the problem $O \mid p_{ij} = 1, prec \mid C_{\max}$ was shown to be strongly NP-hard by Timkovsky [14]; when the precedence graph is an out-tree, then the problem $O \mid p_{ij} = 1, outtree \mid C_{\max}$ becomes polynomially solvable [1]; for a more general objective of minimizing the maximum lateness, Timkovsky proved that $O \mid p_{ij} = 1, outtree \mid L_{\max}$ is weakly NP-hard [14], while the problem $O \mid p_{ij} = 1, intree \mid L_{\max}$ is polynomial solvable [2,3]. We note that, however, there are polynomial time algorithms for $O2 \mid p_{ij} = 1, prec \mid L_{\max}$, even if the jobs have different release times [2,3].

The problem we study in this paper is the m-machine open-shop for unit jobs under arbitrary precedence constraints, $Om \mid p_{ij} = 1, prec \mid C_{\max}$, where $m \geq 3$. For this fundamental problem in scheduling theory, there is no known computational complexity result in the literature. In fact, even when $m = 3$, whether or not $O3 \mid p_{ij} = 1, prec \mid C_{\max}$ is NP-hard is an open question explicitly listed in the websites maintained by Brucker and Knust [4] and Dürr [5], and in the survey paper by Prot and Bellenguez-Morinea [10].

We first introduce a natural layered representation for the precedence graph in Sect. 2, based on which we can construct a partition of the job set into agreeable subsets. We then construct a schedule using the partition and show that it is a 5/3-approximation for the problem $O3 \mid p_{ij} = 1, prec \mid C_{\max}$. In Sect. 3, we propose a greedy algorithm to reduce the number of singleton-job subsets in the earlier partition, resulting in an improved partition, which leads to a 4/3-approximation. We also show that both approximation algorithms apply to the general m-machine open-shops.

2   Preliminaries

We study the problem $O3 \mid p_{ij} = 1, prec \mid C_{\max}$, in which the unit jobs should be processed under the given precedence constraints. These precedence constraints are described as a directed acyclic graph (DAG), the *precedence graph*, in which a vertex corresponds to a job and a directed edge represents a precedence relationship between a pair of jobs. In the rest of the paper, we use a job and a vertex interchangeably. Due to all jobs having unit processing times, we assume without loss of generality that in any feasible schedule the starting processing time of every job is an integer.

Let $V = \{v_1, v_2, \ldots, v_n\}$ be the given set of unit jobs. If v_i precedes v_j, that is, we can start processing the job v_j only if the job v_i is finished by the three-machine openshop $O3$, then there is a directed path beginning from v_i and ending at v_j. Such a directed path is a directed edge (v_i, v_j) in the simplest case, in the DAG $G = (V, E)$.

A subset $X \subseteq V$ of jobs is *agreeable* if none of the jobs of X precedes another. In particular, two jobs are *agreeable* if none of them precedes the other, and thus they can be processed concurrently on different machines in a feasible schedule.

Lemma 1. *An agreeable subset $X \subseteq V$ of jobs can be processed by the three-machine openshop $O3$ in $|X|$ units of time if $|X| \geq 3$, or in 3 units of time if $|X| = 1, 2$.*

Proof. Let the jobs of X be $v_1, v_2, \ldots, v_k$. When $k = 1$, at any time point T, v_1 can be processed on the first machine M_1 (the second machine M_2, the third machine M_3, respectively) starting at T ($T + 1$, $T + 2$, respectively), and thus finished within 3 units of time.

When $k = 2$, at any time point T, v_1 can be processed on the first machine M_1 (the second machine M_2, the third machine M_3, respectively) starting at T ($T + 1$, $T + 2$, respectively); v_2 can be processed on the third machine M_3 (the first machine M_1, the second machine M_2, respectively) starting at T ($T + 1$, $T + 2$, respectively). Thus both of them are finished within 3 units of time.

When $k \geq 3$, at any time point T, for $j = 1, 2, \ldots, k - 2$, v_j can be processed on the first machine M_1 (the second machine M_2, the third machine M_3, respectively) starting at $T + j - 1$ ($T + j$, $T + j + 1$, respectively); v_{k-1} can be processed on the third machine M_3 (the first machine M_1, the second machine M_2, respectively) starting at T ($T + k - 2$, $T + k - 1$, respectively); v_k can be processed on the second machine M_2 (the third machine M_3, the first machine M_1, respectively) starting at T ($T+1, T+k-1$, respectively). See Fig. 1 for an illustration. Thus all of them are finished within k units of time.   □

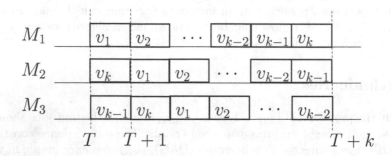

Fig. 1. An sub-schedule to process an agreeable subset $X \subseteq V$ of jobs in $|X|$ units of time when $k = |X| \geq 3$.

Given two disjoint agreeable subsets X_1 and X_2, if a job of X_1 precedes a job of X_2, then we say X_1 *precedes* X_2. A collection of mutual disjoint agreeable subsets is *acyclic* if the precedence relations among the subsets do not contain any cycle. A subset of k jobs is called a k-subset, for $k = 1, 2, \ldots$. For simplicity, a 1-subset is also called a *singleton*.

Corollary 1. *Let C be an acyclic partition of V into agreeable subsets, in which there are b 2-subsets and c singletons. Then a schedule π can be constructed to achieve the makespan $C_{\max}^\pi = n + b + 2c$, where $n = |V|$.*

Proof. Using Lemma 1, all the $n - 2b - c$ jobs outside of those 2-subsets and singletons can be finished in $n - 2b - c$ units of time, and each 2-subset and each singleton can be finished in 3 units of time, respectively. Putting them together, we have a schedule π of makespan $C_{\max}^\pi = (n - 2b - c) + 3b + 3c = n + b + 2c$. □

By Corollary 1, we wish to solve the problem $O3 \mid p_{ij} = 1, prec \mid C_{\max}$ by partitioning the jobs into acyclic agreeable subsets such that the quantity $b + 2c$ is minimized. Our main contribution is an algorithm that produces an acyclic partition achieving a number of singletons no more than the number of isolated jobs (to be defined) in the optimal schedule.

In the rest of the section, we introduce a representation for the DAGs which is used in our algorithm design and analysis.

2.1   A DAG Representation

Let $G = (V, E)$ be the precedence graph describing all the given precedence constraints, where a directed path from v_i to v_j suggests that the job v_i precedes the job v_j (that is, v_j cannot be processed unless v_i is finished by the three-machine openshop). Through out the paper, we let $n = |V|$ and $m = |E|$.

If $(v_i, v_j) \in E$ and there exists a path from v_i to v_j not involving the edge (v_i, v_j), then we call (v_i, v_j) a *redundant* edge, in the sense that the precedence constraint between every pair of jobs is still there after we remove the edge (v_i, v_j) from the graph. We may thus simplify the graph G by removing all redundant edges, which can be executed in $O(m)$ time by a *breadth-first-search* (BFS). Afterwards, for each edge $(v_i, v_j) \in E$, we call v_i a *parent* of v_j and v_j a *child* of v_i. Note that a job can have multiple parents, and multiple children as well.

In the following layered representation of the graph $G = (V, E)$, each job will be associated with a level (a positive integer). The first layer consists of all the jobs with in-degree 0, and these are the level-1 jobs. Iteratively, after the level-ℓ jobs are determined, they and the edges (these are out-edges) incident at them are removed from the graph; then the $(\ell + 1)$-st layer consists of all the jobs with in-degree 0 in the remainder graph, and these are the level-$(\ell + 1)$ jobs. The process terminates when all the jobs of the original graph G have been partitioned into their respective layers. We assume that there are $\ell_{\max}$ layers in total. The entire layer partitioning process is executed in $O(m)$ time. In the sequel, without loss of generality, a DAG $G = (V, E)$ is always represented in this way, in which every job is associated with a level and L_i denotes the subset of all the level-i jobs, for $i = 1, 2, \ldots, \ell_{\max}$. See Fig. 2 for an illustration.

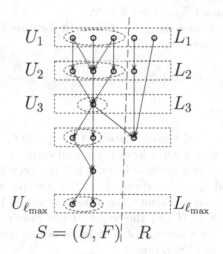

Fig. 2. A layered representation of the precedence graph $G = (V, E)$, in which there are $\ell_{\max}$ layers (each as a dashed rectangle) in total, $L_1, L_2, \ldots, L_{\ell_{\max}}$. U denotes the subset of all the vertices on the longest paths in G, $U_i = L_i \cap U$, for $i = 1, 2, \ldots, \ell_{\max}$ (each as a dashed oval), and $S = (U, F)$ denotes the induced subgraph on U.

Lemma 2. *Given a DAG $G = (V, E)$, L_i is agreeable for every i, and a level-i job has at least one level-$(i-1)$ parent $(i \geq 2)$.*

Proof. By how the layers are constructed.   □

Lemma 3. *Given a DAG $G = (V, E)$, the partition $\mathcal{C} = \{L_1, L_2, \ldots, L_{\ell_{max}}\}$ is an acyclic collection of agreeable subsets.*

Proof. By how the layers are constructed and Lemma 2, L_i precedes L_j if and only if $i < j$.   □

Lemma 4. *Given a DAG $G = (V, E)$, the minimum makespan $C^*_{max} \geq \max\{n, 3\ell_{max}\}$.*

Proof. Since we are dealing with unit jobs, $C^*_{max} \geq n$. Select one job v_i from L_i, for every i, such that v_i is a child of the job v_{i-1}. One clearly sees that in any feasible schedule, the job v_i starts processing after the job v_{i-1} is finished by the three-machine openshop; the makespan of the schedule is thus at least $3\ell_{max}$. This proves the lemma.   □

Theorem 1. *A schedule π can be constructed from the partition $\mathcal{C} = \{L_1, L_2, \ldots, L_{\ell_{max}}\}$ to achieve the makespan $C^\pi_{max} \leq \frac{5}{3}C^*_{max}$.*

Proof. Let b and c denote the number of 2-subsets and the number of singletons among $L_1, L_2, \ldots, L_{\ell_{max}}$. By Corollary 1 a schedule π can be constructed from $\mathcal{C}$ to achieve the makespan $C^\pi_{max} = n + b + 2c$.

Using the trivial bound $\ell_{max} \geq b + c$ in Lemma 4, we have $C^*_{max} \geq \max\{n, 3(b+c)\}$. It follows that

$$C^\pi_{max} = n + b + 2c \leq C^*_{max} + \frac{2}{3}C^*_{max} = \frac{5}{3}C^*_{max}.$$

This proves the theorem.   □

Clearly, from the layered representation of the graph $G = (V, E)$, we see that every longest path begins with a level-1 job and ends at a level-ℓ_{max} job, and it passes through every intermediate layer. That is, every longest path contains exactly ℓ_{max} jobs (and $\ell_{max} - 1$ edges). Let U denote the subset of all the jobs on the longest paths and F denote the subset of edges inherited by U (*i.e.*, $F = E[U]$). We call $S = (U, F)$ the *spine* of the graph $G = (V, E)$, and let $H = G[V - U]$ denote the subgraph of G induced on the remaining subset $V - U$ of jobs. See Fig. 2 for an illustration.

We define a connected component in a DAG in the usual way by ignoring the direction of the edges. If the spine $S = (U, F)$ has more than one connected component, then we can safely conclude that every layer of the graph $G = (V, E)$ contains at least two jobs, that is, $|L_i| \geq 2$ for $i = 1, 2, \ldots, \ell_{max}$. Recall that our goal is to partition all the jobs into acyclic agreeable subsets to minimize the number of singletons. We call such partitions the *optimal partitions* or *optimal collections* of acyclic agreeable subsets. We assume in the rest of the paper that the spine $S = (U, F)$ of the input graph $G = (V, E)$ is connected and there are singleton layers in $S = (U, F)$, as otherwise we trivially achieve an optimal partition without any singletons. Let U_i denote the subset of level-i jobs of U, for $i = 1, 2, \ldots, \ell_{\max}$. If $|U_i| = 1$, then the job of U_i, denoted as s_i, is called a singleton job of U.

Lemma 5. *Given a DAG $G = (V, E)$ and its spine $S = (U, F)$, any acyclic partition of agreeable subsets contains at least $\ell_{\max}$ subsets.*

Proof. Select one job u_i from U_i, for every i, such that u_i is a child of the job u_{i-1}. (For example, these can be the jobs on a single longest path.) One clearly sees that in acyclic partition of agreeable subsets, the jobs u_i and u_j do not belong to a common subset when $i \neq j$. This suggests there are at least $\ell_{\max}$ subsets in the partition. This proves the lemma. $\square$

Lemma 6. *Given a DAG $G = (V, E)$ and its spine $S = (U, F)$, a singleton job of U cannot be processed concurrently with any other job of U in any feasible schedule.*

Proof. Because the singleton job is not agreeable with any other job of U. $\square$

Assume there are in total k singleton jobs in U, which are $s_{i_1}, s_{i_2}, \ldots, s_{i_k}$, where $s_{i_j} \in U_{i_j}$ (that is, $|U_{i_j}| = 1$) and $1 \leq i_1 < i_2 < \ldots < i_k \leq \ell_{\max}$. Let v_i be a level-i job outside of U, i.e., $v_i \in L_i - U_i$. If $i > i_j$ and v_i is agreeable with s_{i_j}, then none of the jobs of U_{i-1} can be a parent of v_i; it follows from Lemma 2 that v_i has a parent $v_{i-1} \in L_{i-1} - U_{i-1}$. When $i - 1 > i_j$, v_{i-1} must also be agreeable with s_{i_j}, and we may repeat the above argument to conclude that there is a job v_{i_j} of $L_{i_j} - s_{i_j}$ which is a predecessor of v_i. Since both s_{i_j} and v_{i_j} are in L_{i_j}, they are agreeable (Lemma 2). We thus have proved the following lemma.

Lemma 7. *Given a DAG $G = (V, E)$ and its spine $S = (U, F)$, for a singleton job $s_{i_j} \in U$ if there is a job of $V - U$ agreeable with s_{i_j}, then there is a level-i job of $L_i - U_i$ with $i \geq i_j$ which is agreeable with s_{i_j}.*

3   A 4/3-Approximation for $O3 \mid prec, p_{ij} = 1 \mid C_{\max}$

We have shown in Theorem 1 that we can construct a schedule π from the partition $\mathcal{C} = \{L_1, L_2, \ldots, L_{\ell_{\max}}\}$ to achieve the makespan $C_{\max}^{\pi} \leq \frac{5}{3} C_{\max}^{*}$, suggesting that the $O3 \mid prec, p_{ij} = 1 \mid C_{\max}$ problem admits a linear time 5/3-approximation. In this section, we present an improved 4/3-approximation algorithm.

3.1   Algorithm Description

Our algorithm is mostly based on the above Lemma 7, for each singleton job s_{i_j} of U, to find a job of $V - U$ which is agreeable with s_{i_j} such that they can be processed concurrently. The algorithm is greedy and iterative, and is denoted as APPROX.

Recall that there are in total k singleton jobs in U, which are $s_{i_1}, s_{i_2}, \ldots, s_{i_k}$ (that is, $U_{i_j} = \{s_{i_j}\}$), with $1 \leq i_1 < i_2 < \ldots < i_k \leq \ell_{\max}$. There are $k+1$ iterations in the algorithm APPROX, which together construct an acyclic partition $\mathcal{D} = \{D_{\ell_{\max}}, D_{\ell_{\max}-1}, \ldots, D_2, D_1\}$. We initialize $R = V - U$.

In the first iteration, sequentially for $i = \ell_{\max}, \ell_{\max} - 1, \ldots, i_k + 1$, we simply let $D_i = L_i$ and remove the jobs of $L_i - U_i$ from R. If $|L_{i_k}| \geq 2$, then we let $D_{i_k} = L_{i_k}$ and remove the jobs of $L_{i_k} - s_{i_k}$ from R. Otherwise, among all the jobs of R, we pick one job that is agreeable with s_{i_k} (i.e., not a predecessor of s_{i_k}) and has the maximum level. Assume this job is $v_i \in L_i - U_i$ such that $i > i_k$. We let $D_{i_k} = \{s_{i_k}, v_i\}$ and remove the job v_i from R. If no job of R is agreeable with s_{i_k}, then we let $D_{i_k} = \{s_{i_k}\}$ and say that s_{i_k} remains as a singleton job in the partition $\mathcal{D}$. This ends the iteration.

In general, in the j-th iteration ($j = 2, 3, \ldots, k$), sequentially for $i = i_{k+2-j} - 1, i_{k+2-j} - 2, \ldots, i_{k+1-j} + 1$, we simply let $D_i = L_i$ and remove the jobs of $L_i - U_i$ from R. We remark that here the set L_i might not be the original L_i, since some of its jobs might be picked in earlier iterations and thus have been removed. Nevertheless, since $|U_i| \geq 2$, we conclude that $|D_i| \geq 2$ too. If $|L_{i_{k+1-j}}| \geq 2$, then we let $D_{i_{k+1-j}} = L_{i_{k+1-j}}$ and remove the jobs of $L_{i_{k+1-j}} - s_{i_{k+1-j}}$ from R. Otherwise, among all the jobs of R, we pick one job that is agreeable with $s_{i_{k+1-j}}$ (i.e., not a predecessor of $s_{i_{k+1-j}}$) and has the maximum level. Assume this job is $v_i \in L_i - U_i$ such that $i > i_{k+1-j}$. We let $D_{i_{k+1-j}} = \{s_{i_{k+1-j}}, v_i\}$ and remove the job v_i from R. If no job of R is agreeable with $s_{i_{k+1-j}}$, then we let $D_{i_{k+1-j}} = \{s_{i_{k+1-j}}\}$ and say that $s_{i_{k+1-j}}$ remains as a singleton job in the partition $\mathcal{D}$. This ends the iteration. A high-level description of such a typical iteration of the algorithm APPROX is depicted in Fig. 3.

The j-th iteration of the algorithm APPROX ($j = 2, 3, \ldots, k$):

1. for $i = i_{k+2-j} - 1, i_{k+2-j} - 2, \ldots, i_{k+1-j} + 1$,
   1.1. set $D_i = L_i$;
   1.2. remove the jobs of $L_i - U_i$ from R;
2. if $|L_{i_{k+1-j}}| \geq 2$,
   2.1. set $D_{i_{k+1-j}} = L_{i_{k+1-j}}$;
   2.2. remove the jobs of $L_{i_{k+1-j}} - s_{i_{k+1-j}}$ from R;
   2.3. end the iteration.
3. if exists $v_i \in R$ (maximum level possible) agreeable with $s_{i_{k+1-j}}$,
   3.1. set $D_{i_{k+1-j}} = \{s_{i_{k+1-j}}, v_i\}$;
   3.3. remove the job v_i from R;
   3.3. end the iteration.
4. 4.1. set $D_{i_{k+1-j}} = \{s_{i_{k+1-j}}\}$;
   4.2. end the iteration.

Fig. 3. A high-level description of a typical iteration of the algorithm APPROX.

In the last (the $(k+1)$-st) iteration, sequentially for $i = i_1 - 1, i_1 - 2, \ldots, 2, 1$, we simply let $D_i = L_i$ and remove the jobs of $L_i - U_i$ from R. Again, we know that here the set L_i might not be the original L_i, since some of its jobs might be picked in earlier iterations. Nevertheless, since $|U_i| \geq 2$, we conclude that $|D_i| \geq 2$ too. This ends the last iteration and the construction of $\mathcal{D}$ is complete. See Fig. 4 for an illustration on $\mathcal{D}$ achieved on the graph $G = (V, E)$ shown in Fig. 2.

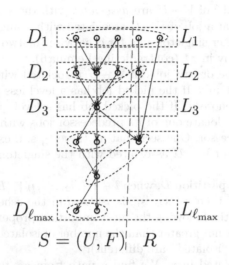

Fig. 4. An illustration on the acyclic partition $\mathcal{D} = \{D_{\ell_{\max}}, D_{\ell_{\max}-1}, \ldots, D_2, D_1\}$ achieved on the precedence graph $G = (V, E)$ shown in Fig. 2. The $\ell_{\max}$ layers $L_1, L_2, \ldots, L_{\ell_{\max}}$ are shown as dashed rectangles and each subset D_i is shown as a dashed oval.

3.2   Performance Analysis

The main result in this section is the following theorem.

Theorem 2. *The schedule* π *constructed from the partition* $\mathcal{D} = \{D_1, D_2, \ldots, D_{\ell_{\max}}\}$ *has a makespan* $C^{\pi}_{\max} \leq \frac{4}{3} C^*_{\max}$.

Proof. We prove first that the partition $\mathcal{D}$ is acyclic, in a way that D_i precedes D_{i+1} for $i = 1, 2, \ldots, \ell_{\max} - 1$. Suppose to the contrary D_i precedes D_j but $i > j$; then D_i precedes D_{i-1}. Note that D_i (D_{i-1}, respectively) consists of a subset of jobs of L_i (L_{i-1}, respectively) and possibly a job v_r with a smaller level $r \leq i - 1$. It follows that $i = i_j$ for some j (that is, s_{i_j} is a singleton job of U), and v_r precedes a job of D_{i-1}, denoted as v_t of level t. Thus we have $r < t \leq i - 1$. If v_t is agreeable with s_{i_j}, then by the algorithm description v_t should be picked into D_{i_j}, a contradiction. Hence v_t precedes s_{i_j}, which implies that v_r precedes s_{i_j} too, again a contradiction. These contradictions together prove that for any $i > j$, D_i doesn't precede D_j.

Next consider an optimal schedule π^* that achieves the minimum makespan $C^*_{\max}$, and assume without loss of generality that the makespan is achieved at the first machine M_1. For a singleton job s_{i_j} of U, Lemma 6 states that it cannot be processed concurrently with any other job of U in π^*. Therefore, there are at most two distinct jobs of $V - U$, such that for each of them, when the machine M_1 is processing it, one of the other machines M_2 and M_3 is processing s_{i_j}. We say that these two jobs of $V - U$ are *associated* with the singleton job s_{i_j}. It is important to note that a job of $V - U$ associated with a singleton job cannot be associated with another singleton job, for otherwise the two singleton jobs were processed concurrently in π^* (contradicting Lemma 6).

Either there is one or two jobs of $V - U$ associated with the singleton job s_{i_j}, we pick one randomly. If the picked job has a level less than or equal to i_j, then we use t_{i_j} to denote it. If the picked job has a level greater than i_j, then we apply Lemma 7 to locate one of its predecessor jobs with level i_j and use t_{i_j} to denote this predecessor. One sees that all these t_{i_j}'s, if exist, are distinct.

If there is no job of $V - U$ associated with the singleton job s_{i_j}, we say s_{i_j} is *isolated* in π^*.

Recall that in the partition $\mathcal{D}$, when $i \notin \{i_1, i_2, \ldots, i_k\}$, $|D_i| \geq 2$. If $|D_{i_j}| = 1$, that is, $D_{i_j} = \{s_{i_j}\}$, then we say s_{i_j} is *isolated* in the schedule π constructed from $\mathcal{D}$. We prove in the following the most important property that the number of isolated jobs in π is not greater than the number of isolated jobs in π^* (though the two meanings of "isolated" are different).

Assume s_{i_j} is isolated in π. We find a path from s_{i_j} to an isolated job in π^* as follows: If s_{i_j} is isolated in π^*, then the path has length 0. If s_{i_j} is not isolated in π^*, that is, we have a job t_{i_j} associated with s_{i_j}, then t_{i_j} should have been picked by the algorithm APPROX in an earlier iteration, since otherwise in this $(k + 1 - j)$-th iteration the singleton job s_{i_j} wouldn't be left alone in the set D_{i_j}. Therefore, we identify another singleton job $s_{i_{j'}}$, where $j' > j$, which is not isolated in π because in the $(k + 1 - j')$-th iteration the algorithm APPROX picked up t_{i_j} to accompany the singleton job $s_{i_{j'}}$. Our path extends from s_{i_j} to $s_{i_{j'}}$. If $s_{i_{j'}}$ happens to be isolated in π^*, then our path ends; otherwise, we continue to use its associated job $t_{i_{j'}}$ to locate a third singleton job $s_{i_{j''}}$, where $j'' > j$ too, which is not isolated in π, and our path extends to $s_{i_{j''}}$. Due to the finitely many singleton jobs, our path ends at a singleton job $s_{i_{j*}}$, which is isolated in π^*.

One sees that we have used the associated jobs t_{i_j}'s, which are distinct from each other, to locate an isolated job in π^* for each isolated job in π. Therefore, an isolated job in π^* wouldn't be discovered by multiple isolated jobs in π. In other words, the number of isolated jobs in π is not greater than the number of isolated jobs in π^*, denoted as c^*. Suppose there are b 2-subsets and c singletons in the partition $\mathcal{D}$; then there are c isolated jobs in π. We have

$$c \leq c^*. \tag{1}$$

In the optimal schedule π^*, the machine M_1 processes nothing while each of the other two machines is processing an isolated job. That is, the machine M_1 idles for at least $2c^*$ units of time before the makespan. Since the load of M_1 is n, we have

$$C_{\max}^* \geq n + 2c^*. \tag{2}$$

On the other hand, we still have $\ell_{\max} \geq b + c$ and $C_{\max}^* \geq 3\ell_{\max}$; therefore,

$$C_{\max}^* \geq \max\{n + 2c^*, 3(b + c)\}, \tag{3}$$

which is a better lower bound than the one in Lemma 4. It follows that

$$C_{\max}^\pi = n + b + 2c = (n + 2c) + b \leq C_{\max}^* + \frac{1}{3}C_{\max}^* = \frac{4}{3}C_{\max}^*.$$

This proves that the performance ratio for the algorithm APPROX is $4/3$.

For the running time, the algorithm APPROX maintains the precedence relationships and updates the subsets L_i's for constructing the partition $\mathcal{D}$. The most time is spent for locating an agreeable job for accompanying a singleton job of U, which might take $O(n)$ time. Therefore, it is safe to conclude that the total running time of the algorithm APPROX is $O(n^2)$. This finishes the proof of the theorem. $\qquad\square$

Corollary 2. *The problem $Om \mid p_{ij} = 1, prec \mid C_{\max}$ admits an $O(n^2)$-time $(2 - \frac{2}{m})$-approximation algorithm.*

Proof. Basically we can construct from the acyclic partition $\mathcal{D}$ a schedule with makespan $C_{\max} \leq n + (m - 2)b + (m - 1)c$. While the lower bounds in Eq. (3) are updated as $C_{\max}^* \geq \max\{n + (m - 1)c^*, m(b + c)\}$. Since we still have $c \leq c^*$, these two inequalities imply that $C_{\max} \leq (1 + (m - 2)/m)C_{\max}^* = (2 - \frac{2}{m})C_{\max}^*$. $\qquad\square$

4   Concluding Remarks

We studied the open-shop scheduling problem for unit jobs under precedence constraints. The problem has been shown to be strongly NP-hard when the number of machines is part of the input [14], but left as an open problem when the number m of machines is a fixed constant greater than 2, since 1978 [8]. We approached this problem by proposing a $(2 - \frac{2}{m})$-approximation algorithm, for $m \geq 3$. Addressing the complexity and designing better approximations are both challenging and exciting.

Acknowledgements. This research is partially supported by the NSFC Grants 11571252, 11771114 and 61672323, the China Scholarship Council Grant 201508330054, and the NSERC Canada.

References

1. Bräsel, H., Kluge, D., Werner, F.: A polynomial algorithm for the $[n/m/0, t_{ij} = 1, tree/C_{max}]$ open shop problem. Eur. J. Oper. Res. **72**, 125–134 (1994)
2. Brucker, P.: Scheduling Algorithms. Springer, New York (2007). https://doi.org/10.1007/978-3-540-69516-5
3. Brucker, P., Jurisch, B., Jurisch, M.Z.: Open shop problems with unit time operations. Oper. Res. **37**, 59–73 (1993)
4. Brucker, P., Knust, S.: Complexity results for scheduling problems (2009). http://www2.informatik.uni-osnabrueck.de/knust/class/
5. Dürr, C.: The scheduling zoo (2016). http://schedulingzoo.lip6.fr
6. Gonzalez, T., Sahni, S.: Open shop scheduling to minimize finish time. J. ACM **23**, 665–679 (1976)
7. Graham, R.L.: Combinatorial scheduling theory. In: Steen, L.A. (ed.) Mathematics Today Twelve Informal Essays. Springer, New York (1978). https://doi.org/10.1007/978-1-4613-9435-8_8
8. Lenstra, J.K., Rinnooy Kan, A.H.G.: Complexity of scheduling under precedence constraints. Oper. Res. **26**, 22–35 (1978)
9. Pinedo, M.L.: Scheduling: Theory, Algorithm and Systems. Springer, New York (2016). https://doi.org/10.1007/978-3-319-26580-3
10. Prot, D., Bellenguez-Morinea, O.: A survey on how the structure of precedence constraints may change the complexity class of scheduling problems. J. Sched. **21**, 3–16 (2018)
11. Sevastianov, S.V., Woeginger, G.J.: Makespan minimization in open shops: a polynomial time approximation scheme. Math. Program. **82**, 191–198 (1998)
12. Sevastianov, S.V., Woeginger, G.J.: Linear time approximation scheme for the multiprocessor open shop problem. Discrete Appl. Math. **114**, 273–288 (2001)
13. Tanaev, V.S., Sotskov, Y.N., Strusevich, V.A.: Scheduling Theory: Multi-Stage Systems. Springer, Heidelberg (1994). https://doi.org/10.1007/978-94-011-1192-8
14. Timkovsky, V.G.: Identical parallel machines vs. unit-time shops and preemptions vs. chains in scheduling complexity. Eur. J. Oper. Res. **149**, 355–376 (2003)
15. Williamson, D.P., et al.: Short shop schedules. Oper. Res. **45**, 288–294 (1997)

Makespan Minimization on Unrelated Parallel Machines with Simple Job-Intersection Structure and Bounded Job Assignments

Daniel R. Page[1]([✉]) [iD], Roberto Solis-Oba[1], and Marten Maack[2]

[1] Department of Computer Science, Western University, London, Canada
dpage6@uwo.ca, solis@csd.uwo.ca
[2] Department of Computer Science, University of Kiel, Kiel, Germany
mmaa@informatik.uni-kiel.de

Abstract. Let there be a set J of n jobs and a set M of m parallel machines, where each job j takes $p_{i,j} \in \mathbb{Z}^+$ time units on machine i and assume $p_{i,j} = \infty$ implies job j cannot be scheduled on machine i. In makespan minimization on unrelated parallel machines ($R||C_{max}$), the goal is to schedule each job non-preemptively on a machine so as to minimize the makespan. A job-intersection graph $G_J = (J, E_J)$ is an unweighted undirected graph where there is an edge $\{j, j'\} \in E_J$ if there is a machine i such that both $p_{i,j} \neq \infty$ and $p_{i,j'} \neq \infty$. In this paper we consider two variants of $R||C_{max}$ where there are a small number of eligible jobs per machine. First, we prove that there is no approximation algorithm with approximation ratio better than $3/2$ for $R||C_{max}$ when restricted to instances where the job-intersection graph contains no diamonds, unless P = NP. Second, we match this lower bound by presenting a $3/2$-approximation algorithm for this special case of $R||C_{max}$, and furthermore show that when G_J is triangle free $R||C_{max}$ is solvable in polynomial time. For $R||C_{max}$ restricted to instances when every machine can process at most ℓ jobs, we give approximation algorithms with approximation ratios $3/2$ and $5/3$ for $\ell = 3$ and $\ell = 4$ respectively, a polynomial-time algorithm when $\ell = 2$, and prove that it is NP-hard to approximate the optimum solution within a factor less than $3/2$ when $\ell \geq 3$. In the special case where every $p_{i,j} \in \{p_j, \infty\}$, called the restricted assignment problem, and there are only two job lengths $p_j \in \{\alpha, \beta\}$ we present a $(2 - 1/(\ell - 1))$-approximation algorithm when $\ell \geq 3$.

Keywords: Makespan minimization · Unrelated parallel machines
Approximation algorithms · Restricted assignment
Bounded job assignments · Job-intersection graphs

Daniel Page is supported by an Ontario Graduate Scholarship. Roberto Solis-Oba was partially supported by the Natural Sciences and Engineering Research Council of Canada, grant 04667-2015 RGPIN. Marten Maack was supported by the DAAD (Deutscher Akademischer Austauschdienst).

D. Kim et al. (Eds.): COCOA 2018, LNCS 11346, pp. 341–356, 2018.
https://doi.org/10.1007/978-3-030-04651-4_23

1   Introduction

Let J be a set of n jobs and M a set of m parallel machines, where a job j takes $p_{i,j} \in \mathbb{Z}^+$ time units on machine i. The goal in *makespan minimization on unrelated parallel machines* is to produce a schedule where each job is scheduled non-preemptively on a machine so as to minimize the length of the schedule or *makespan*. Makespan minimization on unrelated parallel machines is a classic NP-hard scheduling problem, and is denoted as $R||C_{max}$ in Graham's notation (see [12]). Note that when the processing time $p_{i,j} = \infty$, we say that job j cannot be scheduled on machine i, and assume the processing times are given as an $m \times n$ *processing matrix* $P = (p_{i,j})$. In this paper we investigate two versions of $R||C_{max}$:

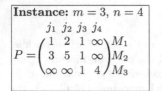 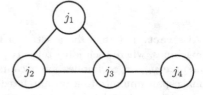

Fig. 1. An instance of $R||C_{max}$ (left), and its job-intersection graph G_J (right).

- $R||C_{max}$ *with simple job-intersection structure.* A *job-intersection graph* $G_J = (J, E_J)$ has a *job vertex* for each job $j \in J$, and for any two jobs $j, j' \in J$, there is an edge $\{j, j'\} \in E_J$ if there is a machine i such that $p_{i,j} \neq \infty$ and $p_{i,j'} \neq \infty$. A set of restrictions on which machines can process a job can be represented as a job-intersection graph. We study $R||C_{max}$ restricted to particular classes of job-intersection graphs. We give an example of a job-intersection graph in Fig. 1.
- $R||C_{max}$ *with bounded job assignments.* Let $\mathcal{J}_i$ be the set of jobs that can be processed by machine i, i.e., $\mathcal{J}_i = \{j \in J \mid p_{i,j} \neq \infty\}$. Let $\ell > 0$. We consider $R||C_{max}$ restricted to instances when, for each machine i, $|\mathcal{J}_i| \leq \ell$. Clearly when $\ell = n$, it is $R||C_{max}$.

Currently the best-known approximation algorithms for $R||C_{max}$ have approximation ratio 2 [10,18,24], and there is no approximation algorithm for $R||C_{max}$ with approximation ratio less than 3/2, unless P = NP [18]. Despite much intensive study, finding an approximation algorithm with approximation ratio strictly less than 2 still remains an open problem and is regarded as one of the most challenging open problems in the study of approximation algorithms today [25]. A natural question is whether there are any "well-structured" and effi-cient to recognize classes of job-intersection graphs, for which the corresponding instances of $R||C_{max}$ can be efficiently solved or for which there are approxima-tion algorithms with approximation ratio less than 2. As we show, both problems given above are closely related from a hardness of approximation standpoint and we present algorithms for both. Furthermore, we establish that there is no approximation algorithm with approximation ratio less than $3/2$ for $R||C_{max}$ restricted to diamondless job-intersection graphs or for $R||C_{max}$ when every machine can process at most $\ell = 3$ jobs, unless $\mathsf{P} = \mathsf{NP}$. However, in both of these cases we can formulate a relatively simple combinatorial algorithm that has the approximation ratio $3/2$, matching the lower bound.

2 Preliminaries

One NP-hard special case of $R||C_{max}$ of recent interest in the literature is the graph balancing problem. In the *graph balancing problem*, every job takes $p_{i,j} = p_j$ time units and can only be scheduled on one of at most two possible machines. This problem can be described as an edge orientation problem: given a weighted multigraph $G = (V, E)$ with weights p_e for each edge $e \in E$, orient all the edges in G such that the maximum load of the vertices is minimized, where the load of a vertex is the sum of all the weights of edges oriented toward that vertex. In this formulation the edges are the jobs, and the vertices are the machines. We would like to remark that another well-known and intensely studied special case of $R||C_{max}$ is the *restricted assignment problem*, which is a general case of the graph balancing problem where every job has a subset of machines on which it can be scheduled.

A graph is *triangle free* if it does not contain any simple cycles of length 3— triangles. Note that all bipartite graphs contain no odd-length cycles, thus all bipartite graphs are triangle free. The *diamond graph* consists of four vertices and five edges, so it is K_4 less one edge. We call a graph *diamondless* if it does not contain the diamond graph as a subgraph. In contrast, a *diamond-free graph* is defined as not having the diamond graph as an induced subgraph. An *induced subgraph* $H = (V', E')$ of a graph $G = (V, E)$ is such that $V' \subseteq V$ and an edge $e = \{u, v\} \in E'$ if both $u, v \in V'$ and $e \in E$; all diamondless graphs are diamond-free, but not all diamond-free graphs are diamondless. For example, the graph K_4 is diamond free but is not diamondless. In Fig. 2 we give an instance of the graph balancing problem where its job-intersection graph is both diamondless and diamond free.

3 Related Work

The best-known approximation algorithms for $R||C_{max}$ and the restricted assignment problem have approximation ratio 2 [10,18,24]. However, for the restricted assignment problem with two job lengths, $\alpha < \beta$, Chakrabarty *et al.* [5] gave a $(2 - \delta)$-approximation algorithm for some small value $\delta > 0$ and a $(2 - \alpha/\beta)$-approximation algorithm. Ebenlendr *et al.* [8] presented a $7/4$-approximation algorithm for the graph balancing problem, and in [13,22] $3/2$-approximation algorithms are presented for the problem when there are only two job lengths.

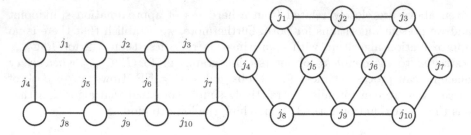

Fig. 2. An instance of the graph balancing problem (left) and its job-intersection graph (right).

The concept of the job-intersection graph goes back to at least Glass and Kellerer [11] with the study of so-called nested-structures and the restricted assignment problem. Research on the restricted assignment problem when instances satisfy certain structural properties is extensive and has grown in interest in recent years [19]. In addition, there has been investigation of scheduling problems on machine-intersection graphs where the machines are the vertices and an edge exists between two vertices when a job can be scheduled on the two corresponding machines [3,14,17]. Jansen *et al.* [14][1] proved that $R||C_{max}$ is fixed-parameter tractable (FPT) in the treewidth tw of the job-intersection graph. That is, if the job-intersection graph G_J has constant treewidth, $R||C_{max}$ can be solved in polynomial time. So when the job-intersection graph belongs to graph classes such as trees ($tw = 1$), cactus graphs ($tw \leq 2$), outerplanar graphs ($tw \leq 2$), and series-parallel graphs ($tw \leq 2$), $R||C_{max}$ is solvable in polynomial time. In this paper we study $R||C_{max}$ restricted to classes of job-intersection graphs that do not have constant treewidth. In Fig. 3 we summarize both computational complexity results found in this paper and presently in the literature for $R||C_{max}$ on job-intersection graphs.

To the best of our knowledge $R||C_{max}$ with bounded job assignments has not been previously studied. Bounded job assignments have been considered in other types of scheduling problems, such as in batch scheduling where a batch size bounds the number of jobs simultaneously processed by a batching machine [6,20]. A generalization of $R||C_{max}$ where every machine has a positive integer called a *machine capacity* that bounds the maximum number of jobs each machine can process has also been studied. For this generalization there is a 2-approximation algorithm [23], and there exists an efficient polynomial-time approximation scheme when the machines are identical [7].

It is important to discuss recognition of the instances for which we design algorithms. For any $0 \leq \ell \leq n$, it is trivial to determine if the set $\mathcal{J}_i$ of jobs that every machine i can process has size at most ℓ. Alon *et al.* [1] gave an algorithm that can test if a graph (V, E) is triangle free in $O(|E|^{1.41})$ time. We note that diamondless graphs can be recognized in $O(|V|^3)$ time by a simple algorithm

---

[1] In this paper the authors refer to the job-intersection graph as the primal graph.

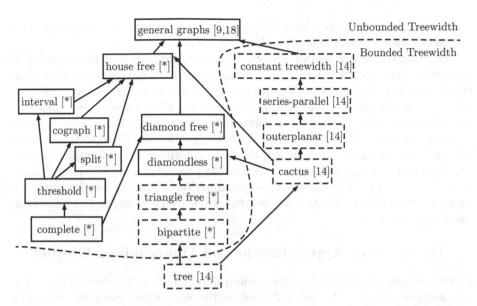

Fig. 3. Summary of results for $R||C_{max}$ with simple job-intersection structure. The job-intersection graphs restricting the machine assignments are grouped by graph class. For two graph classes A and B, "$A \rightarrow B$" in the diagram means that any graph in A is in graph class B. Problems boxed with dashed lines are polynomial-time solvable, and problems with boxed solid lines are strongly NP-hard. The number(s) in brackets are reference numbers, and graph classes with [*] beside them refer to computational complexity results found in this paper.

that looks for a pair of triangles with a common edge. Kloks *et al.* [16] showed that one can recognize if a graph is diamond-free (and give a diamond in the graph if it is not) in $O(|V|^c + |E|^{3/2})$ time, where $O(|V|^c)$ is the time complexity to compute the square of a $|V| \times |V|$ 0-1 adjacency matrix.

4   Our Results

First, we establish that once one admits triangles but forbids diamonds in the job-intersection graph, there is no k-approximation algorithm for $R||C_{max}$ with $k < 3/2$, unless $P \neq NP$. This matches the same inapproximability bounds as those that exist for the restricted assignment problem with two job lengths [18] and the graph balancing problem with two job lengths [2,9], both special cases of $R||C_{max}$. To do this, we strengthen the inapproximability result of Ebenlendr *et al.* [9] for the graph balancing problem with two job lengths. Employing this result we can also prove that for $R||C_{max}$ when every machine i satisfies $|\mathcal{J}_i| \leq 3$, the inapproximability lower bound of 3/2 holds. In Sect. 8, we show that $R||C_{max}$ restricted to job-intersection graphs belonging to several well-studied graph classes such as complete graphs, threshold graphs, interval graphs, cographs, split graphs, and house-free graphs do not have any k-approximation algorithm with $k < 3/2$, unless $\mathsf{P} = \mathsf{NP}$.

To complement our inapproximability results, we present a flow-based $3/2$-approximation algorithm for $R||C_{max}$ when every machine can process at most three jobs. As we will later justify, this problem contains as special cases $R||C_{max}$ when G_J is triangle free and $R||C_{max}$ when G_J is diamondless. Our algorithm can also be used to exactly solve in polynomial time $R||C_{max}$ when every machine can process at most two jobs, as well as $R||C_{max}$ when restricted to triangle-free job-intersection graphs. In addition, the same algorithm is a $5/3$-approximation algorithm for $R||C_{max}$ when every machine can process at most four jobs. Finally, in Sect. 7 we give a $(2 - 1/(\ell - 1))$-approximation algorithm for the restricted assignment problem with two job lengths when every machine can process at most $\ell \geq 3$ jobs.

5   Hardness of Approximation on Diamondless Graphs

In this section we prove under the assumption that $\mathsf{P} \neq \mathsf{NP}$ that $R||C_{max}$ has the inapproximability bound $3/2$ even in the case when instances have job-intersection graphs that are diamondless. To do this, we employ a similar reduction as that used by Ebenlendr et al. [9]. Since Ebenlendr et al. showed this reduction yields the inapproximability bound we desire, we must show that the job-intersection graphs from graph-balancing instances produced by the reduction are diamondless.

The reduction by Ebenlendr et al. [9] uses a variant of the satisfiability problem we will denote as At-most-3-SAT(2L). Let there be n' boolean variables $x_1, \ldots, x_{n'}$, and m' clauses $\alpha_1, \ldots, \alpha_{m'}$. Given a propositional logic formula ϕ in conjunctive normal form (CNF) where there are at most three literals per clause, each variable appears at most three times in ϕ, and each literal (a variable or its negation) appears at most twice in ϕ, the problem is to decide whether there is an assignment of values to the variables $x_1, \ldots, x_{n'}$ so that ϕ is satisfied. At-most-3-SAT(2L) is known to be NP-complete [2]. Without loss of generality we assume that no clause contains a tautology, and that no clause contains duplicate literals.

Now we describe how to construct the graph balancing instance I' from At-most-3-SAT(2L) instance $I = (\phi, n', m')$. Introduce two types of vertices: literal vertices, and clause vertices. Given a variable x_i, a *literal vertex* corresponds to a literal x_i or $\neg x_i$. For each clause α_j, a *clause vertex* is created that corresponds to clause α_j. There will be two types of edges: tautologous edges, and clause edges. For each variable x_i, a *tautologous edge* $\{x_i, \neg x_i\}$ has weight 2. For each clause α_j and literal λ that appears in α_j, a *clause edge* $\{\lambda, \alpha_j\}$ has weight 1. Finally, for clause α_j add $3 - |\alpha_j|$ self-loops with weight 1 on its clause vertex, where $|\alpha_j|$ is the number of literals in clause α_j. The idea is that the orientation of the tautologous edges will determine the assignment of values to the variables of ϕ. Instance I' can be built from I in polynomial time.

To illustrate the reduction, we give an example. Let the propositional logic formula $\phi = (x_1 \lor \neg x_2) \land (\neg x_1 \lor \neg x_2 \lor \neg x_3)$, where $n' = 3$ and $m' = 2$. Then

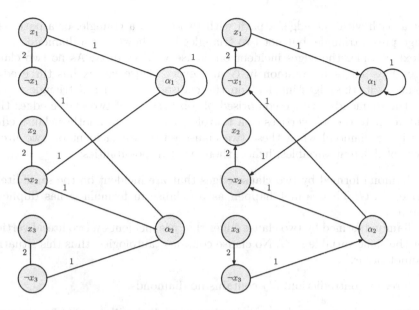

Fig. 4. Given the formula $\phi = (x_1 \vee \neg x_2) \wedge (\neg x_1 \vee \neg x_2 \vee \neg x_3)$, the resulting graph balancing instance applying the construction is shown on the left. Its optimal orientation is given on the right.

$\alpha_1 = (x_1 \vee \neg x_2)$ and $\alpha_2 = (\neg x_1 \vee \neg x_2 \vee \neg x_3)$. Applying the reduction we obtain the graph-balancing instance shown in Fig. 4. The formula ϕ can be satisfied, and the resulting instance has an optimal orientation with makespan 2.

Ebenlendr *et al.* [9] proved that I' has a schedule with makespan at most two if ϕ is satisfied, but the makespan is at least three otherwise. Hence, if there were a k-approximation algorithm with $k < 3/2$, one could apply the above reduction, apply said k-approximation algorithm, then correctly decide whether ϕ is satisfiable or not in polynomial time: if the makespan is less than three return "yes"; and if the makespan is at least three, return "no".

Lemma 1. *The job-intersection graph G_J of the weighted multigraph G produced by the above reduction contains no diamonds.*

Proof. Assume that G_J has at least one diamond. Observe that every vertex in G has at most three incident edges, so G_J can only be comprised of isolated job vertices, paths, or triangles. Then, there must be two triangles that share two job vertices to form a diamond.

First, consider the edges incident on literal vertices in G. Recall that each variable appears in at most three clauses in formula ϕ and each literal for that variable appears at most twice. So the job vertex corresponding to the tautologous edge $\{x_i, \neg x_i\}$ has degree at most three in G_J. Furthermore, this job vertex is only adjacent to job vertices that are clause edges in G, and at most two clause edges may have the same literal vertex as an endpoint in G. Thus, any job vertex $\{x_i, \neg x_i\}$ along with its adjacent job vertices for clause edges form in G_J

either a path with one edge, a path with two edges, a triangle, or a path with an edge plus a triangle, but not two triangles i.e. a bowtie or a diamond.

Next consider the edges incident on clause vertices in G. As no two clause vertices have edges in common in G and every clause vertex has three edges incident on it, the edges incident on the clause vertex form a triangle in G_J. Thus, the diamond must be comprised of job vertices of two clause edges that are adjacent to the job vertices of a tautologous edge and another clause edge. There is no diamond when these two clause edges are incident on two literal vertices of different variables, hence, there are two possibilities:

1. A diamond formed by two clause edges that are incident on the same literal vertex in G. This cannot happen as no clause in formula ϕ has duplicate literals.
2. A diamond formed by two clause edges that are incident on two literal vertices for the same variable in G. No clause contains tautologies, thus this situation cannot occur.

Therefore, by contradiction, G_J contains no diamonds.   □

Theorem 1. *There is no k-approximation algorithm with $k < 3/2$ for the graph balancing problem with two job lengths when the job-intersection graph G_J contains no diamonds, unless $\mathsf{P} = \mathsf{NP}$.*

Corollary 1. *There is no k-approximation algorithm with $k < 3/2$ for $R||C_{max}$ restricted to diamondless job-intersection graphs, unless $\mathsf{P} = \mathsf{NP}$.*

If $|\mathcal{J}_i| > 3$ for some machine i, then there are at least four jobs j_1, j_2, j_3, j_4 such that $p_{i,j_1} \neq \infty$, $p_{i,j_2} \neq \infty$, $p_{i,j_3} \neq \infty$, and $p_{i,j_4} \neq \infty$. This would imply G_J contains a diamond; thus, for any machine i, $|\mathcal{J}_i| \leq 3$ is satisfied if G_J is diamondless. Hence, the diamondless case is a special case of when, for each machine i, $|\mathcal{J}_i| \leq 3$. Thus our inapproximability results carry over to the special case where every machine can process at most three jobs. Do note that proving this special case has the inapproximability bound stated in the corollary can also be made trivially by simply observing that every vertex has at most three incident edges in the graph-balancing instance constructed in the above reduction.

Corollary 2. *There is no k-approximation algorithm for the graph balancing problem with two job lengths when every machine can process at most three jobs where $k < 3/2$, unless $\mathsf{P} = \mathsf{NP}$.*

6   Approximation Results for Unrelated Scheduling with Bounded Job Assignments

As we stated at the end of the previous section, $R||C_{max}$ restricted to diamondless job-intersection graphs is a special case of $R||C_{max}$ when every machine i satisfies $|\mathcal{J}_i| \leq 3$. We present a 5/3-approximation algorithm for $R||C_{max}$ when every machine can process at most four jobs. In our analysis we show the same approximation algorithm has approximation ratio $3/2$ in the case when every machine can process at most three jobs. Note that for $R||C_{max}$ restricted to triangle-free job-intersection graphs, no machine can process three jobs as doing so implies three jobs share a common machine where they can be scheduled, so every machine i satisfies $|\mathcal{J}_i| \leq 2$ in this particular situation.

Let OPT be the value of an optimal solution for $R||C_{max}$ when every machine i satisfies $|\mathcal{J}_i| \leq 4$. Similar to [18], we perform a binary search procedure to find the smallest value T over the interval $[0, \sum_{i \in M, j \in J} (p_{i,j})]$ such that the algorithm given below produces a schedule with makespan at most $(5/3)T$. If a schedule is produced then we decrease the value of T in the binary search, and if REJECT is reported, then there is no schedule with makespan at most T and thus T is increased in the binary search. At the end of the binary search the smallest value for T is found, so it must be the case that $T \leq OPT$ and so the approximation ratio is $5/3$. We say a job j is *small* on machine i if its processing time is $p_{i,j} \leq T/2$, and is *big* on machine i if $T/2 < p_{i,j} \leq T$. Observe that if $OPT \leq T$, at most one big job can be scheduled on a machine. Note that by our definitions, there can be a job j that is neither big nor small with respect to its processing time on some machine i, if $p_{i,j} > T$.

1. First, there may be machines for which some jobs are neither big nor small. For each machine i, if any job j has a processing time $p_{i,j} > T$ on machine i, we remove job j from job set $\mathcal{J}_i$. As a result, every job j in each job set $\mathcal{J}_i$ has processing time $p_{i,j} \leq T$.
2. Build a single-source single-sink flow network N with source s^* and sink t^*. In this network, create a job node for each job, and add arcs from s^* to each job node with capacity 1. Now, for each machine i, we create a machine node and a buffer node with arcs according to the *Machine Plans* given in Fig. 5, which we describe now. Let disjoint sets $\mathcal{S}_i, \mathcal{B}_i \subseteq \mathcal{J}_i$, where by default it is assumed that $\mathcal{S}_i$ and $\mathcal{B}_i$ are the small jobs and big jobs in $\mathcal{J}_i$, respectively. Consider the following cases in the order provided:
   (a) If $|\mathcal{J}_i| = 0$, no arcs are added for machine node i.
   (b) If $\sum_{j \in \mathcal{J}_i} p_{i,j} \leq T$, then every job $j \in \mathcal{J}_i$ can be scheduled on machine i, so we add arcs according to the Machine Plan with $d = |\mathcal{J}_i|$ and set $\mathcal{S}_i = \mathcal{J}_i$ and $\mathcal{B}_i = \varnothing$.
   (c) If $|\mathcal{J}_i| \leq 3$, then use the Machine Plan with $d = |\mathcal{J}_i| - 1$.
   In the last set of cases $|\mathcal{J}_i| = 4$. Sort the jobs of each $\mathcal{J}_i$ in non-increasing order by processing time; let these jobs be denoted as $j_1^{(i)}, j_2^{(i)}, j_3^{(i)}, j_4^{(i)}$.
   (d) If $\sum_{k=2}^{4} p_{i,j_k^{(i)}} > T$, add arcs according to the Machine Plan with $d = 2$.
   (e) If $\sum_{k=2}^{4} p_{i,j_k^{(i)}} \leq T$ and $p_{i,j_1^{(i)}} + p_{i,j_2^{(i)}} > T$, put $j_1^{(i)}$ and $j_2^{(i)}$ (if either is not already) into $\mathcal{B}_i$ and use the Machine Plan with $d = 3$.
   (f) If $\sum_{k=2}^{4} p_{i,j_k^{(i)}} \leq T$ and $p_{i,j_1^{(i)}} + p_{i,j_2^{(i)}} \leq T$, use the Machine Plan with $d = 3$.
3. Now that N is constructed, the algorithm computes an integral maximum flow f on N. If any arc leaving the source does not send one unit of flow then

there is a job node that receives no flow, report REJECT if this is the case. If all the job nodes receive one unit of flow, we build the schedule as follows: for each job node j, if machine node i receives 1 unit of flow from j, schedule job j on machine i.

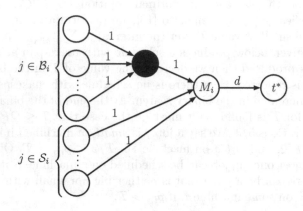

Fig. 5. Flow network N is built in part by determining the appropriate machine plan for each machine. Assume an integer value d is provided along with each plan, and unless otherwise stated, let $\mathcal{S}_i, \mathcal{B}_i \subseteq \mathcal{J}_i$ be the set of small jobs and set of big jobs, respectively. The machine plan for machine node i shows the arcs and capacities of the arcs included. Unlabelled white nodes are job nodes, the black node is a *buffer* node of machine i that only allows one unit of flow to be sent from job nodes in $\mathcal{B}_i$, and t^* is the sink of N.

By the way we designed the flow network, it is not hard to see that if $OPT \leq T$, all the arcs leaving the source are saturated, and as a result, a schedule is produced.

Now we analyze the load of each machine. First, it is trivial to observe that the load of any machine i is at most T if either $\sum_{j \in \mathcal{J}_i} p_{i,j} \leq T$ (case (b)) or all the jobs in $\mathcal{J}_i$ are big (case (c) if $|\mathcal{J}_i| \leq 3$, case (d) if $|\mathcal{J}_i| = 4$). Thus, we consider each machine i when there is at least one small job and $\sum_{j \in \mathcal{J}} p_{i,j} > T$ based on the number of jobs in job set $\mathcal{J}_i$:

- $|\mathcal{J}_i| \leq 2$. If $|\mathcal{J}_i| \leq 1$, then either case (a) or case (b) occurs, which we already considered above. If $|\mathcal{J}_i| = 2$ and all the jobs are small, then $\sum_{j \in \mathcal{J}_i} p_{i,j} \leq T/2 + T/2 = T$ and falls under case (b). If $|\mathcal{J}_i| = 2$ and $\sum_{j \in \mathcal{J}_i} p_{i,j} > T$, then the only remaining case is when there is one big job and one small job that cannot be scheduled together. The algorithm applies case (c), which permits only $|\mathcal{J}_i| - 1 = 1$ job to be scheduled on machine i, and the load of machine i is at most T. Therefore, the load of any machine with $|\mathcal{J}_i| \leq 2$ is at most T.

- $|\mathcal{J}_i| = 3$. Since $\sum_{j \in \mathcal{J}_i} p_{i,j} > T$, at most two jobs can be scheduled on machine i; case (c) is applied here, and the Machine Plan will allow at most $|\mathcal{J}_i| - 1 = 2$ jobs to be scheduled on machine i. If all three jobs in $\mathcal{J}_i$ are small, then at most two jobs are scheduled on machine i and the load is at most T. Otherwise at least one job is big and at most two jobs are small in $\mathcal{J}_i$, and at most one big job will be scheduled with a small job and so the load is at most $T + T/2 = (3/2)T$. Therefore, the load of any machine with $|\mathcal{J}_i| = 3$ is at most $(3/2)T$.

- $|\mathcal{J}_i| = 4$. First, we identify a few key observations that will simplify our analysis. First, if ever case (d) is applied, $d = 2$ in the Machine Plan, so at most one big job is scheduled with one small job and the load is at most $T + T/2 = (3/2)T$. Thus we only need to consider the algorithm in situations when it applies case (e) or case (f). In either of these two cases, $d = 3$, so at most one big job is scheduled with two small jobs, as when three small jobs are scheduled on machine i the load is at most $(3/2)T$. Recall that the jobs in $\mathcal{J}_i$ are sorted in non-increasing order $j_1^{(i)}$, $j_2^{(i)}$, $j_3^{(i)}$, $j_4^{(i)}$. If there are at least three big jobs and at most one small job, then $\sum_{k=2}^{4} p_{i,j_k^{(i)}} > T$ and this falls under case (d); thus we only need to consider below when there are at most two big jobs and at least one small job in $\mathcal{J}_i$.

  - If all four jobs are small, then only case (f) applies as the sum of any two small jobs on machine i cannot exceed T. Again, at most three small jobs can be scheduled on machine i and the load is at most $(3/2)T$.

  - If three jobs are small and one job is big, then either case (e) or case (f) is applied by the algorithm. In case (e), if $p_{i,j_2^{(i)}} \leq T/3$ then the sorting of the jobs implies that the load is at most $T + 2(T/3) = (5/3)T$. Then, observe that if $p_{i,j_2^{(i)}} > T/3$ and $\sum_{k=2}^{4} p_{i,j_k^{(i)}} \leq T$, then $\sum_{k=3}^{4} p_{i,j_k^{(i)}} < T - T/3 = (2/3)T$, and the load on machine i is at most $p_{i,j_1^{(i)}} + p_{i,j_3^{(i)}} + p_{i,j_4^{(i)}} \leq T + (2/3)T = (5/3)T$. Next if case (f) is applied, then $p_{i,j_1^{(i)}} + p_{i,j_2^{(i)}} \leq T$ implies the load of machine i is at most $p_{i,j_1^{(i)}} + p_{i,j_2^{(i)}} + p_{i,j_3^{(i)}} \leq T + T/2 = (3/2)T$.

  - If two jobs are small and two jobs are big, only case (e) applies as the sum of any two big jobs exceeds T. Job $j_2^{(i)}$ is big, so observe that $\sum_{k=2}^{4} p_{i,j_k^{(i)}} \leq T \Rightarrow \sum_{k=3}^{4} p_{i,j_k^{(i)}} < T - (T/2) = T/2$. Thus, the load of machine i is at most $T + T/2 = (3/2)T$.

  Hence, the maximum load of a machine with $|\mathcal{J}_i| = 4$ is at most $(5/3)T$.

Therefore, we obtain the following results that match the inapproximability bounds given by Corollary 1 and Corollary 2.

Theorem 2. *There is a polynomial-time algorithm for $R||C_{max}$ when every machine can process at most two, three, or four jobs with approximation ratio 1, 3/2, or 5/3, respectively.*

Corollary 3. *There is a polynomial-time algorithm for $R||C_{max}$ restricted to job-intersection graphs that are either triangle free or diamondless with approximation ratio 1 or 3/2, respectively. Furthermore, there is a polynomial-time algorithm for $R||C_{max}$ restricted to bipartite job-intersection graphs.*

7   A $(2 - 1/(\ell - 1))$-Approximation Algorithm for Restricted Assignment with Two Job Lengths and Bounded Job Assignments

Let $\alpha, \beta \in \mathbb{Z}^+$, where $\alpha < \beta$. Recall that the restricted assignment problem with two job lengths is a special case of $R||C_{max}$ where every processing time $p_{i,j} \in \{p_j, \infty\}$ and job length $p_j \in \{\alpha, \beta\}$. Note that if every job has the same job length, this is equivalent to the restricted assignment problem with unit job lengths and can be solved in polynomial time [21]. So below we consider instances where at least one job differs in length, and every machine can process at most $\ell \geq 3$ jobs. By modifying the algorithm we gave in Sect. 6 along with using some known results, we obtain an approximation algorithm with approximation ratio $2 - 1/(\ell - 1)$. Like in Sect. 6, there is an estimate T of the optimal makespan where binary search is performed to find the smallest value for T such that the algorithm below produces a schedule with makespan at most $(2 - 1/(\ell - 1))T$. Below we assume if not all of the jobs are scheduled, the algorithm reports REJECT. Given estimate T, consider the following cases in the order provided.

1. If there is a job $j \in J$ with no machine i where $p_{i,j} = p_j \leq T$, report REJECT.
2. $\alpha > T/(\ell - 1)$ and $\beta \leq T$. Apply the $(2 - \alpha/\beta)$-approximation algorithm of Chakrabarty *et al.* [5] for estimate T. If a schedule exists with makespan T, this algorithm will compute a schedule with makespan at most

$$\left(2 - \frac{\alpha}{\beta}\right)T < \left(2 - \frac{\frac{T}{\ell-1}}{T}\right)T = \left(2 - \frac{1}{\ell - 1}\right)T.$$

3. $\alpha \leq T/(\ell - 1)$ and $\beta \leq T/2$. Use the algorithm of Lenstra *et al.* [18]. In this algorithm a fractional solution is computed using linear programming, and then a rounding is performed to integrally assign the remaining fractionally assigned jobs. If $OPT \leq T$, then solving the linear program guarantees the load of each machine is at most T, and the rounding step schedules at most one additional job per machine. Thus, the makespan is at most $T + \max\{\alpha, \beta\} \leq T + T/2 = (3/2)T$.

4. $\alpha \leq T/(\ell - 1)$ and $T/2 < \beta \leq T$. Use the algorithm given in Sect. 6 except than in Step 2, for every machine i proceed as follows:
   - If every job in $\mathcal{J}_i$ is small, it is possible for every job in $\mathcal{J}_i$ to be scheduled on machine i, so use the Machine Plan with $d = |\mathcal{J}_i|$. The load of the machine i is at most $|\mathcal{J}_i|\alpha \leq |\mathcal{J}_i|(T/(\ell - 1)) \leq \ell(T/(\ell - 1)) \leq (2 - 1/(\ell - 1))T$ as $\ell \geq 3$.

– There is at least one big job in job set $\mathcal{J}_i$. If $\beta + \sum_{j \in \mathcal{S}_i} p_j \leq T$, use the Machine Plan with $d = |\mathcal{S}_i| + 1$, where $\mathcal{S}_i$ is the set of small jobs of job set $\mathcal{J}_i$. At most one big job can be scheduled with every job in $\mathcal{S}_i$, so the load of a machine i is at most $\beta + \sum_{j \in \mathcal{S}_i} p_j \leq T$.

If $\beta + \sum_{j \in \mathcal{S}_i} p_j > T$, then either at most one big job can be scheduled with $|\mathcal{S}_i| - 1$ small jobs or at most all $|\mathcal{S}_i|$ small jobs are scheduled together. Add arcs according to the Machine Plan with $d = \max\{|\mathcal{S}_i|, 1\}$. Since at least one job in job set $\mathcal{J}_i$ is big, $|\mathcal{S}_i| \leq |\mathcal{J}_i| - 1 \leq \ell - 1$. If every job that is scheduled on machine i is small, then the load is at most $|\mathcal{S}_i| \alpha \leq (\ell - 1)(T/(\ell - 1)) = T$. Otherwise, at most one big job can be scheduled with $|\mathcal{S}_i| - 1$ small jobs and the load of machine i is at most

$$\beta + (|\mathcal{S}_i| - 1)\alpha \leq T + ((\ell - 1) - 1)\left(\frac{T}{\ell - 1}\right) = \left(2 - \frac{1}{\ell - 1}\right)T.$$

Theorem 3. *There is a $(2 - 1/(\ell - 1))$-approximation algorithm for the restricted assignment problem with two job lengths when every machine can process at most $\ell \geq 3$ jobs.*

8 Inapproximability Results for Job-Intersection Graphs with Cliques

For any instance $I = (P = (p_{i,j}), m, n)$ of $R||C_{max}$ with some $p_{i,j} = \infty$, there is another instance $I' = (P' = (p'_{i,j}), m, n)$ of $R||C_{max}$ with the same optimal solution but every $p'_{i,j} \neq \infty$: set $p'_{i,j} = p_{i,j}$ for any $p_{i,j} \neq \infty$; and if $p_{i,j} = \infty$, set $p'_{i,j}$ to some prohibitively large number, for example, $p'_{i,j} = np_{max} + 1$ where p_{max} is the largest processing time that is not ∞ in P. For $T \leq np_{max}$, there is a schedule for instance I with makespan T if and only if there is a schedule for instance I' with makespan T. Every job in instance I' can be scheduled on any of the machines, so the job-intersection graph G_J for I' is the complete graph K_n. We note that an alternate construction to arrive at the complete job-intersection graph is given at the start of Section 4 in [15]. Therefore, we can carry forward the inapproximability lower bound $3/2$ from the graph balancing problem with two job lengths given in Sect. 5.

Corollary 4. *There is no k-approximation algorithm with $k < 3/2$ for $R||C_{max}$ restricted to instances where the job-intersection graph is the complete graph K_n, unless P = NP.*

From Corollary 4, $R||C_{max}$ restricted to any superclass[2] of the complete job-intersection graphs inherits the $3/2$-inapproximability lower bound of $R||C_{max}$. We name some of these graph classes as they are of interest from a graph-theoretic standpoint. To begin, define a job-intersection graph as a *threshold*

---

[2] For a comprehensive list of superclasses, we recommend the Java application at http://www.graphclasses.org.

graph if it can be constructed by repeatedly performing the following two operations: insert an isolated vertex; or insert a vertex and add edges from this vertex to every other vertex presently in the graph, this vertex is called a *dominating* vertex. All complete graphs are threshold graphs, and three superclasses of threshold graphs are interval graphs, cographs, and split graphs [4, Corollary 7.1.1]. Note that all these graphs belong to the house-free graphs. A graph is called *house free* if the graph does not contain as an induced subgraph the *house graph*, shown in Fig. 6.

Fig. 6. The house graph.

Corollary 5. *There is no k-approximation algorithm with $k < 3/2$ for $R||C_{max}$ restricted to instances where the job-intersection graphs belong to either the threshold graphs, interval graphs, cographs, split graphs, or house-free graphs, unless* P = NP.

9   Conclusion

In this paper we have established several graph classes where $R||C_{max}$ with simple job-intersection structure is either polynomial-time solvable or 3/2-inapproximable. For $R||C_{max}$ with bounded job assignments we have shown that there are polynomial-time algorithms with approximation ratios less than two when the bounds are small. As we have demonstrated, the structure of a job-intersection graph presents another way of investigating the complexity of $R||C_{max}$. However, our work does not address planar job-intersection graphs. $R||C_{max}$ restricted to planar job-intersection graphs seems like it might not be polynomial-time solvable nor 3/2-inapproximable, we would be interested in its complexity.

References

1. Alon, N., Yuster, R., Zwick, U.: Finding and counting given length cycles. Algorithmica **17**(3), 209–223 (1997)
2. Asahiro, Y., Jansson, J., Miyano, E., Ono, H., Zenmyo, K.: Approximation algorithms for the graph orientation minimizing the maximum weighted outdegree. J. Comb. Optim. **22**(1), 78–96 (2011)

3. Asahiro, Y., Miyano, E., Ono, H.: Graph classes and the complexity of the graph orientation minimizing the maximum weighted outdegree. Discret. Appl. Math. **159**(7), 498–508 (2011)
4. Brandstädt, A., Le, V., Spinrad, J.: Graph Classes: A Survey. SIAM, Philadelphia (1999)
5. Chakrabarty, D., Khanna, S., Li, S.: On $(1, \varepsilon)$-restricted assignment makespan minimization. In: 26th Annual ACM-SIAM Symposium on Discrete Algorithms, pp. 1087–1101 (2015)
6. Chang, P.Y., Damodaran, P., Melouk, S.: Minimizing makespan on parallel batch processing machines. Int. J. Prod. Res. **42**(19), 4211–4220 (2004)
7. Chen, L., Jansen, K., Luo, W., Zhang, G.: An efficient PTAS for parallel machine scheduling with capacity constraints. In: Chan, T.-H.H., Li, M., Wang, L. (eds.) COCOA 2016. LNCS, vol. 10043, pp. 608–623. Springer, Cham (2016). https://doi.org/10.1007/978-3-319-48749-6_44
8. Ebenlendr, T., Krčál, M., Sgall, J.: Graph balancing: a special case of scheduling unrelated parallel machines. Algorithmica **68**(1), 62–80 (2014)
9. Ebenlendr, T., Krčál, M., Sgall, J.: Graph balancing: a special case of scheduling unrelated parallel machines. In: 19th Annual ACM-SIAM Symposium on Discrete Algorithms, pp. 483–490 (2008)
10. Gairing, M., Monien, B., Woclaw, A.: A faster combinatorial approximation algorithm for scheduling unrelated parallel machines. Theor. Comput. Sci. **380**(1), 87–99 (2007)
11. Glass, C., Kellerer, H.: Parallel machine scheduling with job assignment restrictions. Nav. Res. Logist. (NRL) **54**(3), 250–257 (2007)
12. Graham, R., Lawler, E., Lenstra, J., Rinnooy, K.: Optimization and approximation in deterministic sequencing and scheduling: a survey. Ann. Discret. Math. **5**, 287–326 (1979)
13. Huang, C., Ott, S.: A combinatorial approximation algorithm for graph balancing with light hyper edges. In: 24th Annual European Symposium on Algorithms. LIPIcs, vol. 57, pp. 49:1–49:15 (2016)
14. Jansen, K., Maack, M., Solis-Oba, R.: Structural parameters for scheduling with assignment restrictions. In: Fotakis, D., Pagourtzis, A., Paschos, V.T. (eds.) CIAC 2017. LNCS, vol. 10236, pp. 357–368. Springer, Cham (2017). https://doi.org/10.1007/978-3-319-57586-5_30
15. Jansen, K., Maack, M., Solis-Oba, R.: Structural parameters for scheduling with assignment restrictions. CoRR abs/1701.07242 (2017). http://arxiv.org/abs/1701.07242
16. Kloks, T., Kratsch, D., Müller, H.: Finding and counting small induced subgraphs efficiently. Inf. Process. Lett. **74**(3–4), 115–121 (2000)
17. Lee, K., Leung, J.Y.T., Pinedo, M.: A note on graph balancing problems with restrictions. Inf. Process. Lett. **110**(1), 24–29 (2009)
18. Lenstra, J., Shmoys, D., Tardos, E.: Approximation algorithms for scheduling unrelated parallel machines. Math. Program. **46**(1–3), 259–271 (1990)
19. Leung, J.Y.T., Li, C.L.: Scheduling with processing set restrictions: a literature update. Int. J. Prod. Econ. **175**, 1–11 (2016)
20. Li, S., Li, G., Zhang, S.: Minimizing makespan with release times on identical parallel batching machines. Discret. Appl. Math. **148**(1), 127–134 (2005)
21. Lin, Y., Li, W.: Parallel machine scheduling of machine-dependent jobs with unit-length. Eur. J. Oper. Res. **156**(1), 261–266 (2004)
22. Page, D.R., Solis-Oba, R.: A 3/2-approximation algorithm for the graph balancing problem with two weights. Algorithms **9**(2), 38 (2016)

23. Saha, B., Srinivasan, A.: A new approximation technique for resource-allocation problems. In: 1st Annual Symposium on Innovations in Computer Science, pp. 342–357 (2010)
24. Shchepin, E., Vakhania, N.: An optimal rounding gives a better approximation for scheduling unrelated machines. Oper. Res. Lett. **33**, 127–133 (2005)
25. Williamson, D., Shmoys, D.: The Design of Approximation Algorithms. Cambridge University Press, Cambridge (2011)

Super-Stability in the Student-Project Allocation Problem with Ties

Sofiat Olaosebikan$^{(\boxtimes)}$ and David Manlove

School of Computing Science, University of Glasgow, Glasgow, Scotland
s.olaosebikan.1@research.gla.ac.uk, David.Manlove@glasgow.ac.uk

Abstract. The *Student-Project Allocation problem with lecturer preferences over Students* (SPA-S) involves assigning students to projects based on student preferences over projects, lecturer preferences over students, and the maximum number of students that each project and lecturer can accommodate. This classical model assumes that preference lists are strictly ordered. Here, we study a generalisation of SPA-S where ties are allowed in the preference lists of students and lecturers, which we refer to as the *Student-Project Allocation problem with lecturer preferences over Students with Ties* (SPA-ST). We investigate stable matchings under the most robust definition of stability in this context, namely *super-stability*. We describe the first polynomial-time algorithm to find a super-stable matching or to report that no such matching exists, given an instance of SPA-ST. Our algorithm runs in $O(L)$ time, where L is the total length of all the preference lists. Finally, we present results obtained from an empirical evaluation of the linear-time algorithm based on randomly-generated SPA-ST instances. Our main finding is that, whilst super-stable matchings can be elusive, the probability of such a matching existing is significantly higher if ties are restricted to the lecturers' preference lists.

1 Introduction

The *Student-Project Allocation problem* (SPA) [4,15] involves sets of students, projects and lecturers, where students are to be assigned to projects offered by lecturers. Applications of SPA can be found in many university departments, for example, the School of Computing Science, University of Glasgow [14], the Faculty of Science, University of Southern Denmark [5], the Department of Computing Science, University of York [13], and elsewhere [3,4,7]. In this setting, lecturers provide a list of projects, and students are required to rank a subset of these projects that they find acceptable, in order of preference. Typically there may be upper bounds on the number of students that each project and lecturer can accommodate. Considering the preferences and the capacities of projects

S. Olaosebikan—Supported by a College of Science and Engineering Scholarship, University of Glasgow.

D. Manlove—Supported by grant EP/P028306/1 from the Engineering and Physical Sciences Research Council.

D. Kim et al. (Eds.): COCOA 2018, LNCS 11346, pp. 357–371, 2018.
https://doi.org/10.1007/978-3-030-04651-4_24

and lecturers, the problem then is to find a *matching* (i.e., an assignment of students to projects such that each student is assigned at most one project, and the capacity constraints on projects and lecturers are not violated), which is optimal in some sense according to the stated preferences.

In this work, we will concern ourselves with a variant of SPA that involves lecturer preferences over students, which is known as the *Student-Project Allocation problem with lecturer preferences over Students* (SPA-S). In this context, it has been argued [21] that a natural property for a matching to satisfy is that of *stability*. Informally, a *stable matching* ensures that no student and lecturer who are not matched together would rather be assigned to each other than remain with their current assignees. Such a pair would have an incentive to form a private arrangement outside of the matching, undermining its integrity. Other variants of SPA in the literature involve lecturer preferences over their proposed projects [12,17,18], lecturer preferences over (student, project) pairs [2], and no lecturer preferences at all [14]. See [5] for a recent survey.

The classical SPA-S model assumes that preferences are strictly ordered. However, this might not be achievable in practice. For instance, a lecturer may be unable or unwilling to provide a strict ordering of all the students who find her projects acceptable. Such a lecturer may be happier to rank two or more students equally in a tie, which indicates that the lecturer is indifferent between the students concerned. This leads to a generalisation of SPA-S which we refer to as the *Student-Project Allocation problem with lecture preferences over Students with Ties* (SPA-ST). If we allow ties in the preference lists of students and lecturers, different stability definitions naturally arise. Suppose M is a matching in an instance of SPA-ST. Informally, we say M is *weakly stable, strongly stable* or *super-stable* if there is no student and lecturer such that if they decide to form an arrangement outside the matching, respectively,

(i) both of them would be better off,
(ii) one of them would be better off and the other no worse off,
(iii) neither of them would be worse off.

With respect to this informal definition, clearly a super-stable matching is strongly stable, and a strongly stable matching is weakly stable. These concepts were first defined and studied by Irving [8] in the context of the Stable Marriage problem with Ties, and subsequently extended to the Hospitals/Residents problem with Ties (HRT) [9,10] (where HRT is the special case of SPA-ST in which each lecturer offers only one project, and the capacity of each project is the same as the capacity of the lecturer offering the project).

Considering the weakest of the three stability concepts mentioned above, every instance of SPA-ST admits a weakly stable matching (this follows by breaking the ties in an arbitrary fashion and applying the stable matching algorithm described in [1] to the resulting SPA-S instance). However, such matchings could be of different sizes [16]. Thus opting for weak stability leads to the problem of finding a weakly stable matching that matches as many students to projects as possible – a problem that is known to be NP-hard [11,16], even for the so-called *Stable Marriage problem with Ties and Incomplete lists*, which is the special case of HRT in which each project (hospital) has capacity 1. Further, a $\frac{3}{2}$-approximation algorithm was described in [6] for the problem of finding a maximum weakly stable matching in an instance of SPA-ST.

Choosing super-stability avoids the problem of finding a weakly stable matching with optimal cardinality, because (i) analogous to the HRT case, all super-stable matchings have the same size [9], (ii) finding one or reporting that none exists can be accomplished in linear-time (as we will see in this paper), and (iii) if a super-stable matching M exists then all weakly stable matchings are of the same size (equal to the size of M), and match exactly the same set of students (see [19] for proof). Furthermore, Irving *et al.* argued in [9] that super-stability is a very natural solution concept in cases where agents have incomplete information. Central to their argument is the following proposition, stated for HRT in [9, Proposition 2], which extends naturally to SPA-ST as follows (see [19] for proof).

Proposition 1. *Let* I *be an instance of SPA-ST, and let M be a matching in* I. *Then M is super-stable in* I *if and only if M is stable in every instance of SPA-S obtained from* I *by breaking the ties in some way.*

In a practical setting, suppose that a student s_i has incomplete information about two or more projects and decides to rank them equally in a tie T, and a super-stable matching M exists in the corresponding SPA-ST instance I, where s_i is assigned to a project in T. Then M is stable in every instance of SPA-S (obtained from I by breaking the ties) that represents the true preferences of s_i. Consequently, we will focus on the concept of super-stability in the SPA-ST context.

Unfortunately not every instance of SPA-ST admits a super-stable matching. This is true, for example, in the case where there are two students, two projects and one lecturer, where the capacity of each project is 1, capacity of the lecturer is 2, and every preference list is a single tie of length 2. Nonetheless, it should be clear from the discussion above that a super-stable matching should be preferred in practical applications when one does exist.

Irving *et al.* [9] described an algorithm to find a super-stable matching given an instance of HRT, or to report that no such matching exists. However, merely reducing an instance of SPA-ST to an instance of HRT and applying the algorithm described in [9] to the resulting HRT instance does not work in general (see [19] for a further explanation).

Our Contribution. In this paper, we describe the first polynomial-time algorithm to find a super-stable matching or to report that no such matching exists, given an instance of SPA-ST – thus solving an open problem given in [1, 15]. Our algorithm runs in time linear in the size of the problem instance. The remaining sections of this paper are structured as follows. We give a formal definition of the SPA-S problem, the SPA-ST variant, and the super-stability concept in Sect. 2. We describe our algorithm for SPA-ST under super-stability in Sect. 3. Further, Sect. 3 also presents our algorithm's correctness results and some structural properties satisfied by the set of super-stable matchings in an instance of SPA-ST. In Sect. 4, we present results arising from an empirical evaluation that investigates how the nature of the preference lists would affect the likelihood of a super-stable matching existing, with respect to randomly-generated SPA-ST instances. Our main finding is that the probability of a super-stable matching existing is significantly higher if ties are restricted to the lecturers' preference lists. Finally, Sect. 5 presents some concluding remarks and potential direction for future work.

2   Preliminary Definitions

2.1   Formal Definition of SPA-S

An instance I of SPA-S involves a set $\mathcal{S} = \{s_1, s_2, \ldots, s_{n_1}\}$ of *students*, a set $\mathcal{P} = \{p_1, p_2, \ldots, p_{n_2}\}$ of *projects* and a set $\mathcal{L} = \{l_1, l_2, \ldots, l_{n_3}\}$ of *lecturers*. Each student s_i ranks a subset of $\mathcal{P}$ in strict order. We denote by A_i the ranked set of projects that s_i finds acceptable. We say that s_i finds p_j *acceptable* if $p_j \in A_i$.

Each lecturer $l_k \in \mathcal{L}$ offers a non-empty set of projects P_k, where $P_1, P_2, \ldots, P_{n_3}$ partitions $\mathcal{P}$, and l_k provides a preference list, denoted by $\mathcal{L}_k$, ranking in strict order of preference those students who find at least one project in P_k acceptable. Also l_k has a capacity $d_k \in \mathbb{Z}^+$, indicating the maximum number of students she is willing to supervise. Similarly each project $p_j \in \mathcal{P}$ has a capacity $c_j \in \mathbb{Z}^+$ indicating the maximum number of students that it can accommodate. We assume that for any lecturer l_k, $\max\{c_j : p_j \in P_k\} \leq d_k \leq \sum\{c_j : p_j \in P_k\}$ (i.e., the capacity of l_k is (i) at least the highest capacity of the projects offered by l_k, and (ii) at most the sum of the capacities of all the projects l_k is offering). We denote by $\mathcal{L}_k^j$, the *projected preference list* of lecturer l_k for p_j, which can be obtained from $\mathcal{L}_k$ by removing those students that do not find p_j acceptable (thereby retaining the order of the remaining students from $\mathcal{L}_k$).

An *assignment* M is a subset of $\mathcal{S} \times \mathcal{P}$ such that $(s_i, p_j) \in M$ implies that s_i finds p_j acceptable. If $(s_i, p_j) \in M$, we say that s_i *is assigned to* p_j, and p_j *is assigned* s_i. For convenience, if s_i is assigned in M to p_j, where p_j is offered by l_k, we may also say that s_i is assigned to l_k, and l_k is assigned s_i.

For any student $s_i \in \mathcal{S}$, we let $M(s_i)$ denote the set of projects assigned to s_i in M. For any project $p_j \in \mathcal{P}$, we denote by $M(p_j)$ the set of students assigned to p_j in M. Project p_j is *undersubscribed*, *full* or *oversubscribed* according as $|M(p_j)|$ is less than, equal to, or greater than c_j, respectively. Similarly, for any lecturer $l_k \in \mathcal{L}$, we denote by $M(l_k)$ the set of students assigned to l_k in M. Lecturer l_k is *undersubscribed*, *full* or *oversubscribed* according as $|M(l_k)|$ is less than, equal to, or greater than d_k, respectively.

A *matching* M is an assignment such that each student is assigned to at most one project in M, each project is assigned at most c_j students in M, and each lecturer is assigned at most d_k students in M. If s_i is assigned to some project in M, for convenience we let $M(s_i)$ denote that project.

In what follows, l_k is the lecturer who offers project p_j.

Definition 1 (stability). Let I' be an instance of SPA-S, and let M be a matching in I'. We say M is *stable* if it admits no blocking pair, where a *blocking pair* is an acceptable pair $(s_i, p_j) \in (\mathcal{S} \times \mathcal{P}) \setminus M$ such that (a) s_i is either unassigned in M or prefers p_j to $M(s_i)$, and (b) either

(i) p_j is undersubscribed and l_k is undersubscribed, or

(ii) p_j is undersubscribed, l_k is full and either $s_i \in M(l_k)$, or l_k prefers s_i to the worst student in $M(l_k)$, or

(iii) p_j is full and l_k prefers s_i to the worst student in $M(p_j)$.

For a full description of an algorithm to find a stable matching in this setting, we refer the interested reader to [1, 15].

2.2   Ties in the Preference Lists

We now define formally the generalisation of SPA-S in which preference lists can include ties. In the preference list of lecturer $l_k \in \mathcal{L}$, a set T of r students forms a *tie of length r* if, for any $s_i, s_{i'} \in T$, l_k does not prefer s_i to $s_{i'}$ (i.e., l_k is *indifferent* between s_i and $s_{i'}$). A tie in a student's preference list is defined similarly. For convenience, in what follows we consider a non-tied entry in a preference list as a tie of length one. We denote by SPA-ST the generalisation of SPA-S in which the preference list of each student (respectively lecturer) comprises a strict ranking of ties, each comprising one or more projects (respectively students).

An example SPA-ST instance I_1 is given in Fig. 1, which involves the set of students $\mathcal{S} = \{s_1, s_2, s_3, s_4, s_5\}$, the set of projects $\mathcal{P} = \{p_1, p_2, p_3\}$ and the set of lecturers $\mathcal{L} = \{l_1, l_2\}$. Ties in the preference lists are indicated by round brackets.

Student preferences	Lecturer preferences	
s_1: p_1	l_1: s_5 $(s_1$ $s_2)$ s_3 s_4	l_1 offers p_1, p_2
s_2: $(p_1$ $p_3)$	l_2: s_4 s_5 s_2	l_2 offers p_3
s_3: p_2		
s_4: p_2 p_3	Project capacities: $c_1 = c_3 = 1$, $c_2 = 2$	
s_5: p_3 p_1	Lecturer capacities: $d_1 = 2$, $d_2 = 1$	

Fig. 1. An example instance I_1 of SPA-ST.

In the context of SPA-ST, we assume that all notation and terminology carries over from Sect. 2.1 as defined for SPA-S with the exception of stability, which we now define. When ties appear in the preference lists, three levels of stability arise (as in the HRT context [9, 10]), namely *weak stability, strong stability and super-stability*. The formal definition for weak stability in SPA-ST follows from the definition for stability in SPA-S (see Definition 1). Moreover, the existence of a weakly stable matching in an instance I of SPA-ST is guaranteed by breaking the ties in I arbitrarily, thus giving rise to an instance I' of SPA-S. Clearly, a stable matching in I' is weakly stable in I. Indeed a converse of sorts holds, which gives rise to the following proposition (see [19] for proof).

Proposition 2. *Let I be an instance of SPA-ST, and let M be a matching in I. Then M is weakly stable in I if and only if M is stable in some instance I' of SPA-S obtained from I by breaking the ties in some way.*

As mentioned earlier, super-stability is the most robust concept to seek in a practical setting. Only if no super-stable matching exists in the underlying problem instance should other forms of stability be sought.

Definition 2 (super-stability). Let I be an instance of SPA-ST, and let M be a matching in I. We say M is *super-stable* if it admits no blocking pair, where a *blocking pair* is an acceptable pair $(s_i, p_j) \in (\mathcal{S} \times \mathcal{P}) \setminus M$ such that (a) either s_i is unassigned in M or s_i prefers p_j to $M(s_i)$ or is indifferent between them; and (b) either

(i) p_j is undersubscribed and l_k is undersubscribed, or
(ii) p_j is undersubscribed, l_k is full, and either $s_i \in M(l_k)$ or l_k prefers s_i to the worst student in $M(l_k)$ or is indifferent between them, or
(iii) p_j is full and l_k prefers s_i to the worst student in $M(p_j)$ or is indifferent between them.

It may be verified that the matching $M = \{(s_3, p_2), (s_4, p_3), (s_5, p_1)\}$ is super-stable in the SPA-ST instance shown in Fig. 1. Clearly, a super-stable matching is weakly stable.

3   An Algorithm for SPA-ST Under Super-stability

In this section we present our algorithm for SPA-ST under super-stability, which we will refer to as Algorithm SPA-ST-super. First, we note that our algorithm is a non-trivial extension of Algorithm SPA-student for SPA-S from [1] and Algorithm HRT-Super-Res for HRT from [9]. Due to the more general setting of SPA-ST, Algorithm SPA-ST-super requires some new ideas, and the proofs of the correctness results are more complex than for the aforementioned algorithms for SPA-S and HRT. In Sect. 3.1, we give a description of our algorithm, before presenting it in pseudocode form. We present the algorithm's correctness results in Sect. 3.2.

3.1   Description of the Algorithm

First, we present some definitions relating to the algorithm. In what follows, I is an instance of SPA-ST, (s_i, p_j) is an acceptable pair in I and l_k is the lecturer who offers p_j. Further, if (s_i, p_j) belongs to some super-stable matching in I, we call (s_i, p_j) a *super-stable pair* and s_i a *super-stable partner* of p_j (and vice-versa).

During the execution of the algorithm, students become *provisionally assigned* to projects. It is possible for a project to be provisionally assigned a number of students that exceed its capacity. This holds analogously for a lecturer. The algorithm proceeds by deleting from the preference lists certain (s_i, p_j) pairs that cannot be super-stable. By the term *delete* (s_i, p_j), we mean the removal of p_j from s_i's preference list and the removal of s_i from $\mathcal{L}_k^j$ (the projected preference list of lecturer l_k for p_j). In addition, if s_i is provisionally assigned to p_j at this point, we break the assignment. By the *head* of a student's preference list at a given point, we mean the set of one or more projects, tied in her preference list after any deletions might have occurred, that she prefers to all other projects in her list.

For project p_j, we define the *tail* of $\mathcal{L}_k^j$ as the least-preferred tie in $\mathcal{L}_k^j$ after any deletions might have occurred (recalling that a tie can be of length one). In the same fashion, we define the *tail* of $\mathcal{L}_k$ (preference list of lecturer l_k) as the least-preferred tie in $\mathcal{L}_k$ after any deletions might have occurred. If s_i is provisionally assigned to p_j, we define the *successors* of s_i in $\mathcal{L}_k^j$ as those students that are worse than s_i in $\mathcal{L}_k^j$. An analogous definition holds for the successors of s_i in $\mathcal{L}_k$.

We now describe our algorithm. **Algorithm SPA-ST-super** begins by initialising an empty set M which will contain the provisional assignments of students to projects (and intuitively to lecturers). We remark that such assignments can subsequently be broken during the algorithm's execution. Also, each project is initially assigned to be empty (i.e., not assigned to any student).

The **while** loop of the algorithm involves each student s_i who is not provisionally assigned to any project in M and who has a non-empty list applying in turn to each project p_j at the head of her list. Immediately, s_i becomes provisionally assigned to p_j in M (and to l_k). If, by gaining a new student, p_j becomes oversubscribed, it turns out that none of the students s_t at the tail of $\mathcal{L}_k^j$ can be assigned to p_j in any super-stable matching – such pairs (s_t, p_j) are deleted. Similarly, if by gaining a new student, l_k becomes oversubscribed, none of the students s_t at the tail of $\mathcal{L}_k$ can be assigned to any project offered by l_k in any super-stable matching – such pairs (s_t, p_u), for each project $p_u \in P_k$ that s_t finds acceptable, are deleted.

Regardless of whether any deletions occurred as a result of the two conditionals described in the previous paragraph, we have two further (possibly non-disjoint) cases in which deletions may occur. If p_j becomes full, we let s_r be any worst student provisionally assigned to p_j (according to $\mathcal{L}_k^j$), and we delete (s_t, p_j) for each successor s_t of s_r in $\mathcal{L}_k^j$. Similarly if l_k becomes full, we let s_r be any worst student provisionally assigned to l_k, and we delete (s_t, p_u), for each successor s_t of s_r in $\mathcal{L}_k$ and for each project $p_u \in P_k$ that s_t finds acceptable. As we will prove later, none of the (student, project) pairs deleted when a project or a lecturer becomes full can be a super-stable pair.

At the point where the **while** loop terminates (i.e., when every student is provisionally assigned to one or more projects or has an empty list), if some project p_j that was previously full ends up undersubscribed, we let s_r be any one of the most preferred students (according to $\mathcal{L}_k^j$) who was provisionally assigned to p_j during some iteration of the algorithm but is not assigned to p_j at this point (for convenience, we henceforth refer to such s_r as the most preferred student rejected from p_j according to $\mathcal{L}_k^j$). If the students at the tail of $\mathcal{L}_k$ (recalling that the tail of $\mathcal{L}_k$ is the least-preferred tie in $\mathcal{L}_k$ after any deletions might have occurred) are no better than s_r, it turns out that none of these students s_t can be assigned to any project offered by l_k in any super-stable matching – such pairs (s_t, p_u), for each project $p_u \in P_k$ that s_t finds acceptable, are deleted. The while loop is then potentially reactivated, and the entire process continues until every student is provisionally assigned to a project or has an empty list.

At the termination of the repeat-until loop, if the set M, containing the provisional assignments of students to projects, is super-stable relative to the given instance I then M is output as a super-stable matching in I. Otherwise, the algorithm reports that no super-stable matching exists in I. We present Algorithm SPA-ST-super in pseudocode form in Algorithm 1.

3.2   Correctness of Algorithm SPA-ST-super

We now present the following results regarding the correctness of Algorithm SPA-ST-super. For several lemmas in this section, we either omit the proof or provide a sketch proof; see [19] for the full proofs. The first of these results deals with the fact that no super-stable pair is ever deleted during the execution of the algorithm. In what follows, I is an instance of SPA-ST, (s_i, p_j) is an acceptable pair in I and l_k is the lecturer who offers p_j.

Lemma 1. *If a pair (s_i, p_j) is deleted during the execution of Algorithm SPA-ST-super, then (s_i, p_j) does not belong to any super-stable matching in I.*

Proof. (Sketch). Suppose (s_i, p_j) is the first super-stable pair to be deleted during some execution of the algorithm, which belongs to some super-stable matching, say M^*. Let l_k be the lecturer who offers p_j. We consider five points in the algorithm at which (s_i, p_j) could be deleted. If (s_i, p_j) is deleted because s_i is in the tail of $\mathcal{L}_k^j$ when p_j became oversubscribed, we show that one of the students provisionally assigned to p_j at this point must form a blocking pair for M^* with p_j, a contradiction. Similarly, if (s_i, p_j) is deleted because s_i is in the tail of $\mathcal{L}_k$ when l_k became oversubscribed, we show that there is some project $p_{j'} \in P_k$ and some student s_r, with s_r provisionally assigned to $p_{j'}$ in M^* at this point, such that $(s_r, p_{j'})$ must form a blocking pair for M^*, a contradiction. Further, if (s_i, p_j) is deleted because s_i is a successor of a worst student provisionally assigned to p_j when p_j became full, we show that one of the students provisionally assigned to p_j at this point must form a blocking pair for M^* with p_j, a contradiction. Similarly, if (s_i, p_j) is deleted because s_i is a successor of a worst student provisionally assigned to l_k when l_k became full, we show that there is some project $p_{j'} \in P_k$ and some student s_r, with s_r provisionally assigned to $p_{j'}$ in M^* at this point, such that $(s_r, p_{j'})$ must form a blocking pair for M^*, a contradiction. Finally, suppose (s_i, p_j) is deleted at line 33 of the algorithm, suppose $p_{j'}$ is a project offered by l_k which was previously full but ended up undersubscribed in line 27 of the algorithm, and suppose s_r is the most preferred student rejected from $p_{j'}$ according to $\mathcal{L}_k^{j'}$. We show that, in order to avoid the pair $(s_{i'}, p_{j'})$ from blocking M^*, we can construct an infinite sequence of distinct students, a contradiction to the finite size of the instance. □

The next three lemmas deal with the case that Algorithm SPA-ST-super reports the non-existence of a super-stable matching in I.

Algorithm 1. Algorithm SPA-ST-super

Input: SPA-ST instance I
Output: a super-stable matching M in I or "no super-stable matching exists in I"
1: $M \leftarrow \emptyset$
2: **for each** $p_j \in \mathcal{P}$ **do**
3: $\quad$ full$(p_j) =$ false
4: **repeat**
5: $\quad$ **while** some student s_i is unassigned and has a non-empty list **do**
6: $\quad\quad$ **for each** project p_j at the head of s_i's list **do**
7: $\quad\quad\quad$ $l_k \leftarrow$ lecturer who offers p_j
8: $\quad\quad\quad$ /* s_i applies to p_j */
9: $\quad\quad\quad$ $M \leftarrow M \cup \{(s_i, p_j)\}$ / *provisionally assign s_i to p_j (and to l_k) */
10: $\quad\quad\quad$ **if** p_j is oversubscribed **then**
11: $\quad\quad\quad\quad$ **for each** student s_t at the tail of $\mathcal{L}_k^j$ **do**
12: $\quad\quad\quad\quad\quad$ delete (s_t, p_j)
13: $\quad\quad\quad$ **else if** l_k is oversubscribed **then**
14: $\quad\quad\quad\quad$ **for each** student s_t at the tail of $\mathcal{L}_k$ **do**
15: $\quad\quad\quad\quad\quad$ **for each** project $p_u \in P_k \cap A_t$ **do**
16: $\quad\quad\quad\quad\quad\quad$ delete (s_t, p_u)
17: $\quad\quad\quad$ **if** p_j is full **then**
18: $\quad\quad\quad\quad$ full$(p_j) =$ true
19: $\quad\quad\quad\quad$ $s_r \leftarrow$ worst student assigned to p_j according to $\mathcal{L}_k^j$ {any if > 1}
20: $\quad\quad\quad\quad$ **for each** successor s_t of s_r on $\mathcal{L}_k^j$ **do**
21: $\quad\quad\quad\quad\quad$ delete (s_t, p_j)
22: $\quad\quad\quad$ **if** l_k is full **then**
23: $\quad\quad\quad\quad$ $s_r \leftarrow$ worst student assigned to l_k according to $\mathcal{L}_k$ {any if > 1}
24: $\quad\quad\quad\quad$ **for each** successor s_t of s_r on $\mathcal{L}_k$ **do**
25: $\quad\quad\quad\quad\quad$ **for each** project $p_u \in P_k \cap A_t$ **do**
26: $\quad\quad\quad\quad\quad\quad$ delete (s_t, p_u)
27: $\quad$ **if** some project p_j is undersubscribed and full(p_j) is true **then**
28: $\quad\quad$ $l_k \leftarrow$ lecturer who offers p_j
29: $\quad\quad$ $s_r \leftarrow$ most preferred student rejected from p_j according to $\mathcal{L}_k^j$ {any if > 1}
30: $\quad\quad$ **if** the students at the tail of $\mathcal{L}_k$ are no better than s_r **then**
31: $\quad\quad\quad$ **for each** student s_t at the tail of $\mathcal{L}_k$ **do**
32: $\quad\quad\quad\quad$ **for each** project $p_u \in P_k \cap A_t$ **do**
33: $\quad\quad\quad\quad\quad$ delete (s_t, p_u)
34: **until** every unassigned student has an empty list
35: **if** M is a super-stable matching in I **then**
36: $\quad$ **return** M
37: **else**
38: $\quad$ **return** "no super-stable matching exists in I"

Lemma 2. *If any student is multiply assigned at the termination of Algorithm SPA-ST-super, then I admits no super-stable matching.*

Lemma 3. *If some lecturer l_k becomes full during some execution of Algorithm SPA-ST-super and l_k subsequently ends up undersubscribed at the termination of the algorithm, then I admits no super-stable matching.*

Lemma 4. *If the pair (s_i, p_j) was deleted during some execution of Algorithm SPA-ST-super, and at the termination of the algorithm s_i is not* *assigned to a project better than p_j, and each of p_j and l_k is undersubscribed, then I admits no super-stable matching.*

The next lemma shows that the final assignment may be used to determine the existence, or otherwise, of a super-stable matching in I.

Lemma 5. *If at the termination of Algorithm spa-st-super, the assignment M is not super-stable in I, then no super-stable matching exists in I.*

Proof. Suppose M is not super-stable in I. If some student s_i is multiply assigned in M, then I admits no super-stable matching, by Lemma 2. Hence suppose no student is multiply assigned in M. Then M is a matching, since no project or lecturer is oversubscribed in M. Let (s_i, p_j) be a blocking pair of M in I, then s_i is either unassigned in M or prefers p_j to $M(s_i)$ or is indifferent between them. Whichever of these is the case, (s_i, p_j) has been deleted and therefore does not belong to any super-stable matching, by Lemma 1. Let l_k be the lecturer who offers p_j. If (s_i, p_j) was deleted as a result of l_k being full or oversubscribed, (s_i, p_j) could only block M if l_k ends up undersubscribed in M. If this is the case, then I admits no super-stable matching, by Lemma 3.

Now suppose (s_i, p_j) was deleted as a result of p_j being full or oversubscribed. Suppose firstly that p_j ends up full in M. Then (s_i, p_j) cannot block M irrespective of whether l_k is undersubscribed or full in M, since l_k prefers the worst assigned student(s) in $M(p_j)$ to s_i. Hence suppose p_j ended up undersubscribed in M. As p_j was previously full, each pair (s_t, p_u), for each s_t that is no better than s_i at the tail of $\mathcal{L}_k$ and each $p_u \in P_k \cap A_t$, would have been deleted in line 33 of the algorithm. Thus if l_k is full in M, then (s_i, p_j) does not block M. Suppose l_k is undersubscribed in M. If l_k was full at some point during the execution of the algorithm, then I admits no super-stable matching, by Lemma 3. Suppose l_k was never full during the algorithm's execution. As (s_i, p_j) is a blocking pair of M, s_i cannot be assigned in M a project that she prefers to p_j. Hence I admits no super-stable matching, by Lemma 4.

Finally suppose (s_i, p_j) was deleted (at line 33) because some other project $p_{j'}$ offered by l_k was previously full and ended up undersubscribed at line 27. Then l_k must have identified her most preferred student, say s_r, rejected from $p_{j'}$ such that s_i is at the tail of $\mathcal{L}_k$ and s_i is no better than s_r in $\mathcal{L}_k$. Moreover, every project offered by l_k that s_i finds acceptable would have been deleted from s_i's preference list at the for loop iteration in line 33. If p_j ends up full in M, then (s_i, p_j) does not block M. Suppose p_j ends up undersubscribed in M. If l_k is full in M, then (s_i, p_j) does not block M, since $s_i \notin M(l_k)$ and l_k prefers the worst student in $M(l_k)$ to s_i. Suppose l_k is undersubscribed in M, again by Lemma 4, I admits no super-stable matching.   □

The next lemma shows that Algorithm spa-st-super may be implemented to run in linear time.

Lemma 6. *Algorithm spa-st-super may be implemented to run in $O(L)$ time and $O(n_1 n_2)$ space, where n_1, n_2, and L are the number of students, number of projects, and the total length of the preference lists, respectively, in I.*

Lemma 1 shows that there is an optimality property for each assigned student in any super-stable matching found by the algorithm, whilst Lemma 5 establishes the correctness of Algorithm SPA-ST-super. The following theorem collects together Lemmas 1, 5 and 6.

Theorem 1. *For a given instance I of SPA-ST, Algorithm SPA-ST-super determines, in $O(L)$ time and $O(n_1 n_2)$ space, whether or not a super-stable matching exists in I. If such a matching does exist, all possible executions of the algorithm find one in which each assigned student is assigned to the best project that she could obtain in any super-stable matching, and each unassigned student is unassigned in all super-stable matchings.*

3.3   Properties of Super-Stable Matchings in SPA-ST

The following theorem, which we will refer to as the *Unpopular Projects Theorem*, gives some properties of super-stable matchings in an instance of SPA-ST that are analogous to those satisfied by stable matchings in SPA-S [1, Theorem 4.1].

Theorem 1. *For a given instance I of SPA-ST, the following holds.*

1. *Each lecturer is assigned the same number of students in all super-stable matchings.*
2. *Exactly the same students are unassigned in all super-stable matchings.*
3. *A project offered by an undersubscribed lecturer has the same number of students in all super-stable matchings.*

4   Empirical Evaluation

In this section, we evaluate an implementation of Algorithm SPA-ST-super. We implemented our algorithm in Python[1], and performed our experiments on a system with dual Intel Xeon CPU E5-2640 processors with 64GB of RAM, running Ubuntu 14.04. For our experiment, we were primarily concerned with the following question: how does the nature of the preference lists in a given SPA-ST instance affect the existence of a super-stable matching?

4.1   Datasets

For datasets, there are clearly several parameters that can be varied, such as the number of students, projects and lecturers; the lengths of the students' preference list as well as a measure of the density of ties present in the preference lists. We denote by t_d, the measure of the density of ties present in the preference lists. In each student's preference list, the tie density t_{d_s} $(0 \le t_{d_s} \le 1)$ is the probability that some project is tied to its successor. The tie density t_{d_l} in each lecturer's preference list is defined similarly. At $t_{d_s} = t_{d_l} = 1$, each preference lists would be contained in a single tie while at $t_{d_s} = t_{d_l} = 0$, no tie would exist in the preference lists, thus reducing the problem to an instance of SPA.

---

[1] https://github.com/sofiatolaosebikan/spa-st-super-stability.

4.2   Experimental Setup

For each range of values for the aforementioned parameters, we randomly-generated a set of SPA-ST instances, involving n_1 students (which we will henceforth refer to as the size of the instance), $0.5n_1$ projects, $0.2n_1$ lecturers and $1.5n_1$ total project capacity which was randomly distributed amongst the projects. The capacity for each lecturer l_k was chosen randomly to lie between the highest capacity of the projects offered by l_k and the sum of the capacities of the projects that l_k offers. In each set, we measured the proportion of instances that admit a super-stable matching. It is worth mentioning that when we varied the tie density on both the students' and lecturers' preference lists between 0.1 and 0.5, super-stable matchings were very elusive, even with an instance size of 100 students. Thus, for the purpose of our experiment, we decided to choose a low tie density.

Experiment 1. In the first experiment, we increased the number of students n_1 while maintaining a constant ratio of projects, lecturers, project capacities and lecturer capacities as described above. For various values of n_1 ($100 \leq n_1 \leq 1000$) in increments of 100, we created 1000 randomly-generated instances. The length of each student's preference list was fixed at 50. For all the preference lists, we set $t_{d_s} = t_{d_l} = 0.005$ (on average, 1 out of 5 students has a single tie of length 2 in their preference list, and this holds similarly for the lecturers). The result displayed in Fig. 2 shows that the proportion of instances that admit a super-stable matching decreases as the number of students increases.

Experiment 2. In our second experiment, we varied the length of each student's preference list while maintaining the number of students, projects, lecturers, project capacities and lecturer capacities, and tie density in the students' and lecturers' preference lists, as in Experiment 1. For various values of n_1 ($100 \leq n_1 \leq 1000$) in increments of 100, we created 1000 randomly-generated instances. In the first part of this experiment, the length of each student's preference list was set to 10, and in the second part, the length of each student's preference list was chosen randomly to lie between $0.25n_1$ and $0.5n_1$. The result is displayed in Fig. 3. Although this result shows that more instances admitted a super-stable matching when the preference list was longer compared to when the preference list was fixed at 10, but this difference is not very significant.

Experiment 3. In our last experiment, we investigated how the variation in tie density in both the students' and lecturers' preference lists affects the existence of a super-stable matching. To achieve this, we varied the tie density in the students' preference lists t_{d_s} ($0 \leq t_d \leq 0.05$) and the tie density in the lecturers' preference lists t_{d_l} ($0 \leq t_d \leq 0.05$), both in increments of 0.005. For each pair of tie densities in $t_{d_s} \times t_{d_l}$ we randomly-generated 1000 SPA-ST instances for various values of n_1 ($100 \leq n_1 \leq 1000$) in increments of 100. For each of these instances, we maintained the same ratio of projects, lecturers, project capacities and lecturer capacities as in Experiment 1. We also fixed the length of each student's preference list at 50. The result displayed in Fig. 4 shows that increasing the tie density in both the students' and lecturers' preference lists reduces the proportion of instances that admit a super-stable matching. In fact, this proportion reduces further as the size of the instance increases. However, it was interesting to see that when we fixed the tie density in the students' preference lists at 0 and varied the tie density in the lecturers' preference lists, about 75% of the randomly-generated SPA-ST instances involving 1000 students admitted a super-stable matching.

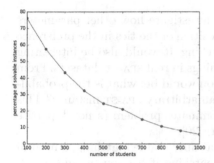

Fig. 2. Result for Experiment 1.   **Fig. 3.** Result for Experiment 2.

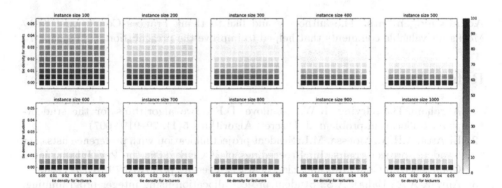

Fig. 4. Each of the coloured square boxes represent the proportion of the 1000 randomly-generated SPA-ST instances that admits a super-stable matching, with respect to the tie density in the students' and lecturers' preference lists. See the colour bar transition, as this proportion ranges from dark (100%) to light (0%).

5   Discussions and Concluding Remarks

Based on the instances we generated randomly, the experimental results suggest that as we increase the size of the instance and the density of ties in the preference lists, the likelihood of a super-stable matching existing decreases. There was no significant uplift in this likelihood even as we increased the lengths of the students' preference lists. Moreover, when the ties occur only in the lecturers' preference lists, we found that a significantly higher proportion of instances admit a super-stable matching. Given that there are typically more students than lecturers in practical applications, it could be that only lecturers are permitted to have some form of indifference over the students that they find acceptable, whilst each student might be able to provide a strict ordering over what may be a small number of projects that she finds acceptable. On the other hand we did not find the same uplift for the case where ties belong to the students' preference lists only.

Further evaluation of our algorithm could investigate how other parameters (e.g., the popularity of some projects, or the position of the ties in the preference lists) affect the existence of a super-stable matching. It would also be interesting to examine the existence of super-stable matchings in real SPA-ST datasets. From a theoretical perspective, an interesting question would be: what is the probability of a super-stable matching existing, given an arbitrary SPA-ST instance? This question has been explored for the Stable Roommates problem (a non-bipartite generalisation of the Stable Marriage problem) [20].

To cope with the possible non-existence of a super-stable matching, a natural strategy would be to seek a strongly stable matching if one exists, and if not, settle for a weakly stable matching. As noted in Sect. 1, every instance of SPA-ST admits a weakly stable matching. We leave open the problem of constructing an algorithm for SPA-ST under the strong stability criterion.

Acknowledgements. The authors would like to thank Frances Cooper and Kitty Meeks for valuable comments that helped to improve the presentation of this paper.

References

1. Abraham, D.J., Irving, R.W., Manlove, D.F.: Two algorithms for the student-project allocation problem. J. Discret. Algorithms **5**(1), 79–91 (2007)
2. El-Atta, A.H.A., Moussa, M.I.: Student project allocation with preference lists over (student, project) pairs. In: Proceedings of ICCEE 2009: the 2nd International Conference on Computer and Electrical Engineering, pp. 375–379. IEEE (2009)
3. Anwar, A.A., Bahaj, A.S.: Student project allocation using integer programming. IEEE Trans. Educ. **46**(3), 359–367 (2003)
4. Calvo-Serrano, R., Guillén-Gosálbez, G., Kohn, S., Masters, A.: Mathematical programming approach for optimally allocating students' projects to academics in large cohorts. Educ. Chem. Eng. **20**, 11–21 (2017)
5. Chiarandini, M., Fagerberg, R., Gualandi, S.: Handling preferences in student-project allocation. Ann. Oper. Res. (2017). https://doi.org/10.1007/s10479-017-2710-1
6. Cooper, F., Manlove, D.F.: A 3/2-approximation algorithm for the student-project allocation problem. In: Proceedings of SEA 2018. LIPIcs, vol. 103 pp. 8:1–8:13 (2018)

7. Harper, P.R., de Senna, V., Vieira, I.T., Shahani, A.K.: A genetic algorithm for the project assignment problem. Comput. Oper. Res. **32**, 1255–1265 (2005)
8. Irving, R.W.: Stable marriage and indifference. Discret. Appl. Math. **48**, 261–272 (1994)
9. Irving, R.W., Manlove, D.F., Scott, S.: The hospitals/residents problem with ties. SWAT 2000. LNCS, vol. 1851, pp. 259–271. Springer, Heidelberg (2000). https://doi.org/10.1007/3-540-44985-X_24
10. Irving, R.W., Manlove, D.F., Scott, S.: Strong stability in the hospitals/residents problem. In: Alt, H., Habib, M. (eds.) STACS 2003. LNCS, vol. 2607, pp. 439–450. Springer, Heidelberg (2003). https://doi.org/10.1007/3-540-36494-3_39
11. Iwama, K., Miyazaki, S., Morita, Y., Manlove, D.: Stable marriage with incomplete lists and ties. In: Wiedermann, J., van Emde Boas, P., Nielsen, M. (eds.) ICALP 1999. LNCS, vol. 1644, pp. 443–452. Springer, Heidelberg (1999). https://doi.org/10.1007/3-540-48523-6_41
12. Iwama, K., Miyazaki, S., Yanagisawa, H.: Improved approximation bounds for the student-project allocation problem with preferences over projects. J. Discret. Algorithms **13**, 59–66 (2012)
13. Kazakov, D.: Co-ordination of student-project allocation. Manuscript, University of York, Department of Computer Science (2001). http://www-users.cs.york.ac.uk/kazakov/papers/proj.pdf. Accessed 8 Mar 2018
14. Kwanashie, A., Irving, R.W., Manlove, D.F., Sng, C.T.S.: Profile-based optimal matchings in the student/project allocation problem. In: Kratochvíl, J., Miller, M., Froncek, D. (eds.) IWOCA 2014. LNCS, vol. 8986, pp. 213–225. Springer, Cham (2015). https://doi.org/10.1007/978-3-319-19315-1_19
15. Manlove, D.F.: Algorithmics of Matching Under Preferences. World Scientific, Singapore (2013)
16. Manlove, D.F., Irving, R.W., Iwama, K., Miyazaki, S., Morita, Y.: Hard variants of stable marriage. Theor. Comput. Sci. **276**(1–2), 261–279 (2002)
17. Manlove, D., Milne, D., Olaosebikan, S.: An integer programming approach to the student-project allocation problem with preferences over projects. In: Lee, J., Rinaldi, G., Mahjoub, A.R. (eds.) ISCO 2018. LNCS, vol. 10856, pp. 313–325. Springer, Cham (2018). https://doi.org/10.1007/978-3-319-96151-4_27
18. Manlove, D.F., O'Malley, G.: Student project allocation with preferences over projects. J. Discret. Algorithms **6**, 553–560 (2008)
19. Olaosebikan, S., Manlove, D.F.: Super-stability in the student-project allocation problem with ties. http://arxiv.org/abs/1805.09887
20. Pittel, B.G., Irving, R.W.: An upper bound for the solvability probability of a random stable roommates instance. Random Struct. Algorithms **5**, 465–486 (1994)
21. Roth, A.E.: The evolution of the labor market for medical interns and residents: a case study in game theory. J. Polit. Econ. **92**(6), 991–1016 (1984)

Primal Dual Algorithm for Partial Set Multi-cover

Yingli Ran[1], Yishuo Shi[2], and Zhao Zhang[1]([✉])

[1] College of Mathematics, Physics, and Information Engineering,
Zhejiang Normal University, Jinhua 321004, Zhejiang, China
`724609171@qq.com`, `hxhzz@sina.com`
[2] Institute of Information Science, Academia Sinica, Taipei 11529, Taiwan
`80491471@qq.com`

Abstract. In a minimum partial set multi-cover problem (MinPSMC), given an element set E, a collection of subsets $\mathcal{S} \subseteq 2^E$, a cost w_S on each set $S \in \mathcal{S}$, a covering requirement r_e for each element $e \in E$, and an integer k, the goal is to find a sub-collection $\mathcal{F} \subseteq \mathcal{S}$ to fully cover at least k elements such that the cost of $\mathcal{F}$ is as small as possible, where element e is fully covered by $\mathcal{F}$ if it belongs to at least r_e sets of $\mathcal{F}$. On the application side, the problem has its background in the seed selection problem in a social network. On the theoretical side, it is a natural combination of the minimum partial (single) set cover problem (MinPSC) and the minimum set multi-cover problem (MinSMC). Although both MinPSC and MinSMC admit good approximations whose performance ratios match those lower bounds for the classic set cover problem, previous studies show that theoretical study on MinPSMC is quite challenging. In this paper, we prove that MinPSMC cannot be approximated within factor $O(n^{\frac{1}{2(\log\log n)^c}})$ under the ETH assumption. A primal dual algorithm for MinPSMC is presented with a guaranteed performance ratio $O(\sqrt{n})$ when $r_{\max}$ and f are constants, where $r_{\max} = \max_{e \in E} r_e$ is the maximum covering requirement and f is the maximum frequency of elements (that is the maximum number of sets containing a common element). We also improve the ratio for a restricted version of MinPSMC which possesses a graph-type structure.

Keywords: Positive influence seeding problem
Partial set multi-cover problem · Densest l-subgraph problem
Approximation algorithm

1 Introduction

The study of this paper is motivated by a *seed selection problem* in a social network. Social network is an important medium for the spread of information and opinions. How information is spread depends on the structure of the network and how opinions are spread depends on the mechanism of influence. One of the most important topics people concern about is to which extent an opinion can

D. Kim et al. (Eds.): COCOA 2018, LNCS 11346, pp. 372–385, 2018.
https://doi.org/10.1007/978-3-030-04651-4_25

be accepted. Following the seminal work of Kempe *et al.* [1] on the influence maximization problem, there are a huge body of studies in this field. Most studies are on probabilistic spreading models such as the *linear threshold model* or the *independent cascade model*. Good performance ratios were achieved by exploring the submodularity of the influence function [1,2]. On the other hand, study on a deterministic model is extremely hard, in which the influence mechanism is such that a node is activated only when at least a predetermined fraction of its neighbors are in the seed set. In fact, Chen proved in [3] that for this model, the minimum seeding problem does not admit an $O(2^{\log^{1-\epsilon} n})$-approximation unless $NP \subseteq DTIME(n^{polylog(n)})$, where n is the number of nodes in the network. However, if one only considers one-step of influence (the goal of which is to select the minimum number of seeds to influence all people *in one time slot*), then the problem is a special case of the minimum set multi-cover problem (which will be explained in the related work section), and thus admits good approximation.

In the real world, because of economic reasons, it is often more cost-effective to influence only a fraction of people. Such a consideration leads to the *minimum partial seeding problem*, which is a special case of the *minimum partial set multi-cover problem* (MinPSMC): given an element set E consisting of n elements, a collection of subsets $\mathcal{S} : 2^E \mapsto \mathbb{R}^+$, a nonnegative weight w_S for each subset $S \in \mathcal{S}$, a covering requirement r_e for each element $e \in E$, and an integer $k \leq n$, the MinPSMC problem is to find a minimum weight sub-collection $\mathcal{F} \subseteq \mathcal{S}$ such that at least k elements are fully covered by $\mathcal{F}$, where an element e is fully covered by $\mathcal{F}$ means that e is contained in at least r_e sets of $\mathcal{F}$, and the weight of sub-collection $\mathcal{F}$ is $w(\mathcal{F}) = \sum_{S \in \mathcal{F}} w_S$. An instance of MinPSMC is denoted as $(E, \mathcal{S}, w, r, k)$. The *minimum partial (single) set cover problem* (MinPSC) and the *minimum set multi-cover problem* (MinSMC) are special cases of the MinPSMC problem, with $r_e \equiv 1$ for MinPSC and $k = n$ for MinSMC.

There are a lot of studies on MinPSC and MinSMC, achieving tight performance ratios matching those lower bounds for the classic set cover problem. However, the study on MinPSMC is very rare. According to recent studies [4,5], this problem is quite challenging, and it is guessed that MinPSMC cannot be approximated within a factor of $O(n^c)$ for some constant c.

1.1   Related Works

In a *one-step minimum seeding problem*, the goal is to select the minimum number of seeds to influence all people in one time slot. This problem is also known as *minimum positive dominating set problem* (MinPDS) in [6] which can be defined as follows: given a graph $G = (V, E)$, a constant $0 < \rho \leq 1$, the goal is to find a node set $D \subseteq V$ with the minimum size such that every vertex v in V has at least $\lceil \rho d(v) \rceil$ neighbors in D, where $d(v)$ is the degree of node v in G. It can be viewed as a special case of MinSMC by setting $E = V(G)$, $\mathcal{S} = \{N_G(v) : v \in V\}$ where $N_G(v)$ is the set of neighbors of v in G, and $r_v = \lceil \rho d(v) \rceil$. Wang et al. [7,8] proved that the MinPDS problem is APX-hard and proposed a greedy algorithm with performance ratio $H(\Delta)$, where Δ is the maximum degree of the graph and

$H(\Delta) = \sum_{i=1}^{\Delta} 1/i$ is the Harmonic number. The same ratio was obtained by Dinh $et\ al.$ [6] by observing the relation between MinPDS and MinSMC.

The minimum set cover problem (MinSC) was one of the 21 problems shown to be NP-hard in Karp's seminal paper [9]. In [10], Feige $et\ al.$ proved that unless $NP \subseteq DTIME\left(n^{O(\log \log n)}\right)$, MinSC does not admit performance ratio $\rho \ln n$ for any $0 < \rho < 1$, where n is the number of elements. Dinur and Steurer proved in [11] that this lower bound holds if $P \neq NP$. For the cardinality version of MinSC, Johnson [12] and Lovász [13] obtained a greedy $H(\Delta)$-approximation algorithm, where Δ is the maximum cardinality of a set and $H(\Delta) = \sum_{i=1}^{\Delta} 1/i$ is the Harmonic number. The same performance ratio was obtained for the weighted version of MinSC by Chvatal $et\ al.$ [14]. Another well-known performance ratio for MinSC is f, the maximum number of sets containing a common element [15], which can be achieved by either an LP rounding algorithm [16] or a local ratio method [17]. By [18], ratio f is also best possible.

For MinSMC, Vazirani $et\ al.$ [15] showed that a greedy algorithm achieves performance ratio $H(n)$ using $dual\ fitting$ analysis. The same performance ratio was obtained by a primal-dual algorithm presented by Rajagopalan $et\ al.$ [19].

The MinPSC problem was first studied by Kearns [20], and a greedy algorithm was presented with performance ratio at most $2H(n) + 3$. Slavík $et\ al.$ [21] improved the algorithm, obtaining performance ratio $\min\{H(\Delta), H(\lceil pn \rceil)\}$. Using primal-dual method, Gandhi $et\ al.$ [22] gave an approximation algorithm with performance ratio f. The same performance ratio f was also obtained by Bar-Yehuda [23] using local ratio method.

It can be seen from the above related work that both MinSMC and MinPSC have approximation algorithms with the best possible performance ratios, matching those for the classic set cover problem. On the contrary, study on MinPSMC seems very difficult. Ran $et\ al.$ [4] were the first to obtain a guaranteed performance ratio for the MinPSMC problem. However, their ratio is meaningful only when the covering percentage $p = k/n$ is very close to 1. Afterwards, in [5], Ran $et\ al.$ presented a simple greedy algorithm for MinPSMC achieving performance ratio Δ. Notice that Δ can be as large as n, and in terms of Δ, the performance ratio for MinSMC and MinPSC is of order $\ln \Delta$. In [5], the authors presented a local ratio algorithm for MinPSMC, which reveals a "shock wave" phenomenon: their performance ratio is f for both MinPSC and MinSMC (which is best possible), but for MinPSMC, the ratio jumps abruptly to $O(n)$ even when the covering percentage p is smaller than 1 by a very small constant. In view of these results, the study of MinPSMC seems to be very challenging.

1.2 Our Contribution

In this paper, we study MinPSMC obtaining the following results.

(i) We prove a lower bound for MinPSMC. The proof makes use of the $densest$ l-$subgraph\ problem$ (DlS), the goal of which is to find a subgraph on l vertices which has the maximum number of edges among all subgraphs on l vertices.

We show that if MinPSMC has a γ-approximation, then DlS has a $2\gamma^2$-approximation. It is shown in [24] that under the ETH assumption, DlS can not be approximated within factor $O(n^{\frac{1}{(\log \log n)^c}})$ for some constant c. So, the same lower bound holds for MinPSMC.

(ii) Under the assumption that the maximum covering requirement $r_{\max} = \max\{r_e : e \in E\}$ is upper bounded by a constant, we present a primal dual algorithm for MinPSMC, obtaining performance ratio $B + \sqrt{B \cdot n}$, where $B = \max\{\binom{f_e}{r_e} : e \in E\}$ and f_e is the number of sets containing element e. To use the primal-dual method, how to design a linear program based on which a good approximation can be achieved is a crucial step. We propose a novel integer program the relaxation of which (using Lovász extension [25]) is a convex program. Using the fact that for a submodular function, its Lovász extension coincides with its convex closure, we modify it into a linear program. Although the linear program has exponential number of variables, we show that our primal-dual algorithm can be executed in polynomial time, making use of an efficient algorithm for minimizing a submodular function divided by a modular function [26]. Our algorithm consists of two stages. The first stage is a primal dual algorithm. After the first stage, the sub-collection of sets selected by the last iteration may fully cover much more elements than required by the remaining covering requirement. Hence the second stage refines the solution by iteratively implementing submodular minimization algorithms [26].

(iii) We improve the performance ratio for a restricted version of MinPSMC, in which $w_S \equiv 1$, $r_e \equiv 2$, and $f_e \equiv 2$, where $f_e = |\{S \in \mathcal{S} : e \in S\}|$ is the frequency of element e. Denote such a restricted problem as MinRPSMC. This restricted version looks more like an optimization problem on a graph. Making use of structural properties of graphs, the performance ratio can be improved to $1 + \sqrt{2}n^{1/4}$.

For limited space, detailed proofs will be contained in a full version.

2 Lower Bound for MinPSMC

A lower bound for MinPSMC can be proved by a reduction from the *densest l-subgraph problem* (DlS). Given a graph $G = (V, E)$ with $|V| = n$ and an integer $l \leq n$, the DlS problem asks for a vertex subset C on l vertices such that the subgraph of G induced by C, denoted as $G[C]$, has the maximum number of edges among all subgraphs of G on l vertices.

Theorem 1. *If MinPSMC has a polynomial-time γ-approximation, then DlS has a polynomial-time $2\gamma^2$-approximation.*

Sketch of the Proof. First, it can be shown that for any vertex set C, we can construct in polynomial time a vertex set $C' \subseteq C$ on $|C|/\gamma$ vertices such that $G[C']$ has at least $|E(G[C'])|/(2\gamma^2)$ edges. Second, for any DlS instance (G, l), we can construct a restricted instance of MinPSMC in which $w_S \equiv 1$, $r_e \equiv 2$, and $f_e \equiv 2$, and there is a one-to-one correspondence between a vertex set C of G and a subcollection of sets $\mathcal{F}$ in the restricted MinPSMC instance, such that the number of edges in $G[C]$ equals the number of elements fully covered by $\mathcal{F}$. Making use of such a relation, it can be proved that if we know the optimal value opt_D of the DlS instance (which is the number of edges in a densest subgraph of G on l vertices), then a γ-approximation algorithm for MinPSMC can be implemented to yield a $2\gamma^2$-approximation for the DlS instance. As to opt_D, it can be guessed by executing the above process for $|E|$ rounds. □

Up to now, the best known performance ratio for DlS is $O(n^{\frac{1}{4}+\varepsilon})$, where $\varepsilon > 0$ is an arbitrary real number [27]. Very recently, Manurangsi [24] showed that DlS has no $n^{\frac{1}{(\log\log n)^c}}$-approximation assuming the exponential time hypothesis (ETH), where $c > 0$ is a constant independent of n. Hence we have the following corollary.

Corollary 1. *MinPSMC cannot be approximated within factor $O(n^{\frac{1}{2(\log\log n)^c}})$ under the ETH assumption.*

3   Primal Dual Algorithm

The following notations and assumptions will be used in this paper. For a sub-collection $\mathcal{F} \subseteq \mathcal{S}$, denote by $\mathcal{C}(\mathcal{F})$ the set of elements fully covered by $\mathcal{F}$. For an element $e \in E$, let f_e be the number of sets containing e, and let $f = \max\{f_e : e \in E\}$. We assume that $r_e \leq f_e$ holds for every $e \in E$, since an element e with $r_e > f_e$ can be removed from consideration. Denote by

$$B = \max\{\binom{f_e}{r_e} : e \in E\}. \tag{1}$$

Notice that $B \leq \binom{f}{r_{\max}}$. This paper studies the MinPSMC problem under the assumption that $r_{\max}$ is a constant.

3.1   Linear Program for MinPSMC

Before we give the integer program formulation for MinPSMC, we introduce some notations. For an element $e \in E$, an r_e-*cover set* is a sub-collection $\mathcal{A} \subseteq \mathcal{S}$ with $|\mathcal{A}| = r_e$ such that $e \in S$ for every $S \in \mathcal{A}$. Denote by Ω_e the family of all r_e-cover sets and $\Omega = \cup_{e \in E}\Omega_e$. The following example illustrates theses concepts.

Example 1. Let $E = \{e_1, e_2, e_3\}$, $\mathcal{S} = \{S_1, S_2, S_3\}$ with $S_1 = \{e_1, e_2\}$, $S_2 = \{e_1, e_3\}$, $S_3 = \{e_1, e_2, e_3\}$, $r(e_i) = 2$ for $i = 1, 2, 3$. For this example, $\Omega_{e_1} = \{\mathcal{A}_1, \mathcal{A}_2, \mathcal{A}_3\}$ with $\mathcal{A}_1 = \{S_1, S_2\}^{e_1}, \mathcal{A}_2 = \{S_1, S_3\}^{e_1}, \mathcal{A}_3 = \{S_2, S_3\}^{e_1}$, $\Omega_{e_2} = \{\mathcal{A}_4\}$ with $\mathcal{A}_4 = \{S_1, S_3\}^{e_2}$, $\Omega_{e_3} = \{\mathcal{A}_5\}$ with $\mathcal{A}_5 = \{S_2, S_3\}^{e_3}$, and $\Omega = \{\mathcal{A}_1, \ldots, \mathcal{A}_5\}$.

Remark 1. Notice that different elements may have a same collection of sets as an r_e-cover set. For the above example, $\{S_1, S_3\}$ is an r_{e_1}-cover set as well as an r_{e_2}-cover set. In this case, this collection of sets should be viewed as different r_e-cover sets. We use superscript in the above example to distinguish them. The idea behind this definition is that if an r_e-cover set $\mathcal{A} \in \Omega$ is taken, then e is fully covered by those sets in $\mathcal{A}$.

For a sub-family $\Omega' \subseteq \Omega$, let

$$\mathcal{S}_{\Omega'} = \bigcup_{\mathcal{A} \in \Omega'} \mathcal{A}$$

be the sub-collection of $\mathcal{S}$ consisting of those sets which belong to some cover set of Ω'. For an instance, in the above example, if $\Omega' = \{\mathcal{A}_3, \mathcal{A}_5\}$, then $\mathcal{S}_{\Omega'} = \{S_2, S_3\}$ (the superscripts are ignored while taking the union).

Let $\rho : 2^\Omega \mapsto \mathbb{R}$ be a function on sub-families of Ω defined by

$$\rho(\Omega') = \sum_{S \in \mathcal{S}_{\Omega'}} w_S \tag{2}$$

for $\Omega' \subseteq \Omega$. It can be proved that ρ is submodular.

The MinPSMC problem can be formulated as an integer program in which a binary variable $x_\mathcal{A} \in \{0,1\}$ is used to indicate whether cover set $\mathcal{A} \in \Omega$ is selected, and a binary variable $y_e \in \{0,1\}$ is used to indicate whether element $e \in E$ is *not* fully covered. By relaxing it into a convex program using Lovász extension of ρ, and linearizing it using the fact that the Lovász extension of a submodular function coincides with its convex closure, we have the following relaxed linear program for MinPSMC:

$$\min \quad \sum_{\Omega': \ \Omega' \subseteq \Omega} \rho(\Omega') \xi_{\Omega'}$$

$$\text{s.t.} \quad \begin{cases} \displaystyle\sum_{\Omega': \ \mathcal{A} \in \Omega'} \xi_{\Omega'} = x_\mathcal{A}, & \forall \mathcal{A} \in \Omega, \\[2mm] \displaystyle\sum_{\mathcal{A}: \ \mathcal{A} \in \Omega_e} x_\mathcal{A} + y_e \geq 1, & \forall e \in E, \\[2mm] \displaystyle\sum_{e \in E} y_e \leq n - k \\[2mm] \xi_{\Omega'} \geq 0, & \forall \Omega' \subset \Omega, \\[1mm] x_\mathcal{A} \geq 0, & \forall \mathcal{A} \in \Omega, \\[1mm] y_e \geq 0, & \forall e \in E. \end{cases} \tag{3}$$

The dual program of (3) is:

$$\max \sum_{e \in E} u_e - (n-k)t$$

$$s.t. \begin{cases} \sum_{\mathcal{A}: \mathcal{A} \in \Omega'} z_{\mathcal{A}} \le \rho(\Omega'), & \forall \Omega' \subseteq \Omega, \\ \sum_{e: \mathcal{A} \in \Omega_e} u_e \le z_{\mathcal{A}}, & \forall \mathcal{A} \in \Omega, \\ u_e \le t, & \forall e \in E, \\ u_e \ge 0, & \forall e \in E, \\ t \ge 0 \end{cases} \tag{4}$$

3.2 Primal-Dual Algorithm

The algorithm is formally described in Algorithm 1. The first step is a standard primal-dual step. Suppose the while loop is executed g rounds. One problem is that the last sub-family might fully cover too many elements the number of which is much more than required by the remaining covering requirement after the $(g-1)$-th iteration. Then the second step uses a submodular minimization algorithm to prune it.

Algorithm 1. $PD(E, \Omega, \rho, r, k)$

Input: An instance (E, Ω, ρ, r, k) of MinPSMC.
Output: A sub-collection $\mathcal{T}$ fully covering at least k elements.
1: $j \leftarrow 0$, $\Omega_0 \leftarrow \Omega$, $E_0 \leftarrow E$, $\Gamma_0 \leftarrow \emptyset$.
   $t \leftarrow 0$, $z_{\mathcal{A}} \leftarrow 0$ for every $\mathcal{A} \in \Omega_0$ and $u_e \leftarrow 0$ for every $e \in E$.
2: **while** $|\mathcal{C}(\Gamma_j)| < k$ **do**
3:      $j \leftarrow j + 1$.
4:      $\alpha_j \leftarrow \min\{\frac{\rho(\Omega') - z(\Omega')}{|\Omega'|} : |\Omega'| \ge 1, \Omega' \subseteq \Omega_{j-1}\}$.
5:      $\Omega^{(j)} \leftarrow \arg\min\{\frac{\rho(\Omega') - z(\Omega')}{|\Omega'|} : |\Omega'| \ge 1, \Omega' \subseteq \Omega_{j-1}\}$.
6:      $t \leftarrow t + \alpha_j$
7:      For each $\mathcal{A} \in \Omega_{j-1}$, $z_{\mathcal{A}} \leftarrow z_{\mathcal{A}} + \alpha_j$
8:      For each $e \in E_{j-1}$, $u_e \leftarrow u_e + \alpha_j$.
9:      $\Gamma_j \leftarrow \Gamma_{j-1} \cup \Omega^{(j)}$.
10:     $E_j \leftarrow E_{j-1} \setminus \{e: e \text{ is newly fully covered}\}$
11:     $\Omega_j \leftarrow \Omega_{j-1} \setminus \{\Omega_e: e \text{ is newly fully covered}\}$
12: **end while**
13: $\widetilde{\Omega} \leftarrow \emptyset$.
14: **while** $|\mathcal{C}(\Gamma_{j-1} \cup \widetilde{\Omega})| < k$ **do**
15:     $\mathcal{A} \leftarrow \arg\min_{\mathcal{A} \in \Omega_{j-1}} \rho(\mathcal{A})$
16:     $\widetilde{\Omega} \leftarrow \widetilde{\Omega} \cup \mathcal{A}$
17:     $\Omega_{j-1} \leftarrow \Omega_{j-1} \setminus \{\Omega_e: e \text{ is newly fully covered}\}$
18: **end while**
19: $\Lambda \leftarrow \arg\min\{\rho(\Gamma_{j-1} \cup \Omega^{(j)}), \rho(\Gamma_{j-1} \cup \widetilde{\Omega})\}$.
20: Output $\mathcal{T} \leftarrow \mathcal{S}_\Lambda$

Lemma 1. *The running time of the above algorithm is polynomial.*

Proof. Notice that both $|\Omega|, |E|$ and the number of dual variables are polynomial. So, to prove the lemma, it suffices to show that line 4 and line 5 of Algorithm 1 can be accomplished in polynomial time (notice that the number of $\Omega' \subseteq \Omega$ is exponential). Notice that the objective to be minimized is a submodular function divided by a modular function. Such an objective can be minimized in polynomial time [26].

The following is an important observation.

Remark 2. For a sub-family Ω, denote by $E(\Omega')$ the set of elements indexing those cover sets of Ω'. Notice that $|\mathcal{C}(\Omega')|$ might be larger than $|E(\Omega')|$. For example, in Example 1, if $\Omega' = \{\mathcal{A}_3\}$, then $E(\Omega') = \{e_1\}$ while $\mathcal{C}(\Omega') = \{e_1, e_3\}$. However, the sub-family $\Omega^{(j)}$ found in line 5 of Algorithm 1 always satisfies

$$|\mathcal{C}(\Omega^{(j)})| = |E(\Omega^{(j)})|. \tag{5}$$

In fact, notice that for any cover set $\mathcal{A} \in \Omega_{j-1}$, the dual variable $z_{\mathcal{A}}$ is active. Hence $z_{\mathcal{A}} = t$ for any $\mathcal{A} \in \Omega_{j-1}$. It follows that

$$\frac{\rho(\Omega') - z(\Omega')}{|\Omega'|} = \frac{\rho(\Omega')}{|\Omega'|} - t. \tag{6}$$

So, for two sub-families Ω' and Ω'' with $\rho(\Omega') = \rho(\Omega'')$, the one with larger cardinality will be preferred (since line 5 chooses the one with the minimum ratio). Hence

$$\Omega^{(j)} \text{ must include all cover sets which are subsets of } \mathcal{S}_{\Omega^{(j)}}. \tag{7}$$

In order words, if e is newly covered by $\mathcal{S}_{\Omega^{(j)}}$, then there is an r_e-cover set which is a subset of $\mathcal{S}_{\Omega^{(j)}}$, adding such an r_e-cover set into $\Omega^{(j)}$ will not change the value of $\rho(\Omega^{(j)})$ while the cardinality of $|\Omega^{(j)}|$ will be larger. Consider Example 1, if we choose $\Omega' = \{\mathcal{A}_3\}$, then $\mathcal{S}_{\Omega'} = \{S_2, S_3\}$, which can fully cover e_1 and e_3. In this case, adding $\mathcal{A}_5 = \{S_2, S_3\}^{e_3}$ into Ω' will lead to a smaller ratio. Then claim (5) follows from (7).

3.3 Performance Analysis

The following lemma shows that throughout the algorithm, a dual feasible solution is maintained.

Lemma 2. *Algorithm 1 maintains the feasibility of* (4).

Proof. Notice that when a sub-family Ω' becomes **z**-tight, for every $\mathcal{A} \in \Omega'$, $z_{\mathcal{A}}$ is deactivated. Hence

$$\text{a } \mathbf{z}\text{-tight set will remain to be } \mathbf{z}\text{-tight to the end of the algorithm.} \tag{8}$$

For those sub-families which are not z-tight, the choice of α in line 4 guarantees that the first constraint of (4) is not violated.

Since all active dual variables increase at the same rate, and for any element e which is fully covered in some iteration, u_e and every $\mathcal{A} \in \Omega_e$ are deactivated at the same time (namely, the time when e is newly fully covered), hence

$$u_e = z_{\mathcal{A}} \text{ holds for } every \ \mathcal{A} \in \Omega_e \text{ throughout of the algorithm.} \qquad (9)$$

Then the second constraint of (4) holds.

Since u_e increases at the same rate as t until it is deactivated, hence $u_e \leq t$ holds for every element e, and

$$u_e = t \text{ for every } e \text{ which is not fully covered yet.} \qquad (10)$$

The third constraint of (4) is maintained. The lemma is proved.

Denote by $\left(\{u_e^{(j)}\}_{e \in E}, \{z_{\mathcal{A}}^{(j)}\}_{\mathcal{A} \in \Omega}, t^{(j)} \right)$ the dual feasible solution after the j-th iteration. Using the submodularity of ρ and the modularity of z, the next lemma can be proved by induction on the number of iterations.

Lemma 3. *For any index j, the sub-family Γ_j satisfies $z^{(j)}(\Gamma_j) = \rho(\Gamma_j)$.*

Proof. We prove the lemma by induction on j. This is obvious for $j = 0$ since $\Gamma_0 = \emptyset$ and $\rho(\emptyset) = z^{(0)}(\emptyset) = 0$. Assume that $j \geq 1$ and the lemma is true for $j - 1$. In the j-th iteration, a sub-family $\Omega^{(j)}$ is found, and Γ_j is set to be $\Gamma_{j-1} \cup \Omega^{(j)}$. Notice that $\Gamma_{j-1} \cap \Omega^{(j)} = \emptyset$. By the choice of α_j in line 4 and the update of z in line 7, we have

$$\rho(\Omega^{(j)}) = z^{(j-1)}(\Omega^{(j)}) + \alpha_j |\Omega^{(j)}| = \sum_{\mathcal{A} \in \Omega^{(j)}} \left(z_{\mathcal{A}}^{(j-1)} + \alpha_j \right) = \sum_{\mathcal{A} \in \Omega^{(j)}} z_{\mathcal{A}}^{(j)} = z^{(j)}(\Omega^{(j)}).$$

Since $z_{\mathcal{A}}$ is deactivated when $\mathcal{A}$ is chosen into Γ_{j-1}, we have $z_{\mathcal{A}}^{(j)} = z_{\mathcal{A}}^{(j-1)}$ for every $\mathcal{A} \in \Gamma_{j-1}$. Hence

$$z^{(j)}(\Gamma_{j-1}) = z^{(j-1)}(\Gamma_{j-1}) = \rho(\Gamma_{j-1}),$$

where the second equality comes from induction hypothesis. It follows that

$$\rho(\Gamma_{j-1}) + \rho(\Omega^{(j)}) = z^{(j)}(\Gamma_{j-1}) + z^{(j)}(\Omega^{(j)}) = z^{(j)}(\Gamma_j). \qquad (11)$$

By the feasibility of $z^{(j)}$, we have

$$z^{(j)}(\Gamma_j) \leq \rho(\Gamma_j). \qquad (12)$$

By the submodularity of ρ and the observation that $\Gamma_{j-1} \cap \Omega^{(j)} = \emptyset$ (and thus $\rho(\Gamma_{j-1} \cap \Omega^{(j)}) = 0$), we have

$$\rho(\Gamma_j) = \rho(\Gamma_{j-1} \cup \Omega^{(j)}) + \rho(\Gamma_{j-1} \cap \Omega^{(j)}) \leq \rho(\Gamma_{j-1}) + \rho(\Omega^{(j)}). \qquad (13)$$

Combining (11), (12), and (13), we have $z^{(j)}(\Gamma_j) = \rho(\Gamma_j)$. The induction step is finished and the lemma is proved.

In order to compare an optimal solution with the output of the algorithm, we have the following observation.

Remark 3. For any sub-collection of sets $\mathcal{F} \subseteq \mathcal{S}$, we can rewrite $\mathcal{F}$ as a family $\Psi_{\mathcal{F}}$ of cover sets. The following example illustrates how this can be done. Let $E = \{e_1, e_2, e_3, e_4\}$, $\mathcal{S} = \{S_1, S_2, S_3, S_4\}$ with $S_1 = \{e_1, e_2, e_4\}$, $S_2 = \{e_2, e_3\}$, $S_3 = \{e_1, e_2, e_3\}$, $S_4 = \{e_3, e_4\}$, and $r(e_i) = 2$ for $i = 1, 2, 3, 4$. For this example, $\mathcal{A}_1 = \{S_1, S_3\}^{e_1}, \mathcal{A}_2 = \{S_1, S_2\}^{e_2}, \mathcal{A}_3 = \{S_1, S_3\}^{e_2}, \mathcal{A}_4 = \{S_2, S_3\}^{e_2}, \mathcal{A}_5 = \{S_2, S_3\}^{e_3}, \mathcal{A}_6 = \{S_2, S_4\}^{e_3}, \mathcal{A}_7 = \{S_3, S_4\}^{e_3}, \mathcal{A}_8 = \{S_1, S_4\}^{e_4}$. If $\mathcal{F} = \{S_2, S_3, S_4\}$, then $\Psi_{\mathcal{F}} = \{\mathcal{A}_4, \mathcal{A}_5, \mathcal{A}_6, \mathcal{A}_7\}$. In general, $\Psi_{\mathcal{F}}$ consists of *all* those cover sets which are subsets of $\mathcal{F}$.

Now, we present the main theorem of this paper.

Theorem 2. *Algorithm 1 has performance ratio at most $B + \sqrt{n \cdot B}$. In particular, if f is a constant, then the performance ratio is $O(\sqrt{n})$.*

Proof. Let $\Omega^{(1)}, \ldots, \Omega^{(g)}$ be the sequence of sub-families of cover sets found by Algorithm 1. Then $\Gamma_j = \cup_{l=1}^{j} \Omega^{(l)}$ for $j = 1, \ldots, g$. Denote by $\mathcal{U}(\Gamma_j)$ the set of elements which are not fully covered by $\mathcal{S}_{\Gamma_j}$. Denote by *opt* the optimal value of the MinPSMC instance. By a standard primal dual analysis and Lemma 3, we can prove the following claim.

Claim 1. $\rho(\Gamma_{g-1}) \leq B \cdot opt$.

For any index j, it can be calculated that

$$
\begin{aligned}
\rho(\Gamma_j) = z^{(j)}(\Gamma_j) &= \sum_{\mathcal{A} \in \Gamma_j} z_{\mathcal{A}}^{(j)} \\
&= \sum_{\mathcal{A} \in \Gamma_j} \sum_{e:\, \mathcal{A} \vdash e} u_e^{(j)} \\
&= \sum_{e:\, e \in \mathcal{C}(\Gamma_j)} u_e^{(j)} \cdot |\{\mathcal{A} \in \Gamma_j : \mathcal{A} \vdash e\}| \\
&\leq B \sum_{e:\, e \in \mathcal{C}(\Gamma_j)} u_e^{(j)} \\
&= B \left(\sum_{e:\, e \in E} u_e^{(j)} - \sum_{e:\, e \in \mathcal{U}(\Gamma_j)} u_e^{(j)} \right) \\
&= B \left(\sum_{e:\, e \in E} u_e^{(j)} - |\mathcal{U}(\Gamma_j)| t^{(j)} \right),
\end{aligned}
\tag{14}
$$

where the first equality comes from Lemma 3; the third equality comes property (9); the inequality holds because the number of sets belonging to e is at most B; the last equality comes from (10).

In particular, taking $j = g - 1$ in inequality (14),

$$\rho(\Gamma_{g-1}) \leq B \left(\sum_{e:\, e \in E} u_e^{(g-1)} - |\mathcal{U}(\Gamma_{g-1})| t^{(g-1)} \right). \tag{15}$$

Since the algorithm does not jump out of the while loop at the $(g-1)$-th iteration, $|\mathcal{C}(\Gamma_{g-1})| < k$ and thus $|\mathcal{U}(\Gamma_{g-1})| > n - k$. Combining this with inequality (15) and the weak duality theorem for linear programs, we have

$$\rho(\Gamma_{g-1}) \leq B \left(\sum_{e:\, e \in E} u_e^{(g-1)} - (n-k) t^{(g-1)} \right) = B \cdot obj_D^{(g-1)} \leq B \cdot opt,$$

where $obj_D^{(g-1)}$ is the objective value of dual program (4) for those dual variables after the $(g-1)$-th iteration. Claim 1 is proved.

Claim 2. $\rho(\Omega^{(g)}) \leq \dfrac{(n - |\mathcal{C}(\Gamma_{g-1})|)}{k - |\mathcal{C}(\Gamma_{g-1})|} \cdot B \cdot opt.$

Let Ψ be the family of cover sets constructed from an optimal solution by the method described in Remark 3. Let

$$\Psi_{g-1} = \Psi \setminus \Gamma_{g-1}. \tag{16}$$

Notice that $\Psi_{g-1} \neq \emptyset$, since otherwise $\Psi \subseteq \Gamma_{g-1}$ which contradicts $|\mathcal{C}(\Gamma_{g-1})| < k$. In the g-th iteration, $\Omega^{(g)}$ is chosen, which means that

$$\frac{\rho(\Omega^{(g)}) - z^{(g-1)}(\Omega^{(g)})}{|\Omega^{(g)}|} \leq \frac{\rho(\Psi_{g-1}) - z^{(g-1)}(\Psi_{g-1})}{|\Psi_{g-1}|}. \tag{17}$$

By an inequality in the proof of Remark 2, we have

$$\rho(\Omega^{(g)}) \leq \frac{|\Omega^{(g)}|}{|\Psi_{g-1}|} \rho(\Psi_{g-1}). \tag{18}$$

Then Claim 2 follows from the observation that

$$\begin{aligned} |\Omega^{(g)}| &\leq B \cdot (n - |\mathcal{C}(\Gamma_{g-1})|), \\ |\Psi_{g-1}| &\geq k - |\mathcal{C}(\Gamma_{g-1})|, \\ \rho(\Psi_{g-1}) &\leq \rho(\Psi) = opt, \end{aligned} \tag{19}$$

where the first inequality holds because there are $n - |\mathcal{C}(\Gamma_{g-1})|$ elements which are not fully covered by Γ_{g-1}, and each element is contained in at most B cover sets; the second inequality holds because Ψ_{g-1} fully covers at least $k - |\mathcal{C}(\Omega_{g-1})|$ elements and each element has at least one cover set in Ψ_{g-1}.

Claim 3. $\rho(\widetilde{\Omega}) \leq (k - |\mathcal{C}(\Gamma_{g-1})|) \cdot opt.$

This claim is obvious by observing that the second while loop picks at most $k - |\mathcal{C}(\Gamma_{g-1})|$ cover sets and every cover set picked has cost upper bounded by opt.

Combining Claims 2 and 3, the last sub-family has cost at most

$$\min \left\{ \frac{(n - |\mathcal{C}(\Gamma_{g-1})|)}{k - |\mathcal{C}(\Gamma_{g-1})|} \cdot B \cdot opt, (k - |\mathcal{C}(\Gamma_{g-1})|) \cdot opt \right\} \leq \sqrt{n \cdot B} \cdot opt. \quad (20)$$

Then the theorem follows from combination of inequality (20) and Claim 1.

4 Improvement on Restricted MinPSMC

In the restricted problem MinRPSMC, $f_e \equiv 2$ and $r_e \equiv 2$, so $B = 1$ by the definition of B in (1). Then it follows from Theorem 2 that Algorithm 1 has performance ratio $1 + \sqrt{n}$ for MinRPSMC. By the proof of Theorem 1, one may have seen a similarity between the restricted version and a graph. Exploring such a similarity and making use of properties of a graph, we can improved the performance ratio for MinRPSMC to $1 + \sqrt{2}n^{1/4}$.

The algorithm for MinRPSMC is modified from Algorithm 1 with the following difference. Instead of outputting the better solution of $\Gamma_{g-1} \cup \Omega^{(g)}$ and $\Gamma_{g-1} \cup \widetilde{\Omega}$, the modified algorithm outputs the better one of $\Gamma_{g-1} \cup \widehat{\Omega}$ and $\Gamma_{g-1} \cup \widetilde{\Omega}$, where $\widehat{\Omega}$ is constructed as follows. If $|\mathcal{C}(\Omega^{(g)})| \leq 2(k - |\mathcal{C}(\Gamma_{g-1})|)$, let $\widehat{\Omega}$ be $\Omega^{(g)}$. Otherwise, find a sub-family $\widehat{\Omega} \subseteq \Omega^{(g)}$ with $\rho(\widehat{\Omega}) \leq \rho(\Omega^{(g)})/\gamma$ and $|\mathcal{C}(\widehat{\Omega})| \geq k - |\mathcal{C}(\Gamma_{j-1})|$, where

$$\gamma = \sqrt{\frac{|\mathcal{C}(\Omega^{(g)})|}{2(k - |\mathcal{C}(\Gamma_{g-1})|)}}. \quad (21)$$

We can prove the following lemma showing the existence of $\widehat{\Omega}$ when $|\mathcal{C}(\Omega^{(g)})| \geq 2(k - |\mathcal{C}(\Gamma_{g-1})|)$.

Lemma 4. In the case $|\mathcal{C}(\Omega^{(g)})| > 2(k - |\mathcal{C}(\Gamma_{g-1})|)$, one can find a sub-family of cover sets $\widehat{\Omega}$ in polynomial time with $\rho(\widehat{\Omega}) \leq \rho(\Omega^{(g)})/\gamma$ and $|\mathcal{C}(\widehat{\Omega})| \geq k - |\mathcal{C}(\Gamma_{g-1})|$.

Making use of the relation between a graph and MinRPSMC, and by properties of a graph, the claimed performance ratio can be proved.

Theorem 3. MinRPSMC problem has performance ratio at most $1 + \sqrt{2}n^{1/4}$.

5 Conclusion and Discussion

This paper proves a lower bound for the minimum partial set multi-cover problem MinPSMC by a reduction from the densest l-subgraph problem. Then, under the assumption that the maximum covering requirement $r_{\max}$ is upper bounded by a constant, this paper gives a $B + \sqrt{n \cdot B}$-approximation algorithm for MinPSMC, where $B \le \binom{f}{r_{\max}}$ and f is the maximum number of sets containing a common element. So, the ratio is $O(\sqrt{n})$ if f is a constant. For a restricted version of Min-RPSMC, the performance ratio can be improved to $1 + \sqrt{2}n^{1/4}$. From Theorem 1 and the fact that the current best known performance ratio for DlS is $O(n^{\frac{1}{4}+\varepsilon})$, where $\varepsilon > 0$ is an arbitrary real number [27], it is natural to ask whether the performance ratio for the restricted version can be improved to $O(n^{\frac{1}{8}+\varepsilon})$? Our algorithm depends on the assumption that r_{max} is upper bounded by a constant. How to obtain an approximation algorithm for the problem without assuming a constant upper bound on $r_{\max}$ is a topic deserving further exploration.

Acknowledgements. This research is supported by NSFC (11771013, 11531011).

References

1. Kempe, D., Kleinberg, J., Tardos, E.: Maximizing the spread of influence through a social network. In: Proceedings of the Ninth ACM SIGKDD International Conference on Knowledge Discovery and Data Mining, pp. 137–146 (2003)
2. Kempe, D., Kleinberg, J., Tardos, É.: Influential nodes in a diffusion model for social networks. In: Caires, L., Italiano, G.F., Monteiro, L., Palamidessi, C., Yung, M. (eds.) ICALP 2005. LNCS, vol. 3580, pp. 1127–1138. Springer, Heidelberg (2005). https://doi.org/10.1007/11523468_91
3. Chen, N.: On the approximability of influence in social networks. SIAM J. Discrete Math. **23**(3), 1400–1415 (2008). A preliminary version appears in SODA'08, pp. 1029–1037
4. Ran, Y., Zhang, Z., Du, H., Zhu, Y.: Approximation algorithm for partial positive influence problem in social network. J. Comb. Optim. **33**(2), 791–802 (2017)
5. Ran, Y., Shi, Y., Zhang, Z.: Local ratio method on partial set multi-cover. J. Comb. Optim. **34**, 302–313 (2017)
6. Dinh, T.N., Shen, Y., Nguyen, D.T., Thai, M.T.: On the approximability of positive influence dominating set in social networks. J. Comb. Optim. **27**, 487–503 (2014)
7. Wang, F., Camacho, E., Xu, K.: Positive influence dominating set in online social networks. In: Du, D.-Z., Hu, X., Pardalos, P.M. (eds.) COCOA 2009. LNCS, vol. 5573, pp. 313–321. Springer, Heidelberg (2009). https://doi.org/10.1007/978-3-642-02026-1_29
8. Wang, F., et al.: On positive influence dominating sets in social networks. Theoret. Comput. Sci. **412**, 265–269 (2011)
9. Karp, R.M.: Reducibility among combinatorial problems. In: Miller, R.E., Thatcher, J.W. (eds.) Complexity of Computer Computations, pp. 85–103. Plenum Press, New York (1972)
10. Feige, U.: A threshold of ln n for approximating set cover. In: Proceedings of 28th ACM Symposium on the Theory of Computing, pp. 312–318 (1996)
11. Dinur, I., Steurer, D.: Analytical approach to parallel repetition. In: STOC, pp. 624–633 (2014)
12. Johnson, D.S.: Approximation algorithms for combinatorial problems. J. Comput **9**, 256–278 (1974)

13. Lovász, L.: On the ratio of optimal integral and fractional covers. Discrete Math. **13**, 383–390 (1975)
14. Chvatal, V.: A greedy heuristic for the set-covering problem. Math. Oper. Res. 233–235 (1979)
15. Vazirani, V.V.: Approximation Algorithms. Springer, Heidelberg (2001)
16. Hochbaum, D.S.: Approximation algorithms for the set covering and vertex cover problems. SIAM J. Comput. **11**, 555–556 (1982)
17. Bar-Yehuda, R., Even, S.: A local-ratio theorem for approximating the weighted vertex cover problem. North-Holland Math. Stud. **109**, 27–45 (1985)
18. Khot, S., Regev, O.: Vertex cover might be hard to approximate to within $2 - \varepsilon$. J. Comput. Syst. Sci. **74**(3), 335–349 (2008)
19. Rajagopalan, S., Vazirani, V.V.: Primal-dual RNC approximation algorithms for (multi)-set (multi)-cover and covering integer programs. In: Proceedings of the 34th Annual IEEE Symposium on Foundations of Computer Science, pp. 322–331 (1993)
20. Kearns, M.: The Computational Complexity of Machine Learning. MIT Press, Cambridge (1990)
21. Slavík, P.: Improved performance of the greedy algorithm for partial cover. Inf. Process. Lett. **64**(5), 251–254 (1997)
22. Gandhi, R., Khuller, S., Srinivasan, A.: Approximation algorithms for partial covering problems. J. Algorithms **53**(1), 55–84 (2004)
23. Bar-Yehuda, R.: Using homogeneous weights for approximating the partial cover problem. J. Algorithms **39**, 137–144 (2001)
24. Manurangsi, P.: Almost-polynomial ratio ETH-hardness of approximating densest k-subgraph. In: STOC, pp. 19–23 (2017)
25. Lovász, L.: Submodular functions and convexity. In: Bachem, A., Korte, B., Grötschel, M. (eds.) Mathematical Programming the State of the Art, pp. 235–257. Springer, Heidelberg (1983). https://doi.org/10.1007/978-3-642-68874-4_10
26. Fleisher, L., Iwata, S.: A push-relabel framework for submodular function minimization and applications to parametric optimization. Discrete Appl. Math. 131, 311–322 (2003)
27. Bhaskara, A., Charikar, M., Chlamtac, E., Feige, U., Vijayaraghavan, A.: Detecting high log-densities: an $O(n^{1/4})$ approximation for densest k-subgraph. In: STOC, pp. 201–210 (2010)
28. Edmonds, J.: Submodular functions, matroids, and certain polyhedra. In: Guy, R., Hanani, H., Sauer, N., Schönheim, J. (eds.) Combinatorial Structures and Their Applications, pp. 69–87. Gordon and Breach, New York (1970)

Reducing Extension Edges of Concurrent Programs for Reachability Analysis

Cong Tian[⊠], Jiaying Wang, Zhenhua Duan, and Liang Zhao

ICTT and ISN Lab, Xidian University, Xi'an 710071, People's Republic of China
`ctian@mail.xidian.edu.cn`

Abstract. Predicate abstraction technique makes boolean programs a simple and popular model for program verification, of which the state reachability problem is decidable. However, the existing approach to reachability analysis of a concurrent boolean program, by applying the backward search (BWS) algorithm to the thread transition diagram (TTD) of the program, is of high complexity. To accelerate this approach, a method that expands the TTD with a kind of expansion edges and summarizes each path in the expanded TTD into a set of Presburger formulas has been proposed, so that the reachability problem is reduced to the satisfiability of the summary formulas. In this paper, we present a method for reachability analysis of concurrent boolean programs which improves the existing work in two aspects. First, with refined constraints on edge expansion, only a small number of expansion edges are required to be added to the TTD, which reduces the space consumption. Second, with optimized algorithm of path summarization using counter abstraction, less local state counters are dealt with and less summary formulas are generated. We have implemented the method and evaluated it on a large set of benchmark concurrent boolean programs. Experimental results show its efficiency on summarization and verification.

Keywords: Concurrent programs · Thread transition diagram
Reachability · Boolean programs · Backward search

1 Introduction

With the rapid development of multi-core hardware and concurrent techniques, multi-thread programming has become one of the prevailing paradigms for software development. Interleaved execution among threads may produce lots of subtle bugs, which are hard to be found with manual inspection. Therefore, it is crucial to verify properties of concurrent programs. However, straightforward implementation of formal verification techniques [12], such as model checking [7], for verifying concurrent programs usually suffer from the problem of state space explosion: the number of reachable states grows tremendously fast when

This research is supported by the NSFC Grant No. 61751207, 61732013 and 61420106004.

the number of concurrent threads increases. As a result of excessive usage of memory and computation resources, the verification efficiency for concurrent programs is impractical in practice.

Abstraction techniques [8] can effectively alleviate the state explosion problem. It only focuses on information related to the attributes which need to be verified and obtains a smaller model, so as to improve the efficiency of formal verification. As one of the most successful abstraction techniques, *predicate abstraction* [5,11] transforms a standard concurrent program into a boolean program that contains only boolean variables. With boolean variables, the state space of a concurrent program is relatively small, and the reachability question is decidable. Further, boolean programs are the same as standard programs in that they can explicitly capture the relationship between the data and the control flow. Successful application of predicate abstraction involves the SLAM project [3] developed by Microsoft Research and the SATABS tool [6,9] supported by Engineering and Physical Sciences Research Council (EPSRC). It is worth pointing out that, although predicate abstraction reduces the state space, the problem of state explosion still exists. In fact, the efficiency in verifying multi-thread boolean programs has become the bottleneck in the verification of concurrent programs and systems.

In a concurrent system that contains many identical threads, we can use *counter abstraction* [10] to reduce the verification complexity. The idea is to introduce a counter c_i for each local state l_i to record the number of threads at the state l_i, so that the global state of a system is represented by a vector of counters $\langle c_1, ..., c_k \rangle$, where k is the number of possible local states. In this way, the size of the state space is greatly reduced. However, even with counter abstraction, the number of local states k is still very large. Specifically, k grows exponentially with respect to the number of local variables. As a result, introducing counters for local states is only suitable for programs with a small quantity of local variables.

A sound and complete method for reachability analysis for well-quasi-ordered transition system (WQOS) is the *backward search* (BWS) algorithm proposed by Abdulla [2]. Starting from a target state whose reachability is under investigation, the algorithm proceeds backwardly by computing the *minimal cover predecessors* of the currently visited state, until either an initial state or a fixpoint is reached. However, the BWS algorithm only works for simple programs with few variables in practice, since the number of states in a WQOS is extremely large with the increase of the program variables.

Using BWS, the work in [14] first propose a method that analyzes the reachability of a boolean program with unbounded number of threads based on the control flow graph of the program. It also introduces local-state counters to record the number of threads at different local states, which can reduce the size of the state space. Then, it provides a more efficient version of the method that operates on a model of *thread transition diagram* (TTD) [13] of boolean programs, and develop a tool CUTR [15] as its implementation. Specifically, the method consists of two phases. The first phase constructs an *expanded TTD* D^+, which is a WQOS, by adding a set of expansion edges to the TTD D, and then execute BWS on D^+. The second phase summarizes the effect of thread transitions on local-state counters along paths from initial states to a given target state in D^+, obtaining a set of Presburger formulas such that the target state is reachable if and only if these formulas are satisfiable. However, (1) there are too many unreachable states and thus unreachable state transitions in the original TTD; and (2) the number of expansion edges that need to be added grows rapidly when the number of states increases. As a result, the number of paths from an initial state to the target state is likely to be very large, which makes it inefficient to summarize the effect of each path in the second phase.

In this paper, we propose an improved method of verifying concurrent boolean programs with multi threads based on predicate abstraction and counter abstraction. On the one hand, we add constraints to the generation of expansion edges, so that the number of expansion edges, as well as the scale of the TTS, is greatly reduced. On the other hand, we optimize the approach of path summarization by maintaining counters only for local states where expansion edges are added. In this way, less summary formulas need to be generated and checked. Experiments are carried out which investigate the performance of our method. The results show that the method reduces the overall computational cost significantly.

The rest of the paper is organized as follows. The next section briefly presents preliminaries, including thread transition diagrams of concurrent Boolean programs and backward search algorithm. Section 3 presents the improved expanded TTD, and Sect. 4 discusses presburger summary of loop-free paths in expanded TTD. Section 5 shows the empirical evaluation results. Finally, the paper is concluded in Sect. 6.

2   Preliminaries

2.1   Concurrent Boolean Programs and Thread Transition Diagrams

This paper focuses on concurrent boolean programs where all variables are boolean, which can be obtained from a standard concurrent program by predicate abstraction. Such a program may contain the following four statements for concurrency [4]:

- **start_thread label**: creates a new thread that starts execution at the program location label. It gets a copy of the local variables of the current thread, which continues execution at the proceeding statement.
- **end_thread**: terminates the current thread.
- **atomic_begin**: begins an atomic section that does not allow a context switch to another thread.
- **atomic_end**: ends an atomic section.

We assume a concurrent program with multi-threads is given in the form of an abstract state machine called *thread transition diagram* (TTD). Specifically, a TTD is a tuple $D = (S, L, T_s, T_t, I)$, where

- S is a finite set of *shared states*, an elements of which is a valuation of the shared variables of the boolean program;

- L is a finite set of *local states*, an elements of which is a valuation of the local variables and the program counter pc;
- $T_s, T_t \subseteq (S \times L) \times (S \times L)$ are two sets of edges, called *standard transitions* and *thread transitions*, respectively; and
- $I \subseteq (S \times L)$ is a set of *initial states*.

An element of $S \times L$ is called a *thread state*. It records the valuation of shared and local variables of an executing thread. We write $(s_1, l_1) \to (s_2, l_2)$ for a standard transition $((s_1, l_1),(s_2, l_2)) \in T_s$, and $(s_1, l_1) \nrightarrow (s_2, l_2)$ for a thread transition $((s_1, l_1),(s_2, l_2)) \in T_t$. TTD of the boolean program P shown in the left hand side of Fig. 2 is depicted in the right hand side of Fig. 2 where each of the initial states $\{((0,0),0),((0,1),0)\}$ is presented as $\circledcirc$. In this example, an extra boolean variable b is considered which indicates whether a new thread is just created, so a shared state s is a pair which is the value of (b, t). The assertion failure state $((0,1),4)$ is represented as $\otimes$ and we are concerned with whether it is reachable (Fig. 1).

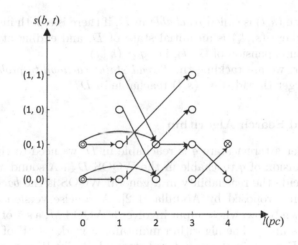

Fig. 1. TTD of program P

A TTD D gives rise to a well-quasi-ordered transition system (WQOS) D_∞. A state in D_∞ is of the form $(s \mid l_1, \ldots, l_n)$ $(n \in \mathbb{N})$ representing a global system state consisting of a shared state s and n threads at local states $l_1, \ldots, l_n$, which are not necessarily distinct. The number n of local states is unbounded and their order is irrelevant. For $n \in \mathbb{N}$, $s \in S$ and $l_1, \ldots, l_n, l_{n+1} \in L$, the global state $g = (s \mid l_1, \ldots, l_n, l_{n+1})$ is called a *unit expansion* of the global state $h = (s \mid l_1, \ldots, l_n)$, denoted as $g \succeq_1 h$. The quasi-order $\succeq$ of the WQOS, called the *expansion* relation, is defined as the smallest quasi-order including $\succeq_1$. The WQOS has two kinds of transitions:

(i) $(s \mid l_1, \ldots, l_n, l) \to (s' \mid l_1, \ldots, l_n, l')$, if $(s, l) \to (s', l')$ is a standard transition of D, and

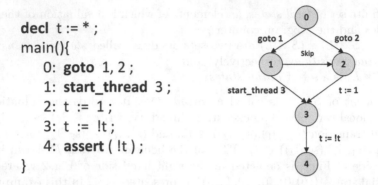

```
decl t := * ;
main(){
    0: goto 1, 2 ;
    1: start_thread 3 ;
    2: t := 1 ;
    3: t := !t ;
    4: assert ( !t ) ;
}
```

Fig. 2. A boolean program P and its Control flow graph

(ii) $(s \mid l_1, \ldots, l_n) \rightarrow (s' \mid l_1, \ldots, l_n, l')$, if $(s, l) \nrightarrow (s', l')$ is a thread transition of D.

A thread state (s, l) is called *reachable* in D, if there is a path in D_∞ starting from $(s_i \mid l_i)$, where (s_i, l_i) is an initial state of D, and ending at some global state g that is an expansion of $(s \mid l)$, i.e. $g \succeq (s \mid l)$.

In this paper, we are tackling the *thread state reachability problem*: given a TTD D, is a target thread state (s, l) reachable in D?

2.2   Backward Search Algorithm

To verify whether a target state q is reachable in D, we need to check whether there is an expansion of q reachable in the WQOS D_∞. A sound and complete algorithm to decide the reachability in a general WQOS is the *backward search* (BWS) algorithm proposed by Abdulla [1,2]. A concise version of the BWS algorithm is shown below, whose input involves a WQOS M, a set of initial states I, and a target state q. The algorithm maintains a work set W of unprocessed states and a set U of minimal encountered states. Initially, W is assigned by $\{q\}$. The algorithm iteratively computes the *minimal cover predecessors* (MCP):

$$MCP(w) = min\{p : \exists w' \succeq w : p \rightarrow w'\} \tag{1}$$

of each state $w \in W$ and updates W accordingly. If an initial state is reached, q is reachable. Otherwise, if a fix point of W is reached without any initial state, q is unreachable.

If a WQOS is finite, the BWS algorithm can be executed directly to decide the reachability of a given target state. However, the state space of the WQOS D_∞ derived from a TTD D may be infinite, since a global state of D_∞ may contain an arbitrary number of threads. To solve the thread state reachability problem in a TTD D, we do not actually generate the whole D_∞, which is of course impossible. Instead, we simulate the execution of the BWS algorithm on D_∞, based on information of D only.

Algorithm 1. BWS(M, I, q)

Input: WQOS M, initial states I, final state $q \notin I$
1: $W := \{q\}; U := \{q\}$
2: **while** $\exists w \in W$ **do**
3:     $W := W \setminus \{w\}$
4:     **for** $p \in MCP(w) \setminus \uparrow U$ **do**
5:         // $\uparrow U$ *stands for the upward closure of* U : $\uparrow U = \{u' : \exists u \in U : u' \succeq u\}$
6:         **if** $p \in I$ **then**
7:             "q is reachable"
8:         **end if**
9:         $W := min(W \cup \{p\})$
10:       $U := min(U \cup \{p\})$
11:    **end for**
12: **end while**
13: "q is unreachable"

Consider a TTD $D = (S, L, T_s, T_t, I)$. For each global state $w = (s \mid l_1, \ldots, l_n)$, $i \in \{1, \ldots, n\}$, we simulate the computation of the MCPs of w in two steps:

(i) compute the set W of direct predecessors of w, i.e., $W = \{(s' \mid l_1, \ldots, l_{n-1}, l'_n) \mid (s', l'_n) \rightarrow (s, l_n) \in T_s\}$ or $W = \{(s' \mid l_1, \ldots, l_{n-1}) \mid (s', l_{n-1}) \twoheadrightarrow (s, l_n) \in T_t\}$,

(ii) if W is non-empty, return W; otherwise, compute the set W of direct predecessors of unit expansions of w, i.e., return $W = \{(s' \mid l_1, \ldots, l_n, l') \mid \exists l \in L : (s', l') \rightarrow (s, l) \in T_s)\}$.

We explain it with an example. Consider the boolean program P in the left hand side of Fig. 2 whose TTD D is shown in the right hand side of Fig. 2. We are concerned with whether the target state $((0, 1), 4)$ is reachable in D, i.e., whether $((0, 1) \mid 4)$ is reachable from initial states $\{((0, 0) \mid 0), ((0, 1) \mid 0)\}$ in D_∞. The simulation of the BWS algorithm is visualized in Fig. 3.

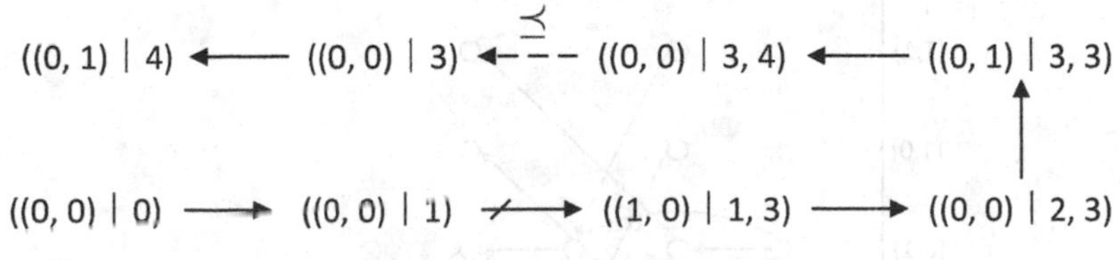

$$((0, 1) \mid 4) \longleftarrow ((0, 0) \mid 3) \overset{\preceq}{\longleftarrow} - - ((0, 0) \mid 3, 4) \longleftarrow ((0, 1) \mid 3, 3)$$

$$((0, 0) \mid 0) \longrightarrow ((0, 0) \mid 1) \not\longrightarrow ((1, 0) \mid 1, 3) \longrightarrow ((0, 0) \mid 2, 3)$$

Fig. 3. Reachability analysis using BWS, applied to the TTD in Fig. 2.

Starting from the target global state $g_0 = ((0, 1) \mid 4)$, we compute an MCP of g_0. Since $((0, 1), 4)$ has exactly one direct predecessor $((0, 0), 3)$ in D, g_0 has exactly one MCP $g_1 = ((0, 0) \mid 3)$. Then we compute an MCP of the global state g_1. Since $((0, 0), 3)$ has no predecessor in D, we need to search for direct predecessors of unit expansions of g_1. For the same reason, the unit expansion

$((0,0) \mid 3,3)$ has no predecessor. We only need to consider unit expansions $((0,0) \mid 3,l)$ of g_1 with $l = 1,2,4$. Notice that $((0,0),4)$ has a direct predecessor $((0,1),3)$ in D, thus $g_2 = ((0,1) \mid 3,3)$ is a direct predecessor of the unit expansion $((0,0) \mid 3,4)$ of g_1. We repeat the step of MCP calculation, arriving at an initial global state $g_6 = ((0,0) \mid 0)$. As a result, the target state $((0,1),4)$ is reachable in D.

3   Improved Expanded TTD

3.1   Constraints on Expanded TTD

There is an alternative approach to simulating the BWS algorithm. Notice that a key operation during BWS is the generation of unit expansions of a global state. Recall that in the previous example, a unit expansion $((0,0) \mid 3,4)$ of $((0,0) \mid 3)$ is generated. It is pointed out that this operation can be simulated in the TTD by adding an *expansion edge* $((0,0),4) \rightarrow ((0,0),3)$. Specifically, they add a set of expansion edges into a TTD D according to the following constraints, obtaining an *expanded TTD* (ETTD) D^+. They show that if a target state (s,l) is reachable in D_∞, it is also reachable in D^+.

According to the existing work, given a TTD D, an expansion edge of D is of the form $(s,l) \dashrightarrow (s,l')$ $(l \neq l')$ such that:

- there exists a transition ending at (s,l) in D, and
- either there exists a transition starting from (s,l') in D, or (s,l') is the target state.

Intuitively, expansion edges fill the gap between two original transitions whose target and source, respectively, differ only in the local state. Figure 4 shows the ETTD generated from the TTD in Fig. 2. According to the above constraints, this ETTD contains as much as 22 expansion edges.

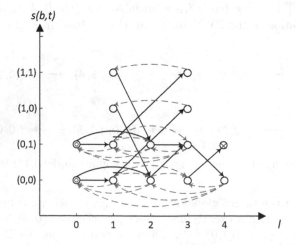

Fig. 4. ETTD generated by the original constraints

The number of expansion edges increases extremely fast with the increase of the number of local states, which consumes massive space and requires large computational effort. Therefore, a smaller ETTD with less expansion edges is vital for efficient verification of concurrent boolean programs.

By investigating the above constraints carefully, we find out that a lot of expansion edges added to the ETTD are redundant, having no effect on the result of BWS.

- For thread states (s, l) with direct predecessors in the TTD, the calculation of MCPs of them does not need to consider their unit expansions. Correspondingly, expansion edges of the form ... $--\rightarrow (s, l)$ are not necessary.
- For all $s \in S$ thread states (s, l) have no direct predecessor in the TTD, they are actually unreachable from other states. Therefore, expansion edges of the form ... $--\rightarrow (s, l)$ are not necessary.

According to the above concern, we add constraints on edge expansion and give a modified definition of expansion edges as follows.

Definition 1. *Given a TTD D, an* expansion edge *is an edge of the form* $(s, l) --\rightarrow (s, l')$ *$(l \neq l')$ such that:*

- *there exists a transition ending at (s, l) in D,*
- *there is no transition ending at (s, l') in D, and*
- *there exists a transition ending at (s', l') in D for some shared state s'.*

The second constraint makes sure that we only add expansion edges for those states with no predecessor transition states, and the last constraint guarantees that the expansion edges added would not contain thread states which are actually unreachable. The modified ETTD generated satisfying our constraints is shown in Fig. 5. Compared with the original ETTD presented in Fig. 4, the modified ETTD is much simpler with a smaller number of expansion edges.

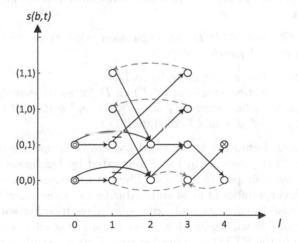

Fig. 5. ETTD generated by the new constraints

3.2 Optimization

A transition $(s, l) \rightarrow (s', l')$ in a TTD means that if a thread is at the state (s, l), it will reach the state (s', l') in finite time. And it is possible that a non-initial state (s, l) has no direct predecessor in a TTD, e.g. $((1, 0), 1)$ or $(0, 0), 3)$ in Fig. 2, due to the modification of the shared variables by other threads.

If a thread state (s', l_1) has no direct predecessor, the corresponding global state $g = (s' \mid l_1)$ has no direct predecessor, either. The unit expansions $(s' \mid l_1, l')$ of g during the simulation aim to find a new thread at local state $l' \neq l_1$ such that the shared state can be modified followed by this thread backwardly. If there exists a transition $(s, l) \rightarrow (s', l')$ in the TTD, we add an expansion edge $(s', l') \dashrightarrow (s', l_1)$ and get an MCP $g' = (s \mid l_1, l)$, i.e. a direct predecessor of a unit expansion, of g.

The constraints of expansion edges can be further optimized. Consider again the example TTD D presented in Fig. 2. The global state $g_0 = ((0, 0) \mid 3)$ has no direct predecessor and 3 unit expansions $g_l = ((0, 0) \mid 3, l)$ with an additional thread at local state $l = 1, 2, 4$, respectively. Notice that $g_1 = ((0, 0) \mid 3, 1)$ has (exactly) one direct predecessor $((0, 0) \mid 3, 0)$, which is an MCP of g_0. However, since there is no transition in D ending at the thread state $((0, 0), 1)$ that modifies the shared state $(0, 0)$, the local state 3 of g_1 cannot transform into a different local state during the BWS, including an initial local state 0. As a result, g_0 is not reachable from an initial global state via its unit expansion g_1 and we only need to consider the other unit expansions g_2 and g_4. In another word, the expansion edge $((0, 0), 1) \dashrightarrow ((0, 0), 3)$ is not necessary. By contrast, g_2 has a direct predecessor $((1, 0) \mid 3, 1)$ with a different shared state $(1, 0)$, and g_4 has a direct predecessor $((0, 1) \mid 3, 3)$ with a different shared state $(0, 1)$. From these MCPs of g_0, the execution of BWS is possible to reach an initial global state. Therefore, the expansion edges $((0, 0), 2) \dashrightarrow ((0, 0), 3)$ and $((0, 0), 4) \dashrightarrow ((0, 0), 3)$ are necessary.

Inspired by the above idea, we provide an optimized definition of expansion edges as follows.

Definition 2. *Given a TTD D, an* expansion edge *is an edge of the form* $(s, l) \dashrightarrow (s, l')$ *($l \neq l'$) such that:*

- *there is no transition ending at (s, l') in D,*
- *there exists a transition ending at (s', l') in D for some shared state s', and*
- *there exists a transition starting from (s'', l'') and ending at (s, l) in D for some shared state $s'' \neq s$ and local state l''.*

Compared with Definition 1, the third constraint is optimized. The ETTD of the example TTD presented in Fig. 2 generated by Definition 2 is shown in Fig. 6, and it is even simpler than Fig. 5. The optimization is more effective for TTDs with a larger number of local states during our experiments.

So far, we finish the definition of expansion edges together with the ETTD. In the next section, we will discuss how to summarize the effect of executing the BWS algorithm in the ETTD with local-state counters recording the number of threads in certain local states.

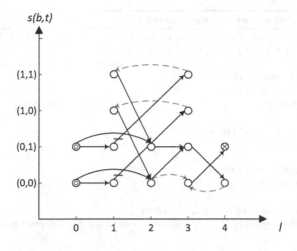

Fig. 6. Optimized ETTD

4   Presburger Summary of Loop-Free Paths in ETTD

We have shown that edges of an ETTD simulate operations of the BWS algorithm. More precisely, an original transition corresponds to a state change for a single thread, while an expansion edge corresponds to a unit expansion of the current global state. By counter abstraction, these operations can be expressed in terms of updates to local-state counters, which record the numbers of threads executing in different local states. The idea is that a transition from a local state A to a local state B can be expressed as decreasing the A counter by 1 while increasing the B counter by 1.

Consider an edge $e = ((s, l), (s', l'))$ of an ETTD:

- Suppose $e = (s, l) \rightarrow (s', l') \in T_s$. If the current global state $g = (s' \mid l'_1, \ldots, l'_n)$ contains a thread at the local state l', i.e. $l'_n = l'$, firing e backwardly amounts to decreasing the counter $c_{l'}$ of l' by 1 and increasing the counter c_l of the local state l by 1.
- Suppose $e = (s, l) \dashrightarrow (s', l') \in T_t$. If the current global state $g = (s' \mid l'_1, \ldots, l'_n)$ contains a thread at the local state l', i.e. $l'_n = l'$, firing e backwardly amounts to decreasing the counter $c_{l'}$ of l' by 1.
- Otherwise, if the current global state g does not contain a thread at l', we first generate a unit expansion $(s' \mid l'_1, \ldots, l'_n, l')$ of g, and then fire e backwardly. Together, the steps amount to an increment of c_l by 1.

The detailed summary of loop-free path segment according to above is shown in Algorithm 2.

A local state l that no expansion edge starts from must be in a path segment of the form $(s_1, l_1) \rightarrow (s, l) \rightarrow (s_2, l_2)$. The backward execution of $(s, l) \rightarrow (s_2, l_2)$ would increase the counter c_l of l by 1, while the backward execution of $(s_1, l_1) \rightarrow (s, l)$ would decrease c_l by 1. Therefore, the value of the counter c_l

---

Algorithm 2. Pathwise summary of a loop-free path with local-state counters in ETTD

Input: ETTD D^+, path $\bar{\sigma} = t_1, \ldots, t_k$, i.e. (t_i, t_{i+1}) in D^+ for $1 \leq i$ ¡ k; local state l.
1: $e_i := (t_i, t_{i+1})$ for $1 \leq i$ ¡ k, $(s_i, l_i) := t_i$ for $1 \leq i$ ¡ k
2: sum := "c_l"   // sum is a string recording the summary
3: **for** $i := k - 1$ downto 1 **do**
4:     **if** $e_i \in T_s$ and $l_i = l$ **then**
5:         sum := sum"+1"   // "." means string concatenation
6:     **end if**
7:     **if** $e_i \in T_s$ and $l_{i+1} = l$ **then**
8:         sum := sum."-1"
9:     **end if**
10:     **if** $e_i \in T_t$ and $l_{i+1} = l$ **then**
11:         sum := sum."-1"
12:     **end if**
13:     **if** e_i is an expansion edge and $l_i = l$ **then**
14:         sum := sum."$\bigoplus$(-1)+1"   // $x \bigoplus_b y = max\{x+y, b\}$, subscript omitted when $b = 0$
15:     **end if**
16: **end for**
17: **return** sum

---

would remain the same. According to the technique of on-the-fly, we only keep counters for those local states that expansion edges start from, as well as the initial local state. In this way, the state space is reduced while the result of the summarization is not affected.

Example. We show how the reachability of the thread state $((0,1), 4)$ for the ETTD in Fig. 5 is established. For each local state $l \in \{0, \ldots, 4\}$, the following formulas are obtained by summarizing the loop-free path segment from the target state $((0,1),4)$ to the initial state $((0,0),0)$ in Fig. 7 with constraints, respectively. The constraints is that, when backward-reaching the initial state t_I along that path, no thread resides in any local state other than l_I.

$$c_0 : 0 + 1 \geq 1$$

$$c_3 : 0 + 1 \bigoplus (-1) + 1 - 1 = 0$$

$$c_4 : 1 - 1 \bigoplus (-1) + 1 - 1 = 0$$

Since the formulas for all local states can be satisfied, we conclude that the target state $t_F = ((0,1), 4)$ is reachable. Notice that the summarization of local states $l = 1, 2$ shown below has no effect on the result.

$$c_1 : 0 + 1 - 1 = 0$$

$$c_2 : 0 + 1 - 1 = 0$$

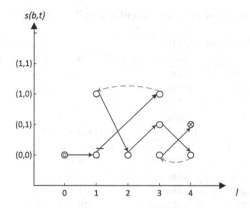

Fig. 7. A path from (0,0) to (4,4) in Fig. 6

5   Empirical Evaluation

We have implemented our method and made experiments on benchmark programs. The experiments aim at measuring the space consumption and performance impact of our implementation compared with the most relative tool CUTR.

We collect an extensive set of benchmarks, 97 TTDs [13] in total obtained from boolean programs, which are in turn obtained from C source programs by SATABS. All experiments are performed on a 64-bit Linux virtual machine with 2 GB memory, and the execution time is limited to 30 min. We use Z3 [16] as the Presburger formula solver. Details of the experimental results on part of the benchmarks are provided in Table 1. The runtime comparison results are shown in Figs. 8 and 9.

The results demonstrate that our method performs much better than CUTR in that the average time of verification is greatly reduced. Especially, the time is reduced by an order of magnitude in certain cases, such as TTDs of 02_inc_c and 09_Stack-L. There are only 4 examples that CUTR performs better than our method. The main reason is that with more constraints on edge expansion we need more time to check whether the expansion edges satisfy the constraints. In another word, more time is needed to preprocess the TTD, especially when the TTD is large. Nevertheless, the time spent for summarization is much less than CUTR.

In addition, the average number of expansion edges drops from 7746 to 92, which greatly reduces the state space and thus the memory consumption. It is also beneficial to the summarization in that less paths from the target state to an initial state need to be summarized.

Table 1. Summary of our experimental results compared with CUTR

Boolean program	Target state	CUTR		Our method	
		Expansion edges	Time (ms)	Expansion edges	Time (ms)
01_inc_l	(4,23)	231	18.766	5	6.222
	(32,70)	16598	331657.000	688	311323.000
02_inc_c	(4,20)	182	13.590	3	5.411
	(4,46)	956	67.403	12	25.681
	(4,448)	101864	1144390.000	480	11809.500
	(4,448)	96562	1027860.000	528	10524.600
03_PrngSimp-C	(8,20)	335	18.938	13	38.179
04_PrngSimp-L	(4,46)	1048	89.463	12	27.839
05_maxsim_l	(4,20)	186	17.451	4	5.632
	(8,24)	475	79.235	18	74.759
06_maxsim_c	(4,24)	301	11.688	3	6.404
	(4,58)	1901	63.764	10	32.891
07_max_opt_l	(4,54)	1542	193.348	16	34.952
	(8,62)	3802	3537.420	60	1563.020
08_maxopt_c	(4,62)	2212	65.827	12	35.277
	(4,136)	10782	585.001	40	167.347
09_Stack-L	(8,33)	887	1056.640	16	165.285
	(8,72)	4332	312586.000	45	2298.380
10_Stack-C	(8,31)	754	257.490	19	134.480
15_Boop	(4,31)	659	11.960	10	10.907
18_Unverif	(4,16)	128	5.660	3	4.066
	(64,23)	4197	669.417	132	605.202
	(64,23)	4205	666.829	132	599.595
19_spin	(8,19)	269	69.279	18	28.180
	(32,24)	1647	3483.500	151	8611.470
	(8,19)	269	77.098	18	58.840
	(32,24)	1647	3536.520	151	8590.500
20_BS_loop	(4,3968)	0	106.246	0	106.671
21_cond	(4,44)	0	3.482	0	3.537
25_tas_l	(16,30)	1829	85.957	66	1762.560
	(32,70)	18618	94279.800	508	372880.000
29_ticket_hc	(8,35)	1331	799.005	32	257.595
30_ticket_lo	(8.27)	715	294.867	26	88.239
dekker	(16,23)	1425	87.767	19	203.169
lu-fig2_fixed	(4,33)	434	37.672	1	6.463
	(16,43)	2535	1606.960	56	1750.870
peterson	(16,13)	303	50.961	42	62.683
pthread5	(32,46)	21244	308310.000	226	76474.300
rand_cas	(4,42)	814	42.801	16	23.965
rand_lock_p0	(8,19)	277	21.700	18	64.848
	(8,100)	10096	40440.700	150	11545.300
Average		**7746**	**79933.103**	**92**	**20049.215**

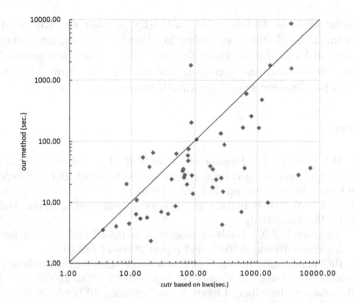

Fig. 8. Time for verification

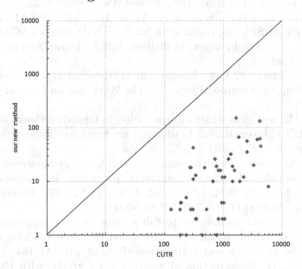

Fig. 9. Number of expansion edges added

6   Conclusion

In this paper, we present an approach to accelerating the verification of reachability properties for concurrent systems with unbounded numbers of threads. A key ingredient is the constraints on edge expansion according to the BWS algorithm, which can prune redundant expansion edges and in turn reduce the scale of the ETTD effectively. In addition, we optimize the summarization of loop-free paths with the technique of on-the-fly, so that we only need to keep counters for initial local states and other local states that expansion edges start from, but not all local states, which further reduces the state space. We have implemented our method and demonstrated its efficiency through experiments on a set of TTDs obtained from benchmark boolean programs.

References

1. Abdulla, P.A., Cerans, K., Jonsson, B., Tsay, Y.K.: General decidability theorems for infinite-state systems. In: Proceedings of the Eleventh IEEE Symposium on Logic in Computer Science, pp. 313–321 (1996)
2. Abdulla, P.A.: Well (and better) quasi-ordered transition systems. Bull. Symb. Log. **16**(4), 457–515 (2010)
3. Ball, T., Rajamani, S.K.: Boolean programs: a model and process for software analysis. Microsoft Research Technical report 2000–14 (2000)
4. Ball, T., Rajamani, S.K.: Bebop: a symbolic model checker for Boolean programs. In: Havelund, K., Penix, J., Visser, W. (eds.) SPIN 2000. LNCS, vol. 1885, pp. 113–130. Springer, Heidelberg (2000). https://doi.org/10.1007/10722468_7
5. Clarke, E., Kroening, D., Sharygina, N., Yorav, K.: Predicate abstraction of ANSI-C programs using SAT. Form. Methods Syst. Des. (FMSD) **25**, 105–127 (2004)
6. Clarke, E., Kroening, D., Sharygina, N., Yorav, K.: SATABS: SAT-based predicate abstraction for ANSI-C. In: Halbwachs, N., Zuck, L.D. (eds.) TACAS 2005. LNCS, vol. 3440, pp. 570–574. Springer, Heidelberg (2005). https://doi.org/10.1007/978-3-540-31980-1_40
7. Clarke, E.M., Emerson, E.A.: Design and synthesis of synchronization skeletons using branching-time temporal logic. In: The Workshop on Logic of Programs, pp. 52–71 (1981)
8. Cousot, P.: The role of abstract interpretation in formal methods. In: Proceedings of the 5th IEEE International Conference on Software Engineering and Formal Methods, pp. 135–137 (2007)
9. Donaldson, A., Kaiser, A., Kroening, D., Wahl, T.: Symmetry-aware predicate abstraction for shared-variable concurrent programs. In: Gopalakrishnan, G., Qadeer, S. (eds.) CAV 2011. LNCS, vol. 6806, pp. 356–371. Springer, Heidelberg (2011). https://doi.org/10.1007/978-3-642-22110-1_28
10. Emerson, E.A.: From asymmetry to full symmetry: new techniques for symmetry reduction in model checking. In: Advanced Research Working Conference on Correct Hardware Design and Verification Methods, pp. 142–157 (1999)
11. Graf, S., Saidi, H.: Construction of abstract state graphs with PVS. In: Grumberg, O. (ed.) CAV 1997. LNCS, vol. 1254, pp. 72–83. Springer, Heidelberg (1997). https://doi.org/10.1007/3-540-63166-6_10
12. Hoare, C.A.R., He, J.: Unifying theories of programming. In: Participants Copies for Relational Methods in Logic, Algebra and Computer Science, International Seminar Relmics, Warsaw, Poland, Septermber, pp. 97–99 (1998)
13. Kaiser, A., Kroening, D., Wahl, T.: A widening approach to multithreaded program verification. ACM Trans. Program. Lang. Syst. **36**(4), 1–29 (2014)

14. Liu, P., Wahl, T.: Infinite-state backward exploration of Boolean broadcast programs. In: Formal Methods in Computer-Aided Design, pp. 155–162 (2014)
15. Liu, P., Wahl, T.: Concolic unbounded-thread reachability via loop summaries. In: International Conference on Formal Engineering Methods, pp. 346–362 (2016)
16. de Moura, L., Bjørner, N.: Z3: an efficient SMT solver. In: Ramakrishnan, C.R., Rehof, J. (eds.) TACAS 2008. LNCS, vol. 4963, pp. 337–340. Springer, Heidelberg (2008). https://doi.org/10.1007/978-3-540-78800-3_24

Robustly Assigning Unstable Items

Ananya Christman[1(✉)], Christine Chung[2(✉)], Nicholas Jaczko[1],
Scott Westvold[1], and David S. Yuen[3]

[1] Department of Computer Science, Middlebury College, Middlebury, VT, USA
{achristman,njaczko,swestvold}@middlebury.edu
[2] Department of Computer Science, Connecticut College, New London, CT, USA
cchung@conncoll.edu
[3] Department of Mathematics, University of Hawaii, Honolulu, HI, USA
yuen@math.hawaii.edu

Abstract. We study the Robust Assignment Problem where the goal
is to assign items of various types to containers without exceeding container capacity. We seek an assignment that uses the fewest number of
containers and is robust, that is, if any item of type t_i becomes corrupt
causing the containers with type t_i to become unstable, every other item
type $t_j \neq t_i$ is still assigned to a stable container. We begin by presenting an optimal polynomial-time algorithm that finds a robust assignment
using the minimum number of containers for the case when the containers have infinite capacity. Then we consider the case where all containers
have some fixed capacity and give an optimal polynomial-time algorithm
for the special case where each type of item has the same size. When the
sizes of the item types are nonuniform, we provide a polynomial-time
2-approximation for the problem. We also prove that the approximation ratio of our algorithm is no lower than 1.813. We conclude with an
experimental evaluation of our algorithm.

1 Introduction

We study the Robust Assignment Problem (RAP) where we are given various
types of items, each with a weight, where items of the same type have the same
weight. We must assign the items to a set of containers with the constraint that
if an item is found to be corrupt (we assume that there may be at most one such
item), then every container containing an item of that type becomes unstable.
Therefore, we would like at least one item of every other type to remain in
at least one stable container. Such an assignment is considered *robust* and we
would like a robust assignment that uses the fewest number of containers while
satisfying their weight limit.

More formally, the input is n item types $t_1, \ldots, t_n$ with sizes (or weights)
$w_1, \ldots, w_n$, respectively, and container capacity C. The output is an assignment
of types to subsets of containers, which uses the lowest number of containers
and satisfies the following constraints: (1) Each type is assigned to k containers,
for some $k \geq 1$ (2) Each container is assigned at most C total weight (3) The

© Springer Nature Switzerland AG 2018
D. Kim et al. (Eds.): COCOA 2018, LNCS 11346, pp. 402–420, 2018.
https://doi.org/10.1007/978-3-030-04651-4_27

assignment is *robust*, that is, for any type t_i, if all containers having an item of type t_i become unstable, for all other types $t_j \neq t_i$, there is a stable container that contains t_j. Formally, let $S_i = \{s_{i,1}, s_{i,2}, \ldots, s_{i,k}\}$ for $1 \leq i \leq n$ denote the set of k containers to which an item of type t_i was assigned. Then for every type $t_j \neq t_i$ such that an item of type t_j is also assigned to any container of S_i, an item of type t_j will exist on some container that is not in S_i. The goal is to find a robust assignment that uses the fewest containers.

RAP has many practical applications. For example, in distributed systems, multiple applications, including instances of the same app, are hosted on a cluster of servers. If a failure occurs in an app (and may therefore possibly occur in the other instances of the faulty app), then the app, all of its hosting servers, and hence all other app instances on those servers, are temporarily suspended. Therefore, the system would like an assignment of app instances to the minimal number of servers such that if a failure occurs in an app and therefore all its hosting servers are temporarily suspended, there is still a running instance of every other app hosted on some unaffected server in the system. RAP can be used to find such an assignment - the apps correspond to the items and the servers correspond to the containers. The goals of our work were in fact motivated by a conversation with industry colleagues who encountered this problem in their company's hosting platforms.

Ad placement on webpages is another application of the Robust Assignment Problem. Ad companies often have ads from multiple clients that must be displayed throughout various webpages of a website. If an ad crashes or slows down, it may affect the entire webpage and hence, the other ads displayed on that webpage as well. Other webpages displaying the faulty ad may need to be temporarily suspended to repair or check the faulty ad. Therefore ad companies would like an assignment of ads to webpages such that if a faulty ad temporarily suspends all of the webpages it is displayed on, there is still a running instance of every other ad on some webpage on the website. Here, the ads and webpages correspond to the items and containers, respectively.

RAP can also be presented as an application to gardening/agriculture. Avid gardeners often grow multiple plants of different varieties in several garden beds. Suppose that during the growing season, it becomes known that a particular plant variety has become disease prone. Therefore, all plants that are planted in the same bed as a disease-prone plant may become infected with the disease. Therefore, gardeners would like to find a way to plan their garden such that if a plant variety becomes prone to disease, then at least one plant of each variety still grows.

Our Results. For RAP we first give an optimal polynomial-time algorithm for finding the minimum number of containers needed to robustly assign the given set of item types, ignoring capacity constraints on the containers (Sect. 3.1). We then introduce the constraint of capacitated containers and give an optimal polynomial-time algorithm for the special case where each type of item has the same size (Sect. 4). For the general case of nonuniform sizes, we provide a polynomial-time 2-approximation for the problem (Sect. 4.2). I.e., our algorithm uses no more than twice the number of containers of the optimal robust assignment. We also prove that the approximation ratio of our algorithm is at least 1.813. We conclude with an experimental evaluation of our algorithm (Sect. 5).

2   Related Work

To the best of our knowledge, our specific model for a robust assignment has not been previously studied. However, our solution ideas draw on those used for the bin-packing problem and some assignment problems, so we first discuss literature related to both problems. As mentioned above, in the context of distributed computing, our work applies to the problem of assigning replicas of applications to servers on a hosting platform, so we also discuss some literature on variations of this problem.

Our problem model has similarities to the problem of bin-packing with conflicts (or constraints) [2,5,6]. In the most general form of this problem, there are conflicts among the items to be packed and these conflicts are captured by a *conflict graph*, where the nodes represent the items and an edge exists between two items that are in a conflict [5]. The goal is to pack the items in the fewest number of bins while satisfying the capacity constraints on the bins and ensuring that no two items in a conflict are packed in the same bin. Jansen proposed an asymptotic approximation scheme for this problem for d-inductive graphs (i.e. where the vertices can be assigned distinct numbers $1 \ldots n$ in such a way that each vertex is adjacent to at most d lower numbered vertices) including trees, grid graphs, planar graphs and graphs with constant treewidth [5]. For all $\epsilon > 0$, Jansen and Öhring [6] presented a $(2 + \epsilon)$-approximation algorithm for the problem on cographs and partial K-trees, and a 2-approximation algorithm for bipartite graphs. Epstein and Levin [2] improved on the 2.7-approximation of [6] on perfect graphs by presenting a 2.5-approximation. They also presented a 7/3-approximation for a sub-class of perfect graphs and a 1.75-approximation for bipartite graphs.

Our problem differs from these previous problems in at least two important ways. First, the conflicts among our items cannot be easily captured by a conflict graph as they do not pertain to specific pairs of items, but rather to *all* pairs of items. Second, for our problem, the total number of items that are packed into bins is not predefined, so an algorithm may create more or less if doing so yields fewer bins.

The wide variety of problems that address the task of assigning items to containers while satisfying constraints and minimizing or maximizing some optimization objective are typically classified as Generalized Assignment Problems [1,11]. While (to our knowledge) no previous works have considered the requirement of a robust assignment as in our model, a few works have had some similarities to ours. Fleischer et al. [3] studied a general class of maximizing assignment problems with packing constraints. In particular, they studied the Separable Assignment Problems (SAP), where the input is a set of n bins, a set of m items, values $f_{i,j}$ for assigning item j to bin i; and a separate packing constraint for each bin – i.e. for bin i, a family of subsets of items that fit in bin i. The goal is to find an assignment of items to bins with the maximum aggregate value. For all examples of SAP that admit an approximation scheme for the single-bin problem, they present an LP-based algorithm with approximation ratio $(1 - \frac{1}{e} - \epsilon)$ and a local search algorithm with ratio $(\frac{1}{2} - \epsilon)$. Korupolu et al. [8] studied the Coupled Placement problem, in which jobs must be assigned to computation and storage nodes with capacity constraints. Each job may prefer some computation-storage node pairs more than others, and may also consume different resources at different nodes. The goal is to find an assignment of jobs to computation nodes and storage nodes that minimizes placement cost and incurs a minimum blowup in the capacity of the individual nodes. The authors present a 3-approximation algorithm for the problem.

One application of our work is the problem of assigning replicas of applications to servers on a hosting platform so that the system is fault-tolerant to a single application failure. There have been a wide variety of studies on related problems and here we discuss a few. Rahman et al. [10] considered the related Replica Placement Problem where copies of data are stored in different locations on the grid such that if one instance at one location becomes unavailable due to failure, the data can be quickly recovered. They present extensive experimental results for this problem. Mills et al. [9] also studied a variation of this problem in the setting where dependencies exist among the failures and the general goal is to find a placement of instances that does not induce a large number of failures. They give two exact algorithms for dependency models represented by trees. Urgaonkar et al. [15] also studied the problem of placing apps on servers, but do not consider fault tolerance and focus instead on satisfying each application's resource requirement. The authors study the usefulness of traditional bin-packing heuristics such as First-Fit and present several approximation algorithms for variations of the problem.

More recently, Korupolu and Rajaraman [7] studied the problem of placing tasks of a parallel job on servers with the goal of increasing availability under two models of failures: adversarial and probabilistic. In the adversarial model, each server has a weight and the adversary can remove any subset of servers of total weight at most a given bound; the goal is to find a placement that incurs the least disruption against such an adversary. For this problem they present a PTAS. In the probabilistic model, each node has a probability of failure and the goal is to find a placement that maximizes the probability that at least a certain minimum number of tasks survive at any time. For the most basic version of the problem they study they give an algorithm that achieves an additive ϵ-approximation. Stein and Zhong [13] studied a related problem of processing jobs on machines to minimize makespan. The jobs must be grouped into sets before the number of machines is known and these sets must then be scheduled on machines without being separated. They present an algorithm that is guaranteed to return a schedule on any number of machines that is within a factor of $(\frac{5}{3} + \epsilon)$ of the optimal schedule, where the optimum is not subject to the restriction that the sets cannot be separated.

3   Preliminaries

As a concrete example, Figs. 1 and 2 show two assignments and the corresponding states of the containers for $n = 6$ item types. In Fig. 1, the assignment is not robust – if type 2 is found to be corrupt, then no items of type 5 will exist in any other containers since all items of type 5 are in the same set of containers as items of type 2 (a similar problem occurs with types 3 and 4). Figure 2 depicts one robust assignment using the optimal number of containers: 4. Note that if an item of type 2 fails, then all item types contained in B and D exist in some other container. Further note that in this assignment if *any* of the item types are found to be corrupt, this robustness property holds.

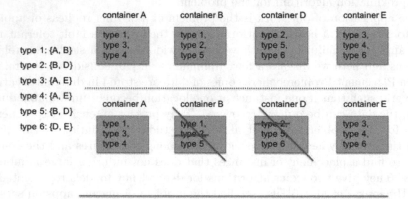

Fig. 1. Example of a non-robust assignment. If type 2 is corrupt, no items of type 5 exist.

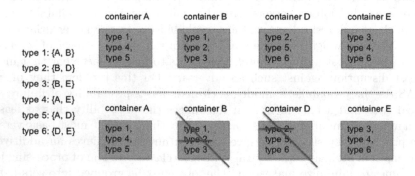

Fig. 2. An optimal robust assignment. If any type is corrupt, all other types still exist.

A *robust* assignment is characterized by whether each type is assigned to a set of containers that is not a subset of the set of containers assigned to any other type. We present this characterization formally as our first Lemma.

Lemma 1. *Let* $S_i = \{s_{i,1}, s_{i,2}, \ldots, s_{i,k}\}$ *for* $1 \leq i \leq n$ *denote the set of* k *containers to which an item of type* t_i *was assigned. An assignment of item types to containers is robust if and only if there is no pair of item types* t_i, t_j *such that* $S_i \subseteq S_j{}^1$.

Proof. First we show that a robust assignment implies no pair of types t_i, t_j will be such that $S_i \subseteq S_j$. Suppose by way of contradiction that there is some pair of types t_i, t_j where $S_i \subseteq S_j$. This means if type t_j is found to be corrupt, and all the containers in S_j become unstable, then all the containers in S_i also become unstable. In this case there are no items of type t_i in stable containers so the assignment was not robust. We now prove the other direction of the lemma. Suppose for contradiction we have no pair of types t_i, t_j such that $S_i \subseteq S_j$, but the assignment is not robust. If it is not robust, there is some type k such that removing the containers in S_k will leave another type k' in no remaining containers. But for this to be true, it must be that $S_{k'} \subseteq S_k$ which is a contradiction.

3.1   Uncapacitated Robust Assignment Problem

In this section we begin by considering the special case of the RAP where the containers have infinite capacity. To tackle the Uncapacitated Robust Assignment Problem we first consider the inverse problem: given m containers, what is the maximum number of item types we can assign robustly? Due to Lemma 1 this problem can be modeled as the combinatorics problem of finding the maximum cardinality *antichain* of a set. Specifically, let P denote the set of subsets of m elements $\{1, 2, \ldots, m\}$. An antichain of P is a set $\bar{P} = \{s_1, s_2, \ldots, s_k\} \subseteq P$ such that for any pair of subsets s_i, s_j in $\bar{P}$, $s_i \not\subseteq s_j$. For a table of all antichains for $m = 1, 2,$ and 3, please refer to the full version of the paper.

Sperner's Theorem [12] states that the maximum cardinality of an antichain $\bar{P}$ of an m-sized set is $\binom{m}{\lfloor m/2 \rfloor}$ and each subset of $\bar{P}$ has size $m/2$. (If m is odd then there will be two maximum cardinality antichains whose subsets will have size $\lfloor m/2 \rfloor$ and $\lceil m/2 \rceil$, respectively.) Therefore, Sperner's Theorem yields the maximum number of item types that can be assigned to m containers as well as the number of containers to which each type is assigned. The values in Table 1 were derived from Sperner's Theorem.

We can thus use this theorem in conjunction with Lemma 1 to solve our original assignment problem – that is, given n types, find the minimum number of containers required to assign these types. Specifically, given n types, we would like to find the smallest m such that $\binom{m}{\lfloor m/2 \rfloor} \geq n$. See Algorithm 1 for further details.

Theorem 1. *The Uncapacitated Robust Assignment Problem is solvable in time polynomial in* n, *the number of item types.*

---

[1] Note that $S_i \subseteq S_j$ is the general condition for nonuniform k; for uniform k the condition is $S_i = S_j$.

Table 1. The maximum number of types that can be robustly assigned to $1 \leq m \leq 10$ containers.

m (# of containers)	Maximum number of item types that can be robustly assigned
1	1
2	2
3	3
4	6
5	10
6	20
7	35
8	70
9	126
10	252

Algorithm 1. Input is n item types.

1: Use Sperner's Theorem to find the minimum integer m such that $\binom{m}{\lfloor m/2 \rfloor} \geq n$.
2: Set up m empty containers.
3: Generate all $\binom{m}{\lfloor m/2 \rfloor}$ of the $\lfloor m/2 \rfloor$-combinations of the m containers.
4: Assign each item type one of the $\lfloor m/2 \rfloor$-combinations, i.e. for each type, assign an item of that type to each of the $\lfloor m/2 \rfloor$ containers in the $\lfloor m/2 \rfloor$-combination that this type was assigned to.

Proof. Due to Sperner's Theorem and Lemma 1, Algorithm 1 correctly returns the minimum number of containers required. Steps 1 and 2 take no more than time linear in the number of types as n serves as a trivial upperbound on the value of m that satisfies the Sperner's Theorem condition. (Furthermore, we can potentially find the solution more quickly by computing upper and lower bounds on m using Stirling's approximation which states that $\binom{m}{\lfloor m/2 \rfloor} \approx m + \frac{1}{2} - \frac{1}{2} \log_2(m\pi)$ [14].) Steps 3 and 4 of the algorithm require enumeration of the $\binom{m}{\lfloor m/2 \rfloor}$ combinations; the number of combinations is exponential in m, but since the chosen m will be $O(\log(n))$, the composite run time is still polynomial in n.

4   The Robust Assignment Problem with Capacity Constraints

In Sect. 3.1, we implicitly assumed that any number of items can be assigned to any one container. However, in practical settings, constraints such as storage space, memory, or other demands will impose limits on the number of items a container may hold. We therefore consider a model where there is one such constraint. We will use the example of a storage constraint for expository purposes.

The problem now becomes: given n types $t_1, t_2, \ldots, t_n$ with integer-valued sizes, $w_1, w_2, \ldots, w_n$, respectively, where items of type t_i have size w_i; and an integer-valued container capacity C, where $1 \leq w_i \leq C$, find an assignment of items to the minimal number of containers such that the assignment is both (1) robust and (2) satisfies the following *capacity constraint*: if A_j is the set of items assigned to container s_j, then for all containers $j = 1 \ldots m$, where m is the number of containers used in the assignment, $\sum_{a \in A_j} w(a) \leq C$, where $w(a)$ is the size of item a. We refer to this variant as the *Capacitated Robust Assignment Problem*.

As a small example, suppose in Fig. 2, types $1, 2, \ldots, 6$ have sizes $1, 2, \ldots 6$, respectively. Then if $C = 12$, the assignment shown in the figure would not satisfy the capacity constraint since both containers D and E currently use 13 units of size. Figure 3 shows an assignment that satisfies both the robustness and capacity constraints.

container A	container B	container D	container E
type 3, type 4, type 5	type 1, type 2, type 3	type 1, type 5, type 6	type 2, type 4, type 6

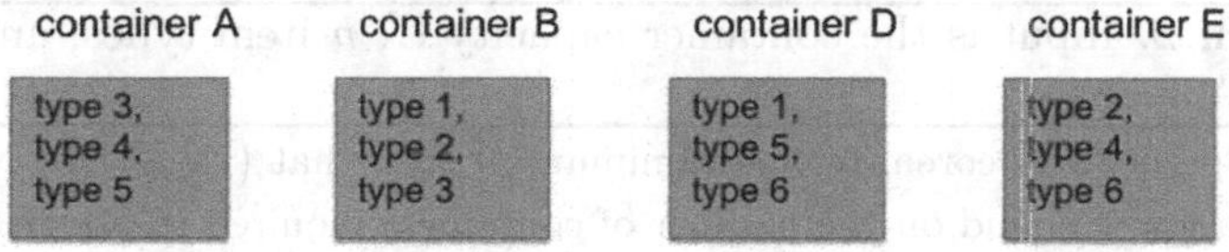

Fig. 3. An assignment that satisfies both the robustness and capacity constraints for capacity equal to 12.

4.1 Uniform Sizes

We first consider the special case where each type, and therefore each item, has the same size w. Given n such types and containers of capacity $C \geq w$, the problem is to find an assignment of types to the minimal number of containers such that the assignment is robust and also satisfies the capacity constraints. In this case the capacity constraint is that if $|A_j|$ denotes the number of items assigned to container s_j, and m is the number of containers in the assignment, then for all $j = 1 \ldots m$, $|A_j| w \leq C$.

Theorem 2. *The Capacitated Robust Assignment Problem with uniform sizes is solvable in time polynomial in n, the number of item types.*

Proof. Algorithm 2 solves this problem optimally. Recall that k denotes the number of items of each type. The algorithm effectively performs an exhaustive search to find the minimum m over all possible k for which the robustness and capacity constraints are satisfied. Specifically, the algorithm starts with the lower bound for m (given by Sperner's Theorem) and searches every possible integral value of k given this m (i.e. starting from $k = \lfloor m/2 \rfloor$ down to $k = 1$) that will satisfy both the robustness and capacity constraints. Robustness is satisfied if $\binom{m}{k} \geq n$ and the capacity constraint is satisfied if $\lfloor \frac{C}{w} \rfloor \geq \frac{nk}{m}$. If no value for k for the given m satisfies both constraints, the algorithm increments m and repeats the search for k.

Note that the algorithm will eventually terminate: if eventually m is incremented to n and k is decremented to 1 both conditions of the while loop will be true. There will be $O(n^2)$ iterations of the while loop. Each iteration takes constant time so the runtime of the loop is $O(n^2)$. The polynomial run-time and correctness of step 11 is addressed in the full version of the paper. So the overall run time of Algorithm 2 is polynomial in n.

The procedure for assigning the n types robustly to the m containers computed by Algorithm 2 is rather technical, so we refer the reader to the full version of the paper for the details on this procedure. We also note that while there are ways to optimize the run-time of our algorithm, the exhaustive-search version we present here is for the sake of simplicity and clarity.

Algorithm 2. Input is the container capacity C, n item types, and item size $w \leq C$.

1: Use Sperner's Theorem to find minimum m such that $\binom{m}{\lfloor \frac{m}{2} \rfloor} \geq n$. Note that m is a lower bound on the number of containers required to assign the types.
2: $k = \lfloor \frac{m}{2} \rfloor$
3: **while** not $(\binom{m}{k} \geq n$ and $\lfloor \frac{C}{w} \rfloor \geq \frac{nk}{m})$ **do**
4:     **if** $k > 1$ **then**
5:         $k--$ //decrease the number of items per type
6:     **else**
7:         $m++$ //add another container
8:         $k = \lfloor \frac{m}{2} \rfloor$ //re-initialize k for the new m
9:     **end if**
10: **end while**
11: For details on how to assign k items of each type to a distinct subset of the m containers, refer to the full version of the paper.

We note that if the problem is simply to find the minimum number of containers needed for the robust assignment, without also requiring the robust assignment itself, one can do so in polylog(n) time by formulating the problem as a fixed-dimension integer program. Namely, given inputs (n, C) where for simplicity we assume $w = 1$, then we want to solve the system $1 \leq k \leq m/2$, $kn \leq mC$, $\binom{m}{k} \geq n$ for k and m with m minimal. The key observation is that for any fixed m, the k satisfying the first two equations that yield the largest $\binom{m}{k}$ is $k = \lfloor \min(m/2, mC/n) \rfloor$. Thus whether or not an m has a corresponding k that satisfies the three equations is equivalent to whether or not $\binom{m}{\lfloor \min(m/2, mC/n) \rfloor} \geq n$. Because we can show $\binom{m}{\lfloor \min(m/2, mC/n) \rfloor}$ is an increasing function in m, then we can do a binary search for the minimal m that satisfies $\binom{m}{\lfloor \min(m/2, mC/n) \rfloor} \geq n$ in the interval $[1, n]$. This would be far more efficient than the brute force while-loop that we give for simplicity in Algorithm 2.

4.2   Nonuniform Sizes

In this section we consider the variant of the problem where there may be a different size $1 \leq w_i \leq C$ for each type t_i. In this case, if we ignore the robustness constraint, the problem would be NP-hard due to its equivalence to bin-packing [4]. We first present our algorithm, Robust First Fit (RFF), for this problem and prove that its approximation ratio is at most 2. We then show that the approximation ratio of RFF is not lower than 1.813.

4.2.1   The Robust First Fit Algorithm

RFF begins by sorting the types in descending order by size. When finding an assignment for type t_i the algorithm first finds the set S of all the containers that have enough empty space to fit an item of type t_i. It then assigns an item of type t_i to the (lexicographically) first container assignment that can be created from the containers in S that has not already been used by a previous type.

Algorithm 3. ROBUST FIRST FIT (RFF). Input is the container capacity C and a set T of n types where all items of type t_i have size $w_i \leq C$ for $1 \leq i \leq n$.

1: Sort the types in descending size order.
2: Use Sperner's Theorem to find minimum m such that $\binom{m}{\lfloor \frac{m}{2} \rfloor} \geq n$. Note that m is a lower bound on the number of containers required to hold the items.
3: **for** $k = \lfloor \frac{m}{2} \rfloor$ to 1 **do**
4:    Set up $m' = m$ empty containers.
5:    **for** each type t_i in T **do**
6:       Let S denote the subset of the m' containers that still have sufficient space to fit an item of type t_i
7:       Assign k items of type t_i to the lexicographically first k-combination of containers in S that is still available (i.e. no type has already been assigned to it)
8:       **if** t_i is still unassigned **then** {there were no available k-combinations in S}
9:          Add a new container to S, maintaining lexicographic order and increment m'. (Note that k does not change.)
10:          Go back to Step 7.
11:       **end if**
12:    **end for**
13:    Store m' along with the corresponding assignment.
14: **end for**
15: Return the assignment from the iteration of the outer-most **for**-loop (Step 3) that used the fewest containers.

Whenever no suitable assignment can be created for type t_i with the existing containers, a new empty container is created. An assignment with a storage constraint will never require fewer containers than an assignment without a storage constraint, so the number of containers (m) and number of items of each type (k) are initialized to the values given by Sperner's Theorem. The **for**-loop in step 3 accounts for the fact that decreasing the number of items of each type might decrease the number of containers required. Therefore, we start with $k = \lfloor m/2 \rfloor$, as given by Sperner's Theorem, and try decreasing the number of items from there.

RFF runs in polynomial time. Since the initial value of m from step 2 can be at most n, the **for**-loop in Step 3 will run for at most $O(n)$ iterations. Each iteration of the loop finds an assignment for n types. To find an assignment for each type t_i, the algorithm searches for an available k-combination whose containers have sufficient space for an item of type t_i. This can be done in poly-time as there will never be more than $O(n)$ k-combinations to check before finding one that is available. Since the number of containers will be no more than $n \cdot \frac{n}{2} = O(n^2)$ (i.e. if one item of each type was assigned to its own dedicated container), the algorithm may reach Step 10 (which causes a new iteration from Step 7) $O(n^2)$ times. Hence the overall run-time of RFF is polynomial.

We note that there are clearly ways to optimize the run-time of our algorithm if one wishes to implement it on a real-world system (for example, using binary search instead of linear search). The version we present here is for the sake of simplicity and clarity.

4.2.2   Upper Bound

We now show that RFF has an approximation ratio of no worse than 2.

Theorem 3. RFF *is a 2-approximation for the Capacitated Robust Assignment Problem with nonuniform sizes. I.e.,* RFF *will use at most $2m^*$ containers to robustly assign all n types, where m^* is the number of containers that an optimal solution uses.*

Proof. Consider any input instance. Let n denote the number of item types to be assigned, let OPT be an optimal robust assignment for them, and let k be the number of items assigned per type in OPT. It suffices to show that for the optimal k RFF uses at most $2m^*$ number of containers since RFF tries each potential value of k and chooses the value of k that minimizes the number of containers required. Hence, if RFF uses no more than $2m^*$ containers when it assigns k items per type, it must ultimately not use more than $2m^*$ containers. We also assume $k \geq 2$, since in the case of $k = 1$ RFF and OPT will both use a dedicated assignment (one item of each type per container) so RFF will return an optimal assignment, using $m = m^*$ containers.

Suppose for contradiction that OPT uses m^* containers while RFF uses strictly more than $2m^*$ containers when assigning the n types. Consider the moment during the execution of the RFF algorithm that container number $2m^* + 1$ was opened and added to S. Let t_i be the type that was being assigned when RFF opened this $(2m^* + 1)$th container. Let w_i be the size of type t_i. Let S_{-i} be the set of $2m^*$ containers already in use by the algorithm when it tried to assign t_i, but before it added container number $2m^* + 1$. Note that RFF has sorted and re-indexed the types in descending size order.

Case 1: $2 \leq k \leq \lfloor m^*/2 \rfloor$, $w_i > C/3$. Let B denote the set of all types t_j for whom $w_j > C/3$. Note that type $t_i \in B$. Due to their size, no more than two items of types in B can fit on a single container, so there can be no more than $2m^*$ such items in total, i.e., $k|B| \leq 2m^*$. Note however, that RFF must be able to assign the $k|B| \leq 2m^*$ items to at most $2m^*$ containers because $2m^*$ containers would indeed be sufficient for even a dedicated assignment: one item of each type per container. This contradicts the assumption that type t_i required RFF to open a $(2m^* + 1)$th container.

Case 2: $2 \leq k \leq \lfloor m^*/2 \rfloor$, $w_i \leq C/3$. In this case, we consider two sub-cases. Subcase 1: there are at least m^* containers in S_{-i} with available space at least w_i (i.e. enough space for an item of type t_i). In this case, we would then have a robust assignment from the set S_{-i} for t_i because OPT needed only m^* containers total to assign all n types, so having m^* containers must provide enough k-combinations to have at least one left for t_i.

Subcase 2: there are fewer than m^* containers in S_{-i} with available space at least w_i. So, in this case there must be $m^* + x$ containers $S_f \subseteq S_{-i}$, where $x > 0$, that have less than w_i available space. We can say that each of these containers in S_f already has filled capacity $C_f > C - w_i$. So if $w(S_f)$ is the total size of all of the items in the containers in S_f, then $w(S_f) > (m^* + x)(C - w_i)$. Since OPT used m^* containers of capacity C to assign all k items of each of the n types robustly, we have $(m^* + x)(C - w_i) < m^*C$. Recalling that we are in the case where $w_i \leq C/3$, we then have

$$(m^* + x)\left(C - \frac{C}{3}\right) = (m^* + x)\left(\frac{2C}{3}\right) < m^*C.$$

This implies $2xC/3 < m^*C/3$, which implies

$$x < m^*/2. \tag{1}$$

Let $S_g = S_{-i} - S_f$ be the set of containers in S_{-i} that still have enough remaining capacity to store an item of type t_i. For type t_i to be unable to be assigned to these $|S_g| = |S_{-i}| - |S_f| = m^* - x$ containers, it must be due to robustness: they must have no remaining available unique combinations of containers. We will show however, that if this were true, it would also lead to a contradiction.

If there are no unique combinations of containers remaining in S_g to assign t_i to, there must be at least $\binom{m^* - x}{k}$ distinct types that are already assigned to those containers. In other words, if

$$T_g = \{t_j \in T : \text{ an item of type } t_j \text{ is assigned to some container in } S_g\},$$

then $|T_g| \geq \binom{m^* - x}{k}$. This is true because if $|T_g| < \binom{m^* - x}{k}$ then there would be at least one remaining available k-combination of the containers in S_g on which to assign t_i.

RFF considers types in descending order by size so each item of the $|T_g| \geq \binom{m^*-x}{k}$ types must take up at least as much space as w_i. Thus, $w(S_g) \geq \binom{m^*-x}{k} k w_i$, where $w(S_g)$ is the total size of all the items on the m^*-x containers of S_g.

The size of all the items which are assigned to the $2m^*$ containers of S_{-i} is $w(S_{-i}) = w(S_f) + w(S_g) \geq (m^* + x)(C - w_i) + \binom{m^*-x}{k} k w_i$. Again, OPT used m^* containers of capacity C so we know the total size of all the items cannot be more than m^*C. Thus,

$$(m^* + x)(C - w_i) + \binom{m^* - x}{k} k w_i \leq m^*C \tag{2}$$

By expanding the left hand side of (2) we get

$$m^*C - m^*w_i + xC - xw_i + \binom{m^* - x}{k} k w_i \leq m^*C$$

and rearranging terms gives us:

$$xC + \binom{m^* - x}{k} k w_i \leq m^*w_i + xw_i \tag{3}$$

By combining Eqs. 3 and 1 with $w_i \leq C/3$ we get

$$3xw_i + \binom{m^* - x}{k} k w_i \leq m^*w_i + \frac{m^*}{2}w_i$$

which implies $\binom{m^*-x}{k} 2k + 6x \leq 3m^*$. Using the fact that $2 \leq k \leq \lfloor \frac{m^*}{2} \rfloor$ yields

$$\binom{m^* - x}{k} 4 + 6x \leq 3m^*. \tag{4}$$

It is a fact for any integers $a, b > 0$, where $b < a$, that $\binom{a}{b} \geq a$; and we know $m^* - x \geq k$ (since by Eq. (1) we know $x < m^*/2$ and we are currently in the case where $k \leq m^*/2$). Hence we can say from Eq. 4 that $4(m^* - x) + 6x \leq 3m^*$, which is a contradiction.

Both cases resulted in contradiction. So, RFF will never use more than $2m^*$ containers.

4.2.3   Lower Bound

We now provide a family of examples that give a lower bound on the approximation ratio of RFF. The family of examples is parameterized by a positive integer $d \geq 3$. We refer to the following instance as $I(d)$. There are $n = \binom{2d+3}{d}$ types, of which $\ell = 2d - 1$ are "large" types and $s = n - \ell$ are "small" types. The small types have size 1, while the large types have size $L = s$. Suppose the containers each have capacity $C = dL$. We first give an optimal assignment for this family.

Proposition 1. *For instance $I(d)$, an optimal assignment uses $m^* = 2d + 3$ containers.*

Proof. First we note that since $d \geq 2$, we have $\frac{1}{d+1} < \frac{2d+3}{(d+3)(d+2)}$. Then

$$\frac{(2d+2)!}{(d+1)d!(d+1)!} < \frac{(2d+3)(2d+2)!}{d!(d+1)!(d+3)(d+2)},$$

from which we get

$$\binom{2d+2}{d+1} < \binom{2d+3}{d} = n.$$

By Sperner's Theorem, this says that instance $I(d)$ requires at least $m = 2d + 3$ containers and this number of containers is possible when $k = d$. Now, letting $k = d$, since

$$k\ell = d(2d - 1) = 2d^2 - d \leq 2d^2 + d - 3 = (d - 1)(2d + 3) = (d - 1)m,$$

we can store k items of each of the ℓ large types on the m containers with at most $d - 1$ items on each container (the full version of the paper describes how). Since the capacity of each container is $C = dL$, each container will have at least capacity L remaining. We will then use the remaining $\binom{2d+3}{d} - \ell$ combinations, which is exactly s, the number of small types, to assign the small items. By design, $Lm \geq ds$ and so there is enough remaining capacity to do this. Therefore $2d + 3$ is the optimal number of containers for the instance $I(d)$.

Given an instance $I(d)$, we now establish the number of containers returned by RFF.

Proposition 2. *For an instance $I(d)$, for each integer $1 \leq k \leq d+1$, we define*

$$J(k) = \min\{j : \binom{j + 2d - 2k + 4}{j} \geq d - 1\}$$

$$z(k) = \min\{j : \binom{j}{k} \geq s\}$$

While using k items of each type, RFF will return the number of containers equal to:

$$m(k) = 2k - 1 - J(k) + z(k).$$

Then RFF will return the number of containers such that $m(k)$ is minimal over $1 \leq k \leq d+1$.

Proof. (Please refer to Table 2 for example values of $J(k)$ and $z(k)$). RFF begins by calculating that at least $2d + 3$ containers are needed, and so RFF will loop from $k = d + 1$ down to $k = 1$ in search of the minimum number of containers needed. In what follows, we index both the containers and item types starting from 0. Consider a fixed k for $1 \leq k \leq d + 1$. RFF will assign each large item type t_j, for each $j = 0, \ldots, d-1$, to containers $\{0, \ldots, k-2, k-1+j\}$. Then the other remaining $d - 1$ large types are assigned to containers $k - 1, \ldots, 2k - 2 - J$ and some order J subset of $\{2k - 1 - J, \ldots, 2d + 2\}$, which has cardinality $J + 2d - 2k + 4$. We need $\binom{J + 2d - 2k + 4}{J} \geq d - 1$. For any such J, the containers numbered 0 through $2k - 2 - J$ would be filled to capacity with large types. Thus taking the minimum such J, calling it $J(k)$, exactly the first $2k - 1 - J(k)$ containers are filled; the other containers have at least capacity L remaining.

Let $z(k)$ be the smallest positive integer such that $\binom{z(k)}{k} \geq s$. To assign the s small types, it is clear we need at least $z(k)$ containers beyond the $2k - 1 - J(k)$. For $d \geq 3$, we can prove by induction that

$$s = \binom{2d + 3}{d} - (2d - 1) > \binom{2d + 2}{d + 1}. \tag{5}$$

This is true for $d = 3$, and assuming it is true for a particular d, then we multiply the lefthand side by $\frac{(2d+4)(2d+5)}{(d+1)(d+4)}$ and the righthand side by $\frac{(2d+3)(2d+4)}{(d+2)^2}$, the latter of which we can prove is smaller by cross-multiplying. We then get

$$\binom{2d + 5}{d + 1} - (2d - 1)\frac{(2d + 4)(2d + 5)}{(d + 1)(d + 4)} > \binom{2d + 4}{d + 2}.$$

Now we can check by cross-multiplication that

$$(2d - 1)\frac{(2d + 4)(2d + 5)}{(d + 1)(d + 4)} > (2d + 1).$$

Then $\binom{2d+5}{d+1} - (2d + 1) > \binom{2d+4}{d+2}$, which is Eq. 5 with d replaced by $d + 1$, completing the induction. Finally,

$$\binom{z(k)}{k} \geq \binom{2d + 3}{d} - (2d - 1) > \binom{2d + 2}{d + 1}$$

implies $z(k) \geq 2d + 3$.

Now note that by definition of $z(k)$ that

$$\binom{z(k) - 1}{k} < s.$$

Because $z(k) \geq 2d + 3$ and $k \leq d + 1 < z(k)/2$, then

$$\binom{z(k) - 1}{k - 1} < \binom{z(k) - 1}{k} < s = L.$$

Thus

$$L z(k) \geq k\binom{z(k)}{k}.$$

We can robustly assign each of $\binom{z(k)}{k}$ small types to k out of $z(k)$ containers each with capacity at least L (the full version describes how) and in particular RFF would naturally do this because every combination of k out of $z(k)$ containers

Table 2. The number of containers output by RFF and OPT for different values of d. RFF outputs the minimal $m(k)$ over $1 \le k \le d+1$ while OPT outputs $m^* = 2d + 3$

d	k	$J(k)$	$z(k)$	RFF output	OPT output	RFF/OPT
5	5	1	13	21	13	1.615
8	8	2	19	32	19	1.684
9	8	2	22	35	21	1.666
15000	10611	2	33185	54404	30003	1.81328
25000	17663	2	55348	90671	50003	1.81331
35000	24710	2	77521	126938	70003	1.81332

is used. Since $s \le \binom{z(k)}{k}$, then RFF would successfully use $z(k)$ containers to robustly assign the s small types. Thus we have shown that RFF with $k \le d+1$ items of each type uses $2k - 1 - J(k) + z(k)$ containers. Thus RFF uses the number of containers equal to the minimum of $2k - 1 - J(k) + z(k)$ for $1 \le k \le d+1$.

Theorem 4. *The approximation ratio of* RFF *is no better (lower) than 1.813.*

Proof. Let $d = 15000$, and consider the instance $I(d)$ as defined above. By Proposition 2, RFF ends up using $k = 10611$ and $J = 2$, $z = 33185$ and $m = 54404$ for this instance, while (by Proposition 1) an optimal assignment requires only $m^* = 2d + 3 = 30003$. (Please see Table 2.)

5   Experimental Results

As described in Sect. 1, RAP can be applied to assigning app instances to the minimal number of servers on a hosting platform while ensuring that if a failure occurs in an app and therefore all its hosting servers are temporarily suspended, there is still a running instance of every other app hosted on some unaffected server. Formally, we are given n apps where app i has size d_i and server capacity C. We would like to find an assignment of app instances to the minimal number of servers m, such that the assignment is robust and satisfies the capacity constraint.

To evaluate the performance of the RFF algorithm, we simulated a hosting platform and measured the number of servers used by the algorithm. Specifically, we tested four values for server capacity C (64 GB, 128 GB, 256 GB, and 512 GB), varied the number of apps from $n = 25$ to $n = 250$ apps (at increments of 25) and set app sizes d_i to be normally distributed between 4 GB and 16 GB. We compared RFF to a dedicated system (i.e. where the number of servers is simply the number of apps) and an "ideal" assignment, which does not correspond to any feasible robust assignment, but serves as a lower bound on the minimally required number of servers. (Recall that it even without the robustness constraint, it

Table 3. Number of servers used when server capacity is 64 GB, 128 GB, 256 GB, and 512 GB.

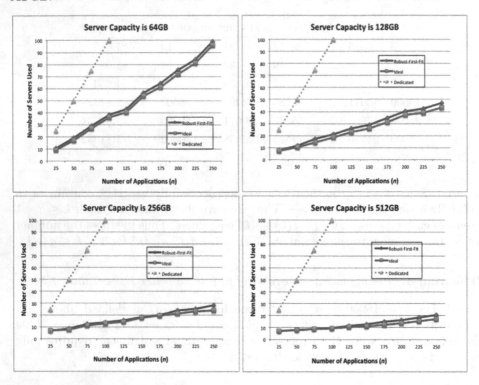

is NP-hard to compute OPT so we did not compute it for the experiments.) We computed the "ideal" assignment by determining the minimum number of servers needed to satisfy robustness alone and the minimum number of servers to satisfy the storage constraints alone and taking the maximum of these two values. I.e., the "ideal" number of servers is defined as: $\min_k \max\{m_r, m_c\}$ where $m_c = \min\{m : mC \geq k \sum_{i=1}^{n} d_i\}$ and $m_r = \min\{m : \binom{m}{k} \geq n\}$.

We tested each setting for 10 iterations and took the average of the results. The graphs in Table 3 show the results. The graphs show that for all settings, RFF performs significantly better than the dedicated system and almost as well as the ideal assignment. Specifically, the worst (minimum) ratio (over all values of n) of servers used by the dedicated system and RFF is 2.40, 3.25, 3.57, and 3.57 for 64 GB, 128 GB, 256 GB, and 512 GB, so RFF always assigned apps more than twice as efficiently as a dedicated system. Note that as the server capacity increases, these ratios either increase or stay the same. The average ratio of servers used by the dedicated system and RFF always increase: 2.64, 4.72, 7.34, and 9.89 for 64 GB, 128 GB, 256 GB, and 512 GB, respectively, so RFF on average performed as much as 9 times as efficiently as a dedicated hosting.

Comparing RFF with the lower bound on optimal, we find that the worst (maximum) ratio (over all values of n) of servers used by RFF and the ideal assignment is 1.17, 1.21, 1.13, and 1.20 for 64 GB, 128 GB, 256 GB, and 512 GB, respectively. So RFF always used close to the same number of servers as an optimal solution. (The average ratios are similar to these values.) The results indicate that when apps sizes are more realistic than those described in Theorem 4 of Sect. 4.2.3, RFF performs close to optimally.

6 Discussion and Conclusions

We proposed a new model for assigning items of various types to containers such that the system is *robust*. We presented an optimal poly-time algorithm in the setting without capacity constraints on the containers. We also presented an optimal poly-time algorithm when item sizes are uniform. Our main algorithm RFF is a poly-time 2-approximation algorithm for the setting where item sizes are nonuniform. Our experimental results suggest that when run on a simulated hosting platform, RFF performs well not only in the worst-case, but even more so on average.

In the lower bound instance, as d increases, it is not clear whether the corresponding ratio is converging (very slowly) to 2 or to a number less than 2, or whether the ratio converges at all; if the ratio does not converge, one can still ask for the limit supremum of the sequence of ratios. If the limit supremum is 2, then the upper bound of 2 is tight.

One direction for future work is to determine whether there is an algorithm with an approximation ratio better than 2. Also, our problem model assumes that the number of items is uniform over all types. A natural extension of this work would be to consider the case where this number is not required to be uniform.

References

1. Chekuri, C., Khanna, S.: A PTAS for the multiple knapsack problem. In: Symposium on Discrete Algorithms (SODA) (2000)
2. Epstein, L., Levin, A.: On bin packing with conflicts. In: Erlebach, T., Kaklamanis, C. (eds.) WAOA 2006. LNCS, vol. 4368, pp. 160–173. Springer, Heidelberg (2007). https://doi.org/10.1007/11970125_13
3. Fleischer, L., Goemans, M.X., Mirrokni, V.S., Sviridenko, M.: Tight approximation algorithms for maximum general assignment problems. In: Proceedings of the Symposium on Discrete Algorithms (2006)
4. Garey, M.R., Johnson, D.S.: Computers and Intractability: A Guide to the Theory of NP-Completeness. Freeman, New York (1979)
5. Jansen, K.: An approximation scheme for bin packing with conflicts. J. Comb. Optim. 3(4), 363–377 (1999)
6. Jansen, K., Öhring, S.: Approximation algorithms for time constrained scheduling. Inf. Comput. 132(2), 85–108 (1997)

7. Korupolu, M., Rajaraman, R.: Robust and probabilistic failure-aware placement. In: Proceedings of the Symposium on Parallelism in Algorithms and Architectures (SPAA), pp. 213–224 (2016)
8. Korupolu, M., Meyerson, A., Rajaraman, R., Tagiku, B.: Robust and probabilistic failure-aware placement. Math. Program. **154**(1–2), 493–514 (2015)
9. Mills, K., Chandrasekaran, R., Mittal, N.: Algorithms for optimal replica placement under correlated failure in hierarchical failure domains. Theor. Comput. Sci. (2017, pre-print)
10. Rahman, R., Barker, K., Alhajj, R.: Replica placement strategies in data grid. J. Grid Comput. **6**(1), 103–123 (2008)
11. Shmoys, D., Tardos, E.: An approximation algorithm for the generalized assignment problem. Math. Program. **62**(3), 461–474 (1993)
12. Sperner, E.: Ein Satz über Untermengen einer endlichen Menge. Mathematische Zeitschrift **27**(1), 544–548 (1928)
13. Stein, C., Zhong, M.: Scheduling when you don't know the number of machines. In: Proceedings of the Symposium on Discrete Algorithms (SODA) (2018)
14. Stirling, J.: Methodus differentialis, sive tractatus de summation et interpolation serierum infinitarium, London (1730)
15. Urgaonkar, B., Rosenberg, A., Shenoy, P.: Application placement on a cluster of servers. Int. J. Found. Comput. Sci. **18**(5), 1023–1041 (2007)

Hardness Results and Approximation Schemes for Discrete Packing and Domination Problems

Raghunath Reddy Madireddy[1]([⊠]), Apurva Mudgal[1], and Supantha Pandit[2]([⊠])

[1] Department of Computer Science and Engineering,
Indian Institute of Technology Ropar, Rupnagar, Punjab, India
{raghunath.reddy,apurva}@iitrpr.ac.in
[2] Stony Brook University, Stony Brook, NY, USA
pantha.pandit@gmail.com

Abstract. The *Maximum Independent Set (MIS)* and *Minimum Dominating Set (MDS)* problems are well-known problems in computer science. In this paper, we consider *discrete* versions of both of these problems - *Maximum Discrete Independent Set (MDIS)* and *Minimum Discrete Dominating Set (MDDS)*. For both problems, the input is a set of geometric objects $\mathcal{O}$ and a set of points $\mathcal{P}$ in the plane. In the *MDIS* problem, the objective is to find a maximum size subset $\mathcal{O}' \subseteq \mathcal{O}$ of objects such that no two objects in $\mathcal{O}'$ have a point in common from $\mathcal{P}$. On the other hand, in the *MDDS* problem, the objective is to find a minimum size subset $\mathcal{O}' \subseteq \mathcal{O}$ such that for every object $O \in \mathcal{O} \setminus \mathcal{O}'$ there exists at least one object $O' \in \mathcal{O}'$ such that $O \cap O'$ contains a point from $\mathcal{P}$.

In this paper, we present PTASes based on *local search* technique for both *MDIS* and *MDDS* problems, where the objects are arbitrary radii disks and arbitrary side length axis-parallel squares. Further, we show that the *MDDS* problem is APX-hard for axis-parallel rectangles, ellipses, axis-parallel strips, downward shadows of line segments, etc. in $\mathbb{R}^2$ and for cubes and spheres in $\mathbb{R}^3$. Finally, we prove that both *MDIS* and *MDDS* problems are NP-hard for unit disks intersecting a horizontal line and for axis-parallel unit squares intersecting a straight line with slope -1.

Keywords: Discrete Independent Set · Discrete Dominating Set
Local search · PTAS · NP-hard · APX-hard · Disks
Axis-parallel squares · Axis-parallel rectangles

1 Introduction

The *Maximum Independent Set (MIS)* and the *Minimum Dominating Set (MDS)* problems attract researchers due to their numerous applications in vari-

S. Pandit—The author is partially supported by the Indo-US Science & Technology Forum (IUSSTF) under the SERB Indo-US Postdoctoral Fellowship scheme with grant number 2017/94, Department of Science and Technology, Government of India.

D. Kim et al. (Eds.): COCOA 2018, LNCS 11346, pp. 421–435, 2018.
https://doi.org/10.1007/978-3-030-04651-4_28

ous fields of computer science like VLSI design, network routing, etc. The input to both problems consists of a set of geometric objects $\mathcal{O}$ in the plane. In the *MIS* problem, we need to find a maximum size sub-collection of objects $\mathcal{O}' \subseteq \mathcal{O}$ such that no two objects in $\mathcal{O}'$ intersect. In the *MDS* problem, we need to find a minimum size sub-collection of objects $\mathcal{O}' \subseteq \mathcal{O}$ such that for every object $O \in (\mathcal{O} \setminus \mathcal{O}')$ there exists at least one object $O' \in \mathcal{O}'$ such that O and O' intersect.

The problems considered in this paper are discrete variants of the *MIS* and *MDS* problems. We formally define these problems as follows.

Maximum Discrete Independent Set (*MDIS*). Let $\mathcal{O}$ be a set of objects and $\mathcal{P}$ be a set of points in the plane. Find a maximum size subset $\mathcal{O}' \subseteq \mathcal{O}$ such that no two objects in $\mathcal{O}'$ cover the same point from $\mathcal{P}$.

Minimum Discrete Dominating Set (*MDDS*). Let $\mathcal{O}$ be a set of objects and $\mathcal{P}$ be a set of points in the plane. Find a minimum size subset $\mathcal{O}' \subseteq \mathcal{O}$ such that for every object $O \in \mathcal{O} \setminus \mathcal{O}'$, $O \cap O' \cap \mathcal{P} \neq \emptyset$ for some $O' \in \mathcal{O}'$.

In this paper, we study the hardness results and polynomial time approximation schemes of the *MDIS* and *MDDS* problems for various geometric objects such as disks, axis-parallel squares, axis-parallel rectangles, etc.

We note that the *MDIS* and *MDDS* problems are at least as hard as *MIS* and *MDS* problems respectively. This can be established by placing a point in each of the intersection regions formed by the given objects in the corresponding instances of the *MIS* and *MDS* problems.

1.1   Previous Work

The *MIS* problem is known to be NP-hard for several classes of objects like unit disks [9], unit squares [12], etc. Further, PTASes are also known for unit squares and unit disks [10,18,21]. On the other hand, Chan and Har-Peled [7] gave a PTAS for the *MIS* problem with pseudo-disks based on the local search algorithm. For axis-parallel rectangles, Adamaszek and Wiese [1] gave a break-through by providing a QPTAS. Very recently, Chuzhoy and Ene [8] also have provided a QPTAS with improved running time. However, a PTAS or a constant factor approximation algorithm is still open.

The *MDIS* problem was first studied by Chan and Har-Peled [7]. They show that an LP-based algorithm gives an $O(1)$-approximation for pseudo-disks. To the best of our knowledge, this is the best approximation factor known till now for the *MDIS* problem even for special classes of pseudo-disks like disks, squares, etc. On the other hand, Chan and Grant [6] have shown that the *MDIS* problem is APX-hard for various classes of objects like axis-parallel rectangles containing a common point, axis-parallel strips, ellipses sharing a common point, downward shadows of line segments, unit balls in $\mathbb{R}^3$ containing the origin, etc. (see Theorem 1.5 in [6]).

The *MDS* problem is NP-complete for unit disk graphs [9] and a PTAS is known for the same [18]. Recently, Gibson and Pirwani [16] obtained a PTAS for *MDS* problem for arbitrary radii disks by local search method first used in [7] and [23]. However, Erlebach and van Leeuwen [11] have shown that the *MDS* problem is APX-hard for several intersection graphs of objects like axis-parallel rectangles, ellipses, etc. Recently, by using local search method, Bandyapadhyay et al. [5] gave a $(2 + \epsilon)$ approximation algorithm for the *MDS* problem with diagonal-anchored axis-parallel rectangles[1], for any $\epsilon > 0$. Actually, they studied L-types of objects which are essentially rectangles when the L-shapes are diagonal-anchored. They gave a local search based PTAS for a special case where the rectangles are anchored from the same side of the diagonal.

1.2 Our Contributions

In [7], Chan and Har-Peled noted that, *"Unlike in the original independent set (MIS) problem, it is not clear if local search yields a good approximation for MDIS problem, even in the unweighted case"*. In this paper, we first answer this partially affirmatively by providing PTASes for the *MDIS* problem with disks and axis-parallel squares. More specifically, we prove the following theorems.

Theorem 1. *There exists a PTAS for the MDIS problem with arbitrary radii disks.*

Theorem 2. *There exists a PTAS for the MDIS problem with arbitrary side length axis-parallel squares.*

The above PTASes are obtained by extending the local search algorithm given in [7]. In addition to the above results, for disks and axis-parallel squares, we present PTASes for the *MDDS* problem by extending the local search method of Gibson and Pirwani [16].

Theorem 3. *There exists a PTAS for the MDDS problem with arbitrary radii disks.*

Theorem 4. *There exists a PTAS for the MDDS problem with arbitrary side length axis-parallel squares.*

To prove the hardness results for the *MDDS* problem, we first define a special case of the *MDS* problem with set systems, the SPECIAL-3DS problem (see Definition 2) and show that it is APX-hard. The proof is inspired from the definition and APX-hardness of the SPECIAL-3SC problem studied by Chan and Grant [6]. Next, by using the SPECIAL-3DS problem, we prove that the following theorem (the classes of objects in this theorem are essentially given in [6]).

---

[1] A set of axis-parallel rectangles is said to be diagonal-anchored, if given a diagonal with slope -1 then either the lower-left or the upper-right corner of each rectangle is on the diagonal.

Theorem 5. *The MDDS problem is* APX-*hard for the following classes of objects.*

A1. *Axis-parallel rectangles in* $\mathbb{R}^2$, *even when all rectangles have upper-left corner inside a square with side length* ϵ *and lower-right corner inside a square with side length* ϵ *for an arbitrary small* $\epsilon > 0$.
A2. *Axis-parallel ellipses in* $\mathbb{R}^2$, *even when all the ellipses contain the origin.*
A3. *Axis-parallel strips in* $\mathbb{R}^2$.
A4. *Axis-parallel rectangles in* $\mathbb{R}^2$, *even when every pair of rectangles intersect either zero or four times.*
A5. *Downward shadows of line segments in* $\mathbb{R}^2$.
A6. *Downward shadows of cubic polynomials in* $\mathbb{R}^2$.
A7. *Unit ball in* $\mathbb{R}^3$, *even when the origin is inside every unit ball.*
A8. *Axis-parallel cubes of similar size in* $\mathbb{R}^3$ *sharing a common point.*
A9. *Half-spaces in* $\mathbb{R}^4$.
A10. *Fat semi-infinite wedges in* $\mathbb{R}^2$ *with apices near the origin.*

We note that for classes **A1-A10**, the *MDIS* problem is known to be APX-hard [6]. Further, in [6], authors also have proved that the set cover problem is APX-hard for all classes of objects **A1-A10** and hitting set is APX-hard for four classes of objects **A3**, **A4**, **A7**, and **A9**. Recently, in [20], the authors have shown that the hitting set problem is APX-hard for the remaining classes of objects.

We also show that both *MDIS* and *MDDS* problems are NP-hard for unit disks intersecting a horizontal line and axis-parallel unit squares intersecting a straight line of slope -1. Our NP-hardness results are inspired by results of Fraser and López-Ortiz [13] and Mudgal and Pandit [22]. We note that in these restricted cases, *MIS* problem can be solved in polynomial time for unit disks [24] and unit squares [22]. Further, the *MDS* problem can also be solved in polynomial-time for unit squares [25]. Our NP-hardness results show the gradation of the complexity between continuous and discrete versions of the problems.

1.3 Organization of the Paper

The rest of the paper is organized as follows. In Sect. 2, we present PTASes for the *MDIS* problem with arbitrary radii disks and arbitrary side length squares. For the same set of objects, we give PTASes for the *MDDS* problem in Sect. 3. The APX-hardness results (proof of Theorem 5 and other related problems) are presented in Sect. 4. Finally, we give a proof sketch of the NP-hardness results for both *MDIS* and *MDDS* problems in Sect. 5.

2 PTAS: Maximum Discrete Independent Set Problem

In this section, we present PTASes for the *MDIS* problem with arbitrary radii disks and arbitrary side length squares. These PTASes are obtained by extending the local search technique of Chan and Har-Peled [7] for the *MIS* problem with pseudo-disks.

Let $(\mathcal{P}, \mathcal{O})$ be the input to the *MDIS* problem where $\mathcal{P}$ is a set of points and $\mathcal{O}$ is a set of objects in the plane. Further, let $m = |\mathcal{O}|$ and $n = |\mathcal{P}|$. Without loss of generality, we assume that no object completely covers another object in $\mathcal{O}$. A set $\mathsf{L} \subseteq \mathcal{O}$ is said to be a *feasible solution* to the *MDIS* problem, if no two objects in L cover the same point from $\mathcal{P}$. For a given integer $t > 1$, we say that a feasible solution L is *t-locally optimal* if we cannot obtain another feasible solution $\mathsf{L}' \subseteq \mathcal{O}$ of larger size, by replacing at most t objects from L with at most $t + 1$ objects from $\mathcal{O}$.

We now describe the procedure to obtain a t-locally optimal solution to the *MDIS* problem in Algorithm 1. Note that, in every local exchange (step 5), the size of L is increased by at least one. Hence, the local exchange can be possible at most m times. However, every such step needs to go over all possible sets $\mathcal{O}'$ and L'. Since $|\mathcal{O}'| \leq t + 1$, there will be at most $O(m^{t+1})$ possibilities for $\mathcal{O}'$ and for every such $\mathcal{O}'$ at most $O(m^t)$ number of different L' are possible. Further, to check whether $(\mathsf{L} \setminus \mathsf{L}') \cup \mathcal{O}'$ is a feasible solution or not, one needs $O(nm)$-time. Hence, Algorithm 1 returns a t-locally optimal solution $\mathsf{L} \subseteq \mathcal{O}$ in $O(nm^{2t+3})$-time.

Algorithm 1. t-level local search for *MDIS* problem

1: Let $\mathsf{L} \leftarrow \emptyset$.
2: **for** $\mathcal{O}' \subseteq \mathcal{O} \setminus \mathsf{L}$ of size at most $t + 1$ **do**
3:     **for** $\mathsf{L}' \subseteq \mathsf{L}$ of size at most t **do**
4:         **if** $(\mathsf{L} \setminus \mathsf{L}') \cup \mathcal{O}'$ is a feasible solution and $|(\mathsf{L} \setminus \mathsf{L}') \cup \mathcal{O}'| \geq |\mathsf{L}| + 1$ **then**
5:             $\mathsf{L} \leftarrow (\mathsf{L} \setminus \mathsf{L}') \cup \mathcal{O}'$ ▷ local exchange step
6:         **end if**
7:     **end for**
8: **end for**

In the following, we first show that Algorithm 1 returns a t-locally optimal solution which has size at least $(1 - O(\frac{1}{\sqrt{t}}))$ times of the size of the optimal solution to *MDIS* problem when the objects are arbitrary radii disks and later we show that the same is also true for arbitrary side length axis-parallel squares.

Consider that $\mathcal{O}$ is a set of arbitrary radii disks. Without loss of generality, we assume that no three disk centers are collinear, and no more than three disks are tangent to a circle [16,27]. For a disk D, let $\mathsf{cen}(D)$ and $\mathsf{radius}(D)$ denote the center and radius of D respectively. Let $\mathsf{dist}(x, y)$ denote the euclidean distance between points x and y in the plane. For a disk D and a point p in the plane, let $\Phi(D, p)$ be the distance between boundary of D and point p i.e., $\Phi(D, p) = \mathsf{dist}(\mathsf{cen}(D), p) - \mathsf{radius}(D)$.

For the given instance $(\mathcal{P}, \mathcal{O})$ of the *MDIS* problem, let $\mathsf{L} \subseteq \mathcal{O}$ be the t-locally optimal solution return by Algorithm 1 and let $\mathsf{OPT} \subseteq \mathcal{O}$ be an optimal solution. For a disk $D \in \mathsf{L} \cup \mathsf{OPT}$, let $\mathsf{cell}(D)$ be the set of points in the plane which are closer to the boundary of D with respect to all other disks in $\mathsf{L} \cup \mathsf{OPT}$ i.e., $\mathsf{cell}(D) = \{p \mid \Phi(D, p) \leq \Phi(D', p) \ \forall D' \in \mathsf{L} \cup \mathsf{OPT}\}$. The collection of all cells of disks in $L \cup OPT$ defines the *Weighed Voronoi Diagram (WVD)* i.e., $WVD = \bigcup_{D \in L \cup OPT} \text{cell}(D)$. We now mention two properties of cells in the WVD.

Lemma 1 ([16]). *For each disk $D \in L \cup OPT$, the following two properties are true.*

1. $\text{cell}(D)$ *is non-empty. In particular,* $\text{cell}(D)$ *contains* $\text{cen}(D)$.
2. $\text{cell}(D)$ *is star-shaped i.e., for any point $p \in \text{cell}(D)$, every point on the line segment $p\, \text{cen}(D)$ is in* $\text{cell}(D)$.

Lemma 2 ([27]). *Let D_1 and D_2 be two disks in $L \cup OPT$. Let x be a point in the plane such that $\Phi(D_1, x) \leq \Phi(D_2, x)$. If D_2 covers x, then D_1 also covers x.*

Let $G = (V, E)$ be a given graph. For a vertex $v \in V$, let $\text{NBH}(v)$ be the set of adjacent vertices of v in V. For a subset $V' \subseteq V$ of vertices, let $\text{NBH}(V')$ be the set of all adjacent vertices of vertices in V' i.e., $\text{NBH}(V') = \bigcup_{v \in V'} \text{NBH}(v)$. Further, let $\text{NBH}(V')^c = \text{NBH}(V') \cup V'$. We now note a planar separator theorem from [14] which is required in proving the performance of the local search algorithm.

Lemma 3 ([14]). *For any given planar graph $G = (V, E)$ and a parameter $r \geq 1$, there exists a subset $X \subseteq V$ of size at most $c_1 |V|/\sqrt{r}$, and a partition of $V \setminus X$ into $|V|/r$ sets $V_1, V_2, \ldots, V_{|V|/r}$ such that (i) $|V_i| \leq c_2 r$, (ii) $\text{NBH}(V_i) \cap V_j = \emptyset$ for $i \neq j$, and (iii) $|\text{NBH}(V_i) \cap X| \leq c_3 \sqrt{r}$ for some constants c_1, c_2, and c_3.*

***Proof of Theorem* 1.** We first define a graph $G = (V, E)$ which can be viewed as the dual of WVD of disks in $L \cup OPT$.

1. For every disk $D \in L \cup OPT$, we place a vertex in G at $\text{cen}(D)$.
2. For every $L \in L$ and $O \in OPT$, we place an edge between $\text{cen}(L)$ and $\text{cen}(O)$ if and only if there exists a point x in the plane such that $\Phi(L, x) = \Phi(O, x)$.

By using the star-shaped property of cells, one can draw G such that no two edges intersect [4]. Thus, graph $G = (V, E)$ is planar bipartite.

As in [7], we apply Lemma 3 on graph $G = (V, E)$ with $r = t/(c_2 + c_3)$. Then, $|\text{NBH}(V_i)^c| \leq |V_i| + |\text{NBH}(V_i)| \leq c_2 r + c_3 \sqrt{r} \leq (c_2 + c_3)r \leq t$. Let $OPT_i = V_i \cap OPT$, $L_i = V_i \cap L$, and $X_i = \text{NBH}(V_i) \cap X$.

We now prove that $(L \setminus (L \cap \text{NBH}(V_i)^c)) \cup OPT_i$ is a feasible solution for the *MDIS* problem. We note that any subset of a feasible solution is also a feasible solution of the *MDIS* problem. Hence, $(L \setminus (L \cap \text{NBH}(V_i)^c))$ and OPT_i are also feasible solutions of the *MDIS* problem. For the sake of contradiction, assume that $(L \setminus (L \cap \text{NBH}(V_i)^c)) \cup OPT_i$ is not a feasible solution. Hence, there exists two disks $O \in OPT_i$ and $L \in (L \setminus (L \cap \text{NBH}(V_i)^c))$ such that both O and L cover the same point $p \in \mathcal{P}$. One can note that, O and L are the unique disks in OPT_i and $L \setminus (L \cap \text{NBH}(V_i)^c)$ respectively, which cover the point p. Without loss of generality, assume that $p \in \text{cell}(O)$. Hence, $\Phi(O, p) \leq \Phi(L, p)$. There are two possible cases. (The arguments in the two cases are on the same lines of the proof of Lemma 3 in [16].)

Case 1: Suppose $\Phi(O, p) = \Phi(L, p)$. Then $p \in \text{cell}(L)$. Hence, $\text{cell}(O)$ and $\text{cell}(L)$ share a common boundary in WVD and further, $O \in \text{OPT}$ and $L \in \text{L}$. Thus, there exists an edge between $\text{cen}(O)$ and $\text{cen}(L)$ in graph G. Hence $L \in \text{NBH}(O)$ which implies $L \notin \text{L} \setminus (\text{L} \cap \text{NBH}(V_i)^c)$.

Case 2: Suppose $\Phi(O, p) < \Phi(L, p)$. Take a walk from p to $\text{cen}(L)$ along line segment $p\,\text{cen}(L)$. Note that we may go through several cells along the walk. Let q be the point at which we enter into $\text{cell}(L)$ along this walk. Therefore, $\Phi(D, q) = \Phi(L, q)$ for some $D \in \text{L} \cup \text{OPT}$ and $D \neq O$. We now prove that D covers p. Consider, $\text{dist}(\text{cen}(D), p) < \text{dist}(p, q) + \text{dist}(\text{cen}(D), q)$ which implies $\Phi(D, p) < \text{dist}(p, q) + \Phi(D, q) = \text{dist}(p, q) + \Phi(L, q) = \Phi(L, p)$. Since $\Phi(D, p) < \Phi(L, p)$ and L covers p, by Lemma 2 disk D also covers p. Suppose $D \in \text{OPT}$. Then, p is covered by two disks O and D in OPT, which is not true. Suppose $D \in \text{L}$. In this case also, p is covered by two disks D and L in L, which is not true.

Thus, $(\text{L} \setminus (\text{L} \cap \text{NBH}(V_i)^c)) \cup \text{OPT}_i$ is a feasible solution for the *MDIS* problem. We now proceed as in [7]. If $|\text{OPT}_i| > |L_i| + |X_i|$, then by replacing disks of $\text{L} \cap \text{NBH}(V_i)^c$ in L with disks in OPT_i, we get a better solution. It contradicts the fact that L is t-locally optimal. Hence, $|\text{OPT}_i| \leq |L_i| + |X_i|$. Thus,

$$|\text{OPT}| \leq \Sigma_i |\text{OPT}_i| + |X| \leq \Sigma_i |L_i| + \Sigma_i |X_i| + |X|$$
$$\leq |\text{L}| + c_3 \sqrt{r} \frac{|V|}{r} + c_1 \frac{|V|}{\sqrt{r}} \leq |\text{L}| + (c_1 + c_3) \frac{|V|}{\sqrt{r}} = |\text{L}| + (c_1 + c_3) \frac{|\text{OPT}| + |\text{L}|}{\sqrt{r}}$$

This implies that $|\text{OPT}| \leq (1 + O(\frac{1}{\sqrt{t}}))|\text{L}|$. Hence, the theorem is proved. □

Proof sketch of Theorem 2. Let $\mathcal{O}$ be the set of axis-parallel squares with arbitrary side lengths. Apply t-level local search given in Algorithm 1 with $t = O(1/\epsilon^2)$. The analysis is similar to the analysis of arbitrary radii disks, except that for any two points p and q in the plane, $\text{dist}(p, q)$ is defined under infinity norm L_∞ instead of L_2-norm as in [3]. □

3   PTAS: Minimum Discrete Dominating Set Problem

In this section, we first give a PTAS for the *MDDS* problem with arbitrary radii disks by using a local search algorithm similar in [16]. Further, we show that the same local search algorithm will give a PTAS for the *MDDS* problem with arbitrary side length axis-parallel squares.

Let $\mathcal{P}$ be a set of n points and $\mathcal{O}$ be a set of m disks in the plane. As in Sect. 2, we assume that no three disk centers are collinear and no more than three disks are tangent to a circle [16,27]. Further, assume that no point in $\mathcal{P}$ lies on the boundary of a disk in $\mathcal{O}$. Also, no three points in $\mathcal{P}$ are collinear.

Let D and D' be two disks in $\mathcal{O}$ such that both D and D' cover a point $p \in \mathcal{P}$, then we say that D is a *dominator* of D' and vice versa. A set $\mathcal{O}' \subseteq \mathcal{O}$ of disks is said to be a *feasible solution* to the *MDDS* problem, if for every disk $O \in (\mathcal{O} \setminus \mathcal{O}')$, there exists at least one dominator in $\mathcal{O}'$. For a given integer

$t > 1$, we say a feasible solution $L \subseteq \mathcal{O}$ is *t-locally optimal* if one cannot obtain a smaller size feasible solution $L' \subseteq \mathcal{O}$ by replacing at most t disks from L with at most $t - 1$ disks from $\mathcal{O}$. One can obtain a t-locally optimal solution to the *MDDS* problem by using a similar local search method in Algorithm 1. Set $L \leftarrow \mathcal{O}$. For $L' \subseteq L$ of size at most t and for every $\mathcal{O}' \subseteq \mathcal{O} \setminus L$ of size at most $t-1$, verify whether $(L \setminus L') \cup \mathcal{O}'$ is a feasible solution and $|(L \setminus L') \cup \mathcal{O}'| \leq |L| - 1$. If yes, replace L with $(L \setminus L') \cup \mathcal{O}'$ (local exchange). Repeat this procedure till no further local exchange is possible. Further, we note that the procedure returns t-locally optimal solution in $O(nm^{2t+3})$-time.

Let $L \subseteq \mathcal{O}$ be a t-locally optimal solution returned by the local search algorithm and let $\mathsf{OPT} \subseteq \mathcal{O}$ be the optimal solution for the *MDDS* problem. Without loss of generality, assume that $L \cap \mathsf{OPT} = \emptyset$ (see [23] for the details). Further, as in [16], we assume that no disk in $L \cup \mathsf{OPT}$ is fully contained in other disk in $\mathcal{O}$.

Definition 1 *(locality condition) [16]. There exists a planar-bipartite graph* $G = (L \cup \mathsf{OPT}, E)$ *such that for every object* $X \in \mathcal{O}$, *there exists an edge between* $L \in L$ *and* $O \in \mathsf{OPT}$ *where both* L *and* O *are dominators of* X. *(Note that the definition of dominator is essentially different in [16]).*

Lemma 4 ([16]). *If the set of disks* $L \cup \mathsf{OPT}$ *satisfies the locality condition then* $|\mathsf{OPT}| \leq (1 + \epsilon)|L|$ *for* $\epsilon = O(\frac{1}{\sqrt{t}})$.

In the following, we construct a graph $G = (V, E)$ which satisfies the locality condition given in Definition 1. Our construction of G is inspired by the results in [19]. Partition the set $\mathcal{O}$ into two sets $\mathcal{O}_1$ and $\mathcal{O}_2$ as follows:

1. $\mathcal{O}_1$ is the collection of disks in $\mathcal{O}$ such that for every disk $D \in \mathcal{O}_1$ there exists at least one point $p \in \mathcal{P}$ that is covered by D and is also covered by at least one disk in L as well as at least one disk in OPT.
2. $\mathcal{O}_2 = \mathcal{O} \setminus \mathcal{O}_1$.

For every disk $D \in L \cup \mathsf{OPT}$, we consider a vertex in graph G at $\mathrm{cen}(D)$. The edge set E is constructed in two phases, i.e., $E = E_1 \cup E_2$. The edge set E_i (for $i = 1$ and 2) make sure that the locality condition is satisfied for the disks in $\mathcal{O}_i$.

Phase I (Construction of the edge set E_1): We first construct the *WVD* (weighted Voronoi diagram) of disks in $L \cup \mathsf{OPT}$ as in Sect. 2. Then for every disk $L \in L$ and every disk $O \in \mathsf{OPT}$, we place an edge in E_1 with end points $\mathrm{cen}(L)$ and $\mathrm{cen}(O)$ if and only if there exists a point q in the plane such that q is on the boundary of both $\mathrm{cell}(L)$ and $\mathrm{cell}(O)$. In particular, the edge is $\overline{\mathrm{cen}(L)q} \cup \overline{\mathrm{cen}(O)q}$.

Lemma 5. *The graph* $G = (V, E_1)$ *satisfies locality condition for disks in* $\mathcal{O}_1$.

(We omit the proof as one can prove this by using the arguments given in Lemma 3 in [16].)

Phase II (Construction of the edge set E_2): For every disk $D \in \mathcal{O}_2$, let $\mathcal{P}^D \subseteq \mathcal{P}$ be the set of points which are covered by disk D. Further, let $\mathcal{P}_L^D \subseteq \mathcal{P}^D$

be the set of points covered by at least one disk in L and $\mathcal{P}^D_{\mathsf{OPT}} \subseteq \mathcal{P}^D$ be the set of points covered by at least one disk in OPT. Note that sets $\mathcal{P}^D_{\mathsf{L}}$ and $\mathcal{P}^D_{\mathsf{OPT}}$ are non-empty. Further, $\mathcal{P}^D_{\mathsf{L}} \cap \mathcal{P}^D_{\mathsf{OPT}} = \emptyset$.

Lemma 6. *Let $p_1 \in \mathcal{P}^D_{\mathsf{L}}$ and $p_2 \in \mathcal{P}^D_{\mathsf{OPT}}$ be two points for some $D \in \mathcal{O}_2$. Further, let $p'_1 \in \mathcal{P}^{D'}_{\mathsf{L}}$ and $p'_2 \in \mathcal{P}^{D'}_{\mathsf{OPT}}$ be two points such that $p_1 \neq p'_1$ and $p_2 \neq p'_2$ for some $D' \in \mathcal{O}_2$. If line segments $\overline{p_1 p_2}$ and $\overline{p'_1 p'_2}$ intersect then at least one of the following is true: (i) p_1 or p_2 is covered by disk D' and (ii) p'_1 or p'_2 is covered by disk D.*

Proof. We note that segment $\overline{p_1 p_2}$ is completely inside D and $\overline{p'_1 p'_2}$ is completely inside D'. If both p_1 and p_2 are not covered by D', then $\overline{p_1 p_2}$ intersects twice the boundary of disk D'. Similarly, if both p'_1 and p'_2 are not covered by disk D, then $\overline{p'_1 p'_2}$ intersects the boundary of the disk D twice. Thus, if the claim is not true then boundary of the both disks D and D' intersect four times, which is not true.   □

Let $\overline{p_1 p_2}$ be a line segment such that $p_1 \in \mathcal{P}^D_{\mathsf{L}}$ and $p_2 \in \mathcal{P}^D_{\mathsf{OPT}}$ for some disk $D \in \mathcal{O}_2$. We say $\overline{p_1 p_2}$ is a *dominator crossing segment*, if there exists a dominator $X \in \mathsf{L} \cup \mathsf{OPT}$ of D such that $\overline{p_1 p_2}$ intersect the boundary of X twice.

We construct a set $\mathcal{S}$ of non-intersecting line segments, except possibly at endpoints. For every disk $D \in \mathcal{O}_2$, we do the following: add line segment $\overline{p_1 p_2}$ in $\mathcal{S}$ for each $p_1 \in \mathcal{P}^D_{\mathsf{L}}, p_2 \in \mathcal{P}^D_{\mathsf{OPT}}$ only if the following two conditions are satisfied.

1. $\overline{p_1 p_2}$ is a non-dominator crossing segment.
2. $\overline{p_1 p_2}$ does not intersect any other non-dominator crossing segment $\overline{p'_1 p'_2}$ where $p'_1 \in \mathcal{P}^{D'}_{\mathsf{L}}$ and $p'_2 \in \mathcal{P}^{D'}_{\mathsf{OPT}}$ for some $D' \in \mathcal{O}_2$. Note that D and D' need not be distinct.

Hence, we have the following lemma.

Lemma 7. *For every disk $D \in \mathcal{O}_2$, the set $\mathcal{S}$ contains at least one line segment which is completely covered by D.*

We now place an edge in E_2 for every segment in $\mathcal{S}$ as follows.

Step 1. Let $\overline{p_1 p_2}$ be segment in $\mathcal{S}$ such that $p_1 \in \mathcal{P}^D_{\mathsf{L}}$ and $p_2 \in \mathcal{P}^D_{\mathsf{OPT}}$ for some disk $D \in \mathcal{O}_2$.
Step 2. Let $\mathsf{OPT}_D \subseteq \mathsf{OPT}$ and $\mathsf{L}_D \subseteq \mathsf{L}$ be the set of dominators of D. Let $\overline{x_1 x_2}$ be the minimal portion of $\overline{p_1 p_2}$ such that $x_1 \in \mathrm{cell}(L)$ and $x_2 \in \mathrm{cell}(O)$ for some $L \in \mathsf{L}_D$ and $O \in \mathsf{OPT}_D$. We further note that, $x_1 \neq p_1$ and $x_2 \neq p_2$ since we assume that no point in $\mathcal{P}$ is on the boundary of any disks in $\mathcal{O}$.
Step 3. Place an edge $\overline{\mathrm{cen}(L)x_1} \cup \overline{x_1 x_2} \cup \overline{\mathrm{cen}(O)x_2}$ in edge set E_2. Note that $\overline{\mathrm{cen}(L)x_1}$ is completely inside $\mathrm{cell}(L)$, $\overline{\mathrm{cen}(O)x_2}$ is completely inside $\mathrm{cell}(O)$, and $\overline{x_1 x_2}$ is completely inside disk D.

We can note that in graph $G = (V, E_1 \cup E_2)$, for every disk $D \in \mathcal{O}$, there exists an edge between a dominator of D in L and a dominator of D in OPT. However, the graph $G = (V, E_1 \cup E_2)$ need not be planar. In the following, we show that one can obtain a planar graph by either *edge perturbation* (slight bend) of some edges or *edge removal* without violating the above property.

Let $e_1 = \overline{\text{cen}(L_1)x_1} \cup \overline{x_1 x_1'} \cup \overline{\text{cen}(O_1)x_1'}$ and $e_2 = \overline{\text{cen}(L_2)x_2} \cup \overline{x_2 x_2'} \cup \overline{\text{cen}(O_2)x_2'}$ be two edges in $E_1 \cup E_2$ for some $L_1, L_2 \in \mathsf{L}$ and $O_1, O_2 \in \mathsf{OPT}$. Assume that L_1 and O_1 dominate disk $D_1 \in \mathcal{O}_2$ and disks L_2 and O_2 dominate disk $D_2 \in \mathcal{O}_2$. Further, assume that $p_1 \in L_1 \cap D_1$, $p_1' \in O_1 \cap D_1$, $p_2 \in L_2 \cap D_2$, and $p_2' \in D_2 \cap O_2$ for some $p_1, p_2, p_1', p_2' \in \mathcal{P}$ such that $\overline{x_1 x_1'}$ is a portion of $\overline{p_1 p_1'}$ and $\overline{x_2 x_2'}$ is a portion of $\overline{p_2 p_2'}$. In particular, if $e_i \in E_1$ for $i = 1, 2$, the line segment $\overline{x_i x_i'}$ is just a point and it is on the boundary of both cell(L_i) and cell(O_i).

We note that both $\overline{x_1 x_1'}$ and $\overline{x_2 x_2'}$ also do not intersect since no two segments in $\mathcal{S}$ intersect.

Lemma 8. *The following pairs of segments do not intersect: (i)* $\overline{\text{cen}(L_1)x_1}$ *and* $\overline{\text{cen}(L_2)x_2}$, *(ii)* $\overline{\text{cen}(L_1)x_1}$ *and* $\overline{\text{cen}(O_2)x_2'}$, *(iii)* $\overline{\text{cen}(O_1)x_1'}$ *and* $\overline{\text{cen}(L_2)x_2}$, *and (iv)* $\overline{\text{cen}(O_1)x_1'}$ *and* $\overline{\text{cen}(O_2)x_2'}$.

Proof. Suppose $\overline{\text{cen}(L_1)x_1}$ and $\overline{\text{cen}(L_2)x_2}$ intersect at point x. Since $\overline{\text{cen}(L_1)x_1}$ is completely inside cell(L_1) and $\overline{\text{cen}(L_2)x_2}$ is completely inside cell(L_2), point x cannot be an interior point to both cell(L_1) and cell(L_2). Hence, $x = x_1 = x_2$. Thus, both line segments $\overline{p_1 p_1'}$ and $\overline{p_2 p_2'}$ intersect at x which is not the common endpoint of the both segments. This is a contradiction to the fact, no two segments in $\mathcal{S}$ intersect, except possibly at endpoints.

The other three cases are similar. Hence the lemma.   □

Thus, if both edges e_1 and e_2 intersect then it must be the case that exactly one of the following four pairs of segments intersect. (i) $\overline{\text{cen}(L_1)x_1}$ and $\overline{x_2 x_2'}$, (ii) $\overline{\text{cen}(O_1)x_1'}$ and $\overline{x_2 x_2'}$, (iii) $\overline{\text{cen}(L_2)x_2}$ and $\overline{x_1 x_1'}$, and (iv) $\overline{\text{cen}(O_2)x_2'}$ and $\overline{x_1 x_1'}$. We now describe the edge perturbation and removal step for the first case and other cases are similar.

Suppose $\overline{\text{cen}(L_1)x_1}$ and $\overline{x_2 x_2'}$ intersect at point x in the plane. There are two possible cases:

Case (a): *Suppose* $\overline{p_2 p_2'}$ *intersect disk* L_1 *twice.* In this case, L_1 cannot be a dominator of D_2, otherwise $\overline{p_2 p_2'}$ is a dominator crossing segment. Partition the disk L_1 into two connected regions which are on the both sides of $\overline{p_2 p_2'}$. In particular, one region contains point p_1 and other region contains no point from $\mathcal{P}$. Further, the latter region is completely covered by disk D_2 and contains cen(L_1). We now slightly bend (*Edge Perturbation*) $\overline{p_2 p_2'}$ as follows: let $\overline{t_1 t_2}$ be the maximal portion of $\overline{x_2 x_2'}$ such that $\overline{t_1 t_2}$ is completely in cell(L_1). We replace $\overline{t_1 t_2}$ with an arc $arc(t_1, t_2)$ such all points on this arc are inside cell(L_1) and goes from outside cen(L_1) such it will not intersect $\overline{\text{cen}(L_1)x_1}$. One can observe that such bend will not intersect any other edge in graph G.

Case (b): *Suppose* $\overline{p_2 p_2'}$ *intersect disk* L_1 *exactly once.* We note that p_2' cannot be inside L_1 otherwise $L_1 \in \mathsf{OPT}$ which is a contradiction to assumption that $\mathsf{L} \cap \mathsf{OPT} = \emptyset$. Thus, p_2 is inside L_1 and hence L_1 is also a dominator of D_2.

Then, segment $\overline{p_1 p_2'}$ is in $\mathcal{S}$ and there exists an edge in E_2 corresponding to segment $\overline{p_1 p_2'}$. Hence, we remove edge e_2 from E_2 (*Edge Removal*). One can note that, removal of e_2 will not violate the locality condition for D_2.

Thus, after applying edge perturbation and removal step on every pair of intersecting edges in G, the resultant graph G becomes planar. Hence, we conclude the following lemma.

Lemma 9. *The disks in* L $\cup$ OPT *satisfy the locality condition.*

Proof of Theorem 3. The Lemma 4 together with Lemma 9 completes the proof of the theorem. $\qquad\square$

Proof Sketch of Theorem 4. A similar argument of the Lemma 9 shows that the locality condition can also be satisfied for the arbitrary side length squares. However, here the distance function is defined with respect to infinity norm instead of the Euclidean norm. $\qquad\square$

4   APX-Hardness Results

In this section, we present APX-hardness results for the *MDIS* and *MDDS* problems. First, we define a restricted version of the *MDS* problem with set systems, the SPECIAL-3DS problem and show that it is APX-hard. We use the SPECIAL-3DS to prove Theorem 5. The work is inspired by the results in [6].

Definition 2 (SPECIAL-3DS). *Let* $(\mathcal{U}, \mathcal{S})$ *be a range space where* $\mathcal{U} = \mathcal{A} \cup \mathcal{B}$, $\mathcal{A} = \{a_1, a_2, \ldots, a_n\}$, $\mathcal{B} = \mathcal{B}^1 \cup \mathcal{B}^2 \cup \cdots \cup \mathcal{B}^6$, *and* $\mathcal{B}^i = \{b_1^i, b_2^i, \ldots, b_m^i\}$ *for* $1 \le i \le 6$ *such that* $3m = 2n$. *Further,* $\mathcal{S}$ *is a collection of* $7m$ *subsets of* $\mathcal{U}$ *such that*

1. *Every element in* $\mathcal{U}$ *is in exactly two sets in* $\mathcal{S}$.
2. *For every* t, $(1 \le t \le m)$, *there exists three integers* $1 \le i < j < k \le n$ *such that the sets* $\{a_i, b_t^1\}$, $\{b_t^1, b_t^2\}$, $\{b_t^2, b_t^3\}$, $\{b_t^3, b_t^4, a_j\}$, $\{b_t^4, b_t^5\}$, $\{b_t^5, b_t^6\}$, *and* $\{b_t^6, a_k\}$ *are in the collection* $\mathcal{S}$.

The objective is to find a minimum size sub-collection $\mathcal{S}' \subseteq \mathcal{S}$ *such that for every* $S \in \mathcal{S}$ *either* $S \in \mathcal{S}'$ *or there exists a set* $S' \in \mathcal{S}'$ *such that* $S \cap S' \ne \emptyset$.

We use the *L-reduction* [26] to prove that the SPECIAL-3DS is APX-hard. Let X and Y be two optimization problems. A polynomial-time computable function f from X to Y is an *L-reduction* if there exist two positive constants α and β (usually 1) such that for each instance x of X the following two conditions hold:

C1: $OPT(f(x)) \le \alpha \cdot OPT(x)$ where $OPT(x)$ and $OPT(f(x))$ are the size of the optimal solutions of x and $f(x)$ respectively.
C2: For any given solution of $f(x)$ with cost $C_{f(x)}$, there exists a polynomial-time algorithm which finds a feasible solution of x with cost C_x such that $|C_x - OPT(x)| \le \beta \cdot |C_{f(x)} - OPT(f(x))|$.

Lemma 10. SPECIAL-3DS *is* APX-*hard.*

Proof. We prove the lemma by giving an L-reduction from an APX-hard problem, dominating set on cubic graphs [2]. Let I_1 be an instance of dominating set problem on a graph $G = (V, E)$ with $V = \{v_1, v_2, \ldots, v_m\}$ and $E = \{e_1, e_2, \ldots, e_n\}$ such that the degree of every vertex in V is exactly three. We now generate an instance I_2 of SPECIAL-3DS from I_1 as follows:

1. Let $\mathcal{A} = \{a_1, a_2, \ldots, a_n\}$ and $\mathcal{B} = \mathcal{B}^1 \cup \mathcal{B}^2 \cup \cdots \cup \mathcal{B}^6$ where $\mathcal{B}^i = \{b_1^i, b_2^i, \ldots, b_m^i\}$ for $i = 1, 2, \ldots, 6$.
2. For a vertex v_t in V ($1 \leq t \leq m$), let e_i, e_j, and e_k ($1 \leq i < j < k \leq n$) be the edges incident on v_t. Then add seven sets $\{a_i, b_t^1\}$, $\{b_t^1, b_t^2\}$, $\{b_t^2, b_t^3\}$, $\{b_t^3, b_t^4, a_j\}$, $\{b_t^4, b_t^5\}$, $\{b_t^5, b_t^6\}$, and $\{b_t^6, a_k\}$ into $\mathcal{S}$. Do the same for every vertex in V.

Let $\mathsf{OPT}(I_1) \subseteq V$ be an optimal dominating set for the instance I_1. We now give a polynomial time algorithm to find an optimal solution $\mathsf{OPT}(I_2)$ for instance I_2 of the SPECIAL-3DS problem from $\mathsf{OPT}(I_1)$. For every vertex $v_t \in V(G)$, do the following:

1. If v_t is in $\mathsf{OPT}(I_1)$ then take sets $\{a_i, b_t^1\}$, $\{b_t^3, a_j, b_t^4\}$, $\{b_t^6, a_k\}$ in $\mathsf{OPT}(I_2)$.
2. If v_t is not in $\mathsf{OPT}(I_1)$ then take sets $\{b_t^2, b_t^3\}$, $\{b_t^4, b_t^5\}$ in $\mathsf{OPT}(I_2)$.

One can easily see that $\mathsf{OPT}(I_2)$ is an optimal dominating set for I_2 and $|\mathsf{OPT}(I_2)| = |\mathsf{OPT}(I_1)| + 2m$. Since $|\mathsf{OPT}(I_1)| \geq m/4$ we get $|\mathsf{OPT}(I_2)| \leq 9 \cdot |\mathsf{OPT}(I_1)|$. Similarly, for any given feasible solution $\mathsf{F}_2 \subseteq \mathsf{S}$ of I_2, one can obtain a feasible solution $\mathsf{F}_1 \subseteq V(G)$ of I_1 such that $|\mathsf{F}_1| \leq |\mathsf{F}_2| - 2m$.

Thus, we conclude that the above reduction is an L-reduction [26] with $\alpha = 9$ and $\beta = 1$. Therefore, the SPECIAL-3DS problem is APX-hard. $\quad\square$

Proof of Theorem 5. The proof is essentially similar to the results in [6]. Similar to the SPECIAL-3SC problem, in the SPECIAL-3DS problem we can order the elements in $\mathcal{B}$ such that every set in $\mathcal{S}$ contains either two consecutive elements from $\mathcal{B}$, one element from $\mathcal{A}$ and one element from $\mathcal{B}$, or one element from $\mathcal{A}$ and two consecutive elements from $\mathcal{B}$. In particular, $(b_t^1, b_t^2, b_t^3, b_t^4, b_t^5, b_t^6)$ must be in the same order for every t ($1 \leq t \leq m$). One can easily encode the instance of the SPECIAL-3DS problem into the *MDDS* problem instances for classes of objects **A1-A10** by following the procedure for obtaining set cover from the SPECIAL-3SC problem given in [6]. In Fig. 1, we depict the encoding of classes **A1**, **A3**, and **A4**. $\quad\square$

We now sketch some additional APX-hardness results.

Theorem 6. *Both MDIS and MDDS problems are* APX-*hard for the classes of objects (i) triangles of similar size and (ii) similar circles.*

Proof. The proof is on the similar lines to the results of Har-Peled [17], who show that set cover problem is APX-hard for rectangles of similar size and similar size

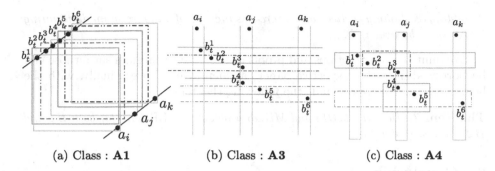

(a) Class : **A1**   (b) Class : **A3**   (c) Class : **A4**

Fig. 1. Encoding of SPECIAL-3DS instance into the *MDDS* problem instances with various classes of geometric objects.

circles by giving a reduction from a known APX-hard problem, vertex cover problem on cubic graphs [2].

We note that one can extend these reductions for our problems. For the *MDIS* problem, a reduction can be given from an APX-hard problem, maximum independent set on cubic graphs [2] and for the *MDDS* problem, the reduction is from minimum dominating set problem on cubic graphs which is known to be APX-hard [2]. The rest of the proofs are same as in [17]. □

5   NP-Hardness Results

In this section, we show that both *MDIS* and *MDDS* problems are NP-hard for the following two classes of geometric objects:

B1: Unit disks intersecting a horizontal line.
B2: Axis-parallel unit squares intersecting a straight line with slope -1.

For **B1**, the reduction is similar to the reduction of covering points by unit disks where the points and disk centers are constrained to be inside a horizontal strip (the *within strip discrete unit disk cover (WSDUDC)* problem) [13]. On the other hand, for **B2**, the reduction is similar to the reduction of the set cover problem with unit squares where the squares intersect a line with slope -1 [22]. However, for our *MDIS* problem, we give a reduction from a known NP hard problem *maximum independent set* on planar graphs where the degree of each vertex of the graph is at most 3 [15] and for the *MDDS* problem we give a reduction from the NP-hard problem *minimum dominating set* on planar graphs such that every vertex is of degree at most 3 [15]. The correctness of the reductions depend on the following lemma.

Lemma 11. *Let G be a graph and e be an edge of G, then*

1. *adding $2k$ dummy vertices on e increases the size of any maximum independent set in G by exactly k.*

2. *adding $3k$ dummy vertices on e increases the size of any minimum dominating set in G by exactly k.*

We omit the constructions and proofs of the hardness results since those can be borrowed from the respective papers as mentioned above. Finally, with the help of Lemma 11, we conclude the following theorem.

Theorem 7. *Both the MDIS and MDDS problems are NP-hard for both **B1** and **B2** classes of objects.*

6   Conclusion

In this paper, for both *MDIS* and *MDDS* problems we design local search based PTASes when the given objects are arbitrary radii disks and arbitrary side length axis-parallel squares. These results partially address the question posed by Chan and Har-Peled [7] about designing a PTAS for the *MDIS* problem with pseudo-disks. Further, we show that the *MDDS* problem is APX-hard for various types of geometric objects in $\mathbb{R}^2$ as well as in $\mathbb{R}^3$. Finally, we prove that both *MDIS* and *MDDS* problems are NP-hard for unit disks intersecting a horizontal line and for axis-parallel unit squares intersecting a straight line with slope -1. A natural open question is the existence of PTASes for the *MDIS* and *MDDS* problems with pseudo-disks.

References

1. Adamaszek, A., Wiese, A.: A QPTAS for maximum weight independent set of polygons with polylogarithmically many vertices. In: Symposium on Discrete Algorithms SODA, pp. 645–656 (2014)
2. Alimonti, P., Kann, V.: Some APX-completeness results for cubic graphs. Theor. Comput. Sci. **237**(1), 123–134 (2000)
3. Aschner, R., Katz, M.J., Morgenstern, G., Yuditsky, Y.: Approximation schemes for covering and packing. In: Ghosh, S.K., Tokuyama, T. (eds.) WALCOM 2013. LNCS, vol. 7748, pp. 89–100. Springer, Heidelberg (2013). https://doi.org/10.1007/978-3-642-36065-7_10
4. Aurenhammer, F.: Voronoi diagrams—a survey of a fundamental geometric data structure. ACM Comput. Surv. **23**(3), 345–405 (1991)
5. Bandyapadhyay, S., Maheshwari, A., Mehrabi, S., Suri, S.: Approximating dominating set on intersection graphs of rectangles and L-frames. In: 43rd International Symposium on Mathematical Foundations of Computer Science, MFCS 2018, Liverpool, UK, 27–31 August 2018, pp. 37:1–37:15 (2018)
6. Chan, T.M., Grant, E.: Exact algorithms and APX-hardness results for geometric packing and covering problems. Comput. Geom. **47**(2), 112–124 (2014)
7. Chan, T.M., Har-Peled, S.: Approximation algorithms for maximum independent set of pseudo-disks. Discret. Comput. Geom. **48**(2), 373–392 (2012)
8. Chuzhoy, J., Ene, A.: On approximating maximum independent set of rectangles. In: Symposium on Foundations of Computer Science, FOCS, pp. 820–829 (2016)
9. Clark, B.N., Colbourn, C.J., Johnson, D.S.: Unit disk graphs. Discret. Math. **86**(1), 165–177 (1990)

10. Das, G.K., De, M., Kolay, S., Nandy, S.C., Sur-Kolay, S.: Approximation algorithms for maximum independent set of a unit disk graph. Inf. Process. Lett. **115**(3), 439–446 (2015)

11. Erlebach, T., van Leeuwen, E.J.: Domination in geometric intersection graphs. In: Laber, E.S., Bornstein, C., Nogueira, L.T., Faria, L. (eds.) LATIN 2008. LNCS, vol. 4957, pp. 747–758. Springer, Heidelberg (2008). https://doi.org/10.1007/978-3-540-78773-0_64

12. Fowler, R.J., Paterson, M.S., Tanimoto, S.L.: Optimal packing and covering in the plane are NP-complete. Inf. Process. Lett. **12**(3), 133–137 (1981)

13. Fraser, R., Lòpez-Ortiz, A.: The within-strip discrete unit disk cover problem. Theor. Comput. Sci. **674**, 99–115 (2017)

14. Frederickson, G.N.: Fast algorithms for shortest paths in planar graphs, with applications. SIAM J. Comput. **16**(6), 1004–1022 (1987)

15. Garey, M., Johnson, D.: The rectilinear Steiner tree problem is NP-complete. SIAM J. Appl. Math. **32**(4), 826–834 (1977)

16. Gibson, M., Pirwani, I.A.: Algorithms for dominating set in disk graphs: breaking the $\log n$ barrier. In: Algorithms - ESA 2010, pp. 243–254 (2010)

17. Har-Peled, S.: Being fat and friendly is not enough. CoRR abs/0908.2369 (2009)

18. Hunt, H.B., Marathe, M.V., Radhakrishnan, V., Ravi, S.S., Rosenkrantz, D.J., Stearns, R.E.: NC-approximation schemes for NP- and PSPACE-hard problems for geometric graphs. J. Algorithms **26**(2), 238–274 (1998)

19. Madireddy, R.R., Mudgal, A.: Stabbing line segments with disks and related problems. In: Canadian Conference on Computational Geometry, CCCG, pp. 201–207 (2016)

20. Madireddy, R.R., Mudgal, A.: Approximability and hardness of geometric hitting set with axis-parallel rectangles. Inf. Process. Lett. **141**, 9–15 (2019)

21. Matsui, T.: Approximation algorithms for maximum independent set problems and fractional coloring problems on unit disk graphs. In: Akiyama, J., Kano, M., Urabe, M. (eds.) JCDCG 1998. LNCS, vol. 1763, pp. 194–200. Springer, Heidelberg (2000). https://doi.org/10.1007/978-3-540-46515-7_16

22. Mudgal, A., Pandit, S.: Covering, hitting, piercing and packing rectangles intersecting an inclined line. In: Lu, Z., Kim, D., Wu, W., Li, W., Du, D.-Z. (eds.) COCOA 2015. LNCS, vol. 9486, pp. 126–137. Springer, Cham (2015). https://doi.org/10.1007/978-3-319-26626-8_10

23. Mustafa, N.H., Ray, S.: Improved results on geometric hitting set problems. Discret. Comput. Geom. **44**(4), 883–895 (2010)

24. Nandy, S.C., Pandit, S., Roy, S.: Faster approximation for maximum independent set on unit disk graph. Inf. Process. Lett. **127**, 58–61 (2017)

25. Pandit, S.: Dominating set of rectangles intersecting a straight line. In: Canadian Conference on Computational Geometry CCCG, pp. 144–140 (2017)

26. Papadimitriou, C.H., Yannakakis, M.: Optimization, approximation, and complexity classes. J. Comput. Syst. Sci. **43**(3), 425–440 (1991)

27. Wan, P., Xu, X., Wang, Z.: Wireless coverage with disparate ranges. In: Symposium on Mobile Ad Hoc Networking and Computing, MobiHoc, p. 11 (2011)

Approximability of Covering Cells
with Line Segments

Paz Carmi[1], Anil Maheshwari[2], Saeed Mehrabi[2(✉)], Luís Fernando Schultz[3],
and Xavier da Silveira[3]

[1] Department of Computer Science, Ben-Gurion University of the Negev,
Beer-Sheva, Israel
carmip@cs.bgu.ac.il
[2] School of Computer Science, Carleton University, Ottawa, Canada
anil@scs.carleton.ca, saeed.mehrabi@carleton.ca
[3] School of Computer Science and Electrical Engineering, University of Ottawa,
Ottawa, Canada
schultz@ime.usp.br

Abstract. In COCOA 2015, Korman et al. studied the following geometric covering problem: given a set S of n line segments in the plane,
find a minimum number of line segments such that every cell in the
arrangement of the line segments is covered. Here, a line segment s covers a cell f if s is incident to f. The problem was shown to be NP-hard,
even if the line segments in S are axis-parallel, and it remains NP-hard
when the goal is cover the "rectangular" cells (i.e., cells that are defined
by exactly four axis-parallel line segments).

In this paper, we consider the approximability of the problem. We
first give a PTAS for the problem when the line segments in S are in
any orientation, but we can only select the covering line segments from
one orientation. Then, we show that when the goal is to cover the rectangular cells using line segments from both horizontal and vertical line
segments, then the problem is APX-hard. We also consider the parameterized complexity of the problem and prove that the problem is FPT
when parameterized by the size of an optimal solution. Our FPT algorithm works when the line segments in S have two orientations and the
goal is to cover *all* cells, complementing that of Korman et al. [9] in
which the goal is to cover the "rectangular" cells.

1 Introduction

Set Cover is a well-studied problem in computer science. The input to the prob-
lem is a ground set $\mathcal{G}$ of n elements and a set $\mathcal{S}$ of m subsets of $\mathcal{G}$; that is,
$\mathcal{S} = \{S_1, S_2, \ldots, S_m\}$ such that $S_i \subseteq \mathcal{G}$ for all $1 \leq i \leq m$. The objective is to

Paz Carmi is supported by Grant 2016116 from the United States-Israel Binational
Science Foundation. Anil Maheshwari is supported in part by Natural Sciences and
Engineering Research Council of Canada (NSERC). Saeed Mehrabi is supported by a
Carleton-Fields postdoctoral fellowship.

© Springer Nature Switzerland AG 2018
D. Kim et al. (Eds.): COCOA 2018, LNCS 11346, pp. 436–448, 2018.
https://doi.org/10.1007/978-3-030-04651-4_29

find a minimum-cardinality subset of S whose union is $\mathcal{G}$. Set Cover is known to be NP-hard [5] and even hard to approximate [8].

In this paper, we consider a geometric variant of the set cover problem that was first studied by Korman et al. [9]. A set of line segments in the plane is said to be *non-overlapping* if any two line segments from the set intersect in at most one point. Given a set S of n non-overlapping line segments in the plane, a *cell* in the arrangement of S is a maximally connected region that is not intersected by any line segment in S [9]. Then, the objective of the *Line Segment Covering* (LSC) problem is to select a minimum number of line segments such that every cell in the arrangement of the line segments is covered. Here, a cell is *covered* by a line segment if it is incident to the line segment (i.e., the line segment is in the set of line segments defining the boundary of the cell). We assume that at most two line segments may share a fixed point in the plane.

Related Work. Korman et al. [9] proved that when the line segments are only horizontal and vertical, the LSC problem is NP-hard and it remains NP-hard when the goal is to cover the "rectangular" cells. By a closer look at their hardness proof, one can see that the problem is NP-hard even if we are only allowed to select the line segments from one orientation (they only select vertical line segments when constructing a solution from a given truth assignment for the corresponding 3SAT problem). Moreover, the authors gave an $O(n \log n)$-time FPT algorithm for covering the rectangular cells when parameterized by k, the size of an optimal solution. However, the algorithm does not work when the goal is to cover *all* cells of the arrangement. The authors leave open studying the approximability of the problem.

The LSC problem is closely related to a guarding problem studied by Bose et al. [3]. Given a set of lines in the plane, they studied the problems of guarding cells of the arrangement by selecting a minimum number of lines, or guarding the lines by selecting a minimum number of cells. Here, "guarding" has the same meaning as "covering" in the LSC problem. However, their results do not extend to the LSC problem, because (as also noted by Korman et al. [9]) they use some properties of lines that are not true for the case of line segments.

Our Results. In this paper, we prove the following results.

- We give a PTAS for the LSC problem when the line segments in S can have any arbitrarily orientations, but we are allowed to select the covering line segments from only one orientation. Given the NP-hardness of the problem [9], this settles the complexity of this variant of the problem.
- When we allow selecting the covering line segments from more than one direction, we show that the LSC problem is APX-hard when the line segments in S have two orientations and the goal is to cover the rectangular cells.
- We give an FPT algorithm for the LSC problem when the line segments in S have only two orientations and the goal is to cover *all* cells of the arrangement. This complements the FPT algorithm of Korman et al. [9] as we do not restrict the covering only to rectangular cells.

Organization. In Sect. 2, we give some definitions and revisit some necessary background. We show our PTAS in Sect. 3 and the APX-hardness result in Sect. 4. Finally, the FPT algorithm is given in Sect. 5 and we conclude the paper in Sect. 6.

2   Preliminaries

In the following, we revisit some techniques and background that are used throughout this paper.

Local Search. Our PTAS for the LSC problem is based on the local search technique, which was introduced independently by Mustafa and Ray [11], and Chan and Har-Peled [4]. Consider an optimization problem in which the objective is to compute a feasible subset S' of a ground set S whose cardinality is minimum over all such feasible subsets of S. Moreover, it is assumed that computing some initial feasible solution and determining whether a subset $S' \subseteq S$ is a feasible solution can be done in polynomial time. The local search algorithm for a minimization problem is as follows. Fix some fixed parameter k, and let A be some initial feasible solution for the problem. In each iteration, if there are $A' \subseteq A$ and $M \subseteq S \setminus A$ such that $|A'| \leq k$, $|M| < |A'|$ and $(A \setminus A') \cup M$ is a feasible solution, then set $A = (A \setminus A') \cup M$ and re-iterate. The algorithm returns A and terminates when no such local improvement is possible.

Clearly, the local search algorithm runs in polynomial time. Let $\mathcal{B}$ and $\mathcal{R}$ be the solutions returned by the algorithm and an optimal solution, respectively. The following result establishes the connection between local search technique and obtaining a PTAS.

Theorem 1 ([4,11]). *Consider the solutions $\mathcal{B}$ and $\mathcal{R}$ for a minimization problem, and suppose that there exists a planar bipartite graph $H = (\mathcal{B} \cup \mathcal{R}, E)$ that satisfies the local exchange property: for any subset $\mathcal{B}' \subseteq \mathcal{B}$, $(\mathcal{B} \setminus \mathcal{B}') \cup N_H(\mathcal{B}')$ is a feasible solution, where $N_H(\mathcal{B}')$ denotes the set of neighbours of $\mathcal{B}'$ in H. Then, the local search algorithm yields a PTAS for the problem.*

The local search was used by Mustafa and Ray [11] to obtain a PTAS for geometric hitting set problem and by Chan and Har-Peled [4] to obtain a PTAS for geometric independent set problem. Since then, the technique has been used to get a PTAS for several other geometric problems, such as geometric dominating set [2] and unique covering [1].

Fixed-Parameter Tractability. The theory of parameterized complexity was developed by Downey and Fellows [6]. Let Σ be a finite alphabet. Then, a *parameterized* problem is a language $L \subseteq \Sigma^* \times \Sigma^*$ in which the second component is called the *parameter* of the problem. A parameterized problem L is said to be *fixed-parameter tractable* or FPT, if the question "$(x_1, x_2) \in L$?" can be decided in time $f(|x_2|) \cdot |x_1|^{O(1)}$, where f is an arbitrary function. We call an algorithm with such running time $f(|x_2|) \cdot |x_1|^{O(1)}$, an FPT algorithm.

For the rest of this paper, we denote a set of n line segments in the plane by S (i.e., $|S| = n$) and the resulting arrangement by $\mathcal{A}(S)$.

3   PTAS

In this section, we show that the LSC problem admits a PTAS when the line segments in S are in any orientations, but we can select line segments from only one orientation to cover the cells. To this end, we run the local search algorithm with parameter $k = c/\epsilon^2$ for some $\epsilon > 0$, where c is a constant. Let $\mathcal{B}$ be the solution returned by the algorithm and let $\mathcal{R}$ be an optimal solution. We can assume that $\mathcal{B} \cap \mathcal{R} = \emptyset$. This is because if $\mathcal{B} \cap \mathcal{R} \neq \emptyset$, then we can consider the sets $\mathcal{B}' = \mathcal{B} \setminus (\mathcal{B} \cap \mathcal{R})$ and $\mathcal{R}' = \mathcal{R} \setminus (\mathcal{B} \cap \mathcal{R})$, and analyze the algorithm with $\mathcal{B}'$ and $\mathcal{R}'$. Here, we mark the faces covered by a line segment in $\mathcal{B} \cap \mathcal{R}$ as "covered" so as we do not need to cover then in the new variant of the problem. This guarantees that the approximation factor of the original instance is upper bounded by that of the new instance with these two new sets $\mathcal{B}'$ and $\mathcal{R}'$.

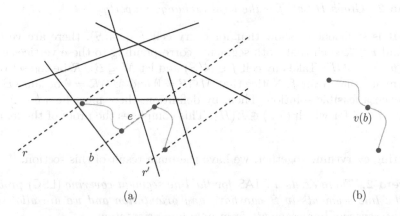

(a)   (b)

Fig. 1. (a) An example of three edges of graph H'. Here, $b \in \mathcal{B}$ and $r, r' \in \mathcal{R}$. (b) Two edges of graph H by contracting the edge e and obtaining one vertex $v(b)$ corresponding to line segment b.

We now construct a planar bipartite graph $H = (\mathcal{B} \cup \mathcal{R}, E)$ that satisfies the local exchange property, hence proving that the problem admits a PTAS by Theorem 1. To this end, we first construct an auxiliary planar graph H' and then show how to obtain H from H' by edge contraction. For each cell $f \subset \mathcal{A}(S)$, let $b \in \mathcal{B}$ and $r \in \mathcal{R}$ be two line segments that cover f; we select a point $p \in b \cap f$ and $q \in r \cap f$ and connect them by a curve c that lies in the interior of f (except its endpoints p and q). Notice that since both $\mathcal{B}$ and $\mathcal{R}$ are feasible solutions, we know that $\mathcal{B}$ contains at least one line segment that covers f and $\mathcal{R}$ also contains at least one line segment that covers f, for all $f \in \mathcal{A}(S)$. We add p and q to $V(H')$ and c to $E(H')$. We complete the definition of H' by connecting every pair of consecutive points in $s \cap V(H')$, for all $s \in S$, by an edge that is exactly the portion of s that lies between the pair of points. See Fig. 1(a) for an example. Clearly, H' is planar because the first set of edges are drawn in the interior of cells and each cell contains at most one edge. Moreover, the second set of edges are aligned with the line segments in $\mathcal{B} \cup \mathcal{R}$. Since the line segments in $\mathcal{B} \cup \mathcal{R}$ are non-overlapping and all have the same orientation, the second set of edges are also non-crossing. To obtain the graph H, for each segment $s \in \mathcal{B} \cup \mathcal{R}$, we contract the edges of H' that are contained in s such that we get a single point $v(s)$ corresponding to s; see Fig. 1(b). So, $V(H) = \{v(s)|s \in \mathcal{B} \cup \mathcal{R}\}$. Graph H is planar since H' remains planar after this edge contraction. Moreover, H is a bipartite graph as the edges of H' with both endpoints belonging to a line segment in $\mathcal{B}$ or both endpoints belonging to a line segment in $\mathcal{R}$ are collapsed into a single point (i.e., $v(s)$).

Lemma 1. *Graph H is planar and bipartite.*

We next show that H satisfies the exchange property.

Lemma 2. *Graph H satisfies the local exchange property.*

Proof. It is sufficient to show that for every cell $f \in \mathcal{A}(S)$, there are vertices $b \in \mathcal{B}$ and $r \in \mathcal{R}$ such that both segments corresponding to these vertices cover f and $(b, r) \in E(H)$. Take any cell $f \in \mathcal{A}(S)$ and let $M \subseteq \mathcal{B} \cup \mathcal{R}$ be the set of all line segments that cover f. Notice that $M \cap \mathcal{B} \neq \emptyset$ and $M \cap \mathcal{R} \neq \emptyset$ because $\mathcal{B}$ and $\mathcal{R}$ are each a feasible solution. Then, by definition, there must be a $b \in M \cap \mathcal{B}$ and $r \in M \cap \mathcal{R}$ for which $(b, r) \in E(H)$. This completes the proof of the lemma. $\square$

Putting everything together, we have the main result of this section.

Theorem 2. *There exists a PTAS for the line segment covering (LSC) problem when the line segments in S can have any orientation and we are allowed to select the covering line segments from only one orientation.*

4   APX-Hardness

In this section, we show that the LSC problem is APX-hard when the line segments in S have only two orientations and the goal is to cover the rectangular cells. To this end, we give an L-reduction from the Minimum Vertex Cover (MVC) problem on graphs with maximum-degree three to this variant of the LSC problem. Our reduction is inspired by the construction of Mehrabi [10]. As a reminder, we first give a formal definition of L-reduction [12], which is one of the gap-preserving reductions. Let Π and Π' be two optimization problems with the cost functions $c_\Pi(.)$ and $c_{\Pi'}(.)$, respectively. We say that Π L-*reduces* to Π' if there are two polynomial-time computable functions f and g such that the followings hold.

1. For any instance x of Π, $f(x)$ is an instance of Π'.
2. If y is a solution to $f(x)$, then $g(y)$ is a solution to x.

3. There exists a constant $\alpha > 0$ such that

$$OPT_{\Pi'}(f(x)) \leq \alpha OPT_{\Pi}(x),$$

where $OPT_Y(x)$ denotes the cost of an optimal solution for problem Y on its instance x.

4. There exists a constant $\beta > 0$ such that for every solution y for $f(x)$,

$$|OPT_{\Pi}(x) - c_{\Pi}(g(y))| \leq \beta |OPT_{\Pi'}(f(x)) - c_{\Pi}(y)|,$$

where $|x|$ denotes the absolute value of x.

Lemma 3. *The minimum vertex cover* (MVC) *problem on graphs with maximum-degree three is* L-*reducible to the* LSC *problem, where S is a set of horizontal and vertical line segments and the goal is to cover the rectangular cells of $\mathcal{A}(S)$.*

Proof. Let I be an instance of MVC on graphs of maximum-degree three; let $G = (V, E)$ be the graph corresponding to I and let k be the size of the smallest vertex cover in G. First, let $u_1, \ldots, u_n$ be an arbitrary ordering of the vertices of G, where $n = |V|$. In the following, we give a polynomial-time computable function f that takes I as input and outputs an instance $f(I)$ of the LSC problem.

We first describe the vertex gadgets. For each vertex u_i, $1 \leq i \leq n$, construct a horizontal line segment H_i and a vertical line segment V_i, and connect them as shown in Fig. 2. We call the (blue) horizontal line segment used in the connection of H_i and V_i the *horizontal connector* C_i of i. Moreover, there are four (small, dashed) line segments used in the connection of H_i and V_i that we call the *small connectors* of i. Notice that these five "connectors" along with H_i and V_i form exactly two rectangular cells. For each edge $(u_i, u_j) \in E$, where $i < j$, we add two small line segments, one horizontal and one vertical, at the intersection point of V_i and H_j such that they intersect each other as well as each intersects one of V_i and H_j, hence forming a rectangular cell; see the two (red, dashed) line segments at the intersection of V_1 and H_2 in Fig. 2 for an example. We call such a pair *edge line segments* and denote them by $E_{i,j}$. Finally, for every rectangular cell whose four sides are *all* defined by the line segments corresponding to a 4-subset of $\{H_i, V_i | 1 \leq i \leq n\}$ (i.e., the cell is not covered by a horizontal connector or edge line segments), we insert a vertical line segment into the cell so as to make it non-rectangular; see the vertical (red) line segment in Fig. 2. This ensures that every rectangular cell is incident either to a horizontal connector or to edge line segments $E_{i,j}$ for some i and j. This gives the instance $f(I)$ of the LSC problem. Notice that f is a polynomial-time computable function. In the following, we denote an optimal solution for the instance X of a problem by $s^*(X)$. We now prove that all the four conditions of L-reduction hold.

First, let M be a vertex cover of G of size k. Denote by $H[M] = \{H_i | u_i \in M\}$ the set of horizontal line segments induced by M and define $V[M]$ analogously. Moreover, let $C[M] = \{C_i | u_i \notin M\}$ be the set of horizontal connectors whose corresponding vertex is not in M. We show that $F = H[M] \cup V[M] \cup C[M]$

is a feasible solution for covering all the rectangular cells of $f(I)$. Let f be a rectangular cell. Then, f must be incident either to a horizontal connector or to edge line segments $E_{i,j}$ for some i and j. First, if f is incident to a horizontal connector C_i, then either $C_i \in F$ or $H_i \in F$ and $V_i \in F$ by the construction of F and so f is covered either way. Next, if f is incident to edge line segments $E_{i,j}$ for some i and j, where w.l.o.g. $i < j$, then either $V_i \in F$ or $H_j \in F$ because we know that $u_i \in M$ or $u_j \in M$. So, f is again covered in this case. Therefore, F is a feasible solution.

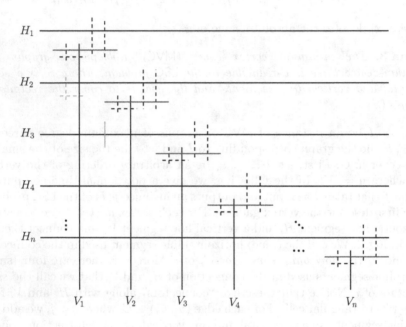

Fig. 2. An illustration in support of the construction in the proof of Lemma 3. (Color figure online)

Second, let F be any feasible solution for $f(I)$. Notice that we can construct a feasible solution F' for $f(I)$ such that $|F'| \leq |F|$ and F' consists of only H_i and V_i for some i, or a horizontal connector. This is because (i) any rectangular cell covered by a small connector is also covered by a horizontal connector, and (ii) any cell covered by a pair of edge line segments $E_{i,j}$ (for some i and j) is also covered by V_i and H_j. For (ii), if exactly one of the line segments in $E_{i,j}$ is in F, then we replace it with exactly one of V_i or H_j. Otherwise, if both line segments of $E_{i,j}$ are in F, then we replace both of them with V_i and H_j. So, $|F'| \leq |F|$ and F' is a feasible solution for $f(I)$. Now, let $M = \{u_i | H_i \in F' \text{ or } V_i \in F'\}$. To show that M is a vertex cover for G, consider any edge $(u_i, u_j) \in E$, where $i < j$. Then, we know that there exists a rectangular cell at the intersection of V_i and H_j that must be covered by F'. Since none of the two edge line segments of $E_{i,j}$ are in F', we conclude that at least one of V_i and H_j is in F', which means that $u_i \in M$ or $u_j \in M$. Hence, M is a vertex cover.

Third, observe that $|H[M]| = |V[M]| = |M| = k$ and also $|C[M]| = n - k$. Given that G has degree three, $k \geq n/4$ and so $|s^*(f(I))| \leq n - k + k + k \leq 5k \leq 5|s^*(I)|$.

We now prove the last condition of L-reduction. First, define $\mathsf{Both}[F'] = \{H_i, V_i | H_i, V_i \in F'\}$; that is, the paths of a vertex u_i, where *both* its horizontal and vertical line segments appear in F'. Also, define $\mathsf{One}[F']$ to be the remaining line segments corresponding to either H_i or V_i for some i; i.e., those of u_i, where *only* one of its line segments appears in F'. Take any vertex i. To cover the two rectangular cells incident to the horizontal connector of i, we must have $C_i \in F'$ or $H_i, V_i \in F'$; this is true for all i. Then, $|C[F']| + |\mathsf{Both}[F']|/2 \geq n$. Moreover, $|\mathsf{One}[F']| + |\mathsf{Both}[F']|/2 \geq k$ since M is a vertex cover of G. Therefore, $|F'| \geq |\mathsf{Both}[F']| + |\mathsf{One}[F']| + |C[F']| \geq |\mathsf{One}[F']| + |\mathsf{Both}[F']|/2 + n \geq k + n$. By this and our earlier inequality $|s^*(f(I))| \leq n - k + k + k$, we have $|s^*(f(I))| = n + k$. Now, suppose that $|F| = |s^*(f(I))| + c$ for some $c \geq 0$. Then,

$$|F| - |s^*(f(I))| = c$$
$$\Rightarrow |F| - (n + k) = c$$
$$\Rightarrow |F'| - (n + k) \leq c$$
$$\Rightarrow |\mathsf{One}[F']| + |\mathsf{Both}[F']|/2 + n - (n + k) \leq c$$
$$\Rightarrow |\mathsf{One}[F']| + |\mathsf{Both}[F']|/2 - k \leq c$$
$$\Rightarrow |M| - |s^*(I)| \leq c.$$

That is, $|M| - |s^*(I)| \leq |F| - |s^*(f(I))|$. This concludes our L-reduction from MVC on graphs of maximum-degree three to LSC with $\alpha = 5$ and $\beta = 1$.   □

Theorem 3. *The line segment covering* (LSC) *problem is* APX-*hard when the line segments in S are either horizontal or vertical and the goal is to cover the rectangular cells of $\mathcal{A}(S)$.*

5   FPT

In this section, we show that the LSC problem is fixed-parameter tractable (parametrized by the size of an optimal solution) when the line segments in S are either horizontal or vertical, and the goal is to cover all the cells in $\mathcal{A}(S)$. This complements the FPT result of Korman et al. [9], where the goal is to cover the rectangular cells. Throughout this section, let k be the size of an optimal solution.

Our FPT follows the framework of Korman et al. [9]. That is, we formulate the LSC problem as a hitting set problem and argue that we only need to hit an $O(k^3)$ number of sets; hence, obtaining a kernel of size $O(k^3)$ for the problem. The FPT of Korman et al. [9] is based on the fact that any three orthogonal line segments can cover at most two "rectangular" cells (i.e., at most two rectangular cells can be incident to all the three line segments). As an analogous result, we prove in Lemma 4 that the number of such cells can be at most six when the goal is to cover *all* cells, including non-rectangular ones. We will then apply this result to obtain the desired kernel.

Lemma 4. *Let S be a set of n axis-parallel line segments in the plane. Then, for any three line segments $s_1, s_2, s_3 \in S$, there are at most six cells in $\mathcal{A}(S)$ that can be covered by all three line segments s_1, s_2 and s_3.*

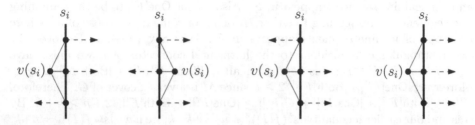

Fig. 3. Placing a new vertex $v(s_i)$ close to s_i and connecting it to the vertices corresponding to the incident cells. The arrows indicate the side on which the cell lies.

Proof. Take any three line segments s_1, s_2 and s_3 in $\mathcal{A}(S)$ and let F be the set of all cells in $\mathcal{A}(S)$ that are covered by all three line segments s_1, s_2 and s_3. We need to show that $|F| \leq 6$. To this end, we construct a planar graph H corresponding to s_1, s_2, s_3 and the cells in F and will then argue that this graph must contain a subdivision of $K_{3,3}$ if $|F| > 6$. We next give the details. Let f be a cell in F. Consider a point $p(f)$ in the interior of f as well as a distinct point $p(s_i, f)$ in $s_i \cap f$, for all $i = 1, 2, 3$ (notice that $p(s_i, f)$ is on the boundary of f). These points together form the set of vertices of H; that is, $V(H) = \{p(f) : f \in F\} \cup \{p(s_i, f) : i = 1, 2, 3, f \in F\}$. Now, for each $i = 1, 2, 3$, consider an ordering of the points $p(s_i, f)$ on s_i, $f \in F$, and connect every two consecutive points by an edge, which is exactly the portion of s_i that lies between the two points. Moreover, for each cell f, we connect $p(f)$ to $p(s_i, f)$ by a curve that lies strictly in the interior of f (except at its endpoints) for all $i = 1, 2, 3$. Then, the edge set $E(H)$ of H consists of the set of all edges connecting the consecutive points as we as the curves $(p(f), p(s_i, f))$, for all $f \in F$ and $i = 1, 2, 3$. Clearly, H is a planar graph. In the following, we consider several cases depending on whether the line segments s_1, s_2 and s_3 intersect each other; observe that there can be at most two intersection points between them.

Case 1. There is no intersection point; that is, the line segments s_1, s_2 and s_3 are pairwise disjoint. In this case, we show that in fact $|F| \leq 2$. To this end, suppose for a contradiction that $|F| > 2$. Take any three cells $f_1, f_2, f_3 \in F$ and consider the subgraph H' of H induced by $\{p(f_i) : i = 1, 2, 3\} \cup \{p(s_i, f_j) : i, j = 1, 2, 3\}$. Now, consider the graph G constructed from H' as follows. For each s_i, $i = 1, 2, 3$, we place a new vertex $v(s_i)$ close to s_i and connect it to the three vertices $p(s_i, f_j)$ for all $j = 1, 2, 3$ such that the resulting graph remains planar. One can easily verify that this is doable since the three line segments are disjoint and so there are a few cases for where to place $v(s_i)$ depending on which side of s_i the three cells lie; see Fig. 3. Observe that the resulting graph G is a planar drawing of a subdivision of $K_{3,3}$, which is not possible. So, $|F| \leq 2$.

Case 2. There is exactly one intersection point; assume w.o.l.g. that s_1 is horizontal, s_2 and s_3 are vertical and s_1 intersects s_2. Here, we show that $|F| \leq 4$. Again, suppose for a contradiction that $|F| > 4$. Then, considering the graph H, there must be at least three vertices in $\{p(s_1, f) : f \in F\}$ that lie w.l.o.g. to the right of s_2. Take any three such vertices and denote the corresponding cells by f_1, f_2, f_3. We can now construct the graph G analogous to the one in Case 1 with these three cells and so obtain a planar drawing of a subdivision of $K_{3,3}$, which is a contradiction.

Case 3. There are two intersection points. Here, we show that $|F| \leq 6$ and we use a similar argument to those in the previous cases. Denote the endpoints of s_1 by a and b, and let p and q be the intersection points of s_1 with s_2 and s_3; assume w.l.o.g. that a is the left endpoint of s_1 and that p lies to the left of q. If $|F| > 7$, then at least one of the line segments ap, pq and qb must contain three vertices of $\{p(s_1, f) : f \in F\}$; assume w.l.o.g. that it is pq. Then, take any three such vertices on pq and consider the three cells f_1, f_2 and f_3 corresponding to these vertices. We can now construct the graph G analogous to the one in Case 1 and so obtain a planar drawing of a subdivision of $K_{3,3}$, which is a contradiction. As such, $|F| \leq 6$.

By the three cases described above, we conclude that $|F| \leq 6$. □

We note that the upper bound in Lemma 4 is tight as Fig. 4 shows an example with three line segments that cover six cell. We now apply Lemma 4 to obtain our FPT. We first formulate the LSC problem as a hitting set problem as follows. The ground set is S and for each cell in $\mathcal{A}(S)$, there exists a set that contains the line segments that cover the cell. Let $\mathcal{C}$ be the resulting set of subsets of S. Then, the LSC problem is equivalent to selecting a minimum number of elements from S such that each set in $\mathcal{C}$ is hit by at least one selected element.

We first reduce the set $\mathcal{C}$ to a set $\mathcal{C}_1$ as follows. For every pair of line segments $\{s_i, s_j\} \in S$, if they appear in more than $6k$ sets $\mathcal{C}$, then we remove all such sets form $\mathcal{C}$ and add the set $\{s_i, s_j\}$ to $\mathcal{C}$. Let $\mathcal{C}_1$ be the resulting set.

Lemma 5. *A set $S' \subseteq S$ with $|S'| \leq k$ is a minimum-size cover of $\mathcal{C}$ if and only if it is a minimum-size cover of $\mathcal{C}_1$.*

Proof. We prove the lemma by an argument similar to the one by Korman et al. [9]. The lemma clearly follows if $\mathcal{C} = \mathcal{C}_1$. So, assume that $X = \mathcal{C}_1 \setminus \mathcal{C}$ and $Y = \mathcal{C} \setminus \mathcal{C}_1$ are two non-empty sets. Let S' with $|S'| \leq k$ be a minimum-size cover for $\mathcal{C}_1$. First, S' is also a cover for $\mathcal{C}$ because for every set $M \in Y$ there exists a pair of line segment s_i and s_j such that both s_i and s_j are in M and we have $\{s_i, s_j\} \in X$. We now prove that S' is also a cover of minimum size for $\mathcal{C}$.

Suppose for a contradiction that there exists a cover S'' for $\mathcal{C}$ such that $|S''| < |S'|$. Then, S'' cannot be a cover $\mathcal{C}_1$ because S' is a cover of minimum size for $\mathcal{C}_1$. Since S'' covers $\mathcal{C} \cap \mathcal{C}_1$, there must exist $\{s_i, s_j\} \in X$ such that neither s_i

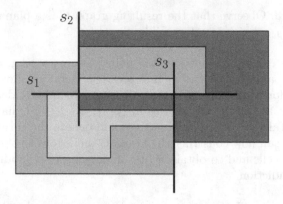

Fig. 4. Three line segments s_1, s_2 and s_3 cover six cells.

nor s_j is in S''. But, we introduced the set $\{s_i, s_j\}$ into $\mathcal{C}_1$ because there were more than $6k$ sets containing both s_i and s_j. If neither s_i nor s_j is in S'', then every other line segment can cover at most six of such subsets by Lemma 4. Therefore, $|S''| > k$ — a contradiction. By a similar argument, we can show that a minimum-size cover of $\mathcal{C}$ is also a minimum-size cover for $\mathcal{C}_1$. This completes the proof of the lemma. □

Next, we reduce $\mathcal{C}_1$ to a new set $\mathcal{C}_2$ as follows. For each line segment $s \in S$, we count how many sets in $\mathcal{C}_1$ contain s. If s appears in more than $6k^2$, then we remove all those sets and add the set $\{s\}$ to $\mathcal{C}_1$. Let $\mathcal{C}_2$ denote the resulting set.

Lemma 6. *A set* $S' \subseteq S$ *with* $|S'| \leq k$ *is a minimum-size cover for* $\mathcal{C}_1$ *if and only if it is a minimum-size cover for* $\mathcal{C}_2$.

Proof. The lemma follows if $\mathcal{C}_1 = \mathcal{C}_2$. So, assume that $X' = \mathcal{C}_2 \setminus \mathcal{C}_1$ and $Y' = \mathcal{C}_1 \setminus \mathcal{C}_2$ are two non-empty sets. Let S' with $|S'| \leq k$ be a minimum-size cover for $\mathcal{C}_2$. For any set $M \in Y'$, there exists a singleton set in X' whose member is in M. This means that S' is also a cover for $\mathcal{C}_1$. We next show that S' is also a minimum-size cover for $\mathcal{C}_1$.

Suppose for a contradiction that there exists a cover S'' for $\mathcal{C}_1$ such that $|S''| < |S'|$. Therefore, S'' is not a cover of $\mathcal{C}_2$. Since S'' covers $\mathcal{C}_1 \cap \mathcal{C}_2$, there must exist a set in X' that is not covered by S''. Notice that this set must be of size 1 from the construction of $\mathcal{C}_2$; let $\{s\}$ be such a set, where $s \in S$. The reason we have the set $\{s\}$ in $\mathcal{C}_2$ is that because there were more than $6k^2$ sets in $\mathcal{C}_1$ containing s. If s is not in S'', then all such sets of S_1 must be covered by other line segments. But, from the construction of $\mathcal{C}_1$, every pair of line segments can appear in at most $6k$ sets. So, $|S''|$ must be greater than k, which is a contradiction. A similar argument can be used to show that a minimum-size cover for $\mathcal{C}_1$ is also is minimum-size cover for $\mathcal{C}_2$. This completes the proof of the lemma. □

Consider the set $\mathcal{C}_2$. By Lemma 6, no line segment of S appears in more than $6k^2$ sets in $\mathcal{C}_2$. Therefore, if $|\mathcal{C}_2| > 6k^3$, then the problem does not have a cover of size at most k. Since the construction of $\mathcal{C}_2$ can be done in polynomial time, we have the following result.

Lemma 7. *For the* LSC *problem on a set of axis-parallel line segments, in polynomial time, we can either obtain a kernel of size $O(k^3)$ or conclude that the problem does not have a cover of size at most k, where k is the size of an optimal cover.*

Since having a kernel of size $O(k^3)$ implies that the problem is FPT [7], we have the main result of this section.

Theorem 4. *The line segment covering* (LSC) *problem on a set of axis-parallel line segments is* FPT *with respect to the size of an optimal cover.*

6 Conclusion

In this paper, we considered the problem of covering the cells in the arrangement of a set of line segments in the plane. We proved that the problem admits a PTAS when the covering line segments can be selected from only one orientation. We then showed that if we allow selecting the covering line segments from more than one orientation, then the problem is APX-hard when we are interested in covering the rectangular cells. Finally, we gave an FPT algorithm for the problem when the line segments have only two orientations, but the goal is to cover all the cells. Our APX-hardness rules out the possibility of a PTAS for "covering rectangular faces" variant of the problem, but is there a 2-approximation algorithm for the problem? For the more general variant, where the line segments are in any orientation, covering line segments can be selected from any orientation and the goal is to cover all the cells, can we obtain a c-approximation algorithm for some small constant c?

References

1. Ashok, P., Kolay, S., Misra, N., Saurabh, S.: Unique covering problems with geometric sets. In: Xu, D., Du, D., Du, D. (eds.) COCOON 2015. LNCS, vol. 9198, pp. 548–558. Springer, Cham (2015). https://doi.org/10.1007/978-3-319-21398-9_43
2. Bandyapadhyay, S., Maheshwari, A., Mehrabi, S., Suri, S.: Approximating dominating set on intersection graphs of rectangles and L-frames. In: MFCS 2018 (2018, to appear). arXiv version https://arxiv.org/abs/1803.06216
3. Bose, P., et al.: Coloring and guarding arrangements. Discrete Math. Theor. Comput. Sci. **15**(3), 139–154 (2013)
4. Chan, T.M., Har-Peled, S.: Approximation algorithms for maximum independent set of pseudo-disks. Discrete Comput. Geom. **48**(2), 373–392 (2012)
5. Cormen, T.H., Leiserson, C.E., Rivest, R.L., Stein, C.: Introduction to Algorithms, 2nd edn. The MIT Press, McGraw-Hill Book Company, Cambridge, New York (2001)
6. Downey, R.G., Fellows, M.R.: Parameterized Complexity. Springer, Heidelberg (1999). https://doi.org/10.1007/978-1-4612-0515-9

7. Downey, R.G., Fellows, M.R.: Fundamentals of Parameterized Complexity. TCS. Springer, Heidelberg (2013). https://doi.org/10.1007/978-1-4471-5559-1
8. Feige, U.: A threshold of ln n for approximating set cover. J. ACM **45**(4), 634–652 (1998)
9. Korman, M., Poon, S.-H., Roeloffzen, M.: Line segment covering of cells in arrangements. Inf. Process. Lett. **129**, 25–30 (2018)
10. Mehrabi, S.: Approximating domination on intersection graphs of paths on a grid. In: Solis-Oba, R., Fleischer, R. (eds.) WAOA 2017. LNCS, vol. 10787, pp. 76–89. Springer, Cham (2018). https://doi.org/10.1007/978-3-319-89441-6_7
11. Mustafa, N.H., Ray, S.: Improved results on geometric hitting set problems. Discrete Comput. Geom. **44**(4), 883–895 (2010)
12. Papadimitriou, C.H., Yannakakis, M.: Optimization, approximation, and complexity classes. J. Comput. Syst. Sci. **43**(3), 425–440 (1991)

Heuristics for the Score-Constrained Strip-Packing Problem

Asyl L. Hawa[✉], Rhyd Lewis, and Jonathan M. Thompson

School of Mathematics, Cardiff University, Senghennydd Road, Cardiff, UK
{hawaa,lewisr9,thompsonjm1}@cardiff.ac.uk

Abstract. This paper investigates the Score-Constrained Strip-Packing Problem (SCSPP), a combinatorial optimisation problem that generalises the one-dimensional bin-packing problem. In the construction of cardboard boxes, rectangular items are packed onto strips to be scored by knives prior to being folded. The order and orientation of the items on the strips determine whether the knives are able to score the items correctly. Initially, we detail an exact polynomial-time algorithm for finding a feasible alignment of items on a single strip. We then integrate this algorithm with a packing heuristic to address the multi-strip problem and compare with two other greedy heuristics, discussing the circumstances in which each method is superior.

Keywords: Strip-packing · Heuristics · Graphs and networks

1 Introduction

The Constrained Ordering Problem (COP) is defined as follows:

Definition 1. *Let $\mathcal{M}$ be a multiset of unordered pairs of positive integers $\mathcal{M} = \{\{a_1, b_1\}, \{a_2, b_2\}, ..., \{a_n, b_n\}\}$, and let $\mathcal{T}$ be an ordering of the elements of $\mathcal{M}$ such that each element is a tuple. The* Constrained Ordering Problem (COP) *consists of finding a solution $\mathcal{T}$ such that, given a fixed value $\tau \in \mathbb{Z}^+$,*

$$\mathbf{rhs}(i) + \mathbf{lhs}(i+1) \geq \tau \quad \forall\, i \in \{1, 2, ..., n-1\}, \tag{1}$$

where $\mathbf{lhs}(i)$ and $\mathbf{rhs}(i)$ denote the left- and right-hand values of the ith tuple. The inequality is referred to as the vicinal sum constraint.

For example, given $\mathcal{M} = \{\{1, 2\}, \{1, 7\}, \{2, 4\}, \{3, 5\}, \{3, 6\}, \{4, 4\}\}$ and $\tau = 7$, one possible solution is $\mathcal{T} = \langle (1, 2), (6, 3), (5, 3), (4, 4), (4, 2), (7, 1) \rangle$.

A prominent application of the COP is in a strip-packing problem brought to light as an open-combinatorial problem by Goulimis [4]. Here, a set $\mathcal{I}$ of rectangular items (where $|\mathcal{I}| = n$) of equal height H made from cardboard are to be packed onto a strip of height H from left to right. Each item $i \in \mathcal{I}$ has width $w_i \in \mathbb{Z}^+$ and possesses two vertical score lines, marked in predetermined

© Springer Nature Switzerland AG 2018
D. Kim et al. (Eds.): COCOA 2018, LNCS 11346, pp. 449–462, 2018.
https://doi.org/10.1007/978-3-030-04651-4_30

places. A pair of knives mounted onto a bar cuts along the score lines of two adjacent items simultaneously, which allows the items to be folded with ease (see Fig. 1). However, by design, the scoring knives cannot be placed too close to one another and, as such, have a "minimum scoring distance" (around 70 mm in industry). The distances between each score line and the nearest edge on an item $i \in \mathcal{I}$ are the score widths, $a_i, b_i \in \mathbb{Z}^+$ where $a_i + b_i < w_i$, assigned such that $a_i \leq b_i$. Since these score widths are not necessarily equal, an item i can be positioned in one of two orientations: "regular", denoted (a_i, b_i), or "rotated", denoted (b_i, a_i), where the smaller of the two score widths a_i is on the left- and right- hand side respectively.

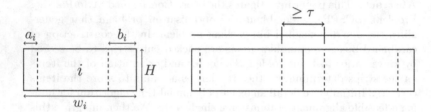

Fig. 1. Dimensions of an item $i \in \mathcal{I}$ in a regular orientation (a_i, b_i), and a feasible alignment of two items whose adjacent score lines can be scored simultaneously.

Clearly, for two items to be feasibly placed alongside one another on a strip, the distance between the two score lines must be equal to or exceed the minimum scoring distance, else the knives will not be able to score the items in the correct locations. Thus, the problem consists of finding a suitable ordering and orientation of the items such that the sum of every pair of adjacent score widths is greater than or equal to the minimum scoring distance.[1] It follows that there are $\frac{n!}{2} 2^n$ distinct arrangements, making complete enumeration infeasible for non-trivial values of n.

It can be seen that, when using a single strip, this packing problem is equivalent to the COP, where each unordered pair in an instance $\mathcal{M}$ contains the score widths of an item and τ is the minimum scoring distance. It then follows that the vicinal sum constraint (1) corresponds to the requirement for the sum of adjacent score widths to exceed τ. Figure 2 shows the same instance $\mathcal{M}$ mentioned earlier depicted as a packing problem.

Observe that in this particular strip-packing problem the widths of the individual items are disregarded, since the aim is to arrange the items onto a single strip of seemingly infinite width. However, in industrial applications, strips of material will often be provided in fixed finite widths. Given a large problem instance, multiple strips may therefore be required to feasibly accommodate all of the items. For this reason, a more generalised problem can also be formulated as follows:

---

[1] Note that the left-hand score width of the first item and the right-hand score width of the last item on the strip are not adjacent to any other item, and can therefore be ignored.

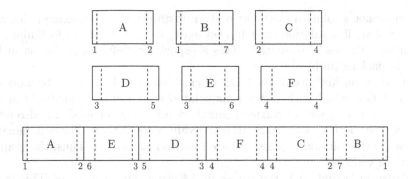

Fig. 2. Example of a single strip-packing problem and a corresponding solution with $\tau = 7$.

Definition 2. *Let $\mathcal{I}$ be a set of n rectangular items of height H with varying widths $w_i \in \mathbb{Z}^+$ and score widths $a_i, b_i \in \mathbb{Z}^+$ for each item $i \in \mathcal{I}$. Given a minimum scoring distance $\tau \in \mathbb{Z}^+$, the Score-Constrained Strip-Packing Problem (SCSPP) consists of finding the minimum number of strips of height H and width W required to pack all items in $\mathcal{I}$ such that the sum of every pair of adjacent score widths is greater than or equal to τ and no strip is overfilled.*

Note that in the special case of $\tau = 0$, the SCSPP is equivalent to the classical one-dimensional bin-packing problem (BPP). When $\tau > 0$, the problem also involves deciding the order in which the items are packed from left to right, and whether each item should be placed in a regular or rotated orientation. Thus, we define the following sub-problem associated with the SCSPP.

Definition 3. *Let $\mathcal{I}' \subseteq \mathcal{I}$ be a set of items whose total width is less than or equal to the capacity of a strip, (i.e. $\sum_{i \in \mathcal{I}'} w_i \leq W$). Given a minimum scoring distance τ, the Score-Constrained Packing Sub-Problem (SCPSP) involves finding an arrangement of the items in $\mathcal{I}'$ such that the sum of every pair of adjacent score widths is greater than or equal to τ.*

The remainder of this article is structured as follows: In Sect. 2, we will detail an exact polynomial-time algorithm for the COP and show how it is applicable to the SCPSP. Section 3 then introduces three heuristics that can be used to find feasible solutions to the SCSPP, one of which makes particular use of the exact algorithm from the previous section, and their associated advantages and disadvantages. An analysis of results gained from extensive experiments and a comparison of the heuristics is provided in Sect. 4, and finally Sect. 5 concludes the paper and proposes some potential directions for future work.

2   Solving the COP

We now present an exact polynomial-time algorithm for the COP. The underlying algorithm was originally proposed by Becker in [1], and is used to determine

whether or not a solution exists for a given instance. Here, we extend this algorithm so that, if a solution does indeed exist, it is also able to formulate and
present us with this final solution. This is especially useful for problems such as
the strip-packing problem.

Let $\mathcal{M}$ be an instance of the COP of cardinality n. It is useful to model $\mathcal{M}$
as a graph G in which each vertex is associated with a single value in $\mathcal{M}$ in nondecreasing order. A pair of vertices, called "dominating vertices" are also added
to the graph, both of which are assigned values equal to τ. These dominating
vertices aid the solution process and are removed at the end. Thus, the graph G
has $2n + 2$ vertices.

As seen in Definition 1, the values in $\mathcal{M}$ are arranged in pairs. This is represented in G by adding a set of "blue" edges, B, between vertices that are
"partners", i.e. whose values make up a pair in $\mathcal{M}$. By introducing a bijective
function $p : V \to V$ that associates each vertex with its partner, $p(v_i) = v_j$,
we can denote this set of edges as $B = \{(v_i, p(v_i)) : v_i \in V\}$. Note that B is
a perfect matching in G, with $|B| = n + 1$. Next, a set of "red" edges R is
added to G that consists of edges between vertices whose sum equals or exceeds
τ, provided they are not partners. It can be seen that the edges in R represent
all possible orderings of values from different pairs in $\mathcal{M}$ that fulfil the vicinal
sum constraint (1). Thus, we have a simple, undirected graph G with vertex set
$V = \{v_1, ..., v_{2n+2}\}$ and two distinct edge sets B and R such that $B \cap R = \emptyset$.
Figure 3 illustrates an example construction of G.

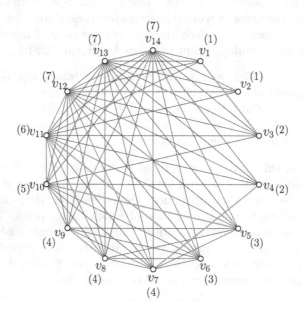

Fig. 3. $G = (V, B \cup R)$ using $\mathcal{M} = \{\{1,2\}, \{1,7\}, \{2,4\}, \{3,5\}, \{3,6\}, \{4,4\}\}$ and $\tau = 7$, where the thinner red edges are in R, and the thicker blue edges are in B. The corresponding values of each vertex are show in parentheses. (Color figure online)

Recall that a Hamiltonian cycle in a graph G is a cycle that visits every vertex exactly once. Now, consider the following definition describing a variant of the Hamiltonian cycle problem involving two edge sets.

Definition 4. *Let $G = (V, B \cup R)$ be a simple, undirected graph where each edge is a member of one of two sets, B or R. G contains an alternating Hamiltonian cycle if there exists a Hamiltonian cycle such that successive edges alternate between sets B and R.*

It is clear that an alternating Hamiltonian cycle in G, if one exists, corresponds to a feasible solution $\mathcal{T}$, as each "blue" edge from B represents a pair of values in $\mathcal{M}$, and each "red" edge from R indicates the values that meet the vicinal sum constraint, and thus can be placed alongside one another feasibly. Also, note that every edge in B must be present in the alternating Hamiltonian cycle. Consequently, the task can also be seen as finding a matching $R' \subseteq R$ of cardinality $n + 1$ such that the edge sets B and R' form an alternating Hamiltonian cycle as described in Definition 4.

The problem of finding an alternating Hamiltonian cycle in general graphs is NP-hard as it generalises the classical Hamiltonian cycle problem [5]. However, the special structure of graphs derived from instances of the COP allows them to be solved in polynomial-time using the following method.

To find a suitable matching $R' \subseteq R$, a Maximum Cardinality Matching (MCM) algorithm is executed, which is based on [1] and also the earlier work of Mahadev and Peled [10]. First, take each vertex $v_1, v_2, ..., v_{2n+2}$ in turn and select the edge in R connecting v_i to the highest-indexed vertex v_j that is not already incident to an edge in R'. Now, add this edge (v_i, v_j) to R' and proceed to the next vertex until all the vertices have been assessed. The two vertices incident to each matching edge in R' are now referred to as being "matched". Similarly to partners, let $m : V \to V$ be a bijective function that assigns each vertex with its match, $m(v_i) = v_j$. Then, we can denote this matching set as $R' = \{(v_i, m(v_i)) : v_i \in V\}$.

During MCM, if a vertex v_i is not adjacent to any other unmatched vertex except its partner $p(v_i)$ via a blue edge, the preceding vertex v_{i-1} can be "rematched", provided that (a) v_i is not the first vertex; (b) the previous vertex v_{i-1} has been matched; and (c) v_{i-1} and $p(v_i)$ are adjacent via a red edge in R. Then, v_i is matched with the vertex that is currently matched with v_{i-1}, and v_{i-1} is rematched with $p(v_i)$. Due to the initial order of the vertices, MCM is guaranteed to produce a maximum cardinality matching.

Clearly, if $|R'| < n + 1$ after MCM has completed, there are an insufficient number of red edges to form an alternating Hamiltonian cycle, and therefore no feasible solution exists for the given instance $\mathcal{M}$. Otherwise, R' is a perfect matching of cardinality $n + 1$, and the spanning subgraph $G' = (V, B \cup R')$ is a 2-regular graph where each vertex $v_i \in V$ is incident to one blue edge and one red edge, as can be seen in Fig. 4. G' therefore consists of one or more cyclic components $C_1, C_2, ..., C_l$.

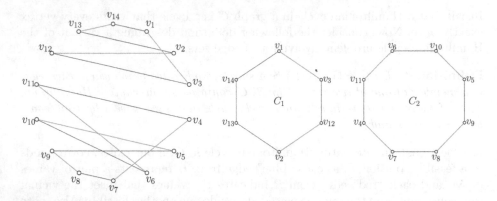

Fig. 4. Subgraph $G' = (V, B \cup R')$ created after MCM. When in planar form, it is clear that $l = 2$. (Color figure online)

In the case where G' comprises one component, i.e. $l = 1$, then this alternating cycle is in fact Hamiltonian and therefore specifies a solution to the COP. However, if $l > 1$, the components of G' must be combined to form a single alternating Hamiltonian cycle. To do this, a Bridge Recognition (BR) procedure is executed that selects suitable edges from $R \backslash R'$ to replace edges in R' to connect the components of G'.

BR operates by first ordering the edges in R' such that the lower-indexed vertices of each edge are in increasing order and the higher-indexed vertices are in decreasing order. Any edges that cannot be placed in such an order are then removed from this list. For instance, in the example in Fig. 4, the edges would be sorted as follows: $(v_1, v_{14}), (v_2, v_{13}), (v_3, v_{12}), (v_4, v_{11}), (v_6, v_{10}), (v_7, v_8)$. Note that, since v_5 was not matched with the highest-indexed vertex possible during MCM, the edge (v_5, v_9) does not adhere to the required structure of the list and is therefore omitted.

Starting from the first edge in the list, BR then searches through the list to find an edge that meets the following conditions: (a) the lower-indexed vertex of the current edge and the higher-indexed vertex of the next edge in the list are adjacent via an edge in R; and (b) the current edge and the next edge are members of different components of G'. If these conditions are met, BR adds the current edge to a set R_1, and continues to add all succeeding edges in the list to R_1 provided that, for each edge, both conditions hold and the succeeding edge is not a member of a component of G' that already has an edge in R_1. Once there are no more valid edges available to add to R_1, BR resumes its search through the remaining edges in the list to find another edge that meets the conditions, and can begin a new set R_2. The procedure terminates once the penultimate edge in the list has been assessed.

Now, one of the following three cases occurs:

1. In the event that BR has produced no sets, there are no suitable edges that can combine the components of G', and therefore an alternating Hamiltonian cycle in G cannot be created. Consequently, no feasible solution exists for the corresponding COP.
2. If there exists a set R_i such that $|R_i| = l$, then all components of G' can be merged together to form a single alternating Hamiltonian cycle. This is achieved by adding the red edge from $R \backslash R'$ connecting the lower-indexed vertex of each edge in R_i to the higher-indexed vertex of the next edge to R' (for the final edge in R_i, add the red edge connecting its lower-indexed vertex to the higher-indexed vertex of the first edge). Edges that appear in both R_i and R' are then removed from R' so that R' remains a perfect matching and $R_i \cap R' = \emptyset$. G' then consists of a single alternating Hamiltonian cycle; hence a solution has been found.
3. It may be that multiple sets need to be used to connect the components of G'. For two edge sets R_i and R_j to "overlap", each set must have exactly one edge from the same component in G', with the other edges in each set being from different components. A collection of sets $\mathcal{R}^*$ therefore needs to be found such that each set in $\mathcal{R}^*$ overlaps with at least one other set, and each component of G' has at least one edge in one of the sets.

In the final case above, a Modified Bridge Recognition (MBR) algorithm is used to find suitable overlapping sets. Firstly, a copy of the set R_i with the highest cardinality[2] generated by BR is created, called R_1^*, and added to $\mathcal{R}^*$. Then, MBR takes the sorted list used in BR and removes edges from the list that are in $\mathcal{R}^*$. Similarly to BR, MBR proceeds through the list to find an edge that meets the following conditions: (a) the lower-indexed vertex of the current edge and the higher-indexed vertex of the next edge in the list are adjacent via an edge in R; and (b) the current edge and an edge in $\mathcal{R}^*$ are members of the same component of G' and the next edge is a member of a component that does not have an edge in $\mathcal{R}^*$, or vice versa. If both conditions hold, the current edge is added to a new set R_2^* which is then added to $\mathcal{R}^*$. MBR continues to add succeeding edges to R_2^*, provided (a) holds and the succeeding edge is a member of a component that does not have an edge in $\mathcal{R}^*$. Then, if every component of G' has an edge in one of the sets in $\mathcal{R}^*$, these sets are able to connect all the components of G' together, and so MBR terminates. Otherwise, the edges in R_2^* are removed from the sorted list, and MBR repeats the search for suitable edges to start a new set R_3^*. This procedure continues until either $\mathcal{R}^*$ contains overlapping sets that cover all components of G', or until there are no more suitable edges in the list to start a new set. If MBR has produced a feasible collection of sets then the components of G' can be merged to create an alternating Hamiltonian cycle by applying the connecting procedure above to every set in $\mathcal{R}^*$.

Using the instance illustrated in Fig. 5 as an example, the edges (v_3, v_{12}) and (v_4, v_{11}) are in the set R_1 formed by BR, as (a) $(v_3, v_{11}) \in R$, i.e. the lower-indexed vertex of the first edge is adjacent to the higher-indexed vertex of the

---

[2] In the event of a tie, MBR selects the set with the lowest index.

next edge; and (b) the edges are members of different components ($(v_3, v_{12}) \in C_1$ and $(v_4, v_{11}) \in C_2$). Note that since v_4 and v_{11} are adjacent, v_4 must also be adjacent to v_{12}, i.e. $(v_4, v_{12}) \in R$, since the value associated with v_{12} is greater than or equal to the value associated with v_{11}. Then, because $|R_1| = l$, the edges (v_3, v_{12}) and (v_4, v_{11}) are removed from R' and replaced by the edges (v_3, v_{11}) and (v_4, v_{12}) from R_1. Removing the dominating vertices and any incident edges results in an alternating Hamiltonian path, which corresponds to a feasible solution $\mathcal{T}$.

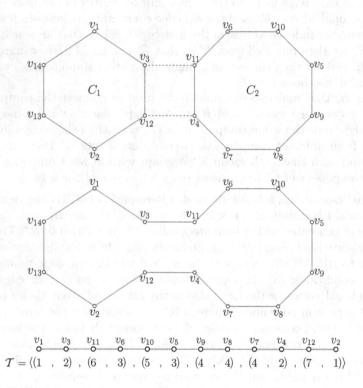

Fig. 5. Specific edges found using BR merge the components of G' together, forming an alternating Hamiltonian cycle which corresponds to a solution.

This entire algorithm that has been described will be referred to as the Alternating Hamiltonian Construction (AHC) algorithm. The time complexity of AHC is quadratic, as stated in the following theorem:

Theorem 1. *Let $G = (V, B \cup R)$ be a graph created from an instance $\mathcal{M}$ of cardinality n of the COP. Then, AHC terminates in at most $O(n^2)$ time.*

Proof. We assess each subprocedure of AHC in turn. Firstly, MCM produces a matching set $R' \subseteq R$ in at most $O(n \lg n)$ time due to the sorting of the vertices in lexicographical order. In BR, sorting the $n + 1$ edges of R' and removing unsuitable edges also requires $O(n \lg n)$ time. The sets R_i can be created in $O(n)$ time, as each edge in the list is considered once. As each set R_i contains at least two edges, BR can produce up to $\frac{n+1}{2}$ sets. Thus, examining each set to find one such that $|R_i| = l$ takes at most $O(n)$ time. The same method is also used to find the set with the highest cardinality in MBR. Since G' comprises a maximum of $\frac{n+1}{2}$ components, it follows that the number of edge sets in $\mathcal{R}^*$ needed to connect all the components is bounded by $\frac{n+1}{2} - 1$. The initial sorted list consists of at most $n - 1$ edges, and therefore MBR is of quadratic complexity $O(n^2)$. Finally, the connecting procedure replaces up to $n+1$ edges, and so can be executed in $O(n)$ time. Consequently, AHC has an overall worst case complexity of $O(n^2)$. $\qquad\qquad\qquad\qquad\qquad\qquad\qquad\qquad\qquad\qquad\qquad\qquad\qquad\square$

3   Heuristics for the SCSPP

In this section, we now turn our attention to the multi-strip version of the problem. As mentioned in the introduction, the SCSPP is a generalisation of the one-dimensional BPP in that we also require the sum of every pair of adjacent score widths to be greater than or equal to a minimum scoring distance τ. It follows that the SCSPP is at least as hard as the BPP, which is known to be NP-hard [3], and so (under the assumption that $P \neq NP$) there is no known algorithm that is able to find an optimal solution for every instance of the SCSPP in polynomial time. Instead, heuristics can be used to find near-optimal solutions in a shorter amount of time.

For an instance of the SCSPP, a feasible solution is represented by the set $\mathcal{S} = \{S_1, S_2, ..., S_k\}$ such that

$$\bigcup_{i=1}^{|\mathcal{S}|} S_i = \mathcal{I}, \tag{2a}$$

$$S_i \cap S_j = \emptyset \qquad \forall\, i, j \in \{1, 2, ..., |\mathcal{S}|\}, \quad i \neq j, \tag{2b}$$

$$\sum_{i=1}^{|S_j|} w_i \leq W \qquad \forall\, S_j \in \mathcal{S}, \tag{2c}$$

$$\mathbf{rhs}(i) + \mathbf{lhs}(i+1) \geq \tau \qquad \forall\, i \in \{1, 2, ..., |S_j| - 1\}, \quad \forall\, S_j \in \mathcal{S}. \tag{2d}$$

That is, all items must be packed onto a strip (2a), each item can only be placed on one strip (2b), the strips cannot be overfilled (2c), and the items on each strip S_j in the solution must be arranged such that the vicinal sum constraint is fulfilled (2d). Note that constraints (2a)–(2c) are the necessary conditions for the BPP. An optimal solution $\mathcal{S}$ for the SCSPP is a solution that consists of the fewest number of strips k needed to feasibly contain the n items in the given problem instance. A simple lower bound for k is the theoretical minimum $t = \lceil \sum_{i=1}^{n} w_i / W \rceil$ which can be computed in $O(n)$ time [12].

Perhaps the simplest and most well-known heuristic for one-dimensional bin packing problems is First-Fit (FF), a greedy online algorithm that places each item, presented in some arbitrary order, onto the lowest-indexed strip such that the capacity of the strip is not exceeded. It is known that there always exists at least one ordering of the items such that FF produces an optimal solution [8]. A minor modification to FF yields the First-Fit Decreasing (FFD) heuristic, which sorts the items in non-increasing order of size prior to performing FF. In 1973, Johnson [7] showed that FFD is guaranteed to return a solution that uses no more than $\frac{11}{9}k + 4$ strips. More recently, Dosa [2] has proven that the worst case for FFD is in fact $\frac{11}{9}k + \frac{6}{9}$, and that this bound is tight. Due to the initial sorting of the items in non-increasing order of sizes, the time complexity of FFD is $O(n \log n)$.

As mentioned in the introduction, the SCSPP shares many similarities with the BPP, however the addition of (2d) brings complications. One obvious difference is the order in which the items appear on the strips. The order of the items in the BPP is unimportant; however in the SCSPP the items must be ordered in a way that meets the vicinal sum constraint. In addition, removing an item from a bin in the BPP retains the feasibility of the bin, whereas in the SCSPP this is not guaranteed, as it may leave a subset of items for which the vicinal sum constraint is not satisfied. Furthermore, the theoretical minimum t has the potential to be less accurate for the SCSPP, as the minimum scoring distance τ is not considered. For example, if the minimum scoring distance is greater than twice the largest score width, then it is clear that n strips will be required, regardless of the items' widths.

To gain an understanding of this problem, three heuristics for the SCSPP have been developed: a basic FFD heuristic with a simple modification; a heuristic that packs strips individually and prioritises score widths; and a more advanced version of FFD that incorporates the polynomial-time AHC algorithm.

The first heuristic is the Modified First-Fit Decreasing (MFFD) heuristic which acts in the same manner as the original FFD, attempting to place each item onto the end of the lowest-indexed strip. If an item is able to be packed onto a strip without exceeding the strip's capacity, MFFD then checks to see if the vicinal sum constraint is met between the right-most score on the strip and one of the score widths on the current item. If the constraint is met, MFFD places the item on the end of the strip in the appropriate orientation, otherwise the next strip is considered. The most prominent issue with this heuristic is due to the items being placed on the end of the strips. Although an item could potentially be packed in a different location on the strip other than the end, it may end up being placed on another strip, or perhaps even begin a new strip, thus increasing the number of strips in the final solution.

The next heuristic is the Pair-Smallest (PS) heuristic, which is an extension of an inexact procedure defined by Lewis et al. [9]. Unlike MFFD, which packs each item in turn, PS focusses on packing one strip at a time, only starting a new strip once the current strip is unable to accommodate any further items. Each strip is initialised by choosing the item from $\mathcal{I}$ with the smallest score width, and packing it in a regular orientation. PS then continues to fill the strip by selecting the item with the smallest score width that meets the vicinal sum constraint with the right-most score width on the strip, and whose width will not cause the strip to be overfilled. This heuristic aligns the smallest score widths with the largest ones, eliminating the possibility of placing larger score widths together unnecessarily. Note that PS therefore prioritises the vicinal sum constraint over the item widths, choosing to fulfil this constraint first before considering whether the item can actually be accommodated. Consequently, there is no use for a procedure such as AHC to find a feasible arrangement of the items.

The last heuristic, Modified First-Fit Decreasing with AHC (MFFD$^+$), incorporates the AHC algorithm from Sect. 2. It operates in a similar fashion to the MFFD, placing items sorted in decreasing order onto the lowest-indexed strip. However, rather than attempting to place the item onto the right-most side of the strip, MFFD$^+$ executes AHC on all items on the strip. If AHC finds a feasible solution, the items are placed on the strip in the order of the solution, which includes the current item, else MFFD$^+$ attempts to pack the current item on the next strip. Using AHC means that if a feasible alignment of the items exists, there is a guarantee that it will be found. Unlike MFFD, where the current item can only be placed on the end of the strip, MFFD$^+$ allows the items to be entirely rearranged (see Fig. 6). This reduces the possibility of having to start a new strip for an item, thus preventing increasing the number of strips in the final solution.

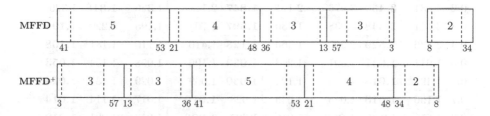

Fig. 6. Example instance of the sub-problem, with $W = 20$ and $\tau = 70$. In MFFD the constraint is not fulfilled in either orientation, however MFFD$^+$ is able find a feasible arrangement.

4   Experimental Results

Benchmark instances currently do not exist for the SCSPP, and so artificial problem instances were produced to compare the performance of the three heuristics. For our experiments, we generated 1000 problem instances for $|\mathcal{I}| \in \{500, 1000\}$. In each problem instance the items have varying widths $w_i \in [150, 1000]$ and score widths $a_i, b_i \in [1, 70]$ to ensure that each item has exactly two score lines, all selected randomly from a uniform distribution. Strips of widths $W = 5000$, 2500 and 1250 were used to influence the number of items per strip. As the width of the strips decrease, the average number of items per strip also decreases, making the problem more constrained. Both the items and the strips have equal height of $H = 1$. Similarly to experiments performed in [9], we also introduced a parameter δ to denote the proportion of pairs of score widths from different items that meet the vicinal sum constraint, i.e. whose sum is greater than or equal to τ. Values of δ from 0.0 to 1.0 were created by changing the value of τ. Clearly, when $\delta = 0.0$, there are no items that can be packed together feasibly, and so n strips will be required (one for each item), whereas if $\delta = 1.0$ all pairs of score widths meet the constraint, and the problem is equivalent to the BPP.

The heuristics were implemented in C++ and executed on a computer with an Intel Core i3-2120 3.30 GHz processor. Our source code and all data is provided at [6]. Since optimal solutions are not available, in our case solution quality q is estimated by comparing each solution to the theoretical minimum, $q = |\mathcal{S}|/t$. For each heuristic, we calculated the average solution quality for every combination of n, δ, and W from 1000 instances. All individual trials were seen to complete in under 160 ms.

Table 1. Average solution quality q for $n = 500$.

δ	$W = 5000, t = 58.039$			$W = 2500, t = 115.571$			$W = 1250, t = 230.648$		
	MFFD	PS	MFFD$^+$	MFFD	PS	MFFD$^+$	MFFD	PS	MFFD$^+$
0.0	8.618	8.618	8.618	4.328	4.328	4.327	2.169	2.169	2.169
0.1	5.214	**4.842**	5.161	2.659	**2.477**	2.657	1.515	**1.479**	1.515
0.2	4.031	**3.459**	3.976	2.121	**1.847**	2.118	1.326	**1.318**	1.326
0.3	3.111	**2.348**	3.038	1.730	**1.397**	1.708	1.195	1.229	**1.194**
0.4	2.436	**1.529**	2.297	1.460	**1.128**	1.410	1.110	1.181	**1.108**
0.5	1.911	**1.041**	1.691	1.263	**1.033**	1.196	1.058	1.154	**1.053**
0.6	1.491	**1.013**	1.246	1.124	**1.030**	1.069	1.029	1.135	**1.024**
0.7	1.196	**1.013**	1.045	1.049	1.028	**1.019**	1.017	1.114	**1.014**
0.8	1.050	1.012	**1.008**	1.016	1.027	**1.008**	1.013	1.091	**1.012**
0.9	1.009	1.012	**1.005**	1.007	1.026	**1.006**	1.012	1.073	**1.011**
1.0	**1.004**	1.012	**1.004**	**1.006**	1.026	**1.006**	**1.011**	1.065	**1.011**

Tables 1 and 2 compare the results obtained from the three heuristics using the different values of δ and W, for $n = 500$ and 1000 respectively. Figures in bold indicate the best average solution quality for the given combination of parameters. We see that q tends towards 1 as δ increases since the proportion of score widths meeting the vicinal sum constraint increases, consequently permissing more items to be packed on each individual strip[3], thus reducing the number of strips required. Note that when $\delta = 1.0$ MFFD and MFFD$^+$ have identical solution qualities, as the instances are equivalent to the original BPP ($\tau = 0$); hence they operate in the same fashion as the original FFD heuristic.

---

[3] The average number of items per strip when $n = 500$ for $W = 5000$, 2500 and 1250 are 8.475, 4.310, and 2.165 respectively, and the average number of items per strip when $n = 1000$ for $W = 5000$, 2500 and 1250 are 8.621, 4.329, and 2.169 respectively.

Table 2. Average solution quality q for $n = 1000$.

δ	$W = 5000, t = 115.534$			$W = 2500, t = 230.563$			$W = 1250, t = 460.623$		
	MFFD	PS	MFFD$^+$	MFFD	PS	MFFD$^+$	MFFD	PS	MFFD$^+$
0.0	8.657	8.657	8.657	4.338	4.338	4.338	2.171	2.171	2.171
0.1	5.173	**4.842**	5.140	2.636	**2.481**	2.643	1.511	**1.467**	1.511
0.2	3.976	**3.462**	3.952	2.093	**1.858**	2.102	1.318	**1.311**	1.319
0.3	3.047	**2.350**	3.013	1.698	**1.409**	1.688	**1.183**	1.221	**1.183**
0.4	2.374	**1.520**	2.266	1.426	**1.131**	1.384	1.097	1.171	**1.096**
0.5	1.847	**1.026**	1.642	1.230	**1.030**	1.170	1.044	1.144	**1.040**
0.6	1.433	**1.012**	1.203	1.099	**1.027**	1.050	1.020	1.124	**1.015**
0.7	1.155	**1.012**	1.026	1.034	1.026	**1.012**	1.012	1.102	**1.009**
0.8	1.032	1.012	**1.006**	1.011	1.025	**1.007**	1.009	1.078	**1.008**
0.9	1.007	1.012	**1.004**	1.007	1.024	**1.006**	**1.008**	1.059	**1.008**
1.0	**1.004**	1.012	**1.004**	**1.006**	1.024	**1.006**	**1.008**	1.054	**1.008**

Looking at results for both $n = 500$ and 1000, we see there is a clear pattern with respect to q across all widths and proportion levels. PS has the best solution quality for a wider range of δ when the strips are wider, and a much smaller range when the strip width decreases. Conversely, MFFD$^+$ obtains better solution qualities for a wider range of δ when $W = 1250$. Although PS does have the best solution quality of the three heuristics for $\delta = 0.1$ and 0.2 using the smallest strip width, we can see that it only marginally superior to MFFD$^+$. For example, take $\delta = 0.2$ in Table 1 for $W = 1250$. The difference between q for PS and MFFD$^+$ is 0.008, which translates to fewer than 2 strips difference between the average number of strips generated by each heuristic. In this particular instance, PS and MFFD$^+$ produced 303 and 305 strips on average, respectively.

Although using the average solution quality from 1000 instances provides a useful overview of the efficiency of a heuristic, there are other characteristics that we can consider. Take, for example, the results obtained with parameters $\delta = 0.7$ and $W = 5000$ in Table 2. Clearly PS obtains solutions with the fewest strips on average, however, we noted that out of the 1000 instances, PS did not produce a single solution containing t strips. On the other hand, there were 152 instances in which MFFD$^+$ was able to generate a solution S such that $|S| = t$, thus implying that there are at least 152 instances that can be solved to optimality. Despite this, MFFD$^+$ has a higher average solution quality than PS, suggesting that the variance in the number of strips required is higher.

5   Conclusions and Further Work

This paper has investigated the Score-Constrained Strip-Packing Problem (SCSPP), a packing problem in which the order and orientation of the items is crucial to the feasibility of the solution. We begun by introducing the Constrained Ordering Problem (COP), and described the Alternating Hamiltonian Construction (AHC) algorithm, an exact polynomial-time algorithm that operates by modelling the problem graphically and using the concept of Hamiltonian cycles. We then showed how AHC can be used to find a solution for the Score-Constrained Packing Sub-Problem (SCPSP), the single-strip version of the SCSPP. Thus, the main problem was to tackle the multi-strip problem. Three heuristics, one of which included the exact AHC algorithm, were described, and thorough experiments using a variety of parameters were executed.

A potential avenue for further research would be to produce an evolutionary algorithm (EA) for the SCSPP which incorporates AHC during local search. One possible addition would be to introduce an approach similar to one used in [11]. During each iteration of the evolutionary algorithm, high quality strips are chosen from each offspring solution and stored in separate set. On completion of the EA, a postoptimisation procedure based on the exact cover problem is executed. The procedure aims to find the fewest number of strips from the set of high quality strips that contains every item exactly once.

References

1. Becker, K.H.: Twin-constrained Hamiltonian paths on threshold graphs - an approach to the minimum score separation problem. Ph.D. thesis, London School of Economics (2010)
2. Dósa, G.: The tight bound of first fit decreasing bin-packing algorithm is $FFD(I) \leq 11/9OPT(I)+6/9$. In: Combinatorics, Algorithms, Probabilistic and Experimental Methodologies, pp. 1–11 (2007)
3. Garey, M.R., Johnson, D.S.: Computers and Intractability: A Guide to the Theory of NP-Completeness, pp. 90–91. WH Freeman Co., San Francisco (1979)
4. Goulimis, C.: Minimum score separation - an open combinatorial problem associated with the cutting stock problem. J. Oper. Res. Soc. **55**(12), 1367–1368 (2004)
5. Häggkvist, R.: On F-Hamiltonian graphs. University of Umeå, Department of Mathematics (1977)
6. Hawa, A.L., Lewis, R., Thompson, J.: Source code for algorithms in the article "Heuristics for the Score-Constrained Strip-Packing Problem" (2018). https://doi.org/10.5281/zenodo.1311857
7. Johnson, D.S.: Near-optimal bin packing algorithms. Ph.D. thesis, Massachusetts Institute of Technology (1973)
8. Lewis, R.: A general-purpose hill-climbing method for order independent minimum grouping problems: a case study in graph colouring and bin packing. Comput. Oper. Res. **36**(7), 2295–2310 (2009)
9. Lewis, R., Song, X., Dowsland, K., Thompson, J.: An investigation into two bin packing problems with ordering and orientation implications. Eur. J. Oper. Res. **213**(1), 52–65 (2011)
10. Mahadev, N.V., Peled, U.N.: Longest cycles in threshold graphs. Discrete Math. **135**(1–3), 169–176 (1994)
11. Malaguti, E., Monaci, M., Toth, P.: A metaheuristic approach for the vertex coloring problem. INFORMS J. Comput. **20**(2), 302–316 (2008)
12. Martello, S., Toth, P.: Lower bounds and reduction procedures for the bin packing problem. Discrete Appl. Math. **28**(1), 59–70 (1990)

Computational Geometry and Combinatorial Optimization

An Algorithm for Reducing Approximate Nearest Neighbor to Approximate Near Neighbor with $O(\log n)$ Query Time

Hengzhao Ma and Jianzhong Li$^{(\boxtimes)}$

Harbin Institute of Technology, Harbin 150001, Heilongjiang, China
hzma@stu.hit.edu.cn, lijzh@hit.edu.cn

Abstract. This paper proposes a new algorithm for reducing Approximate Nearest Neighbor problem to Approximate Near Neighbor problem. The advantage of this algorithm is that it achieves $O(\log n)$ query time. As a reduction problem, the query time complexity is the times of invoking the algorithm for Approximate Near Neighbor problem. All former algorithms for the same reduction need $polylog(n)$ query time. A box split method proposed by Vaidya is used in our paper to achieve the $O(\log n)$ query time complexity.

Keywords: Computation geometry · Approximate nearest neighbor
Reduction

1 Introduction

The approximate nearest neighbor problem, ϵ-NN for short, can be defined as follows: given a set P of points in a metric space S equipped with a distance function D, and a query point $q \in S$, find a point $p \in P$ such that $D(p,q) \leq (1+\epsilon)D(p^*,q)$, where p^* has the minimal distance to q in P. ϵ-NN is one of the most important proximity problems in computation geometry. Many proximity problems in computation geometry can be reduced to ϵ-NN [11], such as approximate diameter, approximate furthest neighbor, and so on. ϵ-NN is also important in many other areas, such as databases, data mining, information retrieval and machine learning.

Due to its importance, ϵ-NN has been the subject of substantial research efforts. Many algorithms for solving ϵ-NN have been discovered. These works can be summarized into four classes.

The first class of the algorithms tries to build data structures that support solving ϵ-NN efficiently. Arya et al. [4] give a such algorithm with query time $1/\epsilon^{O(d)} \cdot \log n$, space $1/\epsilon^{O(d)} \cdot n$ and preprocessing time $1/\epsilon^{O(d)} \cdot n \log n$. Another work [5] gives an algorithm requiring $O(dn)$ space and $O(dn \log n)$ preprocessing

This work was supported by the National Natural Science Foundation of China under grant 61832003.

D. Kim et al. (Eds.): COCOA 2018, LNCS 11346, pp. 465–479, 2018.
https://doi.org/10.1007/978-3-030-04651-4_31

time but query time as high as $(d/\epsilon)^{O(d)} \cdot \log n$. Kleinberg proposes two algorithms in [15]. The first algorithm is deterministic and achieves query time of $O(d \log^2 d(d + \log n))$, using a data structure that requires $O((n \log d)^{2d})$ space and $O((n \log d)^{2d})$ preprocessing time. The second algorithm is a randomized version of the first one. By a preprocessing procedure that takes $O(d^2 \log^2 d \cdot n \log^2 n)$ time, it reduces the storage requirement to $O(dn \cdot \log^3 n)$, but raises the query time up to $O(n + d \log^3 n)$.

The second class of the algorithms considers the situation of $\epsilon = d^{O(1)}$. One such algorithm is given in [6]. It can answer $O(\sqrt{d})$-NN in $O(2^d \log n)$ time with $O(d 8^d n \log n)$ preprocessing time and $O(d 2^d n)$ space. Chan [8] improves this result by giving an algorithm that can answer $O(d^{3/2})$-NN in $O(d^2 \log n)$ query time with $O(d^2 n \log n)$ preprocessing time and $O(dn \log n)$ space.

The third interesting class of work tries to solve ϵ-NN by inspecting some intrinsic dimension of the input point set P. An exemplar work is in [16]. The paper gives an algorithm whose query time is bounded by $2^{O(dim(P))} \log \Delta + (1/\epsilon)^{O(dim(P))}$, where $dim(P)$ is the intrinsic dimension of the input point set P, and Δ is the diameter of P.

Besides these algorithms mentioned above, Indyk et al. [14] initiate the work on the fourth class of algorithms. The key idea is to define an Approximate *Near* Neighbor problem, denoted as (c, r)-NN, and reduce ϵ-NN to it. The (c, r)-NN problem can be viewed as a decisive version of ϵ-NN. The formal definition of (c, r)-NN is give in Definition 2 in the next section.

To use this method to solve ϵ-NN, two parts of problem must be considered. One is how to solve (c, r)-NN, and the other is how to reduce ϵ-NN to (c, r)-NN. Some works about the two parts of problem are discussed below. Our study focuses on the latter part.

Algorithms to Solve (c,r)-NN. The existing algorithms for (c, r)-NN mainly consider the specific situation of d-dimensional Euclidean space with 1-order and 2-order Minkowski distance metrics. Each input point x is given in the form of $(x_1, \cdots, x_d)$. And q-order Minkowski L_q distance between points x and y is given by $D(x, y) = \left(\sum_{i=1}^{d} |x_i - y_i|^q \right)^{\frac{1}{q}}$. The 1-order and 2-order Minkowski distance are well-known Manhattan distance and Euclidean distance, respectively. Another simpler situation, which is the (c, r)-NN problem under Hamming cube $\{0, 1\}^d$ equipped with Hamming distance, is usually considered in theoretical studies.

Table 1 summarizes the complexities of the existing algorithms for (c, r)-NN under Euclidean space and L_1 distance. These papers also give solutions under L_2 distance, but we omit these results due to space limitation. Usually the complexities under L_2 distance is higher than that under L_1 distance. It is a key characteristic of the existing algorithms for (c, r)-NN that they usually have different complexities for problems under different order of Minkowski distance metrics.

The listed solutions in Table 1 can be divided into three groups. The first group includes the one given in [14], which is deterministic, and the other groups are randomized. The advantage of randomization is that the exponential

Table 1. Solutions to (c,r)-NN under Euclidean space and L_1 distance.

Source	Data structure building		Query		Space	Update time
	Time	Failure probability	Time	Failure probability		
[14] $(\epsilon = c - 1)$	$O(n \cdot \frac{1}{\epsilon^d})$	0	$O(1)$	0	$O(n \cdot \frac{1}{\epsilon^d})$	$O(\frac{1}{\epsilon^d})$
[17] $(\epsilon = c - 1)$	$O\left(n\frac{d^3}{\epsilon^2}(n \log d)^{O(\frac{1}{\epsilon^4})}\right)$	$O(1)$	$O\left(\frac{d}{\epsilon^2}polylog(dn)\cdot\log\frac{1}{f}\right)$	f	$O\left(\frac{d^3}{\epsilon^2}(n\log d)^{O(\frac{1}{\epsilon^4})}\right)$	$O(n^{O(\frac{1}{\epsilon^2})})$
[19]	$O(n^{\left(\frac{\epsilon}{2c-1}\right)^2}\log n)$	0	$O(dn^{o(1)})$	$O(1)$	$O(n^{\left(\frac{\epsilon}{2c-1}\right)^2})$	$O(n^{\left(\frac{\epsilon}{2c-1}\right)^2})$
[9,3]	$O(dn^{1+\frac{1}{2c-1}}\log n)$	0	$O(dn^{\frac{1}{2c-1}})$	$O(1)$	$O(dn + n^{1+\frac{1}{2c-1}})$	$O(dn^{\frac{1}{2c-1}+o(1)})$
[1]	$O(dn^{1+o(1)}\log n)$	0	$O(n^{\frac{2c-1}{c^2}})$	$O(1)$	$O(dn^{1+o(1)})$	$O(dn^{o(1)})$

complexity about d is freed. The second group includes the one given in [17], which is based on a random projection method proposed in [15]. One distinguished characteristic of the method is that the data structure building stage is also randomized. The last group includes a long line of research work based on Locality Sensitive Hashing (LSH), which is first proposed in [14]. These works are summarized into three terms in Table 1, which can be viewed as the space-time trade-off under LSH framework.

Finally, comparing the five results in Table 1, it can be seen that the query time grows and the space requirement drops from the first one to the last. The five results form a general space-time trade-off about the solution to (c, r)-NN.

Reducing ϵ-NN to (c,r)-NN. So far there are three different algorithms for such a reduction. Two of the three algorithms are deterministic [12,14], and the other one is randomized [13]. The complexities of the three reduction algorithms are summarized in Table 2. Note that query time in Table 2 is the number of invocations of (c, r)-NN algorithm. And the preprocessing time about [14] is not given because there is no such analysis in that paper.

Table 2. Comparison of three reductions.

Source	Approximation factor	Preprocessing		Query		Space
		Time	Failure probability	Time (# of (c,r)-NN invoked)	Failure probability	
[13]	$c(1+\gamma)^2$ $(\gamma\in(\frac{1}{n},\frac{1}{2}))$ $(c = 1 + \epsilon)$	$O\left(\frac{T(n,c,f)}{\gamma\log^2 n} + n\log n[Q(n,c,f) + D(n,c,f)]\right)$	$f\log n$	$O(\log^{O(1)} n)$	$f\log n$	$O(\frac{S(n,c,f)}{\gamma\log^2 n})$
[14]	$1+\epsilon$	-	-	$O(\log^2 n)$	0	$O(n\cdot polylog(n))$
[12]	$1+\epsilon$	$O(d\cdot n^{\frac{\log n}{\epsilon}}\log\frac{n}{\epsilon})$	0	$O(\log\frac{n}{\epsilon})$	0	$O(d\cdot n^{\frac{\log n}{\epsilon}}\log\frac{n}{\epsilon})$

Among the three reduction algorithms, the one proposed in [13] need to be explained in detail. First, the algorithm outputs a point p' such that $D(q, p') \leq c(1 + \gamma)^2 D(q, p^*)$, where $c = 1 + \epsilon$ and p^* is the exact NN of q. Second, the $T(n, c, f)$, $Q(n, c, f)$, $D(n, c, f)$ and $S(n, c, f)$ functions represent the complexity functions of the data structure building time, query time, update time and storage usage for (c, r)-NN, respectively. Third, the parameter f is the failure probability of one (c, r)-NN invocation, and is selected so that $f \log n$ is a constant less that 1.

The fourth and the most important point about [13] is the $O(\log^{O(1)} n)$ query time. The algorithm given in [13] explicitly invokes $O(\log n)$ times of (c, r)-NN, and each invocation needs $T(n, c, f)$ time. As explained above, the parameter f, which is the failure probability of one (c, r)-NN invocation, is set to $O(\frac{1}{\log n})$. Note that the algorithms for (c, r)-NN given in Table 1 all have constant failure probability[1]. In order to satisfy the requirement of $O(\frac{1}{\log n})$ failure probability of one (c, r)-NN invocation, each time the algorithm in [13] invokes (c, r)-NN, the algorithms for (c, r)-NN with constant failure probability must be executed multiple times, which is $O(\log^{O(1)} n)$ times in expectation. Multiplying $O(\log n)$ invocations of (c, r)-NN and $O(\log^{O(1)} n)$ executions of (c, r)-NN algorithm for each invocation, we obtain that the algorithm in [13] actually invokes $O(\log^{O(1)} n)$ times of (c, r)-NN algorithm. This observation is confirmed in [2].

Our Method. We propose a new algorithm in this paper for reducing ϵ-NN to (c, r)-NN. Comparing with the former works [12–14], our algorithm has the following characteristics:

(1) It achieves $O(\log n)$ query time, counted in the number of invocations of (c, r)-NN algorithm. It is superior to all the other three works. This is the most distinguished contribution of this paper.
(2) Its preprocessing time is $O((\frac{d}{\epsilon})^d \cdot n \log n)$, and the space complexity is $O((\frac{d}{\epsilon})^d \cdot n)$. Our method has better complexity than the other three works in terms of n, so that it is much suitable to big data with low or fixed dimensionality. This situation is plausible in many applications like road-networks and so on.
(3) In terms of the parameterized complexity treating d as a constant, our result is the closest to the well recognized *optimal* complexity claimed in [5], which requires $O(n \log n)$ preprocessing time, $O(n)$ space and $O(\log n)$ query time.

Note that there is an $O((d/\epsilon)^d)$ factor in our preprocessing and space complexity. This factor originates from a lemma we used in [20]. We point out that the upper bound $O((d/\epsilon)^d)$ is actually very loose. There really is possibility to reduce the upper bound, and thus make our result more close to optimal. In this sense, our work is more promising than all the other three works. However, reducing the upper bound $O((d/\epsilon)^d)$ is out of this paper's scope, and is left as our future work.

2   Problem Definitions and Mathematical Preparations

2.1   Problem Definitions

We focus on ϵ-NN in euclidean space R^d. The input is a set P of n points extracted from R^d and a distance metric L_q. Each point x is given as the form

---

[1] The deterministic one has exponential dependence on d, so it is rarely used in theory and practice.

$(x_1, \cdots, x_d)$. L_q distance metric between points x and y is given by $D(x, y) = \left(\sum\limits_{i=1}^{d} |x_i - y_i|^q \right)^{\frac{1}{q}}$.

Denote $B(p, r)$ to be the d-dimensional ball centered at p and with radius r. And let $p' \in B(p, r)$ be equivalent to $D(p', p) \leq r$. We first give the definitions of ϵ-NN and (c, r)-NN problems.

Definition 1 (ϵ-NN). *Given a set P of points extracted from R^d, a query point $q \in R^d$, and an approximation factor ϵ, find a point $p' \in P$ such that $D(p', q) \leq (1 + \epsilon)D(p^*, q)$ where $D(p^*, q) = \min\limits_{p \in P}\{D(p, q)\}$.*

Remark 1. p^* is called the nearest neighbor (NN), or exact NN to q, and p' is called an ϵ-NN to q.

Definition 2 ((c, r)-NN). *Given a set P of points extracted from R^d, a query point $q \in R^d$, a query range r, and an approximation factor $c > 1$, (c, r)-NN problem is to design an algorithm satisfying these:*

1. *if there is a point $p_0 \in P$ satisfying $p_0 \in B(q, r)$, then return a point $p' \in P$ such that $p' \in B(q, c \cdot r)$;*
2. *if $D(p, q) > c \cdot r$ for $\forall p \in P$, then return No.*

Remark 2. There are multiple names referring to the same problem defined above. In the papers related to LSH, it is referred as (c, r)-NN. In [14], it is called approximate Point Location in Equal Balls, which is denoted as ϵ-PLEB where $\epsilon = c - 1$. In more recent papers like [12], it is called Approximate Near Neighbor problem.

Next we give the definition of the reduction problem to be solved in this paper, i.e., the problem of reducing ϵ-NN to (c, r)-NN.

Definition 3 (Reduction Problem). *Given a set P of points extracted from R^d, a query point $q \in R^d$, an approximation factor ϵ, and an algorithm $\mathcal{A}$ for (c, r)-NN, the reduction problem is to find an ϵ-NN to q by invoking the algorithm $\mathcal{A}$ as an oracle.*

Remark 3. To solve the reduction problem, a preprocessing phase is usually needed, which is to devise a data structure $\mathcal{D}$ based on P. Thus the problem of reducing ϵ-NN to (c, r)-NN is divided into two phases. The first is data structure building phase, or preprocessing phase. The second is ϵ-NN searching phase, or query phase. The (c, r)-NN algorithm $\mathcal{A}$ is invoked in query phase as an oracle, which characterizes the algorithm as a Turing reduction from ϵ-NN to (c, r)-NN.

The time complexity of the algorithm for the reduction problem consists of two parts, namely, preprocessing time complexity and query time complexity. An important note is that the query time complexity is the number of invocations of (c, r)-NN algorithm $\mathcal{A}$. This is the well recognized method for analyze the time complexity of a Turing reduction.

2.2 Mathematical Preparations

In this section we introduce some denotations and lemmas to support the idea
of our algorithm for reducing ϵ-NN to (c, r)-NN.

Denotations. Define a box $\mathfrak{b}$ in R^d to be the product of d intervals, i.e., $I_1 \times I_2 \times \cdots \times I_d$ where I_i is either open, closed or semi-closed interval, $1 \leq i \leq m$.
A box is cubical iff all the d intervals defining the box are of the same length.
The side length a cubical box, which is the length of any interval defining the
cubical box, is denoted as $len(\mathfrak{b})$. A minimal cubical box (MCB) for a point set
P, denoted as $MCB(P)$, is a cubical box containing all the points in P and has
the minimal side length. Note that $MCB(P)$ may not be unique.

Given a point set P and a box $\mathfrak{b}$, let $p \in \mathfrak{b}$ denote that a point $p \in P$ falls
inside box $\mathfrak{b}$, and let $|\mathfrak{b} \cap P|$ denote the number of points in P that falls inside
$\mathfrak{b}$. We will use $|\mathfrak{b}|$ for short, if not causing ambiguity.

Given a collection of MCBs $\mathcal{B} = \{\mathfrak{b}_1, \cdots, \mathfrak{b}_m\}$, define $D_{max}(\mathfrak{b})$, $D_{min}(\mathfrak{b}, \mathfrak{b}')$,
$D_{max}(\mathfrak{b}, \mathfrak{b}')$ as follows:

$$D_{max}(\mathfrak{b}) = \max_{p_1, p_2 \in \mathfrak{b}} D(p_1, p_2), \forall \mathfrak{b} \in \mathcal{B}$$

$$D_{min}(\mathfrak{b}, \mathfrak{b}') = \min_{p \in \mathfrak{b}, p' \in \mathfrak{b}'} \{D(p, p')\}, \forall \mathfrak{b}, \mathfrak{b}' \in \mathcal{B}$$

$$D_{max}(\mathfrak{b}, \mathfrak{b}') = \max_{p \in \mathfrak{b}, p' \in \mathfrak{b}'} \{D(p, p')\}, \forall \mathfrak{b}, \mathfrak{b}' \in \mathcal{B}$$

With the above denotations, define $Est(\mathfrak{b})$ as follows:

$$Est(\mathfrak{b}) = \begin{cases} D_{max}(\mathfrak{b}), & if \ |\mathfrak{b} \cap P| \geq 2 \\ \min_{\mathfrak{b}' \in Nbr(\mathfrak{b})} \{D_{max}(\mathfrak{b}, \mathfrak{b}')\}, & otherwise \end{cases} \tag{1}$$

where $Nbr(\mathfrak{b}) = \{\mathfrak{b}' \mid D_{min}(\mathfrak{b}, \mathfrak{b}') \leq r\}$, and the parameter r should satisfy $r \geq Est(\mathfrak{b})$.[2]

For an MCB $\mathfrak{b}$, we associate a ball with it. Pick an arbitrary point $c_\mathfrak{b} \in \mathfrak{b}$,
and let $r_\mathfrak{b} = Est(\mathfrak{b})$, then we have a ball $B(c_\mathfrak{b}, r_\mathfrak{b})$. It is easily verified that every
point in $\mathfrak{b}$ is within a distance of $Est(\mathfrak{b})$ from $c_\mathfrak{b}$, in another way to say, the ball
$B(c_\mathfrak{b}, r_\mathfrak{b})$ encloses every point in $\mathfrak{b}$. We call $B(c_\mathfrak{b}, r_\mathfrak{b})$ the *enclosing ball* for box
$\mathfrak{b}$.

Next we start to introduce the lemmas while discussing different situations of
ϵ-NN search. In the following discussion, we will assume that we have an MCB $\mathfrak{b}$
of the input point set P, an enclosing ball $B(c_\mathfrak{b}, r_\mathfrak{b})$ of the MCB $\mathfrak{b}$, and a query
point q.

---

[2] It can be verified that, as long as $r \geq Est(\mathfrak{b})$ is satisfied, the value of r doesn't
influence the value of $Est(\mathfrak{b})$. The specific value of r will be shown latter.

Situation 1. The first and an easy situation is that, if q is far enough from $c_\mathfrak{b}$ then every point in $\mathfrak{b}$ is an ϵ-NN to q. The following value $T_1(\mathfrak{b})$ explains the threshold for *far enough*, and Lemma 1 depicts the situation discussed above.

Definition 4. *For an MCB of a point set P, define $T_1(\mathfrak{b}) = (1 + 2/\epsilon)r_\mathfrak{b}$.*

Lemma 1. *If $D(q, c_\mathfrak{b}) \geq T_1(\mathfrak{b})$, then every point in $\mathfrak{b}$ is an ϵ-NN to q.*

Proof. See the full paper [18] for the proof.   □

Situation 2. If q is not as far from $c_\mathfrak{b}$ as a distance of $T_1(\mathfrak{b})$, i.e., $D(q, c_\mathfrak{b}) < T_1(\mathfrak{b})$, then we split $\mathfrak{b}$ into a set of sub-boxes $\{\mathfrak{b}_1, \cdots, \mathfrak{b}_m\}$, and calculate the enclosing balls $B(c_{\mathfrak{b}_i}, r_{\mathfrak{b}_i})$ for each box $\mathfrak{b}_i, 1 \leq i \leq m$. The next situation is that if q is still far enough from each point in $\{c_{\mathfrak{b}_1}, \cdots, c_{\mathfrak{b}_m}\}$, i.e., the centers of the enclosing balls, then we can still tell that every point in $\mathfrak{b}$ is an ϵ-NN to q. We give another threshold $T_2(\mathfrak{b})$ based on this idea, and formalize the idea into Lemma 2. This lemma also discusses the quantitative relationship between $T_2(\mathfrak{b})$ and $T_1(\mathfrak{b})$.

Definition 5. *For an MCB of a point set P, split $\mathfrak{b}$ into a set of sub-boxes $\{\mathfrak{b}_1, \cdots, \mathfrak{b}_m\}$. Each of these sub-boxes is an MCB of a point set $P' \subset P$. Then let $B(c_{\mathfrak{b}_i}, r_{\mathfrak{b}_i})$ be the enclosing ball of sub-box $\mathfrak{b}_i$, $1 \leq i \leq m$. Define $rmax_\mathfrak{b} = \max\limits_i\{r_{\mathfrak{b}_i}\}$. In case of $|\mathfrak{b}| = 1$, let $rmax_\mathfrak{b} = 0$.*

Definition 6. *Define $T_2(\mathfrak{b}) = r_\mathfrak{b} + (1 + 2/\epsilon)rmax_\mathfrak{b}$.*

Lemma 2. *We have the following statements:*

1. if $D(q, c_\mathfrak{b}) \geq T_2(\mathfrak{b})$, then every point in $\mathfrak{b}$ is ϵ-NN to q;
2. if $rmax_\mathfrak{b} < \frac{2}{2+\epsilon}r_\mathfrak{b}$, then $T_2(\mathfrak{b}) < T_1(\mathfrak{b})$.

Proof. See the full paper [18] for the proof.   □

Situation 3. If q is still not as far from $c_\mathfrak{b}$ as a distance of $T_2(\mathfrak{b})$, it is time to ask the algorithm of (c, r)-NN for help. Let $\mathcal{A}(Q, \mathfrak{q}, c, r)$ be any algorithm for solving (c, r)-NN, where Q is the input point set, $\mathfrak{q}$ is the query point, r is the query range, and c is the approximation factor. The meanings of these four parameters are already given in Definition 2. The goal of invoking $\mathcal{A}$ is that, if $\mathcal{A}$ answers No then still every point in $\mathfrak{b}$ is an ϵ-NN to q. The following lemma shows how to set the four input parameters to fulfill the goal.

Lemma 3. *Let $\mathcal{A}(Q, \mathfrak{q}, c, r)$ be any algorithm for (c, r)-NN. We have the following statements:*

1. if we set $Q = \{c_{\mathfrak{b}_1}, \cdots c_{\mathfrak{b}_m}\}$, $\mathfrak{q} = q$, $r = \max\limits_i\{T_2(\mathfrak{b}_i)\}$, and let c satisfy $c \cdot r = \max\limits_i\{T_1(\mathfrak{b}_i)\}$, and invoke $\mathcal{A}(Q, \mathfrak{q}, c, r)$, then if $\mathcal{A}$ returns No, we can pick any point in $\mathfrak{b}$ as the answer of ϵ-NN to q;
2. if $rmax_{\mathfrak{b}_i} < \frac{2}{2+\epsilon}r_{\mathfrak{b}_i}$ holds for each $\mathfrak{b}_i$, $1 \leq i \leq m$, then our settings for c and r satisfy the requirement of (c, r)-NN problem definition. i.e. $c > 1$.

Proof. See the full paper [18] for the proof.   □

Situation 4. As what is said in Lemma 3, if the algorithm $\mathcal{A}$ returns No then the search of ϵ-NN terminates with returning an arbitrary point in $\mathfrak{b}$. According to Definition 2, $\mathcal{A}$ can also return some point $c_{\mathfrak{b}_i} \in Q$ other than No. In that case the search must continues. At first glance, the same procedure should be recursively carried out, by applying Lemmas 1, 2, 3 one by one on box $\mathfrak{b}_i$, where the point $c_{\mathfrak{b}_i}$ returned by $\mathcal{A}$ is the center of the enclosing ball of box $\mathfrak{b}_i$. However, to guarantee that the algorithm returns a correct ϵ-NN, the box considered by the algorithm must encloses the exact NN p^*. But the box $\mathfrak{b}_i$ may not enclose p^*, which would ruin the correctness of the algorithm. Thus, we need to expand the search range to the boxes near to $\mathfrak{b}_i$. The following Lemma 4 gives the bounds of the search range and ensures that p^* lies in the range.

Definition 7. *For a collection of MCBs $\mathcal{B} = \{\mathfrak{b}_1, \cdots, \mathfrak{b}_m\}$, let $rmax_{\mathfrak{b}_i}$ be defined as Definition 5 for each $\mathfrak{b}_i$, $1 \leq i \leq m$. Then define $rmax_{\mathcal{B}} = \max\limits_{\mathfrak{b}_i \in \mathcal{B}}\{rmax_{\mathfrak{b}_i}\}$.*

Definition 8. *Define $Nbr(\mathfrak{b})$ as*

$$\{\mathfrak{b}' \in \mathcal{B} \mid D(c_{\mathfrak{b}'}, c_{\mathfrak{b}}) \leq (3 + 4\epsilon)rmax_{\mathcal{B}_{\mathfrak{b}_s}}\}$$

where $\mathcal{B}_{\mathfrak{b}_s} = Nbr(\mathfrak{b}_s)$ and $\mathfrak{b}_s$ is the super box of $\mathfrak{b}$.

Remark 4. The definition of Nbr sets is a *recursive definition*. For a box $\mathfrak{b}$, its $Nbr(\mathfrak{b})$ set is defined based on the $Nbr(\mathfrak{b}_s)$ set of its super box $\mathfrak{b}_s$. It requires that the boxes are recursively split, which can be represented as a tree structure. The formal description of the tree structure is given in Sect. 3.1.

Lemma 4. *Given the query point q, and a collection of boxes $\{\mathfrak{b}_1, \cdots, \mathfrak{b}_m\}$, if we find a box $\mathfrak{b}_i$ satisfying $D(q, c_{\mathfrak{b}_i}) \leq \max\limits_{i}\{T_1(\mathfrak{b}_i)\}$, then the nearest neighbor of q lies in and can only lie in $Nbr(\mathfrak{b}_i)$, i.e., $p^* \in Nbr(\mathfrak{b}_i)$.*

Proof. See the full paper [18] for the proof. □

We are done introducing the mathematical preparations. In the next section we will propose our algorithm based on the lemmas given above.

3   Algorithms

In this section we propose our algorithm for reducing ϵ-NN to (c, r)-NN, including the preprocessing and query algorithm.

3.1   Preprocessing

Our preprocessing algorithm mainly consists of two sub-procedures. One is to build the box split tree T, and the other is to construct the Nbr sets.

Building the Box Split Tree. We first give the definition of the box split tree.

Definition 9 (Box split tree). *Given a point set P and its MCB $\mathfrak{b}_P$, a tree T is a box split tree iff:*

1. *the root of T is $\mathfrak{b}_P$;*
2. *each non-root node of T is an MCB of a point set $P' \subset P$;*
3. *if box $\mathfrak{b}'$ is a sub-box of $\mathfrak{b}$, then there is an edge between the node for $\mathfrak{b}$ and the node for $\mathfrak{b}'$ in T;*
4. *each node has at least 2 child nodes, and at most $|P|$ child nodes;*
5. *$rmax_{\mathfrak{b}} < \frac{2}{2+\epsilon} r_{\mathfrak{b}}$ holds for each box $\mathfrak{b}$ in T.*

Further, T is fully built iff each box at the leaf nodes of T contains only one point.

Remark 5. The fifth term is required by the second statement of Lemma 3.

We use a box split method to build the box split tree. This method is originally proposed in [20], and also used in several other papers [7,10]. It starts from the MCB $\mathfrak{b}_P$ of the point set P, then continuously splits $\mathfrak{b}_P$ into a collection $\mathcal{B}$ of cubical boxes until each box in $\mathcal{B}$ contains only one point. The method proceeds in a series of split steps. In each split step, the box $\mathfrak{b}_L$ with the largest side length in the current collection $\mathcal{B}$ is chosen and split. Define $h_i(\mathfrak{b})$ to be the hyperplane orthogonal to the i-th coordinate axis and passing through the center of $\mathfrak{b}$. One split step will split $\mathfrak{b}_L$ into at most 2^d sub-boxes using all $h_i(\mathfrak{b}_L)$, each of which is an MCB. The set of non-empty sub-boxes generated by conducting one split step on $\mathfrak{b}$ is denoted as $Succ(\mathfrak{b})$. The details of the box splitting method can be found in [20].

Next we describe how to use this method to build the box split tree T. The main obstacle is to satisfy the fifth term in Definition 9, i.e., $rmax_{\mathfrak{b}} < \frac{2}{2+\epsilon} r_{\mathfrak{b}}$ for each box $\mathfrak{b}$ in T. We use the following techniques to solve this problem.

When a split step is executed and a box $\mathfrak{b}$ is split, we temporarily store the sub-boxes of $\mathfrak{b}$ in a max-heap $H_{\mathfrak{b}}$, which is ordered on the side length of the boxes in the heap. Recall the definitions in Sect. 2.2, the side length of a box $\mathfrak{b}$ is denoted as $len(\mathfrak{b})$. When box $\mathfrak{b}$ is split fine enough so that $rmax_{\mathfrak{b}} < \frac{2}{2+\epsilon} r_{\mathfrak{b}}$ is satisfied, the algorithm will create a node for each $\mathfrak{b}' \in H_{\mathfrak{b}}$, and hang it under the node for box $\mathfrak{b}$ in the box split tree T. Then for each $\mathfrak{b}'$ at these newly created leaf nodes, a max-heap is created to store its sub-boxes. In an overview, a max-heap is maintained for each box at the leaf nodes of the box split tree.

In each split step, the box with the largest volume is split. To efficiently pick out this box, a secondary heap H_2 is maintained. The heaps for the leaf nodes are called the primary heaps in contrast. The elements in H_2 is just the top elements in each primary heap, together with a pointer to its corresponding primary heap. Apparently the top element $\mathfrak{b}_{top}$ in H_2 is the box with largest volume. When $\mathfrak{b}_{top}$ is picked, the primary and secondary heap will pop it out simultaneously. Then $\mathfrak{b}_{top}$ is split by conducting one split step on it, generating

---

Algorithm 1. Preprocessing

Input: a point set P, and an approximation factor ϵ
Output: a box split tree T
`// Initialization`

1 Compute $\mathfrak{b}_0 = MCB(P)$;
2 Compute the enclosing ball $B(c_{\mathfrak{b}_0}, r_{\mathfrak{b}_0})$ of $\mathfrak{b}_0$;
3 Set $\mathfrak{b}_0$ to be the root of T;
4 Initialize the primary heap for $\mathfrak{b}_0$ with one key-value pair $(len(\mathfrak{b}_0), \mathfrak{b}_0)$;
5 Initialize the secondary heap H_2 with one key-value pair $(len(\mathfrak{b}_0), \mathfrak{b}_0)$;

`// Main loop`

6 **while** $|\mathcal{B}| < n$ **do**
7 $\quad$ Pop out the top element $\mathfrak{b}_{top}$ from H_2 and the corresponding primary heap $H_{\mathfrak{b}_s}$;
8 $\quad$ Split $\mathfrak{b}_{top}$ by conducting one split step on $\mathfrak{b}$, generating $Succ(\mathfrak{b}_{top})$;
9 $\quad$ **foreach** $\mathfrak{b} \in Succ(\mathfrak{b}_{top})$ **do**
10 $\quad\quad$ Add $\mathfrak{b}$ into $H_{\mathfrak{b}_s}$, and maintain the heap;
11 $\quad$ **end**
12 $\quad$ Let the current top element of $H_{\mathfrak{b}_s}$ to be $\mathfrak{b}_t$;
13 $\quad$ Let $Flag = false$;
14 $\quad$ **if** $len(\mathfrak{b}_t) < \frac{2}{(2+\epsilon)d} len(\mathfrak{b}_s)$ **then** `// Applying Lemma 6`
15 $\quad\quad$ Let $Flag = true$;
16 $\quad\quad$ **foreach** $\mathfrak{b} \in H_{\mathfrak{b}_s}$ **do**
17 $\quad\quad\quad$ Create a node and hang it under the node of $\mathfrak{b}_s$;
18 $\quad\quad\quad$ Initialize the primary heap for $\mathfrak{b}$ with one key-value pair $(len(\mathfrak{b}), \mathfrak{b})$;
19 $\quad\quad$ **end**
20 $\quad$ **else**
21 $\quad\quad$ Add $\mathfrak{b}_t$ into H_2.
22 $\quad$ **end**
23 $\quad$ Invoke Algorithm 2, taking $\mathfrak{b}, Succ(\mathfrak{b}), rmax_{\mathcal{B}_{\mathfrak{b}_s}}$, and the boolean value $Flag$ as the input of this invocation;
24 **end**

$Succ(\mathfrak{b}_{top})$. These sub-boxes in $Succ(\mathfrak{b}_{top})$ will be added into the primary heap where $\mathfrak{b}_{top}$ formerly resides. When this primary heap finishes maintaining, its top element is inserted into the secondary heap. And then the iteration continues.

We point out the last problem to solve in order to satisfy the fifth term in Definition 9. The heaps, including the primary heaps and the secondary heap, are organized according to the len value of the boxes, in order to retrieve the box with the largest volume. On the other hand, the condition of $rmax_{\mathfrak{b}} < \frac{2}{2+\epsilon} r_{\mathfrak{b}}$ is based on the Est value of the boxes, because here we have $rmax_{\mathfrak{b}} = \max\limits_{\mathfrak{b}' \in H_{\mathfrak{b}}} \{Est(\mathfrak{b}')\}$.

Notice that the top element $\mathfrak{b}_{top}$ in the primary heap have the largest len value, but may not has the largest Est value. So we can not directly check $r_{\mathfrak{b}_{top}} < \frac{2}{2+\epsilon} r_{\mathfrak{b}}$ to decide whether $\mathfrak{b}$ is split fine enough. Fortunately, the len and Est value of a box have certain quantity relationships, which is formalized into the following lemma.

Lemma 5. *For the MCB $\mathfrak{b}$ of any point set P where $|\mathfrak{b}| \geq 2$, we have $len(\mathfrak{b}) \leq Est(\mathfrak{b}) \leq d \cdot len(\mathfrak{b})$. In the situation that $|\mathfrak{b}| = 1$, we redefine $len(\mathfrak{b})$ as $len(\mathfrak{b}) = Est(\mathfrak{b})$ to make this inequality consistent.*

Proof. See the full paper [18] for the proof.   □

With the help of Lemma 5, we have the following Lemma 6 about the criteria for deciding whether a box is split fine enough.

Lemma 6. *For box $\mathfrak{b}$ and its primary heap $H_\mathfrak{b}$, if the top element $\mathfrak{b}_{top}$ satisfies $len(\mathfrak{b}_{top}) < \frac{2}{(2+\epsilon)d} len(\mathfrak{b})$, then $rmax_\mathfrak{b} < \frac{2}{2+\epsilon} r_\mathfrak{b}$.*

Proof. See the full paper [18] for the proof.   □

The pseudo codes for building the box split tree are given in Algorithm 1. The algorithm also includes the invocation of Algorithm 2 aimed to maintain the Nbr sets, which will be introduced in the next section.

Nbr Sets Maintaining. Algorithm 2 for maintaining $Nbr(\mathfrak{b})$ is given below. It is invoked each time the main loop of Algorithm 1 is executed, as shown above.

Algorithm 2. Maintaining $Nbr(b)$

Input: box $\mathfrak{b}$, $Succ(\mathfrak{b})$, the neighbor range parameter $rmax_{\mathcal{B}_{b_s}}$, and a
        boolean value $Flag$.

1   **foreach** $\mathfrak{b}' \in Succ(\mathfrak{b})$ **do**
2 | $Nbr(\mathfrak{b}') \leftarrow Nbr(\mathfrak{b}) \cup Succ(\mathfrak{b}) - \{\mathfrak{b}\}$;
3 | Set $Est(\mathfrak{b}')$ according to Equation 1;
4 | Update $rmax_{\mathfrak{b}_s}$;
5   **end**
6   **foreach** $\mathfrak{b}' \in Nbr(\mathfrak{b})$ **do**
7 | **if** $Flag = ture$ and $\mathfrak{b}'$ is in a higher level that $\mathfrak{b}$ **then**
8 | | $Nbr(\mathfrak{b}') \leftarrow Nbr(\mathfrak{b}') \cup Succ(\mathfrak{b})$
9 | **else**
10 | | $Nbr(\mathfrak{b}') \leftarrow Nbr(\mathfrak{b}') \cup Succ(\mathfrak{b}) - \{\mathfrak{b}\}$
11 | **end**
12   **end**
13   **foreach** $\mathfrak{b}' \in Succ(\mathfrak{b})$ **do**
14 | **foreach** $\mathfrak{b}'' \in Nbr(\mathfrak{b}')$ **do**
15 | | **if** $D(c_{\mathfrak{b}''}, c_{\mathfrak{b}'}) > (3 + 4/\epsilon)rmax_{\mathcal{B}_{b_s}}$ **then**
16 | | | Delete $\mathfrak{b}''$ from $Nbr(\mathfrak{b}')$;
17 | | | Delete $\mathfrak{b}'$ from $Nbr(\mathfrak{b}'')$;
18 | | **end**
19 | **end**
20   **end**

There are two parts of Algorithm 2 that need to be explained in detail.

The first is Line 4. From Definition 8 for $Nbr(\mathfrak{b})$, we can see that the maintaining of $Nbr(\mathfrak{b})$ is based on the value $rmax_{\mathcal{B}_{b_s}}$ passed down by its super-box $\mathfrak{b}_s$. In Algorithm 2, Line 4 is aimed for updating $rmax_{\mathfrak{b}_s}$ when the set of sub-boxes of $\mathfrak{b}_s$ is changed. If $Nbr(\mathfrak{b})$ is implemented as a heap, then whenever any sub-box of $\mathfrak{b}_s$ needs $rmax_{\mathcal{B}_{\mathfrak{b}_s}}$, this value can be retrieved from the heap in constant time.

The other part is the second *foreach* loop in Algorithm 2. The functionality of the loop is explained in the following Lemma 7.

Lemma 7. *The second foreach loop ensures that for all box $\mathfrak{b}$ in the box split tree T, each $\mathfrak{b}' \in Nbr(\mathfrak{b})$ is either in the same level with $\mathfrak{b}$, or a degenerated box containing only one point.*

Proof. See the full paper [18] for the proof.   □

3.2   Query

The query algorithm goes down the tree T returned by Algorithm 1 level by level. At each level of T, the algorithm $\mathcal{A}$ for (c, r)-NN will be invoked, and the input parameters of $\mathcal{A}$ are set according to Lemma 3. The pseudo codes are given in Algorithm 3.

Algorithm 3. Query

Input: query point q, data set P, box split tree T, and algorithm $\mathcal{A}$ for
      (c, r)-NN
Output: ϵ-NN of q in P
1 set $\mathfrak{b}_c = root(T)$;
2 **if** $D(q, c_{\mathfrak{b}_c}) \geq T_2(\mathfrak{b})$ **then**
3 | pick any point $p' \in \mathfrak{b}_c \cap P$;
4 | **return** p';
5 **end**
6 **while** $|\mathfrak{b}_c| > 1$ **do**
7 | $\mathcal{B}_c \leftarrow Nbr(\mathfrak{b}_c)$;
8 | $P_c \leftarrow \bigcup_{\mathfrak{b} \in \mathcal{B}_c} \mathfrak{b} \cap P$;
9 | invoke $\mathcal{A}$, where the input of $\mathcal{A}$ is set according to Lemma 3;
10 | **if** *the query returns NO* **then**
11 | | pick any point $p' \in P_c$;
12 | | **return** p';
13 | **else** // the query returns the center $c_{\mathfrak{b}'}$ of box $\mathfrak{b}'$
14 | | set $\mathfrak{b}_c = \mathfrak{b}'$;
15 | | **continue**;
16 | **end**
17 **end**
18 $P_c \leftarrow Nbr(\mathfrak{b}_c) \cap P$;
19 Conduct brute-force search in P_c to find the exact NN;

We should spend some efforts to explain the termination condition in Algorithm 3. We introduce a lemma about $Nbr(\mathfrak{b})$ when $|\mathfrak{b}| = 1$.

Lemma 8. *For a box* $\mathfrak{b}$ *satisfying* $|\mathfrak{b}| = 1$, *all the boxes* $\mathfrak{b}' \in Nbr(\mathfrak{b})$ *contain only one point, i.e.,* $|\mathfrak{b}'| = 1$.

Proof. See the full paper [18] for the proof. □

According to the above lemma, when the WHILE loop breaks, all boxes in $Nbr(\mathfrak{b}_c)$ contains only one point. Thus the brute-force search takes $O(|Nbr(\mathfrak{b}_c)|)$ time. We will bound this complexity in the next section.

4 Analysis

4.1 Correctness

We prove the correctness of our algorithm by introducing the following lemma.

Lemma 9. *In every execution of the loop body, Algorithm 3 ensures that the exact nearest neighbor* $p^* \in P_c$ *after the assignment of* P_c *(Line 8).*

Proof. See the full paper [18] for the proof. □

Theorem 1 (Correctness). *The point* p' *returned by Algorithm 3 is an* ϵ-NN *to* q *in* P, *i.e., if* p^* *is the exact NN to* q *in* P, *then* $D(q, p') \leq (1 + \epsilon)D(q, p^*)$.

Proof. See the full paper [18] for the proof. □

4.2 Complexities

Before we bound the complexity of our algorithm, we should first bound the size of $Nbr(\mathfrak{b})$ for any box $\mathfrak{b}$ by introducing a lemma from [20].

Lemma 10 ([20]). *Let* r *be a positive number. During the execution of the split method described in Sect. 3.1, at each time before splitting a box, let* $\mathcal{B}$ *be the current box collection, and let* $\mathfrak{b}_L$ *be the box with the largest volume in* $\mathcal{B}$. *For any box* $\mathfrak{b} \in \mathcal{B}$, *the size of the set* $\{\mathfrak{b}' \in \mathcal{B} \mid D_{min}(\mathfrak{b}, \mathfrak{b}') \leq r \cdot Est(\mathfrak{b}_L)\}$ *is at most* $2^d(2d\lceil r \rceil + 3)^d$.

Based on the lemma above, we can bound the size of $Nbr(\mathfrak{b})$ for any box $\mathfrak{b}$ in the box split tree T constructed in Algorithm 1.

Lemma 11. *The size of* $Nbr(\mathfrak{b})$ *defined in Definition 8 and constructed in Algorithm 2 is* $O((\frac{d}{\epsilon})^d)$.

Proof. See the full paper [18] for the proof. □

We introduce and prove another lemma which is about the property of the box split tree T constructed in preprocessing phase.

Lemma 12. *For a point set* P *where* $|P| = n$, *the fully built split tree* T *constructed based on* P *has the following properties:*

1. *There are at most $2n$ nodes in T.*
2. *The total time to build T is $O(dn \log n)$.*

Proof. See the full paper [18] for the proof.   □

Now we start to prove the complexities of our algorithm, including preprocessing time, space and query time complexities.

Theorem 2 (Preprocessing Time Complexity). *The complexity of Algorithm 1 for preprocessing is $O(O((\frac{d}{\epsilon})^d \cdot n \log n))$.*

Proof. See the full paper [18] for the proof.   □

Theorem 3 (Space Complexity). *The space complexity of Algorithm 1 is $O((\frac{d}{\epsilon})^d \cdot n)$.*

Proof. See the full paper [18] for the proof.   □

Theorem 4 (Query Time Complexity). *Algorithm 3 invokes $O(\log n)$ times of the algorithm $\mathcal{A}$ for (c, r)-NN problem.*

Proof. See the full paper [18] for the proof.   □

5   Conclusion

In this paper we proposed a new algorithm for reducing ϵ-NN problem to (c, r)-NN problem. Compared to the former works for the same reduction problem, our algorithm achieves the lowest query time complexity, which is $O(\log n)$ times of invocations of the algorithm for (c, r)-NN problem. We elaborately designed the input parameters of each of the invocation, and built a dedicated data structure in preprocessing phase to support the query procedure. A box split method proposed in [20] is used as a building block for the algorithm of preprocessing phase. Our paper also raises a problem which is to reduce the exponential complexity on d introduced by the box split method. This is left as our future work.

References

1. Andoni, A., Indyk, P.: Near-optimal hashing algorithms for approximate nearest neighbor in high dimensions. In: 47th Annual IEEE Symposium on Foundations of Computer Science, vol. 51, pp. 459–468 (2006)
2. Andoni, A., Indyk, P.: Nearest neighbors in high-dimensional spaces. In: Handbook of Discrete and Computational Geometry, 3rd edn, chap. 43, pp. 1135–1155. CRC Press Inc., Boca Raton (2017)
3. Andoni, A., Razenshteyn, I.: Optimal data-dependent hashing for approximate near neighbors. In: Proceedings of the Forty-Seventh Annual ACM Symposium on Theory of Computing, pp. 793–801 (2015)
4. Arya, S., Mount, D.M.: Approximate nearest neighbor queries in fixed dimensions. In: Proceedings of the Fourth Annual Symposium on Discrete Algorithms, pp. 271–280 (1993)

5. Arya, S., Mount, D.M., Netanyahu, N.S., Silverman, R., Wu, A.Y.: An optimal algorithm for approximate nearest neighbor searching fixed dimensions. J. ACM **45**(6), 891–923 (1998)
6. Bern, M.W.: Approximate closest-point queries in high dimensions. Inf. Process. Lett. **45**(2), 95–99 (1993)
7. Callahan, P.B., Kosaraju, S.R.: A decomposition of multidimensional point sets with applications to k-nearest-neighbors and n-body potential fields. J. ACM **42**(1), 67–90 (1995)
8. Chan, T.M.: Approximate nearest neighbor queries revisited. In: Proceedings of the Thirteenth Annual Symposium on Computational Geometry, vol. 20, pp. 352–358 (1997)
9. Datar, M., Immorlica, N., Indyk, P., Mirrokni, V.S.: Locality-sensitive hashing scheme based on p-table distributions. In: Proceedings of the Twentieth annual Symposium on Computational Geometry, pp. 253–262 (2004)
10. Feder, T., Greene, D.: Optimal algorithms for approximate clustering. In: Proceedings of the Twentieth Annual ACM Symposium on Theory of Computing, pp. 434–444 (1988)
11. Goel, A., Indyk, P., Varadarajan, K.: Reductions among high dimensional proximity problems. In: Proceedings of the Twelfth Annual ACM-SIAM Symposium on Discrete Algorithms, pp. 769–778 (2001)
12. Har-Peled, S.: A replacement for voronoi diagrams of near linear size. In: 42nd Annual IEEE Symposium on Foundations of Computer Science, pp. 94–103 (2001)
13. Har-Peled, S., Indyk, P., Motwani, R.: Approximate nearest neighbor: towards removing the curse of dimensionality. Theory Comput. **8**(1), 321–350 (2012)
14. Indyk, P., Motwani, R.: Approximate nearest neighbors: towards removing the curse of dimensionality. In: Proceedings of the Thirteenth Annual ACM Symposium on Theory of Computing, pp. 604–613 (1998)
15. Kleinberg, J.M.: Two algorithms for nearest-neighbor search in high dimensions. In: Proceedings of the Twenty-ninth Annual ACM Symposium on Theory of Computing, pp. 599–608 (1997)
16. Krauthgamer, R., Lee, J.R.: Navigating nets: simple algorithms for proximity search. In: Proceedings of the Fifteenth Annual ACM-SIAM Symposium on Discrete Algorithms, pp. 798–807 (2004)
17. Kushilevitz, E., Ostrovsky, R., Rabani, Y.: Efficient search for approximate nearest neighbor in high dimensional spaces. SIAM J. Comput. **30**(2), 457–474 (2000)
18. Ma, H., Li, J.: An Algorithm for Reducing Approximate Nearest Neighbor to Approximate Near Neighbor with O(logn) Query Time (2018). http://arxiv.org/abs/1809.09776
19. Panigrahy, R.: Entropy based nearest neighbor search in high dimensions. In: Proceedings of the Seventeenth Annual ACM-SIAM Symposium on Discrete Algorithms, pp. 1186–1195 (2006)
20. Vaidya, P.M.: An optimal algorithm for the all-nearest-neighbors problem. In: 27th Annual Symposium on Foundations of Computer Science, pp. 117–122 (1986)

Exact and Approximate Map-Reduce Algorithms for Convex Hull

Anirban Ghosh$^{(\boxtimes)}$ and Samuel Schwartz

School of Computing, University of North Florida, Jacksonville, FL, USA
{anirban.ghosh,n00448518}@unf.edu

Abstract. Given a set of points P in the Euclidean plane, the classic problem of convex hull in computational geometry asks to compute the smallest convex polygon C with the vertex set $X \subseteq P$, such that every point in P belongs to C.

In our knowledge, only two map-reduce convex hull algorithms have been designed so far. The exact map-reduce algorithm designed by Goodrich et al. (2011) is intricate and runs in constant number of rounds when the mappers and reducers have a memory of $\Theta(|P|^\epsilon)$, for a small constant $\epsilon > 0$. Otherwise, their algorithm runs in logarithmic number of rounds with high probability. In Big Data, easy-to-implement constant-round map-reduce algorithms are highly preferred. The other exact map-reduce algorithm, designed by Eldawy et al. (2011), does not perform efficiently when X contains sufficiently high number of points from P.

In this paper, we design two new simple constant-round map-reduce algorithms along with map-reduce implementable pruning heuristics to address the above shortcomings. Our first algorithm CH-MR is exact and outperforms Eldawy et al.'s algorithm when reasonable computing resources are available, and the heuristics are able to prune away sufficient number of points. The second algorithm, named APXCH-MR, can run efficiently on any point set to return an approximate convex hull, when the input parameters are sub-linear in $|P|$.

The designed algorithms are theoretically analyzed in the light of the popular $\mathcal{MRC}$ model. Our algorithms are easy to implement and do not use any complicated data structure.

Keywords: Convex hull · Parallel computing · Map-reduce

1 Introduction

Given a set P of n points $p_1, \ldots, p_n$ in the Euclidean plane, the convex hull of P is defined to be the smallest convex polygon C containing P. Clearly, the vertex set of $C \subseteq P$. If P is convex, then P itself is the vertex set of C. A popular analogy is to imagine the points as nails on a wall, then the convex hull consists

Research supported by the University of North Florida start-up fund.

D. Kim et al. (Eds.): COCOA 2018, LNCS 11346, pp. 480–494, 2018.
https://doi.org/10.1007/978-3-030-04651-4_32

of the nails touched by a rubber band when put around them. Mathematically, convex hull of P, denoted by $\mathrm{CH}(P)$, is given by the following expression:

$$\mathrm{CH}(P) = \Big\{ \sum_{i=1}^{|P|} \lambda_i p_i \text{ such that } \forall i \; \lambda_i \geq 0 \text{ and } \sum_{i=1}^{|P|} \lambda_i = 1 \Big\}.$$

In computational geometry, convex hull algorithms are used extensively. These algorithms have innumerable applications in image processing, robotics, geographic information systems, pattern matching and several others. The convex hull of a given point set P consists of the extreme points in P. Indeed, in the era of Big Data, convex hull of a point set has turned out to be an important tool from data analytic point of view.

Convex hull is a well-researched problem. Several sequential algorithms have been obtained till date. Some of them are even output-sensitive, i.e. the asymptotic runtime is a function of n and the size of the convex hull, $h := |CH(P)|$. Refer to [4,6,7,9,14,16,18,21] for a list of well-known exact sequential convex hull algorithms.

In modern times, data sizes are in gigabytes, terabytes and even in petabytes. Naturally, we have taken recourse to parallel computing to process the massive data sets. The map-reduce framework [8] has become a de-facto standard in parallel computing for Big Data. Hadoop [15] is a popular software framework for map-reduce. First, let us understand the computational model of map-reduce algorithms.

A map-reduce algorithm consists of multiple rounds where each round is composed of three sequential stages: *map* (M), *shuffle* (S) and *reduce* (R). The map function is commonly known as the **mapper**, the shuffle function as the **shuffler**, and the reduce function as the **reducer**. Mappers and reducers can be deterministic as well as randomized.

An algorithm in this framework can be represented as a sequence of rounds $[M_1, S_1, R_1], \ldots, [M_r, S_r, R_r]$, where r is the number of rounds and the round i is denoted by $[M_i, S_i, R_i]$, $1 \leq i \leq r$. In a certain round of execution, depending on the environment, there can be many mappers and reducers running in parallel. Given a cluster setup, many mappers/reducers can be allocated to a machine. However, it must be noted that the three steps cannot be run in parallel. In other words, in every round, first the mapping phase must complete, then the shuffling, and finally the reduction. Moreover, although the mappers and reducers run in parallel, a particular mapper or reducer is sequential in nature.

The basic unit of information in a map-reduce computation is a `<key,value>` pair. The input to round j, or specifically the map phase, is a set of `<key,value>` pairs which we denote by $\mathcal{I}_j$. Clearly, $\mathcal{I}_1$ is the input to the algorithm. Note that the input to a round can be the output from the previous round. Now, to understand the idea of a map-reduce computation, consider the round j.

1. The *mapper* takes one ordered `<key,value>` pair from $\mathcal{I}_j$ and outputs a finite multiset of new `<key,value>` pairs. This means, if the input to a mapper is $\langle k; v \rangle \in \mathcal{I}_j$, then the output from the mapper is a multiset of pairs

$\mathcal{U}(k, v) := \{\langle k_1; v_1 \rangle, \langle k_2; v_2 \rangle, \ldots\}$. Hence, the output from the map phase is $\mathcal{U} := \cup_{<k,v> \in \mathcal{I}_j} \mathcal{U}(k, v)$. Since a mapper works with one <key,value> pair at a time, the map phase can be parallelized. In other words, many mappers can run in parallel to generate the pool of <key,value> pairs $\mathcal{U}$. The pool $\mathcal{U}$ is then fed to the shuffler which in turn does the job of sending the pairs to appropriate reducers, as explained next.

2. The *shuffler* sends all of the values that are associated with an individual key to a particular reducer. To understand this better, let $\mathcal{K}$ denote the set of distinct keys in $\mathcal{U}$. For every $k \in \mathcal{K}$, let $\mathcal{P}(k)$ denote the pairs in $\mathcal{U}$ with the key k. Then, all the values in $\mathcal{P}(k)$ are sent to a single reducer since they all share the same key k. In particular, the reducer receives $\langle k; \mathcal{V} \rangle$ where $\mathcal{V}$ is the multiset of values which have k as the corresponding key. The underlying system that implements map-reduce does this task of shuffling by itself, and is seamless to the algorithm designer/programmer. This implies that we do not need to supply a shuffler function. Henceforth, in our discussion we will not consider shufflers anymore.

3. The *reducer* takes an ordered pair $\langle k, \mathcal{V} \rangle$ as input, where $\mathcal{V}$ is a multiset of values associated with the key k and outputs an another multiset $\mathcal{X}(k, \mathcal{V})$ of <key,value> pairs. Hence, the output from this round j is $\mathcal{X} := \cup_{k \in \mathcal{K}} \mathcal{X}(k, \mathcal{V})$. The pairs in $\mathcal{X}$ may be the final output of the algorithm or an intermediate result which can be fed as an input to round $j + 1$ of the algorithm.

A great advantage of map-reduce is that we do not need to worry about the various low level aspects such as, inter-system communication, data transfer, task scheduling, fault tolerance etc., unlike other models of parallel computing. The underlying system which implements map-reduce takes care of the above low-level issues. Here, the only task of the programmer is to specify the map and reduce functions for every round of map-reduce. But it is not a silver bullet. There are several non-traditional challenges of efficiency in map-reduce algorithms. These are addressed by the following widely accepted model.

The $\mathcal{MRC}$ Model. The first theoretical model for map-reduce was introduced by Karloff et al. [17]. Following is a description of their model. For a given problem, consider a r-round map-reduce algorithm $\mathcal{A} := \{[M_1, R_1], \ldots, [M_r, R_r]\}$. Let n denote the size of the input. Then $\mathcal{A} \in \mathcal{MRC}$ if the following are true:

1. $\mathcal{A}$ outputs the correct answer with probability at least 0.75.
2. Runtime of any mapper or reducer is $\mathcal{O}(n^k)$, for some fixed k.
3. Memory used by any mapper or reducer is $o(n)$.
4. Total space required by $\mathcal{A}$ is $o(n^2)$. This implies that the number of distinct keys $|\mathcal{K}|$ can also be $o(n^2)$.
5. The number of rounds $r = \mathcal{O}(\log^j n)$, for some fixed $j \geq 0$.

For some popular $\mathcal{MRC}$ algorithms in the literature, refer to [11,12,17,19,26].

In our knowledge, there are exactly two map-reduce convex hull algorithms in the literature. One of these is designed by Eldawy et al. [10], and the other by Goodrich et al. [13]. We start with a brief description of the Eldawy et al.'s algorithm. First, a grid of a certain size is overlayed on P. After this step, we obtain a cell partition of P where every cell is either empty or contains at least one point. Next, the points in the innermost cells are discarded since those do not belong to the convex hull. Finally, the points in the outermost cells are sent to a single reducer where the final exact convex hull is computed using a traditional sequential algorithm. We recommend the reader to refer to their paper [10] for further details. The algorithm has the following limitations:

1. The grid-based approach used in their algorithm may not be able to prune sufficient number of points for certain point sets. For instance, one may consider point sets where $|h| = \Theta(n)$. Eventually, it may happen that $\Theta(n)$ points are sent to the single reducer. This clearly shows that their algorithm does not belong to the $\mathcal{MRC}$ class.
2. There is no load-balancing in the reduce phase since only one reducer is always used. If there are m machines, only one among them will be used and the remaining $m - 1$ machines will stay idle. Certainly, this is not desirable in map-reduce algorithms.

The other algorithm designed by Goodrich et al. [13] is intricate for implementation and runs in logarithmic rounds with high probability. However, the authors pointed out that their algorithm runs in constant number of rounds when the mappers and reducers have memory size of $\Theta(n^\epsilon)$, for a small constant $\epsilon > 0$.

Our Results. In this paper, we design new simple map-reduce convex hull algorithms to address the above concerns. In Sect. 2, we present a naive exact algorithm. Next, we design and analyze the following two new algorithms in the light of the $\mathcal{MRC}$ model.

1. An exact map-reduce algorithm, named **CH-MR**, which runs in constant number of rounds on any point set. CH-MR $\in \mathcal{MRC}$ when the designed map-reduce heuristic pruners succeeds in pruning away sufficient number of points, leaving $o(n)$ points for the main round of convex hull computation. We show that CH-MR asymptotically outperforms Eldawy et al.'s algorithm when the points in P are chosen uniformly and independently from a disk or a polygon, and sufficient number of mappers/reducers are allowed to run in parallel. See Sect. 3.
2. An approximate constant-round map reduce algorithm, named **APXCH-MR**, which returns an approximate convex hull of P depending on two input parameters k_1, k_2. When $k_1, k_2 = o(n)$ (which is quite reasonable in practice), APXCH-MR $\in \mathcal{MRC}$. See Sect. 4 for the definition of approximate convex hull and the algorithm. APXCH-MR runs efficiently on any type of point set, as long as k_1, k_2 are sub-linear.

Related Works. For some recent map-reduce algorithms in computational geometry refer to [1,2,5,27].

Notation. We denote generation of a `<key,value>` pair using emit$\langle$key, value$\rangle$. The notation is used uniformly both in the map and reduce phases of our algorithms.

It must be noted that our algorithms do not output the points in convex hull in particular order (clockwise or anticlockwise). If required, a map-reduce sorting algorithm can be run on the output to obtain an ordered sequence.

2   A Naive Exact Algorithm

In this section, we present an exact map-reduce algorithm. Let $p \in P$. Then, $p \in \mathrm{CH}(P)$, if and only if, for every triplet $\{q, r, s\} \in P \setminus p$, $p \notin \Delta qrs$. Clearly, the number of such triplets is $\binom{n-1}{3}$. Refer to Algorithm 1 for a map-reduce implementation of this idea.

Algorithm 1. A naive exact algorithm

Map 1: input $\langle p; \emptyset \rangle$
1: **for each** triplet $\{q, r, s\} \in P \setminus p$ **do**
2:     **if** $p \notin \Delta qrs$ **then**
3:         emit $\langle p; 1 \rangle$;
4:     **end if**
5: **end for**

Reduce 1: input $\langle p; \mathcal{V} = \{1, 1, \ldots, 1\} \rangle$
1: **if** $|\mathcal{V}| = \binom{n-1}{3}$ **then**
2:     emit $\langle p; \emptyset \rangle$;
3: **end if**

It is not hard to see that this algorithm is practically unusable for massive data sets which follows from Theorem 1 (stated without proof on account of its straightforwardness).

Theorem 1. *The following observations are made in order for the Algorithm 1.*

1. *The number of mappers is n.*
2. *Each mapper takes $\mathcal{O}(n^3)$ time since there are $\binom{n-1}{3}$ triplets.*
3. *Total work done by the map stage is $\mathcal{O}(n^4)$.*
4. *Total number of pairs emitted by a single mapper is $\mathcal{O}(n^3)$.*
5. *Total number of pairs generated after the map stage is $\mathcal{O}(n^4)$.*
6. *The number of reducers is at most n.*
7. *Each reducer takes $\mathcal{O}(n^3)$ time since $|\mathcal{V}| \leq \binom{n-1}{3}$.*
8. *Total work done by the reduce stage is $\mathcal{O}(n^4)$.*

Remark. Since the total number of pairs generated after the map stage is $\mathcal{O}(n^4)$, the Algorithm 1 is not in $\mathcal{MRC}$.

3 An Improved Exact Algorithm: CH-MR

Parallel heuristics play a vital role in Big Data. This motivates us to design a number of practical parallel heuristics, which will be used in our algorithms for pruning away points not in CH(P).

Akl-Toussaint+. The classic Akl-Toussaint heuristic was proposed in 1979 [3]. Given a point set P, find out N, W, S, E, where N is the point in P with the largest y-coordinate, W is the point with the smallest x-coordinate, S is the point with the smallest y-coordinate, and E is the point with the largest x-coordinate. In the case of ties, choose arbitrarily. Clearly, any point lying inside the quadrilateral $NWSE$ is not in CH(P), and hence can be pruned right away. See Fig. 1 (Left) for an illustration of the heuristic. The notations N, W, S, E correspond to the cardinal directions.

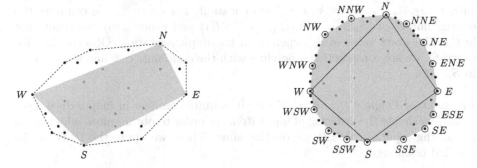

Fig. 1. Left: Points in the quadrilateral $NWSE$ (shown in gray) do not belong to the convex hull and hence can be pruned away. Convex hull of the point set is shown using the dashed polygon. Right: The 16-gon used for the *Akl-Toussaint+* heuristic using the sixteen cardinal points. Here, the 16-gon prunes more points than the quadrilateral $NWSE$.

The four cardinal points N, W, S, E belong to the convex hull of P. While the heuristic works effectively for most regular data sets, but may fail to prune sufficient number of points for massive data sets. The heuristic does not prune a point $p \notin$ CH(P) if it lies outside the quadrilateral. So, we extend it and propose the following *Akl-Toussaint+* heuristic. We also show in this section how to efficiently implement the heuristic in map-reduce.

Earlier, we were interested in the quadrilateral $NWSE$ for pruning points, but here, we extend this idea to a 16-gon. Refer to Fig. 1 (Right) for the sixteen cardinal points. These sixteen points belong to the convex hull of the given point set. Furthermore, observe that it is still possible that this heuristic like before, fails to eliminate any point if every point in P lies outside the aforesaid 16-gon. For instance, if P is convex, this heuristic will not prune away any point.

After the application of this heuristic, we obtain sixteen disjoint subsets having the form $P(\alpha, \beta)$ where $\alpha \neq \beta$ and $(\alpha, \beta) \in \{(E, ENE), (ENE,$

NE), (NE, NNE), (NNE, N), (N, NNW), (NNW, NW), (NW, WNW), (WNW, W), (W, WSW), (WSW, SW), (SW, SSW), (SSW, S), (S, SSE), (SSE, SE), (SE, ESE), $(ESE, E)\}$. For a pair (α, β), let γ denote a direction which is absent in these two where $\gamma \in \{N, W, S, E\}$. For instance, for (E, ENE), γ can be either W or S. We define $P(\alpha, \beta)$ as follows: $P(\alpha, \beta) := \{p \mid p \in P \wedge p \text{ lies to that half-plane of } \ell_{\alpha,\beta} \text{ which does not contain } \gamma\}$. This general definition of $P(\alpha, \beta)$ will be used later in our algorithms.

For certain point sets, there does not exist such sixteen distinct points. For example, the point set where $n - 3$ points are enclosed inside a triangle defined by three points.

Polyline Pruner – PLP. Now, we propose another heuristic which is applied independently for every $P(\alpha, \beta)$. In the following, we show how the pruning works for one set. Application of the heuristic to other sets work in a similar way. Consider the set $P(E, ENE)$. Partition it arbitrarily into $t := \lceil |P(E, ENE)|/k_2 \rceil$ disjoint subsets $S_1, \ldots, S_t$, where k_2 is a fixed constant. For every S_i, we compute the convex hull C_i of $\{S_i \cup (\alpha = E) \cup (\beta = ENE)\}$ and prune away the points not in C_i. The exact value of k_2 depends on the implementation. Observe that for every S_i, we are constructing a polyline with the endpoints α, β and some points in S_i.

Randomized Polyline Pruner – RPLP. It is quite common in map-reduce algorithms to shuffle the ordering of input items in order to obtain more information about the input. In RPLP, we do the same. Then, we apply the PLP on the shuffled point set.

Description of CH-MR. Using the heuristics defined above, we are ready to design the algorithm CH-MR; refer to Algorithm 2.

In round 1, we divide the input point set P into $t := \lceil n/k_1 \rceil$ parts $P_1, \ldots, P_t$, each of constant size k_1 and possibly $|P_t| < k_1$. For every P_i, we compute its convex hull C_i independently using an optimal sequential convex hull algorithm and eliminate the points which do not belong C_i. Clearly, if $p \in P_i$ and $p \notin C_i$, then $p \notin \text{CH}(P)$.

Next, we apply the Akl-Touissant$^+$ heuristic on the point set obtained after round 1. This gives us sixteen disjoint subsets (possibly empty) of the form $P(\alpha, \beta)$, as explained next. Using a constant number of rounds and restricting the input size to any reducer or mapper to $o(n)$, this can be achieved. Note that $\sqrt{n} = o(n)$.

Refer to Algorithm 3. Here, it is shown how to obtain the extreme points N, W, S, E in round A. From Map A, for each of N, W, S, E, $\sqrt{n}$ local extremes are emitted which goes as input to a single reducer. This reducer then computes the corresponding global extreme. Hence in round A, four reducers are used. In round B, we can easily separate out the points emitted from the round 1 into $P(E, N), P(N, W), P(W, S), P(S, E)$. This technique of finding global extremes can be applied repeatedly to obtain the sixteen required extremes and the corresponding sixteen subsets of P.

Algorithm 2. CH-MR(P)

Map 1: input $\langle Q := \{p_{i1}, \ldots, p_{ik_1}\}; \emptyset \rangle$
1: $\mathcal{C} \leftarrow$ Convex-Hull(Q);
2: **for each** $x \in \mathcal{C}$ **do**
3:      emit $\langle x; 1 \rangle$;
4: **end for**

Reduce 1: input $\langle p; 1 \rangle$
1: emit $\langle p; \emptyset \rangle$;

Apply the Akl-Touissant$^+$ heuristic to obtain the sixteen disjoint subsets of the form $P(\alpha, \beta)$. On these sixteen sets, execute the following rounds:

Map 2: input $\langle Q := \{p_{i1}, \ldots, p_{ik_2}\} \subseteq P(\alpha, \beta); \emptyset \rangle$
1: $\mathcal{C} \leftarrow$ Convex-Hull($Q \cup \{\alpha, \beta\}$);
2: **for each** $x \in \mathcal{C} \setminus \{\alpha, \beta\}$ **do**
3:      emit $\langle x; 1 \rangle$;
4: **end for**

Reduce 2: input $\langle p; 1 \rangle$
1: emit $\langle p; \emptyset \rangle$;

Map 3: input $\langle p \in P'(\alpha, \beta); \emptyset \rangle$
1: $\tau \leftarrow$ Random($1, \sqrt{|P'(\alpha, \beta)|}$);
2: emit $\langle \tau; p \rangle$;

Reduce 3: input $\langle \text{key}; Q \rangle$
1: Divide Q into subsets $Q_1, Q_2, \ldots$ each of constant size k_2;
2: **for each** Q_i **do**
3:      $\mathcal{C} \leftarrow$ Convex-Hull($Q_i \cup \{\alpha, \beta\}$;
4:      **for each** $p \in \mathcal{C} \setminus \{\alpha, \beta\}$ **do**
5:          emit $\langle p; \emptyset \rangle$;
6:      **end for**
7: **end for**

Map 4: input $\langle p \in P''(\alpha, \beta); \emptyset \rangle$
1: **for each** $q \in P''(\alpha, \beta) \setminus p$ **do**
2:      **if** $p \in \Delta \alpha \beta q$ **then**
3:          **return**;
4:      **end if**
5: **end for**
6: emit $\langle p; 1 \rangle$;

Reduce 4: input $\langle p; 1 \rangle$
1: emit $\langle p; \emptyset \rangle$;

The final output of the algorithm includes the sixteen cardinal points computed using the Akl-Touissant$^+$ heuristic.

Algorithm 3. FIND-EXTREMES

Map A: input $\langle Q := \{p_{i1}, \ldots, p_{i\sqrt{n}}\}; \emptyset \rangle$ **Map** B: input $\langle p; \emptyset \rangle$
1: local$W \leftarrow$ FINDMIN-X(Q);
2: local$E \leftarrow$ FINDMAX-X(Q);
3: local$S \leftarrow$ FINDMIN-Y(Q);
4: local$N \leftarrow$ FINDMAX-Y(Q);
5: emit $\langle \$W; \text{local}W \rangle$;
6: emit $\langle \$E; \text{local}E \rangle$;
7: emit $\langle \$S; \text{local}S \rangle$;
8: emit $\langle \$N; \text{local}N \rangle$;

1: **if** $p \in P(E, N)$ **then**
2:     emit $\langle [p, \$EN]; \emptyset \rangle$;
3: **else if** $p \in P(N, W)$ **then**
4:     emit $\langle [p, \$NW]; \emptyset \rangle$;
5: **else if** $p \in P(W, S)$ **then**
6:     emit $\langle [p, \$WS]; \emptyset \rangle$;
7: **else**
8:     emit $\langle [p, \$SE]; \emptyset \rangle$;
9: **end if**

Reduce A: input $\langle k \in \{N, E, W, S\}; \mathcal{V} \rangle$ **Reduce** B: input $\langle [p, x]; \emptyset \rangle$
1: **if** $k = \$W$ **then**
2:     global$W \leftarrow$ FINDMIN-X$(\mathcal{V})$;
3: **else if** $k = \$E$ **then**
4:     global$E \leftarrow$ FINDMAX-X$(\mathcal{V})$;
5: **else if** $k = \$S$ **then**
6:     global$S \leftarrow$ FINDMIN-Y$(\mathcal{V})$;
7: **else**
8:     global$N \leftarrow$ FINDMAX-Y$(\mathcal{V})$;
9: **end if**

1: **if** $x = \$EN$ **then**
2:     add p to global$P(E, N)$;
3: **else if** $x = \$NW$ **then**
4:     add p to global$P(N, W)$;
5: **else if** $k = \$WS$ **then**
6:     add p to global$P(W, S)$;
7: **else**
8:     add p to global$P(S, E)$;
9: **end if**

Now, for the sixteen subsets, we execute the following three rounds. Round 2 implements the PLP heuristic and acts somewhat similar to round 1 but with a difference; in this round for each partition we compute its convex hull by the inclusion of α and β.

In round 3, we implement the RPLP heuristic. We denote the output from round 2 using $P'(\alpha, \beta)$. Here, we use $\sqrt{|P'(\alpha, \beta)|}$ reducers in order to shuffle $P'(\alpha, \beta)$, and load balance the reducers simultaneously.

Next, we show in Lemma 1 that with high probability a reducer does not get overloaded using the following version of the well-known Chernoff bound. In our analysis, we say that a reducer R_i, $1 \leq i \leq \sqrt{|P'(\alpha, \beta)|}$ in the round 2 is *overloaded* if it receives at least $6\sqrt{|P'(\alpha, \beta)|}$ points.

Chernoff Bound. Let $X_1, \ldots, X_n$ be independent Bernoulli random variables, and $X = \sum_{i=1}^{n} X_i$, such that $\Pr(X_i = 1) = p$, then for $R \geq 6\mu$, $\Pr(X \geq R) \leq 2^{-R}$ where μ denotes the mean. Refer to [20, Chap. 4] for a proof of this bound.

Lemma 1. *Consider a particular $P'(\alpha, \beta)$ and round 3. Then, the probability that a reducer in round 3 gets overloaded is at most $2^{-6\sqrt{|P'(\alpha, \beta)|}}$. Moreover, the probability that there exists an overloaded reducer is $\leq \sqrt{|P'(\alpha, \beta)|} 2^{-6\sqrt{|P'(\alpha, \beta)|}}$.*

Proof. Fix a reducer R_i, $1 \leq i \leq \sqrt{|P'(\alpha, \beta)|}$. Define 0–1 random variables $X_1, \ldots, X_{|P'(\alpha, \beta)|}$ such that $X_j = 1$ if point $p_j \in P'(\alpha, \beta)$ is sent to R_i,

$X_j = 0$, otherwise. Let $X = \sum_{j=0}^{|P'(\alpha,\beta)|} X_j$. Clearly, $\Pr(p_j$ is received by $R_i)$ $= 1/\sqrt{|P'(\alpha,\beta)|}$. Also, in this case, $\mu = \sqrt{|P'(\alpha,\beta)|}$. By the application of the above version of Chernoff bound, the required probability can be concluded.

By the union bound, we conclude that the probability that there exists a overloaded reducer is at most $\sqrt{|P'(\alpha,\beta)|}2^{-6\sqrt{|P'(\alpha,\beta)|}}$. This completes the proof.

Remark. It can be checked that,

$$\lim_{|P'(\alpha,\beta)|\to\infty} \frac{\sqrt{|P'(\alpha,\beta)|}}{2^{6\sqrt{|P'(\alpha,\beta)|}}} = 0.$$

This implies that for large $|P'(\alpha,\beta)|$, it is very unlikely that a reducer will get overloaded.

Let us denote the output of round 3 by $P''(\alpha,\beta)$. In round 4, for every point p in $P''(\alpha,\beta)$, we check that if there exists a point $q \neq p \in P''(\alpha,\beta)$ such that $p \in \Delta\alpha\beta q$. If yes, we do not emit p, else we emit p since $p \in CH(P)$. This also shows that CH-MR computes the exact convex hull of P. As a part of the final output, the sixteen cardinal points are included.

CH-MR is Better than Algorithm 1. Every reducer and mapper in CH-MR runs in $\mathcal{O}(n)$ time. In each round, there are at most n mappers and n reducers. Thus, in each round, the total work done by the map or the reduce phase is $\mathcal{O}(n^2)$. Also, in every round at most n pairs are generated and the sum of length of the pairs is $\mathcal{O}(n)$. This shows that CH-MR is asymptotically better than the naive version since the later generates $\mathcal{O}(n^4)$ pairs after the map phase.

Now, we come to an important question. Is CH-MR a $\mathcal{MRC}$ algorithm? Our next theorem gives the answer.

Theorem 2. *If the size of every $P''(\alpha,\beta)$ is $o(n)$, CH-MR performs as an exact constant-round $\mathcal{MRC}$ algorithm with probability $\geq 1 - \sqrt{P'(\alpha,\beta)}2^{-6\sqrt{P'(\alpha,\beta)}}$.*

Proof. It can be checked easily from our previous discussion that CH-MR has constant number of rounds. Runtime of every mapper and reducer in CH-MR is polynomial in n. Also, as stated earlier, at each round at most n pairs are generated and the sum of length of those pairs is $\mathcal{O}(n)$.

If the size of every $P''(\alpha,\beta)$ is $o(n)$, it follows from our previous discussion that the mappers in every round always receive $o(n)$ amount of input. The same is true of the reducers except the ones in round 3. Now, it follows from Lemma 1 that the probability that no reducer in round 3 gets overloaded i.e. every reducer receives less than $6\sqrt{P'(\alpha,\beta)}$ points is at least $1 - \sqrt{P'(\alpha,\beta)}2^{-6\sqrt{P'(\alpha,\beta)}}$. To this end, observe that $6\sqrt{P'(\alpha,\beta)}$ is $o(P'(\alpha,\beta))$. This completes the proof.

Remark. It is important to note that

$$\lim_{|P'(\alpha,\beta)|\to\infty} 1 - \frac{\sqrt{|P'(\alpha,\beta)|}}{2^{6\sqrt{|P'(\alpha,\beta)|}}} = 1.$$

This implies for large data sets, when every $|P''(\alpha,\beta)| = o(n)$, CH-MR performs like a $\mathcal{MRC}$ algorithm with high probability (in fact, close to 1). However, if in the algorithm CH-MR we do not include the RPLP heuristic, then the algorithm is indeed a $\mathcal{MRC}$ algorithm for such a class of data sets. Hence, we state the following without a proof.

Corollary 1. *If the size of every $P''(\alpha,\beta)$ is $o(n)$, CH-MR without the RPLP heuristic, is an exact constant-round $\mathcal{MRC}$ algorithm.*

Comparison with Eldawy et al.'s Algorithm. In the following, we refer to their algorithm as E-MR. Now, we will observe the difference in asymptotic runtime between CH-MR and E-MR on random point sets. Assume that at most $m = o(n)$ (as advised by the authors of $\mathcal{MRC}$) mappers or reducers can run in parallel. Let t be the number of points remaining after the execution of the heuristics in both the algorithms. In the following, we focus on the main round where the hull points are computed.

It is shown in [22–24] that the expected number of vertices of the convex hull of n points, chosen uniformly and independently from a disk is $\mathcal{O}(n^{1/3})$. Consider the situation when this many points are sent to the main rounds of both the algorithms where the convex hull points are computed. When $m = t = \mathcal{O}(n^{1/3})$, the map phase of E-MR can complete in $\mathcal{O}(1)$ time and the single reducer in $\mathcal{O}(n^{1/3} \log n)$ time. Thus, the total runtime of the round amounts to $\mathcal{O}(n^{1/3} \log n)$. For CH-MR, it is $\mathcal{O}(n^{1/3})$ since the map phase completes in $\mathcal{O}(n^{1/3})$ time, and the reduce phase in $\mathcal{O}(1)$ time. In this case, CH-MR is faster by a factor of $\mathcal{O}(\log n)$.

It is also shown in [22–24] that the expected number of vertices of the convex hull of n points, chosen uniformly and independently from a polygon having a constant number of sides is $\mathcal{O}(\log n)$. Arguing similarly as above, when $m = t = \mathcal{O}(\log n)$, E-MR takes $\mathcal{O}(\log n \log \log n)$ time and CH-MR takes $\mathcal{O}(\log n)$. Once again, CH-MR is faster by a $\mathcal{O}(\log \log n)$ factor.

Thus, we can conclude that when sufficient number of mappers and reducers are allowed to run in parallel, CH-MR performs better than E-MR on random point sets. The load-balanced aspect of CH-MR gives us this advantage.

4   An Approximate Algorithm: APXCH-MR

In this section, we design an approximate map-reduce algorithm (named APXCH-MR) by adapting the sequential approximate algorithm designed by Soisalon-Soininen [25]. The definition of approximate convex hull is conveyed through Theorem 3. Our objective is to design a $\mathcal{MRC}$ algorithm which is load-balanced and at the same time runs fast for all point sets (including the ones where $h = \Theta(n)$). Note that CH-MR does not perform as a $\mathcal{MRC}$ algorithm for point sets where $h = \Theta(n)$. In the following, we present the original sequential algorithm by Soisalon-Soininen [25].

Description of Soisalon-Soininen's Algorithm. Given two integer constants k_1, k_2 as input, the following algorithm returns an approximate convex hull of P having size $\mathcal{O}(k_1 + k_2)$.

Calculate the points E, W and then divide the area between these two points into k_1 equal-sized vertical strips. For each of these strips, compute the point with largest y-coordinate and the point with smallest y-coordinate. In a strip, if there are multiple points with a minimum or maximum y-coordinate, then two points among these with the maximum and minimum x-coordinates, respectively, are considered. Thus, from this step we obtain at most $2k_1$ points. Denote this set by S_1.

Similarly, we calculate N, S of P and then divide the region between them into k_2 horizontal strips. For each of these strips, compute the point with smallest x-coordinate and the point with largest x-coordinate. Once again from this step, we obtain at most $2k_2$ points. Denote this set by S_2. Return the convex hull of $S = S_1 \cup S_2 \cup \{N, W, S, E\}$. Observe that $|S| = \mathcal{O}(k_1 + k_2)$.

Description of APXCH-MR. Refer to Algorithm 4. In the algorithm by p_x and p_y, we refer to the x-coordinate and y-coordinate of p, respectively. In the beginning, using round A of Algorithm 3, we find out the points N, W, S, E. In round 1, each mapper gets $\sqrt{n}$ points and for each point its vertical and horizontal strip numbers are calculated. The reducer sends all the points in a particular strip (vertical/horizontal) into a single file.

Next, in round 2, from each file obtained from round 1, we send at most $\sqrt{n}$ points to a mapper where depending on the strip alignment (vertical/horizontal), we find out the required two extreme points. A single reducer in this round accepts at most $\sqrt{n}$ candidate points for a particular strip and computes the final two extreme points for that strip.

Let P' be the output from round 2. Then the convex hull for the set $S = P' \cup \{N, W, S, E\}$ is found using the idea of Algorithm 2. But here, we consider the four subsets $P(E, N), P(N, W), P(W, S), P(S, E)$, instead of sixteen. Furthermore, we do not use the PLP and RPLP heuristics to reduce the number of rounds, although one may use them, if required.

Now, we are ready to present our theorem regarding APXCH-MR.

Theorem 3. *Given $k_1 = o(n), k_2 = o(n)$, APXCH-MR is a constant-round $\mathcal{MRC}$ algorithm which returns a set of points $C \subseteq P$, such that if $p \notin C$, then the distance between p and the approximate convex hull polygon is*

$$\rho := \frac{d(E, W)d(N, S)}{\sqrt{(d(E, W)k_1)^2 + (d(N, S)k_2)^2}},$$

where $d(p, q)$ denotes the Euclidean distance between p, q. Furthermore, if $k := \min(k_1, k_2)$ and $D := \max(d(E, W), d(N, S))$, then $\rho \leq D/\sqrt{2}k$.

Proof. The proof of the approximation can be found in [25]. Next, it can be checked that APXCH-MR is a constant-round $\mathcal{MRC}$ algorithm when both k_1 and k_2 are both $o(n)$.

Algorithm 4. APXCH-MR(P, k_1, k_2)

Find N, W, S, E using round A of Algorithm 3;

Map 1: input $\langle Q := \{p_{i1}, \ldots, p_{i\sqrt{n}}\} \subseteq P \setminus \{N, E, W, S\}; \emptyset \rangle$

1: **for each** $p \in Q$ **do**
2:     $v(p) \leftarrow \lfloor (p_x - W_x)k_1/(E_x - W_x) \rfloor$;
3:     $h(p) \leftarrow \lfloor (p_y - S_y)k_2/(N_y - S_y) \rfloor$;
4:     emit $\langle p; \$v(p) \rangle$;
5:     emit $\langle p; \$\$h(p) \rangle$;
6: **end for**

Reduce 1: input $\langle p; \mathcal{V} = \{\$v(p), \$\$h(p)\} \rangle$

1: emit $\langle \$v(p); p \rangle$; {Points in a vertical strip go to a single file}
2: emit $\langle \$\$h(p); p \rangle$; {Points in a horizontal strip go to a single file}

Map 2: input $\langle Q := \{p_{i1}, \ldots, p_{it}\}, t \leq \sqrt{n}; \emptyset \rangle$

1: **if** Q contains points from vertical strip v **then**
2:     $v_T \leftarrow$ FindMax-Y(Q);
3:     $v_B \leftarrow$ FindMin-Y(Q);
4:     emit $\langle [v, \$Y]; v_T \rangle$;
5:     emit $\langle [v, \$Y]; v_B \rangle$;
6: **else**
7:     $h_L \leftarrow$ FindMin-X(Q);
8:     $h_R \leftarrow$ FindMax-X(Q);
9:     emit $\langle [h, \$X]; h_L \rangle$;
10:     emit $\langle [h, \$X]; h_R \rangle$;
11: **end if**

Reduce 2: input $\langle k; \mathcal{V} \rangle$

1: **if** $\$Y \in k$ **then**
2:     $T \leftarrow$ FindMax-Y$(\mathcal{V})$;
3:     $B \leftarrow$ FindMin-Y$(\mathcal{V})$;
4:     emit $\langle T; \emptyset \rangle$;
5:     emit $\langle B; \emptyset \rangle$;
6: **else**
7:     $L \leftarrow$ FindMin-X$(\mathcal{V})$;
8:     $R \leftarrow$ FindMax-X$(\mathcal{V})$;
9:     emit $\langle L; \emptyset \rangle$;
10:     emit $\langle R; \emptyset \rangle$;
11: **end if**

Let P' denote the output of the above round. Find $P(E, N), P(N, W), P(W, S), P(S, E)$ using round B of Algorithm 3 based on P'. Using Algorithm 2 without the PLP and RPLP heuristics and considering the sets $P(E, N), P(N, W), P(W, S), P(S, E)$, compute the convex hull C of $\mathcal{S} = P' \cup \{N, W, S, E\}$;

APXCH-MR is Efficient. In Theorem 3, we have assumed generous upper bounds for k_1, k_2. In practice, k_1, k_2 are much smaller than n. If they are fixed constants, then computation of the approximate hull in the final round takes constant time. Thus, our algorithm is not only load-balanced but also fast for fixed k_1, k_2. In comparison, for point sets where $h = \Theta(n)$, the map phase of round 4 in CH-MR does $\mathcal{O}(n^2)$ amount of work, and the reduce phase does $\mathcal{O}(n)$. This heavy work is avoided in APXCH-MR when k_1, k_2 are small compared to n. Furthermore, APXCH-MR is a $\mathcal{MRC}$ algorithm when k_1, k_2 are sub-linear in n, irrespective of the geometry of P.

Acknowledgment. We are thankful to the anonymous reviewers for their insightful comments that helped us to improve the presentation of the paper.

References

1. Agarwal, P.K., Fox, K., Munagala, K., Nath, A.: Parallel algorithms for constructing range and nearest-neighbor searching data structures. In: Proceedings of the 35th ACM SIGMOD-SIGACT-SIGAI Symposium on Principles of Database Systems, pp. 429–440 (2016)
2. Aghamolaei, S., Baharifard, F., Ghodsi, M.: Geometric spanners in the MapReduce model. In: Wang, L., Zhu, D. (eds.) COCOON 2018. LNCS, vol. 10976, pp. 675–687. Springer, Cham (2018). https://doi.org/10.1007/978-3-319-94776-1_56
3. Akl, S.G., Toussaint, G.T.: Efficient convex hull algorithms for pattern recognition applications. In: Proceedings of the 4th International Joint Conference on Pattern Recognition, pp. 483–487 (1979)
4. Andrew, A.M.: Another efficient algorithm for convex hulls in two dimensions. Inform. Process. Lett. **9**(5), 216–219 (1979)
5. Andoni, A., Nikolov, A., Onak, K., Yaroslavtsev, G.: Parallel algorithms for geometric graph problems. In: Proceedings of the 46th Annual ACM Symposium on Theory of Computing, pp. 574–583 (2014)
6. Bykat, A.: Convex hull of a finite set of points in two dimensions. Inform. Process. Lett. **7**(6), 296–298 (1978)
7. Chan, T.M.: Optimal output-sensitive convex hull algorithms in two and three dimensions. Discrete Comput. Geom. **16**(4), 361–368 (1996)
8. Dean, J., Ghemawat, S.: MapReduce: simplified data processing on large clusters. Commun. ACM **51**(1), 107–113 (2008)
9. Eddy, W.F.: A new convex hull algorithm for planar sets. ACM Trans. Math. Softw. **3**(4), 398–403 (1977)
10. Eldawy, A., Li, Y., Mokbel, M.F., Janardan, R.: CG_Hadoop: computational geometry in MapReduce. In: Proceedings of the 21st ACM SIGSPATIAL International Conference on Advances in Geographic Information Systems, pp. 294–303 (2013)
11. Ene, A., Im, S., Moseley, B.: Fast clustering using MapReduce. In: Proceedings of the 17th ACM SIGKDD International Conference on Knowledge Discovery and Data Mining, pp. 681–689 (2011)
12. Finocchi, I., Finocchi, M., Fusco, E.G.: Clique counting in MapReduce: algorithms and experiments. ACM J. Exp. Algorithmics **20**, 1–7 (2015)
13. Goodrich, M.T., Sitchinava, N., Zhang, Q.: Sorting, searching, and simulation in the MapReduce framework. In: Asano, T., Nakano, S., Okamoto, Y., Watanabe, O. (eds.) ISAAC 2011. LNCS, vol. 7074, pp. 374–383. Springer, Heidelberg (2011). https://doi.org/10.1007/978-3-642-25591-5_39
14. Graham, R.L.: An efficient algorithm for determining the convex hull of a finite planar set. Inform. Process. Lett. **1**(4), 132–133 (1972)
15. hadoop.apache.org

16. Jarvis, R.A.: On the identification of the convex hull of a finite set of points in the plane. Inform. Process. Lett. **2**(1), 18–21 (1973)
17. Karloff, H., Suri, S., Vassilvitskii, S.: A model of computation for MapReduce. In: Proceedings of the Twenty-First Annual ACM-SIAM Symposium on Discrete Algorithms, pp. 938–948. Society for Industrial and Applied Mathematics (2010)
18. Kirkpatrick, D.G., Seidel, R.: The ultimate planar convex hull algorithm? SIAM J. Comput. **15**(1), 18–21 (1986)
19. Lattanzi, S., Moseley, B., Suri, S., Vassilvitskii, S.: Filtering: a method for solving graph problems in MapReduce. In: Proceedings of the Twenty-Third Annual ACM Symposium on Parallelism in Algorithms and Architectures, pp. 85–94 (2011)
20. Mitzenmacher, M., Upfal, E.: Probability and Computing: Randomization and Probabilistic Techniques in Algorithms and Data Analysis. Cambridge University Press, Cambridge (2017)
21. Preparata, F.P., Hong, S.J.: Convex hulls of finite sets of points in two and three dimensions. Commun. ACM **20**(2), 87–93 (1977)
22. Preparata, F.P., Shamos, M.I.: Computational Geometry: An Introduction. Springer, New York (1985). https://doi.org/10.1007/978-1-4612-1098-6
23. Raynaud, H.: Sur l'enveloppe convex des nuages de points aleatoires dans R^n. J. Appl. Probab. **7**, 35–48 (1970)
24. Rényi, A., Sulanke, R.: Über die konvexe Hülle von n zufällig gerwähten Punkten I. Z. Wahrsch. Verw. Gebiete **2**, 75–84 (1963)
25. Soisalon-Soininen, E.: On computing approximate convex hulls. Inform. Process. Lett. **16**(3), 121–126 (1983). A correction to this paper appeared in Inform. Process. Lett. **22**, 55–56 (1986). coauthored by Ivan Stojmenović
26. Suri, S., Vassilvitskii, S.: Counting triangles and the curse of the last reducer. In: Proceedings of the 20th International Conference on World Wide Web, pp. 607–614 (2011)
27. Yaroslavtsev, G., Vadapalli, A.: Massively parallel algorithms and hardness for single-linkage clustering under ℓ_p-distances, arXiv:1710.01431 (2017)

Transmitting Particles in a Polygonal Domain by Repulsion

Amirhossein Mozafari$^{(\boxtimes)}$ and Thomas C. Shermer

School of Computing Science, Simon Fraser University, Burnaby, BC, Canada
{amozafar,shermer}@sfu.ca

Abstract. In this paper, we introduce the problem of transmitting particles to a target point by the effect of a *repulsion actuator (RA)*. In this problem, we are given a polygonal domain P and a target point t inside it. Also, there is a particle at each point of P. The question is which particles can get to the target point t by activating a RA in P. We present the first polynomial time algorithm to solve this problem.

Keywords: Geometric algorithm · Polygonal domain
Repulsion actuator · Polynomial time algorithm

1 Introduction

Studying the behaviour of a set of objects in a polygonal region when they interact each other is an important class of problems in computational geometry and industry such as sensor networks. A natural way that objects can interact each other is by attraction and repulsion forces. The most investigated problems in this context concern the behaviour of objects when they interact each other under the attraction force. The beacon attraction problem [1,2,4] can be considered as an example of such problems. In this problem, we are given a polygonal region P full of point particles (there is a particle at each point in P) and a beacon point inside it. The beacon is an object that can attract particles. So, when we activate the beacon, particles start to move toward the beacon. In [4], Biro *et al.* gave an $O(n)$ time algorithm to find the attraction region of P with respect to a given beacon point. The attraction region is the subset of P that the particles in it get to the beacon after activating it. In [7], Kouhestani *et al.* investigated the inverse beacon attraction problem and gave an $O(n^3)$ algorithm to compute the inverse attraction region of a given point in a simple polygon. The inverse attraction region of a point p is subset of the polygon that if we activate a beacon in it, p gets to the beacon. In contrast to the problems raising from interaction of objects by attraction, problems raising from interaction of objects under the repulsion force have been rarely investigated. In [5], Bose and Shermer studied the effect of putting a repulsion actuator (RA) in a convex polygon full of point particles. They gave an $O(n^2)$ algorithm to compute all RA locations that can gather particles into a point. They also gave an $O(n)$ time algorithms

© Springer Nature Switzerland AG 2018
D. Kim et al. (Eds.): COCOA 2018, LNCS 11346, pp. 495–508, 2018.
https://doi.org/10.1007/978-3-030-04651-4_33

to determine whether such a location exists for a given convex polygon. In this paper, we consider another problem regarding the behaviour of objects under the repulsion force called *transmitting particles to a target point by activating a repulsion actuator*. In this problem, we have a polygonal domain (polygons with holes inside it) P and a target point t in the interior of P. Suppose that initially we have a particle at each point of P. When we activate a repulsion actuator (RA) at a point r inside P, all particles move away from it until either stop at a corner of P or they hit the target point t. Precisely, each particle traverses a path such that at each time it goes in a direction that takes itself farthest from r while it remains inside the polygon. Figure 1 shows the behaviour of two particles when we activate a RA at point r in P.

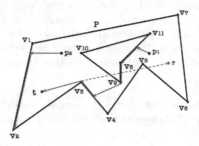

Fig. 1. The behavior of two particles p_1 and p_2 when we activate a RA at r in a polygonal domain P. In this example p_1 can get to the target point t while p_2 stops at the vertex v_2.

A natural question here asks which particles can get to t by activating only one RA in P. Precisely, we say that a point $x \in P$ is *acceptable* if there exists a point $r_x \in P$ such that if we put a RA at r_x, the particle at x gets to the target point t. According to this definition the problem becomes computing all acceptable points of P with respect to a given target point t in its interior.

In Sect. 2, we give some basic definitions and essential properties. In Sect. 3, we present a polynomial time algorithm for the problem and in Sect. 4 we discuss the complexity of the algorithm.

2   Preliminaries and Definitions

Let x, y and r be three points inside a polygonal domain P. We say that r can send x to y if activating a RA at r can cause the particle at x get to the point y. Note that the path that the particle traverses to take itself farthest from r at each time, may not be unique: when the particle hits ∂P (boundary of P), it might get farther from r by moving clockwise or counter-clockwise around the component (a connected part of ∂P) that it has hit. So, r can send x to y if there exist such a path that ends up at y. Henceforth, instead of saying the particle at x traverses a path we simply say x traverses a path.

We say that r is a *repulsion point* for x if r can send x to the target point t. We denote the set of all repulsion points for x by $R(x)$. By this notation, the set $\mathcal{A}$ of all acceptable points of P can be written as follows:

$$\mathcal{A} = \{x \in P \mid R(x) \neq \emptyset\} \tag{1}$$

Let $V = \{v_1, \ldots, v_n\}$ be the set of vertices of P. In order to compute $\mathcal{A}$, we first compute a subset A' of $\mathcal{A}$ which is the set of all points of P that can get to t by a direct path (a line segment). For example, in Fig. 1, the vertex v_3 can get to t by a direct path. Then, we compute subsets $A_1, \ldots, A_n$ of $\mathcal{A}$ in which, A_i is the set of all points $x \in P$ for which there exist a point $r_x \in P$ such that r_x can send x to t by a path having $v_i t$ as its last segment. For example, in Fig. 1, p_1 belongs A_3 because v_3 is the last bend point of the path that p_1 traverses to move away from the RA in r and get to t. It is clear that:

$$\mathcal{A} = A' \cup A_1 \cup \cdots \cup A_n \tag{2}$$

An immediate observation is that if v_i is not visible from t then $A_i = \emptyset$. For a boundary points x and y we say that a path p from x to y has j *jumps* if $p \setminus \partial P$ has j components.

3   The Algorithm

Let $L = \{l_1, \ldots, l_n\}$ where l_i $(1 \leq i \leq n)$ is the half line starting from t and passing through v_i. We assume that there is no line that passes t and two other vertices (we can have this condition by slightly perturbing the vertices). So, $|L| = n$. In order to compute A', we first partition the plane into a set of cones having common tip at t obtained by radially adjacent half lines of L. Figure 2 shows an example of such cones:

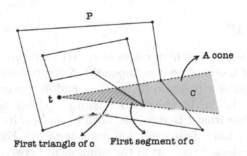

Fig. 2. Partitioning the plane by the set of cones according to L

Let CN be the set of all these cones. Each $c \in CN$ consists of a tip at the point t, two half lines as its boundaries and a set of line segments each of which has an endpoint on each of the boundary half lines of c. Within a cone, these segments are internally disjoint so we can have an order on them according to their closeness to t. Similarly, they partition the cone into a sequence of regions starting with a triangle having t as its vertex and the first segment as its base. Lets call this triangle as the first triangle of the cone. It can be easily seen that A' is the union of all first triangles of the cones in CN that have more than one segment. So, in order to compute A', we only need to compute the subdivision of the plane induced by ∂P and the half lines in L using well-known algorithms [6,8] and then consider the cones with more than one segment.

It remains to compute all acceptable points that follow a path that bends in order to get t. Let $x \in A_i$ and so there must be a point $r_x \in R(x)$ that v_i is the last bend point of a path that r_x sends x to t along. Note that t lies on the interior of P and hence, r_x should lie on l_i otherwise, x can never reach t after leaving v_i. Also, note that we can compute each A_i individually and then consider their union to specify $A \setminus A'$. Henceforth, we fix the index i and assume that the vertex v_i is visible by t (otherwise $A_i = \emptyset$) and discuss how to compute A_i.

Let $\widehat{l_i} = l_i \cap P$. To compute A_i, we first consider a sequence of functions $T^0, \ldots, T^n$ on V such that for a vertex $v \in V$, $T^j(v)$ $(1 \leq j \leq n)$ is defined as the set of all points r on $\widehat{l_i}$ that can send v to v_i (and therefore from v_i to t) using at most j jumps.

Lemma 1: For a boundary point x and a point $r \in P$, by activating a RA at r, x can have at most n jumps before it stops.

This is because if x jumps from an edge of ∂P, it can never back to that edge again. According to the above lemma, T^n is a function that for a vertex v, returns all of its repulsion points on $\widehat{l_i}$. In order to build this sequence, we use a procedure called *expand*. This procedure gets T^{j-1} as its input and builds T^j. Running this procedure n times starting from T^0 gives us T^n. In the next sections, we discuss how to compute T_0 and the procedure *expand*.

3.1   Computing T^0

Lets e_1 and e_2 be the two neighbor edges of v_i. We can consider that e_1 and e_2 lie on a same side of l_i otherwise it is impossible for v_i to be the last bend point when we have a repulsion point on l_i. So, if e_1 and e_2 are on different sides of l_i, we have $A_i = \emptyset$. Let e_1 be the closer edge to t (closer in the sense that if we consider a half-line from t that passes both e_1 and e_2, its intersection point with e_1 is closer to t than its intersection point with e_2). For a point x on the component of v_i, we denote the part of the component between v_i and x starting from e_2 by $[v_i, x]$. We chose this direction because it is impossible for a repulsion point on $\widehat{l_i}$ to send a point to t along e_1. Figure 3 shows an example of $[v_i, x]$.

Let e be an edge of ∂P, then the interior of P should lie on one side of e. We call this side as the P-side of e. Also, the supporting line of e divides the plane into two half planes. Denote the half plane not in the P-side of e by H_e. For a point $x \in \partial P$, we define $J(x) \subseteq \widehat{l_i}$ as the set of points $r \in \widehat{l_i}$ such that when we

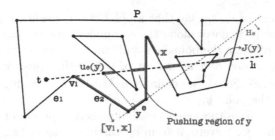

Fig. 3. $[v_i, x]$, $u_e(y)$ and the pushing region of y.

activate a RA at r, x immediately jumps off from the boundary. Figure 3 shows an example of $J(y)$ for a point y on the interior of an edge of ∂P. The following lemma shows the connection between $J(x)$ and the edge(s) containing it:

Lemma 2: For any point $x \in \partial P$, we have:

1. If x lies on the interior of an edge e we have $J(x) = H_e \cap \widehat{l_i}$.
2. If x is a reflex vertex of two edges e' and e'', we have $J(x) = (H_{e'} \cup H_{e''}) \cap \widehat{l_i}$.
3. If x is a convex vertex of two edges e' and e'', we have $J(x) = H_{e'} \cap H_{e''} \cap \widehat{l_i}$.

Lemma 2 says that if x is a reflex vertex, putting a RA at a point in $\widehat{l_i}$ makes x jump if and only if the RA makes it jump from one of the supporting lines of e' and e''. Similarly, if x is a convex vertex, x jumps if and only if a RA make it jump from both supporting lines of e' and e''. Figure 4 shows an example:

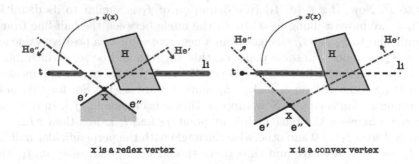

Fig. 4. When we activate a RA on $J(x)$, x jumps immediately into the interior of P.

Let $sp(l_i)$ be the supporting line of l_i (the line that contains l_i). For a point x on an edge e of ∂P, consider the half line from x perpendicular to e towards the P-side of e. If this half line intersects $sp(l_i)$, we denote this intersection point by $u_e(x)$ otherwise $u_e(x)$ is undefined. Note that if e does not intersect $sp(l_i)$, then u_e is defined for either all or none of the points of e. In the first case, we simply say that u_e is defined and in the second case we say u_e is undefined. For a point

x on the component of v_i, we say that a point $r \in \widehat{l_i}$ pushes x into $[v_i, x]$ if by activating a RA at r, x moves along the component toward the inside of $[v_i, x]$ (x does not jump off from the boundary and enters to the interior of $[v_i, x]$). We define the *pushing region* of x as the set of points that push x into $[v_i, x]$ and denote it by $Push(x)$. For v_i, we define $Push(v_i) = \widehat{l_i} \setminus tv_i$. Figure 3 shows the pushing region of the point y.

Let $(f_0, f_1, f_2, \dots)$ be the sequence of intersection points of the component of v_i and $sp(l_i)$ when we traverse it from v_i starting along e_2 (so, $f_0 = v_i$). These f_i points break each edge of the component of v_i passing through $sp(l_i)$ into two parts. So, for simplicity in computing T^0, we can consider each of these parts as a separate edge. By this modification, we have the sequence $(e_2, e_3, \dots, e_1)$ of edges of the component of v_i and the order on this sequence is the order as we traverse the component starting from e_2. Also, each edge lies in one side of $sp(l_i)$. Let x be a point of the component of v_i and e be the edge containing x (if x is a vertex consider the edge in $[v_i, x]$). Then, we have the following lemma:

Lemma 3: If u_e is undefined, then there exists a point $y \in [v_i, x]$ such that $Push(y) = \emptyset$.

Proof. If the angle between e_2 and l_i (the angle that doesn't contain e_1) is greater than $\pi/2$, it is trivial that the pushing region of any point on the interior of e_2 is empty. So, consider the case that the angle between e_2 and l_i is less than $\pi/2$. In this case for any point y on the interior of e_2, $u_{e_2}(y)$ intersects $sp(l_i)$. Let λ be the one third of the length of smallest edge in P. Now, consider the sequence P_k of shapes in which P_k is obtained from P by replacing each vertex v with a circular arc tangent to the incident edges of v such that the distance between the contact points of the arc to v is λ/k. Now, as k goes to infinity, P_k tends to P. Now, If $x \in [v_i, f_1]$ (the definition of f_1 is similar to its definition for P), as we move x along $[v_i, x]$ to v_i, the angle between the half line from x perpendicular to P_k and l_i, changes from a negative value to a positive value and because P_k is smooth, in some point x_0, this angle should be zero which means that the perpendicular line from x_0 is parallel to l_i. In this case it is impossible that putting some RA on l_i makes x_0 move toward $sp(l_i)$. So, for this point the pushing region is empty. Now suppose that x is outside $[v_i, f_1]$. In this case, if the angle between the tangent line at point f_1 and l_i is less than $\pi/2$, then obviously $Push(f_1) = \emptyset$ and otherwise the angle with the perpendicular half line from f_1 and l_i is negative and thus there should be a point $x_0 \in [v_i, f_1]$ that $Push(x_0) = \emptyset$. Because this fact is true for all P_k, this should also be true for P. $\square$

Let $e(f_k)$ be the incident edge of f_k on $[v_i, f_k]$ (consider $e(f_0)$ as e_1). So, $(e_2, e_3, \dots, e(f_1))$ is the sequence of edges in $[v_i, f_1]$. Orient the polygonal domain so that $\overrightarrow{v_i t}$ is directing leftward. For any edge $e_k \in (e_2, \dots, e(f_1))$, we denote its right vertex by $a(e_k)$ and its left vertex by $b(e_k)$. Then we have:

Lemma 4: u_{e_k} is defined if and only if traversing the boundary starting from any point in e_k in the direction $\overrightarrow{a(e_k)b(e_k)}$ goes to v_i via e_2.

Proof. We proceed by induction on k. Trivially the lemma is true for e_2 (this is because e_2 is that farther incident edge of v_i to t). Suppose that the above statements is true up to the edge e_{k-1}. Now, for the connection of e_{k-1} and e_k four cases may occur: $a(e_k) = a(e_{k-1})$, $a(e_k) = b(e_{k-1})$, $b(e_k) = a(e_{k-1})$ and $b(e_k) = b(e_{k-1})$. Also, each of $u_{e_{k-1}}$ and u_e may be defined or undefined which gives us 16 cases. But, because e_{k-1} and e_k are neighbour edges, it is impossible that $a(e_k) = a(e_{k-1})$ or $b(e_k) = b(e_{k-1})$ and both $u_{e_{k-1}}$ and u_e are defined or undefined. Similarly, it is impossible that $a(e_k) = b(e_{k-1})$ or $b(e_k) = a(e_{k-1})$ and one of e_k or e_{k-1} is defined and another isn't. So, eight cases left. Note that in these eight cases, four of them are exactly the mirror of others which exchanges left and right vertices. So, four cases left which is shown in Fig. 5:

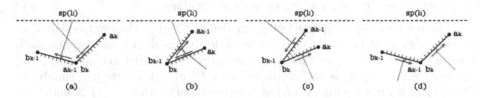

Fig. 5. Consistency of directions to v_i in e_{k-1} and e_k according to the Lemma 4.

In (a) of Fig. 5, both $u_{e_{k-1}}$ and u_{e_k} are defined and in (d) they are undefined. Also, in (b) and (c) u_{e_k} (resp. $u_{e_{k-1}}$) is defined (resp. undefined). As we can see, for all of these cases, the direction that the lemma gives for e_k is consistent with the direction the Lemma gives for e_{k-1} and so this direction should go toward v_i along e_2 which proves the lemma.   □

Corollary 1: We can extend the above lemma for each part $[f_m, f_{m+1}]$ of the component of v_i as follows:

1. If m is even, for $e_k \in (e(f_m) + 1, \ldots, e(f_{m+1}))$, u_{e_k} is defined if and only if the direction $\overrightarrow{a(e_k)b(e_k)}$ goes to v_i via e_2.
2. If m is odd, for $e_k \in (e(f_m)+1, \ldots, e(f_{m+1}))$, u_{e_k} is defined if and only if the direction $\overrightarrow{b(e_k)a(e_k)}$ goes to v_i via e_2.

The proof of the above corollary is obtained from the proof of Lemma 4 by simply replacing e_2 by $e(f_m) + 1$.

For a given point $z \in l_i$, we introduce notations $\dot{z}$ and $\ddot{z}$ as follows: z divides l_i into two parts each in one side of z. We denote the intersection of tz with P by $\dot{z}$, and the intersection of the side that does not contain t with P by $\ddot{z}$.

Corollary 2: Let x be a point of an edge $e \in [f_m, f_{m+1}]$ (if x is a vertex, consider the edge in $[v_i, x]$) such that u_e is defined. Then:

1. If m is even:
   (a) If $u_e(x) \in sp(l_i) \setminus l_i$ then $Push(x) = \widehat{l_i} \setminus J(x)$.

(b) If $u_e(x) \in l_i$, we have $Push(x) = \ddot{u}(x) \setminus J(x)$.
2. If m is odd:
   (a) If $u_e(x) \in sp(l_i) \setminus l_i$ then $Push(x) = \emptyset$.
   (b) If $u_e(x) \in l_i$, we have $Push(x) = \dot{u}(x) \setminus J(x)$.

Proof. Suppose that r is a point of $\widehat{l_i}$. Note that if $r \in J(x)$, then it can not push x into the interior of $[v_i, x]$ so suppose that $r \notin J(x)$. According to the Corollary 1, if e lies on the same side of e_2 with respect to $sp(l_i)$, in order that x goes to the interior of $[v_i, x]$, r should push x to the left. This can happen if and only if r lies on the right side of $u_e(x)$. Similarly, if e lies on the opposite side of e_2 with respect to $sp(l_i)$, r should push x to the right side to sent it into the interior of $[v_i, x]$ and this happens only if r lies on the left side $u_e(x)$. Note that if m is even (resp. odd), e must lie on the same side (resp. opposite side) of e_2 with respect to $sp(l_i)$ which proves the corollary. □

For a vertex v on the component of v_i, we define $T^0(v)$ as the intersection of all pushing regions of points in $[v_i, v]$. According to this definition and Lemma 3, if for a point $x \in [v_i, v]$ and its containing edge e (consider the edge in $[v_i, x]$ if x is a vertex), $u_e(x)$ is undefined, we have $T^0(v) = \emptyset$ so, we can simply set $Push(x) = \emptyset$ for all points x of the component of v_i that $u_e(x)$ is undefined and this assignment does not change the result of T^0 for the vertices of the component of v_i.

Observation 1: $T^0(v)$ is exactly the set of repulsion points of v that sends the particle at v along $[v_i, v]$ to the vertex v_i and then, make the particle jump from v_i to t.

Note that $T^0(v)$ is a set of intervals because it is an intersection of regions each of which consists of a set of disjoint intervals on l_i. According to the above definition, in order to obtain $T^0(v)$ for a given vertex v, we need to have the pushing region of infinitely many points but if we consider the inclusion relation on these pushing regions as a partial order on them, it would be enough to only consider the minimal pushing regions.

Observation 2: For a given vertex v, the minimal pushing regions of points in $[v_i, v]$ are among the following candidate regions:

$$Candidate\ regions\ for\ v\ =\ \{Push(v') \mid v'\ is\ a\ vertex\ in\ [v_i, v]\}$$

To see that why the above observation is correct, first note that for any edge e, $J(x)$ is same for all $x \in e$. So, for any point x on the interior of e, by slightly moving x on e we can get a pushing region not greater than the pushing region of x. So, in order to compute $T^0(v)$ for a given vertex v, we compute these candidate regions according to the Corollary 2 and Lemma 3 then intersect them to obtain $T^0(v)$. Because $T^0(v)$ is a set of disjoint intervals, we can represent it with a sequence of points with even length according to their closeness to t. In this representation, the first and the second elements of the sequence represent the first interval and similarly, the third and forth elements represent the second interval and so on. Also, because all points are on l_i, we can represent each point by its distance to t.

3.2   The Expand Procedure

To explain the procedure *expand*, we assume that we have computed $T^{j-1}(v)$ for all vertices in V and discuss how to compute $T^j(v)$ for a given vertex $v \in V$. Because each point can have at most n jumps when we put a RA in P, we can say $v \in A_i$ if and only if $T^n(v) \neq \emptyset$. For a fixed vertex $v \in V$, we have:

$$T^j(v) = \left(T^j(v) \cap J(v) \right) \cup \left(T^j(v) \cap \overline{J(v)} \right) \tag{3}$$

where $\overline{J(v)}$ is the complement of $J(v)$ with respect to $\widehat{l_i}$. In the above equation, lets call the first intersection by $N_1(v)$ and the second intersection by $N_2(v)$. We first compute $N_1(v)$ for all vertices and then compute $N_2(v)$ for each vertex in V using our information about $N_1(v)$ for the vertices of P. By computing the union of $N_1(v)$ and $N_2(v)$ for each vertex, we can obtain $T^j(v)$.

Computing $N_1(v)$. In order to compute $N_1(v)$, we need a map of $J(v)$ denoted by M_v^1. For each vertex $v' \in V$, there is a corresponding region in M_v^1 denoted by $M_v^1(v')$ such that for all $z \in M_v^1(v')$, z makes v jump off from the boundary and then sends it to v' as its first visiting vertex (v' is the first vertex that the v reaches after jumping). Note that some regions of M_v^1 might be empty. To construct this partition, we build the *visible triangle decomposition* of P with respect to v denoted by $VTD(v)$. This decomposition partitions the region of P visible from v by the set of line segments from v passing all vertices visible from v. Figure 6 shows an example of $VTD(v)$.

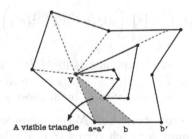

Fig. 6. Visible triangle decomposition of P according to v.

Let vab be a triangle in $VTD(v)$ with base edge $e = ab$ and sides va and vb. Note that a and b may not be vertices of P. Also, let c_{vab} be the opposite cone of the triangle (the cone with vertex v and half line sides along va and vb from v in the directions of $\overrightarrow{av}$ and $\overrightarrow{bv}$ respectively). Also, let $\widetilde{vab}$ be the intersection of c_{vab} with $\widehat{l_i}$. $\widetilde{vab}$ becomes empty if there is no such intersection. In this case, no point on $\widehat{l_i}$ that can make v jump into the triangle and thus, we don't consider this triangle in computation of $N_1(v)$. So, we assume that $\widetilde{vab} \neq \emptyset$. The property of $\widetilde{vab}$ is that any point in this region makes v jump into triangle vab. If we denote the vertices of P next to a and b by a' and b' respectively, we can find the partition of $\widetilde{vab}$ into subsets $\widetilde{vab}_a$ and $\widetilde{vab}_b$ such that the points in $\widetilde{vab}_a$ send v to a' and the points in $\widetilde{vab}_b$ sends v to b' (it is possible that one of these part becomes empty). Figure 7 shows this configuration:

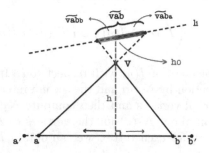

Fig. 7. Obtaining $\widetilde{vab}_a$ and $\widetilde{vab}_b$.

In order to compute $\widetilde{vab}_a$ and $\widetilde{vab}_b$, consider the line h perpendicular to the supporting line of e passing through v. The intersection point h_0 of h and $sp(l_i)$, divides $sp(l_i)$ into two parts. The intersection of these parts with $\widetilde{vab}$ becomes $\widetilde{vab}_a$ and $\widetilde{vab}_b$. In fact, points in the b-side (resp. a-side) of h in $\widetilde{vab}$, sends v to a' (resp. b'). We apply the above method to all triangles in $VTD(v)$ and put all regions on $J(v)$ that send v to v' in $M_v^1(v')$. Having the map M_v^1, we have:

$$N_1(v) = \bigcup_{v' \in V} \left(M_v^1(v') \cap T^{j-1}(v') \right) \tag{4}$$

This is because if for a vertex v', a point r is in $M_v^1(v') \cap T^{j-1}(v')$, r sends v to v' and because r is also in $T^{j-1}(v')$, r can send it from v' to v_i using at most $j-1$ jumps. So, in total r can send v to v_i by at most j jumps.

Computing $N_2(v)$. In order to obtain $N_2(v)$, again we need a map of $\overline{J(v)}$ denoted by M_v^2. In this map, for each vertex $v' \in V$, there is a corresponding region in M_v^2 denoted by $M_v^2(v')$ which is the subset of $\overline{J(v)}$ such that each $z \in M_v^2(v')$ sends v to v' without jumping (v' is not necessarily the first vertex that v reaches). According to this definition, the regions of M_v^2 may overlap each other and some regions may become a subset of another. Instead of directly computing M_v^2, we compute two maps M_v^{21} and M_v^{22} separately such that each $z \in M_v^{21}(v')$ (resp. $z \in M_v^{22}(v')$) sends v to v' without jumping on the clockwise (resp. counter clockwise) path on the component of v. It is clear that:

$$M_v^2(v') = M_v^{21}(v') \cup M_v^{22}(v') \tag{5}$$

We describe how to compute M_v^{21} and computing M_v^{22} is similar. We assume that $M_v^{21}(v) = \overline{J(v)}$. It is trivial that if v' is not in the component of v we have

$M_v^{21}(v') = \emptyset$. Let $e = ab$ be an edge of the component of v. The perpendicular line on the supporting line of e passing from a divides the plane into two half planes. Denote the half plane doesn't include b by H_e^a. Note that any $r \in H_e^a \cap \overline{J(a)}$ sends a to b without jumping. So, we have:

Lemma 5: If $M_v^{21}(v') \neq \emptyset$ and v'' is the neighbour of v' not in the clockwise path vv' on the component. Then we have:

$$M_v^{21}(v'') = M_v^{21}(v') \cap H_{v'v''}^{v'} \cap \overline{J(v')} \tag{6}$$

According to above lemma, we can start from v and traverse the component of v clockwise and build the regions of M_v^{21} (note that there must be a vertex v' on the component with $M_v^{21}(v') = \emptyset$). After computing M_v^2, we can construct $N_2(v)$ as follows:

$$N_2(v) = \bigcup_{v' \in V} \left(M_v^2(v') \cap N_1(v') \right) \tag{7}$$

Note that if $r \in M_v^2(v') \cap N_1(v')$ for a vertex $v' \in V$, r can send v to v' without jumping and because $r \in N_1(v')$, r can send it from v' to v_i using at most j jumps. This means that r can send v to v_i with at most j jumps.

3.3   Building A_i

After computing $N_2(v)$, we have $T^j(v) = N_1(v) \cup N_2(v)$ and we go for the next iteration until computing $T^n(v)$ for all vertices $v \in V$. We include all vertices with $T^n(v) \neq \emptyset$ in A_i. Now, a point $x \in P$ is in A_i if there exist $r \in \widehat{l_i}$ that sends x to a vertex v as its first visiting vertex and $r \in T^n(v)$. To obtain all points in A_i, we consider each pair (v, e) individually where v is a vertex in V and e is an incident edge of v and compute a set $A_i^{(v,e)}$ which is the subset of A_i that can be sent to v_i by reaching v as their first vertex via e. So, we have:

$$A_i = \bigcup_{\text{All pairs } (v,e)} A_i^{(v,e)} \tag{8}$$

Here, suppose that a pair (v, e) is given and we discuss how to compute $A_i^{(v,e)}$. Let $I_1, \ldots, I_q$ be the set of intervals of $T^n(v)$. We denote by $A_i^v(k)$ as the set of all points of P that can be sent to v as their first visiting vertex via e by some point in I_k $(1 \le k \le q)$. So,

$$A_i^{(v,e)} = \bigcup_{k \in \{1, \ldots, q\}} A_i^{(v,e)}(k) \tag{9}$$

Again we just need to compute each $A_i^{(v,e)}(k)$ independently. Let $I_k = [r_1^k, r_2^k]$ where r_1^k and r_2^k are two endpoints of I_k. For two points $r \in [r_1^k, r_2^k]$ and y on e, the segment ry might have some intersections with ∂P and so, these intersection points divide ry into a set of segments. We call the segment incident to e as the first segment of ry and denote it by $FS(ry)$. Note that it is impossible for the points on $ry \setminus FS(ry)$ to reach v as their first visiting vertex. Also, let $p(r)$ be the intersection point of $sp(e)$ and the line perpendicular to $sp(e)$ passing from r. Now, $FS(ry) \subseteq A_i^{(v,e)}(k)$ if and only if $y \in e \cap vp(r)$. So, for a given point $r \in I_k$, we can compute the set of all such $FS(ry)$ as follows: we consider the set of lines passing through r and every vertex inside the triangle obtained by r and the endpoints of $e \cap vp(r)$. These lines and ∂P partition the triangle. The union of parts incident with e are exactly the set of all $FS(ry)$ with $y \in e \cap vp(r)$. Figure 8 shows such configuration:

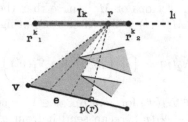

Fig. 8. An example of $Region(r)$.

Lets denote this union by $Region(r)$. So, we have:

$$A_i^{(v,e)}(k) = \bigcup_{r \in I_k} Region(r) \qquad (10)$$

In order to compute the union of infinitely many regions, let $(\alpha_0, \alpha_1, \alpha_2, \ldots, \alpha_{d_k})$ be the sequence of points on I_k such that $\alpha_0 = r_1^k, \alpha_{d_k} = r_2^k$ and for each $0 < w < d_k$, $\alpha_w p(\alpha_w)$ or $\alpha_w v$ passes a vertex of ∂P as we traverse I_k from r_1^k to r_2^k. Now, as r moves from α_w to α_{w+1}, the segments of the boundary of $Region(r)$ changes uniformly. So, to see that which points are covered by $Region(r)$ when r ranges in $[\alpha_w, \alpha_{w+1}]$, it is enough to check these segments at $r = \alpha_w$ and $r = \alpha_{w+1}$. So, $A_i^{(v,e)}(k)$ is computed by considering all intervals $[\alpha_w, \alpha_{w+1}]$.

4   Complexity of the Algorithm

The first part of the algorithm is computing A'. It costs $O(n \log n)$ to obtain the subdivision and build the cones. Because the total number of segments in each cone is linear, we can check in linear time if a cone has more than one segment and store its first triangle. Since there are a linear number of cones, computing A' costs $O(n^2)$. We can also compute the f_m sequence for each vertex v_i using this subdivision.

Computing A_i takes three independent steps: computing T^0, computing T^n using the *expand* procedure and building A_i having T^n. In computing T^0, first we use linear time (using the map we obtained to build cones) to check which vertices are visible from t and find the neighbor edges of the vertices that lie on the same side of the line connecting them to t. Computing the $J(v)$ for a vertex $v \in V$ takes linear time. Also, the intersection and union operations can be done linearly. In order to compute $T^0(v)$ for a given vertex v, we should compute $\dot{u}(v') \setminus J(v')$ or $\ddot{u}(v') \setminus J(v')$ for $O(n)$ vertices which costs $O(n^2)$. So, building T^0 costs $O(n^3)$.

In the procedure *expand*, we need to compute $T^j(v)$ for all $v \in V$ having T^{j-1}. For a given $v \in V$, the maps M_v^1 and M_v^2 are independent of j and so, we can build them once and use them whenever they are needed in the *expand* procedure. In order to compute these maps, we spend $O(n \log n)$ time to build $VTD(v)$. Next, we have $O(n)$ triangles and it take constant time for each triangle to obtain $\widetilde{vab}_a$ and $\widetilde{vab}_b$. So, building M_v^1 costs $O(n \log n)$. Computing each of M_v^{21} and M_v^{22} costs $O(n^2)$ because we need to traverse the component of v and in each step, we should compute an intersection. So, M_v^2 can be computed in $O(n^2)$ and thus, building these maps for all vertices costs $O(n^3)$.

Note that each $T^0(v)$ has at most $O(n)$ endpoints and thus we have at most $O(n^2)$ endpoints in the intervals of $T^0(v)$ for all $v \in V$. On the other hand, the regions of both M_v^1 and M_v^2 have at most $O(n)$ endpoints and so, we have at most $O(n^2)$ endpoints for all maps. Now, because we don't introduce any new endpoint in the expand procedure, $T^n(v)$ should have at most $O(n^2)$ endpoints. For a fixed vertex v, $T^{j-1}(v) \subseteq T^j(v)$. So, in the expand procedure, we can compute the Eqs. (4) and (7) for $T^{j-1}(v') \setminus T^{j-2}(v')$ instead of $T^{j-1}(v')$ and add the results to the $N_1(v)$ and $N_2(v)$ in the previous iteration to obtain new $N_1(v)$ and $N_2(v)$. So, by this modification in obtaining $N_1(v)$ and $N_2(v)$, computing $T^1(v), \ldots, T^n(v)$ costs $O(n^2)$ and because we have n vertices, computing T^n takes $O(n^3)$.

In order to build A_i, for each pair (v, e), we have at most $O(n)$ α_w points in total (for all I_k). We need to spend $O(n \log n)$ time to have these points sorted on each I_k. Now, for each interval $[\alpha_w, \alpha_{w+1}]$, in a constant time we can obtain which points are covered by $Region(r)$ for some r in this interval. Because we have at most $O(n)$ pairs (v, e), Building A_i having T^n costs $O(n^2 \log n)$. So, in total building A_i costs $O(n^3 + n^3 + n^2 \log n) = O(n^3)$ and because i varies between 1 and n, the total complexity of the algorithm is $O(n^4)$.

5   Conclusion

In this paper, we studied the problem of transmitting particles in a polygonal domain to a target point by activating a repulsion actuator and presented an algorithm with $O(n^4)$ time complexity to determine which points of the polygon can be sent to the target point by activating only one repulsion actuator. A natural question here asks is it possible to improve this running time to solve the problem? Another interesting problem is, given an integer $k \geq 1$, to determine which points of the polygonal domain can get to the target point by activating at most k repulsion actuators. One can think of activating repulsion actuators one after the other, or all at the same time.

References

1. Biro, M.: Beacon-based routing and guarding. Ph.D. dissertation. State University of New York at Stony Brook (2013)
2. Biro, M., Gao, J., Iwerks, J., Kostitsyna, I., Mitchell, J.S.B.: Beacon-based routing and coverage. In: 21st Fall Workshop on Computational Geometry, FWCG 2011 (2011)
3. Biro, M., Gao, J., Iwerks, J., Kostitsyna, I., Mitchell,J.S.B.: Combinatorics of beacon-based routing and coverage. In: Proceedings of the 25th Canadian Conference on Computational Geometry, CCCG 2013, vol. 1, p. 3 (2013)
4. Biro, M., Iwerks, J., Kostitsyna, I., Mitchell, J.S.B.: Beacon-based algorithms for geometric routing. In: Dehne, F., Solis-Oba, R., Sack, J.-R. (eds.) WADS 2013. LNCS, vol. 8037, pp. 158–169. Springer, Heidelberg (2013). https://doi.org/10.1007/978-3-642-40104-6_14
5. Bose, P., Shermer, T.C.: Gathering by repulsion. In: LIPIcs-Leibniz International Proceedings in Informatics, vol. 101. Schloss Dagstuhl-Leibniz-Zentrum fuer Informatik (2018)
6. de Berg, M., van Kreveld, M., Overmars, M., Schwarzkopf, O.: Computational geometry. In: Computational Geometry, pp. 1–17. Springer, Heidelberg (1997). https://doi.org/10.1007/978-3-662-03427-9_1
7. Kouhestani, B., Rappaport, D., Salomaa, K.: On the inverse beacon attraction region of a point. In: Proceedings of the 27th Canadian Conference on Computational Geometry, CCCG 2015, pp. 205–212 (2015)
8. Toth, C.D., O'Rourke, J., Goodman, J.E.: Handbook of Discrete and Computational Geometry. Chapman and Hall/CRC, Boca Raton (2017)

Does a Robot Path Have Clearance C?

Ovidiu Daescu$^{(\boxtimes)}$ and Hemant Malik

University of Texas at Dallas, Richardson, TX 75080, USA
{daescu,malik}@utdallas.edu

Abstract. Most path planning problems among polygonal obstacles ask
to find a path that avoids the obstacles and is optimal with respect to
some measure or a combination of measures, for example an u-to-v shortest path of clearance at least c, where u and v are points in the free space
and c is a positive constant. In practical applications, such as emergency
interventions/evacuations and medical treatment planning, a number of
u-to-v paths are suggested by experts and the question is whether such
paths satisfy specific requirements, such as a given clearance from the
obstacles. We address the following path query problem: Given a set S
of m disjoint simple polygons in the plane, with a total of n vertices,
preprocess them so that for a query consisting of a positive constant
c and a simple polygonal path π with k vertices, from a point u to a
point v in free space, where k is much smaller than n, one can quickly
decide whether π has clearance at least c (that is, there is no polygonal
obstacle within distance c of π). To do so, we show how to solve the
following related problem: Given a set S of m simple polygons in $\Re^2$,
preprocess S into a data structure so that the polygon in S closest to a
query line segment s can be reported quickly. We present an $O(t \log n)$
time, $O(t)$ space preprocessing, $O((n/\sqrt{t}) \log^{7/2} n)$ query time solution
for this problem, for any $n^{1+\epsilon} \le t \le n^2$. For a path with k segments, this
results in $O((nk/\sqrt{t}) \log^{7/2} n)$ query time, which is a significant improvement over algorithms that can be derived from existing computational
geometry methods when k is small.

Keywords: Path query · Polygonal obstacles · Clearance
Proximity queries

1 Introduction

Path planning problems among polygonal obstacles in the plane usually ask to
find a path that avoids the obstacles and is optimal with respect to some measure
or a combination of measures, for example a shortest u-to-v path [12,13,18] or
an u-to-v shortest path of clearance at least c [19,20], where u and v are points
in the free space and c is a positive constant. In some practical applications
however, such as emergency interventions/evacuations and medical treatment
planning, a number of u-to-v (polygonal or circular arc) paths are suggested
by experts and the question is whether such paths satisfy specific requirements,

© Springer Nature Switzerland AG 2018
D. Kim et al. (Eds.): COCOA 2018, LNCS 11346, pp. 509–521, 2018.
https://doi.org/10.1007/978-3-030-04651-4_34

Fig. 1. California fire evacuation map, with a 4 feet clearance demand (drone acquired image). The shortest path (in blue) does not have good clearance. The proposed path (in red), should be checked (queried) for the desired clearance. (Color figure online)

such as a given clearance from the polygonal obstacles. This is illustrated in Fig. 1. In this paper we address the following path query problem:

Path-Obstacles Proximity Queries: Given a set S of m disjoint simple polygons in the plane, with a total of n vertices, preprocess it to quickly answer queries of the following type: for a positive constant c and a simple polygonal path π, from a point u to a point v in free space, decide whether π has clearance at least c, that is, there is no polygonal obstacle within distance c of π.

Somehow surprisingly, it seems this problem has not been addressed in computational geometry. To solve it, we show how to solve the following related problem:

Object-Obstacles Proximity Queries: Given a set S of m polygonal obstacles with a total of n vertices, preprocess S into a data structure so that the obstacle in S closest to a query object ρ can be reported quickly.

In this paper we consider the set S as a collection of disjoint simple polygons and the query object corresponds to a line segment (or line).

Once the segment-polygon proximity problem is solved, one can check for each of the segments of the given path π whether the segment has a clearance of c or not, and also report the *minimum clearance of the path*, defined as the minimum of the clearances of the line segments along the path.

1.1 Related Work

A simple, brute force solution to the path clearance problem would be to take each line segment along the path and find its distance (zero in case of intersections) to each of the line segments defining the boundary of the polygonal obstacles, which can be done in constant time per pair of segments. The clearance of the path would be reported as the minimum clearance over its line segments. For a path π with k line segments, among a set of polygonal obstacles with a total of n vertices, this leads to an $O(nk)$ time, $O(n + k)$ space solution that requires no preprocessing. This is linear in n and thus inefficient for a query type problem. It is good however to contrast this with results that can be extracted from using complex data structures, such as the Visibility-Voronoi Complex (VVC) [19]. The Visibility-Voronoi diagram for clearance c, $VV^{(c)}$, introduced in [19], encodes the visibility graph of the obstacles dilated with a disc of radius c and can be used to compute paths of clearance c and other desired properties between two points u and v by a search in this graph. The Visibility-Voronoi complex is a generalization of $VV^{(c)}$, that allows to find u-to-v paths for any given clearance value c without having to first construct the $VV^{(c)}$, by performing a Dijkstra like search on the graph encoding the VVC. $VV^{(c)}$ and the VVC require $O(n \log n + n_1)$ preprocessing time and can report an u-to-v path of clearance at least c in $O(n \log n + n_2)$ time, where n_1 is the number of visibility edges and n_2 is the number of edges of the diagram encountered during the search; both n_1 and n_2 are $O(n^2)$ in the worst case. However, neither $VV^{(c)}$ or VVC can be used directly to check whether a given path has clearance at least c, since the edges of the path are in general not encoded by the underlying graphs.

For finding a closest point to a query line, Cole and Yap [6] and Lee and Ching [9] reported a solution with preprocessing time and space in $O(n^2)$ and query time in $O(\log n)$. Mitra and Chaudhuri [14] presented an algorithm with $O(n \log n)$ preprocessing time, $O(n)$ space, and $O(n^{0.695})$ query time. Mukhopadhyay [16], used the simplicial partition technique of Matousek [10] to improve the query time to $O(n^{1/2+\epsilon}))$ for arbitrary $\epsilon > 0$, with $O(n^{1+\epsilon}))$ preprocessing time and $O(n \log n)$ space.

The problem of locating the nearest point to a query line segment among a set P of n points in the plane was addressed in [4]. If the query line segment is known to lie outside the convex hull of P, an $O(n)$ size data structure can be constructed in $O(n \log n)$ time, which can answer the nearest neighbor of a line segment in $O(\log n)$ time. If k non-intersecting line segments are given at a time, then the nearest neighbors of all these line segments can be reported in $O(k \log^3 n + n \log^2 n + k \log k)$ time using divide and conquer and the data structure for queries outside the convex hull. Later on, in [3], the time was reduced to $O(n \log^2 n)$ when $n = k$. Moreover, given n disjoint red segments and k disjoint blue segments in the plane, the algorithm in [3] can be used to find the closest pair of segments of a different color in $O((n + k) \log^2(n + k))$ time. Thus, with the red segments the edges of polygons in S and the blue segments the segments along the query path, the path-polygon proximity problem we study can be solved within $O((n + k) \log^2(n + k))$ time, without any preprocessing. Our goal is to obtain a query time that is sublinear in n, and thus more efficient for small values of k.

Goswami et al. [7] reported an algorithm for closest point to line segment queries with $O(\log^2 n)$ query time and $O(n^2)$ preprocessing time and space, based on simplex range searching. Segal and Zeitlin [17] provided an algorithm which takes $O(\log^2 n \log \log n)$ query time, using $O(n^2/\log n)$ space and $O(n^2)$ preprocessing time. However, these algorithms do not answer the segment-polygon proximity query problem as described here, as it is not enough to consider only the vertices of the polygons in S.

1.2   Results

For a set S of m disjoint simple polygonal obstacles in the plane, with a total of n vertices, the goal is to preprocess S so that given a positive constant c and an u-to-v polygonal path with k edges, where k is much smaller than n (i.e., $k = o(n)$) one can quickly answer whether the path has clearance at least c. We have the following results.

- We present an $O(t \log n)$ time, $O(t)$ space preprocessing, $O((n/\sqrt{t}) \log^{7/2} n)$ query time solution, for any $n^{1+\epsilon} \leq t \leq n^2$, to report the closest polygon in S to a query line segment.
- For a path π with k segments, we obtain $O((nk/\sqrt{t}) \log^{7/2} n)$ query time, with $O(t \log n)$ time, $O(t)$ space preprocessing, using segment-polygon proximity queries. When $t = n$ this gives $O(\sqrt{n}k \log^{7/2} n)$ query time with linear space and $O(n \log n)$ time preprocessing, improving over previous methods whenever k is small (i.e. $k = o(\sqrt{n})$). When $t = \Theta(n^2)$ it gives a query time of $O(k \log^{7/2} n)$, which is an $O(n)$ time improvement in query time over applying the solution derived from [3,4], when k is small. Moreover, unlike [3,4], our result is easily parallelizable, since queries with line segments along the query path are independent of each other. Thus, with k processors available, a query with a k segment path would take time proportional to the time to answer a line segment query. Assuming k is small, this can be easily implemented by multithreading (JAVA, C++) on modern laptop and desktop computers.

Our solutions differ from algorithms that could possibly be derived from existing visibility graph or Voronoi diagram based methods. They offer a preprocessing-query time trade-off and result in significant improvements when k is much smaller than n.

2   Line Segment Proximity Queries

In a **nearest neighbor query** problem a set S of n geometric objects in $\Re^d$, d a positive integer constant, is preprocessed into a data structure so that the object of S closest to a query object (point, line, line segment, etc.) can be reported quickly. In this section we address nearest-neighbor queries in the plane, where the input S corresponds to a set of disjoint simple polygons and the query object corresponds to a line segment. This is illustrated in Fig. 2.

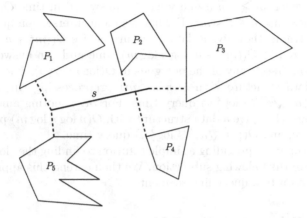

Fig. 2. A set S of polygons, to be preprocessed for closest polygon to a query line segment (or line).

Obviously, if the query object intersects a polygon in S then that polygon is a closest polygon and the closest distance from the query object to S is zero. Following this observation, a query can be divided into two parts, executed in this order:

1. **Emptiness Query:** Query if any polygon of S is intersected by the query object. If there is such polygon, then report it as the answer, with a distance of zero.
2. **Proximity Query:** (No polygon in S intersect the query object) Query for the closest polygon in S.

Thus, one can separately develop data structures for the two steps above, aiming for the best trade-offs on preprocessing-space-query on both structures.

Emptiness queries have been addressed in the context of ray shooting among polygonal obstacles in the plane. Chazelle et al. [5] gave an algorithm for ray shooting queries among m disjoint simple polygons with a total of n edges, with $O(n\sqrt{m} + m^{3/2}\log m + n\log n)$ preprocessing time, $O(\sqrt{m}\log n)$ query time, and $O(n)$ space. Obviously, ray shooting queries can be used to answer emptiness queries for both lines and line segments, within the same time and space bounds, by replacing each such query with two, respectively one, ray shooting queries.

Agarwal and Sharir [2] develop data structures for ray shooting queries by first building data structures for line and line segment intersection queries. They first address line intersection queries and show that a set of m simple polygons with a total of n vertices can be preprocessed in time $O((m^2 + n\log m)\log n)$

into a data structure of size $O(m^2 + n)$ so that an intersection between a query line and the polygons can be detected on $O(\log n)$ time. Alternately, they give a data structure with $O(n \log n)$ preprocessing time, $O(n)$ space, and $O(\lceil m/\sqrt{n} \rceil^{1+\epsilon} \log n)$ query time. When $m \leq \sqrt{n}$ the query time becomes $O(\log n)$ while when $m \geq \sqrt{n}$ a query can be answered in time $O(\lceil m/\sqrt{t} \rceil^{1+\epsilon})$ with space t such that $n \leq t \leq m^2$. For line segment intersection queries for disjoint simple polygons they give a data structure of size $O((m^2 + n) \log m)$, that can be constructed in $O((m^2 + n) \log n \log m)$ time and can answer whether a query segment intersects any of the polygons in $O(\log m \log n)$ time. Alternately, they gave a data structure with $O(n \log^2 m)$ preprocessing time, $O(n \log m)$ space, and $O(\lceil m/\sqrt{n} \rceil^{1+\epsilon} \log^2 n)$ query time. For ray shooting among pairwise disjoint polygons, they give a data structure with $O(n \log n \log m)$ preprocessing time, $O(n)$ space, and $O(\lceil m/\sqrt{n} \rceil^{1+\epsilon} \log^5 n)$ query time.

We first warm up by providing a simple solution to finding the closest polygon to a query line in the following subsection. We then extend this approach to find the closest polygon to a query line segment.

2.1   Closest Polygon to a Query Line

Given a set S of m disjoint simple polygons, with a total of n vertices, to find the closest polygon to a query line l we first perform an emptiness query with l, as described earlier. Using the result in [2], this can be done with $O(n \log n)$ preprocessing time, $O(n)$ space, and $O(\lceil m/\sqrt{n} \rceil^{1+\epsilon} \log n)$ query time.

Observation 1. *Given a simple polygon P and a line l such that l does not intersect P, the closest point of P from l is a vertex of P.*

Assume that none of the polygons in S intersect the query line l. Based on Observation 1 to find the closest polygon in S to the query line l reduces to finding the closest point to a query line problem, where points corresponds to the vertices of the polygons in S. We further preprocess S by computing the convex hull of each polygon in S and taking the vertices of the convex hulls as the set of points. This requires only an additional $O(n)$ time and storage. Thus, we have a set of $n' \leq n$ points, where n' could be much smaller than n in practice.

We can then use the results in [14,16] for closest point to line queries. Putting things together we obtain the following result.

Lemma 1. *A set S of m polygons, with a total of n vertices, can be preprocessed in $O(n^{1+\epsilon})$ time into a data structure of size $O(n \log n)$, that can report the closest polygon to a query line in $O(n^{(1/2)+\epsilon} + \lceil m/\sqrt{n} \rceil^{1+\epsilon} \log n)$ time, for arbitrary $\epsilon > 0$. Alternately, with $O(n \log n)$ time preprocessing one can construct a data structure of size $O(n)$ that can report the closest polygon to a query line in $O(n^{0.695})$ time.*

2.2 Closest Polygon to a Query Line Segment

Given a set S of m polygons, with a total of n vertices, to find the closest polygon to a query line segment s we first perform an emptiness query with s. To facilitate that, we preprocess S into a data structure for planar point location queries, which requires $O(n \log n)$ time and $O(n)$ space [8]. Given a segment s, we locate the endpoints of s in this data structure, in $O(\log n)$ query time; if any of the two endpoints is inside a polygon in S then we are done.

If both endpoints of s are in free space we proceed with a segment intersection query. As described earlier, this can be done with $O((m^2 + n) \log m)$ space, $O((m^2 + n) \log n \log m)$ preprocessing time, and $O(\log m \log n)$ query time or, alternately, with $O(\lceil m/\sqrt{n} \rceil^{1+\epsilon} \log^2 n)$ query time, $O(n \log^2 m)$ preprocessing time and $O(n \log m)$ space.

Assume that the line segment s does not intersect any polygon in S. Obviously, the closest distance from s to S is attained by a point on s and a point on a line segment on the boundary of some polygon in S.

Observation 2. *Consider a line segment e on the boundary of some polygon in S. The closest distance between s and e is either along:*

1. *A line perpendicular to s and passing through an endpoint of e, or*
2. *A line perpendicular to e and passing through an endpoint of s, or*
3. *A line joining an endpoint of s and an endpoint of e.*

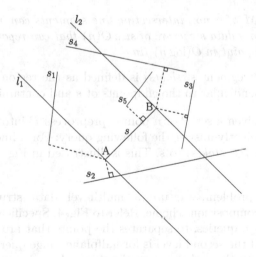

Fig. 3. Illustrating the closest distance between query line segment AB and some other line segments.

This is illustrated in Fig. 3 where AB is the query segment s and s_1, s_2, s_3, s_4 and s_5 are five line segments (see also [19]). Lines l_1 and l_2 are perpendicular to segment AB and passing through points A and B, respectively.

Thus, we can focus on the n line segments on the boundaries of the polygons in S. Given a set M of n non-intersecting line segments, we want to build a data structure so that, for a query segment s with endpoints A and B, the closest segment of M can be quickly determined. From Observation 2, it is clear that the minimum distance involves an end point of at least one line segment. Therefore we can decompose this problem into the following two subproblems:

1. Find the line segment of set M closest to point A or point B.
2. Let l_1 and l_2 be the lines perpendicular to s at its endpoints and consider the endpoints of the line segments in M which lie between l_1 and l_2. Find the closest such endpoint to s.

Notice that for our purpose we could relax the second subproblem, and ask instead for finding the closest endpoint of M to the query segment s.

Subproblem 1: Given a set M of n line segments, preprocess M so that we can efficiently find the closest line segment to a query point q.

To answer Subproblem 1 we construct the Voronoi diagram of the line segments in M and preprocess it for point location queries. Yap [21] provided an $O(n \log n)$ time algorithm to construct the Voronoi diagram of non intersecting line segments. After constructing the Voronoi diagram, preprocessing for point location takes $O(n \log n)$ time with $O(n)$ storage, and a point location query can be answered in $O(\log n)$ time [8].

Lemma 2. *A set M of n non-intersecting line segments can be preprocessed in $O(n \log n)$ time into a data structure of size $O(n)$ that can report the closest line segment to a query point in $O(\log n)$ time.*

For a given line segment s, *slab(s)* is defined as the region bounded by the lines l_1 and l_2 perpendicular to the endpoints of s and containing s (see [4]).

Subproblem 2: Given a set P of n points, preprocess P into a data structure so that one can efficiently answer the following query: For a line segment s, find the closest point in $P \cap slab(s)$ to s. This is illustrated in Fig. 4.

To solve this problem, we use a multilevel data structure based on Matousek's [11] decomposition scheme. Refer to Fig. 4. Specifically, the first level is for halfplane range queries, to separates the points that are on the side of l_1 that contains s, and the second level is for halfplane range queries on the resulting points to separate those that are on the side of l_2 and contain s. These two levels are used to isolate the points in $P \cap slab(s)$. The third level is for halfplane range queries bounded by the line supporting s, to isolate the subsets of $P \cap slab(s)$ that are on either side of s. The subsets on this level are further processed for closest point to line segment queries, when the query segment is outside the convex hull of the points, as in [4].

Using Theorem 6.1 from [11], we obtain the following trade-off.

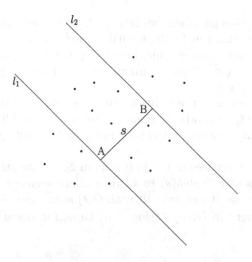

Fig. 4. Slab(s) and the points in $P \in slab(s)$.

Lemma 3. *A set P of n points in the plane can be preprocessed in $O(t \log n)$ time and $O(t)$ space into a data structure that can answer the query in Subproblem 2 in $O((n/\sqrt{t}) \log^{7/2} n)$ time, for any $n^{1+\epsilon} \leq t \leq n^2$ and $\epsilon > 0$.*

Summing up, we have the following result.

Lemma 4. *Given a set S of m disjoint simple polygons, it can be preprocessed in $O(t \log n)$ time into a data structure of size $O(t)$ that can report the closest polygon in S to a query line segment in $O((n/\sqrt{t}) \log^{7/2} n + \lceil m/\sqrt{n} \rceil^{1+\epsilon} \log^2 n)$ time, for any $n^{1+\epsilon} \leq t \leq n^2$ and $\epsilon > 0$.*

3   Closest Polygon to Path Queries

We now turn our attention to finding the closest polygon to a query path. Given a set S of m disjoint simple polygons in the plane, with a total of n vertices, we want to preprocess S into a data structure so that for a query consisting of a positive constant c and a simple polygonal path π with k vertices, from a point u to a point v in free space, one can quickly decide whether there is no polygonal obstacle within distance c of π.

Our solution actually works even if π has self intersections, however we do not see the practical aspect of such paths.

To solve the path-polygons proximity query problem we proceed as follows.

Preprocessing. We build the following data structures.

1. A point location data structure D_1 for the polygons is S. It can be built with $O(n)$ space and $O(n \log n)$ time, and can answer point location queries in $O(\log n)$ time.

2. A segment intersection data structure D_2 for the polygons of S. As described earlier, this can be done with $O((m^2 + n)\log m)$ space, $O((m^2 + n)\log n \log m)$ preprocessing time, and $O(\log m \log n)$ query time or, alternately, with $O(\lceil m/\sqrt{n}\rceil^{1+\epsilon}\log^2 n)$ query time, $O(n\log^2 m)$ preprocessing time and $O(n\log m)$ space.
3. The Voronoi diagram of the polygons in S, enhanced with a point location data structure, D_3. It can be built with $O(n)$ space and $O(n\log n)$ time, and can answer point location and closest polygon to point queries in $O(\log n)$ time.
4. With P the set of vertices of the polygons in S, a data structure D_4 to find the closest point of $P \cap slab(s)$ to s, for a query segment s, as described in the previous section. It can be built with $O(t)$ space and $O(t\log n)$ time, and can answer a query in $O((n/\sqrt{t})\log^{7/2} n)$ time, for any $n^{1+\epsilon} \le t \le n^2$ and $\epsilon > 0$.

Query. Given a simple polygonal path π with k vertices, to answer a query we proceed as follows.

1. Query D_1 with the vertices of π. If any such vertex is inside some polygon of S we stop and report it as the closest polygon to π, with a zero distance (or an *intersection* flag). Otherwise, all vertices of π are in free space and we proceed with the next step. This step takes $O(\log n)$ time per query and thus $O(k\log n)$ time overall.
2. Query D_2 with the line segments on π. If it is found that a line segment intersects a polygon in S then stop and report it as the closest polygon to π, with a zero distance (or an *intersection* flag).
3. Query $D3$ with the vertices of π and keep tract of the closest distance found. That distance gives the closest polygon of S to the vertices of π. This step takes $O(\log n)$ time per query and thus $O(k\log n)$ time overall.
4. Query $D4$ to find the closest obstacle vertex in $slab(s)$, for each segment s of π. This takes $O((n/\sqrt{t})\log^{7/2} n)$ time per query and $O((kn/\sqrt{t})\log^{7/2} n)$ time overall, for any $n^{1+\epsilon} \le t \le n^2$ and $\epsilon > 0$.

Theorem 1. *A set S of m disjoint simple polygons in the plane, with a total of n vertices, can be preprocessed in $O(t\log n)$ time into a data structure of size $O(t)$ so that given a query consisting of a positive value c and a simple polygonal path π with k vertices one can answer if π has clearance at least c in $O(k((n/\sqrt{t})\log^{7/2} n + \lceil m/\sqrt{n}\rceil^{1+\epsilon}\log^2 n))$ time, for any $n^{1+\epsilon} \le t \le n^2$ and $\epsilon > 0$.*

When $t = n^2$ and $m = o(\sqrt{n})$ the query time becomes $o(k\log^{7/2} n)$, which is a linear time faster than what can be obtained from previous algorithms [3,4] when k is much smaller than n. When $t = n^{1+\epsilon}$ the query time is $O(k(\sqrt{n}\log^{7/2} n + \lceil m/\sqrt{n}\rceil^{1+\epsilon}\log^2 n))$, which is asymptotically faster than what can be obtained from previous algorithms [3,4] when $k = o(\sqrt{n})$.

4 Conclusion and Extensions

In this paper, we studied the problem of finding the closest polygon of a set S of disjoint simple polygons to a query line segment or simple polygonal path. We proposed solutions that are significantly better in query time, when k is small relative to n, than what could be obtained from existing, non-query based approaches. Since queries with line segments along the query path are independent of each other, our result is easily parallelizable: with k processors available, a query with a k segment path would take time proportional to the time to answer a line segment query. When k is small, this can be easily implemented by multithreading (JAVA, C++) on modern laptop and desktop computers.

A possible extension of our work, that we leave as an open problem, is to query with paths that are not polygonal, but formed of, or including, circular arcs. This version has direct applications in minimally invasive surgery, for instruments formed of circular tubes [15]. This problem seems significantly harder, even when the clearance c (diameter of the tube) is known in advance. We sketch a possible approach here and underline the missing data structures needed to address this version.

It is easy to see that the minimum distance between a circular arc σ and a line segment s could be achieved by a point $p \in \sigma$ and a point $q \in s$ neither of which is an endpoint of σ or s.

For the general version, with clearance given at query time one would need data structures for the following two types of queries: (1) circular arc intersection queries for disjoint polygons and (2) circular arc proximity queries for disjoint polygons. So far, neither of these data structures have been described in the computational geometry literature. There are however data structures for ray shooting queries among circular arcs [1], so if k is comparable to n one could instead answer ray shooting queries against π at query time. Such method however seems inefficient.

Consider now the case when the clearance c is known at preprocessing time. As before, we have a set S of m disjoint simple polygons in the plane, with a total of n vertices. In addition, we also know the clearance c, given as a positive real value. A query consists of a path π with k circular arcs and asks whether π has clearance at least c. To solve it, one can proceed as follows.

Preprocessing. Build the following data structures.

1. Find the Minkowski sum of the obstacles in S with a disk of radius c and compute the union Γ of the resulting objects, which can be done with $O(n)$ space and $O(n \log^2 n)$ time [19]. The boundary of Γ consists of both line segments and circular arcs. Further process Γ for point location queries, for an additional $O(n)$ space and $O(n \log n)$ time. A point location query in the resulting data structure D_1 can be answered in $O(\log n)$ time.
2. Preporcess Γ for circular ray shooting queries: given a circular arc σ determine the first line segment or arc on the boundary of Γ hit by σ. Notice that the radius for the query circular arc is known in advance.

Thus, in this case, we are dealing only with a special case of circular ray shooting queries among disjoint line segments and circular arcs, where the radius of all circular arcs given at preprocessing time is the same. Still, we are not aware of any data structure that can efficiently handle such queries.

References

1. Agarwal, P.K., van Kreveld, M., Overmars, M.: Intersection queries for curved objects. In: Proceedings of the seventh annual symposium on Computational geometry, pp. 41–50. ACM (1991)
2. Agarwal, P.K., Sharir, M.: Ray shooting amidst convex polygons in 2D. J. Algorithms **21**(3), 508–519 (1996)
3. Bespamyatnikh, S.: Computing closest points for segments. Int. J. Comput. Geom. Appl. **13**(05), 419–438 (2003)
4. Bespamyatnikh, S., Snoeyink, J.: Queries with segments in Voronoi diagrams. In: Proceedings of the Tenth Annual ACM-SIAM Symposium on Discrete Algorithms, pp. 122–129. Society for Industrial and Applied Mathematics (1999)
5. Chazelle, B., et al.: Ray shooting in polygons using geodesic triangulations. Algorithmica **12**(1), 54–68 (1994)
6. Cole, R., Yap, C.K.: Geometric retrieval problems. In: 24th Annual Symposium on Foundations of Computer Science, pp. 112–121. IEEE (1983)
7. Goswami, P.P., Das, S., Nandy, S.C.: Triangular range counting query in 2D and its application in finding k nearest neighbors of a line segment. Comput. Geom. **29**(3), 163–175 (2004)
8. Kirkpatrick, D.: Optimal search in planar subdivisions. SIAM J. Comput. **12**(1), 28–35 (1983)
9. Lee, D., Ching, Y., et al.: The power of geometric duality revisited. Info. Process. Lett. **21**, 117–122 (1985)
10. Matoušek, J.: Efficient partition trees. Discret. Comput. Geom. **8**(3), 315–334 (1992)
11. Matoušek, J.: Range searching with efficient hierarchical cuttings. Discret. Comput. Geom. **10**(2), 157–182 (1993)
12. Mitchell, J.S.: Shortest paths among obstacles in the plane. Int. J. Comput. Geom. Appl. **6**(03), 309–332 (1996)
13. Mitchell, J.S.: Geometric shortest paths and network optimization. Handb. Comput. Geom. **334**, 633–702 (2000)
14. Mitra, P., Chaudhuri, B.: Efficiently computing the closest point to a query line1. Pattern Recognit. Lett. **19**(11), 1027–1035 (1998)
15. Morimoto, T.K., Cerrolaza, J.J., Hsieh, M.H., Cleary, K., Okamura, A.M., Linguraru, M.G.: Design of patient-specific concentric tube robots using path planning from 3-D ultrasound. In: 2017 39th Annual International Conference of the IEEE Engineering in Medicine and Biology Society (EMBC), pp. 165–168. IEEE (2017)
16. Mukhopadhyay, A.: Using simplicial partitions to determine a closest point to a query line. Pattern Recognit. Lett. **24**(12), 1915–1920 (2003)
17. Segal, M., Zeitlin, E.: Computing closest and farthest points for a query segment. Theor. Comput. Sci. **393**(1–3), 294–300 (2008)
18. Storer, J.A., Reif, J.H.: Shortest paths in the plane with polygonal obstacles. J. ACM (JACM) **41**(5), 982–1012 (1994)

19. Wein, R., Van Den Berg, J., Halperin, D.: Planning high-quality paths and corridors amidst obstacles. Int. J. Robot. Res. **27**(11–12), 1213–1231 (2008)
20. Weina, R., van den Bergb, J.P., Halperina, D.: The visibility-voronoi complex and its applications. Comput. Geom. **36**, 66–87 (2007)
21. Yap, C.K.: Ano (n logn) algorithm for the voronoi diagram of a set of simple curve segments. Discret. Comput. Geom. **2**(4), 365–393 (1987)

Star Routing: Between Vehicle Routing and Vertex Cover

Diego Delle Donne[1] and Guido Tagliavini[2(✉)]

[1] Instituto de Ciencias, Universidad Nacional de General Sarmiento, Malvinas
Argentinas, Argentina
ddelledo@ungs.edu.ar
[2] School of Arts and Sciences, Rutgers University, New Brunswick, NJ, USA
guido.tag@rutgers.edu

Abstract. We consider an optimization problem posed by an actual
newspaper company, which consists of computing a minimum length
route for a delivery truck, such that the driver only stops at street crossings, each time delivering copies to all customers adjacent to the crossing.
This can be modeled as an abstract problem that takes an unweighted
simple graph $G = (V, E)$ and a subset of edges X and asks for a shortest
cycle, not necessarily simple, such that every edge of X has an endpoint
in the cycle.

We show that the decision version of the problem is strongly NP-
complete, even if G is a grid graph. Regarding approximate solutions,
we show that the general case of the problem is APX-hard, and thus no
PTAS is possible unless P = NP. Despite the hardness of approximation,
we show that given any α-approximation algorithm for metric TSP, we
can build a 3α-approximation algorithm for our optimization problem,
yielding a concrete $9/2$-approximation algorithm.

The grid case is of particular importance, because it models a city
map or some part of it. A usual scenario is having some neighborhood full of customers, which translates as an instance of the abstract
problem where almost every edge of G is in X. We model this prop-
erty as $|E - X| = o(|E|)$, and for these instances we give a $(3/2 + \varepsilon)$-
approximation algorithm, for any $\varepsilon > 0$, provided that the grid is sufficiently big.

Keywords: Vehicle routing · Vertex cover
Approximation algorithms · Computational complexity

1 Introduction

Every morning, a well-known newspaper[1] in Buenos Aires needs to deliver a copy
to each subscriber by trucks. For now, assume there is only one truck. Tradi-
tionally, the truck stops in front of each customer's house, every time delivering

---

[1] Unfortunately, for confidentiality reasons, we cannot disclose their identity.

© Springer Nature Switzerland AG 2018
D. Kim et al. (Eds.): COCOA 2018, LNCS 11346, pp. 522–536, 2018.
https://doi.org/10.1007/978-3-030-04651-4_35

a single copy of the paper. But now, the company thinks there could be a better (that is, cheaper) way to do it: instead of stopping to make a single delivery, the truck will only stop at street crossings, and each time the driver will pick up a pile of copies and deliver them to all customers located on any of the (typically four) adjacent streets. The goal is to minimize the number of blocks traveled by the truck.

We model the city topology as a simple graph, and the set of customers as a subset of edges. In other words, we distinguish blocks that have at least one customer, but we don't care if there is more than one customer in a single block. If C is a cycle and X is a subset of edges of a simple graph, we say that C *covers* X if every edge of X has an endpoint in C. The formal description of the problem is the following:

STAR ROUTING

INSTANCE: A simple graph $G = (V, E)$, a non-empty subset of edges $X \subseteq E$, and a positive integer K.

QUESTION: Does G have a cycle, not necessarily simple, of length at most K that covers X?

Since all edges can be traversed in both directions, STAR ROUTING (or simply STAR) models all streets as two-way streets. Also, note that STAR doesn't ask about which (or how many) road crossings the truck should stop at and deliver during its journey.

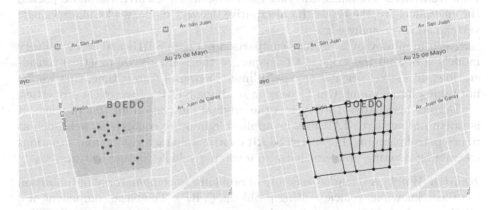

Fig. 1. (a) A possible set of customers, marked as red dots, on a small part of Boedo neighborhood in Buenos Aires. The light blue area is an arbitrary boundary for the truck. (b) A STAR instance that models the problem. Red edges are blocks that contain customers. (Color figure online)

Consider the example of Fig. 1a, which represents a real-life setting with subscribers shown as red dots. This is mapped to the STAR instance shown in Fig. 1b. Each block that contains at least one customer is mapped as a red edge,

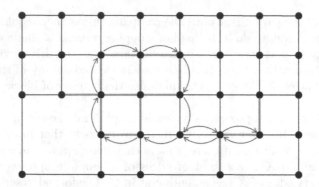

Fig. 2. Cleaned-up version of the STAR instance of Fig. 1b. The arrows show a feasible solution.

and X is the set of red edges. A feasible solution is presented in Fig. 2. Indeed, this cycle of length 12 given by the arrows is a feasible solution because every red edge has one endpoint in the cycle. In contrast, if we wanted to stop precisely at every customer's address, we would need to go through at least 16 edges: one per red edge, plus 4 more edges to move between the two connected components induced by the red edges. Thus, STAR's solution is at least 40% better, in terms of number of blocks traversed. This improvement may (or may not) be at the cost of greater overall time to perform the delivery, since now the driver has to walk from street crossings, carrying the newspapers. Clearly, the more packed the customers are, the better this alternative delivery model works, since a single vertex can cover many edges.

Keep in mind that a rigorous comparison between STAR and other delivery models is beyond the scope of this paper, as there are several practical considerations, like street orientation, speed limits or overall transit time, that we are not taking into account. Our focus is on studying STAR's theoretical properties.

Despite newspaper delivery was the original motivation for this problem, it is worth noting that STAR may be applicable in other contexts as well, such as police patrol planning. In general, STAR captures characteristics from situations resembling covering problems but also involving vehicle routing features.

Related Work. A remarkable family of problems in combinatorial optimization are those known as vehicle routing problems (VRP). The basic component of a VRP are vehicles that move throughout a network, maybe starting and ending at some depot point, and moving between customers located over the network to deliver some sort of merchandise. The goal is usually minimizing some metric related to the total consumed time or the traveled distance. The origin of these problems can be traced back to the 1954 paper of Dantzig, Fulkerson and Johnson [6], in which they considered the TSP, which is a particular case of VRP. This work was followed by several other papers about the TSP. Clarke and Wright [4] added more than one vehicle to the problem, which led to the first proper formulation of VRP, though that name was not coined until the work of Golden, Magnanti and Nguyan [11].

In 1974, Orloff [14] identified a class of routing problems of a single vehicle, which he called GENERAL ROUTING PROBLEM (GRP). The GRP takes a weighted graph $G = (V, E)$, and two sets $W \subseteq V$ and $F \subseteq E$, and asks to find a shortest cycle of G that traverses every vertex in W and every edge in F. This is a generalization of other well-known routing problems, like the CHINESE POSTMAN PROBLEM ($W = \emptyset$ and $F = E$), the RURAL POSTMAN PROBLEM ($W = \emptyset$), and the TSP (G complete, $W = V$ and $F = \emptyset$). Notably, the first can be solved in polynomial time, whereas the decision versions of the latter two are NP-complete [13].

The STAR problem is a simple VRP with a single vehicle fleet, where we want to minimize the delivery cost, which we model as the total distance traveled by the vehicle. However, in contrast with traditional VRPs, the subset X of edges containing customers can be covered just by visiting any of the two adjacent endpoints, rather than traveling along it. There are some variants of TSP with a similar flavor to that of STAR, in which the objective is to cover vertices with a more relaxed criteria than standard TSP. One of them is the COVERING SALESMAN PROBLEM (CSP) [5,16] that takes a directed weighted graph and a positive integer D, and asks to find a minimum-length tour over a subset of vertices of G such that every vertex not in the tour is within distance D of some vertex in the tour. Current and Schilling [5] devised a simple heuristic for this problem, but its performance guarantee cannot be bounded due to the arbitrary weights. Interestingly, our approximation algorithm for the general version of STAR is similar to theirs, but since we assume unit weights we are able to derive a bound on the approximation ratio. Shaelaie et al. [16] presented metaheuristics for the CSP but, once again, they do not provide any theoretical guarantees. Another related problem is the TSP WITH NEIGHBORHOODS (TSPN) [1,2,9], that takes a set of regions in the euclidean plane, and asks for a shortest closed curve that visits each region. We should note that the grid version of STAR, which we will discuss later on, can be reduced to the rectilinear version of TSPN, in which each edge from X is a region, but unfortunately the rectilinear TSPN has not been studied.

To the best of our knowledge, STAR hasn't been considered before, existing literature has little overlap with it, and it's the first VRP based on the notion of vertex cover.

Organization. In Sect. 2 we show that STAR is strongly NP-complete, even when the input graph is a grid. In Sect. 3 we study how well it is possible to approximate the general version of STAR. First, we give a lower bound by showing that STAR is APX-hard. Second, we provide a factory of approximation algorithms, which takes an α-approximation algorithm for metric TSP and produces a 3α-approximation algorithm for STAR. This yields a $9/2$ approximation factor for the general case, and a $3 + \varepsilon$ factor when the graph is planar. In Sect. 4 we develop a $(3/2 + \varepsilon)$-approximation algorithm for grid graphs, assuming there are asymptotically more edges with customers than not and that the grid is large enough. Finally, in Sect. 5 we state some open problems.

Notation. Let Π be an optimization problem. Let I be a valid input of Π. We write $\Pi^*(I)$ the value of an optimal solution of the problem Π for I.

If A is a finite set, $|A|$ is the cardinality of A. We denote $K(A)$ the complete graph whose set of vertices is A.

All graphs we consider in this paper are simple. All cycles and paths we consider are not necessarily simple. If G is a graph, $\tau(G)$ is the cardinality of any minimum vertex cover of G. If X is a subset of edges of G, $G[X]$ is the subgraph of G induced by X. If S is a path of G, $\ell(S)$ is the number of edges of S (counting repetitions). If the edges of G have weights given by a function w, $\ell_w(S)$ is the sum of the weights of the edges of S (counting repetitions). If T is another path of G, that starts where S ends, $S \circ T$ is the path we get from first traversing S and then T. If u and v are two vertices of G, $d_G(u, v)$ is the minimum $\ell(S)$ over every path S between u and v. If p and q are two points in $\mathbb{R}^2$, $d_1(p, q)$ is the Manhattan distance between them.

A grid graph with n rows and m columns is the cartesian product of graphs $P_n \square P_m$, where P_k is the path of k vertices. A star graph is a complete bipartite graph $K_{1,n}$, for some $n \geq 1$.

2   STAR is Hard, Even for Grids

In this section we show that STAR is NP-complete when we restrict G to the class of grid graphs. We call this version of the problem *grid* STAR. These instances are of practical interest, since grids are the most simple way of modelling a city layout. In particular, the problem is hard for planar graphs and for bipartite graphs, among all superclasses of grid graphs.

To prove completeness, we will reduce from the rectilinear variant of TSP. Recall the TSP takes a set of elements S equipped with weights between each pair of elements, and a positive integer L, and asks if there exists a hamiltonian cycle in $K(S)$ with total weight L or less. In the rectilinear version, the input is a set of points P in the plane, with positive integer coordinates, and a positive integer L, and asks if $K(P)$ has a hamiltonian cycle with total Manhattan distance length L or less.

The rectilinear TSP is NP-complete. In 1976, Garey et al. [10] proved this, by reducing from EXACT COVER BY 3-SETS (X3C), which takes a family $\mathcal{F} = \{F_1, \ldots, F_t\}$ of 3-element subsets of a set U of $3n$ elements, and asks if there exists a subfamily $\mathcal{F}' \subseteq \mathcal{F}$ of parwise disjoint subsets such that $\cup_{F \in \mathcal{F}'} = U$. Since X3C has no numerical arguments, it is strongly NP-complete. The rectilinear TSP instance they build is such that both coordinates of every point in the set P, as well as the optimization bound L, are bounded by a polynomial on the size of the X3C instance. Thus, rectilinear TSP is strongly NP-complete.

The transformation we will use has a similar flavor than the one devised by Demaine and Rudoy [7] to show that solving a certain puzzle is NP-complete.

Theorem 1. *Grid* STAR *is strongly* NP-complete.

Proof. Given a cycle of G it's easy to check in polynomial time if it covers all edges in X, and if it has length K or less. Thus, the problem is in NP.

Now we reduce from rectilinear TSP. Let $P = \{p_1, \ldots, p_n\}$ and the bound L be an instance of rectilinear TSP. Let m be the maximum coordinate of any point in P, so that all points lie in the rectangle $[1, m] \times [1, m]$. Let $c = 2(n+1)$. We will build a grid graph G by taking the $m \times m$ rectangular grid of points with lower left corner at $(1, 1)$, and expanding it by a factor of c. Formally, if $G = (V, E)$, then V is the set of all integer coordinates points in $[c, cm] \times [c, cm]$, and E is the natural set of edges we need to produce a grid out of V. Note that $cp_i \in V$ for every i. That is, multiplying by c we map points from P to V.

Let e_i be any adjacent edge to cp_i in G, and let $X = \{e_1, \ldots, e_n\}$. Finally, let $K = cL$.

Polynomial time. Since rectilinear TSP is strongly NP-complete, we can assume m and L are polynomial. The grid G has size $O((cm)^2)$, which is polynomial because both c and m are. The coordinates of every vertex are bounded by $O(cm)$. Computing X is obviously polynomial. Finally, computing $K = cL$ is also polynomial, since L is polynomial. Thus, the reduction takes polynomial time, and every numerical value is bounded by a polynomial in the transformation's input size.

Rectilinear TSP **to grid** STAR. Assume there is a hamiltonian cycle with Manhattan distance length L or less, in $K(P)$. W.l.o.g., suppose $T = \langle p_1, \ldots, p_n, p_{n+1} = p_1 \rangle$ is such a cycle. For each i, let S_i be a shortest path in G from cp_i to cp_{i+1}. Then $S = S_1 \circ \cdots \circ S_n$ is a cycle in G that goes through every vertex cp_i, and thus covers X. Its length is

$$\ell(S) = \sum_{i=1}^{n} \ell(S_i) = \sum_{i=1}^{n} d_1(cp_i, cp_{i+1}) = c \sum_{i=1}^{n} d_1(p_i, p_{i+1}) = c\ell_{d_1}(T) \leq cL = K$$

Grid STAR **to rectilinear** TSP. Suppose there is a cycle $S = \langle s_1, \ldots, s_m, s_{m+1} = s_1 \rangle$ of length K or less that covers X in G. At some point while we traverse S, we must get close to each cp_i, since the cycle covers e_i. More specifically, there exists an index $1 \leq j_i \leq m$ such that either s_{j_i} is exactly cp_i, or $e_i = (cp_i, s_{j_i})$. This implies that $d_1(cp_i, s_{j_i}) \leq 1$. Assume w.l.o.g. that $j_1 \leq \cdots \leq j_n$, since otherwise we can rearrange the indexes of the points cp_i. Consider the hamiltonian cycle $T = \langle p_1, \ldots, p_n, p_{n+1} = p_1 \rangle$ of $K(P)$. (Define, for convenience, $j_{n+1} := j_1$.) We need to show that $\ell_{d_1}(T) \leq L$. Since $\ell_{d_1}(T)$ is an integer, it suffices to prove $\ell_{d_1}(T) < L+1$. We start by rewriting

$$\ell_{d_1}(T) = \sum_{i=1}^{n} d_1(p_i, p_{i+1}) = (1/c) \sum_{i=1}^{n} d_1(cp_i, cp_{i+1})$$

Since d_1 is a metric, we can decompose

$$d_1(cp_i, cp_{i+1}) \leq d_1(cp_i, s_{j_i}) + d_1(s_{j_i}, s_{j_{i+1}}) + d_1(s_{j_{i+1}}, cp_{i+1}) \leq 2 + d_1(s_{j_i}, s_{j_{i+1}})$$

Therefore

$$\ell_{d_1}(T) \le (1/c) \left(2n + \sum_{i=1}^{n} d_1(s_{j_i}, s_{j_{i+1}}) \right)$$

Consider the subpaths $S_i := \langle s_{j_i}, s_{j_i+1}, \ldots, s_{j_{i+1}} \rangle$ (here we are using the fact that the indexes j_i are ordered). Then $d_1(s_{j_i}, s_{j_{i+1}}) \le \ell_{d_1}(S_i)$. Since these subpaths are disjoint pieces of S, we have $\sum_{i=1}^{n} \ell_{d_1}(S_i) \le \ell_{d_1}(S) = \ell_{d_G}(S)$, so

$$\ell_{d_1}(T) \le (1/c)\,(2n + \ell_{d_G}(S)) \le 2n/c + K/c = n/(n+1) + L < 1 + L$$

as desired.   □

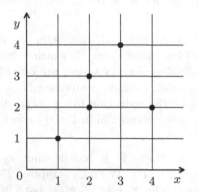
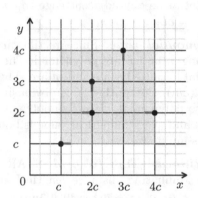

Fig. 3. Mapping an instance of rectilinear TSP (on the left) to grid STAR (on the right). The marked points on the first grid are p_is, which are mapped to the second grid as cp_i. The light blue area denotes the graph G. The red edges make up the set X. (Color figure online)

3   An Approximation Algorithm for the General Case

Since STAR is a hard problem in regards to finding exact solutions, we investigate approximation algorithms. We start by showing that the general version of the problem is hard to approximate within a constant factor arbitrarily close to 1. For this, we reduce from approximating the VERTEX COVER (VC) problem, which is known to be APX-hard [8]. Given a simple graph G, VC asks for a minimum cardinality vertex cover of G.

Theorem 2. *For every α-approximation algorithm for STAR there is an α-approximation algorithm for VC.*

Proof. Let A_{STAR} be an α-approximation algorithm for STAR. Given an input graph $G = (V, E)$, the approximation algorithm for VC proceeds as follows. If $E = \emptyset$, return an empty set. If G is a star graph, return the central vertex. Otherwise, every feasible vertex cover has two or more vertices. Consider the instance $I = (K(V), E)$ of STAR, that is, a complete graph where the set of customers are the edges of G. The algorithm computes $S = A_{\text{STAR}}(I)$ and outputs S as a set.

The algorithm is polynomial, since we can construct I in polynomial time. Note that every cycle in $K(V)$ that covers E induces a vertex cover of G, and therefore S is a feasible vertex cover of G. Reciprocally, every vertex cover of G induces a cycle in $K(V)$ that covers E (by fixing any order among the vertices in the cover), which implies that $\text{STAR}^*(I) \leq \tau(G)$. Since S is an α-approximation, we have $|S| \leq \ell(S) \leq \alpha \, \text{STAR}^*(I) \leq \alpha \, \tau(G)$. $\qquad\square$

Dinur and Safra showed that it's hard to approximate VC within a factor 1.3606 of optimal [8]. Thus, STAR is hard to approximate as well.

Corollary 3. *It's NP-hard to approximate STAR within a factor 1.3606 of optimal.*

Therefore, STAR doesn't admit a PTAS unless P = NP, and thus the best we can hope for is some constant-factor approximation algorithm. Indeed, we now show that STAR admits one.

During the rest of this paper, we denote (G, X) an instance of STAR, and write $\text{OPT} := \text{STAR}^*(G, X)$. Recall that $X \neq \emptyset$.

Lemma 4. *If $G[X]$ is not a star graph, then $\text{OPT} \geq \tau(G[X])$.*

Proof. Let S be an optimal solution of STAR. Since S covers X, we can extract a vertex cover C of $G[X]$ from the set of vertices of S. Since $G[X]$ is not a star, it's easy to see that S has two or more vertices, and thus $\ell(S) \geq |C|$. Hence, $\text{OPT} = \ell(S) \geq |C| \geq \tau(G[X])$. $\qquad\square$

From now on, we assume $G[X]$ is not a star. It's easy to both recognize a star graph and, in that case, return the optimal solution (the central vertex of the star) in polynomial time.

Lemma 5. *Let C be a vertex cover of $G[X]$. Starting from a feasible solution T of TSP for (C, d_G) we can build, in polynomial time in G, a feasible solution S of STAR for (G, X), such that $\ell(S) - \ell_{d_G}(T)$.*

Proof. Let $T = \langle t_1, \ldots, t_n, t_{n+1} = t_1 \rangle$. Let S_i be any shortest path between t_i and t_{i+1}, in G. Consider the path $S = S_1 \circ \cdots \circ S_n$ of G, which covers X, since it traverses every vertex in C. This path can be computed in polynomial time, since it's the union of a polynomial number of shortest paths of G. We have $\ell(S) = \sum_{i=1}^n \ell(S_i) = \sum_{i=1}^n d_G(t_i, t_{i+1}) = \ell_{d_G}(T)$. $\qquad\square$

Recall the classic 2-approximation for VC, shown in Algorithm 1. We will refer to it as the *approximation via matching*.

Algorithm 1. VC 2-approximation via matching

Input: A simple graph G.
1: Compute any maximal matching M of G. Let $M = \{(u_1, v_1), \ldots, (u_m, v_m)\}$.
2: **return** $\{u_1, \ldots, u_m, v_1, \ldots, v_m\}$

Theorem 6. *Let C be a vertex cover of $G[X]$, built with the approximation via matching. Then* $\mathrm{TSP}^*(C, d_G) \le 3\, \mathrm{OPT}$.

Proof. Let $C = \{u_1, \ldots, u_m, v_1, \ldots, v_m\}$, such that each $e_i := (u_i, v_i)$ is an edge of the maximal matching. Let $S = \langle s_1, \ldots, s_n, s_{n+1} = s_1 \rangle$ be an optimal solution of STAR for (G, X). The key observation is that since $e_i \in X$, and S covers X, at least one of u_i or v_i is in S. W.l.o.g., assume u_i is in S. Hence, for each u_i, there exists an index $1 \le j_i \le n$ such that $u_i = s_{j_i}$ (we define $j_{m+1} := j_1$). W.l.o.g., assume that $j_1 \le \cdots \le j_m$, since otherwise we can rearrange the elements of C to satisfy it. Given this ordering, consider $T = \langle u_1, v_1, u_2, v_2, \ldots, u_m, v_m, u_{m+1} = u_1 \rangle$, which is a feasible solution of TSP for (C, d_G). It suffices to show that $\ell_{d_G}(T) \le 3\, \mathrm{OPT}$. Figure 4 shows the sets and cycles defined so far.

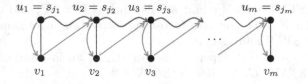

Fig. 4. Relation between C, S, and T. The curly blue arrows denote S, and the green arrows denote T. We do not show the edges that close the cycle. Also, S may contain v_is, but we don't illustrate this. (Color figure online)

We have that

$$\ell_{d_G}(T) = \sum_{i=1}^{m} (d_G(u_i, v_i) + d_G(v_i, u_{i+1})) = \sum_{i=1}^{m} (1 + d_G(v_i, u_{i+1}))$$

Since d_G is a metric, $d_G(v_i, u_{i+1}) \le d_G(v_i, u_i) + d_G(u_i, u_{i+1}) = 1 + d_G(u_i, u_{i+1})$. Hence,

$$\ell_{d_G}(T) \le \sum_{i=1}^{m} (2 + d_G(u_i, u_{i+1})) = 2m + \sum_{i=1}^{m} d_G(u_i, u_{i+1})$$

Recall that $u_i = s_{j_i}$ for each i. Consider the subpaths $S_i := \langle s_{j_i}, s_{j_i+1}, \ldots, s_{j_{i+1}} \rangle$. Then, $d_G(u_i, u_{i+1}) = d_G(s_{j_i}, s_{j_{i+1}}) \le \ell(S_i)$, and therefore

$$\ell_{d_G}(T) \le 2m + \sum_{i=1}^{m} \ell(S_i) \le 2m + \ell(S) = 2m + \mathrm{OPT}$$

Since C is a 2-approximation, $2m = |C| \le 2\, \tau(G[X])$. Finally, we use Lemma 4 to get $|C| \le 2\, \mathrm{OPT}$, and we arrive to the desired bound. $\qquad\square$

The proposed approximation algorithm for STAR is shown in Algorithm 2. Note that the instance (C, d_G) of TSP that A_{TSP} approximates is, indeed, a metric instance, because d_G is a metric.

Algorithm 2. Approximation algorithm for STAR

Input: An instance (G, X) of STAR.

1: Let A_{VC} be the approximation via matching algorithm. Let A_{TSP} be an approximation algorithm for metric TSP.
2: Compute $C = A_{\text{VC}}(G[X])$.
3: Compute $T = A_{\text{TSP}}(C, d_G)$.
4: Using T, build S as in Lemma 5.
5: **return** S

Theorem 7. *If A_{TSP} is an α-approximation algorithm for metric TSP, then Algorithm 2 is a 3α-approximation algorithm for STAR.*

Proof. The algorithm is polynomial, because each step is polynomial. The answer S is a feasible solution of STAR, as stated in Lemma 5. Regarding the performance guarantee,

$$
\begin{aligned}
\ell(S) &= \ell_{d_G}(T) & \text{(Lemma 5)} \\
&\leq \alpha \, \text{TSP}^*(C, d_G) & (A_{\text{TSP}} \text{ is an } \alpha\text{-approximation}) \\
&\leq 3\alpha \, \text{OPT} & \text{(Theorem 6)}
\end{aligned}
$$

□

Using Christofides' 3/2-approximation algorithm for metric TSP [3], we get the following concrete algorithm.

Corollary 8. *There is a 9/2-approximation algorithm for STAR.*

If G is restricted to some subclass of graphs, we could use a more specific approximation algorithm A_{TSP} (one that doesn't work for all metric instances), and get a better approximation guarantee. For example, if G is a planar graph (for instance, if G is a grid graph), then we can use a PTAS [12].

Corollary 0. *For every constant $\varepsilon > 0$, there is a $(3 + \varepsilon)$-approximation algorithm for planar instances of STAR.*

4   An Approximation Algorithm for Grids Full of Customers

A typical and desired case in the newspaper delivery business is having neighborhoods full of customers. We model such a dense neighborhood with a grid graph, where almost every edge is in X. In this section, we propose a method to approximate the optimal solution, tailored for this dense setting.

The key idea is that since almost every edge is in X, any feasible solution will cover almost every edge of E. What if instead of covering just X, we cover the whole set E? We show that if $|E - X| = o(|E|)$, then there is such a naïve tour that is guaranteed to have length at most a factor $3/2 + \varepsilon$ of the optimal, for sufficiently large grids.

A cycle that covers every edge in a graph is somewhat similar to the concept of *space-filling curve*. Mathematically, a space-filling curve is a curve whose range contains a certain 2-dimensional area, for example the unit square. Space-filling curves have been used before to compute tours for the TSP. In 1989, Platzman and Bartholdi [15] proved that if we visit the vertices in the order given by a specific space-filling curve, we get an $O(\log n)$-approximation algorithm. In the graph-theoretical setting of STAR, *filling* means to cover edges, but not necessarily to visit every vertex. Our dense-case approximation can be thought of as a *space-filling cycle*.

Before constructing this particular cycle we prove some auxiliary results that will help us to analyze its performance.

Lemma 10. *Let e be an edge of a graph G. Then $\tau(G) \leq \tau(G - e) + 1$.*

Proof. If we take any vertex cover of $G - e$ and add one of the endpoints of e (if not already in the vertex cover), we get a vertex cover of G.   □

In what follows, we will write $\overline{X} := E - X$.

Lemma 11. *Let (G, X) be an instance of STAR, such that $G[X]$ is not a star graph. Then $\tau(G) \leq \mathrm{OPT} + |\overline{X}|$.*

Proof. If we repeatedly apply the previous lemma, each time subtracting a new vertex of $\overline{X}$, we get $\tau(G) \leq \tau(G - \overline{X}) + |\overline{X}| = \tau(G[X]) + |\overline{X}|$. Using Lemma 4 we arrive to the desired inequality.   □

The proof plan is to construct a space-filling cycle, compare its length with $\tau(G)$ and then use Lemma 11 to bound its performance. It will come in handy to know the exact value of $\tau(G)$ when G is a grid graph.

Lemma 12. *Let G be a grid graph with n rows and m columns. Then $\tau(G) = \lfloor nm/2 \rfloor$.*

Proof. ($\leq$) Note that G is bipartite. Consider any bipartition of its vertices. Both subsets of the partition are vertex covers, and since there are nm vertices in total, one of them must have size at most $\lfloor nm/2 \rfloor$.

($\geq$) We use the fact that the size of any matching is always less than or equal to the size of any vertex cover. It suffices to exhibit a matching of size $\lfloor nm/2 \rfloor$. To build such a matching, we go over every other row, and for each one we take every other horizontal edge. If m is odd, we also take every other vertical edge of the last column. It's clear that this is a matching, and it's a matter of simple algebra to verify that it has $\lfloor nm/2 \rfloor$ edges.   □

We are ready to exhibit and analyze our space-filling cycle.

Theorem 13. *There is an approximation algorithm for grid STAR that computes solutions with length at most* $(3/2 + O(1/n + 1/m))(\text{OPT} + |\overline{X}|)$, *where* n *and* m *are the number of rows and columns, respectively, of the input grid graph.*

Proof. We introduce some terminology to describe the cycle. Enumerate the grid's rows from 1 to n, being 1 the uppermost row and n the lowest one. We divide the grid into horizontal *stripes*, such that the i-th stripe, $i \geq 1$, consists of rows $2i - 1$ and $2i$. If n is odd, the last stripe is formed only by the last row.

First we sketch a high-level description. Starting from the upper left corner, we will visit the stripes in order. Initially we move right, until we get to the right border of the grid, the end of the first stripe. Then we go down to the second stripe, and now move left until we get to the left border. Next we go down to the third stripe. The process continues until we finish visiting the last stripe. If the last one is a single row, we move in a straight line. Finally, we go back to the starting position.

More specifically, on stripe i, for some odd i, we move from left to right following a square wave pattern, which we call *period*. A period is a sequence of the following single-edge moves: down, right, right, up, right, right. This is illustrated in Fig. 5a. We repeat this sequence of moves until it's no longer possible, at the right border of the grid. At this point, we could be anywhere between the beginning and the end of a period. In any case, we stop, and move exactly two edges down. On stripe $i + 1$ we move in the opposite direction, from right to left, repeating the steps we did on stripe i, but in reverse order. When we get to the left border, we go down two edges again, and we are ready to repeat the process. When we reach the end of the grid, we close the cycle by adding a shortest path to the initial vertex. An example of this construction is shown in Fig. 5b.

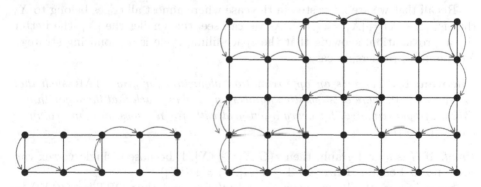

Fig. 5. (a) A period. (b) The tour for a 5×7 grid. The blue arrows show how each stripe is traversed. The gray arrows show how the cycle goes from one stripe to the next one. The path between the lower right corner and the upper left corner, that closes the cycle, is not drawn for clarity. (Color figure online)

Let C be this cycle. It's easy to see that C covers each edge of G, and that it can be computed in polynomial time. Our approximation algorithm simply outputs C. We now show that $\ell(C) \leq (3/2 + O(1/n + 1/m))\tau(G)$. By Lemma 11, this implies the desired bound.

Each of the $\lfloor n/2 \rfloor$ two-rows stripes contains $m - 1$ horizontal and $\lceil m/2 \rceil$ vertical edges of C. To move between two consecutive two-rows stripes, C uses exactly 2 edges. Additionally, if n is odd, the last stripe is a single row, and we account $m - 1$ edges for moving along that row, plus 2 edges to move from the previous stripe. Finally, we have at most $n + m - 2$ extra moves to go from the last stripe to the initial position. Summing everything,

$$\ell(C) \leq \lfloor n/2 \rfloor (m - 1 + \lceil m/2 \rceil) + (\lfloor n/2 \rfloor - 1)\, 2 + (m - 1 + 2) + (n + m - 2)$$

The first term accounts for intra-stripes moves, the second for inter-stripes moves, the third for a potential single-row stripe, and the last one for the cost to go back to the initial position. A sloppy bounding of the floor and ceiling functions yields

$$
\begin{aligned}
\ell(C) &\leq (n/2)(m - 1 + m/2 + 1) + 2(n/2 - 1) + (m + 1) + (n + m - 2)\\
&= (3/2)(nm/2) + 2n + 2m - 3\\
&\leq (3/2)\tau(G) + 2n + 2m - 9/4\\
&= (3/2 + O(1/m + 1/n))\tau(G)
\end{aligned}
$$

$\square$

Corollary 14. *There is an approximation algorithm for grid STAR such that for every $\varepsilon > 0$, there exist positive numbers n_ε and m_ε such that the algorithm computes solutions with length at most $(3/2 + \varepsilon)(\mathrm{OPT} + |\overline{X}|)$, for every input grid with $n \geq n_\varepsilon$ rows and $m \geq m_\varepsilon$ columns.*

Recall that we are interested in the case where almost all edges belong to X, that is, $|E - X| = |\overline{X}| = o(|E|)$. As we can see, the smaller the $|\overline{X}|$, the better the approximation, showing that the space-filling cycle is a promising strategy for the dense readership case.

Theorem 15. *There is an approximation algorithm for grid STAR such that for every $\varepsilon > 0$, there exist positive numbers n_ε and m_ε such that the algorithm is $(3/2 + \varepsilon)$-approximated, for every input grid with $n \geq n_\varepsilon$ rows, $m \geq m_\varepsilon$ columns and $|\overline{X}| = o(|E|)$.*

Proof. If G is a grid graph, then $\tau(G[X]) \geq |X|/4$, because a single vertex can cover up to 4 edges. Hence, $\mathrm{OPT} \geq \tau(G[X]) = \Omega(|X|)$.

Since $|\overline{X}| = o(|E|)$, we have $|X| = \Theta(|E|)$, and thus $\mathrm{OPT} = \Omega(|X|) = \Omega(|E|)$. This in turn implies that $|\overline{X}| = o(\mathrm{OPT})$, which means that for all $\varepsilon > 0$ there exist positive integers n_ε, m_ε such that $|\overline{X}| < \varepsilon\,\mathrm{OPT}$ for every $n \geq n_\varepsilon$ and $m \geq m_\varepsilon$.

Fix any $\varepsilon > 0$. Let $\varepsilon_1, \varepsilon_2 > 0$ be any two positive reals such that $\varepsilon_1 + (3/2)\varepsilon_2 + \varepsilon_1\varepsilon_2 \leq \varepsilon$. Instantiate Corollary 14 with ε_1, and let n_{ε_1} and m_{ε_1} be the minimum numbers of rows and columns, respectively. Let $n_{\varepsilon_2}, m_{\varepsilon_2}$ be such that if $n \geq n_{\varepsilon_2}$ and $m \geq m_{\varepsilon_2}$, then $|\overline{X}| < \varepsilon$ OPT.

Under these choices, if $n \geq n_\varepsilon = \max\{n_{\varepsilon_1}, n_{\varepsilon_2}\}$ and $m \geq m_\varepsilon = \max\{m_{\varepsilon_1}, m_{\varepsilon_2}\}$, the performance guarantee is

$$
\begin{aligned}
(3/2 + \varepsilon_1)(\text{OPT} + |\overline{X}|) &< (3/2 + \varepsilon_1)(\text{OPT} + \varepsilon_2\text{OPT}) \\
&\leq (3/2 + \varepsilon_1)(1 + \varepsilon_2)\text{OPT} \\
&= (3/2 + \varepsilon_1 + (3/2)\varepsilon_2 + \varepsilon_1\varepsilon_2)\text{OPT} \\
&\leq (3/2 + \varepsilon)\text{OPT}
\end{aligned}
$$

$\square$

5 Open Questions

In this paper we only considered the unweighted case of STAR. If the input graph has weights, the problem obviously remains hard, in terms of finding both exact and approximate solutions. Unfortunately, for that case, the approximation strategy we proposed in Theorem 7 is no longer useful, because if the vertex cover is agnostic of the weights, then the constructed cycle may be forced to use heavy edges, and therefore the output can be made arbitrarily longer than an optimal solution. Is it possible to adapt the algorithm for the weighted case, or to devise a different constant-factor approximation algorithm?

On a separate note, we showed that there cannot be a PTAS for STAR unless P = NP. However, this doesn't rule out the possibility of a PTAS for the grid case, for which the best we have achieved is a $(3/2 + \varepsilon)$-approximation algorithm that only works for a proper subset of instances. Since the grid case is of practical interest, it would be worthwhile to investigate this possibility.

Finally, the problem may be extended in natural ways, like using multiple trucks or considering the time it takes the driver to carry newspapers to the households.

Acknowledgements. Thanks to Martín Farach-Colton for useful discussions and suggestions about the presentation.

References

1. Arkin, E.M., Hassin, H.: Approximation algorithms for the geometric covering salesman problem. Discrete Appl. Math. **55**(3), 197–218 (1994). https://doi.org/10. 1016/0166-218X(94)90008-6. http://www.sciencedirect.com/science/article/pii/ 0166218X94900086
2. de Berg, M., Gudmundsson, J., Katz, M.J., Levcopoulos, C., Overmars, M.H., van der Stappen, A.F.: TSP with neighborhoods of varying size. J. Algorithms **57**(1), 22–36 (2005). https://doi.org/10.1016/j.jalgor.2005.01.010. http://www.sciencedirect.com/science/article/pii/S0196677405000246

3. Christofides, N.: Worst-case analysis of a new heuristic for the travelling salesman problem. Technical report 388, Graduate School of Industrial Administration, Carnegie Mellon University (1976)
4. Clarke, G., Wright, J.W.: Scheduling of vehicles from a central depot to a number of delivery points. Oper. Res. **12**(4), 568–581 (1964)
5. Current, J.R., Schilling, D.A.: The covering salesman problem. Transp. Sci. **23**(3), 208–213 (1989). http://www.jstor.org/stable/25768381
6. Dantzig, G., Fulkerson, R., Johnson, S.: Solution of a large-scale traveling-salesman problem. J. Oper. Res. Soc. America **2**(4), 393–410 (1954)
7. Demaine, E.D., Rudoy, M.: A simple proof that the $(n^2 - 1)$-puzzle is hard. Computing Research Repository abs/1707.03146 (2017)
8. Dinur, I., Safra, S.: On the hardness of approximating minimum vertex cover. Ann. Math. **162**, 2005 (2004)
9. Dumitrescu, A., Mitchell, J.S.B.: Approximation algorithms for tsp with neighborhoods in the plane. J. Algorithms **48**(1), 135–159 (2003). https://doi.org/10.1016/S0196-6774(03)00047-6. http://www.sciencedirect.com/science/article/pii/S0196677403000476. Twelfth Annual ACM-SIAM Symposium on Discrete Algorithms
10. Garey, M.R., Graham, R.L., Johnson, D.S.: Some NP-complete geometric problems. In: Proceedings of the Eighth Annual ACM Symposium on Theory of Computing, pp. 10–22. STOC 1976. ACM, New York, NY, USA (1976)
11. Golden, B.L., Magnanti, T.L., Nguyen, H.Q.: Implementing vehicle routing algorithms. Networks **7**(2), 113–148 (1977)
12. Grigni, M., Koutsoupias, E., Papadimitriou, C.: An approximation scheme for planar graph TSP. In: Proceedings of the 36th Annual Symposium on Foundations of Computer Science. p. 640. FOCS 1995. IEEE Computer Society, Washington, DC, USA (1995)
13. Lenstra, J.K., Kan, A.H.G.R.: On general routing problems. Networks **6**(3), 273–280 (1976)
14. Orloff, C.S.: A fundamental problem in vehicle routing. Networks **4**(1), 35–64 (1974)
15. Platzman, L.K., Bartholdi III, J.J.: Spacefilling curves and the planar travelling salesman problem. J. ACM **36**(4), 719–737 (1989)
16. Shaelaie, M.H., Salari, M., Naji-Azimi, Z.: The generalized covering traveling salesman problem. Appl. Soft Comput. **24**(c), 867–878 (2014). https://doi.org/10.1016/j.asoc.2014.08.057

Combinatorial Optimization and Data Structure

Effect of Crowd Composition on the Wisdom of Artificial Crowds Metaheuristic

Christopher J. Lowrance(✉)[iD], Dominic M. Larkin[iD], and Sang M. Yim[iD]

Department of Electrical Engineering and Computer Science, United States Military
Academy, West Point, NY 10996, USA
{christopher.lowrance,dominic.larkin,sang.yim}@usma.edu

Abstract. This paper investigates the impact that task difficulty and
crowd composition have on the success of the *Wisdom of Artificial
Crowds* metaheuristic. The metaheuristic, which is inspired by the *wisdom of crowds* phenomenon, combines the intelligence from a group of
optimization searches to form a new solution. Unfortunately, the aggregate formed by the metaheuristic is not always better than the best
individual solution within the crowd, and little is known about the variables which maximize the metaheuristic's success. Our study offers new
insights into the influential factors of artificial crowds and the collective intelligence of multiple optimization searches performed on the same
problem. The results show that favoring the opinions of experts (i.e., the
better searches) improves the chances of the metaheuristic succeeding by
more than 15% when compared to the traditional means of equal weighting. Furthermore, weighting expertise was found to require smaller crowd
sizes for the metaheuristic to reach its peak chances of success. Finally,
crowd size was discovered to be a critical factor, especially as problem
complexity grows or average crowd expertise declines. However, crowd
size matters only up to a point, after which the probability of success
plateaus.

Keywords: Wisdom of crowds · Combinatorial optimization
Collective intelligence · Metaheuristic optimization · Crowd factors

1 Introduction

The phrase *'wisdom of crowds'* refers to an observation popularized by
Surowiecki [13] who found that groups of people tend to be collectively smarter
than individuals, even experts. As a result, crowds are increasingly being used
to solve complex problems. For instance, crowds have been leveraged to produce
and test software, update Wikipedia content, make stock market predictions,
and generate new ideas for organizations [2,7,9,15].

Wisdom of Artificial Crowds (WoAC) is a metaheuristic that is related to
the original *wisdom of crowds* concept [19]. However, instead of applying to

© Springer Nature Switzerland AG 2018
D. Kim et al. (Eds.): COCOA 2018, LNCS 11346, pp. 539–551, 2018.
https://doi.org/10.1007/978-3-030-04651-4_36

people, WoAC extracts the collective intelligence from a group of stochastic and independent searches (e.g., multiple genetic algorithm searches) performed on the same optimization problem. Because searches on NP-hard problems are incomplete and indeterminate, multiple attempts on the same problem will tend to converge to different solutions [20]. Similar to the approach used in the *wisdom of crowds*, the metaheuristic considers popular choices within the crowd as wiser options which should be incorporated into the group's aggregate solution. The goal of WoAC is to combine the intelligence from the group of independent searches into a new solution that outperforms the best individual within the crowd.

In order to apply WoAC in the most effective manner, it is critical to understand the elements that interplay in its success. Therefore, in this paper, we control and analyze the effect of dominant variables including expertise, crowd size, and task difficulty on the metaheuristic's probability of success.

Crowds in our study consisted of multiple searches generated by a genetic algorithm on a combinatorial optimization problem. The optimization problem used in our study was the Traveling Salesman Problem (TSP) [19]. We selected the TSP and the genetic algorithm for our means of crowd generation and testing because of their familiarity within the research community. Our intent is to abstract the details associated with the TSP and genetic algorithm so that the paper can focus on the crowd variables that influence the success of the WoAC metaheuristic. It is worth emphasizing that our goal is not about how to best solve the TSP, nor trying to beat the best-known solutions to the TSP. The metaheuristic is applicable to other types of hard combinatorial optimization problems where searches are stochastic and incomplete due to complexity. Rather than focusing on the application of WoAC to a particular problem, we narrowed the scope of the paper to identifying and analyzing the factors that affect crowd success so that the metaheuristic can be more effectively applied in a range of combinatorial optimization problems.

2 Related Work

Wisdom of crowds (WoC) has been applied to a number of applications involving humans, and the results of several studies suggest that the input from experts should be weighted more strongly for best performance. For instance, Velic et al. applied WoC to stock market prediction and discovered that less knowledgeable contributors sometimes negatively impact the input from highly experienced experts due to the so-called *groupthink* phenomenon [15]. Moore and Clayton used crowds of computer users to report suspected phishing websites and discovered that mistakes tend to be repeatedly made by novices [10]. Welsh studied WoC for aggregating people's reviews of entertainment films and found that success depends upon crowd diversity and expertise [18]. Especially in smaller crowds, Welsh reported that it is better to weight the advice of experts to avoid bias from the majority.

Other WoC studies investigated the impact of crowd size, expertise, and task difficulty on the success of aggregating people's opinions. Wagner and Suh subjected people to tasks such as guessing the temperatures and weights of substances [17]; they found that group success depends upon task difficulty, and the chances of success were best for tasks of medium difficulty. Additionally, they discovered that crowd size matters, but within limits due to the balance between diversity and similarity. Robert and Romero studied the effect of WoC on the WikiProject community, and they found that larger crowd sizes tend to have better performance when there is more diversity [12]. In a study using crowds to forecast financial and political outcomes, Gracht et al. discovered that weighting the input from each individual equally produced the best results given the difficulty of assessing and forecasting human expertise [16].

In addition to the social settings above, the concept of exploiting collective wisdom has also been applied in computer science. For instance, the outputs from multiple machine learning algorithms are sometimes aggregated to improve prediction accuracy using so-called *ensemble* techniques [5]. In terms of optimization search algorithms, the WoAC metaheuristic has been employed in a number of NP-hard problems, including games [1,3,6,10,11] and combinatorial optimizations [4,8,14,19,20]. However, all of the work dealing with WoAC primarily has been a function of applying the metaheuristic to new types of problems, instead of investigating the factors that make the aggregation method successful. Preliminary research related to weighting expertise and varying the crowd size by Lowrance et al. [8] is expanded in this paper by conducting more experiments over a wider range of crowd sizes, as well as considering the impact of diversity and task complexity.

3   Procedure for Generating and Aggregating Artificial Crowds

3.1   Forming Crowds with Different Compositions

The TSP is an NP-hard problem in combinatorial optimization where the objective is to find the shortest route that allows a person to visit a set of cities once and then return to the origin city [19]. The complexity of the TSP is correlated to the number of cities and generally increases as the city count grows. Although the TSP was used in our experiments, the WoAC metaheuristic can be applied to other types of hard problems in combinatorial optimization.

A total of five TSP datasets were used in the evaluation, and each dataset consisted of one of the following numbers of cities: 50, 75, 100, 125, or 150. The coordinates of our datasets, as well as the best-known solutions, were generated using the Concorde program [21]. A total of 10,000 trials were conducted by testing the metaheuristic 2,000 times on each of the five TSP datasets.

Multiple searches using the same genetic algorithm were conducted during each of the 10,000 trials in order to form crowds of potential solutions. The number of searches (i.e., crowd size) was varied from 10 to 200 in increments of 10, and for each crowd step size, a total of 100 independent trials were executed for statistical purposes. To speed up the generation of each crowd, multiprocessing was leveraged to spawn concurrent searches.

Crowd size was a directly controlled variable in the experiments. The levels of *expertise* and *diversity* within the crowds were also varied, but indirectly as a result of the same genetic algorithm working on different datasets of ranging complexities. The genetic algorithm generally converged to solutions that were less wise and more diverse when faced with increasingly complex datasets as indicated in Table 1. The table shows statistics related to the mean crowd *expertise* and *diversity* for the 2,000 trials conducted at each city size. We quantified crowd expertise as the percent difference in fitness scores between the crowd mean and the best-known solution to the dataset. To quantify the amount of diversity in each crowd, we used a statistic referred to as *Jaccard distance*, which measures dissimilarity among sets.

Table 1. Crowd statistics for each dataset

Cities in the TSP	Mean (μ) percent difference in fitness (crowd mean to best known)	Std. Deviation (σ) of percent difference in fitness (crowd mean to best known)	Mean Jaccard distance
50	13.68	0.72	0.38
75	15.51	0.63	0.40
100	17.26	0.59	0.42
125	19.50	0.60	0.42
150	21.83	0.57	0.45

We adopted the following method for calculating the Jaccard distance of individual crowds. First, we started by finding the Jaccard index (similarity) between every combinational pair of generic searches. The intersection between two combinatorial graphs was found by counting the number of edges they shared in common and then dividing by their union or the total number of edges used by the pair. For a pair of graphs, x and y from two genetic searches, or members within the crowd, we defined their Jaccard index to be

$$J_{x,y} = \frac{|x \cap y|}{|x \cup y|} \tag{1}$$

where the numerator is the total number of matching edges in the graphs x and y, and the denominator is the number of unique edge combinations found in the pair of graphs. A single statistic that captures the crowd's mean diversity (D), or Jaccard distance, was obtained by repeating the process of (1) for every possible pair of graphs, $|P_c|$, within the crowd, and then taking one minus the average Jaccard similarity:

$$D = 1 - \frac{1}{|P_c|} \sum_{(x,y) \epsilon (P_c)} J_{x,y} \tag{2}$$

3.2   Extracting and Combining Wisdom from a Crowd of Searches

The first step of each experimental trial was to launch multiple instances of our genetic algorithm in parallel so that a crowd of varying solutions to a TSP could be formed. Once the searches converged, the WoAC metaheuristic operated on the resulting graphs as a post-processing optimization step; the procedure built a new solution to the problem using the wisdom gained from the crowd by adopting common edge selections to be a part of the aggregate. Once the combined solution was formed, its fitness score was compared to the best individual in the crowd to determine whether the heuristic was successful in surpassing the top expert.

The aggregation process of WoAC is fully explained in [8] and was followed in our experiments. As a synopsis of the process, the metaheuristic involves traversing the graphs of each search (i.e., each crowd member) and updating a master matrix that stores the frequency (count) that each edge combination was selected by the crowd. According to the WoAC metaheuristic, edges with higher counts in the matrix are incorporated into the aggregate that forms the crowd's solution to the problem.

For the TSP, the metaheuristic builds a new solution by choosing a starting node and then using the counts in the matrix to find the crowd's most popular edge choice for this particular node. This process is repeated for every adjacent node until the Hamiltonian cycle is complete. If a node is already in the newly constructed path, then the next highest occurrence is selected and so on. It is worth noting that the objective function (e.g., spatial information in the case of the TSP) is only referenced if all options in the histogram have been exhausted, meaning that the crowd's preferred choices have already been selected as part of the newly constructed graph. In this case, a greedy heuristic is used to find the nearest node as the next destination based on the objective function. Two other optimization heuristics are employed to improve the chances of WoAC success: (1) the metaheuristic tries every node as the starting point, and (2) a new Hamiltonian path is built after each crowd member (i.e., genetic algorithm solution) is added to the edge frequency matrix. The latter step effectively varies the crowd size up to the maximum number of searches collected. Ultimately, the WoAC attempt that yields the lowest cost is chosen as the best crowd's solution.

The traditional means of updating the WoAC matrix is to update the counts in the matrix by one for each edge occurrence noted in the crowd of graphs. We refer to this approach as the *equal weighting method* because each search within the crowd has the same voting power regardless of fitness scores. Other weighting schemes introduced in [8] include the *linear* and *exponential weighting methods*. The general idea of these latter two schemes is to increase the weight carried by the better searches (i.e., those with better relative fitness scores within the crowd) so that their opinions are more strongly favored in the aggregation process.

The weight assigned to a search affects the WoAC aggregation process by changing the value added to each cell of the edge occurrence matrix. We normalized the weights of each crowd member between 0 and 1 using feature scaling in the form of min-max normalization. For instance, assume that we have a crowd of candidate solutions to a particular combinatorial optimization problem

$$\underline{c} = \{c_1, c_2, \ldots, c_n\} \tag{3}$$

where $\underline{c}$ is the cost vector (array) that contains the fitness scores associated with each search and n is the total number of searches performed. We can rescale the search scores using

$$\hat{c}_i = \frac{c_i - \min(\underline{c})}{\max(\underline{c}) - \min(\underline{c})} \tag{4}$$

where $\hat{c}_i$ is the normalized score of the ith search in the crowd. The normalized fitness levels produced by (4) are used in the linear weighting scheme according to

$$w_{l_i} = 1 - \hat{c}_i \tag{5}$$

where w_{l_i} is the linear weight assigned to the ith member in the crowd. The scheme assigns the best search a count value of 1 to all its edge choices and the worst crowd member a 0. All other searches are given weights (i.e., values to be added to the WoAC matrix) that are linearly distributed between 0 and 1.

In the exponential weighting scheme, the values added to cells of the WoAC matrix for a given search are determined by

$$w_{e_i} = e^{-(\hat{c}_i * m)} \tag{6}$$

where w_{e_i} is the exponential weight assigned to the ith search and m is the exponential decay constant that facilitates the mapping of the w_{e_i} to a range of values between 1 to 0. In our study, we let $m = 5$ because it leads to (6) assigning a weight of roughly 0 to the worst-performing search in the crowd. On the other hand, (6) gives the search with the best fitness a weight of 1. The weights of all other searches decay exponentially from 1 to 0 based on their normalized fitness score defined in (4). Due to its exponential nature, this scheme more rapidly diminishes the input of less-fit crowd members than the linear weighting method, but at the cost of accepting less diversity in the aggregate WoAC solution.

4   Experimental Results: Identifying the Factors of Crowd Success

4.1   Impact of Crowd Composition and Problem Complexity

Time and computational power are required to form artificial crowds using multiple instances of an evolutionary search on a complex problem. Therefore, it would be more efficient to apply the metaheuristic using the smallest crowd that yields the maximum likelihood of success. But to achieve this objective, a deeper understanding of the predominant factors that impact the metaheuristic's success is necessary. For this purpose, we analyzed the data from our experiments and found that the optimal crowd size is mostly dependent upon the problem's complexity and the crowd's expertise.

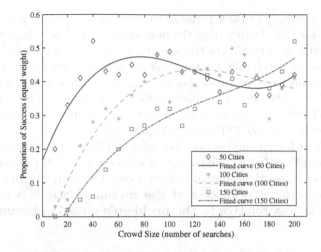

Fig. 1. Comparing metaheuristic success as a function of crowd size and multiple datasets of varying complexity.

Figure 1 is the first example that shows the interplay of these variables on the probability of the metaheuristic succeeding when the input from each search is given equal weight in forming the aggregate. For all of our plots, we defined success to be when the WoAC metaheuristic formed a solution with a fitness score superior to the best individual in the crowd. The left-half of the figure (crowd sizes <100) indicates that more complex problems require larger crowds in order to achieve commensurate levels of success. We attribute this observation to the following two causes: (1) the average expertise within a given crowd size decreased as problem complexity grew (see the second column in Table 1), and (2) the average diversity within a given crowd increased as problem complexity grew (see the last column in Table 1). As a result of these factors, larger crowd sizes were required to mitigate the shortcomings of having less knowledgeable populations and less consensus.

There are other noteworthy observations from Fig. 1. First, there appears to be an optimal crowd size for the city sizes of 50 and 100. For both of these datasets, the average success of the metaheuristic began to plateau and deteriorate slightly. This observation can be attributed to there being a limit to the amount of wisdom that can be extracted from the crowd. Additionally, as a crowd size grows, the novel ideas of the top-performing searches tend to be suppressed by the lesser-fit majority when each member gets an equal vote. We attribute this '*crowding out*' effect as the significant drawback associated with the equal-weighting scheme.

Another observation in Fig. 1 is that the dataset of 150 cities eventually trended to higher levels of success than the smaller, less complex datasets when the crowd size was sufficiently large (>170). We attribute this result to the nature of each crowd's position relative to their respective best-known solutions.

For example, according to Table 1, the crowd average fitness for the TSP with 150 cities was weaker (i.e., farther from the best-known solution) when compared to the smaller TSPs. As a result, the trials operating on the 150 dataset had a somewhat easier task of succeeding because there generally existed more edge combinations that could have led to success than those which worked on the smaller datasets. However, the pool of searches performed on the 150 TSP typically possessed less collective wisdom than the crowds formed using the other datasets, and consequently, the 150 TSP usually required larger crowd sizes to achieve a commensurate or higher success rate. We refer to this crowd phenomenon as the *Crowd Fitness Effect (CFE)*, which can cause more complex problems to yield higher levels of success than simpler problems counterintuitively. In other words, the CFE implies that **the success of the metaheuristic is a function of the crowd's fitness relative to the problem's global optimum.**

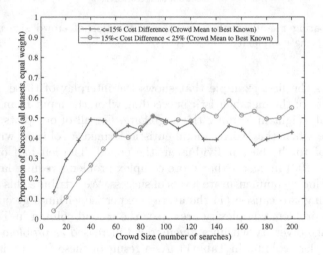

Fig. 2. Comparing metaheuristic success as a function of crowd size and the mean difference between the crowd's average fitness score and the best-known solution. The line plots are based on all dataset trials.

To analyze the CFE further, all 10,000 trials were divided into two groups depending upon whether the crowd's mean fitness level was less or greater than 15% from the best-known solution. Figure 2 shows that knowledgeable crowds closer to the global optimum generally require a smaller crowd size before reaching a peak rate of success and plateauing. For instance, the average success rate of wiser crowds (those $<= 15\%$ from the best-known) is shown in the figure to plateau at a crowd size of approximately 40. On the other hand, less knowledgeable crowds (those $>15\%$ from best-known) required a larger crowd size of roughly 130 before plateauing. Additionally, there is a crossover point in Fig. 2, similar to the one in Fig. 1, where groups that were farther from the global optimum began to achieve higher success rates at sufficiently large crowd sizes. In general, the plot shows that **the metaheuristic requires larger crowds to achieve peak success when the pool of searches are less fit.**

4.2 Impact of Weighting Expertise

The effects of weighting individual solutions based on their fitness scores can be seen graphically in Fig. 3. The plot shows the mean success rate for the aforementioned weighting methods as a function of crowd size. The exponential weighting method consistently achieved the highest levels of success, followed by the linear method, and lastly by the equal voting method. In summary and consistent with the findings in [8], **weighting the input of better searches more strongly increases the likelihood of metaheuristic succeeding,** in part because it mitigates the chances that the expertise from these standout searches will be overshadowed or *'crowded out'* by the majority that tends to be less fit.

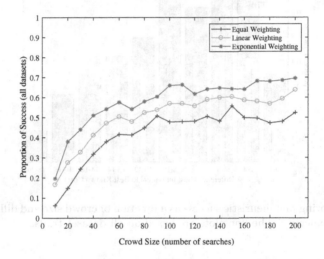

Fig. 3. Comparing metaheuristic success as a function of crowd size and different means of weighting expertise. The line plots are based on all dataset trials.

To statistically quantify the mean difference between the weighting methods, paired t-tests were performed as summarized in Table 2. The statistics are based on all 10,000 trials conducted at different crowd and city sizes. This table shows the mean chances of success for the exponential method to be higher than 6% when compared to the linear method and more than 15% when compared to the equal vote method.

Another aspect affecting the metaheuristic is the cost difference between the *best individual* in the crowd to the *best-known* solution, as opposed to the *crowd's average* fitness relative to the best-known. Intuitively, when the fitness of the best individual in the crowd is near the global optimum, then the task of the

Table 2. Paired t-tests comparing weighting techniques

Null hypothesis	Test decision	95% Confidence	p-value
$\mu_1 - \mu_2 = 0$	Rejected	(0.063, 0.084)	$3.55x10^{-43}$
$\mu_1 - \mu_3 = 0$	Rejected	(0.151, 0.173)	$4.37x10^{-174}$
μ_1 = exponential	μ_2 = linear	μ_3 = equal	

metaheuristic succeeding becomes more difficult. Figure 4 illustrates this effect, as well as the impact of weighting expertise, on the chances of the metaheuristic surpassing the best individual fitness score.

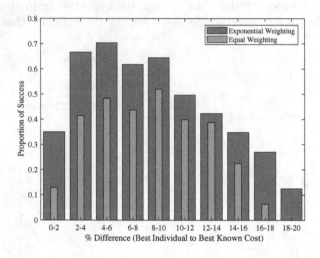

Fig. 4. Comparing metaheuristic success as a function of crowd size and different means of weighting expertise. The line plots are based on all dataset trials.

Both weighting methods shown in Fig. 4 indicate that there is a peak when the metaheuristic is most effective. The chances of succeeding were highest when the fitness of the best individual in the crowd was somewhat farther from the best-known solution. However, away from this point, the probability of success decreased either because the task of surpassing the best individual became more challenging (left side of plot) or the crowd became less knowledgeable (right side of plot).

In terms of the impact of weighting expertise, the following observations can be made: (1) the exponential weighting method shows an earlier peak which implies that it favors the presence of stronger expertise, whereas the equal method works best when the strongest individual is of moderate expertise and (2) the margin of difference between the weighting techniques is largest near the extremes which implies that the exponential method is less sensitive to the expertise level of the best individual.

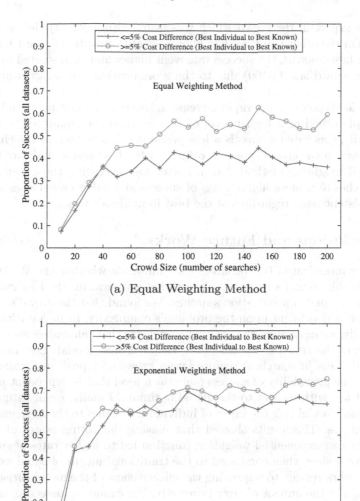

(a) Equal Weighting Method

(b) Exponential Weighting Method

Fig. 5. Comparing metaheuristic success as a function of crowd size, weighting technique, and cost difference to the best-known solution. Both plots are based on all dataset trials.

Figure 5 continues to investigate the impact that the top expert, as well as the weighting technique, has on the metaheuristic's success rate; however, in these plots, crowd size was the independent variable. In the equal-weighting subplot of 5(a), the average success rate plateaued at a smaller crowd size (>40) when there existed an expert in the crowd which was near (within 5% of) the best-known solution. On the other hand, for less knowledgeable crowds (i.e., best individual >5% from best-known), the success rate went higher and then started to plateau at a larger crowd size (>100) due to the aforementioned discussion about the CFE.

Figure 5(b) shows a more rapid increase in success for the exponential method before it plateaued at a higher level of success than the equal-weight method. Additionally, this subplot reveals a less pronounced difference in the chances of success between the crowd types. In other words, these results indicate that the exponential weighting method does a better job leveraging the expertise from wiser searches to achieve higher rates of success at smaller crowd sizes and on a more consistent basis regardless of the best individual's fitness.

5   Conclusions and Future Work

This paper investigated the variables that influence whether the WoAC meta-heuristic is able to find a superior solution that outperforms the best expert in a crowd of incomplete optimization searches. We found that the crowd's expertise level, which is dependent upon the problem's complexity, to be a critical factor. Additionally, we discovered the optimal crowd size that maximizes success to be a function of the crowd's expertise (i.e., search fitness levels). Less knowledge-able crowds (less fit searches) require larger crowds for peak performance, but eventually the probability of success plateaus a level that is dependent upon the crowd's fitness with respect to the global optimum. Finally, we investigated the effectiveness of weighting the input of individual searches to the aggregate based on their fitness. The results showed that biasing the voting power of searches according to an exponential weighting function led to higher rates of success at smaller crowd sizes when compared to the traditional means of equal weighting.

Future work related to improving the effectiveness of the metaheuristic could further explore the impact of crowd diversity. For example, instead of using the same genetic algorithm to form the crowd, properties associated with the algo-rithm, such as its selection and mutation methods, could be modified between each individual search in order to create more diversity within the crowd. Alter-natively, different stochastic search methods could be used as well to produce a broader assortment of search results. Either method should produce a wider range of diversity that could be effectively controlled and analyzed for its impact on the success of the metaheuristic.

References

1. Ashby, L.H., Yampolskiy, R.V.: Genetic algorithm and wisdom of artificial crowds algorithm applied to light up. In: International Conference on Computer Games 2011, pp. 27–32 (2011)
2. Chen, J., Ren, Y., Riedl, J.: The effects of diversity on group productivity and member withdrawal in online volunteer groups. In: Proceedings of the SIGCHI Conference on Human Factors in Computing Systems, pp. 821–830 (2010)

3. Hughes, R., Yampolskiy, R.V.: Solving sudoku puzzles with wisdom of artificial crowds. Int. J. Intell. Games Simul. **7**(1), 24–29 (2012)
4. Hundley, M., Yampolskiy, R.V.: Shortest total path length spanning tree via wisdom of artificial crowds algorithm. In: Proceedings of the 28th Modern Artificial Intelligence and Cognitive Science Conference (2017)
5. Kantardzic, M.: Data Mining: Concepts, Models, Methods, and Algorithms. Wiley, Hoboken (2011)
6. Khalifa, A.B., Yampolskiy, R.V.: Ga with wisdom of artificial crowds for solving mastermind satisfiability problem. Int. J. Intell. Games Simul. **6**(2), 12–17 (2011)
7. Kittur, A., Kraut, R.E.: Harnessing the wisdom of crowds in wikipedia: quality through coordination. In: Proceedings of the 2008 ACM Conference on Computer Supported Cooperative Work, pp. 37–46 (2008)
8. Lowrance, C.J., Abdelwahab, O., Yampolskiy, R.V.: Evolution of a metaheuristic for aggregating wisdom from artificial crowds. In: Pereira, F., Machado, P., Costa, E., Cardoso, A. (eds.) EPIA 2015. LNCS (LNAI), vol. 9273, pp. 238–249. Springer, Cham (2015). https://doi.org/10.1007/978-3-319-23485-4_24
9. Mäntylä, M.V., Itkonen, J.: The effect of crowd size and time restriction in software testing. Inf. Softw. Technol. **55**(6), 986–1003 (2013)
10. Moore, T., Clayton, R.: Evaluating the wisdom of crowds in assessing phishing websites. In: Tsudik, G. (ed.) FC 2008. LNCS, vol. 5143, pp. 16–30. Springer, Heidelberg (2008). https://doi.org/10.1007/978-3-540-85230-8_2
11. Port, A.C., Yampolskiy, R.V.: Using a GA and wisdom of artificial crowds to solve solitaire battleship puzzles. In: International Conference on Computer Games, pp. 25–29 (2012)
12. Robert, L., Romero, D.M.: Crowd size, diversity and performance. In: Proceedings of the 33rd Annual ACM Conference on Human Factors in Computing Systems, pp. 1379–1382 (2015)
13. Surowiecki, J.: The Wisdom of Crowds. Random House LLC, New York City (2005)
14. Trainor, P.J., Yampolskiy, R.V., DeFilippis, A.P.: Wisdom of artificial crowds feature selection in untargeted metabolomics: an application to the development of a blood-based diagnostic test for thrombotic myocardial infarction. J. Biomed. Inform. **81**, 53–60 (2017)
15. Velic, M., Grzinic, T., Padavic, I.: Wisdom of crowds algorithm for stock market predictions. In: Proceedings of the International Conference on Information Technology Interfaces, pp. 137–144 (2013)
16. von der Gracht, H.A., Hommel, U., Prokesch, T., Wohlenberg, H.: Testing weighting approaches for forecasting in a group wisdom support system environment. J. Bus. Res. **69**(10), 4081–4094 (2016)
17. Wagner, C., Suh, A.: The wisdom of crowds - impact of collective size and expertise transfer on collective performance. In: 47th Hawaii International Conference on System Sciences, pp. 594–603 (2014)
18. Welsh, M.: Expertise and the wisdom of crowds - whose judgments to trust and when. In: 34th Annual Meeting of the Cognitive Science Society, pp. 594–603 (2012)
19. Yampolskiy, R.V., Ashby, L., Hassan, L.: Wisdom of artificial crowds—a metaheuristic algorithm for optimization. J. Intell. Learn. Syst. Appl. **4**(02), 98–107 (2012)
20. Yampolskiy, R.V., El-Barkouky, A.: Wisdom of artificial crowds algorithm for solving NP-hard problems. Int. J. Bio-inspired Comput. **3**(6), 358–369 (2011)
21. Concorde TSP Solver. http://www.math.uwaterloo.ca/tsp/concorde/index.html. Accessed 10 July 2018

Analysis of Consensus Sorting via the Cycle Metric

Ivan Avramovic$^{(\boxtimes)}$ (iD) and Dana S. Richards

George Mason University, Fairfax, VA 22030, USA
{iavramo2,richards}@gmu.edu

Abstract. Sorting is studied in this paper as an archetypal example to explore the optimizing power of consensus. In conceptualizing the consensus sort, the classical hill-climbing method of optimization is paired with the modern notion that value and fitness can be judged by data mining. Consensus sorting is a randomized sorting algorithm which is based on randomly selecting pairs of elements within an unsorted list (expressed in this paper as a permutation), and deciding whether to swap them based on appeals to a database of other permutations. The permutations in the database are all scored via some adaptive sorting metric, and the decision to swap depends on whether the database consensus suggests a better score as a result of swapping. This uninformed search process does not require the definition of the concept of sorting, but rather depends on selecting a metric which does a good job of distinguishing a good path to the goal, a sorted list. A previous paper has shown that the ability of the algorithm to converge on the goal depends strongly on the metric which is used, and analyzed the performance of the algorithm when number of inversions was used as a metric. This paper continues by analyzing the performance of a much more efficient metric, the number of cycles in the permutation.

Keywords: Adaptive sorting · Randomized algorithms
Uninformed search · Combinatorics · Simulation and modeling

1 Introduction

Hill-climbing is a well-known, fundamental approach to solving optimization problems. Likewise, a fundamental approach to solving search problems can be realized if value can be expressed in terms of reachability or distance to the goal. This paper studies sorting as a search problem, but sorting is only used here as an archetypal example to explore the optimizing power of consensus. The optimization used here advances the notion of hill-climbing by drawing its fitness scores from comparisons against an external body of data, thus applying the modern notion that data mining can be used to judge value and fitness.

Consensus sorting, introduced in [1], is a randomized sorting algorithm in which pairs of elements are selected at random, and swapped if a certain condition is satisfied. The decision to swap is not based on the definition of sorting,

© Springer Nature Switzerland AG 2018
D. Kim et al. (Eds.): COCOA 2018, LNCS 11346, pp. 552–565, 2018.
https://doi.org/10.1007/978-3-030-04651-4_37

but rather on an appeal to a consensus from a permutation database. Every permutation in the database has an associated score, and depending whether the scores are on average better when the chosen pair of elements is swapped or remains the same (see Fig. 1), the decision is made. The process is repeated for as long as reasonable progress towards the goal state (a sorted array) is being made. Thus, the ability to sort depends on using a scoring metric which is effective in guiding sorting decisions.

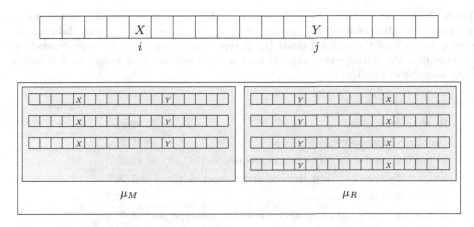

Fig. 1. A consensus swapping decision. Elements X and Y at positions i and j are selected at random. One group in the database is formed from all permutations matching X and Y, and another group is formed from all permutations matching Y and X. The mean scores of the two groups, μ_M and μ_R, are compared to find out whether to swap.

It makes sense that metrics which incorporate some notion of "sortedness" will be more likely to produce a sorted array. It turns out, however, that this is not enough, and that out of a number of sensible scoring metrics taken from the adaptive sorting literature [2], only a couple were found to make reasonable progress towards producing a sorted array [1]. Table 1 shows a comparison of the sorting effectiveness of various metrics, expressed in terms of the average percent improvement in score over the array's initial score. The meaning of each metric in the table is briefly described below.

1. The *cycle count* subtracts the number of cycles from the total number of elements.
2. The *number of inversions* metric counts the number of pairs of elements which are out of order with respect to one another.
3. The *minimum number of ascending subsequences* is the smallest number of subsequences of the sequence such that all subsequences are ascending.
4. The *largest distance between element and correct slot* is the largest magnitude difference between an element's index and it's correct index.

5. The *size of the largest inversion* measures the largest magnitude difference between the indices of a pair of inverted elements.
6. The *minimum number of removals for a sorted subsequence* counts the smallest number of removals it would take to procude a sorted subsequence.
7. The *number of runs* counts the number of elements which are out order relative to the preceeding element in the array.

Table 1. A comparison of the performance of the consensus sorting algorithm under a number of different adaptive sorting method. The score improvement data shown here is taken from the original study [1]. It represents the amount of improvement on average over the initial score, over 50 runs using a permutation length of 100 and a database size of 1,000,000.

Metric	Score improvement
Reverse cycle count	98.8%
Number of inversions	48.2%
Minimum number of ascending subsequences	12.6%
Largest distance between element and correct slot	7.26%
Size of the largest inversion	3.43%
Minimum number of removals for a sorted subsequence	1.41%
Number of runs	−2.00%

At this point, one thing should be noted about the consensus sorting algorithm. It is not by any means an efficient sorting algorithm, and it will require far more iterations than a typical comparison-based sort before converging. Furthermore, as can be seen from the table, the result of sorting is typically not a fully sorted array. The performance depends on several factors, including database size, database quality, permutation size, and sorting metric. Only one of the metrics shown in the list has any likelihood of producing a fully sorted result for a reasonably sized array.

The reason that such an imperfect sorting algorithm may be of interest is that it models a type of uninformed search. Consider a search problem in which one is given a starting state based on currently known information, and a well-defined goal state. One is also provided with a large set of data elements relevant to the problem, which may be used in the search for the goal. However, assume that the understanding of or ability to determine the goal state is limited, and thus progress towards the goal must be inferred from the data. Thus, the steps towards the goal are made through appeals to the data set.

Since the sense of position is derived from the database, it is of interest how close one can come to the goal via this process. Alternatively, a negative phrasing can be used: will successive appeals to the large data set bias the result towards some particular set of values?

As an example, consider the case where one has a photo, and wants to determine the location where the photo was taken. The state is a set of potential location of the photo, which has a starting state that includes the entire world, and a goal state which is the location where the photo was taken. Each movement towards the goal is a refinement of the set of location, by comparing the image for similarity against a large database of images from around the world. This general approach was implemented in the Im2gps algorithm presented by Hays and Efros [5].

The form of the search is evident in web and social media searches, where where a particular user seeks a subset of the available data which is relevant to their interests. They may not know in advance what their goal looks like, but they have access to a large body of data to assist in their search. In general, increased use of information sharing through Cloud computing has led to increased volumes of available data [4]. Data available on the Cloud tend to have decreasing informational value density but increasing overall value [6], and may have restriction on the nature of access to the available data. Thus, a user may find themselves in a position of seeking a method which best leverages the value of the data available on the Cloud. The rapid growth of social media services has made it more complex to retrieve desired data, and social media searches have been studied intensively [8]. Personalized search is a concept which has been introduced due to the fact that, within a broad search category, different users may have differing specific goals [7].

The consensus sorting algorithm considered in this paper is an abstract version of the problem. This is done for the sake of focusing on the mathematics of the analysis of the behavior as the state approaches the goal state. The formulation considered here is a randomized sorting algorithm, where the state is a permutation, and the goal state is the identity permutation, representing a sorted array. Each movement step involves the selection of elements and the decision whether or not to swap by comparing against the database consensus. As stated previously, the quality of the sort depends on the scoring metric used to mediate swapping decisions.

The metrics of interest are the top two from Table 1, the reverse cycle counting metric (which will be referred to simply as the *cycle metric*) and the number of inversions metric, due to their relative effectiveness in sorting compared to the others. The inversion metric is studied in detail in [1]. This paper will focus on the cycle metric.

2   Consensus Sorting Algorithm

From the original paper [1], the consensus sorting algorithm is defined as follows. Assume that the input array is a permutation $\pi \in S_n$ with a length of n. Thus, array and permutation may be used synonymously here. A *permutation database* $\mathcal{D}$ is given as a collection of permutations of length n. Also given is a scoring function

$$\phi : S_n \to \mathbb{N}_0.$$

The function ϕ must satisfy the condition that a result of zero uniquely corresponds to a fully sorted array (an identity permutation), while lower scores should generally correspond to greater degrees of "sortedness."

A *consensus swapping function* is a function

$$F_{\phi,\mathcal{D}} : \mathbb{Z} \times \mathbb{Z} \times S_n \to S_n.$$

The function takes as input indices i and j, as well as a permutation π_a. The output is the subsequent permutation π_{a+1}, as given in (3) below.

Let $\mathcal{C}_M$ be the collection of all permutations from $\mathcal{D}$ which match π_a at positions i and j, and let $\mathcal{C}_R$ be the collection of all permutations from $\mathcal{D}$ which have the same elements as π_a in positions i and j, but in reverse order. Then μ_M and μ_R are the mean scores of $\mathcal{C}_M$ and $\mathcal{C}_R$, respectively. Thus,

$$\mu_M = \frac{1}{|\mathcal{C}_M|} \cdot \sum_{\pi \in \mathcal{C}_M} \phi(\pi) \tag{1}$$

$$\mu_R = \frac{1}{|\mathcal{C}_R|} \cdot \sum_{\pi \in \mathcal{C}_R} \phi(\pi) \tag{2}$$

$$F_{\phi,\mathcal{D}}(i,j,\pi_a) = \begin{cases} \pi_a & \text{if } \mu_M \text{ or } \mu_R \text{ is undefined,} \\ & \text{or } \mu_M \leq \mu_R, \\ & \text{or } i = j, \\ (i\ j) \cdot \pi_a & \text{otherwise.} \end{cases} \tag{3}$$

The $(i\ j) \cdot \pi_a$ in (3) denotes that the elements at indices i and j in π_a are swapped. Let $\pi_1 = \pi$ be the initial permutation. At each iteration a, a pair of distinct indices i, j are selected from the array at random. And the consensus swapping function is used to compute π_{a+1} from π_a. The algorithm terminates when it is deemed that progress is no longer being made.

3   Perfect Sorting

An *ideal database* is a database in which every permutation of length n is represented a nonzero equal number of times. For simplicity's sake, one can assume that an ideal database contains exactly one instance of every possible permutation of length n. An ideal database is denoted by $\mathcal{D}_I$.

When a sorting decision for a given database differs from that of an ideal database, the decision is referred to as a sorting *error*, even if the error results in an improvement in score.

A *perfect sort* is an execution of the consensus sort which reaches the goal state of a fully sorted array. Typically, a perfect sort can only be assured in cases in which an ideal database is used, otherwise sorting errors may arise and prevent the achievement of the goal state. An observation can be made about the sorting capability of a scoring metric, regardless of the nature of the metric.

Theorem 1. *Suppose that*

$$\phi\left(F_{\phi,\mathcal{D}}(i,j,\pi)\right) \le \phi(\pi), \tag{4}$$

for some scoring metric ϕ and database $\mathcal{D}$, and for any legal values of i, j, and π. In other words, the score is monotonically non-increasing as a result of iterating the consensus sorting algorithm. If every non-sorted array has some legal choice of i and j such that

$$\phi\left(F_{\phi,\mathcal{D}}(i,j,\pi)\right) \ne \phi(\pi), \tag{5}$$

then the algorithm will produce a fully sorted array with probability approaching 1 as the number of iterations of the algorithm increases without bound.

Proof. The set of possible scores produced by ϕ is finite, due to the finite number of permutations of size n, and has a lower bound of 0, which is achieved only when the array is fully sorted. Score is non-increasing, which means that once the score has changed, it will never return to the same score again. Therefore, there can be at most a finite number of score changes due to iteration of the algorithm.

If the algorithm does not result in a fully sorted array, then it must eventually yield a member of some subset $M \subset S_n$ of permutations which each have the same score $s > 0$, such that no amount of additional iterations will yield a permutation with a score less than s.

However, regardless of the $\pi \in M$, it is a given that there must exist at least one choice of i and j which will result in a reduction in score, and no choice of elements which will increase the score. If $0 < p < 1$ is the chance of picking one particular i,j pair, it follows that the chance of selecting a score-reducing pair of elements is at least p. Thus, k iterations of the algorithm would have a chance of at least

$$p \cdot \frac{1-(1-p)^{k+1}}{1-(1-p)} = 1 - (1-p)^{k+1}$$

of selecting a score-reducing pair of elements. As k increases without bound, this probability approaches 1, contradicting the supposition that a fully sorted array will not be the eventual result. □

The theoretical guarantee of a sorted array from Theorem 1 applies mostly to situations when an ideal database is used, because if a randomly selected database is used, it is unlikely that the conditions of Theorem 1 will hold. However, if an ideal database is used, the scoring function of the inversion and cycle metrics will produce monotonically non-increasing results, and a score-reducing selection always exists for arrays which are not fully sorted. Thus, both metrics will theoretically produce sorted arrays in the ideal case.

4   Breakdown of Cyclical Cases

Assume that the metric used to score permutations is the reverse cycle metric (which will simply be called the cycle metric), defined as follows. Let π be a permutation of length n containing c cycles. The reverse cycle metric is given by

$$\phi(\pi) = n - c. \tag{6}$$

Thus, when a permutation is in order, all cycles are 1-cycles, so its score is zero. Higher scores correspond to lesser sortedness.

Lemma 1. *If two distinct indices are part of the same cycle, then swapping them will result in them being part of two different cycles, and if they are part of different cycles, swapping them will merge their cycles.*

Proof. Suppose that i and j are part of the same cycle of length k. And suppose the indices that comprise the cycle can be represented in order as $m_0 \rightarrow m_1 \rightarrow \cdots \rightarrow m_k$, where $m_0 = m_k = i$. Thus, since i and j are distinct, there must be some integer x, with $0 < x < k$, such that $m_x = j$. By swapping i and j, m_x would then point to m_1, resulting in the cycle $m_x \rightarrow m_1 \rightarrow \cdots \rightarrow m_x$,which does not include either m_0 or m_k. Index m_0 would point to m_{x+1}, resulting in a cycle containing the remaining elements of the original cycle. Therefore the i and j positions are no longer part of the same cycle, but rather two different cycles.

Now suppose that i is part of a cycle of length k, and that $m_0, \ldots m_k$ are defined as before, but that j is part of a different cycle of length l. Similarly, let j's cycle contain $n_0, \ldots n_l$. Swapping i and j will cause m_0 to point to n_1 and n_0 to point to m_1, while keeping the remaining elements intact. Thus, after swapping they form a single cycle. $\qquad\square$

A corollary to Lemma 1 is that any swap of elements will alter the score of the permutation by 1, using the cycle metric. The change will be an improvement if a cycle is broken in two, and a degradation if a pair of cycles are merged into one.

The algorithm's decision whether or not to swap is slightly more complex. It shall be approached by analyzing cases. For a permutation π of length n, let $s = \phi(\pi)$ be the score, $c(\pi)$ (or simply c) be the number of cycles, and $c_k(\pi)$ (or simply c_k) be the number of cycles of length k in π. By (6), $s = n - c$, and thus $c = n - s$. Given a pair of distinct indices i and j, the possible cases which they may belong to are broken down in Table 2.

Proposition 1. *The number of choices of i and j which result in a case $(1/1)$ selection is*

$$c_1(c_1 - 1). \tag{7}$$

Proof. Since both i and j must be 1-cycles, i must be selected among the c_1 1-cycles, and j must be selected among the remaining $c_1 - 1$ 1-cycles. The total is the product of the two. $\qquad\square$

Proposition 2. *The number of choices of i and j which result in a case $(1/2^+)$ selection is*

$$2c_1(n - c_1). \tag{8}$$

Table 2. A breakdown of the ways to select a pair of distinct indices i and j from a permutation. The set of cases shown here is exhaustive.

Case	Description
Case (1/1)	Both i and j are 1-cycles
Case (1/2$^+$)	Either i or j is a 1-cycle, while the other is part of a longer cycle
Case (2$^+$/2$^+$)	i and j are members of different cycles of length greater than 1
Case (2)	i and j share a 2-cycle
Case (3$^+$)	i and j are consecutive members of a cycle, but not a 2-cycle
Case (2$^+$2$^+$)	Both i and j are members of the same cycle, but not consecutive

Proof. There are c_1 elements which are 1-cycles, and $n - c_1$ elements which are not members of 1-cycles, and are thus members of longer cycles. Either i is the 1-cycle and j is a member of a longer cycle, or j is the 1-cycle and i is a member of a longer cycle. In either case, one is selected from the first group of c_1 elements, and the other is selected from the second group of $n - c_1$ elements. The total, considering both roles of i and j, is

$$c_1(n - c_1) + (n - c_1)c_1 = 2c_1(n - c_1).$$

$\square$

Proposition 3. *The number of choices of i and j which result in a case $(2^+/2^+)$ selection is*

$$\sum_{k=2}^{n}(kc_k)(n - c_1 - k). \tag{9}$$

Proof. Begin by selecting i as a member of a cycle of length greater than 1. That means that for every cycle length $k \geq 2$, there are k elements from each of c_k cycles to select from. After i has been selected, j must be selected from one of the remaining n elements, excluding the c_1 elements which are 1-cycles and the k elements which are members of the same cycle as i. Thus, the total comes from summing the product of number of ways to select i, which is (kc_k), with the number of ways to select j, which is $(n - c_1 - k)$, over all $k \geq 2$. $\square$

Proposition 4. *The number of choices of i and j which result in a case (2) selection is*

$$2c_2. \tag{10}$$

Proof. There are c_2 2-cycles in a permutation. Since each 2-cycle contains two elements, there are a total of $2c_2$ elements which are members of 2-cycles. i can be selected as any of those $2c_2$ indices, but the selection of i uniquely determines the selection of j as the other member of the selected cycle. $\square$

Proposition 5. *The number of choices of i and j which result in a case (3^+) selection is*

$$2(n - c_1 - 2c_2). \tag{11}$$

Proof. Select i as a member of a cycle of length greater than 2. Since there are c_1 elements in 1-cycles and $2c_2$ elements in 2-cycles, the number of possible selections of i is $n - c_1 - 2c_2$. After i has been selected, j can be selected as the index either before or after i in the cycle, resulting in 2 possible selections. The total is the product of the two, $2(n - c_1 - 2c_2)$. □

Proposition 6. *The number of choices of i and j which result in a case (2^+2^+) selection is*

$$\sum_{k=4}^{n}(kc_k)(k-3). \tag{12}$$

Proof. If i and j are members of the same cycle, but not consecutive members, then the length of the cycle must be at least 4. Select i as a member of a cycle whose length is at least 4. For every cycle length $k \geq 4$, there are k elements from each of c_k which may be selected. Once i has been selected, j can be selected as one of the k members of the same cycle, excluding i and the members which immediately precede and succeed i in the cycle. Thus, the total comes from summing the product of number of ways to select i, which is (kc_k), with the number of ways to select j, which is $(k-3)$, over all $k \geq 4$. □

Case $(1/1)$, case $(1/2^+)$, and case $(2^+/2^+)$ shall be referred to as *complements* of case (2), case (3^+), and case (2^+2^+), respectively.

Remark 1. Swapping i and j will swap between a case and its complement case.

Remark 2. Since the set of elements in a permutation is partitioned into cycles, and since each of the c_k k-cycles contains k elements, it follows that

$$\sum_{k=1}^{n} kc_k = n. \tag{13}$$

Remark 3. Since there are $n(n-1)$ ways to select two distinct elements from a permutation of length n, the sum of the number of selections from each of the six cases as given by (7), (8), (9), (10), (11), and (12) will be

$$n(n-1). \tag{14}$$

Remark 4. The number of permutations of length n containing c cycles is given by the *unsigned Stirling number of the first kind*. Since the score of a permutation is defined as $s = n - c$, it follows that the number of permutations with a given score s is:

$$\left[\begin{matrix} n \\ n - s \end{matrix} \right] \tag{15}$$

Proposition 7. *The total number of k-cycles over all permutations of length n with c cycles is given by*

$$\sum_{\pi \in \mathcal{C}_{s,n}} c_k(\pi) = \frac{n!}{k(n-k)!} \left[\begin{matrix} n - k \\ c - 1 \end{matrix} \right] \tag{16}$$

Proof. The proof shall be approached by using *exponential generating functions* (EGFs)[3] to count the number of cycles of a given length over all permutations of size n and total number of cycles c. For a given k, begin by defining a labeled combinatorial class of permutations, $\mathcal{P}_{c,k}$. Every object in class $\mathcal{P}_{c,k}$ is a permutation which has c cycles, and has one particular cycle of length k selected. Thus, the total number of cycles of length k among all permutations with c cycles and n elements is equal to the total number of objects of size n in $\mathcal{P}_{c,k}$.

Since every object in $\mathcal{P}_{c,k}$ has c cycles, the class can be constructed by first selecting a cycle of length k, then selecting a set of $c - 1$ additional arbitrary-sized cycles. The construction uses the atomic object $\mathcal{A}$ which has size 1, the cycle construction CYC which constructs a cycle of arbitrary size, the m-cycle construction CYC_m which constructs a cycle of length m, the m-set construction SET_m which constructs m sets, and the labeled product $*$ which convolves two classes. See Table 3 for how each of these constructions translate into EGFs.

Table 3. Exponential generating functions for the required combinatorial class constructions

Name	Construction	Exponential generating function
Atom	$\mathcal{B} = \mathcal{A}$	$B(z) = z$
Cycle	$\mathcal{B} = CYC(\mathcal{C})$	$B(z) = \ln\frac{1}{1-C(z)}$
m-Cycle	$\mathcal{B} = CYC_m(\mathcal{C})$	$B(z) = \frac{1}{m}C(z)^m$
m-Set	$\mathcal{B} = SET_m(\mathcal{C})$	$B(z) = \frac{1}{m!}C(z)^m$
Labeled product	$\mathcal{B} = \mathcal{C} * \mathcal{D}$	$B(z) = C(z) \cdot D(z)$

Thus, constructing the combinatorial class and using it to determine the corresponding EGF gives

$$\mathcal{P}_{c,k} \cong CYC_k(\mathcal{A}) * SET_{c-1}(CYC(\mathcal{A})) \tag{17}$$

$$P_{c,k}(z) = \frac{z^k}{k} \frac{1}{(c-1)!} \left(\ln\frac{1}{1-z}\right)^{c-1} \tag{18}$$

The EGF in (18) can then be used to sum c_k over all permutations with a given score.

$$\sum_{\pi \in \mathcal{C}_{s,n}} c_k(\pi) = n![z^n]P_{c,k}(z)$$

$$= n![z^n]\frac{z^k}{k}\frac{1}{(c-1)!}\left(\ln\frac{1}{1-z}\right)^{c-1}$$

$$= \frac{n!}{k}[z^{n-k}]\frac{1}{(c-1)!}\left(\ln\frac{1}{1-z}\right)^{c-1}$$

$$= \frac{n!}{k(n-k)!}[z^{n-k}]\frac{(n-k)!}{(c-1)!}\left(ln\frac{1}{1-z}\right)^{c-1}$$

$$= \frac{n!}{k(n-k)!}\begin{bmatrix} n-k \\ c-1 \end{bmatrix} \tag{19}$$

The final step in (19) uses the fact [3] that

$$\begin{bmatrix} m \\ r \end{bmatrix} = \frac{m!}{r!}[z^m]\left(ln\frac{1}{1-z}\right)^r. \tag{20}$$

□

5   Perfect Sorting with the Cycle Metric

Lemma 2. *If indices i and j in permutation π are fixed to give a case (1/1), case (1/2$^+$), case (2), or case (3$^+$) selection, then any permutation π' of the same length, with $\pi(i) = \pi'(i)$ and $\pi(j) = \pi'(j)$, will be the same case.*

Proof. The proof follows trivially from the fact that a 1-cycle at i or j is a 1-cycle in both π and π', and a cycle containing both i and j in consecutive order does so in both π and π'.   □

Lemma 3. *Suppose that indices i and j are selected to give a case (1/1), case (1/2$^+$), case (2), or case (3$^+$) selection. Then the consensus sorting algorithm with an ideal database and cycle metric will make the decision which leads to the greatest improvement in score, which will be to swap in case (2) and case (3$^+$), and to do nothing in case (1/1) and (1/2$^+$).*

Proof. Consider a permutation π' of length n, with $\pi'(i)$ at index i and $\pi'(j)$ at j. Then the selection of i and j in π' represents the same case as the same selection in π, by Lemma 2. By Lemma 1, if π contains a 1-cycle at i or j, then swapping i and j in π' must degrade the score of π' by 1. Likewise, if π does not contain a 1-cycle at either i or j, then swapping i and j in π' must create a 1-cycle, and thus improve the score by 1.

Thus, for case (2) and case (3$^+$), every matching permutation (see Fig. 1) $\pi_M \in \mathcal{D}_I$ has a corresponding permutation $\pi_R \in \mathcal{D}_I$ with i and j reversed, where

$$\phi(\pi_M) - 1 = \phi(\pi_R).$$

Likewise, for case (1/1) and case (1/2$^+$), every matching permutation $\pi_M \in \mathcal{D}_I$ has a corresponding permutation $\pi_R \in \mathcal{D}_I$ with i and j reversed, where

$$\phi(\pi_M) + 1 = \phi(\pi_R).$$

Therefore, case (2) and case (3$^+$) will result in a swap, whereas case (1/1) and case (1/2$^+$) will result in no swap.   □

Lemma 4. *Suppose that indices i and j are selected to give a case $(2^+/2^+)$ or a case (2^+2^+) selection. Then the consensus sorting algorithm with an ideal database and cycle metric will make the decision not to swap, regardless of whether a swap would improve the score.*

Proof. Given any matching permutation $\pi_M \in \mathcal{D}_I$, the members i and j are followed by i' and j', respectively. Construct a permutation $\pi_R \in \mathcal{D}_I$ from π_M as follows. First, swap the elements at i and j, resulting in a reversed permutation, and then swap the elements at i' and j', giving the result

$$\pi_R = (i'j') \cdot (ij) \cdot \pi_M.$$

If i and j were originally part of the same cycle, then the first swap creates a cycle, and the second swap re-merges the cycles. Likewise, if i and j were originally part of different cycles, then the first swap merges the cycles and the second swap creates a cycle.

Thus, there is a one-to-one correspondence between matching permutations $\pi_M \in \mathcal{D}_I$ and reversed permutations $\pi_R \in \mathcal{D}_I$, such that both permutations in the pair have the same score. Therefore, the consensus sorting algorithm would not be able to decide between swapping and not, and will not swap by default.   □

Proposition 8. *The cycle metric together with an ideal database will fully sort an array with probability approaching 1, allowing for an unbounded number of consensus sorting iterations.*

Proof. Lemmas 3 and 4 demonstrate that only case (2) and case (3^+) will take part in a swap, if an ideal database is used, and in either of those two cases, the score will be reduced. Furthermore, any array which is not fully sorted must contain at least one cycle which is not a 1-cycle. Therefore, it is possible to select a pair of indices i and j which are consecutive members of the same cycle, making i and j a case (2) or case (3^+) selection. Therefore there would exist a score-reducing choice of elements.

Since both conditions of Theorem 1 are satisfied, the array is expected to be fully sorted with probability approaching 1.   □

Remark 5. If an error occurs, there will be a relative degradation of score by 1 due to noise in case (1/1), case $(1/2^+)$, case $(2^+/2^+)$, case (2), and case $(3+)$. If an error occurs in case (2^+2^+), the result will be an improvement in score by 1, because of Lemma 4.

6   Empirical Results

The consensus sort algorithm using the cycle metric was run over a range of database sizes and database sizes. Figures 2 and 3 show the average level of score improvement (measured by taking the final score at the point of convergence as a percentage of the initial random permutation score) for fixed permutation sizes over a range of database sizes. The plots show the average over an epoch of 500 runs with a 95% confidence interval.

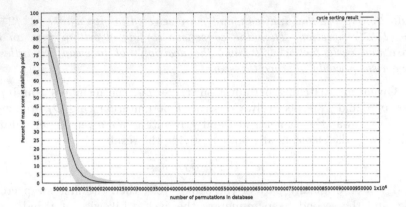

Fig. 2. The empirical equilibrium of a cycle metric consensus sort for an array size of 50 and a range of database sizes, averaged over an epoch of 500.

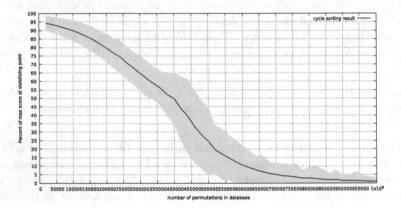

Fig. 3. The empirical equilibrium of a cycle metric consensus sort for an array size of 100 and a range of database sizes, averaged over an epoch of 500.

7   Conclusion

It has been shown that when the cycle metric is used in conjunction with the consensus sort, a perfect sort is expected in the ideal case. However, the same can be said of the inversion metric. Empirical results suggest that if a different database is used, a certain database size is needed before the cycle metric will perform well. Beyond that point, the cycle metric is far superior to the inversion metric, in that it converges to zero far more quickly as database size increases, meaning that it produces arrays which are closer to the fully sorted goal. The inversion metric is far more likely than the cycle metric to produce a sorting error which will result in a significant degradation in score, and thus impede sorting.

Previous work [1] has analyzed score degradation in the inversion metric, and concluded that the score of an array at the point of convergence could be predicted by heuristic. The heuristic used to predict the final score used the theoretical mean rate of improvement in score in each sorting step, as well as the mean rate of score degradation due to sorting errors from using a database which was not ideal, and determined when the two were in steady-state.

The task becomes more complicated in the case of the cycle metric. Indeed, the steady state of the mean rate of improvement and the mean rate of degradation due to error can be computed for the the cycle metric in order to determine a steady-state, using the tools developed in this paper. This has been done, with partial success. The difficulty lies in the fact that using mean rates makes an assumption of a uniform distribution of permutations by score, which is unlikely to hold during the actual sorting process while using the cycle metric. What is needed to improve the analysis is a classification of permutations which is less sensitive to the exact distribution of permutations within a class.

Regardless of the ability to predict convergence point, it is undeniable that in the example studied in this paper, the fitness score derived from data mining worked as an effective optimization tool in finding permutations with scores very close to the goal.

References

1. Avramovic, I., Richards, D.S.: Randomized sorting as a big data search algorithm. In: International Conference on Advances in Big Data Analytics, ABDA 2015, pp. 57–63 (2015)
2. Estivill-Castro, V., Wood, D.: A survey of adaptive sorting algorithms. ACM Comput. Surv. **24**(4), 441–476 (1992)
3. Flajolet, P., Sedgewick, R.: Analytic combinatorics - symbolic combinatorics. Technical report, Algorithms Project, INRIA Rocquencourt (2002)
4. Hashem, I.A.T., Yaqoob, I., Anuar, N.B., Mokhtar, S., Gani, A., Khan, S.U.: The rise of "big data" on cloud computing: Review and open research issues. Inf. Syst. **47**, 98–115 (2015)
5. Hays, J., Efros, A.A.: IM2GPS: estimating geographic information from a single image. In: IEEE Conference on Computer Vision and Pattern Recognition, pp. 1–8 (2008)
6. Tian, W., Zhao, Y.: Optimized Cloud Resource Management and Scheduling, Chap. 2. Morgan Kaufmann, Elsevier Science & Technology Books, Amsterdam (2014)
7. Wang, H., He, X., Chang, M.W., Song, Y., White, R.W., Chu, W.: Personalized ranking model adaptation for web search. In: Proceedings of the 36th International ACM SIGIR Conference on Research and Development in Information Retrieval, SIGIR 2013, pp. 323–332. ACM, New York (2013). https://doi.org/10.1145/2484028.2484068
8. Yang, D., Zhang, D., Yu, Z., Yu, Z., Zeghlache, D.: SESAME: mining user digital footprints for fine-grained preference-aware social media search. ACM Trans. Internet Technol. **14**(4), 281–2824 (2014). https://doi.org/10.1145/2677209

On the Competitiveness of Memoryless Strategies for the k-Canadian Traveller Problem

Pierre Bergé[1(✉)], Julien Hemery[2], Arpad Rimmel[2], and Joanna Tomasik[2]

[1] LRI, Université Paris-Sud, Université Paris-Saclay, Orsay, France
Pierre.Berge@lri.fr
[2] LRI, CentraleSupélec, Université Paris-Saclay, Orsay, France
Julien.Hemery@supelec.fr, {Arpad.Rimmel,Joanna.Tomasik}@lri.fr

Abstract. The k-Canadian Traveller Problem (k-CTP), proven PSPACE-complete by Papadimitriou and Yannakakis, is a generalization of the Shortest Path Problem which admits blocked edges. Its objective is to determine the strategy that makes the traveller traverse graph G between two given nodes s and t with the minimal distance, knowing that at most k edges are blocked. The traveller discovers that an edge is blocked when arriving at one of its endpoints.

We study the competitiveness of randomized memoryless strategies to solve the k-CTP. Memoryless strategies are attractive in practice as a decision made by the strategy for a traveller in node v of G does not depend on his anterior moves. We establish that the competitive ratio of any randomized memoryless strategy cannot be better than $2k + O(1)$. This means that randomized memoryless strategies are asymptotically as competitive as deterministic strategies which achieve a ratio $2k+1$ at best.

Keywords: Online algorithms · Competitive analysis Canadian traveller problem

1 Introduction

The *Canadian Traveller Problem* (CTP), a generalization of the *Shortest Path Problem*, was introduced in [6]. Given an undirected weighted graph $G = (V, E, \omega)$ and two nodes $s, t \in V$, the objective is to design a strategy to make a traveller walk from s to t through G on the shortest path possible. An additional strain comes from set E_*, $E_* \subset E$ of blocked edges. The traveller does not know, however, which edges are blocked. He discovers a blocked edge, also called *blockage*, when arriving at one of its endpoints. This implies that we solve the CTP with online algorithms, called strategies. The k-*Canadian Traveller Problem* (k-CTP) is the parameterized variant of CTP, where an upper bound k for the number of blocked edges is given. Both CTP and k-CTP are PSPACE-complete [2,6].

© Springer Nature Switzerland AG 2018
D. Kim et al. (Eds.): COCOA 2018, LNCS 11346, pp. 566–576, 2018.
https://doi.org/10.1007/978-3-030-04651-4_38

State-of-the-Art. Strategies for the k-CTP are studied through the competitive analysis, which evaluates their quality [4]. The competitive ratio of a strategy is the maximum, over all satisfiable instances, of the ratio of the distance traversed by the traveller following the strategy and the *optimal offline cost*, which is the distance he would traverse if he knew blocked edges from the beginning.

There are two classes of strategies: deterministic and randomized. Westphal [7] proved that there is no deterministic strategy that achieves a competitive ratio better than $2k + 1$. This ratio is reached by REPOSITION and COMPARISON strategies [7,8]. The REPOSITION strategy repeats an attempt to reach t through the shortest (s,t)-path going back to s after the discovery of an obstacle. As in practical cases, such as urban networks, it does not seem realistic, Xu *et al.* [8] introduced the GREEDY algorithm. For grids, it achieves ratio $O(1)$, regardless of k. However, for any graph, this ratio is $O(2^k)$.

We evaluate the competitiveness of the randomized strategies by calculating the maximal ratio of the mean distance traversed by the traveller following the strategy and the optimal offline cost. Westphal [7] proved that no randomized algorithm can attain a ratio smaller than $k + 1$. However, unlike the deterministic case, no $(\alpha k + 1)$-competitive randomized strategy, $\alpha < 2$, was identified, excepted two very particular cases for which randomized strategies have been proposed. Demaine *et al.* [5] designed a strategy with a ratio $\left(1 + \frac{\sqrt{2}}{2}\right) k + 1$, executed in time of $O\left(k\mu^2 |E|^2\right)$, where parameter μ may be exponential. It is dedicated to graphs that can be transformed into apex trees. Bender *et al.* studied in [3] a restriction of k-CTP for graphs composed of node-disjoint (s,t)-paths and proposed a polynomial-time strategy with ratio $(k + 1)$.

Contributions and Paper Plan. We study the competitiveness of memoryless strategies [1,4]. The choice that the strategy makes for the traveller at node v (to decide where he should go next) depends only on the graph deprived of edges already discovered and the current position of the traveller. Memoryless strategies are easy to be implemented as they do not memorize the edges already visited by the traveller to make a decision. The only information they use is the graph $G \backslash E'_*$, which is the graph G deprived of the blocked edges discovered $E'_* \subseteq E_*$. Given that deterministic strategies cannot achieve a ratio better than $2k + 1$, our goal is to prove that randomized memoryless strategies are not more competitive asymptotically and attain the ratio $2k + O(1)$. To do this, we compute a lower bound $c_{l_0} = 2k + O(1)$ on the competitive ratio of any randomized memoryless strategy for a certain set of instances of the k-CTP.

We remind, in Sect. 2, the definitions of k-CTP, memoryless strategies, and the competitive ratio. In Sect. 3, we present sets $\mathscr{R}_k$ of *road maps*, *i.e.* pairs (G, E_*), which are a means to study the performance of memoryless strategies. We prove in Sect. 4 that randomized memoryless strategies cannot drop below a ratio $c_k = 2k + O(1)$ on road maps in $\mathscr{R}_k$, where expression $O(1)$ is made precise. Eventually, we draw conclusions and highlight the future work in Sect. 5.

2   Definitions

We start by introducing the notation. For any graph $G = (V, E, \omega)$, let $G \backslash E'$ denotes its subgraph $(V, E \backslash E', \omega)$.

2.1   Memoryless Strategies for the k-CTP

Let $G = (V, E, \omega)$ be an undirected graph with positive weights. The objective is to make a traveller traverse the graph from a source node s to a target one t, with $s, t \in V$ and a set $E_* \subsetneq E$ of blocked edges. The traveller does not know a priori which edges are blocked. He discovers a blocked edge only when arriving at one of its endpoints. The goal is to design a strategy A with the minimum competitive ratio.

We focus on *memoryless strategies* (MS). Concretely, we suppose that the traveller forgets the nodes which he has already visited. In other words, a decision of an MS is independent of the nodes already visited. Each time when he starts to trace a path to target t, his map is refreshed. In the literature, the term *memoryless* was used in the context of online algorithms (*e.g.* PAGING PROBLEM [4], LIST UPDATE PROBLEM [1]) which make decisions according to the current state, ignoring past events. An MS can be either deterministic or randomized.

Definition 1 (Memoryless Strategies for the k-CTP). *A deterministic strategy A is an MS if and only if (iff) the next node w the traveller visits depends only on graph G deprived of blocked edges already discovered E_*' and the current traveller position v: $w = A(G \backslash E_*', v)$. Similarly, a randomized strategy A is an MS iff node w is the realization of a discrete random variable $X = A(G \backslash E_*', v)$.*

For example, the GREEDY strategy [8] is a deterministic MS. It consists in choosing at each step the first edge of the shortest path between the current node v and the target t. In contrast, the REPOSITION strategy [7] is not an MS as its decision refers to the past moves of the traveller. The polynomial-time strategies proposed in the literature do not use much memory information in the decision-making process. Either they are memoryless or they use a small amount of memory. For example, REPOSITION (deterministic [7], randomized [3]) can be implemented with a one bit memory given that the only information to retain is whether the traveller tries to reach t or returns to s.

The following process allows us to identify whether a deterministic strategy A is an MS. Let us suppose that a traveller T_1 follows strategy A: he has already visited certain nodes of the graph, he is currently at node v but he has not reached target t yet. Let us imagine a second traveller T_2 who is airdropped on node v of graph $G \backslash E_*'$ and is guided by strategy A. If the traveller T_2 always follows the same path as T_1 until reaching t, A is a deterministic MS. If T_1 and T_2 may follow different paths, then A is not an MS. Formally, proving that a strategy is a MS consists in finding the function which transforms the pair $(G \backslash E_*', v)$ into node $w = A(G \backslash E_*', v)$.

2.2   Competitive Ratio

Let (G, E_*) be a *road map*, *i.e.* a pair with graph $G = (V, E, \omega)$ and blocked edges $E_* \subsetneq E$, such that there is an (s, t)-path in graph $G \backslash E_*$ (nodes s and t remain in the same connected component when all blocked edges are discovered). We denote by $\omega_A (G, E_*)$ the distance traversed by the traveller reaching t with strategy A on graph G with blocked edges E_* and $\omega_{\min} (G, E_*)$ the cost of the shortest (s, t)-path in graph $G \backslash E_*$.

The ratio $\omega_A (G, E_*) / \omega_{\min} (G, E_*)$ is abbreviated as $c_A (G, E_*)$. A strategy A is c_A-competitive [4,8] iff for any (G, E_*), $\omega_A (G, E_*) \leq c_A \omega_{\min} (G, E_*)$. Otherwise stated, for any (G, E_*), $c_A (G, E_*) \leq c_A$. If strategy A is randomized, $\omega_A (G, E_*)$ is replaced by $\mathbb{E} (\omega_A (G, E_*))$ which is the expected distance traversed by the traveller to reach t with strategy A. The competitive ratio can also be evaluated on a family $\mathscr{R}$ of road maps, put formally:

$$c_{A, \mathscr{R}} = \max_{(G, E_*) \in \mathscr{R}} c_A (G, E_*). \tag{1}$$

This "local" competitive ratio fulfils $c_{A, \mathscr{R}} \leq c_A$. The definition of the competitive ratio can also be extended to families of strategies. We denote by c_{MS} the competitive ratio of MSes, which is the minimum over competitive ratios of any MSes: $c_{\mathrm{MS}} = \min_{A \, \mathrm{MS}} c_A$.

3   Road Atlas Used to Study Randomized MSes

Before specifying *road atlases* $\mathscr{R}_k$, *i.e.* families of road maps we construct to evaluate the competitiveness of randomized MSes, we need to introduce the concepts used in their definition.

We define recursively a sequence of graphs G_i for $i \geq 1$ with weights from $\{1, \varepsilon\}$, $0 < \varepsilon \ll 1$. Graphs G_1 and G_{i+1} are represented in Fig. 1a and b, graphs G_2 and G_3 are shown in Fig. 1c and d. Edges with weight 1 are thicker than edges with weight ε (weights ε are omitted in Fig. 1c and d). For any graph G_i, axis Δ_{vert} is its vertical axis of symmetry (Fig. 1c and d).

We focus on road maps (G_i, E_*) composed of graph G_i but also at most i blocked edges which are on the right side of axis Δ_{vert}. Indeed, blocking edges on the left side of Δ_{vert} in G_i would affect negligibly the total distance traversed by a traveller. Let us suppose that a traveller traverses graph G_i and has already discovered some blocked edges $E'_* \subseteq E_*$. Then, he considers graph $G_i \backslash E'_*$ and tries to reach t, being ignorant of the identity of the undiscovered blocked edges. We denote by $\mathscr{G}$ the set of all the subgraphs of G_i, *i.e.* graphs $G_i \backslash E'_*$ with at most i edges in E'_* on the right side of Δ_{vert}, for any $i \geq 1$. We call them *diamond graphs* because of their appearance, diamonds joined together. Formally, we write $\mathscr{G} = \bigcup_{i=1}^{+\infty} \{G_i \backslash E'_* : |E'_*| \leq i\}$. For any graph $G \in \mathscr{G}$, we partition its edges, denoted by E_G, into two sets $E_{G, \mathrm{left}}$ (on the left side of axis Δ_{vert}) and $E_{G, \mathrm{right}}$ (on the right side of axis Δ_{vert}).

To any diamond graph G of $\mathscr{G}$, we associate a *diamond binary tree* (DBT), denoted by T_G. Tree T_G, rooted in t, is obtained from the right half of graph G

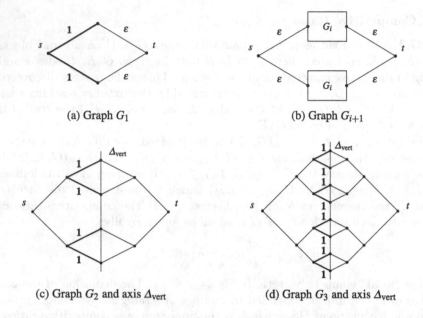

(a) Graph G_1 (b) Graph G_{i+1}

(c) Graph G_2 and axis Δ_{vert} (d) Graph G_3 and axis Δ_{vert}

Fig. 1. Recursive construction of graphs G_i

(on the right side of axis Δ_{vert}) by successive contractions of edges: any node with a single son is merged with its father (in Fig. 2a: edge (t, v_2) is contracted, v_2 merges with t). We denote by $T_\varnothing$ the empty tree. Any nonempty tree is a triplet (v, T_a, T_b) with a root $v \in V$ and trees T_a and T_b. Figures 2a and b illustrate the construction of the DBT.

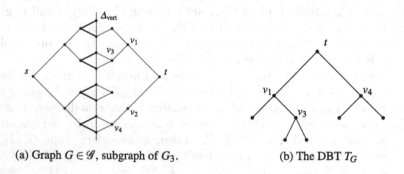

(a) Graph $G \in \mathcal{G}$, subgraph of G_3.   (b) The DBT T_G

Fig. 2. An example of graph $G \in \mathcal{G}$ and its DBT T_G.

To put the definition of DBTs, let $L(v)$ denote the set of sons of node v which is only defined for the nodes on the right side of Δ_{vert}. For all $v \in \Delta_{\text{vert}}$, we have $L(v) = \emptyset$. For graph G of Fig. 2a, $L(t) = \{v_1, v_2\}$, $L(v_2) = \{v_4\}$, for example.

Function BIN-TREE gives the construction of tree T_G, which is BIN-TREE (t):

$$\text{BIN-TREE } (v) = \begin{cases} T_\varnothing \text{ if } L(v) = \emptyset, \\ \text{BIN-TREE } (v_{\text{next}}) \text{ if } L(v) = \{v_{\text{next}}\}, \\ (v, \text{BIN-TREE}(v_{\text{up}}), \text{BIN-TREE}(v_{\text{down}})) \text{ if } L(v) = \{v_{\text{up}}, v_{\text{down}}\}. \end{cases}$$

We say that the depth of a node v in a DBT T, denoted by $d(v)$, is equal to the number of edges separating it from the root. We denote by $d_{\min}(T)$ the minimum depth of all T leaves. For example, for DBT T_G in Fig. 2b, $d_{\min}(T_G) = 2$.

The depth of an edge (u,v), $D(u,v)$, is defined as $D(u,v) = \max\{d(u), d(v)\}$. We say that edge e' is the *mother* of edge e if these two edges share one endpoint and $D(e') = D(e) - 1$, putting it shortly $e' = P(e)$. Conversely, we say e is the *daughter* of $P(e)$. Edge e^* is the *aunt* of edge e if e^* and $P(e)$ share one endpoint and $D(P(e)) = D(e^*)$. We indicate this fact as $e^* = U(e)$. Observe that the aunt of e and its mother share the same ancestor. For example, in Fig. 2b, edge (t, v_4) is the aunt of (v_1, v_3).

Now we define the graphs contained in the road maps of atlas $\mathscr{R}_k$.

Definition 2 (Sets $\mathscr{D}_k$). *Infinite set $\mathscr{D}_k$ contains graphs of $\mathscr{G}$ such that their DBT T_G fulfils $d_{min}(T_G) \geq k$: $\mathscr{D}_k = \{G \in \mathscr{G} : d_{min}(T_G) \geq k\}$.*

In other words, if graph G belongs to $\mathscr{D}_k$, then its DBT T_G induced on nodes of depth less than k forms a complete binary tree. For example, the DBT T_G on Fig. 2b contains a complete binary tree of depth 2, so $G \in \mathscr{D}_2$. Finally, we define road atlases $\mathscr{R}_k$:

Definition 3 (Road atlas $\mathscr{R}_k$). *Road atlas $\mathscr{R}_k$ is composed of road maps (G, E_*), where:*

- *Graph G belongs to $\mathscr{D}_k$: $G \in \mathscr{D}_k$,*
- *Set E_* becomes $\{\hat{e}, U(\hat{e}), U^2(\hat{e}), \ldots, U^{k-1}(\hat{e})\}$ in the DBT T_G, with $D(\hat{e}) = k$.*

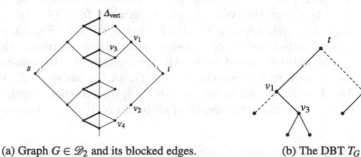

(a) Graph $G \in \mathscr{D}_2$ and its blocked edges.   (b) The DBT T_G

Fig. 3. An example of road map $(G, E_*) \in \mathscr{R}_2$, edges of E_* are dashed and blue. (Color figure online)

In Fig. 3a, we give an example of road map (G, E_*) in $\mathscr{R}_2$ where G is the graph initially drawn in Fig. 2a and edges of E_* are dashed and in blue. In Fig. 3b, we provide the corresponding DBT to see that the road map fulfils Definition 3 for $k = 2$.

For road map $(G, E_*) \in \mathscr{R}_k$, set $E_* \subsetneq E_{G,\text{right}}$ contains k edges and there is no two of them with the same depth in T_G. Moreover, there is a unique node $v_{k,j}$ among all nodes of depth k such that there is an open (s, t)-path containing $v_{k,j}$ in $G \backslash E_*$. In brief, any traveller on road map $(G, E_*) \in \mathscr{R}_k$ must traverse this node in order to reach t directly.

4   Competitiveness of Randomized MSes

We study the competitiveness of randomized MSes for road atlases $\mathscr{R}_k$. The MS performance is determined by properties of the corresponding DBTs T_G. These properties result from relations which exist between DBT edges.

The following theorem states that cutting one edge from $G \in \mathscr{D}_k$ produces a graph $G \backslash \{e\} \in \mathscr{D}_{k-1}$.

Theorem 1. *For any $G \in \mathscr{D}_k$ and edge $e \in E_{G,\text{right}}$, graph $G \backslash \{e\} \in \mathscr{D}_{k-1}$.*

Proof. Let $G \in \mathscr{D}_k$ and e be an edge in $E_{G,\text{right}}$. There is an edge e_T in T_G for which $T_{G \backslash \{e\}}$ is obtained by removing e_T and its descendants from T_G and next applying the edge contraction, if necessary. For example, if $e = (t, v_2)$ in Fig. 2b, then $e_T = (t, v_4)$. Let v be the "shallower" endpoint of edge $e_T = \{u, v\}$, *i.e.* $d(v) < d(u)$. Edge e_T and its mother have this node in common, $v \in P(e_T)$. We distinguish two cases:

- **The depth of node u is greater or equal to $d_{\min}(T_G)$:** If u is the unique leaf of depth $d_{\min}(T_G)$, the depth of leaves of the DBT $T_{G \backslash \{e\}}$ is $d_{\min}(T_G) - 1 \geq k - 1$. Otherwise, in DBT $T_{G \backslash \{e\}}$, the depth of leaves is still equal to $d_{\min}(T_G) \geq k$. In both cases, $G \backslash \{e\} \in \mathscr{D}_{k-1}$.
- **The depth of node u is strictly inferior to $d_{\min}(T_G)$:** Let T_v be the subtree of T_G with root v. We denote by w the brother of node u, *i.e.* the other son of node v (Fig. 4b). When edge e is removed from G, edge e_T and its descendants are withdrawn in the DBT (in the DBT in Fig. 4b, edge e has not been contracted, so $e = e_T$). Consequently, after the contraction, T_v becomes T_w, the subtree rooted in w. All the leaves of T_w have initially a depth greater than k, so by removing e from G, all the leaves of T_w have a depth greater than $k - 1$. All leaves outside T_v, preserve their depth which is greater than k. Therefore, the depth of all leaves of $T_{G \backslash \{e\}}$ is greater than $k - 1$.

After examining these two cases, we conclude that $G \backslash \{e\}$ belongs to $\mathscr{D}_{k-1}$. $\square$

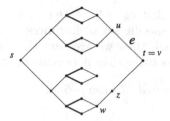

(a) Example of graph $G \in \mathscr{D}_2$. Edge e (red) is about to be removed from this graph.

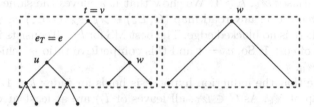

(b) The DBT T_G. Blue frame covers nodes of depth 0 to 2. Two leaves are at depth 2.

(c) The DBT $T_{G \setminus \{e\}}$. Nodes of depth 0 to 1 are framed. Graph $G \setminus \{e\}$ belongs to $\mathscr{D}_1$.

Fig. 4. Illustration of the proof of Theorem 1 on a subgraph of G_3.

Corollary 1. *For any road map* $(G, E_*) \in \mathscr{R}_k$ *and edge* $e \in E_*$, *we have:*

$$(G \setminus \{e\}, E_* \setminus \{e\}) \in \mathscr{R}_{k-1}.$$

Proof. Let $e = (u, v)$ and v be the shallowest endpoint of e. As $e \in E_*$, its depth is less than k. We know that $E_* = \{\hat{e}, U(\hat{e}), \ldots, U^{k-1}(\hat{e})\}$ with $D(\hat{e}) = k$. We denote edge e by $U^j(\hat{e})$ with $0 \le j \le k-1$. As a consequence, the depth of edge e is $k - j$: $D(e) = k - j$.

As $G \in \mathscr{D}_k$, any edge of depth less than $k - 1$ has two daughters. In graph $G \setminus \{e\}$, edge $P(e)$ has only one daughter as edge e disappeared. Consequently, nodes v and the sibling node of u are merged in the DBT of graph $G \setminus \{e\}$.

Now we prove that $U^{j+1}(\hat{e})$ becomes the aunt of $U^{j-1}(\hat{e})$ in the DBT $T_{G \setminus \{e\}}$, *i.e.* after the removal of $e = U^j(\hat{e})$. Indeed, the daughters of the sibling edge of e in T_G are now the daughters of $P(e)$ in $T_{G \setminus \{e\}}$. So, edge $U^{j+1}(\hat{e})$ which used to be the aunt of e is the aunt of $U^{j-1}(\hat{e})$ in $T_{G \setminus \{e\}}$. Therefore, set $E_* \setminus \{e\}$ can be written $\{\hat{e}, U(\hat{e}), \ldots, U^{k-2}(\hat{e})\}$ in $G \setminus \{e\}$. Thanks to Theorem 1, we have $G \setminus \{e\} \in \mathscr{D}_{k-1}$ which terminates the proof. □

We denote by c_k the competitive ratio of the best memoryless strategy for road atlases $\mathscr{R}_k$, $k \ge 1$. Formally:

$$c_k = \min_{A \in \text{MS}} c_{A, \mathscr{R}_k}. \tag{2}$$

Our objective is to show that $c_k = 2k + O(1)$. As value c_k gives the competitiveness of MSes over a specific set of instances, it is a lower bound of c_{MS}. If our objective is achieved, then we are sure that randomized MSes are not asymptotically more competitive than deterministic strategies.

Theorem 2. *Any randomized MS competitive ratio is at least* $2k + O(1)$ *over road atlas* $\mathcal{R}_k$.

Proof. We prove by induction that $c_k = 2k + 1 - \psi(k-1)$, where $\psi(k-1) = \sum_{j=0}^{k-1} \frac{c_j + 1}{2^{j+1}}$ is a convergent series bounded by a constant. Let A be the best MS over all road atlases $\mathcal{R}_k$, $k \geq 1$. We show that it achieves the same competitive ratio for a given k over any road map $(G, E_*) \in \mathcal{R}_k$: $c_A(G, E_*) = c_k$.

If $k = 0$, there is no blocked edge. The best MS for $k = 0$ consists in traversing an (s,t)-path of cost 1. So, $c_0 = 1$ and this competitive ratio is achieved for any road map in $\mathcal{R}_0$.

We assume that the induction hypothesis holds for index $k - 1$. Let (G, E_*) be a road map of $\mathcal{R}_k$. As $G \in \mathcal{D}_k$, all leaves of T_G are at least at depth k. So, T_G is complete up to depth k and has 2^k nodes of depth k. We suppose that the traveller, guided by the most competitive MS A over atlas $\mathcal{R}_k$, is standing at source s and starts his walk on a road map $(G, E_*) \in \mathcal{R}_k$. We focus on value $c_A(G, E_*)$.

As strategy A is the most competitive, the traveller using it either reaches t directly with distance 1 or meets a blocked edge and thus traverses a total distance less than $2 + c_{k-1}$: distance 1 to reach the blockage, distance 1 to go back to a node on the left-hand side of Δ_{vert}, and at most distance c_{k-1} to reach t on the new road map which belongs to $\mathcal{R}_{k-1}$ (Corollary 1).

Indeed, remember that when the traveller meets a blockage e^* for the first time, the only information taken into account by the MS A after this moment is the position of the traveller and the current graph $G \setminus \{e^*\}$. This justifies the use of the inductive term c_{k-1}, as strategy A is not influenced by the past and guides the traveller independently of its previous trips. The traveller, who necessarily returns to s after being blocked in an instance from $\mathcal{R}_k$, faces now an instance of $\mathcal{R}_{k-1}$.

For any $1 \leq j \leq 2^k$, let $p_{k,j}^A$ signify the probability that the traveller visits the jth node at depth k, denoted by $v_{k,j}$ (index j passes from left to right in the DBT representation).

We denote by j^* the index of the only node v_{k,j^*} such that there is an open (s,t)-path containing it. Obviously, the traveller does not know the identity of node v_{k,j^*} as he ignores E_*. If he chooses luckily to walk on a simple (s,t)-path containing v_{k,j^*}, then he reaches t with distance 1. Otherwise, if he chooses an (s,t)-path traversing node $v_{k,j}$ with $j \neq j^*$, he meets a certain blocked edge $e_{k,j}$. We have:

$$c_A(G, E_*) = p_{k,j^*}^A + \sum_{j \neq j^*} p_{k,j}^A \left(2 + c_A(G \setminus \{e_{k,j}\}, E_* \setminus \{e_{k,j}\})\right). \tag{3}$$

According to Yao's principle [9], probabilities $p_{k,j}^A$ necessarily follow the uniform distribution and are all equal to $\frac{1}{2^k}$. From the traveller point of view, all nodes $v_{k,j}$ are indistinguishable. A strategy with a non-uniform distribution necessarily puts some nodes $v_{k,j}$ at a disadvantage, with $p_{k,j}^A < \frac{1}{2^k}$. Moreover, strategy A has to be competitive on any instance of $\mathscr{R}_k$. Applied on a road map of $\mathscr{R}_k$ where one of these penalized nodes is v_{k,j^*}, such a strategy makes the probability to reach t with distance 1 decrease and, therefore, the competitive ratio increases. Consequently, the best MS A fulfils $p_{k,j}^A = \frac{1}{2^k}$ for any node $v_{k,j}$.

According to the induction hypothesis, the best MS for road atlas $\mathscr{R}_{k-1}$ performs ratio c_{k-1} for any road map in $\mathscr{R}_{k-1}$. Thanks to this observation and the previous remark on the probability values $p_{k,j}$, we obtain from Equation (3) that the competitive ratio of A is the same for all road maps of $\mathscr{R}_k$:

$$c_A\left(G, E_*\right) = \frac{1}{2^k} + \left(1 - \frac{1}{2^k}\right)\left(2 + c_{k-1}\right) = c_k.$$

We observe that $c_k - c_{k-1} = 2 - \frac{1}{2^k} - \frac{c_{k-1}}{2^k}$ and we obtain the following iterative formula thanks to the induction hypothesis:

$$c_k = 2 - \frac{1}{2^k} - \frac{c_{k-1}}{2^k} + 2(k-1) + 1 - \sum_{j=0}^{k-2} \frac{c_j + 1}{2^{j+1}} = 2k + 1 - \sum_{j=0}^{k-1} \frac{c_j + 1}{2^{j+1}}.$$

As $c_k \leq 2k + 1$, $\sum_{j=0}^{+\infty} \frac{c_j + 1}{2^{j+1}}$ converges and $c_k = 2k + O(1)$. For $k = 10^4$, the numerical computations give $\psi(10^4) = \sum_{j=0}^{10^4} \frac{c_j + 1}{2^{j+1}} = 3.213$, so value c_k is larger than $2k - 2.22$.   □

As c_k represents the competitive ratio of the best competitive MS over road atlas $\mathscr{R}_k$, no MS can go below $2k + O(1)$ in terms of competitiveness.

5   Conclusion and Further Work

We studied the competitiveness of the MSes for the k-CTP. An MS is a strategy which does not make decisions referring to the anterior moves of the traveller, in other words, the nodes the traveller visited until his current position.

Then, we constructed a series of k-CTP instances, called road atlases and denoted by $\mathscr{R}_k$. We foremost concluded that a randomized MS cannot reach a competitive ratio better than $2k + O(1)$ on road atlas $\mathscr{R}_k$. That is to say that we identified an upper bound on the competitive ratio of randomized MSes which is significantly higher than the existing one $k + 1$. In future research, if we aim at designing a strategy with competitive ratio $\alpha k + O(1)$, $\alpha < 2$, we shall focus on strategies which are not only randomized but use memory as well.

References

1. Albers, S.: Online algorithms: a survey. Math. Program. **97**, 3–26 (2003)
2. Bar-Noy, A., Schieber, B.: The Canadian traveller problem. In: Proceedings of ACM/SIAM SODA, pp. 261–270 (1991)
3. Bender, M., Westphal, S.: An optimal randomized online algorithm for the k-Canadian traveller problem on node-disjoint paths. J. Comb. Optim. **30**(1), 87–96 (2015)
4. Borodin, A., El-Yaniv, R.: Online Computation and Competitive Analysis. Cambridge University Press, Cambridge (1998)
5. Demaine, E.D., Huang, Y., Liao, C.-S., Sadakane, K.: Canadians should travel randomly. In: Esparza, J., Fraigniaud, P., Husfeldt, T., Koutsoupias, E. (eds.) ICALP 2014. LNCS, vol. 8572, pp. 380–391. Springer, Heidelberg (2014). https://doi.org/10.1007/978-3-662-43948-7_32
6. Papadimitriou, C., Yannakakis, M.: Shortest paths without a map. Theor. Comput. Sci. **84**(1), 127–150 (1991)
7. Westphal, S.: A note on the k-Canadian traveller problem. Inform. Proces. Lett. **106**(3), 87–89 (2008)
8. Xu, Y., Hu, M., Su, B., Zhu, B., Zhu, Z.: The Canadian traveller problem and its competitive analysis. J. Comb. Optim. **18**(2), 195–205 (2009)
9. Yao, A.C.: Probabilistic computations: toward a unified measure of complexity. In: Proceedings of FOCS, pp. 222–227 (1977)

Rent Division Among Groups

Mohammad Ghodsi[1,2], Mohamad Latifian[1(✉)], Arman Mohammadi[1],
Sadra Moradian[1], and Masoud Seddighin[1]

[1] Department of Computer Engineering, Sharif University of Technology,
Tehran, Iran
mohamad.latifian@gmail.com
[2] School of Computer Science,
Institute for Research in Fundamental Sciences (IPM), Tehran, Iran

Abstract. In this paper, we extend the *Rent Sharing* problem to the
case that every room must be allocated to a group of agents. In the classic Rent Sharing problem, there are n agents and a house with n rooms.
The goal is to allocate one room to each agent and assign a rent to each
room in a way that no agent envies any other option. Our setting deviates
from the classic Rent Sharing problem in a sense that the rent charged
to each room must be divided among the members of the resident group.

We define three notions to evaluate fairness, namely, *weak envy-
freeness*, *aggregate envy-freeness* and *strong envy-freeness*. We also define
three different policies to divide the cost among the group members,
namely, *equal, proportional,* and *free* cost-sharing policies.

We present several positive and negative results for different combinations of the fairness criteria and rent-division policies. Specifically, when
the groups are pre-determined, we propose a *strong envy-free* solution
that allocates the rooms to the agents, with free cost-sharing policy. In
addition, for the case that the groups are not pre-determined, we propose a strong envy-free allocation algorithm with equal cost-sharing policy. We leverage our results to obtain an algorithm that determines the
maximum total rent along with the proper allocation and rent-division
method.

Keywords: Fairness · Envy-freeness · Rent sharing · House allocation

1 Introduction

Envy-freeness is a famous notion and a central concept studied extensively since
1960's in the literature of economics [1–8]. An allocation is envy-free, if every
agent prefers his allocated share to that of other players.

Envy-free resource allocation is studied for various types of resources. In the
setting we study, the resources are a set of indivisible goods (rooms) along with
one divisible bad (money). Although there are many different real-life applications that fit into this setting, here, we use the terminology of Rent Sharing.
In the well known *Rent Sharing* problem, n agents are willing to rent a house

© Springer Nature Switzerland AG 2018
D. Kim et al. (Eds.): COCOA 2018, LNCS 11346, pp. 577–591, 2018.
https://doi.org/10.1007/978-3-030-04651-4_39

with n rooms, and one seeks to somehow allocate the rooms to the agents and determine the rent of the rooms so that each agent prefers his own option. The challenge in this problem is that the rooms are heterogeneous and the agents have different valuations over the rooms. Thus, to maintain fairness, the rent must be wisely divided.

Formally, let $\mathcal{H}$ be a house with n rooms and let $v_{i,j}$ be the value of room j for agent a_i. The utility of a_i for renting room j at price r_j is $u_{i,j} = v_{i,j} - r_j$. Agent a_i (weakly) prefers room j to k, if $u_{i,j} \geq u_{i,k}$. In the *Rent Sharing* problem, we would like to charge a rent to each room and allocate one room to each agent, such that the resulting allocation preserves envy-freeness, i.e., every agent prefers his option. The most interesting aspect of the Rent Sharing problem is that with mild assumptions, an envy-free solution is always guaranteed to exist.

The existence of an envy-free solution for the Rent Sharing problem is proved by Su [8]. In addition, Aragones [9] proposed a polynomial time algorithm to find an envy-free solution. The solution is not necessarily unique, and there may be several envy-free allocations. Therewith, we can optimize other objectives among the feasible solutions. For example, Gal, Mash, Procaccia, and Zick [6] consider the problem of finding an envy-free solution that maximizes the value of agent with the minimum utility to his room (maximin solution).

The basic assumption in the classic Rent Sharing problem is that every room must be allocated to a single agent. However, there are situations that this assumption is no longer applicable. For example in dormitories, each room is allocated to a group of students. As another example (albeit, not in the terminology of Rent Sharing) think of the following scenario: a set of n tasks that must be performed by a set of workers having different skills and interests. In order to perform each task, we must assign it to a group of workers. How can we fairly assign the tasks to the groups and pay the agents? One can think of tasks as rooms and workers as agents and their salary as their rent share.

In such situations, a new challenge arises: the agents within a group may have diverse valuations, i.e., a room might be acceptable to some of the agents in a group but not by the others. In this paper, our goal is to discuss such situations.

Many previous studies consider the fair division problem among groups of agents [10–17]. The groups can be either pre-determined or be formed by the algorithm. The majority of the article focuses on the case that the groups are pre-determined. In this model, there is a set $\mathcal{G}$ of groups, with each group g_i consisting of m_i agents and each agent having a value function over the rooms. Our goal is to fairly allocate a room to each group and divide the rent among the agents. We name this problem *Group Rent Sharing*.

In Sect. 2.1, we introduce three notions for evaluating fairness: *aggregate envy-freeness*, *weak envy-freeness*, and *strong envy-freeness*. Recall that every solution for the Group Rent Sharing problem must determine a method by which the rent is divided among the residents of a room. We name this dividing method *cost-sharing policy*. In Sect. 2.2, different possible policies and their relations to various fairness criteria are discussed. We define three policies: *equal, proportional and free* and study the consistency of these policies and fairness notions.

After all we consider the case where the groups are not pre-determined. In this case, the allocation algorithm must also partition the agents into groups. We propose a strong envy-free allocation with an equal cost-sharing policy in Sect. 4.

As in the classic rent sharing problem, along with fairness, we can consider optimizing other objectives. Here, we seek to maximize the total rent. Note that the allocation must be individually rational, i.e., the utility of every agent in the final solution must be non-negative. The summary of our results for different fairness criteria and cost-sharing policies can be found in Table 2.

1.1 Related Works

Previous works that are related to ours fall into two categories: a stream of studies on the Rent Sharing problem and the works that consider fair division among groups.

Fair division of resources is widely studied in the context of economics and mathematics where the problem mostly considers cases with either single divisible item (also known as cake-cutting) [4,5,7,18–21], or a set of indivisible items [22–28]. In addition, a combination of these two settings is studied, where there is a set of indivisible items together with a single divisible resource. The rent division problem is, in fact, a combination of a divisible resource (money) and a set of indivisible items (rooms).

The problem of fairly dividing indivisible items with money was firstly introduced by Alkan, Demange, and Gale [22]. They show that for a sufficiently large amount of money, an envy-free allocation exists. They also suggest optimizing other objectives over myriad possible envy-free solutions. Specifically, they introduce the *money Rawlsian* solution in which the goal is to maximize the minimum money taken from every agent. Aragones [9], suggests an algorithm for computing a money Rawlsian envy-free solution in polynomial time. He also shows that every envy-free solution preserves envy-freeness if we re-allocate the rooms by the welfare-maximizing allocation.

Su [8] explains the *Sperner's Lemma* and describes its applications to fair division problems. Especially, he investigates the Rent Sharing problem and used Sperner's Lemma to show the existence of an envy-free allocation. Procaccia, Velez, and Yu [29] extend the classic rent sharing problem by considering a budget for each agent. They study the conditions under which an envy-free allocation with given budget constraints is possible and propose an algorithm to find one. Gal, Mash, Procaccia, and Zick [6] have recently conducted a study on finding equitable and maximin envy-free allocations. The former is the envy-free allocation that minimizes the disparity (the maximum difference) of the agents' utilities and the latter aims to maximize the minimum utility of the agents. They show that a maximin allocation is also equitable. Then, they propose an LP-based method to compute these allocation in polynomial time.

Some works consider the fair allocation problem among groups or families, for example Segal-Halevi and Nitzan [11] consider the proportional allocation for the case that the resource must be divided among families. They introduce three notions to evaluate fairness for this case, namely, *Average, Unanimous* and *Democratic* proportionality and show various results for these notions. Chan et al. [10] consider the Rent Sharing problem for the case where every room must be allocated to 2 agents. In their model, the groups are not known in advance. They define various solution concepts and study the complexity of their corresponding search problem. In contrast to our work, they do not consider any cost-sharing policy.

From a practical point of view, there are a considerable number of empirical studies that consider notion of fairness between groups rather than individuals [12–15,30], mostly in the context of ultimatum games.

2 Model Definition and Preliminaries

We refer to the rooms by their indices and denote the set of Groups by $\mathcal{G} = \{g_1, g_2, \ldots, g_n\}$. Furthermore we suppose that each group g_i consists of m_i agents and denote the j'th agent of g_i by $a_{i,j}$. The value of room k for agent $a_{i,j}$ is denoted by $v_{i,j,k}$. In this paper we suppose that the valuations are normalized, so that for each agent $a_{i,j}$, $\sum_k v_{i,j,k} = \frac{1}{m_i}$. Hence, the total value of each group for the house is 1. In the *Group Rent Sharing* problem, we seek to find a triple $\mathcal{S} = (\mathcal{A}, \mathcal{R}, \mathcal{D})$ where:

- $\mathcal{A} : \mathcal{G} \to [n]^1$ is a bijection that allocates one room to each group.
- $\mathcal{R} : [n] \to \mathbb{R}_{\geq 0}$ is a rent division function that determines the rent of each room.
- $\mathcal{D} : (\mathcal{N}, [n]) \to \mathbb{R}^2$, is a cost-sharing function where $\mathcal{D}(a_{i,j}, k)$ determines the cost assigned to agent $a_{i,j}$ for living in room k. Each cost-sharing function must have the property that the total amount of cost assigned to each group a room must be exactly equal to the rent determined for that room, i.e., for all i, $\sum_{j=1}^{m_i} \mathcal{D}(a_{i,j}, k) = \mathcal{R}(k)$ for every k. We use $d_{i,j,k}$ to refer to the value of $\mathcal{D}(a_{i,j}, k)$.

We refer to such a triple as allocation-triple. Roughly, in every allocation-triple, one should determine the room that must be allocated to each group ($\mathcal{A}$), the rent charged to each room ($\mathcal{R}$) and the way that rent is divided among the agents in each group for each room ($\mathcal{D}$).

When the groups are not pre-determined, we suppose that $\mathcal{N}$ is the set of agents and the allocation algorithm must determine a quadruple $\mathcal{S} = (\mathcal{B}, \mathcal{A}, \mathcal{R}, \mathcal{D})$, where $\mathcal{B}$ is a function that allocates a group to each agent. The allocation function $\mathcal{B}$ has the restriction that each group must contain exactly m agents. Functions $\mathcal{A}$, $\mathcal{R}$, and $\mathcal{D}$ are defined similar to the Group Rent Sharing problem.

---

[1] $[n]$ refers to the set $\{1, 2, \ldots, n\}$.
[2] $\mathcal{N} = \bigcup_{i,j} a_{i,j}$.

2.1   Fairness Criteria

Our goal in this section is to extend the notion of envy-freeness to the case that every room must be allocated to a group of agents. For this, we define three notions to evaluate fairness: aggregate envy-freeness, weak envy-freeness, and strong envy-freeness. Fix an allocation-triple $S = (A, R, D)$ (or quadruple $S = (B, A, R, D)$ when the groups are not pre-determined). Denote by $u_{i,j,k}$, the utility of agent $a_{i,j}$, if room k is allocated to group g_i, regarding cost-sharing function D, i.e., $u_{i,j,k} = v_{i,j,k} - d_{i,j,k}$. In this paper, we consider the allocation triples which are *individual rational*, meaning that every agent receives a non-negative utility.

Definition 1. S is weak envy-free, *if for every k, there exists at least one agent in room k that does not envy any other option.*

In other words, S is weak envy-free if for each group g_i, at least one agent $a_{i,j}$ in g_i does not envy any other room, which means for all $k \neq A(g_i)$ we have $u_{i,j,A(g_i)} \geq u_{i,j,k}$.

One shortcoming of this notion is that it does not consider the preferences of everyone and hence, an allocation may be unfair to all the agents in a group except one. Our next notion seeks to somehow resolve this issue.

Definition 2. S is aggregate envy-free, *if the total utility of agents in each room is at least as large as their total utility for any other room, i.e., for every group g_i,*

$$\sum_j u_{i,j,A(g_i)} \geq \sum_j u_{i,j,k} \quad \forall k.$$

To put it simply, if we consider each group g_i as an agent a_i with value $\sum_j v_{i,j,k}$ for each room k, we want to find an envy-free solution for the classic Rent Sharing problem with agents $\{a_1, a_2, \ldots, a_n\}$ and rooms in $\mathcal{H}$. In fact, this notion takes the aggregate utility of a group into account, instead of considering only one agent from each group. However, the aggregate utility of a group does not capture the utility of every individual, i.e., some of the agent may still be unsatisfied. In the strong envy-free notion, we desire to satisfy all the agents.

Definition 3. S is strong envy-free, *if for each agent $a_{i,j}$ and every $k \neq A(g_i)$,*

$$u_{i,j,A(g_i)} > u_{i,j,k}.$$

2.2   Cost-Sharing

Recall that in the classic *Rent Sharing* problem, we desire to charge a rent to each room. For the Group Rent Sharing problem, in addition to a method for dividing the rent, we need to formulate a policy to split the rent among the agents in each room. As mentioned before, we denote such a policy by D. In this section, we discuss on different possible polices for dividing the rent of a room among the resident agents.

The first and the easiest solution that immediately bears in mind, is to split the rent equally among the agents in each group. We name such a cost-sharing policy, *equal cost-sharing*.

Definition 4. *A cost-sharing policy D is equal, if the cost assigned to each agent for a room equals to his roommates, i.e., for each group g_i and each room k and agents $a_j, a_{j'}$ we have $d_{i,j,k} = d_{i,j',k}$.*

Even though this policy seems natural, considering the fact that the agents in one group may have diverse interests in a room, rent discrimination would be more reasonable. One idea is that each agent pays a price proportional to his valuation for that room.

Definition 5. *A cost-sharing policy D is proportional, if for each group g_i, agent $a_{i,j}$ pays the rent $d_{i,j,k} = \frac{v_{i,j,k}}{V_{i,k}} \mathcal{R}(k)$ for room k, where $V_{i,k} = \sum_{j'} v_{i,j',k}$.*[3]

As we discuss in Sect. 3, the proportional cost-sharing policy is inconsistent to strong envy-freeness, i.e., there are cases that no strong envy-free allocation exists with respect to proportional cost-sharing policy.

The final policy we introduce considers no restriction on the rent charged to each agent for each room. We call such a method *free cost-sharing* policy. The only criterion for a free cost-sharing policy is that the total payment of the agents in each group for a specific room must sum up to the rent fixed for that room. The main results of this paper are concerned with the strong envy-freeness and free cost-sharing policy. Prior to explaining our main results, we will justify the model and subsequently introduce some possible and impossible results in pertaining to our model.

3   Pre-determined Groups

As elaborated earlier, our model comprises two essential ingredients: the fairness criterion and the cost-sharing policy. In this section, we shed light on the relation between these two components. To do so, we define the concept of consistency.

Definition 6. *A fairness criterion $\mathcal{F}$ is consistent with cost-sharing function D, if for every instance of the* Group Rent Sharing *problem, an allocation-triple S with cost-sharing function D exists, such that S preserves $\mathcal{F}$.*

In Lemmas 1 and 2, we show, through counter-examples, that some fairness criteria and some cost-sharing policies are inconsistent.

Lemma 1. *For the case that the groups are pre-determined, the equal cost-sharing policy and strong envy-freeness are inconsistent.*

---

[3] Note that if $V_{i,k} = 0$, by individual rationality, $\mathcal{R}(k) = 0$ and no agent has to pay any cost.

Proof. Consider the following instance: let $|\mathcal{G}| = 2n$, $g_i = \{a_{i,1}, a_{i,2}\}$ and there are $2n$ rooms. Furthermore, suppose that the valuations of the agents in each group g_i are as follows:

$$v_{i,1,k} = \begin{cases} \frac{1}{3n} & k \le n \\ \frac{1}{6n} & k > n \end{cases} \qquad v_{i,2,k} = \begin{cases} \frac{1}{6n} & k \le n \\ \frac{1}{3n} & k > n \end{cases}$$

Consider an arbitrary rent division function $\mathcal{R}$. Due to the symmetric construction of the valuations, we can observe w.l.o.g. that the room 1 is the one with the maximum rent and is assigned to g_1. Let $p = \mathcal{R}(1)$ be the maximum rent. Considering the fact that the cost-sharing policy is equal, the utility of the agents in g_1 would be $u_{1,1,1} = \frac{1}{3n} - \frac{p}{2}$ and $u_{1,2,1} = \frac{1}{6n} - \frac{p}{2}$. Note that the rent assigned to room $i \le n$ must exactly equal p. Otherwise, agent $a_{1,1}$ envies that room. Now, suppose that $\mathcal{R}(2n) = q$. Thus, the utility of the agents in group g_1, when room $2n$ is allocated to g_1 would be $u_{1,1,2n} = \frac{1}{6n} - \frac{q}{2}$ and $u_{1,2,2n} = \frac{1}{3n} - \frac{q}{2}$. Since we intend our allocation to be strong envy-free, we have:

$$u_{1,1,1} \ge u_{1,1,2n} \Rightarrow \frac{1}{3n} - \frac{p}{2} \ge \frac{1}{6n} - \frac{q}{2} \tag{1}$$

$$u_{1,2,1} \ge u_{1,2,2n} \Rightarrow \frac{1}{6n} - \frac{p}{2} \ge \frac{1}{3n} - \frac{q}{2} \tag{2}$$

$$(1), (2) \Rightarrow q \ge p \tag{3}$$

As p is the maximum rent:

$$\Rightarrow q = p$$

$$(2), (4) \Rightarrow \frac{1}{n} \le 0 \tag{4}$$

Which contradicts $n > 0$.

The counter-example described in the proof of Lemma 1 is independent of $\mathcal{R}$ which means even for a very large amount of rent, a strong envy-free allocation-triple with the equal cost-sharing policy is impossible. In Lemma 2, we show that the proportional cost-sharing policy and strong envy-freeness are also inconsistent.

Lemma 2. *For the case that the groups are pre-determined, the proportional cost-sharing policy and strong envy-freeness are inconsistent.*

Due to lack of space, we omit the proof of Lemma 2[4] but, to give an intuition, take into account the following instance: let $|\mathcal{G}| = 3$ and let $g_i = \{a_{i,1}, a_{i,2}\}$. Moreover, suppose that the valuation functions of the agents are as in Table 1. As we illustrate in the proof of Lemma 2, for this instance, no allocation of rooms can guarantee strong envy-freeness with the proportional cost-sharing policy. To show this, we consider different allocation possibilities and show that in each of them, at least one agent envies another choice.

---

[4] We refer the reader to the full version of the paper for this proof.

Table 1. Valuation of the agents

	1	2	3
$a_{1,1}$	0	3/8	1/8
$a_{1,2}$	3/8	0	1/8
$a_{2,1}$	0	0	1/2
$a_{2,2}$	1/2	0	0
$a_{3,1}$	0	1/4	1/4
$a_{3,2}$	0	1/4	1/4

Observation 1. *Aggregate envy-freeness is implied by strong envy-freeness. Furthermore, aggregate envy-freeness implies weak envy-freeness.*

In Lemmas 3 and 4, we show that both equal and proportional cost-sharing policies are consistent to aggregate envy-freeness. Moreover, Considering Observation 1, both the policies are consistent to weak envy-freeness as well.

Lemma 3. *The equal cost-sharing policy and aggregate envy-freeness are consistent.*

Lemma 4. *The proportional cost-sharing policy and aggregate envy-freeness are consistent.*

The general idea behind proving both Lemmas 3 and 4 is to build a classic Rent Sharing problem by aggregating the valuations of the agents in each group and then showing that dividing the rent by each of these two policies preserves aggregate envy-freeness for every group.

3.1   Strong Envy-Freeness and Free Cost-Sharing

This section deals with the results surrounding the strong envy-free allocations with the free cost-sharing policy. As described in Sect. 3, proportional and equal cost-sharing policies are not consistent with strong envy-freeness. Here, we show that with the free cost-sharing policy, one can find a strong envy-free allocation (Lemma 3). Our assumption in this section is that the groups are known in advance. However, the results can be trivially extended to the case that the groups are not pre-determined.

We start this section with Observation 2, which indicates that increasing the rent for all the agents preserves envy-freeness. We use Observation 2 as a basis upon which the proof of Theorem 3 is obtained.

Observation 2. *Let $\mathcal{S} = (\mathcal{A}, \mathcal{R}, \mathcal{D})$ be a strong envy-free allocation-triple and let c be a constant. Furthermore, let $\mathcal{S}^* = (\mathcal{A}, \mathcal{R}^*, \mathcal{D}^*)$ be an allocation-triple such that for all k, $\mathcal{R}^*(k) = \mathcal{R}(k) + c$ and for every agent $a_{i,j}$, $d^*_{i,j,k} = d_{i,j,k} + \frac{c}{m_i}$. Then, $\mathcal{S}^*$ is also strong envy-free.*

Theorem 3. *Strong envy-freeness is consistent with the free cost-sharing policy.*

Proof. Consider a proxy agent a_i for each group g_i and set the valuation of a_i for room k as $v_{i,k} = \sum_{j=1}^{m_i} v_{i,j,k}$. Now, consider the classic *Rent Sharing* problem instance with agents $a_1, a_2, \ldots, a_n$ and house $\mathcal{H}$. We know that the utility of a_i for room k is $u_{i,k} = v_{i,k} - \mathcal{R}(k)$. On the other hand, we know that an envy-free allocation for this instance always exists [8]. Thus, we can find an allocation $\mathcal{A}^*$ and a rent division function $\mathcal{R}^*$ such that for each proxy agent a_i and any $k \neq \mathcal{A}^*(g_i)$, $u_{i,\mathcal{A}^*(g_i)} \geq u_{i,k}$, which means $v_{i,\mathcal{A}^*(g_i)} - \mathcal{R}^*(\mathcal{A}^*(g_i)) \geq v_{i,k} - \mathcal{R}^*(k)$. By definition, for all $k \neq \mathcal{A}^*(g_i)$,

$$\sum_{j=1}^{m_i} v_{i,j,\mathcal{A}^*(g_i)} - \mathcal{R}^*(\mathcal{A}^*(g_i)) \geq \sum_{j=1}^{m_i} v_{i,j,k} - \mathcal{R}^*(k). \tag{5}$$

Now, consider the allocation-triple $\mathcal{S}^* = (\mathcal{A}^*, \mathcal{R}^*, \mathcal{D}^*)$, where the cost-sharing function $\mathcal{D}^*$ is determined as follows:

$$d^*_{i,j,k} = v_{i,j,k} - \frac{\sum_{t=1}^{m_i} v_{i,t,k} - \mathcal{R}^*(k)}{m_i} \tag{6}$$

Note that $\sum_j d^*_{i,j,k} = \mathcal{R}^*(k)$. We claim that $\mathcal{S}^*$ is strong envy-free. To show this, take an arbitrary agent $a_{i,j}$. We have $u_{i,j,k} = v_{i,j,k} - d^*_{i,j,k}$, which means

$$u_{i,j,k} = v_{i,j,k} - (v_{i,j,k} - \frac{\sum_{t=1}^{m_i} v_{i,t,k} - \mathcal{R}^*(k)}{m_i}).$$

Notice that $u_{i,j,\mathcal{A}^*(g_i)} = v_{i,j,\mathcal{A}^*(g_i)} - d^*_{i,j,\mathcal{A}^*(g_i)}$. Regarding Eq. (6),

$$\begin{aligned}
u_{i,j,\mathcal{A}^*(g_i)} &= v_{i,j,\mathcal{A}^*(g_i)} - \Big(v_{i,j,\mathcal{A}^*(g_i)} \\
&\quad - \frac{\sum_{t=1}^{m_i} v_{i,t,\mathcal{A}^*(g_i)} - \mathcal{R}^*(\mathcal{A}^*(g_i))}{m_i}\Big) \\
&= \frac{\sum_{t=1}^{m_i} v_{i,t,\mathcal{A}^*(g_i)} - \mathcal{R}^*(\mathcal{A}^*(g_i))}{m_i} \\
&\geq \frac{\sum_{t=1}^{m_i} v_{i,t,k} - \mathcal{R}^*(k)}{m_i} \\
&> v_{i,j,k} - (v_{i,j,k} - \frac{\sum_{t=1}^{m_i} v_{i,t,k} - \mathcal{R}^*(k)}{m_i}) \\
&\geq v_{i,j,k} - d^*_{i,j,k} = u_{i,j,k}
\end{aligned}$$

Thus far, we've shown that $\mathcal{S}^*$ is strong envy-free. However, in $\mathcal{S}^*$, there may be agents with negative utilities. Let $u_{min} = \min_{i,j} v_{i,j,\mathcal{A}^*(g_i)} - d^*_{i,j,\mathcal{A}^*(g_i)}$ and let $Z = \max_i |g_i|$. Note that if $u_{min} > 0$, then the individual rationality constraint has already fulfilled. Otherwise, let $\mathcal{R}^{**}$ be the rent function such that for every room k, $r^{**}(k) = r^*(k) + u_{min} \cdot Z$ and let $\mathcal{D}^{**}$ be a function such that for all i, j, k, $d^{**}_{i,j,k} = d^*_{i,j,k} + \frac{Z \cdot u_{min}}{|g_i|}$. By Observation 2, $\mathcal{D}^{**}$ is also strong envy-free with nonnegative utilities and hence guarantees individual rationality. In summary, value of $d_{i,j,k}^{**}$ would be

$$v_{i,j,k} - \frac{\sum_{t=1}^{m_i} v_{i,t,k} - \mathcal{R}^*(k)}{m_i} + \frac{Z \cdot (\min_{w,t} v_{w,t,\mathcal{A}^*(g_w)} - d_{w,t,\mathcal{A}^*(g_w)}^*)}{|g_i|}.$$

In light of Theorem 3, we can present an algorithm for computing a strong envy-free allocation-triple. We already know that a solution to the classic rent sharing problem can be found in polynomial time. All the other steps described in Theorem 3 can be easily implemented in polynomial time. Thus, a strong envy-free solution with the free cost-sharing policy can be found in polynomial time.

The idea to find a solution with maximum possible total rent is inspired by [6]. Let $\mathcal{A}^*$ be an allocation, which is welfare-maximizing. In [6], it is shown that if an envy-free solution exists with arbitrary allocation function $\mathcal{A}$ and rent sharing function $\mathcal{R}$, then the pair $\mathcal{A}^*$ and $\mathcal{R}$ is also envy-free. In Theorem 4, we use a generalized form of this statement to obtain a strong envy-free allocation. In fact, we show that if $\mathcal{S} = (\mathcal{A}, \mathcal{R}, \mathcal{D})$ is a strong envy-free allocation-triple, then so is $\mathcal{S}' = (\mathcal{A}^*, \mathcal{R}, \mathcal{D})$.

Theorem 4. *A strong envy-free allocation-triple with the free cost-sharing policy that maximizes total rent can be found in polynomial time.*

Proof. Recall the definition of proxy agent from the proof of Theorem 3. An allocation $\mathcal{A}$ is welfare-maximizing, if it maximizes value of the following expression:

$$\sum_{i=1}^{|\mathcal{G}|} \sum_{j=1}^{m_i} v_{i,j,\mathcal{A}(g_i)} = \sum_{i=1}^{|\mathcal{G}|} v_{i,\mathcal{A}(a_i)}.$$

Such an allocation can be found in polynomial time by finding a maximum weighted matching in the bipartite graph representing the tendency of the proxy agents to the rooms, i.e. the weight of the edge between proxy agent i and room k is $\sum_{j=1}^{m_i} v_{i,j,k}$. Now, consider the pseudo-code described in Algorithm 1. The algorithm begins with computing a welfare-maximizing allocation of the rooms to the proxy agents. Let $\mathcal{A}$ be the welfare-maximizing allocation. we find the desired allocation-triple by solving a linear program described in Algorithm 1, which computes the envy-free allocation with the maximum possible price. In this LP, the first set of constraints ensures that the sum of the costs assigned to the agents in each group is equal to the room rent. The second set of constraints guarantee strong envy-freeness and the third set ensures the individual rationality condition. Theorem 3 ensures that the LP described in Algorithm 1 is feasible. However, we still must overcome a technical hurdle: we did not show that the allocation-triple suggested by Algorithm 1 is the one that maximizes the total rent. In fact, by Algorithm 1 we find a solution that maximizes the total rent among the solutions with welfare-maximizing allocation functions. But the allocation-triple with the maximum possible total rent may be obtained by some other allocation functions. Let $\mathcal{S} = (\mathcal{A}, \mathcal{R}, \mathcal{D})$ be the optimal strong envy-free solution and let $\mathcal{A}^*$ be the welfare-maximizing allocation. By strong envy-freeness we know

---

ALGORITHM 1. Strong envy-free allocation with maximum rent

---

(1) Let $\mathcal{A}$ be a welfare-maximizing allocation
(2) Compute a rent division $\mathcal{R}$ and cost-sharing function $\mathcal{D}$ by the linear program

$$\max \sum_{k=1}^{n} \sum_{j=1}^{m_k} d_{k,j,\mathcal{A}(k)}$$

$$s.t.$$

$$\mathcal{R}_k = \sum_{j=1}^{m_i} d_{i,j,k} \qquad \forall i, k$$

$$v_{i,j,\mathcal{A}(g_i)} - d_{i,j,\mathcal{A}(g_i)} \geq v_{i,j,k} - d_{i,j,k} \qquad \forall i, j, k$$

$$v_{i,j,\mathcal{A}(g_i)} - d_{i,j,\mathcal{A}(g_i)} \geq 0 \qquad \forall i, j$$

$$v_{i,j,\mathcal{A}(g_i)} - d_{i,j,\mathcal{A}(g_i)} \geq v_{i,j,k} - d_{i,j,k} \qquad \forall i, j, k. \tag{7}$$

By summing over all agents in group g_i, for all i, k we have:

$$\sum_{j=1}^{m_i} v_{i,j,\mathcal{A}(g_i)} - \mathcal{R}(\mathcal{A}(g_i)) \geq \sum_{j=1}^{m_i} v_{i,j,k} - \mathcal{R}(h_k). \tag{8}$$

Since Eq. (7) holds for all $k \neq \mathcal{A}(g_i)$, it also holds for room $\mathcal{A}^*(g_i)$. Therefore, for all g_i we have:

$$\sum_{j=1}^{m_i} v_{i,j,\mathcal{A}(g_i)} - \mathcal{R}(\mathcal{A}(g_i)) \geq \sum_{j=1}^{m_i} v_{i,j,\mathcal{A}^*(g_i)} - \mathcal{R}(\mathcal{A}^*(g_i)). \tag{9}$$

Summing Inequality (9) over all the groups yields:

$$\sum_{i=1}^{|\mathcal{G}|} \sum_{j=1}^{m_i} v_{i,j,\mathcal{A}(q_i)} - \sum_{k=1}^{|\mathcal{G}|} \mathcal{R}(\mathcal{A}(q_i))$$

$$\geq \sum_{i=1}^{|\mathcal{G}|} \sum_{j=1}^{m_i} v_{i,j,\mathcal{A}^*(g_i)} - \sum_{k=1}^{|\mathcal{G}|} \mathcal{R}(\mathcal{A}^*(g_i)).$$

Since both $\mathcal{A}$ and $\mathcal{A}^*$ are bijections, we have

$$\sum_{k=1}^{|\mathcal{G}|} \mathcal{R}(\mathcal{A}(g_i)) = \sum_{k=1}^{|\mathcal{G}|} \mathcal{R}(\mathcal{A}^*(g_i)). \tag{10}$$

Hence,

$$\sum_{i=1}^{|\mathcal{G}|}\sum_{j=1}^{m_i} v_{i,j,\mathcal{A}(g_i)} \geq \sum_{i=1}^{|\mathcal{G}|}\sum_{j=1}^{m_i} v_{i,j,\mathcal{A}^*(g_i)}. \tag{11}$$

By definition of welfare-maximizing allocation,

$$\sum_{i=1}^{|\mathcal{G}|}\sum_{j=1}^{m_i} v_{i,j,\mathcal{A}(g_i)} \leq \sum_{i=1}^{|\mathcal{G}|}\sum_{j=1}^{m_i} v_{i,j,\mathcal{A}^*(g_i)},$$

$$\sum_{i=1}^{|\mathcal{G}|}\sum_{j=1}^{m_i} v_{i,j,\mathcal{A}(g_i)} = \sum_{i=1}^{|\mathcal{G}|}\sum_{j=1}^{m_i} v_{i,j,\mathcal{A}^*(g_i)}. \qquad \text{Inequality (11)}$$

Regarding Eq. (10),

$$\sum_{i=1}^{|\mathcal{G}|}\sum_{j=1}^{m_i} v_{i,j,\mathcal{A}(g_i)} - \sum_{i=1}^{|\mathcal{G}|} \mathcal{R}(\mathcal{A}(g_i))$$

$$= \sum_{i=1}^{|\mathcal{G}|}\sum_{j=1}^{m_i} v_{i,j,\mathcal{A}^*(g_i)} - \sum_{i=1}^{|\mathcal{G}|} \mathcal{R}(\mathcal{A}^*(g_i)). \tag{12}$$

Equality (12) together with Inequality (9) results in the following expression for all i:

$$\sum_{j=1}^{m_i} v_{i,j,\mathcal{A}(g_i)} - \mathcal{R}(\mathcal{A}(g_i)) = \sum_{j=1}^{m_i} v_{i,j,\mathcal{A}^*(g_i)} - \mathcal{R}(\mathcal{A}^*(g_i)),$$

$$\sum_{j=1}^{m_i} v_{i,j,\mathcal{A}(g_i)} - \sum_{j=1}^{m_i} d_{i,j,\mathcal{A}(g_i)} = \sum_{j=1}^{m_i} v_{i,j,\mathcal{A}^*(g_i)} - \sum_{j=1}^{m_i} d_{i,j,\mathcal{A}^*(g_i)}. \tag{13}$$

In addition, since Inequality (7) holds for every room k,

$$v_{i,j,\mathcal{A}(g_i)} - d_{i,j,\mathcal{A}(g_i)} \geq v_{i,j,\mathcal{A}^*(g_i)} - d_{i,j,\mathcal{A}^*(g_i)} \qquad \forall i,j. \tag{14}$$

Equation (13) together with Inequality (14) yield:

$$v_{i,j,\mathcal{A}(g_i)} - d_{i,j,\mathcal{A}(g_i)} = v_{i,j,\mathcal{A}^*(g_i)} - d_{i,j,\mathcal{A}^*(g_i)} \qquad \forall i,j.$$

Hence, for all i,j,k, we have

$$v_{i,j,\mathcal{A}^*(g_i)} - d_{i,j,\mathcal{A}^*(g_i)} \geq v_{i,j,k} - d_{i,j,k}.$$

This shows that we can change the allocation of the optimal solution to the welfare-maximizing allocation, without violating the strong envy-freeness condition. Thus, the solution offered by LP maximizes the total rent amongst all admissible allocation-triples.

4   Not Pre-determined Groups

In this section, we consider the case that the groups are not known in advance. As regards the dormitories, for instance, it is more realistic to assume that the students request for a room individually.

For this case, we show that a strong envy-free allocation with the equal cost-sharing policy always exists. Assume that we have a house with n rooms with capacity of m agents per room. In addition, we further suppose that the set of agents is $\mathcal{N} = \{a_1, a_2, \ldots, a_n\}$ and value of i'th room for agent a_j is $v_{i,j}$. The goal is to provide a strong envy free quadruple $\mathcal{S} = (\mathcal{B}, \mathcal{A}, \mathcal{R}, \mathcal{D})$ as defined in Sect. 2.

Theorem 5. *For the case that the groups are not pre-determined, Strong envy-freeness is consistent with equal cost-sharing.*

Proof. First, we construct an instance of the classic Rent Sharing problem as follows: let $\mathcal{H}' = \{r_{1,1}, \ldots, r_{1,m}, r_{2,1}, \ldots, r_{2,m}, \ldots, r_{n,1}, \ldots, r_{n,m}\}$ be a house consisting m copies of every room in $\mathcal{H}$. Furthermore, let $r_{i,j}$ be the j'th copy of the i'th room in $\mathcal{H}$. Now, we solve the classic Rent Sharing problem instance considering $\mathcal{H}$ and $\mathcal{N}$. First note that for this case Observation 6 holds.

Observation 6. *For every i, j, j', the rent charged for room $r_{i,j}$ is the same as $r_{i,j'}$.*

Observation 6 is because of the fact that the agents in rooms $r_{i,j}$ and $r_{i,j'}$ must not envy each other. Let A^* be the allocation function that allocates a room to each agent and let R^* be the function that assigns a rent to each room. Now, let $\mathcal{S}^* = (\mathcal{B}^*, \mathcal{A}^*, \mathcal{R}^*, \mathcal{D}^*)$ be an allocation quadruple where the cost sharing function $\mathcal{D}^*$ is equal and the rent charged for i'th room of $\mathcal{H}$ is $\sum_{j=1}^{m} r_{i,j}$. Moreover, we define the i'th group of $\mathcal{B}^*$ as the agents located in one of the copies of i'th room and the allocation function $\mathcal{A}^*$ allocates i'th room to g_i.

By observation 6, every agent pays the same rent in $\mathcal{S}^*$ as in the classic rent sharing instance. Thus, envy-freeness of the classic instance implies strong envy-freeness of $\mathcal{S}^*$. In addition, the agents in the same group pay equal price for their room. Hence, the allocation quadruple $\mathcal{S}^*$ is strong envy-free with the equal cost-sharing policy.

Table 2. Predetermined groups

	Weak envy-free	Aggregate envy-free	Strong envy-free
Equal	✓ Observation 1	✓ Lemma 3	✗ Lemma 1
Proportional	✓ Observation 1	✓ Lemma 4	✗ Lemma 2
Free	✓ Observation 1	✓ Observation 1	✓ Theorem 3

Recall that in Sect. 3 we proved that when the groups are pre-determined, no allocation-triple can guarantee strong envy-freeness with the equal cost-sharing policy.

Table 3. Not pre-determined groups

	Strong envy-free
Equal	✓ Theorem 5
Free	✓ Lemmas 3, 4

5   Conclusion and Future Works

In this paper, we considered the Group Rent Sharing problem, which is an extension of the classic Rent Sharing problem to the case where each room must be allocated to a group of agents. We generalized the envy-freeness notion for such situations. We also defined the cost-sharing policy, which adopts the method by which the rent is divided among the resident agents of a room.

We defined three fairness criteria (weak, aggregate, and strong envy-free) and three cost-sharing policies (equal, proportional, and free). Our results encompass several positive and negative results regarding the consistency of different fairness notions and cost-sharing policies. You can find a summary of these results in Tables 2 and 3.

We proposed two positive results regarding strong envy-freeness: consistency of this notion with the free cost-sharing policy in the case that the groups are predetermined and consistency with equal cost-sharing in the case that the groups are not pre-determined. For both of these cases, we can find the allocation with the maximum total rent. One interesting open question is to give an upper-bound on the ratio of maximum total rent in these two cases. Another direction would be the analysis of the problem in stochastic settings where the valuation of the houses to the agents are drawn from a given distribution.

References

1. Foley, D.K.: Resource allocation and the public sector (1967)
2. Gamov, G., Stern, M.: Puzzle math. Viking, New York (1958)
3. Procaccia, A.D.: Cake cutting: not just child's play. Commun. ACM **56**(7), 78–87 (2013)
4. Robertson, J., Webb, W.: Cake-Cutting Algorithms: Be Fair If You Can (1998)
5. Segal-Halevi, E., Hassidim, A., Aumann, Y.: Envy-free cake-cutting in two dimensions. In: AAAI, vol. 15, pp. 1021–1028 (2015)
6. Gal, Y.K., Mash, M., Procaccia, A.D., Zick, Y.: Which is the fairest (rent division) of them all? In: Proceedings of the 2016 ACM Conference on Economics and Computation, pp. 67–84. ACM (2016)
7. Cohler, Y.J., Lai, J.K., Parkes, D.C., Procaccia, A.D.: Optimal envy-free cake cutting. In: AAAI (2011)
8. Su, F.E.: Rental harmony: Sperner's lemma in fair division. Am. Math. Mon. **106**(10), 930–942 (1999)
9. Aragones, E.: A derivation of the money Rawlsian solution. Soc. Choice Welf. **12**(3), 267–276 (1995)

10. Chan, P.H., Huang, X., Liu, Z., Zhang, C., Zhang, S.: Assignment and pricing in roommate market. In: AAAI, pp. 446–452 (2016)
11. Segal-Halevi, E., Nitzan, S.: Proportional cake-cutting among families. arXiv preprint arXiv:1510.03903 (2015)
12. Robert, C., Carnevale, P.J.: Group choice in ultimatum bargaining. Organ. Behav. Hum. Decis. Process. **72**(2), 256–279 (1997)
13. Bornstein, G., Yaniv, I.: Individual and group behavior in the ultimatum game: are groups more "rational" players? Exp. Econ. **1**(1), 101–108 (1998)
14. Messick, D.M., Moore, D.A., Bazerman, M.H.: Ultimatum bargaining with a group: underestimating the importance of the decision rule. Organ. Behav. Hum. Decis. Process. **69**(2), 87–101 (1997)
15. Santos, F.P., Santos, F.C., Paiva, A., Pacheco, J.M.: Evolutionary dynamics of group fairness. J. Theor. Biol. **378**, 96–102 (2015)
16. Manurangsi, P., Suksompong, W.: Asymptotic existence of fair divisions for groups. Math. Soc. Sci. **89**, 100–108 (2017)
17. Suksompong, W.: Approximate maximin shares for groups of agents. Math. Soc. Sci. **92**, 40–47 (2018)
18. Chen, Y., Lai, J.K., Parkes, D.C., Procaccia, A.D.: Truth, justice, and cake cutting. Games Econ. Behav. **77**(1), 284–297 (2013)
19. Alijani, R., Farhadi, M., Ghodsi, M., Seddighin, M., Tajik, A.S.: Envy-free mechanisms with minimum number of cuts. In: AAAI, pp. 312–318 (2017)
20. Stromquist, W.: Envy-free cake divisions cannot be found by finite protocols. Electron. J. Comb. **15**(1), 11 (2008)
21. Dehghani, S., Farhadi, A., HajiAghayi, M., Yami, H.: Envy-free chore division for an arbitrary number of agents. In: Proceedings of the Twenty-Ninth Annual ACM-SIAM Symposium on Discrete Algorithms, pp. 2564–2583. SIAM (2018)
22. Alkan, A., Demange, G., Gale, D.: Fair allocation of indivisible goods and criteria of justice. Econ.: J. Econ. Soc. 1023–1039 (1991)
23. Ghodsi, M., HajiAghayi, M., Seddighin, M., Seddighin, S., Yami, H.: Fair allocation of indivisible goods: improvements and generalizations. In: Proceedings of the 2018 ACM Conference on Economics and Computation, pp. 539–556. ACM (2018)
24. Farhadi, A., et al.: Fair allocation of indivisible goods to asymmetric agents. In: Proceedings of the 16th Conference on Autonomous Agents and MultiAgent Systems, pp. 1535–1537. International Foundation for Autonomous Agents and Multiagent Systems (2017)
25. Procaccia, A.D., Wang, J.: Fair enough: guaranteeing approximate maximin shares. In: Proceedings of the Fifteenth ACM Conference on Economics and Computation, pp. 675–692. ACM (2014)
26. Amanatidis, G., Markakis, E., Nikzad, A., Saberi, A.: Approximation algorithms for computing maximin share allocations. ACM Trans. Algorithms (TALG) **13**(4), 52 (2017)
27. Cole, R., Gkatzelis, V.: Approximating the nash social welfare with indivisible items. In: Proceedings of the Forty-Seventh Annual ACM on Symposium on Theory of Computing, pp. 371–380. ACM (2015)
28. Ghodsi, M., Saleh, H., Seddighin, M.: Fair allocation of indivisible items with externalities. arXiv preprint arXiv:1805.06191 (2018)
29. Procaccia, A.D., Velez, R.A., Yu, D.: Fair rent division on a budget. In: AAAI (2018)
30. Azrieli, Y., Shmaya, E.: Rental harmony with roommates. J. Econ. Theory **153**, 128–137 (2014)

Sequence Sentential Decision Diagrams

Shuhei Denzumi(✉)

The University of Tokyo, Hongo 7-3-1, Bunkyo City, Tokyo 113-8656, Japan
denzumi@mist.i.u-tokyo.ac.jp

Abstract. In this paper, we propose a new data structure sequence sentential decision diagram (SSDD) that represents sets of strings. SSDD is a generalized data structure of Sequence Binary Decision Diagram (SeqBDD), that is a similar data structure to a deterministic finite automaton, but the size can be exponentially smaller than the SeqBDD for the same string set. We also provide algorithms to manipulate sets of strings on SSDD. These algorithms allow operations such as intersection, union, and concatenation to be executed on SSDDs under their compressed representations without expanding. We analyzed the size complexity of SSDD and the time complexity of proposed algorithms.

Keywords: Data structure · Compression · Decision diagram
Set of strings

1 Introduction

Discrete structures are fundamental concepts in the field of computer science and discrete mathematics. It is an important technique to represent various types of discrete structures compactly on computers and processing them efficiently. Therefore, improvement of compact representation of discrete structures and efficient manipulation them will give a huge impact on modern society.

Nowadays, binary decision diagrams (BDDs) and its family have been recognized as an important data structure to manipulate discrete structures. Using BDD, we can represent Boolean functions in compact and canonical form. In addition, we can compute the BDD for the result of binary Boolean operations of two BDDs directly without expanding them. There is a variant of a BDD, Zero-suppressed BDD (ZDD) that is specialized for manipulating families of sets. ZDD is proposed by Minato twenty years ago [9]. However, ZDD cannot represent sets of strings efficiently. Loekito et al. proposed SeqBDD (SeqBDD) in 2010 [8]. SeqBDD is almost the same data structure as ZDD, but its restriction on the structure is modified to handle strings [5]. The basic operations of SeqBDD are very similar to those of ZDD. SeqBDD is an efficient representation especially for sets of strings having strings of various length. A SeqBDD is a vertex-labeled graph structure, which resembles an acyclic DFA in binary form. A SeqBDD can be more compact than an equivalent ADFA.

© Springer Nature Switzerland AG 2018
D. Kim et al. (Eds.): COCOA 2018, LNCS 11346, pp. 592–606, 2018.
https://doi.org/10.1007/978-3-030-04651-4_40

Compact string indexes for storing sets of strings are fundamental data structures in computer science, and have been extensively studied for decades [2,6]. Examples of compact string indexes include tries [1,2], finite automata and transducers [3,7]. Because of the rapid increase in the massive amounts of sequence data, such as biological sequences, natural language texts, and event sequences, these compact string indexes have attracted much attention and gained more importance [2,6]. In such applications, an index is required not only to store sets of strings compactly for searching but also to manipulate efficiently them with various set operations.

Darwiche proposed sentential decision diagram (SDD) in 2011 [4]. SDD is a generalization of BDD. SDD represents Boolean functions more compactly than BDD in canonical form, and support Boolean operations in polynomial time. Nishino et al. proposed zero-suppressed SDD (ZSDD) in 2016 [11]. ZSDD is a variant of SDD, and a generalization of ZDD. ZSDD has almost the same features as SDD and more effective for sparse families of sets. In this paper, we introduce a new data structure called *sequence sentential decision diagram (SSDD)*. Likewise the relationship among ZDD, SDD, and ZSDD, SSDD is a variant of SDD, and a generalization of SeqBDD. SDD and ZSDD represent Boolean functions or families of sets on a fixed number of variables. SeqBDD requires a fixed size of an alphabet, but it can represent variable length strings. We modified ZSDD structure so as to deal with sets of strings without restriction on the length. We also provide algorithms to compute string set operations on SSDD, such as union and concatenation, without expanding them. We show that SSDD can be exponentially smaller than SeqBDD, and can be polynomially larger than SeqBDD in the worst case.

2   Preliminary

Let $\Sigma = \{a, b, \ldots\}$ be a countable alphabet of *letters*. We assume that the letters of Σ are ordered by a precedence $\prec_\Sigma$ such as $a \prec_\Sigma b \prec_\Sigma \cdots$ in a standard way.

Let $s = a_1 \cdots a_n$, $n \geq 0$, be a string over Σ. For every $i = 1, \ldots, n$, we denote by $s[i] = a_i$ the i-th letter and by $|s| = n$ the *length* of s. The *empty string* of length zero is denoted by ε. We denote by Σ^* the set of all strings of length $n \geq 0$. For two strings s and t, we denote the *concatenation* of s and t by $s \cdot t$ or st. If $s = pqr$ for some possibly empty strings p, q, and r, we refer to p, q, and r as a *prefix*, *factor*, and *suffix* of s, respectively.

A *language* on an alphabet Σ is a set $L \subseteq \Sigma^*$ of strings on Σ. A *finite language* of size $m \geq 0$ is just a finite set $L = \{s_1, \ldots, s_m\}$ of m strings on Σ. A finite language L is referred to as a *string set*. We define the *cardinality* of L by $|L| = m$, the *total length* of L by $\|L\| = \sum_{i=1}^m |s_i|$, and the *maximal string length* of L by $\mathsf{maxlen}(L) = \max\{ |s| \mid s \in L \}$. The *empty language* of cardinality 0 is denoted by $\emptyset$. The *concatenation* of languages L and M is defined as $L \cdot M = \{ st \mid s \in L, t \in M \}$.

We define *ℓ-prefix decomposition* ($\ell > 0$) of a language L by $L = P_1 \cdot S_1 \cup \cdots \cup P_k \cdot S_k \cup R$, $k \geq 0$, where $P_i \subseteq \Sigma^\ell, 1 \leq i \leq k$ and $\mathsf{maxlen}(R) < l$. If

$P_i \cap P_j = \emptyset$ for all $i \neq j$ and $\bigcup_{i=1}^k P_i = \Sigma^\ell$ and $P_i \neq \emptyset$ for all i, then we say the decomposition is ℓ-prefix partition of a language L. We denote it as $(\{(P_1, S_1), \ldots, (P_k, S_k)\}, R) \in 2^{\Sigma^* \times \Sigma^*} \times \Sigma^*$.

3   Data Structure

In this section, we propose a new data structure to represent a set of strings compactly. Our main idea is to decompose the string set recursively according to a given binary tree and represent the decompositions as a directed acyclic graph.

Before defining our data structure, we prepare *prefix partitioning tree (ptree)* that determines how given languages are prefix partitioned. A ptree $T = (V_T, E_T)$ is a rooted, ordered, full binary tree. We assume that a ptree has at least 2 leaves. For an interior node q of a ptree, we refer to the left/right child of q as *p-child/s-child* of v and denote it by $p(q)/s(q)$, respectively. We call the edges pointing p-child/s-child p-edge/s-edge, respectively. We denote the number of leaves that are contained in the subtree rooted by q by $leaf(q)$, and denote the preorder rank of q in left to right depth-first search by $rank(q)$ where the preorder rank of the root node is 0.

For a language L consisting of strings of length ℓ and a ptree node of $leaf(q) = \ell$, the *recursive prefix partitioning* of a language L with ptree node q is defined as follows: (i) if $leaf(q) > 1$, do $leaf(p(q))$-prefix partition of L as $P_1 \cdot S_1 \cup \cdots \cup P_k \cdot S_k \cup R$, and execute recursive prefix partitioning of $P_1, \ldots, P_k, R$ with $p(q)$, and $S_1, \ldots, S_k$ with $s(q)$. (ii) if $leaf(q) = 1$, stop partitioning. By this recursive prefix partitioning, we can partition a language until each set of strings become set of letters or a set of empty string. However, above partitioning can only deal with languages of fixed length strings. To solve this problem, we make a change of ptree so as to deal with variable length strings. We delete the rightmost node q_r of ptree, the leaf with the largest preorder rank, and make the s-child of the parent of q_r be one of the nodes in the rightmost path of the ptree including root. That is, we make a cycle in the ptree consisting of only s-edges. The cycle is not reachable after we traverse p-edges at least once. Obviously, this ptree is not a tree, but an almost (1) tree. We refer to such a ptree as ptree+. We show how to make a ptree+ from a ptree in Fig. 1. Now, we can define recursive prefix partitioning for any finite languages.

Definition 1 (Recursive prefix partitioning). *For a ptree+ $T = (V_T, E_T)$ and a finite language L that includes at least one string with length more than 1, recursive prefix partitioning of L with the node q of T is $leaf(p(q))$-prefix partition is $L = P_1 \cdot S_1 \cup \cdots \cup P_k \cdot S_k \cup R$. And, recursive prefix partitioning of $P_1, \ldots, P_k$ with ptree+ node $p(q)$ and $S_1, \ldots, S_k$ with ptree+ node $s(q)$.*

We define the *set of prefix partitions* of a language L with ptree+ T as the set consisting of all partitions that are obtained during the recursive prefix partition of L with T. In Fig. 2, an example of recursive prefix partitioning is shown. Each pair of string sets is denoted by a one-source two-destination edge where solid

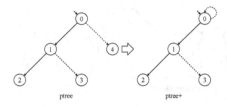

Fig. 1. Examples of ptree and ptree+.

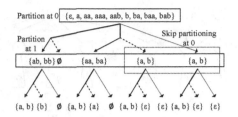

Fig. 2. Recursive prefix partitioning of $\{\varepsilon, a, aa, aaa, aab, b, ba, baa, bab\}$ by ptree+ in Fig. 1.

line head of edge denotes the prefix set and dashed line head of edge denotes the suffix set. Sets consisting of strings less than threshold ℓ is pointed by dotted line edges.

Based on prefix partitioning, we introduce a data structure *sequence sentential decision diagram* (*SSDD*, SeqSDD) as graphical representation of a finite language. A given language is converted into SSDD by recursive prefix partitioning and merging the same languages occur during the partitioning.

Attribute	Terminal	Nonterminal
pt	null	$pt(v)$
part	null	$parts(v)$
rem	null	$rem(v)$
label	$label(v)$	null
val	$value(v)$	null

Fig. 3. The attribute values for a vertex v.

Definition 2 (Sequence Sentential Decision Diagram). *A sequence sentential decision diagram (an SSDD or a SeqSDD) is a multi-rooted, directed graph* $G = (V, E)$, *with a ptree+* $T = (V_T, E_T)$, *satisfying the following:*

- *V is a vertex set containing two types of vertices known as nonterminal and terminal vertices. Each has certain attributes, id, pt, parts, rem, label, and val. The respective attributes are shown in Fig. 3.*
- *There are three types of terminal vertices, called $\top$ (top), $\bot$ (bottom) and letter vertices, respectively. A letter vertex v has a subset of the alphabet as a label $label(v) \subseteq \Sigma$. The top and bot have empty set as their label. An SSDD may have at most one of $\top$ and $\bot$. A terminal vertex v has as an attribute a value $value(v) \in \{0, 1\}$, indicating whether it is a $\top$ or a $\bot$, denoted by $\mathbf{1}$ or $\mathbf{0}$, respectively. A value of a letter vertex is null. A nonterminal vertex v has as attributes a ptree+ node $pt(v) \in V_T$, that is a branching node of T, called the respecting ptree+ node of v, a set of paired SSDD vertices $parts(v)$ called*

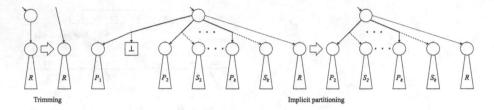

Trimming   Implicit partitioning

Fig. 4. Reduction rules of SSDD.

the ps-pair set, *and a SSDD node* $rem(v)$ *called the* reminder. *For an element* $(u, w) \in parts(v)$, *we call* (u, w) *ps-pair, and call* u/w *p-vertex/s-vertex of the pair, respectively. We refer to the corresponding outgoing edges as the p-edge and s-edge from* v . *We define the attribute triple for* v *by* $triple(v) = \langle pt(v), parts(v), rem(v) \rangle$. *For distinct vertices* u *and* v, $id(u) \neq id(v)$ *holds. A root is any vertex with no parent.*

- *We assume that the graph is* acyclic. *That is, there exists some partial order* $\prec_V$ *on vertices of* V *such that* $v \prec_V u$, $v \prec_V w$ *and* $v \prec_V rem(v)$ *for any nonterminal* v *and* $(u, w) \in parts(v)$.

We define the *size* of the graph, denoted by $|G|$, as the number of its edges. By definition, the graph consisting of a single terminal vertex, $\bot$ or $\top$, is an SSDD of size zero. For any vertex v in an SSDD G, the *subgraph rooted by* v is defined as the graph consisting of v and all its descendants. A graph G is called *single-rooted* if it has exactly one root, and *multi-rooted* otherwise. In this paper, we identify a single-rooted SSDD and its root node name. When we want to deal with multiple sets of strings at the same time, multi-rooted graphs are useful. We call it the shared SSDD environment that is the same idea of shared BDD [10].

We introduce three reduction rules of SSDD as follows: (i) merging: if there are multiple SSDD vertices that have the same attributes except for *id*, these vertices are merged into one vertex. (ii) trimming: if there is a nonterminal vertex v such that $parts(v) = \emptyset$, the vertex is deleted, and the edges that point v are changed so as to point $rem(v)$. (iii) implicit partitioning: if a nonterminal vertex v has a pair of vertices $(P, \bot)$ in $parts(v)$, remove the pair from $parts(v)$. If this change makes vertices referred from no vertices, delete such vertices and their outgoing edges from G except for root vertices. Figure 4 shows the rules trimming and implicit partitioning.

After we applying the reduction rules (i) and (ii), we can obtain a canonical form for a given set of strings. Figure 5 shows an example of SSDD. For each vertex v, $rem(v)$ is denoted by a dotted arrow, and each $(u, v) \in parts(v)$ is denoted by a one-source two-destination arrow that the solid line head of edge points u and the dashed line head of edge points v. We call such SSDD *canonical SSDD*. We do not require implicit partitioning always. This rule can reduce size of SSDD drastically, but we have to restore the deleted vertices when we want to compute the union of two given SSDDs. This problem will be discussed later. Figure 6, shows an example of implicit partitioning.

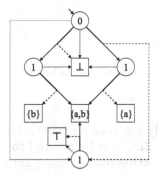
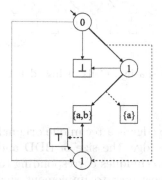
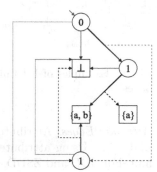

Fig. 5. An SSDD that represents the string set in Fig. 2.

Fig. 6. The SSDD obtained by applying implicit partitioning to the SSDD in Fig. 5

Fig. 7. The SSDD obtained by introducing ε-flag into the SSDD in Fig. 6

3.1 Semantics

Clearly, we see that an SSDD for a language L simulates the recursive prefix partitioning of L with ptree+ T by using two attributes, *parts*, and *rem*, of each vertex. For a vertex $v \in V$, let $L_G(v)$ be the language that is represented by the vertex. Note that the ptree+ node q determines the beginning position of strings included in $L_G(x)$. That is, there can be multiple nodes that represent the same language and respects different ptree+ nodes. Now, we give the first semantics of a sequence BDD.

Definition 3 (The definition of the language). *In an SSDD G, a vertex v in G denotes a finite language $L_G(v)$ on Σ defined recursively as:*

1. *If v is a terminal vertex, $L_G(v)$ is the trivial language defined as: (i) if value(v) = 1, $L_G(v) = \{\varepsilon\}$, and (ii) if value(v) = 0, $L_G(v) = \emptyset$. (ii) if value(v) = null, $L_G(v) = label(v)$.*
2. *If x is a nonterminal vertex, $L_G(x)$ is the finite language*
   $L_G(x) = (\bigcup_{(u,v) \in parts(x)} L_G(u) \cdot L_G(v)) \cup L_G(rem(x)).$

We write $L(v)$ for $L_G(v)$ if the underlying graph G is clearly understood. Moreover, if G is an SSDD with the single root r, we write $L(G)$ for $L_G(r)$. We say that G is an *SSDD for L* if $L = L(G)$.

3.2 Size Reducing Techniques

In this subsection, we show techniques to reduce the size of SSDD.

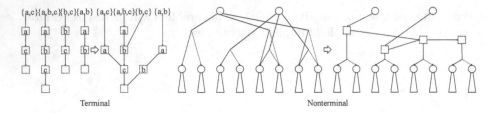

Fig. 8. Examples of list tail sharing. Cells of linked lists are denoted as dashed line boxes.

Attribute Edges. Attribute edge is a technique originally proposed for BDD and ZDD [10]. Using attribute edge, the size of BDD and ZDD can be reduced to half. For SDD and ZSDD, techniques corresponding to attributed edge have not been proposed yet. We show how to implement such a technique for SSDD. The main idea of attribute edge is deleting the $\top$ from a decision diagram and adding 1-bit flag to edges instead of the $\top$. For SSDD, we call such flags ε-flags. When we use ε-flags, we do not create SSDD vertices that represent languages including ε. Since the terminal node $\top$ represents $\{\varepsilon\}$, the $\top$ is not used. Thus, the number of vertices is reduced to half in the best case. For a given SSDD, we can construct an SSDD with ε-flags by the following procedure: (i) Add 1-bit flag to every s-edge of nonterminal vertices, and initialize it to 0. (ii) If an s-edge is pointing vertex v such that $\varepsilon \in L(v)$, create an SSDD vertex v' that represents $L(v) - \{\varepsilon\}$ and change the destination of the s-edge to v'. Then we change the ε-flag of the s-edge to 1 to indicate that the destination vertex do not include ε but we deal with the vertex as if it includes ε. An example of SSDD with ε-flag is shown in Fig. 7.

List Tail Sharing. SSDD vertices have to store set as their attributes. This technique is to reduce the memory usage by such sets by using linked lists and sharing the equivalent contents. The main idea is simple is simple. We sort the set of paired vertices of a nonterminal vertex in increasing order by the id of the p-side vertex of each pair. We also sort the set of letters of terminal vertices in lexicographic order. Next, we store the elements of the sets in linked list. If there are multiple vertices that their lists have the same sequence of elements in their list, we merge the tail of lists into one. We can do this by using a hashtable whose keys are a pair of a list content and pointer to the next list cell, and value is a pointer to such list cell if it exists. Figure 8 shows examples of list tail sharing between terminal vertices and nonterminal vertices.

4   Algorithm

In this section, we provide fundamental algorithms for SSDD. They are similar to the algorithms for ZSDD, that represents families of sets, but not the same because there are some differences between SSDD and ZSDD as shown below:

1. SSDD terminal vertices have subsets of the alphabet as labels. Terminal vertices of ZSDD are only $\top$ and $\bot$.
2. SSDD nonterminal vertices have the attribute *rem*. ZSDD nonterminal vertices only have an attribute corresponding to *parts*.
3. For $(u, v) \in parts(v)$ of an SSDD vertex v, the strings in $L(u)$ are the same length more than 0. On the other hand, ZSDD does not have such a restriction for sizes of sets in a family.
4. SSDD uses a ptree+ that has a cycle so as to deal with variable length strings. ZSDD uses a tree like a ptree to partition sets because the size of the universal set is fixed.

First, we discuss how to construct an SSDD for a string. During construction of an SSDD, we create SSDD vertices. To avoid creating equivalent vertices, we call the procedure **Getnode** that returns a vertex with given attributes. There are two **Getnode** for nonterminal and terminal. These are called **GetnodeN** and **GetnodeT**, respectively. They use hashtables to keep one-to-one correspondence between languages and vertices. The reduction rule merging is realized by these procedures. To construct an SSDD for a given string, we decompose it according to recursive prefix partitioning, and create vertices in a bottom-up manner.

Next, we discuss set operations such as intersection, difference, and union. These are implemented as recursive algorithms. The pseudo code of union is given in Algorithm 1. Intersection and difference can be computed in the same manner as the algorithm of union. During set operations of nonterminal vertices x_1, x_2, we compute the intersection of all combination of $(u_1, v_1) \in parts(x_1)$ and $(u_2, v_2) \in parts(x_2)$ to keep the disjoint condition on p-side. Note that $\{s_1 \cap s_2 | s_1 \in D_1, s_2 \in D_2\}$ is a disjoint set for two disjoint sets D_1 and D_2. After trying all combination, we get a set of paired vertices of size $|parts(x_1)| \times |parts(x_2)|$. It can contain multiple pairs that have the same vertex as their second element. We use the procedure **Compress** to merge such vertices. There is one important difference between union and intersection. If we apply implicit partitioning, we have to restore the vertices deleted by implicit partitioning when we compute union. Without the deleted vertices, we will miss some strings that must be included in the resultant vertex. For example, imagine the union of two nonterminal whose *parts* are $\{(\{a\}, \{a, b\})\}$ and $\{(\{b\}, \{a, b\})\}$, respectively. To restore the deleted vertices, we use the **All** procedure that returns the vertex r consisting of all possible strings respecting a given ptree node. By computing difference between r and all u of $(u, v) \in parts(x)$, we can get the implicit partitioned vertices. Note that the number of restored vertices can be exponentially larger than the original SSDD.

Third, we show the algorithm for concatenation of two SSDD vertices. The pseudo code is shown as Algorithm 2. Since concatenation is an operation for string sets, ZSDD has no corresponding algorithm. An obvious way to realize concatenation is expanding both SSDD, concatenate them explicitly and construct the SSDD for the resultant string set. The difficulty of concatenation

---

Algorithm 1. Union(y, z): Make an SSDD vertex that represents a language $L(y) \cup L(z)$ for two given SSDD vertices y and z.

---

*Input:*Two SSDD vertices y and z
*Output:*The SSDD vertex x such that $L(x) = L(y) \cup L(z)$
Global variable: A hash table *memocache* whose key is in $\langle Op, V, V \rangle$ and value is in V where Op is a set of operation names.

```
 1: if y = ⊥ then
 2:     return z;
 3: end if
 4: if z = ⊥ then
 5:     return y;
 6: end if
 7: if y = z then
 8:     return y;
 9: end if
10: if pt(y) > pt(z) or (pt(y) = pt(z) and id(y) > id(z)) then
11:     swap(y, z);
12: end if
13: if (x ← memocache.search(⟨∪, y, z⟩)) ≠ null then
14:     return x;
15: end if
16: if pt(y) = pt(z) = null then
17:     x ← GetnodeT(label(y) ∪ label(z));
18: else if pt(y) < pt(z) then
19:     x ← GetnodeN(pt(y), parts(y), Union(rem(y), z));
20: else
21:     U ← ∅;
22:     if we employ implicit partitioning then
23:         u_y, u_z ← All(p(pt(y)));
24:         for all (u, v) ∈ parts(y) do
25:             u_y ← Difference(u_y, u);
26:         end for
27:         for all (u, v) ∈ parts(z) do
28:             u_z ← Difference(u_z, u);
29:         end for
30:         temporarily add (u_y, ⊥) and (u_z, ⊥) to part(y) and part(z), respectively.
31:     end if
32:     for all (u_y, v_y) ∈ parts(y) do
33:         for all (u_z, v_z) ∈ parts(z) do
34:             U ← U ∪ {(Intersection(u_y, u_z), Union(v_y, v_z))};
35:         end for
36:     end for
37:     U' ← Compress(U);
38:     w ← Union(rem(y), rem(z));
39:     x ← GetnodeN(pt(y), U', w);
40: end if
41: memocache.insert(⟨∪, y, z⟩, x);
42: return x;
```

Algorithm 2. Concatenation(q, y, z): Make an SSDD vertex that represents a language $L(y) \cdot L(z)$ for two given SSDD vertices y and z.

Input: Two SSDD vertices y and z, and a ptree+ node q where $pt(y)$ and $pt(z)$ are reachable from q by traversing only p-edges.

Output: The SSDD vertex x such that $L(x) = L(y) \cdot L(z)$ and $pt(x)$ is a p-side descendant of q.

Global variable: A hash table *memocache* whose key is in $\langle Op, V, V \rangle$ and value is in V where Op is a set of operation names.

```
 1: if y = ⊥ or z = ⊥ then
 2:     return ⊥;
 3: end if
 4: if (x ← memocache.search(⟨·, y, z⟩)) ≠ null then
 5:     return x;
 6: end if
 7: x ← ⊥, U ← ∅;
 8: if pt(y) = q then
 9:     for all (u, v) ∈ part(x) do
10:         U ← U ∪ {(u, Concatenation(s(q), v, z))};
11:     end for
12:     x ← GetnodeN(q, U, ⊥);
13:     F_p ← Convert0(rem(y), an empty list);
14: else
15:     F_p ← Convert0(y, an empty list);
16: end if
17: for all U_p ∈ F_p do
18:     x ← Union(x, Convert(q, U_p + + a list containing only z));
19:                          ▷ "++" means list concatenation operation.
20: end for
21: memocache.insert(⟨·, y, z⟩, x);
22: return x;
```

comes from that we need to convert a language represented by a vertex respecting a qtree+ node into new vertex respecting other ptree+ nodes.

Our algorithm realizes this with less number of expanding of SSDD vertices. The main idea is to delay expanding vertices until it is required like call-by-need. During conversion for ptree+ node q, we want to know what languages must be represented by vertices respecting $p(q)$ and what languages must be paired for each p child. We can determine it only by the length of prefixes. Therefore, we do not need to expanding all SSDD vertices. We just find sequences of SSDD vertices $u_1, \ldots, u_k$ such that $\sum_{1 \leq i \leq k} maxlen(L(u_i)) = leaf(p(q))$ and $minlen(L(u_i)) = maxlen(L(u_i))$ for all $1 \leq i \leq k$. We divide such sequences into prefix part and suffix part recursively, and construct corresponding new SSDD vertices.

Algorithm 3. Convert(q, U_p)**:** Make an SSDD vertex, respecting a given ptree+ node, that represents the same language that a given sequence of SSDD vertices represent.

Input: A ptree+ node q and a list of SSDD vertices $U_p = U_p[1] \cdots U_p[i]$.

Output: The SSDD vertex x, that respecting ptree+ node q, such that $L(x) = L(U_p[1]) \times \cdots \times L(U_p[i])$

Global variable: A hash table *memocache* whose key is in $\langle V_P, V* \rangle$ and value is in V.

```
 1: if the size of U_p is 1 and U_s[1] = ⊤ then
 2:     return ⊤;
 3: end if
 4: if leaf(q) = 1 then
 5:     return U_p[1];
 6: end if
 7: if (x ← memocache.search(q, U_p) ≠ null then
 8:     return x;
 9: end if
10: F ← Convert1(q, an empty list, U_p, 0);
11: U ← ∅, w ← ⊥;
12: for all (U'_p, U'_s) ∈ F do
13:     if U'_s is empty then
14:         w ← Union(w, Convert(p(q), U'_p));
15:     else
16:         U ← U ∪ {(Convert(p(q), U'_p), Convert(s(q), U'_s))};
17:     end if
18: end for
19: U' ← Compress(U);
20: x ← GetnodeN(q, U, w);
21: memocache.insert(⟨q, U_p, U_s⟩, x);
22: return x;
```

---

5   Analysis

In this section, we show complexity analysis of SSDD.

Size Complexity: We show that the size of an SSDD can be exponentially smaller than the SeqBDD for the same language. SeqBDD is a data structure to represent sets of strings, and the size of a SeqBDD can be $\mathcal{O}(\Sigma)$ times smaller than the equivalent deterministic finite automaton (DFA) [5]. The size of SeqBDD is at most the same as the size of DFA.

Theorem 1. *Size of an SSDD can be $\mathcal{O}(m\Sigma^{\ell/2})$ times smaller than the SeqBDD that represents the same language where ℓ is the number of leaves of ptree+ and m is the length of the longest string.*

Proof. Define a language L_i and ptree T_i as follows: $L_{n+1} = \bigcup_{a \in \Sigma} \{a\} \cdot L_n \cdot \{a\}$, and $L_1 = \Sigma$. That is, L_n is the set of all palindrome of length $2n - 1$. $T_{n+1} =$ a ptree with root q such that $p(q)$ is a leaf, $s(s(p))$ is a leaf, and $p(s(q))$ is

Algorithm 4. Convert0(x, U_p): Make sequences of SSDD vertices such that the language represented by each sequence contains strings of the same length.

Input: An SSDD vertex x

Output: A set of sequences of SSDD vertices F such that $L(u_1) \times \cdots \times L(u_k)$ is a set of string of the same length for $u_1, \ldots, u_k \in F$, and $\bigcup_{u_1, \ldots, u_k \in F} L(u_1) \times \cdots \times L(u_k) = L(x)$ where $\times$ means concatenation of languages.

1: **if** $\mathbf{minlen}(x) = \mathbf{maxlen}(x)$ **then**
2:             ▷ **minlen** and **maxlen** can be computed by simple recursive algorithms.
3:     **return** { $U_p + +$ a list contains only x };
4: **end if**
5: $F \leftarrow \emptyset$;
6: **if** $rem(x) \neq null$ **then**
7:     $F \leftarrow F \cup \mathbf{Convert0}(rem(x), U_p)$;
8: **end if**
9: **for all** $(u, v) \in parts(x)$ **do**
10:     $F \leftarrow F \cup \mathbf{Convert0}(v, U_p + +$a list contains only $u)$;
11: **end for**
12: **return** F;

T_n. T_1 is just a leaf. Figure 9 shows the graphical image of T_{n+1}. We show the SeqBDD and SSDD that represent L_{n+1} in Figs. 10 and 11. The size f_{n+1} of SeqBDD for L_{n+1} is $|\Sigma| \times f_n + 2|\Sigma|$ and $f_1 = |\Sigma|$. The size g_{n+1} of SSDD for L_{n+1} is $g_n + 2|\Sigma|$ and $g_1 = |\Sigma|$. Thus, $f_n = \frac{(|\Sigma|+1)|\Sigma|^n - 2|\Sigma|}{|\Sigma|-1} = \mathcal{O}(|\Sigma|^n)$ and $g_n = (2n-1)|\Sigma|$. Next, we consider the language L_n^m with ptree+ in Fig. 12. Then, the size of the SeqBDD for L_n^m is $\mathcal{O}(m|\Sigma|^n)$ as shown in Fig. 13, and the size of the SSDD for L_n^m is $m + (2n+1)|\Sigma|$ as shown in Fig. 14. As a result, the SSDD for L_n^m is $\mathcal{O}(m\Sigma^{\ell/2})$ times smaller than the SeqBDD for L_n^m.

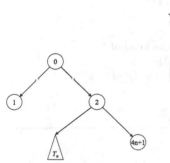

Fig. 9. The ptree T_{n+1}

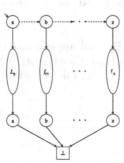

Fig. 10. The SeqBDD represents L_{n+1}

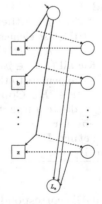

Fig. 11. The SSDD represents L_{n+1}

Algorithm 5. Convert1(q, U_p, U_s, ℓ): Make an SSDD vertex, respecting a given ptree+ node, that represents the same language that a given sequence of SSDD vertices represent.

*Input:*A ptree+ node q and two doubly linked lists of SSDD vertices U_p and U_s. And, the length ℓ of strings represented by U_p.
*Output:*The set of two lists of SSDD vertices that are corresponding $p(q)$ and $s(q)$ respectively.

1: **if** $\ell = leaf(p(q))$ **or** $(\ell < leaf(p(q))$ **and** U_s is empty) **then**
2:      **return** $\{(U_p, U_s)\}$;
3: **end if**
4: **if** $\ell < leaf(p(q))$ **then**
5:      **while** U_s is not empty **and** $leaf(pt(U_s[1])) \le 1$ **and** $\ell < leaf(p(q))$ **do**
6:          Let the first element of U_s be x and delete it from U_s;
7:          Append x to the tail of U_p;
8:          $\ell \leftarrow \ell + leaf(pt(U_s[1]))$;
9:      **end while**
10:     **if** $\ell = leaf(p(q))$ **or** U_s is empty **then**
11:         **return** $\{(U_p, U_s)\}$;
12:     **end if**
13:     let the first element of U_s be x and delete it from U_s;
14:     $F \leftarrow \emptyset$;
15:     **if** $rem(x) \ne null$ **then**
16:         append $rem(x)$ to the head of U_s;
17:         $F \leftarrow F \cup$ **Convert1**(U_p, U_s, ℓ);
18:         delete $rem(x)$ from the head of U_s;
19:     **end if**
20:     **for all** $(u, v) \in parts(x)$ **do**
21:         append u to the tail of U_p, and append v to the head of U_s;
22:         $F \leftarrow F \cup$ **Convert1**$(U_p, U_s, \ell + leaf(p(pt(x))))$
23:         delete u from the tail of U_p, and delete v from the head of U_s;
24:     **end for**
25:     **return** F;
26: **end if**
27: **if** $\ell > leaf(p(q))$ **then**
28:     let the last element of U_p be x and delete it from U_p;
29:     $\ell \leftarrow \ell - leaf(pt(x))$;
30:     **for all** $(u, v) \in parts(x)$ **do**
31:         append u to the tail of U_p, and append v to the head of U_s;
32:         $F \leftarrow F \cup$ **Convert1**$(U_p, U_s, \ell + leaf(p(pt(x))))$
33:         delete u from the tail of U_p, and delete v from the head of U_s;
34:     **end for**
35:     **return** F;
36: **end if**

SeqBDD corresponds SSDD with ptree+ consisting of only one leaf. We show the next theorem.

Theorem 2. *Size of an SSDD with ptree+ consisting of only one leaf is at most* $\mathcal{O}(|\Sigma|)$ *times larger than the SeqBDD for the same language.*

Proof. In the SSDD with such ptree+ T, every nonterminal vertices respect the root node of T. There is no vertices representing the same language but respects different ptree+ nodes. This structure is almost the same as a DFA. The difference between DFA and this SSDD is that DFA's labels are letters but this SSDD's labels are subsets of the alphabet. However, the number of vertices is the same. The number of outgoing edges from this SSDD nonterminal vertices is at most Σ. That is the same as the number of outgoing edges of DFA vertices. Therefore, the size of SSDD is at most $\mathcal{O}(|\Sigma|)$ times larger than the equivalent SeqBDD.

Theorem 3. *Size of an SSDD with implicit partitioning is at most* $\mathcal{O}(\ell m^2)$ *times larger than the equivalent SeqBDD where ℓ is the number of leaves of ptree+ and m is the size of the SeqBDD.*

Proof. Remember the algorithm of **Convert**. This situation is similar to converting an SSDD with a ptree+ consisting of only one leaf into an SSDD with a certain ptree+. For each vertex of SeqBDD, we may create SSDD vertices that respect different ptree+ nodes. And, for ptree+ node q, we will create SSDD vertices for sets of strings of length $leaf(q)$ that connect two different SeqBDD vertices, as a start vertex and a goal vertex. The number of outgoing edges is at most m where m is the size of the original SeqBDD because we have to choose halfway vertices that splits the strings existing between the start and goal vertices. Therefore, the number of SSDD vertices is $\mathcal{O}(\ell m^2 + \ell m^3)$.

These above theorems say that size of an SSDD is polynomially larger than the equivalent SeqBDD in the worst case and exponentially smaller than it in the best case.

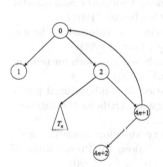

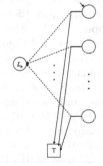

Fig. 12. The ptree+ for L_n^m

Fig. 13. The SeqBDD represents L_n^m

Fig. 14. The SSDD represents L_n^m

Time Complexity: The algorithms for operations such as **Search**, **Intersection**, **Difference**, and **Union** are similar to algorithms on ZSDD. The time complexity of these algorithms is proven in [11]. The **Union** algorithm for SSDD with implicit partitioning is not polytime algorithm because it takes to exponential time to restore deleted vertices. Other above algorithms are polytime. The **Concatenation** algorithm is not polytime because we have to expand all vertices to raw sets of strings in the worst case.

6 Conclusion

In this paper, we propose a sequence sentential decision diagram that is a new data structure to represent and manipulate sets of strings. SSDD is a generalized version of the sequence binary decision diagram and can be exponentially smaller than SeqBDD. For future work, we will evaluate the performance of SSDD for practical data sets and use SSDD for real-life applications. We also address the question of the relationship between SSDD and grammar compression.

References

1. Aho, A.V., Hopcroft, J.E., Ullman, J.D.: The Design and Analysis of Computer Algorithms. Addison-Wesley, Boston (1974)
2. Crochemore, M., Hancart, C., Lecroq, T.: Algorithms on Strings. Cambridge University Press, Cambridge (2007)
3. Daciuk, J., Mihov, S., Watson, B.W., Watson, R.: Incremental construction of minimal acyclic finite state automata. Comput. Linguist. **26**(1), 3–16 (2000)
4. Darwiche, A.: SDD: a new canonical representation of propositional knowledge bases. In: Proceedings of the Twenty-Second International Joint Conference on Artificial Intelligence, IJCAI 2011, vol. 22, pp. 819–826. AAAI Press (2011)
5. Denzumi, S., Yoshinaka, R., Arimura, H., Minato, S.-I.: Sequence binary decision diagram: minimization, relationship to acyclic automata, and complexities of boolean set operations. Discret. Appl. Math. **212**, 61–80 (2016)
6. Gusfield, D.: Algorithms on Strings, Trees, and Sequences: Computer Science and Computational Biology. Cambridge University Press, Cambridge (1997)
7. Hopcroft, J.E., Motwani, R., Ullman, J.D.: Introduction to Automata Theory, Languages, and Computation, 3rd edn. Addison-Wesley, Boston (2006)
8. Loekito, E., Bailey, J., Pei, J.: A binary decision diagram based approach for mining frequent subsequences. Knowl. Inf. Syst. **24**(2), 235–268 (2010)
9. Minato, S.-I.: Zero-suppressed BDDs for set manipulation in combinatorial problems. In: Proceedings of the Thirtieth AAAI Conference on Artificial Intelligence, pp. 1058–1066 (2016)
10. Minato, S.-I., Ishiura, N., Yajima, S.: Shared binary decision diagram with attributed edges for efficient Boolean function manipulation. In: Proceedings of the 27th ACM/IEEE Design Automation Conference, pp. 52–57 (1990)
11. Nishino, M., Yasuda, N., Minato, S.-I., Nagata, M.: Zero-suppressed sentential decision diagrams. In: Proceedings of the 30th ACM/IEEE Design Automation Conference, pp. 272–277 (2016)

Clustering

Online Unit Covering in Euclidean Space

Adrian Dumitrescu[1], Anirban Ghosh[2(✉)], and Csaba D. Tóth[3,4]

[1] University of Wisconsin–Milwaukee, Milwaukee, WI, USA
dumitres@uwm.edu
[2] University of North Florida, Jacksonville, FL, USA
anirban.ghosh@unf.edu
[3] California State University Northridge, Los Angeles, CA, USA
[4] Tufts University, Medford, MA, USA
cdtoth@acm.org

Abstract. We revisit the online UNIT COVERING problem in higher dimensions: Given a set of n points in $\mathbb{R}^d$, that arrive one by one, cover the points by balls of unit radius, so as to minimize the number of balls used. In this paper, we work in $\mathbb{R}^d$ using Euclidean distance. The current best competitive ratio of an online algorithm, $O(2^d d \log d)$, is due to Charikar et al. (2004); their algorithm is deterministic.
(I) We give an online deterministic algorithm with competitive ratio $O(1.321^d)$, thereby improving on the earlier record by an exponential factor. In particular, the competitive ratios are 5 for the plane and 12 for 3-space (the previous ratios were 7 and 21, respectively). For $d = 3$, the ratio of our online algorithm matches the ratio of the current best offline algorithm for the same problem due to Biniaz et al. (2017), which is remarkable (and rather unusual).
(II) We show that the competitive ratio of every deterministic online algorithm (with an adaptive deterministic adversary) for UNIT COVERING in $\mathbb{R}^d$ under the L_2 norm is at least $d + 1$ for every $d \geq 1$. This greatly improves upon the previous best lower bound, $\Omega(\log d / \log \log \log d)$, due to Charikar et al. (2004).
(III) We obtain lower bounds of 4 and 5 for the competitive ratio of any deterministic algorithm for online UNIT COVERING in $\mathbb{R}^2$ and respectively $\mathbb{R}^3$; the previous best lower bounds were both 3.
(IV) When the input points are taken from the square or hexagonal lattices in $\mathbb{R}^2$, we give deterministic online algorithms for UNIT COVERING with an optimal competitive ratio of 3.

Keywords: Online algorithm · Unit covering · Unit clustering
Competitive ratio · Lower bound · Newton number

1 Introduction

Covering and clustering are fundamental problems in the theory of algorithms, computational geometry, optimization, and other areas. They arise in a wide

Research supported, in part, by the University of North Florida start-up fund and by the NSF awards CCF-1422311 and CCF-1423615.

D. Kim et al. (Eds.): COCOA 2018, LNCS 11346, pp. 609–623, 2018.
https://doi.org/10.1007/978-3-030-04651-4_41

range of applications, such as facility location, information retrieval, robotics, and wireless networks. While these problems have been studied in an offline setting for decades, they have been considered only recently in a more dynamic (and thereby realistic) setting. Here we study such problems in a high-dimensional Euclidean space and mostly in the L_2 norm. We first formulate them in the classic *offline* setting.

Problem 1. k-CENTER. Given a set of n points in $\mathbb{R}^d$ and a positive integer k, cover the set by k congruent balls centered at the points so that the diameter of the balls is minimized.

The following two problems are dual to Problem 1.

Problem 2. UNIT COVERING. Given a set of n points in $\mathbb{R}^d$, cover the set by balls of unit diameter so that the number of balls is minimized.

Problem 3. UNIT CLUSTERING. Given a set of n points in $\mathbb{R}^d$, partition the set into clusters of diameter at most one so that the number of clusters is minimized.

Problems 1 and 2 are easily solved in polynomial time for points on the line, i.e., for $d = 1$; but both problems become NP-hard already in Euclidean plane [14,22]. Factor 2 approximations are known for k-CENTER in any metric space (and so for any dimension) [13,15]; see also [25, Chap. 2], while polynomial-time approximation schemes are known for UNIT COVERING for any fixed dimension [17]. However, these algorithms are notoriously inefficient and thereby impractical; see also [2] for a summary of such results and different time vs. ratio trade-offs.

Problems 2 and 3 are identical in the offline setting: indeed, one can go from clusters to balls in a straightforward way; and conversely, one can assign multiply covered points in an arbitrary fashion to unique balls. In this paper we focus on the second problem, namely *online* UNIT COVERING; we however point out key differences between this problem and online UNIT CLUSTERING.

The performance of an online algorithm ALG is measured by comparing it to an optimal offline algorithm OPT using the standard notion of competitive ratio [3, Chap. 1]. The competitive ratio of ALG is defined as $\sup_\sigma \frac{\mathsf{ALG}(\sigma)}{\mathsf{OPT}(\sigma)}$, where σ is an input sequence of points, $\mathsf{OPT}(\sigma)$ is the cost of an optimal offline algorithm for σ and $\mathsf{ALG}(\sigma)$ denotes the cost of the solution produced by ALG for this input. For randomized algorithms, $\mathsf{ALG}(\sigma)$ is replaced by the expectation $E[\mathsf{ALG}(\sigma)]$, and the competitive ratio of ALG is $\sup_\sigma \frac{E[\mathsf{ALG}(\sigma)]}{\mathsf{OPT}(\sigma)}$. If there is no danger of confusion, we use ALG to refer to an algorithm or the cost of its solution, as needed.

Charikar et al. [8, Sect. 6] studied the online version of UNIT COVERING (under the name of "Dual Clustering"). The points arrive one by one and each point needs to be assigned to a new or to an existing unit ball upon arrival; the L_2 norm is used in $\mathbb{R}^d$, $d \in \mathbb{N}$. The location of each new ball is fixed as soon as it is placed. The authors provided a deterministic algorithm of competitive ratio $O(2^d d \log d)$ and gave a lower bound of $\Omega(\log d / \log \log \log d)$ on the competitive ratio of any deterministic algorithm for this problem. For $d = 1$ a tight bound of 2 is folklore; for $d = 2$ the best known upper and lower bounds on the competitive ratio are 7 and 3, respectively, as implied by the results in [8][1].

The online UNIT CLUSTERING problem was introduced by Chan and Zarrabi-Zadeh [7] in 2006. While the input and the objective of this problem are identical to those for UNIT COVERING, UNIT CLUSTERING is more flexible in that the algorithm is not required to produce unit balls at any time, but rather the smallest enclosing ball of each cluster should have diameter *at most* 1; furthermore, a ball may change (grow or shift) in time. In regard to their *online* versions, It is worth emphasizing two properties (shared with UNIT COVERING): (i) a point assigned to a cluster must remain in that cluster; and (ii) two distinct clusters cannot merge into one cluster, i.e., the clusters maintain their identities. The authors showed that several standard approaches for UNIT CLUSTERING, namely the deterministic algorithms Centered, Grid, and Greedy, all have competitive ratio at most 2 for points on the line ($d = 1$). Moreover, the first two algorithms are applicable for UNIT COVERING, with a competitive ratio at most 2 for $d = 1$, as well. These algorithms naturally extend to any higher dimension (including Grid provided the L_∞ norm is used).

> **Algorithm Centered.** For each new point p, if p is covered by an existing unit ball, do nothing; otherwise place a new unit ball centered at p.

> **Algorithm Grid.** Build a uniform grid in $\mathbb{R}^d$ where cells are unit cubes of the form $\prod [i_j, i_j + 1)$, where $i_j \in \mathbb{Z}$ for $j = 1, \ldots, d$. For each new point p, if the grid cell containing p is nonempty, put p in the corresponding cluster; otherwise open a new cluster for the grid cell and put p in it.

Since in $\mathbb{R}^d$ each cluster of OPT can be split into at most 2^d grid-cell clusters created by the algorithm, its competitive ratio is at most 2^d, and this analysis is tight for the L_∞ norm. It is worth noting that there is no direct analogue of this algorithm under the L_2 norm.

Some (easy) remarks are in order. Any lower bound on the competitive ratio of an online algorithm for UNIT CLUSTERING applies to the competitive ratio of the same type of algorithm for UNIT COVERING. Conversely, any upper bound on the competitive ratio of an online algorithm for UNIT COVERING yields an upper bound on the competitive ratio of the same type of algorithm for UNIT CLUSTERING.

Related Work. UNIT COVERING is a variant of SET COVER. Alon et al. [1] gave a deterministic online algorithm of competitive ratio $O(\log m \log n)$ for this problem, where n is the size of the ground set and m is the number of sets in the

---

[1] Charikar et al. [8] claim (on p. 1435) that a lower bound of 4 for $d = 2$ under the L_2 norm follows from their Theorem 6.2; but this claim appears unjustified; only a lower bound of 3 is implied. Unfortunately, this misinformation has been carried over also by [7,9].

family. Buchbinder and Naor [6] obtained sharper results under the assumption that every element appears in at most Δ sets.

Chan and Zarrabi-Zadeh [7] showed that no online algorithm (deterministic or randomized) for UNIT COVERING can have a competitive ratio better than 2 in one dimension ($d = 1$). They also showed that it is possible to get better results for UNIT CLUSTERING than for UNIT COVERING. Specifically, they developed the first algorithm with competitive ratio below 2 for $d = 1$, namely a randomized algorithm with competitive ratio 15/8. This fact has been confirmed by subsequent algorithms designed for this problem; the current best ratio 5/3, for $d = 1$, is due to Ehmsen and Larsen [11], and this gives a ratio of $2^d \cdot \frac{5}{6}$ for every $d \geq 2$ (the L_∞ norm is used); their algorithm is deterministic. The appropriate "lifting" technique has been layed out in [7,28]. From the other direction, the lower bound for deterministic algorithms has evolved from 3/2 in [7] to 8/5 in [12], and then to 13/8 in [20].

Answering a question of Epstein and van Stee [12], Dumitrescu and Tóth [9] showed that the competitive ratio of any algorithm (deterministic or randomized) for UNIT CLUSTERING in $\mathbb{R}^d$ under the L_∞ norm must depend on the dimension d; in particular, it is $\Omega(d)$ for every $d \geq 2$.

Liao and Hu [21] gave a PTAS for a related disk cover problem (another variant of SET COVER): given a set of m disks of arbitrary radii and a set P of n points in $\mathbb{R}^2$, find a minimum-size subset of disks that jointly cover P; see also [23, Corollary 1.1].

Our Results. (i) We show that the competitive ratio of Algorithm Centered for online UNIT COVERING in $\mathbb{R}^d$, $d \in \mathbb{N}$, under the L_2 norm is bounded by the Newton number of the Euclidean ball in the same dimension. In particular, it follows that this ratio is $O(1.321^d)$ (Theorem 1 in Sect. 2). This greatly improves on the ratio of the previous best algorithm due to Charikar et al. [8]. The competitive ratio of their algorithm is at most $f(d) = O(2^d d \log d)$, where $f(d)$ is the number of unit balls needed to cover a ball of radius 2 (i.e., the doubling constant). By a volume argument, $f(d)$ is at least 2^d. In particular $f(2) = 7$ and $f(3) = 21$ [27]; see also [2]. The competitive ratios of our algorithm are 5 in the plane and 12 in 3-space, improving the earlier ratios of 7 and 21, respectively.

(ii) We show that the competitive ratio of every deterministic online algorithm (with an adaptive deterministic adversary[2]) for UNIT COVERING in $\mathbb{R}^d$ under the L_2 norm is at least $d + 1$ for every $d \geq 1$ (Theorem 3 in Sect. 3). This greatly improves the previous best lower bound, $\Omega(\log d / \log \log \log d)$, due to Charikar et al. [8].

(iii) We obtain lower bounds of 4 and 5 for the competitive ratio of any deterministic algorithm (with an adaptive deterministic adversary) for UNIT COVERING

---

[2] An *adaptive adversary* is one that tries to force the algorithm perform extensive work by observing each of its actions and constructing the input accordingly step by step.

in $\mathbb{R}^2$ and respectively $\mathbb{R}^3$ (Theorems 2 and 3 in Sect. 3). The previous best lower bounds were both 3.

(iv) For input point sequences that are subsets of the infinite square or hexagonal lattices, we give deterministic online algorithms for UNIT COVERING with an optimal competitive ratio of 3 (Theorems 4 and 5 in Sect. 4).

Notation and Terminology. For two points $p, q \in \mathbb{R}^d$, let $d(p, q)$ denote the Euclidean distance between them. Throughout this paper the L_2-norm is used. The closed ball of radius r in $\mathbb{R}^d$ centered at point $z = (z_1, \ldots, z_d)$ is $B_d(z, r) = \{x \in \mathbb{R}^d \mid d(z, x) \le r\} = \{(x_1, \ldots, x_d) \mid \sum_{i=1}^d (x_i - z_i)^2 \le r^2\}$. A *unit ball* is a ball of unit radius in $\mathbb{R}^d$. The UNIT COVERING problem is to cover a set of points in $\mathbb{R}^d$ by a minimum number of unit balls.

The *unit sphere* is the surface of the d-dimensional unit ball centered at the origin $\mathbf{0}$, namely, the set of points $\mathbb{S}^{d-1} \subset B_d(\mathbf{0}, 1)$ for which equality holds: $\sum_{i=1}^d x_i^2 = 1$. A *spherical cap* $C(\alpha)$ of angular radius $\alpha \le \pi$ and center P on $\mathbb{S}^{d-1}$ is the set of points Q in $\mathbb{S}^{d-1}$ for which $\angle P0Q \le \alpha$; see [24].

2   Analysis of Algorithm Centered for Online Unit Covering in Euclidean d-Space

For a convex body $C \subset \mathbb{R}^d$, the *Newton number* (a.k.a. *kissing number*) of C is the maximum number of nonoverlapping congruent copies of C that can be arranged around C so that each of them is touching C [5, Sect. 2.4]. Some values $N(B_d)$, where $B_d = B_d(\mathbf{0}, 1)$, are known exactly for small d, while for most dimensions d we only have estimates. For instance, it is easy to see that $N(B_2) = 6$, and it is known that $N(B_3) = 12$ and $N(B_4) = 24$. The problem of estimating $N(B_d)$ in higher dimensions is closely related to the problem of determining the densest sphere packing and the knowledge in this area is largely incomplete with large gaps between lower and upper bounds; see [5, Sect. 2.4] and the references therein; in particular, many upper and lower estimates up to $d = 128$ are given in [4,10]. In this section, we prove the following theorem.

Theorem 1. *Let $\varrho(d)$ be the competitive ratio of Algorithm Centered in $\mathbb{R}^d$ (when using the L_2 norm). Then $\varrho(2) = N(B_2) - 1 = 5$, $\varrho(3) = N(B_3) = 12$, and $\varrho(d) \le N(B_d)$ for every $d \ge 4$. In particular, $\rho(d) = O(1.321^d)$*

A key fact for proving the theorem is the following easy lemma.

Lemma 1. *Let B be a unit ball centered at o, that is part of OPT. Let $p, q \in B$ be any two points in B presented to the online algorithm that forced the algorithm to place new balls centered at p and q; refer to Fig. 1. Then $\angle poq > \pi/3$.*

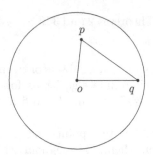

Fig. 1. Lemma 1.

Proof. Assume for contradiction that $\alpha = \angle poq \leq \pi/3$. Assume also, as we may, that p arrives before q. Since $q \notin B(p)$, we have $|pq| > 1$. Consider the triangle $\triangle poq$; we may further assume that $\angle opq \geq \angle oqp$ (if we have the opposite inequality, the argument is symmetric). In particular, we have $\angle opq \geq \pi/3$. Since $\angle poq \leq \pi/3$ and $\angle opq \geq \pi/3$, the law of sines yields that $|oq| \geq |pq| > 1$. However, this contradicts the fact that q is contained in B, and the proof is complete.   □

Corollary 1. *Let B be a unit ball centered at o, that is part of OPT. For every point $p \in B$ presented to the online algorithm that forced the algorithm to place a new ball centered at p, let $\Psi(p)$ denote the cone with apex at o, axis $\overrightarrow{op}$, and angle $\pi/6$ around $\overrightarrow{op}$. Then the cones $\Psi(p)$ are pairwise disjoint in B; hence the corresponding caps on the surface of B are also nonoverlapping.*

Proof of Theorem 1. For every unit ball B of OPT we bound from above the number of unit balls placed by **Algorithm Centered** whose center lies in B. Suppose this number is at most A (for every ball in OPT). Since the center of every unit ball placed by the algorithm is a point of the set and all points in the set are covered by balls in OPT, it follows that the competitive ratio of **Algorithm Centered** is at most A.

By Corollary 1 we are interested in the maximum number $A(\alpha)$ of nonoverlapping caps $C(\alpha)$ that can be placed on $\mathbb{S}^{d-1}$, for $\alpha = \pi/6$. This is precisely the maximum number of nonoverlapping balls that can touch a fixed unit ball externally, which is the Newton number $N(B_d)$ in dimension d.

For $d = 2$ we gain 1 in the bound due to the fact that the inequality in Lemma 1 is strict and we are dealing with the unit circle; the five vertices of a regular pentagon inscribed in a unit circle make a tight example with ratio 5; note that the minimum pairwise distance between points is $2\sin(\pi/5) > 1$, and so the algorithm places a new ball for each point. For $d = 3$ the twelve vertices of a regular icosahedron inscribed in a unit sphere make a tight example with ratio 12; since the minimum pairwise distance between points is $(\sin(2\pi/5))^{-1} > 1$, the same observation applies.   □

Bounds on the Newton Number of the Ball. A classic formula established by Rankin [24] yields that

$$N(B_d) \leq \sqrt{\frac{\pi}{8}} \, d^{3/2} \, 2^{d/2}(1 + o(1)).$$ (1)

More recently, Kabatiansky and Levenshtein [19] have established a sharper upper bound

$$N(B_d) \leq 2^{0.401d(1+o(1))}.$$ (2)

In particular, $N(B_d) = O(1.321^d)$. It is worth noting that the best lower known on the Newton number, due to Jenssen et al. [18] is far apart; see also [26].

$$N(B_d) = \Omega\left(d^{3/2} \cdot \left(\frac{2}{\sqrt{3}}\right)^d\right).$$ (3)

In particular, $N(B_d) = \Omega(1.154^d)$.

3   Lower Bounds on the Competitive Ratio for Online Unit Covering in Euclidean d-Space

Theorem 3 that we prove in this section greatly improves the previous best lower bound on the competitive ratio of a deterministic algorithm, $\Omega(\log d / \log \log \log d)$, due to Charikar et al. [8].

Previous lower bounds for $d = 2, 3$. To clarify matters, we briefly summarize the calculation leading to the previous best lower bounds on the competitive ratio. Charikar et al. [8] claim (on p. 1435) that a lower bound of 4 for $d = 2$ under the L_2 norm follows from their Theorem 6.2; but this claim appears unjustified; only a lower bound of 3 is implied. The proof uses a volume argument. For a given d, the parameters R_t are iteratively computed for $t = 1, 2, \ldots$ by using the recurrence relation

$$R_{t+1} = \frac{R_t + t^{1/d}}{2}, \text{ where } R_1 = 0.$$ (4)

The lower bound on the competitive ratio of any deterministic algorithm given by the argument is the largest t for which $R_t \leq 1$. The values obtained for R_t, for $t = 1, 2, \ldots$ and $d = 2, 3$ are listed in Table 1; as such, both lower bounds are equal to 3.

3.1   A New Lower Bound in the Plane

In this section, we deduce an improved lower bound of 4 (an alternative proof will be provided by Theorem 3).

Theorem 2. *The competitive ratio of any deterministic online algorithm for* UNIT COVERING *in the plane (in the L_2 norm) is at least 4.*

Table 1. Values R_t, for $t = 1, 2, 3, 4$.

d	R_1	R_2	R_3	R_4
2	0	0.5	0.957...	1.344...
3	0	0.5	0.879...	1.161...

Proof. Consider a deterministic online algorithm ALG. We present an input instance σ for ALG and show that the solution $\mathsf{ALG}(\sigma)$ is at least 4 times $\mathsf{OPT}(\sigma)$. Our proof works like a two player game, played by Alice and Bob. Here, Alice is presenting points to Bob, one at a time. Bob (who plays the role of the algorithm) makes the decision whether to place a new disk or not. If a new disk is required, Bob decides where to place it. Alice tries to force Bob to place as many new disks as possible by presenting the points in a smart way. Bob tries to place new disks in a way such that they may cover other points presented by Alice in the future, thereby reducing the need of placing new disks quite often.

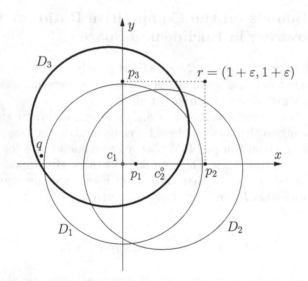

Fig. 2. A lower bound of 4 on the competitive ratio in the plane. The figure illustrates the case $p_4 = r$.

The center of a disk D_i is denoted by c_i, $i = 1, 2, \ldots$; refer to Fig. 2. The point coordinates will depend on a parameter $\varepsilon > 0$; a sufficiently small $\varepsilon \leq 0.01$ is chosen so that the inequalities appearing in the proof hold. First, point p_1 arrives and the algorithm places disk D_1 to cover it. Without loss of generality, it can be assumed that $c_1 = (0,0)$ and $p_1 = (x,0)$, where $0 \leq x \leq 1$. The second point presented is $p_2 = (1+\varepsilon^2, 0)$ and, since $p_2 \notin D_1$, a second disk D_2 is placed to cover it. By symmetry, it can be assumed that $y(c_2) \leq 0$. The third point presented is $p_3 = (0, 1 + \varepsilon)$, and neither D_1 nor D_2 covers it; thus a new disk, D_3, is placed to cover p_3.

Consider two other candidate points, $q = (-1 + \varepsilon, \sqrt{2\varepsilon})$ and $r = (1 + \varepsilon, 1 + \varepsilon)$. Since

$$|qc_1|^2 = (-1 + \varepsilon)^2 + 2\varepsilon = 1 + \varepsilon^2 - 2\varepsilon + 2\varepsilon = 1 + \varepsilon^2 > 1,$$

q is not covered by D_1; and clearly r is not covered by D_1. Since

$$|qc_2|^2 \geq (1 - \varepsilon + \varepsilon^2)^2 + 2\varepsilon = 1 + 3\varepsilon^2 + O(\varepsilon^3) > 1,$$

q is not covered by D_2; and clearly r is not covered by D_2. Note also that the D_3 cannot cover both q and r, since their distance is close to $\sqrt{5} > 2$. We now specify p_4, the fourth point presented to the algorithm. If q is covered by D_3, let $p_4 = r$, otherwise let $p_4 = q$. In either case, a fourth disk, D_4, is required to cover p_4.

To conclude the proof, we verify that p_1, p_2, p_3, p_4 can be covered by a unit disk.

Case 1: $p_4 = r$. It is easily seen that p_1, p_2, p_3, p_4 can be covered by the unit disk D centered at $(\frac{1}{2}, \frac{1}{2})$; indeed, the four points are close to the boundary of the unit square $[0, 1]^2$.

Case 2: $p_4 = q$. Consider the unit disk D centered at the midpoint c of qp_2. We have

$$|qp_2|^2 = (2 - \varepsilon + \varepsilon^2)^2 + 2\varepsilon = 4 - 2\varepsilon + O(\varepsilon^2) < 4.$$

It follows that D covers p_2 and p_4. Note that

$$c = \left(\frac{\varepsilon + \varepsilon^2}{2}, \sqrt{\frac{\varepsilon}{2}} \right).$$

We next check the containment of p_1 and p_3.

$$|cp_1|^2 \leq \left(1 - \frac{\varepsilon + \varepsilon^2}{2} \right)^2 + \frac{\varepsilon}{2} = 1 - \frac{\varepsilon}{2} + O(\varepsilon^2) < 1,$$

thus D also covers p_1. Finally, we have

$$|cp_3|^2 \leq \left(\frac{\varepsilon + \varepsilon^2}{2} \right)^2 + \left(1 + \varepsilon - \sqrt{\frac{\varepsilon}{2}} \right)^2 \leq 1 - \sqrt{2\varepsilon} + O(\varepsilon) < 1,$$

thus D also covers p_3.

We have shown that $\mathsf{ALG}(\sigma)/\mathsf{OPT}(\sigma) \geq 4$, and the proof is complete.   □

3.2   A New Lower Bound in d-Space

We introduce some additional terminology. For every integer k, $0 \leq k < d$, a *k-sphere* of radius r centered at a point $c \in \mathbb{R}^d$ is the locus of points in $\mathbb{R}^d$ at distance r from a center c, and lying in a $(k + 1)$-dimensional affine subspace that contains c. In particular, a $(d - 1)$-sphere of radius r centered at c is the set of *all* points $p \in \mathbb{R}^d$ such that $|cp| = r$; a 1-sphere is a circle lying in a 2-dimensional affine plane; and a 0-sphere is a pair of points whose midpoint is c. A *k-hemisphere* is a k-dimensional manifold with boundary, defined as the intersection $S \cap H$, where S is a k-sphere centered at some point $c \in \mathbb{R}^d$ and H is a halfspace whose boundary ∂H contains c but does not contain S. For $k \geq 1$, the *relative boundary* of the k-hemisphere $S \cap H$ is the $(k-1)$-sphere $S \cap (\partial H)$ concentric with S; and the *pole* of $S \cap H$ is the unique point $p \in H$ such that $\overrightarrow{cp}$ is orthogonal to the k-dimensional affine subspace that contains $S \cap (\partial H)$. For $k = 0$, a 0-hemisphere consists of a single point, and we define the *pole* to be that point. We make use of the following observation.

Observation 1. *Let S be a k-sphere of radius $1+\varepsilon$, where $0 \leq k < d$ and $\varepsilon > 0$; and let B be a unit ball in $\mathbb{R}^d$. Then $S \setminus B$ contains a k-hemisphere.*

Proof. Without loss of generality, S is centered at the origin, and lies in the subspace spanned by the coordinate axes $x_1, \ldots, x_{k+1}$. By symmetry, we may also assume that the center of B is on the nonnegative x_1-axis, say, at $(b, 0, \ldots, 0)$ for some $b \geq 0$. If $b = 0$, then S and B are concentric and B lies in the interior of S, consequently, $S \setminus B = S$. Otherwise, $S \cap B$ lies in the open halfspace $x_1 > 0$, and $S \setminus B$ contains the k-hemisphere $S \cap \{(x_1, \ldots, x_d) \in \mathbb{R}^d : x_1 \leq 0\}$.   □

Theorem 3. *The competitive ratio of every deterministic online algorithm (with an adaptive deterministic adversary) for* UNIT COVERING *in $\mathbb{R}^d$ under the L_2 norm is at least $d+1$ for every $d \geq 1$; and at least $d+2$ for $d = 2, 3$.*

Proof. Consider a deterministic online algorithm ALG. We present an input instance σ for ALG and show that the solution $\mathsf{ALG}(\sigma)$ is at least $d+1$ times $\mathsf{OPT}(\sigma)$. In particular, σ consists of $d+1$ points in $\mathbb{R}^d$ that fit in a unit ball, hence $\mathsf{OPT}(\sigma) = 1$, and we show that ALG is required to place a new unit ball for each point in σ. Similarly to the proof of Theorem 2, our proof works like a two player game between Alice and Bob.

Let the first point $p_0 = o$ be the origin in $\mathbb{R}^d$ (we will use either notation as convenient). For a constant $\varepsilon \in (0, \frac{1}{2d})$, let S_0 be the $(d-1)$-sphere of radius $1 + \varepsilon$ centered at the origin o. Refer to Fig. 3. Next, B_0 is placed to cover p_0. The remaining points $p_1, \ldots, p_d$ in σ are chosen adaptively, depending on Bob's moves. We maintain the following two invariants: For $i = 1, \ldots, d$, when Alice has placed points $p_0, \ldots, p_{i-1}$, and Bob placed unit balls $B_0, \ldots B_{i-1}$,

(I) the vectors $\overrightarrow{op_j}$, for $j = 1, \ldots, i-1$, are pairwise orthogonal and they each have length $1 + \varepsilon$;

(II) there exists a $(d-i)$-hemisphere $H_i \subset S_0$ that lies in the $(d-i+1)$-dimensional subspace orthogonal to $\langle \overrightarrow{op_j} : j = 1, \ldots, i-1 \rangle$ and is disjoint from $\bigcup_{j=0}^{i-1} B_j$.

Both invariants hold for $i = 1$: (I) is vacuously true, and (II) holds by Observation 1 (the first condition of (II) is vacuous in this case).

At the beginning of step i (for $i = 1, \ldots, d$), assume that both invariants hold. Alice chooses p_i to be the pole of the $(d-i)$-hemisphere H_i. By Invariant (II), p_i is not covered by $B_0, \ldots, B_{i-1}$, and Bob has to choose a new unit ball B_i

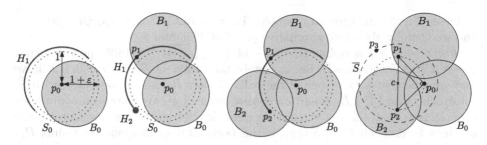

Fig. 3. The first three steps of the game between Alice and Bob in the proof of Theorem 3 for $d = 2$. After the 3rd step, Alice can place a 4th point $p_3 \in \overline{S}$ which is not covered by the balls B_0, B_1, B_2.

that contains p_i. By Invariant (I), $H_i \subset S_0$, so $|op_i| = 1 + \varepsilon$. By Invariant (II), $\overrightarrow{op_i}$ is orthogonal to the vectors $\overrightarrow{op_j}$, for $j = 1, \ldots, i - 1$. Hence Invariant (I) is maintained.

Let S_i be the relative boundary of H_i, which is a $(d-i-1)$-sphere centered at the origin. Since p_i is the pole of H_i, $\overrightarrow{op_i}$ is orthogonal to the $(d-i)$-dimensional subspace spanned by S_i. By Observation 1, S_i contains a $(d-i-1)$-hemisphere that is disjoint from B_i. Denote such a $(d-i-1)$-hemisphere by $H_{i+1} \subset S_i$. Clearly, H_{i+1} is disjoint from the balls $B_0, \ldots, B_{i-1}, B_i$; so Invariant (II) is also maintained.

By construction, p_i $(i = 1, \ldots, d)$ is not covered by the balls $B_0, \ldots, B_{i-1}$, so Bob has to place a unit ball for each of the $d + 1$ points $p_0, p_1, \ldots, p_d$. By Invariant (I), the points $p_1, \ldots, p_d$ span a regular $(d-1)$-dimensional simplex of side length $(1 + \varepsilon)\sqrt{2}$. By Jung's Theorem [16, p. 46], the radius of the smallest enclosing ball of $p_1, \ldots, p_d$ is

$$R = (1 + \varepsilon)\sqrt{2} \cdot \sqrt{\frac{d-1}{2d}} < \left(1 + \frac{1}{2d}\right)\sqrt{\frac{d-1}{d}} = \sqrt{\frac{4d^3 - 3d - 1}{4d^3}} < 1,$$

and this ball contains the origin p_0, as well.

We next show how to adjust the argument to derive a slightly better lower bound of $d + 2$ for $d = 2, 3$. Let B be the smallest enclosing ball of the points $p_0, p_1, \ldots, p_d$, and let c be the center of B. As noted above, the radius of B is $R = (1 + \varepsilon)\sqrt{(d-1)/d}$. Let $\overline{S}$ be the $(d-1)$-sphere of radius $2 - R = 2 - (1 + \varepsilon)\sqrt{(d-1)/d}$ centered at c. Then the smallest enclosing ball of B and an arbitrary point $p_{d+1} \in \overline{S}$ has unit radius. That is, points $p_0, \ldots, p_d, p_{d+1}$ fit in a unit ball. This raises the question whether Alice can choose yet another point $p_{d+1} \in \overline{S}$ outside of the balls $B_0, \ldots, B_d$ placed by Bob.

For $d = 2$, $\overline{S}$ has radius $2 - (1 + \varepsilon)\sqrt{1/2} = 2 - (1 + \varepsilon)(\sqrt{2}/2) \geq 1.2928$ (provided that $\varepsilon > 0$ is sufficiently small). A unit disk can cover a circular arc in $\overline{S}$ of diameter at most 2. If 3 unit disks can cover $\overline{S}$, then $\overline{S}$ would be the smallest enclosing circle of a triangle of diameter at most 2, and its radius would be at most $\frac{2}{3}\sqrt{3} \leq 1.1548$ by Jung's Theorem. Consequently, Alice can place a 4th point $p_3 \in \overline{S}$ outside of B_0, B_1, B_2, and all four points $p_0, \ldots, p_3$ fit in a unit disk; see Fig. 3(right) for an example. That is, $\mathsf{ALG}(\sigma) = 4$ and $\mathsf{OPT}(\sigma) = 1$; and we thereby obtain an alternative proof of Theorem 2.

For $d = 3$, $\overline{S}$ has radius $R_1 = 2 - (1 + \varepsilon)(\sqrt{2/3}) \geq 1.1835$ (provided that $\varepsilon > 0$ is sufficiently small). Let c_i denote the center of B_i, for $i = 0, 1, 2, 3$; we may assume that at least one of the balls B_i, say B_0, is not concentric with $\overline{S}$, since otherwise $\bigcup_{i=0}^{3} B_i$ would cover zero area of $\overline{S}$. We may also assume for concreteness that cc_0 is a vertical segment; let π_0 denote the horizontal plane incident to c. Then $C = \overline{S} \cap \pi_0$ is the horizontal great circle (of radius R_1) of $\overline{C}$, centered at c. Note that $C \cap B_0 = \emptyset$, and so if $\bigcup_{i=0}^{3} B_i$ covers $\overline{S}$, then $\bigcup_{i=1}^{3}(B_i \cap \pi_0)$ covers C. However, the analysis of the planar case ($d = 2$) shows that this is impossible; indeed, we have $R_1 \geq 1.1835 > 1.1548$. Consequently, Alice can place a 5th point $p_4 \in \overline{S}$ outside of B_0, B_1, B_2, B_3, and all five points $p_0, \ldots, p_4$ fit in a unit ball. That is, $\mathsf{ALG}(\sigma) = 5$ and $\mathsf{OPT}(\sigma) = 1$ and a lower bound of 5 on the competitive ratio is implied.   □

4   Unit Covering for Lattice Points in the Plane

In this section, we describe optimal deterministic algorithms for the online UNIT COVERING of points from the infinite unit square and hexagonal lattices. We start with integer lattice $\mathbb{Z}^2$.

Theorem 4. *There exists a deterministic online algorithm for online* UNIT COVERING *of integer points (points in $\mathbb{Z}^2$) with competitive ratio 3. This result is tight: the competitive ratio of any deterministic online algorithm for this problem is at least 3.*

Proof. First, we prove the lower bound; refer to Fig. 4(left). First, point p_1 arrives and disk D_1 covers it. Observe that D_1 misses at least one point from $\{p_2, p_3\}$, since $|p_2 p_3| = 2\sqrt{2} > 2$. We may assume that D_1 missed p_2; and this further implies that D_1 covers neither p_4 nor p_5, otherwise it would also cover p_2, which is a contradiction. Now, p_2 arrives and some disk D_2 is placed to cover it. If D_2 covers p_4, then the next input point is p_5, otherwise it is p_4. In either case, a third disk is needed. To finish the proof, observe that $\{p_1, p_2, p_4\}$ and $\{p_1, p_2, p_5\}$ can each be covered by a single unit disk; hence the competitive ratio of any deterministic algorithm is at least 3.

Next, we present an algorithm of competitive ratio 3. Refer to Fig. 4(right). Partition the lattice points using unit disks as shown in the figure. When a point arrives, cover it with the disk it belongs to in the partition. For the analysis, consider a disk D from an optimal cover. As seen in the figure, D can cover points that belong to at most three disks used for partitioning the lattice. Thus, we conclude that the algorithm has competitive ratio 3.   □

In the following, we state our result for the infinite hexagonal lattice.

Theorem 5. *There exists a deterministic online algorithm for online* UNIT COVERING *of points of the hexagonal lattice with competitive ratio 3. This result is tight: the competitive ratio of any deterministic online algorithm for this problem is at least 3.*

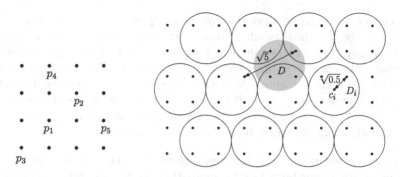

Fig. 4. Left: lower bound for $\mathbb{Z}^2$. Right: illustration of the upper bound; the disk D is shaded.

Proof. The proof is similar to that of Theorem 4. We start by proving the lower bound of 3; refer to Fig. 5(left). The first point, p_1, arrives and D_1 is placed to cover it. D_1 misses at least one of $\{p_2, p_3\}$, since $|p_2 p_3| = 2\sqrt{3} > 2$. By symmetry, we may assume that D_1 misses p_3. Now, point p_3 arrives and D_2 is placed to cover it. We distinguish two cases:

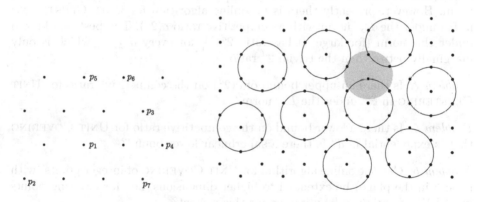

Fig. 5. Left: lower bound for the hexagonal lattice. Right: illustration of the upper bound.

Case 1: D_2 misses p_4. Since D_1 misses p_3, D_1 also misses p_4. Otherwise, if D_1 covers p_4, then D_1 also covers p_3, a contradiction. The algorithm uses D_3 to cover p_4. Thus the ratio is 3 since $\{p_1, p_3, p_4\}$ can be covered by a single disk centered at $(p_1 + p_4)/2$, and the algorithm has used three disks: D_1, D_2, and D_3.

Case 2: D_2 covers p_4. This means that D_2 misses p_5. Two cases may occur:

1. D_1 *misses p_5 too.* Then p_5 is the next input point, and the algorithm uses D_3 to cover it. Here $\{p_1, p_3, p_5\}$ can be covered by a single disk centered at $(p_1 + p_3 + p_5)/3$, but the algorithm has used three disks: D_1, D_2, and D_3.

2. D_1 *covers* p_5. Since D_1 does not cover p_3, D_1 cannot cover p_6. If D_2 misses p_6, let p_6 be the third point presented; the algorithm uses D_3 to cover p_6. Here $\{p_1, p_3, p_6\}$ can be covered by a single disk centered at $(p_1 + p_6)/2$, but the algorithm has used three disks: D_1, D_2, and D_3. If D_2 covers p_6, let p_7 be the third point presented. Note that D_1 cannot cover p_7 since it covers p_5; also, D_2 cannot cover p_7 since it covers p_6. The algorithm uses D_3 to cover p_7. Here $\{p_1, p_3, p_7\}$ can be covered by a single disk centered at $(p_1 + p_3 + p_7)/3$, but the algorithm has used three disks: D_1, D_2, and D_3.

In all cases a lower bound of 3 has been enforced by Alice, as required.

To prove the upper bound of 3 we partition the lattice points using disks as shown in Fig. 5(right). It can be concluded that the algorithm has competitive ratio 3 in this case.   □

5   Conclusion

Our results suggest several directions for future study. We summarize a few specific questions of interest.

By Theorem 1 and a remark in the Introduction, `Algorithm Centered` has a competitive ratio $O(1.321^d)$ also for UNIT CLUSTERING in $\mathbb{R}^d$ under the L_2 norm. However, presently there is no online algorithm for UNIT CLUSTERING in $\mathbb{R}^d$ under the L_∞ norm with a competitive ratio $o(2^d)$. The best one known under this norm (for large d) has ratio $2^d \cdot \frac{5}{6}$ for every $d \geq 2$, which is only marginally better than the trivial 2^d ratio.

Problem 4. Is there an upper bound of $o(2^d)$ on the competitive ratio for UNIT CLUSTERING in $\mathbb{R}^d$ under the L_∞ norm?

Problem 5. Is there a lower bound on the competitive ratio for UNIT COVERING that is exponential in d? Is there a superlinear lower bound?

Problem 6. Can the online algorithm for UNIT COVERING of integer points (with ratio 3 in the plane) be extended to higher dimensions, i.e., for covering points in $\mathbb{Z}^d$? What ratio can be obtained for this variant?

References

1. Alon, N., Awerbuch, B., Azar, Y., Buchbinder, N., Naor, J.: The online set cover problem. SIAM J. Comput. **39**(2), 361–370 (2009)
2. Biniaz, A., Liu, P., Maheshwari, A., Smid, M.H.M.: Approximation algorithms for the unit disk cover problem in 2D and 3D. Comput. Geom. **60**, 8–18 (2017)
3. Borodin, A., El-Yaniv, R.: Online Computation and Competitive Analysis. Cambridge University Press, Cambridge (1998)
4. Boyvalenkov, P., Dodunekov, S., Musin, O.R.: A survey on the kissing numbers. Serdica Math. J. **38**, 507–522 (2012)
5. Brass, P., Moser, W.O.J., Pach, J.: Research Problems in Discrete Geometry. Springer, New York (2005). https://doi.org/10.1007/0-387-29929-7

6. Buchbinder, N., Naor, J.: Online primal-dual algorithms for covering and packing. Math. Oper. Res. **34**(2), 270–286 (2009)

7. Chan, T.M., Zarrabi-Zadeh, H.: A randomized algorithm for online unit clustering. Theory Comput. Syst. **45**(3), 486–496 (2009)

8. Charikar, M., Chekuri, C., Feder, T., Motwani, R.: Incremental clustering and dynamic information retrieval. SIAM J. Comput. **33**(6), 1417–1440 (2004)

9. Dumitrescu, A., Tóth, C.D.: Online unit clustering in higher dimensions. In: Solis-Oba, R., Fleischer, R. (eds.) WAOA 2017. LNCS, vol. 10787, pp. 238–252. Springer, Cham (2018). https://doi.org/10.1007/978-3-319-89441-6_18

10. Edel, Y., Rains, E.M., Sloane, N.J.A.: On kissing numbers in dimensions 32 to 128. Electr. J. Comb. **5**, R22 (1998)

11. Ehmsen, M.R., Larsen, K.S.: Better bounds on online unit clustering. Theor. Comput. Sci. **500**, 1–24 (2013)

12. Epstein, L., van Stee, R.: On the online unit clustering problem. ACM Trans. Algorithms **7**(1), 7:1–7:18 (2010)

13. Feder, T., Greene, D.H.: Optimal algorithms for approximate clustering. In: Proceedings of 20th ACM Symposium on Theory of Computing (STOC), pp. 434–444 (1988)

14. Fowler, R.J., Paterson, M., Tanimoto, S.L.: Optimal packing and covering in the plane are NP-complete. Inf. Process. Lett. **12**(3), 133–137 (1981)

15. Gonzalez, T.F.: Clustering to minimize the maximum intercluster distance. Theor. Comput. Sci. **38**, 293–306 (1985)

16. Hadwiger, H., Debrunner, H.: Combinatorial Geometry in the Plane. Holt, Rinehart and Winston, New York (1964). (English translation by Victor Klee)

17. Hochbaum, D.S., Maass, W.: Approximation schemes for covering and packing problems in image processing and VLSI. J. ACM **32**(1), 130–136 (1985)

18. Jenssen, M., Joos, F., Perkins, W.: On kissing numbers and spherical codes in high dimensions. Adv. Math. **335**, 307–321 (2018)

19. Kabatiansky, G.A., Levenshtein, V.I.: On bounds for packings on the sphere and in space. Probl. Inform. Transm. **14**(1), 1–17 (1978)

20. Kawahara, J., Kobayashi, K.M.: An improved lower bound for one-dimensional online unit clustering. Theor. Comput. Sci. **600**, 171–173 (2015)

21. Liao, C., Hu, S.: Polynomial time approximation schemes for minimum disk cover problems. J. Comb. Optim. **20**(4), 399–412 (2010)

22. Megiddo, N., Supowit, K.J.: On the complexity of some common geometric location problems. SIAM J. Comput. **13**(1), 182–196 (1984)

23. Mustafa, N.H., Ray, S.: Improved results on geometric hitting set problems. Discrete Comput. Geom. **44**(4), 883–895 (2010)

24. Rankin, R.A.: The closest packing of spherical caps in n dimensions. Proc. Glasgow Math. Assoc. **2**(3), 139–144 (1955)

25. Williamson, D.P., Shmoys, D.B.: The Design of Approximation Algorithms. Cambridge University Press, Cambridge (2011)

26. Wyner, A.D.: Capabilities of bounded discrepancy decoding. Bell Syst. Tech. J. **44**(6), 1061–1122 (1965)

27. Wynn, E.: Covering a unit ball with balls half the radius (2012). https://mathoverflow.net/questions/98007/covering-a-unit-ball-with-balls-half-the-radius

28. Zarrabi-Zadeh, H., Chan, T.M.: An improved algorithm for online unit clustering. Algorithmica **54**(4), 490–500 (2009)

Isolation Branching: A Branch and Bound Algorithm for the k-Terminal Cut Problem

Mark Velednitsky[(✉)] and Dorit S. Hochbaum

University of California, Berkeley, Berkeley, USA
{marvel,dhochbaum}@berkeley.edu

Abstract. In the k-terminal cut problem, we are given a graph with edge weights and k distinct vertices called "terminals." The goal is to remove a minimum weight collection of edges from the graph such that there is no path between any pair of terminals. The k-terminal cut problem is NP-hard.

The k-terminal cut problem has been extensively studied and a number of algorithms have been devised for it. Most of the algorithms devised for the problem are approximation algorithms or heuristic algorithms. There are also fixed-parameter tractable algorithms that solve the problem optimally in time that is polynomial when the value of the optimum is fixed, but none have been shown empirically practical. It is possible to apply implicit enumeration using any integer programming formulation of the problem and solve it with a branch-and-bound algorithm.

Here, we present a branch-and-bound algorithm for the k-terminal cut problem which does not rely on an integer programming formulation. Our algorithm employs "isolating cuts" and, for this reason, we call our branch-and-bound algorithm *Isolation Branching*.

In an empirical experiment, we compare the performance of the Isolation Branching algorithm to that of a branch-and-bound applied to the strongest known integer programming formulation of k-terminal cut. The integer programming branch-and-bound procedure is implemented with Gurobi, a commercial mixed-integer programming solver. We compare the performance of the two approaches for real-world instances and synthetic data. The results on real data indicate that Isolation Branching, coded in Python, runs an order of magnitude faster than Gurobi for problems of sizes of up to tens of thousands of vertices and hundreds of thousands of edges. Our results on synthetic data also indicate that Isolation Branching scales more effectively.

Though we primarily focus on creating a practical tool for k-terminal cut, as a byproduct of our algorithm we prove that the complexity of Isolation Branching is also fixed-parameter tractable with respect to the size of the optimal solution, thus providing an alternative, constructive, and somewhat simpler, proof of this fact.

The first author's work is supported by the NPSC Fellowship. The second author's work has been funded in part by NSF award CMMI-1760102.

© Springer Nature Switzerland AG 2018
D. Kim et al. (Eds.): COCOA 2018, LNCS 11346, pp. 624–639, 2018.
https://doi.org/10.1007/978-3-030-04651-4_42

Keywords: k-terminal cut · Optimization · Branch-and-bound
Isolating cut · Clustering

1   Introduction

In the k-terminal cut problem, we are given an graph with edge weights and k distinct vertices called "terminals." The goal is to remove a minimum weight collection of edges from the graph such that there is no path between any pair of terminals. The k-terminal cut problem is NP-hard [5]. The problem has been studied extensively in the literature and has also been referred to as the multiterminal cut problem or the multiway cut or multicut problem with k terminals.

The k-terminal cut problem has a number of applications. Specific application areas include distributing computational jobs in a parallel computing system [7], partitioning elements of a circuit into sub-circuits that will be put on different chips [5], scheduling tasks [12], understanding transportation bottlenecks [12], planning the "divide" step in divide-and-conquer algorithms [9], and even partitioning Markov Random Fields for computer vision [1]. More generally, graph cut problems, including k-terminal cut, have applications to graph clustering [6]. Minimizing the weight of edges between clusters is equivalent to maximizing the weight within clusters. In a setting where the weights measure similarity between vertices, the result is a graph clustering procedure. Thus, k-terminal cut gives an explicit combinatorial objective function for supervised graph clustering.

The k-terminal cut problem has been extensively studied and a number of algorithms have been devised for it. Most of the algorithms devised for the problem are approximation algorithms or heuristic algorithms which provide good feasible, but typically non-optimal, solutions. There are also fixed-parameter tractable algorithms that solve the problem optimally in time that is polynomial when the value of the optimum is fixed, but none have been shown empirically practical.

The first approximation algorithm for k-terminal cut gave an approximation ratio of 2.0 [5]. Improved approximation algorithms are based on the linear programming relaxation of the integer programming formulation of the problem known in the literature as the *geometric IP* [3]. A sequence of improved approximation algorithms delivered an approximation factor of 1.5 [3], followed by 1.3438 [11], followed by 1.32388 [2]. The best-to-date approximation factor is 1.2965 [15].

The popular graph partitioning library METIS offers heuristics for k-terminal cut which are used extensively in practice [12]. METIS uses a "top-down" strategy. The procedure starts by finding a feasible solution for k-terminal cut which is likely to have cut value close to the optimal cut value. Then it refines the solution with various heuristics.

It is possible to apply implicit enumeration using any integer programming (IP) formulation of the problem and solve it with a branch-and-bound algorithm. Indeed, we compare the performance of our algorithm to a branch-and-bound procedure based on the geometric IP formulation. The geometric IP is the IP that was proved to have the smallest integrality gap, assuming the Unique Games Conjecture [13].

There are several fixed-parameter tractable optimization algorithms for k-terminal cut. It was first proven in 2004 that k-terminal cut is fixed-parameter tractable with respect to the value of the optimal solution [14]. That proof was not constructive. A constructive proof in the form of an algorithm was given in [4] with running time $O(|OPT|4^{|OPT|}n^3)$, where $|OPT|$ is the weight of the optimal cut and n is the number of vertices in the graph. The algorithm of [4] is of theoretical value and has never been implemented. A by-product of our algorithm is an alternative, constructive proof of the fixed-parameter tractability of k-terminal cut.

We note that k-terminal cut can be solved in polynomial time on graphs which are $(2-2/k)$-stable [17]. A graph is $(2-2/k)$-stable if the optimal solution remains optimal even when the weights of the edges in the solution are multiplied by $(2-2/k)$.

The algorithm devised here applies concepts used in the closely related k-cut problem. In the k-cut problem, there are no terminals. The goal in the k-cut problem is to remove a minimum weight collection of edges such that the resulting graph consists of k non-empty, disjoint connected components. The k-cut problem was proved to be easier than k-terminal cut: in [8], the authors proved that k-cut is polynomial for fixed k, whereas [5] showed that k-terminal cut is NP-hard even for $k = 3$.

The polynomial-time algorithm for k-cut introduced in [8] relies on two building blocks which we will also use here: *seed sets* and *minimum isolating cuts*. A *seed set* is a set of vertices in the graph which we assume belong to the same component in an optimal solution. Given a set of seed sets, a *minimum isolating cut* is the smallest (s, t)-cut which separates one seed set from the rest. For the k-cut problem, [8] shows that if the "correct" set of $2k$ vertices are chosen as seeds, then the source set of the minimum isolating cut recovers one of the components in the optimal k-cut. It is not possible to know in advance which $2k$ vertices are the "correct" ones to choose, so all $\sim \binom{n}{2k}$ possibilities must be enumerated. Ultimately, for k-cut, the exponent of the running time polynomial depends quadratically on k and is prohibitively large in practice even for small values of k.

Isolating cuts have also been used in the 2-approximation algorithm for k-terminal cut presented in [5]. For each of the k terminals, consider the minimum isolating cut which separates the chosen terminal from the rest of the terminals. If we take the union of the edges which appear in all k of these minimum isolating cuts, then the result is a feasible k-terminal cut. It is shown in [5] that the value of this solution is at most twice the value of the optimal k-terminal cut.

Our contributions in this paper are as follows:

1. We devise a branch-and-bound algorithm for the k-terminal cut problem, which does not rely on an integer programming formulation, and is demon-

strated to be practical and scalable. Our algorithm employs isolating cuts and, for this reason, we call our branch-and-bound algorithm *Isolation Branching*.

2. We conduct an empirical study of optimization procedures for k-terminal cut, in which the performance of Isolation Branching is compared to branch-and-bound on the geometric IP formulation of k-terminal cut. The IP branch-and-bound procedure is implemented with Gurobi, a commercial mixed-integer programming solver. The performance is evaluated for real-world instances and synthetic data. The results on real data indicate that Isolation Branching, coded in Python, runs an order of magnitude faster than Gurobi on graphs with up to tens of thousands of vertices and hundreds of thousands of edges. The results on synthetic data indicate that Isolation Branching scales more effectively.

3. Though our primary motivation is developing a practical branch-and-bound algorithm for the k-terminal cut problem, we also prove that the running time of our algorithm is fixed-parameter tractable with respect to the size of the optimal solution. Thus, a byproduct of our algorithm is an alternative, constructive, and somewhat simpler proof of an already-known result: that the k-terminal cut problem is fixed-parameter tractable.

2   Preliminaries

The input to the k-terminal cut problem is an undirected graph $G = (V, E, w)$ with vertex set V, edge set E, weights w, and k terminals $s_1, \ldots, s_k \in V$. The graph G is assumed to have positive integer edge weights $w_{ij} \in \mathbb{Z}^+$ for $(i, j) \in E$. Throughout our algorithm, we maintain k sets, one for each terminal, which we call the *seed* sets. The ith seed set, S_i, is a subset of V which contains the terminal s_i and none of the other s_j $(j \neq i)$.

A collection of k subsets of V is a k-terminal cut if and only if the k sets partition the vertex set V: an edge is cut if and only if its endpoints are in two different sets. Our algorithm initializes with the smallest possible seed sets, $S_i = \{s_i\}$, and adds vertices as seeds until the seed sets form a partition. The tool needed to add new vertices to the seed sets is a *minimum isolating cut*.

Definition 1 (Isolating Cut). *Given a collection of seed sets $\{S_1, \ldots, S_k\}$, an isolating cut for S_i is a cut which separates all the vertices in S_i from all the vertices in $\cup_{j \neq i} S_j$.*

To describe an isolating cut for seed set S_i, we use the term *source set* to denote the set of vertices which remain connected to S_i and *sink set* to denote the set of vertices which remain connected to $\cup_{j \neq i} S_j$ when the edges of the cut are removed.

Definition 2 (Minimum Isolating Cut). *Given a collection of seed sets $\{S_1, \ldots, S_k\}$, the notation $\mathcal{I}(S_i)$ denotes the source set of the isolating cut for S_i with minimum cut weight. If the minimum isolating cut for S_i is not unique, then $\mathcal{I}(S_i)$ denotes the minimum isolating cut with maximal source set.*

In the definition above, *maximal* source set means that the source set of this minimum isolating cut is not contained in the source set of another minimum isolating cut. A minimum isolating cut with maximal source set can be computed in polynomial time. First, connect all the vertices in S_i to a single source vertex s and all the vertices $\cup_{j \neq i} S_j$ to a single sink vertex t. Now, solve a 2-terminal cut problem (an (s, t)-cut problem) using an algorithm which computes the residual flow graph, such as Ford-Fulkerson. The set of vertices reachable from vertex t via unsaturated edges is the sink set of the desired minimum isolating cut with maximal source set.

Next, we develop an understanding of how these seed sets relate to the optimal **k-terminal cut**:

Definition 3 (Containment Property). *A collection of seed sets* $(S_1, \ldots, S_k)$ *is said to have the containment property if there exists an optimal* **k-terminal cut** $(S_1^*, \ldots, S_k^*)$ *such that* $S_i \subseteq S_i^*$ $\forall i$.

In [5], they prove the following lemma, which we have rephrased here:

Lemma 1 (Isolation Lemma). *Consider the collection of seed sets* $S_i = \{s_i\}$. *For any* i,

$$(\{s_1\}, \ldots, \{s_{i-1}\}, \mathcal{I}(\{s_i\}), \{s_{i+1}\}, \ldots, \{s_k\})$$

has the containment property in G.

As an example, consider Fig. 1. The optimal k-terminal cut has weight 8 (cutting the four edges that form the central square) while the four minimum isolating cuts for the terminals each have weight 3. The source sets of the four minimum isolating cuts are subsets of the source sets of the optimal k-terminal cut. The isolation lemma proves that this is always the case. In our analysis, we rely on a simple generalization:

Lemma 2 (Seed Set Isolation Lemma). *Consider a collection of seed sets* $(S_1, \ldots, S_k)$ *with the containment property in* G. *Then*

$$(S_1, \ldots, S_{i-1}, \mathcal{I}(S_i), S_{i+1}, \ldots, S_k)$$

has the containment property in G.

Proof. Let $(S_1^*, S_2^*, \ldots, S_k^*)$ be an optimal **k-terminal cut** in G and let E_{OPT} be the edges of that cut. Merge the vertices of each S_i into their respective s_i to create the new graph G' with terminals s_i'. By the containment property, $S_i \subseteq S_i^*$, so none of the edges in E_{OPT} have both endpoints in the same S_i, so all of the edges in E_{OPT} still connect two distinct vertices in G'. Thus, E_{OPT} is still an optimal solution in G' to the **k-terminal cut** problem.

We apply the isolation Lemma (1) in G'. The minimum isolating cut $\mathcal{I}(s_i')$ in G' adds the same vertices as $\mathcal{I}(S_i)$ in G. That is,

$$\mathcal{I}(s_i') \setminus s_i' = \mathcal{I}(S_i) \setminus S_i.$$

From the isolation lemma, we have that

$$(s_1', \ldots, \mathcal{I}(s_i'), \ldots, s_k')$$

has the containment property in G'.

$$\mathcal{I}(s_i') \setminus s_i' \subseteq S_i^* \setminus S_i$$
$$\implies \mathcal{I}(S_i) \setminus S_i \subseteq S_i^* \setminus S_i$$
$$\implies \mathcal{I}(S_i) \subseteq S_i^*.$$

We conclude that

$$(S_1, \ldots, \mathcal{I}(S_i), \ldots, S_k)$$

has the containment property in G.   □

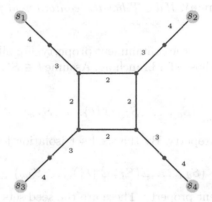

Fig. 1. Isolation lemma in a small graph

3   Branch and Bound

In our algorithm for k-terminal cut, we take a "bottom-up" approach of assigning vertices to seed sets until all the vertices have been assigned. When possible, we try to use isolating cuts to add unassigned vertices to seed sets. Otherwise, we "branch" by considering assigning a certain unassigned vertex to all possible seed sets. Following the branches where we make the "correct" assignment (where the vertex is assigned to the seed set to which it belongs in the optimal solution), we will reach the optimal solution. In branches where we make the "wrong" assignment, we will eventually arrive at a sub-optimal solution. Using a bound we derive for this purpose, based on the 2-approximation from [5], we can ignore many of these branches by proving that they are sub-optimal before arriving at a leaf node.

3.1   Branching

Let T be the branch and bound tree. At each node of the tree, $d \in T$, we will store a collection of pairwise disjoint seed sets $S_{d,i} \subset V$ ($i \in \{1, \ldots, k\}$, where k is the number of terminals). For convenience, we will use *nodes* when referring to the branch and bound tree T and *vertices* when referring to V in the original graph G.

We will say that a vertex $\ell \in V$ is *unassigned* in $d \in T$ if it is not in any of the $S_{d,i}$. In our branching step, we choose an unassigned vertex ℓ in d and create k children in T by assigning ℓ to each of the $S_{d,i}$ in turn and computing the new isolating cuts (with maximal source sets). If d does not have any unassigned vertices then it is a leaf node in T. Algorithm 1 is the pseudo-code of the Isolation Branching algorithm. Figure 2 provides an illustration.

The following lemma shows that the containment property propagates down the tree:

Lemma 3 (Inheritance). *If $d \in T$ has the containment property, then at least one child of d has it.*

Proof. Assume node d has the containment property. For all i, $S_{d,i} \subseteq S_i^*$. Let ℓ be the unassigned vertex chosen for branching. Assume $\ell \in S_j^*$. Then $S_{d,j} \cup \{\ell\} \subseteq S_j^*$, so the collection of sets

$$(S_{d,1}, \ldots, S_{d,j} \cup \{\ell\}, \ldots, S_{d,k})$$

has the containment property. By the seed set isolation lemma (Lemma 2), the collection of sets

$$(S_{d,1}, \ldots, \mathcal{I}(S_{d,j} \cup \{\ell\}), \ldots, S_{d,k})$$

also has the containment property. These are the seed sets for the jth child of d. □

3.2   Bounding

We would prefer to only expand nodes which have a chance of having the optimal solution as one of their leaves. At each node $d \in T$, we will consider the value of the function

$$L(d) = \frac{1}{2} \sum_{i=1}^{k} w(S_{d,i}, \bar{S}_{d,i}).$$

In words, this is half the sum of the weights of the isolating cuts. If the collection of seed sets at d has the containment property, then the sum of isolating cuts is known to be a 2-approximation to the optimal solution, so $L(d)$ must be less than the value of the optimal cut. The next two lemmas show that $L(d)$ can be used to cull branches even when the collection of seed sets does *not* have the containment property.

Algorithm 1. Isolation Branching (IB)

> *# initialization loop*
> **for** $i = 1 \ldots k$ **do**
> $S_{0,i} \leftarrow \mathcal{I}(\{s_i\})$.
> **end for**
> $d \leftarrow 0$.
> *# main loop*
> **while** node d has unassigned vertices **do**
> Choose vertex ℓ unassigned in d (see Sect. 4.1).
> $D \leftarrow |T|$.
> **for** $i = 1 \ldots k$ **do**
> $S_{D+i,i} \leftarrow \mathcal{I}(S_{d,i} \cup \{\ell\})$.
>   For $j \neq i$, $S_{D+i,j} \leftarrow S_{d,j}$.
> **end for**
> $d = \operatorname{argmin}_{d' \in T} L(d')$ (see Sect. 3.2).
> **end while**
> *# output step*
> Return cut $(S_{d,1}, \ldots, S_{d,k})$.

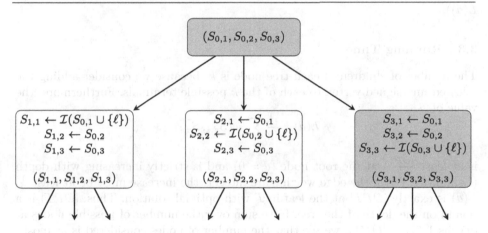

Fig. 2. Branch and bound tree for $k = 3$.

Lemma 4 (Leaf Nodes). *If $d \in T$ is a leaf node, then $L(d)$ is the value of the feasible k-terminal cut at node d.*

Proof. Each edge in the feasible k-terminal cut is exactly double-counted inside the sum $L(d)$ since it appears in exactly two isolating cuts. Multiplying by one-half returns the weight of the feasible k-terminal cut. $\square$

Lemma 5 (Lower Bound). *If $d_2 \in T$ is a descendant of $d_1 \in T$, then*

$$L(d_2) > L(d_1).$$

Proof. It is sufficient to prove the inequality when d_2 is a child of d_1. Recall that $S_{d_1,i}$ is required to be a *maximal* source set for all i. Assume that from d_1 to d_2

we add our unassigned vertex ℓ to $S_{d_1,j}$ (and then take an isolating cut). The size of the new isolating cut must *strictly* increase, otherwise it contradicts the maximality of the previous source set. Formally,

$$w(S_{d_2,j}, S_{d_2,j}^-) > w(S_{d_1,j}, S_{d_1,j}^-).$$

For the rest ($i \neq j$),

$$w(S_{d_2,i}, S_{d_2,i}^-) = w(S_{d_1,i}, S_{d_1,i}^-).$$

In total, the value of the sum $L(d_2)$ strictly increases from $L(d_1)$.   □

Taken together, these two lemmas give us the desired restriction. If we know that the weight of the optimal k-terminal cut is bounded above by B, then we need not expand nodes where $L(d) \geq B$ since any leaf nodes which descend from these nodes cannot be optimal. If we expand the nodes in order from lowest lower bound, then the first leaf we encounter must be the leaf with the optimal solution (because any leaf node we encounter later will have a larger value of $L(d)$).

3.3   Running Time

The number of children of each tree node is k, because we consider adding the selected unassigned vertex to each of the k possible terminals. Furthermore, the value of

$$L(d) = \frac{1}{2} \sum_i w(S_{d,i}, S_{d,i}^-)$$

is at least $\frac{|OPT|}{2}$ at the root node ($d = 0$) and is strictly increasing with depth (Lemma 5). When the edge weights are integer, the increase must be at least $\frac{1}{2}$. $L(d)$ is exactly $|OPT|$ at the leaf in T with optimal solution. Thus, $|OPT|$ is a bound on the depth of the tree. If we sum over the number of possible nodes at depths $1, 2, \ldots, |OPT|$, we see that the number of nodes considered is at most

$$1 + k + k^2 + \ldots + k^{|OPT|} < 2k^{|OPT|}.$$

Let $\tau(n)$ be the complexity of evaluating a minimum s, t-cut on a graph with n nodes. The complexity of our algorithm is thus $O(2k^{|OPT|}\tau(n))$. From this, we have fixed-parameter tractability.

Corollary 1. *When we can bound $|OPT|$ by a factor that does not depend on n (for example, graphs with terminals of bounded weighted degree), then the algorithm Isolation Branching runs in polynomial time.*

4   Empirical Study

4.1   Isolation Branching (IB) Implementation

Our open-source implementation is available online at https://github.com/marvel2010/k-terminal-cut and works as a Python package (ktcut). It represents graphs using NetworkX. We chose Python for ease of implementation and portability, even though it is not the fastest language in terms of its practical running time [16].

There are a number of hyper-parameters which affect the performance of our branch-and-bound algorithm. The first is the **Branching Node**. That is, after expanding a node in the tree, how do we decide which node to expand next? The second is the **Branching Vertex**. At each tree node, how do we decide which unassigned graph vertex to branch on to create the children nodes?

Branching Node: For choosing the branching node, it can be shown that the strategy of choosing the node with the best bound will explore the fewest number of nodes before reaching the optimal solution and having proof of optimality. The reason to use other strategies would be to quickly reach feasible solutions without optimality guarantees. Since that is a separate problem, here we only consider the best bound strategy.

Branching Vertex: For choosing the branching vertex, we considered a few options. The options included choosing a vertex randomly, choosing the vertex farthest from an existing source set, or choosing the vertex of largest degree. Initial experiments suggested that the last strategy was best (largest degree), so our results use that strategy. The largest degree strategy makes some sense. If a vertex is forced to be in a particular source set, then its neighbors must either join the source set or the edge between them is cut. This means that, when a high-degree node is added to a source set, either the source set grows significantly in the next isolating cut or the weight of the cut grows significantly. Either outcome is good, since it means that either the source sets grow quickly or we create tree nodes that do not need to be explored (large values of $L(d)$). In our implementation, we contract source sets into a single terminal vertex at each node in the branch-and-bound tree. This allows subsequent minimum cuts to be evaluated on a smaller graph.

4.2   Comparison to Integer Programming

To compare our Isolating Branching algorithm to integer programming branch-and-bound, we used Gurobi, a popular commercial software package for integer programming. The formulation to which we applied integer programming branch-and-bound is below. It is referred to in the literature as the *geometric* IP formulation [3]. Assuming the Unique Games Conjecture, it was proved in [13] that no other formulation can have a smaller integrality gap than this one.

The variable x_{it} is a binary variable: it is 1 if vertex i is assigned to terminal t and 0 otherwise. The variable z_{ij} is forced to be 1 if edge (i,j) is cut and can be 0 otherwise. To avoid double-counting edges, we assume $i < j$. In total, if there are n nodes and m edges in G, the IP formulation has $nk + m$ variables.

$$
\begin{aligned}
minimize \; & \sum_{(i,j)\in E} w_{ij}z_{ij} \\
s.t. \; & z_{ij} \geq x_{it} - x_{jt} \quad && \forall 1 \leq i < j \leq n, 1 \leq t \leq k \\
& z_{ij} \geq x_{jt} - x_{it} \quad && \forall 1 \leq i < j \leq n, 1 \leq t \leq k \\
& \sum_{t=1}^{k} x_{it} = 1 \quad && \forall 1 \leq i \leq n \\
& x_{it} \in \{0,1\} \quad && \forall 1 \leq i \leq n, 1 \leq t \leq k \\
& z_{ij} \in \{0,1\} \quad && \forall 1 \leq i < j \leq n \\
& x_{tt} = 1 \quad && \forall 1 \leq t \leq k
\end{aligned}
\tag{IP}
$$

4.3 Data Sets

Real Data Sets: The ten real-world data sets we used were part of the tenth DIMACS Implementation Challenge, a graph clustering challenge that concluded in 2012. According to the website, "These real-world graphs are often used as benchmarks in the graph clustering and community detection communities." The data sets are available online at https://www.cc.gatech.edu/dimacs10/. In most of the data sets, the graphs are already connected. In the rest, we only consider the largest connected component, otherwise the k-terminal cut problem decomposes into smaller problems on each component.

Synthetic Data Sets: To systematically study the running time scaling of our Isolation Branching algorithm, we used synthetic graphs. It has been observed that many real-world graphs, from social networks to computer networks to metabolic networks, exhibit both a power-law degree distribution and high clustering [10]. The POWER CLUSTER model, introduced in [10], generates random graphs which exhibit both of these properties. NetworkX includes a tool for randomly generating graphs according to the POWERLAW CLUSTER model with three parameters: the number of vertices, the number of random edges to add for each vertex, and the probability of creating a triangle. In our scaling experiment, we vary the first parameter (the number of nodes) while leaving the latter two at the fixed values of 10 and 0.1, respectively.

Terminals: In the data sets, terminals are not specified. In order to find suggested terminals, do the following: first, we perform spectral clustering on the graph to get an approximate clustering (by performing k-means clustering on the spectral embedding of the graph). Next, we choose the largest-degree vertex in each approximate cluster and set those vertices to be our k terminals.

Typically, the true number of clusters is quite small. For example, in the "political blogs" data set, there are really only two clusters. For that reason, our experiments consider a small number of terminals: five terminals on all graphs and ten terminals on larger graphs.

4.4 Results

We compare the running time of the Isolation Branching (IB) algorithm to the time it takes Gurobi to solve the Integer Program (IP). We compare on ten real-world test graphs (see Sect. 4.3). Details about the graphs are presented in Table 1.

Table 1. Ten real-world graphs, from the tenth DIMACS challenge, in which we are interested in solving `k-terminal cut`. The graphs are sorted by number of edges.

Graph	Size		Description
	NumVertices	NumEdges	
netscience	1589	2742	Network science collaborations
celegans_metabolic	453	2025	Metabolic network of an organism
jazz	198	2742	Jazz musicians
email	1133	5451	Email exchanges within university
power	4941	6594	US Western States Power Grid
hep-th	8361	15751	High-energy theory collaborations
polblogs	1490	16715	Links between political blogs
PGPgiantcompo	10680	24316	Network of PGP algorithm users
as-22july06	22963	48436	High-level snapshot of the internet
astro-ph	16706	121251	Astrophysics collaborations

First, we compare the running time of IB versus IP on all ten graphs, each with five terminals (chosen as described in Sect. 4.3). The running time and speed-up results can be found in Table 2. In all cases, IB outperforms IP. On graphs with less than 10^4 edges, the performance improvement is modest, typically less than an order of magnitude. On graphs with more than 10^4 edges, the difference is more pronounced and the improvement is at least an order of magnitude. IB solves all instances to optimality in approximately ten seconds or less. On the largest data sets, IP requires three minutes to reach an optimal solution.

Next, we run experiments with ten terminals. For this experiment, we only consider graphs with approximately 5000 vertices or more. The reason for this restriction is that, for small real-world graphs, the `10-terminal-cut` solution tends to be uninteresting: several clusters contain only a singleton terminal. The results can be found in Table 3. With the exception of the "power" graph, where

Table 2. The running time of Isolation Branching (IB) versus Gurobi Integer Programming (IP), measured in CPU seconds, on graphs from Table 1 with 5 terminals.

Graph (5 terminals)	Size		Running time		
	NumVertices	NumEdges	IB	IP	SpeedUp
netscience	379	914	0.21	1.6	7.8
celegans_metabolic	453	2025	0.59	3	5.1
jazz	198	2742	1.9	4.3	2.2
email	1133	5451	5.1	10	2.0
power	4941	6594	11	22	2.0
hep-th	5835	13815	0.8	24	28
polblogs	1222	16714	3.4	28	8.1
PGPgiantcompo	10680	24316	3.5	40	12
as-22july06	22963	48436	10	150	15
astro-ph	14845	119652	5.6	190	33

both algorithms took disproportionately long, IB outperformed IP by a factor of at least twenty.

Table 3. The running time of Isolation Branching (IB) versus Gurobi Integer Programming (IP), measured in CPU seconds, on large graphs from Table 1 with 10 terminals.

Graph (10 terminals)	Size		Running time		
	NumVertices	NumEdges	IB	IP	SpeedUp
power	4941	6594	380	723	1.9
hep-th	5835	13815	2.32	48.7	21
PGPgiantcompo	10680	24316	3.96	80.2	20
as-22july06	22963	48436	41.5	1560	38
astro-ph	14845	119652	15.9	387	24

To explore how the speed-up of IB versus IP scales, we next consider random k-terminal cut instances generated according to the Powerlaw Cluster model where we systematically increase the number of edges from 10000 to 90000 (see Sect. 4.3). The results can be found in Table 4 and Fig. 3. The running time reported is the average running time of each algorithm across fifty randomly generated k-terminal cut instances. The error bars reflect the standard deviation of running time across those instances.

As before, IB consistently outperforms IP. Furthermore, the improvement of IB over IP increases with the size of the graph. With 10000 vertices, the improvement is a factor of five. With 90000 vertices, the improvement is more than an order of magnitude. These results are consistent with our observations in

Table 4. The average running time of Isolation Branching (IB) versus Gurobi Integer Programming (IP), measured in CPU seconds, on fifty synthetic data sets generated according to the Powerlaw Cluster (PC) model with 10 new edges per node and probability 0.1 of creating a triangle.

Graph (5 terminals)	Size		Running time		
	NumVertices	NumEdges	IB	IP	SpeedUp
PC(1000, 10, 0.1)	1000	10000	2.8	16	5.5
PC(2000, 10, 0.1)	2000	20000	5.3	31	5.8
PC(3000, 10, 0.1)	3000	30000	7.2	47	6.5
PC(4000, 10, 0.1)	4000	40000	8.1	63	7.8
PC(5000, 10, 0.1)	5000	50000	10	79	7.9
PC(6000, 10, 0.1)	6000	60000	10	95	9.5
PC(7000, 10, 0.1)	7000	70000	7.7	110	14
PC(8000, 10, 0.1)	8000	80000	11	130	12
PC(9000, 10, 0.1)	9000	90000	10	140	14

the real-world data sets and suggest that the speed-up of IB versus IP increases with the size of the data set.

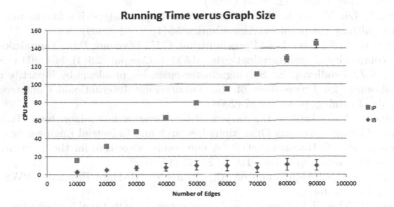

Fig. 3. The average running time of our Isolation Branching (IB) versus Gurobi Integer Programming (IP) on fifty random instances of k-terminal cut generated using the PowerlawCluster model.

5   Conclusions

In this paper, we introduce Isolation Branching, a branch-and-bound algorithm devised specifically for solving the k-terminal cut problem. In the empirical study, we demonstrate that Isolation Branching offers improvements of an order of magnitude over solving the Integer Program with Gurobi, especially on large graphs. Using synthetic data, we demonstrate that the Isolation Branching algorithm scales better from small to large instances. Our open-source code is available online at https://github.com/marvel2010/k-terminal-cut.

An advantage of our algorithm is that it uses only minimum (s, t)-cuts, avoiding linear programming. As a byproduct of our analysis of the running time of Isolation Branching, we offer an alternative proof that `k-terminal cut` is fixed-parameter tractable with respect to the size of the optimal solution.

In future work, we plan to explore additional tools that might be used to speed up the Isolation Branching algorithm. We will also consider modifications to the Isolation Branching algorithm to allow us to solve *balanced* cuts.

References

1. Boykov, Y., Veksler, O., Zabih, R.: Markov random fields with efficient approximations. In: 1998 Proceedings of IEEE Computer Society Conference on Computer Vision and Pattern Recognition, pp. 648–655. IEEE (1998)
2. Buchbinder, N., Naor, J.S., Schwartz, R.: Simplex partitioning via exponential clocks and the multiway cut problem. In: Proceedings of the Forty-Fifth Annual ACM Symposium on Theory of Computing, pp. 535–544. ACM (2013)
3. Călinescu, G., Karloff, H., Rabani, Y.: An improved approximation algorithm for multiway cut. In: Proceedings of the Thirtieth Annual ACM Symposium on Theory of Computing, pp. 48–52. ACM (1998)
4. Chen, J., Liu, Y., Lu, S.: An improved parameterized algorithm for the minimum node multiway cut problem. Algorithmica **55**(1), 1–13 (2009)
5. Dahlhaus, E., Johnson, D.S., Papadimitriou, C.H., Seymour, P.D., Yannakakis, M.: The complexity of multiterminal cuts. SIAM J. Comput. **23**(4), 864–894 (1994)
6. Fern, X.Z., Brodley, C.E.: Solving cluster ensemble problems by bipartite graph partitioning. In: Proceedings of the Twenty-First International Conference on Machine Learning, p. 36. ACM (2004)
7. Goldberg, A.V., Tardos, É., Tarjan, R.E.: Network flow algorithms. Technical report, Cornell University Operations Research and Industrial Engineering (1989)
8. Goldschmidt, O., Hochbaum, D.S.: A polynomial algorithm for the k-cut problem for fixed k. Math. Oper. Res. **19**(1), 24–37 (1994)
9. Hochbaum, D.S.: Approximation Algorithms for NP-Hard Problems. PWS Publishing Co., Boston (1996)
10. Holme, P., Kim, B.J.: Growing scale-free networks with tunable clustering. Phys. Rev. E **65**(2), 026107 (2002)
11. Karger, D.R., Klein, P., Stein, C., Thorup, M., Young, N.E.: Rounding algorithms for a geometric embedding of minimum multiway cut. Math. Oper. Res. **29**(3), 436–461 (2004)
12. Karypis, G., Kumar, V.: A fast and high quality multilevel scheme for partitioning irregular graphs. SIAM J. Sci. Comput. **20**(1), 359–392 (1998)
13. Manokaran, R., Naor, J.S., Raghavendra, P., Schwartz, R.: SDP gaps and UGC hardness for multiway cut, 0-extension, and metric labeling. In: Proceedings of the Fortieth Annual ACM Symposium on Theory of Computing, pp. 11–20. ACM (2008)

14. Marx, D.: Parameterized graph separation problems. In: Downey, R., Fellows, M., Dehne, F. (eds.) IWPEC 2004. LNCS, vol. 3162, pp. 71–82. Springer, Heidelberg (2004). https://doi.org/10.1007/978-3-540-28639-4_7
15. Sharma, A., Vondrák, J.: Multiway cut, pairwise realizable distributions, and descending thresholds. In: Proceedings of the Forty-Sixth Annual ACM Symposium on Theory of Computing, pp. 724–733. ACM (2014)
16. Stein, M., Geyer-Schulz, A.: A comparison of five programming languages in a graph clustering scenario. J. Univ. Comput. Sci. **19**, 428–456 (2013)
17. Velednitsky, M., Hochbaum, D.: A polynomial-time algorithm for 2-stable instances of the k-terminal cut problem. arXiv preprint arXiv:1806.06091 (2018)

Characterizing Cycle-Complete Dissimilarities in Terms of Associated Indexed 2-Hierarchies

Kazutoshi Ando[1][(✉)] and Kazuya Shoji[2]

[1] Faculty of Engineering, Shizuoka University,
Hamamatsu, Shizuoka 432-8561, Japan
ando.kazutoshi@shizuoka.ac.jp
[2] Graduate School of Integrated Science and Technology, Shizuoka University,
Hamamatsu, Shizuoka 432-8561, Japan

Abstract. 2-ultrametrics are a generalization of the ultrametrics and it is known that there is a one-to-one correspondence between the set of 2-ultrametrics and the set of indexed 2-hierarchies (which are a generalization of indexed hierarchies). Cycle-complete dissimilarities, recently introduced by Trudeau, are a generalization of ultrametrics and form a subset of the 2-ultrametrics; therefore the set of cycle-complete dissimilarities corresponds to a subset of the indexed 2-hierarchies. In this study, we characterize this subset as the set of indexed *acyclic* 2-hierarchies, which in turn allows us to characterize the cycle-complete dissimilarities. In addition, we present an $O(n^2 \log n)$ time algorithm that, given an arbitrary cycle-complete dissimilarities of order n, finds the corresponding indexed acyclic 2-hierarchy.

Keywords: Hierarchical classification · Quasi-hierarchy
Quasi-ultrametric · Cluster analysis

1 Introduction

Ultrametrics appear in a wide variety of research fields, including phylogenetics [10], cluster analysis [9], and cooperative game theory [2]. They have, among others, two important properties: there is a one-to-one correspondence between the set of ultrametrics and the set of indexed hierarchies [3,6,8], and every dissimilarity has a corresponding subdominant ultrametric [7].

2-ultrametrics [7] are a generalization of the ultrametrics and maintain their important properties: there is a one-to-one correspondence between the set of the 2-ultrametrics and the set of indexed 2-hierarchies [7] (which are a generalization of indexed hierarchies), and every dissimilarity has a corresponding subdominant 2-ultrametric [7].

This work was supported by JSPS KAKENHI Grant Number 18K11180.

D. Kim et al. (Eds.): COCOA 2018, LNCS 11346, pp. 640–650, 2018.
https://doi.org/10.1007/978-3-030-04651-4_43

Motivated by the work of Trudeau [11], Ando et al. [1] introduced the concept of cycle-complete dissimilarities. These form a subset of the 2-ultrametrics, so there is a corresponding subset of the indexed 2-hierarchies. In this study, we characterize this subset as the set of indexed *acyclic* 2-hierarchies, which in turn allows us to characterize the cycle-complete dissimilarities. In addition, we present an $O(n^2 \log n)$ time algorithm that, given an arbitrary cycle-complete dissimilarity of order n, finds the corresponding indexed acyclic 2-hierarchy.

The rest of this paper is organized as follows. In Sect. 2, we review 2-ultrametrics and 2-hierarchies and the one-to-one correspondence between them. In Sect. 3, we characterize the cycle-complete dissimilarities in terms of indexed 2-hierarchies. In Sect. 4, we present an $O(n^2 \log n)$ time algorithm for finding the indexed 2-hierarchy corresponding to a given cycle-complete dissimilarities. Finally, in Sect. 5, we conclude this paper.

2 2-Ultrametrics and Indexed 2-Hierarchies

Let X be a finite set. A mapping $d: X \times X \to \mathbb{R}_+$ is called a *dissimilarity* on X if for all $x, y \in X$ we have

$$d(x, y) = d(y, x) \text{ and } d(x, x) = 0. \tag{1}$$

A dissimilarity d on X is *proper* if $d(x, y) = 0$ implies $x = y$ for all $x, y \in X$. In addition, it is called a *quasi-ultrametric* [5] if for all $x, y, z, t \in X$ we have

$$\max\{d(x, z), d(y, z)\} \le d(x, y) \implies d(z, t) \le \max\{d(x, t), d(y, t), d(x, y)\}. \tag{2}$$

A family $\mathcal{K}$ of subsets of X is called a *quasi-hierarchy* on X if $\mathcal{K}$ satisfies the following conditions.

 (i) $X \in \mathcal{K}, \emptyset \notin \mathcal{K}$,
 (ii) $\{x\} \in \mathcal{K}$ for all $x \in X$,
(iii) $\forall A, B \in \mathcal{K} : A \cap B \in \mathcal{K} \cup \{\emptyset\}$,
(iv) $\forall A, B, C \in \mathcal{K} : A \cap B \cap C \in \{A \cap B, B \cap C, C \cap A\}$.

For any quasi-hierarchy $\mathcal{K}$ on X, a mapping $f: \mathcal{K} \to \mathbb{R}_+$ satisfying the following two conditions is called an *index* of $\mathcal{K}$ and the pair $(\mathcal{K}, f)$ is called an *indexed quasi-hierarchy* on X.

(1) $\forall x \in X: f(\{x\}) = 0$,
(2) $\forall A, B \in \mathcal{K}: A \subset B \implies f(A) < f(B)$.

A quasi-hierarchy $(X, \mathcal{K})$ is said to be a *2-hierarchy* if it also satisfies

(v) $\forall A, B \in \mathcal{K} : A \cap B \notin \{A, B\} \implies |A \cap B| \le 1$.

Likewise, a dissimilarity d on X is called a *2-ultrametric* [7] if for all $x, y, z, t \in X$, we have

$$d(x, y) \le \max\{d(x, z), d(y, z), d(x, t), d(y, t), d(z, t)\}. \tag{3}$$

Let d be a dissimilarity on X and σ be a positive real number. Then, the undirected graph $G_d^\sigma = (X, E_d^\sigma)$ defined by

$$E_d^\sigma = \{\{x, y\} \mid x, y \in X, x \neq y, d(x, y) \leq \sigma\} \tag{4}$$

is called the *threshold graph* of d at the threshold σ. We denote the set of all the maximal cliques of threshold graphs of d's by $\mathcal{K}_d$, i.e.,

$$\mathcal{K}_d = \bigcup_{\sigma \geq 0} \{K \mid K \text{ is a maximal clique of } G_d^\sigma\}. \tag{5}$$

In addition, for each $K \in \mathcal{K}_d$ we define $\mathrm{diam}_d(K)$ as

$$\mathrm{diam}_d(K) = \max\{d(x, y) \mid x, y \in K\} \tag{6}$$

and call it the *diameter* of K with respect to d.

With these definitions in place, we can now present the following useful lemma, followed by two propositions that clarify the relationships between quasi-ultrametrics and indexed quasi-hierarchies and between 2-ultrametrics and indexed 2-hierarchies.

Lemma 1. *Let d be a dissimilarity on X. If $K \in \mathcal{K}_d$, then K is a maximal clique of G_d^σ for $\sigma = \mathrm{diam}_d(K)$.*

Proof. Let $K \in \mathcal{K}_d$ be arbitrary and $\sigma = \mathrm{diam}_d(K)$. Since $d(x, y) \leq \mathrm{diam}_d(K) = \sigma$ for all $x, y \in K$, K is a clique of G_d^σ. Also, K is not a clique of $G_d^{\sigma'}$ for any σ' such that $\sigma' < \sigma$ since $d(x, y) = \sigma$ for some $x, y \in K$. Therefore, K is a maximal clique of $G_d^{\sigma''}$ for some σ'' such that $\sigma \leq \sigma''$. However, since for such a σ'', every clique of G_d^σ is a clique of $G_d^{\sigma''}$, it follows that K must be a maximal clique of G_d^σ. $\square$

Proposition 1 (Diatta and Fichet [5]). *A proper dissimilarity d on X is a quasi-ultrametric if and only if $(\mathcal{K}_d, \mathrm{diam}_d)$ is an indexed quasi-hierarchy on X.*

Proposition 2 (Jardin and Sibson [7]). *A proper dissimilarity d on X is a 2-ultrametric if and only if $(\mathcal{K}_d, \mathrm{diam}_d)$ is an indexed 2-hierarchy on X.*

3   Characterizing Cycle-Complete Dissimilarities in Terms of their Associated Indexed 2-Hierarchies

Let d be a dissimilarity on X. First, we introduce the complete weighted graph K_X, whose vertex set is X and whose edges $\{x, y\}$ have weight $d(x, y) = d(y, x)$.

We call a sequence

$$F : x_0, x_1, \cdots, x_{l-1}, x_l \tag{7}$$

of elements in X a *cycle* in K_X if all the x_i $(i = 0, \cdots, l-1)$ are distinct and $x_0 = x_l$. A dissimilarity d on X is called *cycle-complete* [1] if for each cycle (7) in K_X and each chord $\{x_p, x_q\}$ of F, we have

$$d(x_p, x_q) \leq \max_{i=1}^{l} d(x_{i-1}, x_i). \tag{8}$$

Proposition 3. *Let d be a dissimilarity on X. If d is cycle-complete, then it is also a 2-ultrametric.*

Proof. Let x, y, z, t be arbitrary distinct elements of X. If d is cycle-complete, then we have

$$d(x, y) \leq \max\{d(x, z), d(z, y), d(y, t), d(t, x)\} \tag{9}$$
$$\leq \max\{d(x, z), d(z, y), d(x, t), d(y, t), d(z, t)\}. \tag{10}$$

$\square$

If a dissimilarity d on X is not cycle-complete, then there must exist a cycle $F : x_0, x_1, \cdots, x_{l-1}, x_l(= x_0)$ of K_X and a chord $\{x_p, x_q\}$ of F such that (8) does not hold. We call such a cycle an *invalid cycle* in K_X.

Lemma 2. *Let d be a dissimilarity on X that is not cycle-complete and*

$$F \colon x_0, x_1, \cdots, x_l(= x_0) \tag{11}$$

be an invalid cycle in K_X of minimum length l. If $l \geq 5$, then for all $0 \leq p \leq l-3$ and $2 \leq q \leq l-1$ such that $2 \leq q - p \leq l-2$, we have

$$\max_{i=1}^{l} d(x_{i-1}, x_i) < d(x_p, x_q) = \text{const.} \tag{12}$$

Proof. Let F be an invalid cycle (11) of minimum length l, where $l \geq 5$. Let

$$\delta = \max\{d(x_p, x_q) \mid \{x_p, x_q\} \text{ is a chord of } F\} \tag{13}$$

and $\delta = d(x_p, x_q)$ for some chord $\{x_p, x_q\}$ of F. We can assume without loss of generality that $0 \leq p$ and $p + 3 \leq q \leq l - 1$. Let

$$Y = \{p, p+1, \cdots, q\},$$
$$W = \{q, q+1, \cdots, l-1, 0, \cdots, p\}.$$

Let $\{x_i, x_j\}$ be a chord of F such that $\{i, j\} \subseteq Y$. If $d(x_i, x_j) < \delta$, then

$$F' \colon x_0, x_1, \cdots, x_{i-1}, x_i, x_j, x_{j+1}, \cdots, x_{l-1}, x_l(= x_0) \tag{14}$$

is an invalid cycle with a length less than l, contradicting the initial choice of F. Hence, we must have $d(x_i, x_j) = \delta$. Similarly, for a chord $\{x_i, x_j\}$ of F such that $\{i, j\} \subseteq W$, we have $d(x_i, x_j) = \delta$.

Next, let $\{x_i, x_j\}$ be a chord of F such that $i \in Y - W$ and $j \in W - Y$. If $i = p + 1$, then, since $d(x_{p+1}, x_q) = \delta$, we have $d(x_{p+1}, x_j) = \delta$ by the same argument as above. If $i > p + 1$, then, since $\{x_p, x_{p+2}\}$ is a chord of F such that $\{p, p+2\} \subseteq Y$, we have $d(p, p+2) = \delta$. Then, we again have that $d(x_i, x_j) = \delta$ by the same argument as above. $\square$

For a family $\mathcal{K}$ of subsets of X, a sequence

$$C_0, C_1, \cdots, C_{l-1}, C_l \tag{15}$$

of subsets in $\mathcal{K}$ is called a *cycle* in $\mathcal{K}$ if we have

(i) $C_{i-1} \cap C_i \notin \{C_{i-1}, C_i, \emptyset\}$ for $i = 1, \cdots, l$,

(ii) $C_i \cap C_j = \emptyset$ for $0 \le i \le l - 3$ and $2 \le j \le l - 1$ with $2 \le j - i \le l - 2$, and

(iii) $C_0 = C_l$,

where $l \ge 3$. If $\mathcal{K}$ has no cycle, we call it *acyclic*.

Theorem 1. *A proper dissimilarity d on X is cycle-complete if and only if $(\mathcal{K}_d, \mathrm{diam}_d)$ is an indexed acyclic 2-hierarchy on X.*

Proof. Here, we treat the "if" and "only if" parts separately.

(The "only if" part:) If we assume d is cycle-complete, that means it is a 2-ultrametric (Proposition 3), and hence, $(\mathcal{K}_d, \mathrm{diam}_d)$ is an indexed 2-hierarchy (Proposition 2). Thus, it only remains to show that $\mathcal{K}_d$ is acyclic. Suppose, to the contrary, that there is a cycle

$$K_0, K_1, \cdots, K_{l-1}, K_l (= K_0) \tag{16}$$

in $\mathcal{K}_d$. Then, let

$$\delta = \max\{\mathrm{diam}_d(K_i) \mid i = 0, \cdots, l - 1\} \tag{17}$$

and $i^* = 0, \cdots, l - 1$ such that $\mathrm{diam}_d(K_{i^*}) = \delta$. If

$$d(x, y) \le \delta \text{ for all } x, y \in \bigcup_{i=0}^{l-1} K_i, \tag{18}$$

then $\cup_{i=0}^{l-1} K_i$ would be a clique of G_d^δ. However, this is impossible since K_{i^*} is a maximal clique of G_d^δ (Lemma 1). Hence, there would have to exist $x, y \in \cup_{i=0}^{l-1} K_i$ such that $d(x, y) > \delta$. Without loss of generality, suppose that $x \in K_a$ and $y \in K_b$ for $0 \le a < b \le l - 1$ and choose $x_i \in K_i \cap K_{i+1}$ for $i = 0, \cdots, l - 1$. For the sake of simplicity, we assume that $x, y \notin K_i \cap K_{i+1}$ for $i = 0, \cdots, l - 1$. Then, we could construct an invalid cycle F in K_X via

$$F : x_0, \cdots, x_{a-1}, x, x_a, \cdots, x_{b-1}, y, x_b, \cdots, x_{l-1}, x_l (= x_0), \tag{19}$$

contradicting the cycle-completeness of d.

(The "if" part:) Here, we assume $(\mathcal{K}_d, \mathrm{diam}_d)$ is an indexed acyclic 2-hierarchy on X and show that the mapping d is cycle-complete. By Proposition 2, d is a 2-ultrametric. If d is not cycle-complete, then there would have to exist an invalid cycle in K_X. Let $F : x_0, x_1, \cdots, x_{l-1}, x_l (= x_0)$ be such a cycle of minimum length l.

First, we consider the case where $l \ge 5$. By Lemma 2, we have

$$d(x_p, x_q) > \max_{i=1}^{l} d(x_{i-1}, x_i) \text{ for all chord } \{x_p, x_q\} \text{ of } F. \tag{20}$$

For each $i = 0, \cdots, l - 1$, let us choose a maximal clique K_i of G_d^σ such that $\{x_i, x_{i+1}\} \subseteq K_i$, where $\sigma = \max_{i=1}^{l} d(x_{i-1}, x_i)$. By (20), we would have

$$K_i \cap \{x_0, x_1, \cdots, x_{l-1}\} = \{x_i, x_{i+1}\} \quad (i = 0, \cdots, l - 1). \tag{21}$$

In particular, all K_i $(i = 0, \cdots, l-1)$ would be pairwise distinct. Also, since each K_i is a maximal clique of G_d^σ, we would have

$$K_i \cap K_{i+1} \notin \{K_i, K_{i+1}, \emptyset\} \quad (i = 0, \cdots, l-1). \tag{22}$$

Let i and j be such that $0 \leq i, j \leq l-1$ and $2 \leq j - i \leq l - 2$. We now show that $K_i \cap K_j = \emptyset$. To the contrary, suppose that $x \in K_i \cap K_j$. Then, we would have

$$d(x_i, x) \leq \sigma \text{ and } d(x, x_{j+1}) \leq \sigma. \tag{23}$$

From this, it would follow that

$$F': x_0, \cdots, x_i, x, x_{j+1}, \cdots, x_l$$

is an invalid cycle of length less than l, contradicting the choice of F. Thus, $K_i \cap K_j = \emptyset$, so we would have shown that $K_0, K_1, \cdots, K_{l-1}, K_l (= K_0)$ is a cycle in $\mathcal{K}_d$, a contradiction.

Next, we consider the case where $l = 4$. Let

$$F: x_0, x_1, x_2, x_3, x_4(= x_0) \tag{24}$$

be an invalid cycle in K_X and $\sigma = \max\{d(x_{i-1}, x_i) \mid i = 1, 2, 3, 4\}$. We assume, without loss of generality, that $d(x_0, x_2) > \sigma$ and show that $d(x_1, x_3) > \sigma$. Suppose, to the contrary, that $d(x_1, x_3) \leq \sigma$. Then, there would exist maximal cliques K and K' of G_d^σ such that $\{x_0, x_1, x_3\} \subseteq K$ and $\{x_1, x_2, x_3\} \subseteq K'$, and hence, $\{x_1, x_3\} \subseteq K \cap K'$. This contradicts the assumption that $\mathcal{K}_d$ is a 2-hierarchy since $K \neq K'$ by $d(x_0, x_2) > \sigma$. Then, by defining K_i as a maximal clique of G_d^σ such that $\{x_i, x_{i+1}\} \subseteq K_i$ for $i = 0, 1, 2, 3$, we would have (21) and (22), similar to the $l \geq 5$ case.

Now, suppose that for some $x \in X - \{x_0, x_1, x_2, x_3\}$ we have $x \in K_0 \cap K_2$. Then, there would have to exist a maximal clique K of G_d^σ such that $\{x_0, x, x_3\} \subseteq K$. It would then follow that $K \cap K_0 \supseteq \{x_0, x\}$ and $K \neq K_0$, contradicting the assumption that $\mathcal{K}_d$ is a 2-hierarchy. Therefore, we have that $K_0 \cap K_2 = \emptyset$ and similarly that $K_1 \cap K_3 = \emptyset$. Then, $K_0, K_1, K_2, K_3, K_4(= K_0)$ would be a cycle in $\mathcal{K}_d$, contradicting the assumption that $\mathcal{K}_d$ is acyclic. $\qquad\square$

Corollary 1. *The mapping* $d \mapsto (\mathcal{K}_d, \text{diam}_d)$ *is a one-to-one correspondence between the set of proper cycle-complete dissimilarities on X and the set of indexed acyclic 2-hierarchies on X.*

4   Algorithm

A vertex v of a connected graph G is called a *cut vertex* if $G - v$ is not connected. A graph is called *2-connected* if it is connected and has no cut vertex. Note that a graph with only one vertex is 2-connected. A maximal 2-connected subgraph of a graph G is called a *2-connected component* of G.

Input : Proper cycle-complete dissimilarity d on X.

Output: Indexed acyclic 2-hierarchy $(\mathcal{K}_d, \text{diam}_d)$.

1 Let
$$0 < \sigma_1 < \cdots < \sigma_l$$
be the distinct values of $d(x, y)$ $(x, y \in X, x \neq y)$;

2 $\mathcal{K} \leftarrow \mathcal{K}^{(0)} \leftarrow \{\{x\} \mid x \in X\}$;

3 $f(\{x\}) \leftarrow 0$ $(x \in X)$;

4 **for** $p = 1$ **to** l **do**

5   Let $\mathcal{K}^{(p)}$ be the vertex sets of the 2-connected components of $G_d^{\sigma_p}$;

6      $\mathcal{L} \leftarrow \mathcal{K}^{(p)} - \mathcal{K}^{(p-1)}$;

7      $\text{diam}_d(K) \leftarrow \sigma_p$ $(K \in \mathcal{L})$;

8      $\mathcal{K} \leftarrow \mathcal{K} \cup \mathcal{L}$;

9 **end**

10 **return** $(\mathcal{K}, f)$;

Algorithm 1: Outline of the algorithm for computing $(\mathcal{K}_d, \text{diam}_d)$.

Lemma 3. *Let d be a cycle-complete dissimilarity on X. Then, for all $\sigma \geq 0$, the vertex set of a 2-connected component of G_d^{σ} is a clique of G_d^{σ}.*

Proof. Let $Q \subseteq X$ be the vertex set of a 2-connected component of G_d^{σ}. If $|Q| \leq 2$, then Q is a clique of G_d^{σ} by the definition of a 2-connected component, so we assume $|Q| \geq 3$. Suppose, to the contrary, that there exist distinct vertices $x, y \in Q$ such that $\{x, y\} \notin E_d^{\sigma}$. By the definition of Q, there are two openly disjoint paths P_1 and P_2 in G_d^{σ} connecting x and y. By concatenating P_1 and P_2, we can create a cycle in K_X, where all the edges have weights of at most σ. Since $\{x, y\}$ is a chord of this cycle, it follows from the cycle-completeness of d that $d(x, y) \leq \sigma$, and hence $\{x, y\} \in E_d^{\sigma}$, a contradiction. $\square$

The set of maximal cliques of the threshold graph of a cycle-complete dissimilarity is characterized as follows.

Lemma 4. *Let d be a cycle-complete dissimilarity on X and $\sigma \geq 0$. Then, $K \subseteq X$ is a maximal clique of $G_d^{\sigma} = (X, E_d^{\sigma})$ if and only if K is the vertex set of some 2-connected component of G_d^{σ}.*

Proof. Assume that $K \subseteq X$ is a maximal clique of $G_d^{\sigma} = (X, E_d^{\sigma})$. Since K corresponds to a 2-connected subgraph of G_d^{σ}, it is a subset of the vertex set Q of some 2-connected component of G_d^{σ}. However, since Q is a clique (Lemma 3), we must have $K = Q$ by the maximality of K. Conversely, if $Q \subseteq X$ is the vertex set of a 2-connected component of G_d^{σ}, then Q is a clique of G_d^{σ} (Lemma 3). If this clique is not maximal, then there must exist a vertex $x \in X - Q$ such that $\{x, y\} \in E_d^{\sigma}$ for all $y \in Q$, contradicting the assumption that Q is the vertex set of a 2-connected component of G_d^{σ}. $\square$

Based on Lemma 4, we have designed an algorithm for constructing the indexed acyclic 2-hierarchy $(\mathcal{K}_d, \mathrm{diam}_d)$ for a given proper cycle-complete dissimilarity d, as outlined in Algorithm 1. The validity of the algorithm follows straightforwardly from the propositions presented above.

Input : Proper cycle-complete dissimilarity d on X.
Output: Indexed acyclic 2-hierarchy $(\mathcal{K}_d, \mathrm{diam}_d)$.
1   Let $e_1, \ldots, e_m$ be the edges of K_X ordered in nondecreasing order of d, where $m = \frac{n(n-1)}{2}$;
2   $\mathcal{K} \leftarrow \{\{x\} \mid x \in X\}$;
3   $f(\{x\}) \leftarrow 0 \ (x \in X)$;
4   $\mathcal{L} \leftarrow \emptyset$;
5   **for** $i = 1$ **to** m **do**
6     $\{x, y\} \leftarrow e_i$;
7   **if** x and y are in different 2-connected components of G_{i-1} **then**
8   **if** x and y are in the same component **then**
9   Let P be a path connecting x and y in G_{i-1};
10   Let $Q_1, \ldots, Q_l$ be the vertex sets of the 2-connected components of G_{i-1} which contain at least two vertices of P;
11        $Q \leftarrow \bigcup_{k=1}^{l} Q_k$;
12        $\mathcal{L} \leftarrow \mathcal{L} \cup \{Q\} - \{Q_1, \ldots, Q_l\}$;
13   **else**
14        $Q \leftarrow \{x, y\}$;
15        $\mathcal{L} \leftarrow \mathcal{L} \cup \{Q\}$;
16   **end**
17   **end**
18   **if** $d(e_i) < d(e_{i+1})$ or $i = m$ **then**
19       $\mathcal{K} \leftarrow \mathcal{K} \cup \mathcal{L}$;
20       $f(K) \leftarrow d(e_i) \ (K \in \mathcal{L})$;
21       $\mathcal{L} \leftarrow \emptyset$;
22   **end**
23 **end**
24 **return** $(\mathcal{K}, f)$;

Algorithm 2: More detailed description of the algorithm for computing $(\mathcal{K}_d, \mathrm{diam}_d)$.

It is not immediately clear how to implement Algorithm 1 efficiently, however. To achieve this, we need to able to identify the 2-connected components of a threshold graph efficiently. Let $e_1, \ldots, e_m$ be the edges of K_X arranged in nondecreasing order of d, where $m = \frac{n(n-1)}{2}$. Then, we construct the vertex sets of the 2-connected components of the undirected graph $G_i = (X, E_i)$ incrementally for $i = 0, 1 \cdots, m$, where E_i is defined by $E_i = \{e_1, \cdots, e_i\}$. A more detailed description of the algorithm is given in Algorithm 2.

Let $G = (X, E)$ be an undirected graph whose vertex set is X. Let A and $\mathcal{Q}$ be the set consisting of all the cut vertices and the set of the 2-connected components of G, respectively. The block forest (cf. [4]) of G is the bipartite graph $B = (A, \mathcal{Q}; F)$ defined by $F = \{(a, Q) \mid a \in A, Q \in \mathcal{Q}, a \in Q\}$, as shown in Fig. 1.

Theorem 2. *Given a proper cycle-complete dissimilarity d on X, Algorithm 2 correctly produces the indexed acyclic 2-hierarchy $(\mathcal{K}_d, \mathrm{diam}_d)$ and terminates in $O(n^2 \log n)$ time, where $n = |X|$.*

Proof. First, we show that the algorithm is valid. In Lines 6–17, it finds the vertex set Q of the 2-connected component of G_i formed by adding the edge $e_i = \{x, y\}$ to G_{i-1}, if it exists. This set is either $Q_1 \cup \cdots \cup Q_l$ or $e_i = \{x, y\}$, depending on whether or not x and y are in the same component. Then, the algorithm adds Q to the list $\mathcal{L}$, removing $Q_1, \cdots, Q_l$ in the first case. Then, the collection $\mathcal{L}$ of vertex sets in Line 19 is exactly the same as $\mathcal{K}^{(p)} - \mathcal{K}^{(p-1)}$ in Line 6 of Algorithm 1, where $d(e_i) = \sigma_p$.

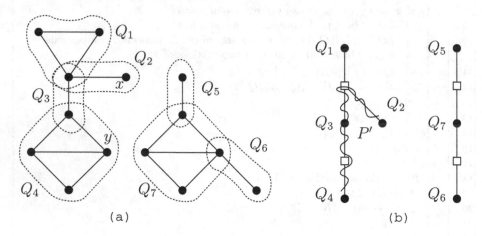

Fig. 1. (a) All 2-connected components of a graph G. (b) Block forest of G, where the cut vertices are indicated by rectangles, and the path P' between Q_2 and Q_4 is indicated by a wavy line.

Next, we consider the algorithm's time complexity. It takes $O(n^2 \log n)$ time to sort the edges of K_X using any standard sorting algorithm, so the complexity must be at least that. Here, we show that the other operations in Algorithm 2 only require $O(n^2)$ time. To achieve this bound, we represent the 2-connected components of G_i as block forest B_i, and assume that each of the trees in the forest B_i is rooted at some vertex for $i = 0, 1 \cdots, m$. In addition, we use a mapping $q \colon X - A \to \mathcal{Q}$ that associates each $x \in X - A$ with the unique 2-connected component $q(x)$ of G_{i-1} to which x belongs. With this, given arbitrary $x, y \in X$, we can determine whether or not x and y are in the same 2-connected component of G_{i-1} in $O(1)$ time. We can also find the 2-connected components $Q_1, \cdots, Q_l$ (Line 10) in $O(n)$ time by searching for the path P' in the forest B_i

connecting the nodes corresponding to x and y, as shown in Fig. 1(b). The block forest can be updated in $O(n)$ time by reducing the 2-connected components $Q_1, \cdots, Q_l$ on the path P' to a single 2-connected component Q. See Fig. 2(b). The mapping q can also be updated in $O(n)$ time. Since the number of i's for which x and y are in different 2-connected components is $O(n)$ [1, Lemma 3.5], it follows that the total time taken to compute Lines 8–16 is $O(n^2)$. □

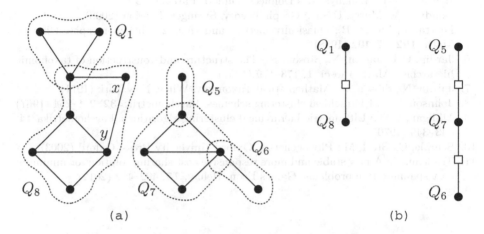

(a)   (b)

Fig. 2. (a) All 2-connected components of the graph $G + \{x, y\}$, where G is the graph in Fig. 1(a). (b) Block forest of $G + \{x, y\}$, where the cut vertices are indicated by rectangles. Here, Q_2, Q_3 and Q_4 in Fig. 1(b) have been reduced to form Q_8.

5   Conclusions

It is known [5] that the mapping $d \mapsto (\mathcal{K}_d, \mathrm{diam}_d)$ gives a one-to-one correspondence between the set of quasi-ultrametrics and the set of indexed quasi-hierarchies on X, where $\mathcal{K}_d$ is the set of all the maximal cliques of threshold graphs of d and the function $\mathrm{diam}_d \colon \mathcal{K}_d \to \mathbb{R}_+$ gives the diameter of each clique in $\mathcal{K}_d$. This leads to a similar one-to-one correspondence between the set of 2-ultrametrics and the set of indexed 2-hierarchies on X [7]. The cycle-complete dissimilarities [1] form a subset of the 2-ultrametrics, so the mapping $d \mapsto (\mathcal{K}_d, \mathrm{diam}_d)$ gives a correspondence between these and a subset of the indexed 2-hierarchies on X. In this paper, we have characterized this subset as the set of indexed acyclic 2-hierarchies on X, which has then allowed us to characterize the cycle-complete dissimilarities. In addition, we have presented an algorithm for finding the indexed acyclic 2-hierarchy $(\mathcal{K}_d, \mathrm{diam}_d)$ on X corresponding to a cycle-complete dissimilarity d on X and shown that runs in $O(n^2 \log n)$ time, where $n = |X|$.

Acknowledgments. The authors are grateful to the anonymous referees for useful comments which improved the presentation of the original version of this paper.

References

1. Ando, K., Inagaki, R., Shoji, K.: Efficient algorithms for subdominant cycle-complete cost functions and cycle-complete solutions. Discrete Appl. Math. **225**, 1–10 (2017)
2. Ando, K., Kato, S.: Reduction of ultrametric minimum cost spanning tree games to cost allocation games on rooted trees. J. Oper. Res. Soc. Japan **53**, 62–68 (2010)
3. Benzécri, J.-P.: L'analyse des Données. Dunod, Paris (1973)
4. Bondy, J.A., Murty, U.S.R.: Graph Theory. Springer, London (2008)
5. Diatta, J., Fichet, B.: Quasi-ultrametrics and their 2-ball hypergraphs. Discrete Math. **192**, 87–102 (1998)
6. Jardin, C.J., Jardin, N., Sibson, R.: The structure and construction of taxonomic hierarchies. Math. Biosci. **1**, 173–179 (1967)
7. Jardin, N., Sibson, R.: Mathematical Taxonomy. Wiley, New York (1971)
8. Johnson, S.C.: Hierarchical clustering schemes. Psychometrika **32**, 241–254 (1967)
9. Milligan, G.W.: Ultrametric hierarchical clustering algorithms. Psychometrika **44**, 343–346 (1979)
10. Semple, C., Steel, M.: Phylogenetics. Oxford University Press, Oxford (2003)
11. Trudeau, C.: A new stable and more responsive cost sharing solution for minimum cost spanning tree problems. Games Econ. Behav. **75**, 402–412 (2012)

Making Multiple RNA Interaction Practical

Syed Ali Ahmed[1,3], Saman Farhat[1], and Saad Mneimneh[1,2(✉)]

[1] The Graduate Center, City University of New York, New York, NY 10016, USA
sahmed@gradcenter.cuny.edu, samanfarhat@yahoo.com
[2] Hunter College, City University of New York, New York, NY 10065, USA
saad@hunter.cuny.edu
[3] Google LLC, 1600 Amphitheatre Parkway, Mountain View, CA 94043, USA

Abstract. Multiple RNA interaction can be modeled as a problem in combinatorial optimization, where the "optimal" structure is driven by an energy-minimization-like algorithm. However, the actual structure may not be optimal in this computational sense. Moreover, it is not necessarily unique. Therefore, alternative sub-optimal solutions are needed to cover the biological ground.

We extend a recent combinatorial formulation for the Multiple RNA Interaction problem with approximation algorithms to handle more elaborate interaction patterns, which when combined with Gibbs sampling and MCMC (Markov Chain Monte Carlo), can efficiently generate a reasonable number of optimal and sub-optimal solutions. When viable structures are far from an optimal solution, exploring dependence among different parts of the interaction can increase their score and boost their candidacy for the sampling algorithm. By clustering the solutions, we identify few representatives that are distinct enough to suggest possible alternative structures.

Keywords: Multiple RNA interaction · NP-hardness
Approximation algorithms · Gibbs sampling · MCMC · Clustering

1 Introduction

The role of interaction between two or more RNA molecules has been increasingly recognized in biological mechanisms, including the regulation of gene expression, methylation, and splicing. Pairwise interaction has been noted for regulating gene expression whereby an anti-sense RNA blocks the ribosomal binding site of the messenger RNA, e.g. [25]. Typical scenarios of multiple (more than two) RNA interaction involve the interaction of small nucleolar RNAs (snoRNAs) and ribosomal RNAs (rRNAs) in the process of methylation [29], small nuclear

---

Supported by a Research Starter Award in Informatics from the PhRMA Foundation www.phrmafoundation.org.

D. Kim et al. (Eds.): COCOA 2018, LNCS 11346, pp. 651–671, 2018.
https://doi.org/10.1007/978-3-030-04651-4_44

RNAs (snRNA) and messenger RNAs in the splicing of introns [40], and several ribozyme complexes of small RNAs as catalytic RNA complexes [22,36,39].

The prediction of structures resulting from pairwise interactions is now somewhat understood; for instance, due to successful efforts in generalizing the partition function of a single RNA to the case of two. Algorithms for pairwise interaction of RNAs based on a generalized partition function and other methods appear in [3,4,6,10,11,15,19,26,29,30,33,35,37], but they do not scale when carried over to multiple RNAs (more than two). The de facto treatment of multiple RNAs has been to account for their interaction by concatenating the RNAs into a single long RNA, which is then folded in order to predict the structure [5,13]. Most folding algorithms prevent the formation of pseudoknots due to their increased computational complexity. While pseudoknots are rare in folded structures, they form legitimate patterns when spanning multiple RNAs, e.g. kissing loops. There are a few attempts that introduce pseudoknots into the concatenation model, e.g. [9], but advances in pairwise interaction algorithms based on a generalized partition function suggest that the latter are more adequate, so they remain the state-of-the-art for two RNAs.

Therefore, a promising approach is to exploit pairwise interaction in the context of multiple RNAs. Indeed, we have recently proposed in a series of works [1,2,31,32] a formulation where multiple RNAs interact along a chain driven by the pairwise interactions of consecutive RNAs (Fig. 1b). This formulation can produce optimal or near optimal solutions as it admits a Polynomial Time Approximation Scheme PTAS. However, correct biological structures are not necessarily "optimal" in any given computational framework and often are not unique. Therefore, some realistic solutions ought to be sub-optimal. But it is challenging to pick up the desired sub-optimal solutions, especially when far from optimal. For instance, many artifact interactions can easily arise when the RNAs are exact complement of each other (they bind perfectly). The CopA-CopT complex represents such an example [25]. It is known as the perfect couple, and has been problematic since the inception of pairwise interaction algorithms in 2005. The correct solution must drop many of these artifacts and, therefore, is typically very far from optimal.

In this paper, we extend our formulation for the Multiple RNA Interaction problem to:

1. Handle more elaborate interaction patterns (not just a chain) guided by what we call bipartite interaction graphs; we provide new approximation algorithms for special cases of the bipartite graph
2. Conform to a sampling algorithm that uses Gibbs sampling and MCMC to produce multiple (sub-optimal) solutions, which are then clustered to reveal several candidate structures, and
3. Explore dependence among interactions to better score sub-optimal solutions and boost their candidacy for sampling; this dependence renders the problem hard to approximate (but remains useful and practical in sampling).

2   The Model: Pegs and Rubber Bands

We describe a combinatorial optimization problem called *Pegs and Rubber Bands* as a framework for multiple RNA interaction. The link between the two will be made shortly following a formal description of *Pegs and Rubber Bands*.

Consider m levels numbered 1 to m with n_l pegs in level l numbered 1 to n_l. We define $n = \max_{l=1}^{m} n_l$. There is an infinite supply of rubber bands, and a rubber band can be placed around pegs in two levels. For instance, we may choose to place a rubber band around pegs $[i_1, i_2]$ (i.e., the set of pegs from i_1 to i_2, where $i_1 \leq i_2$), in level l_1, and pegs $[j_1, j_2]$ in level l_2, where $l_1 < l_2$. In this case, the rubber band defines a window with a given weight $w(l_1, l_2, i_2, j_2, u, v)$, where $u = i_2 - i_1 + 1$ and $v = j_2 - j_1 + 1$ represent the lengths of the intervals covered by the window in levels l_1 and l_2, respectively (as in Fig. 1a). For convenience, we will use $w(l_1, l_2, i, j, u, v)$ interchangeably to denote both the window and its weight, depending on context. As such, each window $w(l_1, l_2, i, j, u, v)$ defines two intervals, $[i - u + 1, i]$ in level l_1 and $[j - v + 1, j]$ in level l_2.

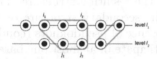

(a) A rubber band around pegs defines a window. The lengths $u = i_2 - i_1 + 1$ and $v = j_2 - j_1 + 1$ of the corresponding intervals may be different.

```
I1    3' UGUAUG 5'
         ||||
U6    5' AUAC...GAUU...GUGAAGCGU 3'
             ||||  |||||||||
U2    3' UAUGAU...CUAG...CACUUCGCA 5'
         |||||
I2    5' UACUAAC 3'
```

(b) Multiple RNA interaction within the eukaryotic spliceosome, showing the predicted structure of spliceosomal U2-U6 snRNA and two introns I1 and I2, which is consistent with biological experiments [40, 43].

Fig. 1. The Pegs and Rubber Bands formulation and an example of multiple RNA interaction.

Assume window w defines interval $[a, b]$ in level l and another w' defines interval $[c, d]$ in the same level l. We say that w' *follows* w if $b < c$. We require **two** conditions:

1. $[a, b] \cap [c, d] = \emptyset$ (disjoint base pairs)
2. *follow* can be extended by transitivity to a **partial order** relation (no pseudoknots).

We refer to the above two requirements as the *no overlap* condition. In particular, there is overlap if some windows define overlapping intervals in the same level, or if there is a sequence of windows (two or more) that follow one another in a cycle.

The *Pegs and Rubber Bands* problem is to maximize the total weight by placing rubber bands around pegs in such a way that none of their corresponding windows make an overlap. Therefore, the goal is to find a set of windows S (a solution) that maximizes $\sum_{w(l_1,l_2,i,j,u,v) \in S} w(l_1, l_2, i, j, u, v)$ subject to the no overlap condition.

The Connection to Multiple RNA Interaction: RNAs are mapped as levels, the ordered pegs in each level represent RNA bases $\{A, C, G, U\}$ in the order of occurrence in their sequence, a window $w(l_1, l_2, i, j, u, v)$ is an interaction between bases $[i - u + 1, i]$ in RNA l_1 and bases $[j - v + 1, j]$ in RNA l_2 (windows are then converted to base pairs), and the weight $w(l_1, l_2, i, j, u, v)$ is chosen based on the energy of that interaction. The energies are obtained using a generalized partition function for pairwise interaction, and account for both intra- and inter- molecular energies; for instance,

$$w(l_1, l_2, i, j, u, v) = RT \ln \frac{P_{l_1}(free[i - u + 1, i]) P_{l_2}(free[j - v + 1, j])}{Z^I_{l_1,l_2}(i - u + 1, i, j - v + 1, j)]}$$

where R is the Boltzman constant and T is temperature, $P_l(free[i, j])$ is the probability that subsequence $[i, j]$ is free (does not fold) in RNA l, and $Z^I_{l_1,l_2}(i_1, i_2, j_1, j_2)$ is the generalized partition function of the interaction of subsequences $[i_1, i_2]$ in RNA l_1 and $[j_1, j_2]$ in RNA l_2 (subject to no folding within the subsequences) [33]. This reflects a two step process in which the RNAs are first freed to interact, and then the interaction takes place. Therefore, RNA folding is not ignored even if it's not explicit in the resulting structure. The no overlap condition reflects a typical nature of the secondary structure of RNA interactions, which may be interpreted as the absence of pseudoknots (condition 2) in addition to the fact that each base can participate in at most one base pair (condition 1). The maximization problem corresponds to energy minimization, which leads to favorably stable structures.

Figure 1b shows an example of a structure predicted using the *Pegs and Rubber Bands* formulation. In the figure, windows are replaced by base pairs in their corresponding intervals. Observe that if the RNAs were handled pairwise, as in [41] for instance, the best interacting pair of RNAs will dominate the solution, and since the pair is required to fully interact before incorporating any further interactions, this will "lock" the interaction pattern of the whole ensemble into a sub-optimal state; thus preventing the correct structure from presenting itself as a solution. Our formulation avoids this "locking" problem since the pairwise interaction would have favored to include the binding of the 5' end of U6 and the 3' end of U2 in Fig. 1b, leaving I1 and I2 detached.

3 RNA Interaction Pattern, Bipartite Graphs, and Approximations

So far, we made an implicit assumption that every pair of RNAs can interact; hence, $w(l_1, l_2, i, j, u, v)$ represents a legitimate interaction for any pair (l_1, l_2). However, for this interaction to take place, RNA l_1 and RNA l_2 have to act as sense and anti-sense. Therefore, we envision the existence of a bipartite graph

$G = (V_1, V_2, E)$, called the interaction graph, where $(u, v) \in E$ (with either $u \in V_1$ and $v \in V_2$, or $u \in V_2$ and $v \in V_1$) iff RNA u and RNA v may interact. For convenience, we say $(u, v) \in G$. The set V_1 represents the sense RNAs (5' to 3') and the set V_2 represents the anti-sense RNAs (3' to 5'). Therefore, $w(l_1, l_2, i, j, u, v)$ may be part of the solution for *Pegs and Rubber Bands* if $(l_1, l_2) \in G$. This will prevent the formation of circular interactions with odd cycles; the shortest cycle will have length four, e.g. RNA 1 interacts with RNA 2, RNA 2 interacts with RNA 3, RNA 3 interacts with RNA 4, and RNA 4 interacts with RNA 1.[1] An exponential time $O((m + |E|)n^m)$ algorithm for *Pegs and Rubber Bands* based on dynamic programming is described in Fig. 2; it consists of decomposing the solution by successively adding windows that define disjoint intervals and preserve *follow* as a partial order relation.

$$W(i_1, \ldots, i_m) = \max \begin{cases} W(i_1, \ldots, i_{l-1}, \underline{i_l - 1}, i_{l+1}, \ldots, i_m) \ \{\text{skip a peg in level } l\} \qquad l = 1, \ldots, m \\ W(i_1, \ldots, i_{l_1-1}, \underline{i_{l_1} - u}, i_{l_1+1}, \ldots, i_{l_2-1}, \underline{i_{l_2} - v}, i_{l_2+1}, \ldots, i_m) \\ \quad + w(l_1, l_2, i_{l_1}, i_{l_2}, u, v) \ \{\text{use } u \text{ and } v \text{ pegs in levels } l_1 \text{ and } l_2\} \ (l_1, l_2) \in G \end{cases}$$

where $w(l_1, l_2, i, j, u, v) = -\infty$ if $u > i$ or $v > j$, $0 < u, v \leq w$ (the maximum window size), and $W(0, 0, \ldots, 0) = 0$.

Fig. 2. Dynamic programming algorithm for *Pegs and Rubber Bands* and a bipartite graph G. The maximum is given by $W(n_1, n_2, \ldots, n_m)$ and the solution can be obtained by standard DP backtracking. This algorithm stands behind the approximation results when it is applied to sub-problems as presented in the following theorems.

3.1 Approximation Algorithms

Theorem 1 (Hardness). *Pegs and Rubber Bands is NP-hard.*

Proof: The NP-hardness was established in [2] for the special case when the bipartite graph is a path (RNAs interact in a chain). Therefore, this result holds for the cases of a cycle and a tree (even when bounded in degree), and for general bipartite graphs. □

Theorem 2 (Path and Cycle Interactions). *Pegs and Rubber Bands admits for every fixed $\epsilon > 0$ a polynomial time algorithm that achieves a total weight within $(1 - \epsilon)$-factor of optimal if the bipartite interaction graph is a path or a cycle.*

Proof: The proof for the case of a path appears in [2]. We present the proof for the case of a cycle. For simplicity of illustration, and since the bipartite graph is a cycle, we assume that the levels are numbered 0 to $m - 1$ modulo

---

[1] Circular interactions with odd cycles (where the interaction graph G is not restricted to being bipartite) can be achieved by allowing *inverted windows* in which the interaction given by $w(l_1, l_2, i, j, u, v)$ occurs between bases $[i - u + 1, i]$ on RNA l_1 and bases $[j, j - v + 1]$ (inverted sequence) on RNA l_2, but we do not explore this direction here.

m. Let OPT be the weight of the optimal solution and denote by $W[i \ldots j]$ the weight of the optimal solution when the problem is a path restricted to levels $i, i+1, \ldots, j$ mod m, i.e. a sub-problem with interaction graph containing edges $(l, l+1)$ for $l = i, i+1, \ldots, j-1$ mod m. Let k be an integer that is a function of ϵ and suppose $m = ka + b$, where $0 \le b < k$ (a and b are the quotient and remainder in the division of m by k, respectively). Consider the following m solutions (weights) obtained by circular shifts, each a concatenation of $\lceil m/k \rceil$ optimal solutions for sub-problems consisting of at most k levels (a of them have k levels and one has b levels).

$$W_0 = W[0 \ldots k-1] + W[k \ldots 2k-1] + \ldots + W[m-b \ldots m-1]$$

$$W_1 = W[1 \ldots k] + W[k+1 \ldots 2k] + \ldots + W[m-b+1 \ldots m]$$

$$\vdots$$

$$W_{m-1} = W[m-1 \ldots m+k-2] + W[m+k-1 \ldots m+2k-2] + \ldots$$
$$+ W[2m-b-1 \ldots 2m-2]$$

While each $W_i \le OPT$, it is easy to verify that every pair of consecutive levels (modulo m) is missing in exactly a of the above m sub-problems if $b = 0$, and $a+1$ otherwise; that's $\lceil m/k \rceil$ in both cases. Therefore,

$$\sum_{i=0}^{m-1} W_i \ge (m - \lceil m/k \rceil)OPT \Rightarrow \max_i W_i \ge \left(1 - \frac{\lceil m/k \rceil}{m}\right)OPT$$

We can achieve the desired $(1-\epsilon)$ factor approximation by making $\lceil m/k \rceil / m \le \epsilon$, which when m is large enough, can be done if $k = O(\frac{1}{\epsilon})$.

There are $O(m)$ sub-problems of at most k levels each. A sub-problem requires a time polynomial in n for a fixed k, $O(kn^k)$, as shown in Fig. 2 using dynamic programming. Furthermore, each of the m solutions has $\lceil m/k \rceil$ sub-problems, so the additional running time required to find all W_i given the sub-problems is $O(m^2/k)$. $\square$

Theorem 3 (Tree Interaction). *Pegs and Rubber Bands admits for every fixed $\epsilon > 0$ a polynomial time algorithm that achieves a total weight within $(1 - \epsilon)$-factor of optimal if the bipartite interaction graph is a tree with bounded degree.*

Proof: Start with an arbitrary vertex (RNA) in G, say v_0, and visit all vertices of the tree using a Breadth First Search traversal. This assigns a "layer" for each vertex, i.e. v_0 will be in layer 0, all neighbors of v_0 will be in layer 1, and so on. Let V_i be the set of all vertices in layer i. Let $G_{i,j}$ for $i \le j$ be the induced graph of G defined by vertices in $V_i \cup V_{i+1}, \ldots \cup V_j$. Now for a given k, consider the following partitioning (sub-problems), given by $G_{i,i+k-1}$:

$$\text{partition } 0 : G_{0,0}, G_{1,k}, G_{k+1,2k}, G_{2k+1,3k}, \ldots$$

$$\text{partition } 1 : G_{0,1}, G_{2,k+1}, G_{k+2,2k+1}, G_{2k+2,3k+1}, \ldots$$

$$\vdots$$

$$\text{partition } k - 1 : G_{0,k-1}, G_{k,2k-1}, G_{2k,3k-1}, \ldots$$

where the first and last sub-problems have k or fewer layers. It is not hard to see that each sub-problem $G_{i,i+k-1}$ consists of disjoint sub-trees with at most $1+d+d(d-1)+\ldots+d(d-1)^{k-2} = O(d^k)$ vertices each, where d is the maximum degree. Using the algorithm of Fig. 2, the optimal solution given each sub-tree can be found in polynomial time $O(d^k n^{d^k})$, for the fixed k and d. The total number of sub-trees after shifting the partition $k - 1$ times is $O(m)$.

The k partitions give us k solutions. By observing that every edge in G appears in exactly $k-1$ of the above k partitions, and using an argument similar to the proof of Theorem 2, we can show that the best of the k solutions is a $(k - 1)/k$-factor approximation. We choose $k = O(\frac{1}{\epsilon})$. □

Theorem 4 (Star Interaction). *Peg and Rubber Bands is also NP-hard if the bipartite graph is a star, and admits there a 1/2-factor approximation algorithm if all windows have $u = v = 1$ and weight 1.*

Proof: We can now think of a window $w(1, l, i, j, 1, 1)$ as an *edge* connecting peg i in level 1 and peg j in level l (all RNAs interact only with RNA 1). We then need to maximize the number of edges with no crossings. A reduction from 3SAT (the special satisfiability problem where each clause in the conjunctive normal form has exactly three variables) with variables $x_1, x_2, \ldots, x_n$ and clauses $C_1, C_2, \ldots, C_k$ to a star instance of *Pegs and Rubber Bands* with $m = k + 2$ levels (RNAs) is done as follows: Each variable is mapped to two sets (representing True or False) of k edges between level 1 and level 2 such that every edge in the first set crosses every edge in the second set. Each clause C_l is mapped to three crossing edges between level 1 and level $l + 2$ representing the state of its variables. Edges for clause C_l and clause $C_{l'}$ are node disjoint on level 1 (there are enough pegs on level 1 for each variable to ensure that). Figure 3 illustrates this construction.

Each variable can contribute at most k edges (depending on its setting of True or False), and each clause at most one edge (indicating the variable that satisfies it). The 3SAT instance is satisfiable iff we can find $nk + k = k(n + 1)$ edges with no crossings.

The 1/2-factor approximation: Given a peg j in level l, list all i in **decreasing** order such that $w(1, l, i, j, 1, 1)$ exists. By repeating this process for $j = 1, 2, \ldots, n_l$ in order, we obtain a sequence S_l for level l. The optimal solution for the star *Pegs and Rubber Bands* corresponds to $m - 1$ **disjoint** sub-sequences $A_2, A_3, \ldots, A_m$ each **increasing** in $S_2, S_3, \ldots, S_m$, respectively, such that their union[2], a subset of $X = \{1, 2, \ldots, n_1\}$, is the largest possible. The increasing

---

[2] For ease of notation, we are thinking of A_l as a sequence and a set at the same time.

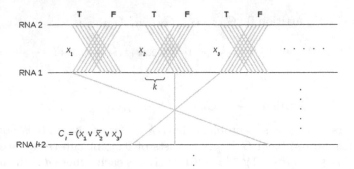

Fig. 3. Reduction from 3SAT to star *Pegs and Rubber Bands*, showing the clause $(x_1 \vee \overline{x}_2 \vee x_3)$.

(and disjoint) sub-sequences ensure that no edges cross. This transformation gives a special case of a coverage problem where the claimed approximation is achieved by a greedy algorithm that repeats the following $m - 1$ times [8]: choose an $l \neq 1$ that has not yet been chosen and contributes the longest A_l, add A_l to the solution, and update X to $X - A_l$ and each S_i to $S_i - A_l$. $\square$

3.2   Finding the Bipartite Interaction Graph

We do not impose a specific bipartite graph to start with, but we infer it in the following way: Given m RNAs, we start with a random permutation over $\{0, 1, \ldots, m\}$. All indices to the left of 0 belong to V_1, and all indices to the right of 0 belong to V_2. We find the optimal solution for *Pegs and Rubber Bands* given the complete bipartite interaction graph, i.e. $G = K_{|V_1|,|V_2|}$, and using the algorithm of Fig. 2, which is exponential in the number of RNAs m, but practical when $m \leq 4$ (when $m > 4$, one could explore the approximation algorithms presented above with the corresponding appropriate bipartite graphs). Afterwards, a random search generates some neighboring permutations, e.g. using a 2-opt strategy in which the permutation $(\sigma_0, \ldots, \sigma_i, \ldots, \sigma_j, \ldots, \sigma_m)$ produces neighbors of the form $(\sigma_0, \ldots, \sigma_{i-1}, \sigma_j, \ldots, \sigma_i, \sigma_{j+1}, \ldots, \sigma_m)$ for several values of i and j, and the same is repeated. When a better solution is revealed, the permutation is updated. When no better solution is found, we stop. The obtained solution represents a locally optimal one, and possibly the global optimal given all possible permutations. If RNA u and RNA v do not interact in this solution, (u, v) is dropped from the complete bipartite graph. We then generate sub-optimal solutions for the given bipartite graph by sampling (Sect. 5).

4   Weight Dependent Pegs and Rubber Bands

We extend the *Pegs and Rubber Bands* formulation by allowing windows to be either *single* or *dependent*. Single windows contribute a weight $w(l_1, l_2, i, j, u, v)$ as before. Recall that each window $w(l_1, l_2, i, j, u, v)$ defines two intervals, $[i -$

$u+1, i]$ in level l_1 and $[j-v+1, j]$ in level l_2. If a solution contains two windows that define intervals $[a, b]$ and $[c, d]$ in level l with $b < c$, then we may consider them dependent in level l (windows can be dependent in one or two levels), and thus add to their single weight contribution a new positive term for level l given by $\Delta(l, a, b, c, d)$. To motivate this idea, imagine that in the folding of RNA l, whenever $[a, b]$ is free, it is energetically favorable that $[c, d]$ is also free; for instance, due to the breaking of a stem in the original folding. One could then use

$$\Delta(l, a, b, c, d) = RT \ln P_l(free[c, d] \,|\, free[a, b]) - RT \ln P_l(free[c, d])$$

when it's positive as a possible term. Therefore, in addition to the single contribution of windows, a solution where both $[a, b]$ and $[c, d]$ of RNA l interact will acquire more weight, due to the net positive effect (since $\Delta(l, a, b, c, d) > 0$) of replacing the individual probabilities with the joint probability (conditioning).

Based on the above motivation, we also require that if $\Delta(l, a, b, c, d)$ and $\Delta(l, e, f, g, h)$ both contribute to the total weight of a given solution, then either $[a, d] \cap [e, h] = \emptyset$ or one is contained in the other; thus mimicking the typical nesting property of folding in RNA l.

Given a solution S, let $I_l(S)$ be the set of all intervals in level l defined by windows in S, i.e. either $l_1 = l$ or $l_2 = l$ in $w(l_1, l_2, i, j, u, v) \in S$. Let $M_l(S)$ be the set of all no-overlap matchings in $I_l(S)$; in other words, if $([a, b], [c, d])$ and $([e, f], [g, h])$ belong to a matching (with $b < c$ and $f < g$), then either $[a, d] \cap [e, h] = \emptyset$ or one is contained in the other. The modified weight of solution S is defined as:

$$w(S) = \sum_{w(l_1, l_2, i, j, u, v) \in S} w(l_1, l_2, i, j, u, v) + \sum_l \max_{M \in M_l(S)} \sum_{([a,b],[c,d]) \in M} \Delta(l, a, b, c, d)$$

We then seek a solution that maximizes the above. We will call this variant the *Weight Dependent Pegs and Rubber Bands*, which remains to be NP-hard. However, this variant of the problem is even hard to approximate.

Theorem 5. *The Weight Dependent Pegs and Rubber Bands has no constant factor approximation unless $P = NP$, even when $m = 2$ (two RNAs).*

Proof: We make a reduction from the Longest Common Subsequence problem, which is known to be hard to approximate [23]. Given n strings $s_1, \ldots, s_n$, let LCS be the length of their longest common subsequence. We show how to construct an instance of *Weight Dependent Pegs and Rubber Bands* with $m = 2$ that has an optimal weight $OPT = LCS[nx - (n+1)]$, where x is chosen such that

$$\frac{n+1}{n} < x < \frac{n}{n-1}$$

Furthermore, we show that any approximation to OPT consists of an integer multiple of $[nx - (n+1)]$, say k, and reveals a common subsequence of length k.

We define s_i' to be string s_i reversed. If string s_i has length $|s_i|$, then we call $s_i[j]$ and $s_i'[|s_i| - j + 1]$ *duplicates* (they represent two copies of the same character due to the reversal). We also define $s_0 = s_1$ and $s_{n+1} = s_n$. We then construct two levels of pegs, where each peg i in level $l = 1, 2$ is represented by a character of some string. In the first level, we lay out the pegs given by the concatenated string $s_1 s_1' s_3 s_3' \ldots$. In the second level, we lay out the pegs given by the concatenated string $s_0 s_2' s_2 s_4' s_4 \ldots$. Figure 4 shows this construction for $s_1 = 0010111$, $s_2 = 01010$, and $s_3 = 100101$ (the choice of a binary alphabet is made for ease of illustration).

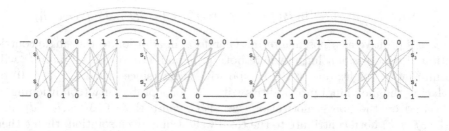

Fig. 4. A construction of a *Weight Dependent Pegs and Rubber Bands* instance for $s_1 = 0010111$, $s_2 = 01010$, and $s_3 = 100101$, with $LCS(s_1, s_2, s_3) = 4$ (showing one possible solution).

Every window in this instance has the form $w(1, 2, i, j, 1, 1)$ where peg i in level 1 and peg j in level 2 represent characters of s_{k+1} and s_k respectively for even k (s_k' and s_{k+1}' respectively for odd k) and the two characters are equal. We set $w(1, 2, i, j, 1, 1) = -1$. In addition, we define $\Delta(l, a, a, b, b) = x$ if pegs a and b of level l correspond to duplicate characters. If we represent windows as edges going across between level 1 and level 2, and dependence among windows (the Δ terms) as arcs connecting duplicate characters in level 1 and in level 2, then every edge has weight -1 and every arc has weight x.

Since an arc (dependence) can only contribute to a solution when two corresponding edges (windows) touching its left and right are also included, the solution must contain chains that alternate in arcs and edges by starting and ending with an edge. Therefore, the only way to achieve a positive weight is by a chain of length $n + (n + 1)$, consisting of an alternation of n arcs and $n + 1$ edges, for a weight of $nx - (n + 1) > 0$. By the choice of x, any shorter such chain will have negative weight. Furthermore, this chain represents one character common to all strings. The optimal solution will consist of LCS such chains that are nested as shown in Fig. 4; this nesting guarantees that the common characters occur in the same order in all strings. Any approximation must contain k chains, for some integer $k \leq LCS$. Therefore, any constant approximation to OPT, say α, must score $k[nx - (n + 1)] \geq \alpha LCS[nx - (n + 1)]$; resulting in $k/LCS \geq \alpha$. This in turn means that we have a constant factor approximation for the Longest Common Subsequence problem, a contradiction unless P = NP. $\qquad \square$

While it is hard to even approximate the optimal solution, it is easy to determine the weight of a given solution S as described by $w(S)$ above using

$\Delta(l, a, b, c, d)$. This is useful in the context of sampling (see following section), when S has already been sampled, and can be done by computing a maximum no-overlap matching in each level l. Nevertheless, we do not fully implement this idea, instead we consider an easier approach.

To avoid the computation of conditional probabilities in $\Delta(l, a, b, c, d)$, we adopt for $a \leq b < c \leq d$ the following more practical definition (Theorem 2 is no longer true for this definition).

$$\Delta(l, a, b, c, d) = RT \ln P_l(free[a, d]) - RT \ln P_l(free[a, b]) - RT \ln P_l(free[c, d])$$

when it's positive and **no other intervals** in level l lie between $[a, b]$ and $[c, d]$. Thus we replace the individual free probabilities by one pertaining to the entire range. This will still capture dependence of two regions that separate from the stem of a hairpin loop of moderate size, as in the case of CopA-CopT with a window split in Fig. 5a. In general, when $\Delta(l, a, b, c, d)$ becomes sufficiently positive, it prevents such window splits from being detrimental to the total weight of the solution (the two windows of the split are no longer scored independently).

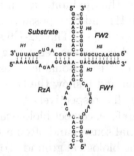

(i)
CopA 5' CGGUUUAAGUGGG...UUUCGUACUCGCCAAAGUUGAAGA...UUUUGCUU 3'
         |||||||||||| |||||||||||||||||||||||| |||||||
CopT 3' GCCAAAUUCACCC...AAAGCAUGAGCGGUUUCAACUUCU...AAAACGAA 5'

(ii)
CopA 5' CGGUUUAAGUGGG...UUUCGUACUCGCCAAAGUUGAAGA...UUUUGCUU 3'
         |||||||||||| ||||||||| |||||||
CopT 3' GCCAAAUUCACCC...AAAGCAUGAGCGGUUUCAACUUCU...AAAACGAA 5'

(a) The pairwise interaction of CopA-CopT: (i) computational prediction with artifact interactions due to the maximization nature of the problem, and (ii) the actual biologically known interaction [25], where the last window is dropped and the middle window is split (reversible kissing loop). (b) A circular 4-way junction construct of a hairpin ribozyme complex [36].

Fig. 5. Some structures that are not "optimal" in the computational sense.

Assume solution S has n intervals in level l given by $[a_i, b_i]$, sorted from left to right for $i = 1, \ldots, n$. The weight of the maximum no-overlap matching in level l can now be computed in linear time as $W_l(n)$ by dynamic programming, as shown below. The matching itself can be obtained by standard DP backtracking.

$$W_l(i) = \max \left[W_l(i-1), W_l(i-2) + \Delta(l, a_{i-1}, b_{i-1}, a_i, b_i) \right]$$

where $W_l(0) = W_l(1) = 0$, and $\Delta(l, a_{i-1}, b_{i-1}, a_i, b_i) = -\infty$ when not positive.

5 Realistic Biological Factors (The Need for Sampling)

Many biological factors affect the observed structure of interacting RNA molecules. For instance, reversible kissing loops (where some hydrogen bonds

(a)
```
  I1 3' UGUAUG
        |||
  U6 5' ACAGAGAUGAUC--AGC
            |||||  |||
  U2 3' AUGA-UGUGAACUAGAUUCG
            |||| ||||
  I2 5' UACUAACACC
```

(b)
```
  I1 3' UGUAUG
        |||
  U6 5' ACAGAGAUGAUCAGC
            |||||
  U2 3' AUGA-UGUGAACUAGAUUCG
            |||| ||||
  I2 5' UACUAACACC
```

(c)
```
  I1 3' UGUAUG
        |||
  U6 5' ACAGAGAUGAUC--AGC
            |||||  |||
  U2 3' AUGA-UGUGAACUAGAUUCG
            |||| ||||
  I2 5' UACUAACACC
```

(d)
```
  I1 3' UGUAUG
        |||
  U6 5' ACAGAGAUGAUCAGC
            |||||
  U2 3' AUGA-UGUGAACUAGAUUCG
            |||| ||||
  I2 5' UACUAACACC
```

Fig. 6. The yeast spliceosome with 4 RNAs (I1 and I2 are functionally independent stretches of the same much longer messenger RNA). (a) Helix Ia and helix Ib with both introns attached. (b) Helix Ia with both introns attached. (c) Helix Ia and helix Ib with I1 detached. (d) Helix Ia with I1 detached. Both (a) and (b) represent biologically correct structures. The actual folding within RNAs (thus the 3-way/4-way junction) is not shown.

of the interaction between hairpins unwind) [24] are generally not captured by energy minimization since a kissing loop is energetically more favorable than a partial one. We observe such artifacts within the pairwise interaction of CopA-CopT in *E. Coli*, as shown in Fig. 5a.

Figure 5b shows a circular 4-way junction construct of a hairpin ribozyme complex that exhibits a similar unwinding of two hydrogen bonds (located between H1 and H2), possibly due to higher order interactions with other parts of the structure [36]. The U2-U6 snRNA complex is a classical example where there is lack of consensus regarding whether the complex forms a 3-way or a 4-way junction (it is reasonable to assume that both structures co-exist [7,34,38,43]). Figure 6a and b show the two possibilities given by the formation of two helices or one helix, respectively.

Therefore, correct biological structures are not always "optimal" (in the computational sense), and often are not unique. Sub-optimal solutions are needed to cover the biological ground. To that end, once the bipartite interaction graph has been fixed, as described in Sect. 3.2, we generate alternative sub-optimal solutions using a sampling approach, and then cluster similar structures to obtain a set of distinct representative. Sampling has been successfully used for single RNAs (folding) and pairs of RNAs, e.g. [12,20,28,42], and we have previously explored it for multiple RNAs when the bipartite interaction graph is a path [32]. We extend the ideas of our sampling techniques to the general case below.

5.1 Gibbs Sampling with Metropolis Hastings (MCMC)

Gibbs sampling has been described in [17]. As a random variable, let S_{l_1,l_2} be a set of non-overlapping windows of the form $w(l_1,l_2,i,j,u,v)$, so S_{l_1,l_2} represents a valid interaction between RNA l_1 and RNA l_2. A Gibbs sampler works by sampling each random variable individually in order, conditioned on the current values of the other variables. In other words, we work with $P(S_{l_1,l_2}| \cup_{(l,l') \in G-(l_1,l_2)} S_{l,l'})$. Given a total order on the pairs $(l_1,l_2) \in G$, a new sample $\cup_{(l,l') \in G} S_{l,l'}$ is declared every time S_{l_1,l_2} has been sampled for all pairs $(l_1,l_2) \in G$ in order. The process repeats until we obtain the desired number of samples. We can assume that we start with $S_{l_1,l_2} = \emptyset$ for all pairs (l_1,l_2).

Under typical conditions of ergodicity [14], the Gibbs guarantee is that $\cup_{(l,l')\in G}S_{l,l'}$ is a sample from $P(\cup_{(l,l')\in G}S_{l,l'})$, which is not necessarily a known distribution, in contrast to $P(S_{l_1,l_2}|\cup_{(l,l')\in G-(l_1,l_2)}S_{l,l'})$, which may be reasonably constructed.

This is convenient because, conditioned on $\cup_{(l,l')\in G-(l_1,l_2)}S_{l,l'}$, the permissible windows of the form $w(l_1,l_2,i,j,u,v)$ are exactly those which when added will not make an overlap in $\cup_{(l,l')\in G}S_{l,l'}$. Therefore, we can assume that:

$$P(S_{l_1,l_2}|\cup_{(l,l')\in G-(l_1,l_2)}S_{l,l'}) \propto \begin{cases} 0 & \cup_{(l,l')\in G}S_{l,l'} \text{ contains an overlap} \\ e^{w(S_{l_1,l_2})/RT} & \text{otherwise} \end{cases}$$

where the exponential term is consistent with the standard Boltzman distribution for the interaction of RNAs l_1 and l_2, knowing that $w(S_{l_1,l_2})$ represents the negative of the energy multiplied by RT. We now describe a method to sample from $P(S_{l_1,l_2}|\cup_{(l,l')\in G-(l_1,l_2)}S_{l,l'})$ based the Metropolis-Hastings algorithm (also known as Markov Chain Monte Carlo MCMC).

Metropolis-Hastings Procedure: To sample from $P(S_{l_1,l_2}|\cup_{(l,l')\in G-(l_1,l_2)}S_{l,l'})$, we first make $S_{l_1,l_2} = \emptyset$ and drop all the windows of the form $w(l_1,l_2,i,j,u,v)$ that make an overlap when added to $\cup_{(l,l')\in G-(l_1,l_2)}S_{l,l'}$. We only work with the remaining windows of the form $w(l_1,l_2,i,j,u,v)$. We then construct a random sequence $S_{l_1,l_2}^0, S_{l_1,l_2}^1, \ldots$, where S_{l_1,l_2}^t is a set of non-overlapping windows of the form $w(l_1,l_2,i,j,u,v)$. This can be done with a Metropolis-Hastings strategy [18,27]: Given S_{l_1,l_2}^t, we randomly generate S_{l_1,l_2}^{t+1} with some proposal probability $Q(S_{l_1,l_2}^{t+1}|S_{l_1,l_2}^t)$, and either accept S_{l_1,l_2}^{t+1} with probability

$$\min\left\{1, \frac{Q(S_{l_1,l_2}^t|S_{l_1,l_2}^{t+1})}{Q(S_{l_1,l_2}^{t+1}|S_{l_1,l_2}^t)} \times \frac{e^{w(S_{l_1,l_2}^{t+1})/RT}}{e^{w(S_{l_1,l_2}^t)/RT}}\right\}$$

or reject it and let $S_{l_1,l_2}^{t+1} = S_{l_1,l_2}^t$.

It is well known and easy to show that such a strategy results in a Markov chain which converges to the desired probability distribution if the proposal chain $Q(S_{l_1,l_2}^{t+1}|S_{l_1,l_2}^t)$ satisfies $Q(S_{l_1,l_2}^{t+1} = y|S_{l_1,l_2}^t = x) > 0 \Leftrightarrow Q(S_{l_1,l_2}^{t+1} = x|S_{l_1,l_2}^t = y) > 0$; this also makes it irreducible [16].

A simple strategy is to make $Q(S_{l_1,l_2}^{t+1}|S_{l_1,l_2}^t)$ **uniform** among all the neighbors of S_{l_1,l_2}^t (including S_{l_1,l_2}^t itself), where a neighbor other than S_{l_1,l_2}^t can be obtained by one of the following three operations:

- a window $w(l_1,l_2,i,j,u,v) \in S_{l_1,l_2}^t$ is removed from S_{l_1,l_2}^t
- a window $w(l_1,l_2,i,j,u,v) \notin S_{l_1,l_2}^t$ that does not overlap in S_{l_1,l_2}^t is added to S_{l_1,l_2}^t
- a window $w(l_1,l_2,i,j,u,v) \in S_{l_1,l_2}^t$ is replaced by a window $w(l_1,l_2,i', j',u',v') \notin S_{l_1,l_2}^t$ that only overlaps with $w(l_1,l_2,i,j,u,v)$ in S_{l_1,l_2}^t.

Therefore, for every S_{l_1,l_2}^{t+1} that is a neighbor of S_{l_1,l_2}^t, $Q(S_{l_1,l_2}^{t+1}|S_{l_1,l_2}^t)$ is the inverse of the number of neighbors of S_{l_1,l_2}^t. This proposal probability defines an irreducible Markov chain since every pair of solutions can be reached from one another through a sequence of neighbors. We do not allow two adjacent windows $w(l_1, l_2, i, j, u, v)$ and $w(l_1, l_2, i - u, j - v, u', v')$ to co-exist (since together they represent one bigger window).

We perform 20 iterations of the Metropolis-Hastings algorithm **without** rejection. This allows us to start with some random solution. We then allow 50 iterations (with rejection) for the "burn-in" time of the Metropolis-Hastings algorithm. Finally, we generate 50 samples in 50 iterations and select one uniformly at random. We generate 1000 solutions (Gibbs samples) by repeating the entire procedure for each pair (l_1, l_2) in order.

5.2   Clustering the Samples

The sampled sub-optimal solutions are usually too many. In addition, many of them will be similar. Therefore, we use clustering to reduce their number. To cluster the samples, we first remove duplicates, so we only work with unique samples. We then sort the solutions to make the output of the clustering deterministic. Finally, we use hierarchical agglomerative clustering with complete linkage, and obtain the clusters by "cutting" the tree where distance between clusters is 1 (largest); we used a distance function similar to the one reported in [31], which was shown to be a metric. For completeness, we describe the distance function in the Appendix.

6   Experimental Results

Given the clusters, the solution with the largest weight (the best) in each cluster acts as a "representative" of the cluster. We sort the representatives of the clusters by decreasing weight. We consider the first k representatives, for a given k. To assess our approach, we repeat the experiment 200 times, which was verified to be enough for the percentage hits (defined below) to converge within $\pm 3\%$. Given a set of candidate structures in mind; for instance, Fig. 6 shows four candidates for the yeast spliceosome, we then count for each candidate the number of runs in which it is found among the first k representatives, as a percentage hit. We also compute the "rank" of each candidate, which is the index of it's first[3] representative (according to the sorted order) if found, averaged over the number of runs with a hit. Finally, we compute for each candidate the $F1$-score of its first representative when found, also averaged in the same way. After converting windows to base pairs, the $F1$-score is given by $\frac{2 \times precision \times recall}{precision + recall}$, where *recall* is defined as the number of base pairs in the representative that are also in the biologically correct structure, divided by the total number of base pairs in the latter, and *precision* is defined as the same but divided by the total number of base pairs in the representative.

---

[3] We use "first representative" because many solutions can represent the same candidate; for instance, a window can split in different ways, but we still refer to it as a window split.

We consider three settings with (a) **all** windows included (no filtering), (b) only **symmetric** windows $w(l_1, l_2, i, j, u, v)$ where $u = v$, and finally (c) the 500 windows $w(l_1, l_2, i, j, u, v)$ with the highest $Z^I_{l_1, l_2}(i - u + 1, i, j - v + 1, j)$ among all **bounded size** windows satisfying $\max(u, v) \leq 10$. **All tables show data for $k = 5$**, and each entry lists the percentage hit followed by the average rank and the average $F1$-score.

6.1 Structural Variation

The interaction of the U2-U6 complex in the spliceosome of yeast (shown in Fig. 6) has the pattern I1—U6—U2—I2 (the bipartite interaction graph is a path). The complex has been reported to have two distinct experimental structures, e.g. [38]. In one conformation, U2 and U6 interact to form a helix known as helix Ia. In another conformation, the interaction reveals a structure containing an additional helix, known as helix Ib. Section 5 describes possible underlying mechanisms that are responsible for this conformational switch.

We consider the set of four candidates in Fig. 6. The results are summarized in Table 1; we did not consider bounded size windows here because all given windows are already small in size.

Table 1. Results for the yeast spliceosome, percentage hit followed by rank and F1-score.

Class	All			Symmetric		
Helices Ia+Ib (as 2 symm. windows)	100	1	1	100	1	1
Helices Ia+Ib (as 1 window)	0.9	5	0.882	-	-	-
Helices Ia+Ib (as 2 symm. windows), I1 detached	100	2	0.914	100	2	0.914
Helices Ia+Ib (as 1 window), I1 detached	0	-	-	-	-	-
Helix Ia	100	3	1	100	3	1
Helix Ia, I1 detached	100	4	0.897	100	4	0.897

6.2 Artifact Interactions

The pairwise interaction of CopA-CopT (the bipartite interaction graph is simply an edge) is shown in Fig. 5a. Due to the optimization nature of our problem, it is sometimes possible to pick up interactions that are not biologically real; dropping these interactions from the solution would make it sub-optimal, even when preferred biologically, as described in Sect. 5. The last interaction window of CopA-CopT in Fig. 5a is an example of such an artifact.

For each of the three interaction windows in Fig. 5a, we consider whether the window is present, dropped, or split. We thus identify six classes of candidates based on presence/absence of windows and window splits, as shown in Table 2.

A real interaction given by window $w(l_1, l_2, i, j, u, v)$ is considered present if the solution contains a window $w(l_1, l_2, i', j', u', v')$ "in range" such that $[i' - u' +$

$1, i'] \subseteq [i - u + 1 - 3, i + 3]$ and $[j' - v' + 1, j'] \subseteq [j - v + 1 - 3, j + 3]$. Furthermore, if exactly two windows fall in that range, we consider them as a window split.[4] Typically, though we do not enforce it, such a window split is declared when the two windows happen to be treated as *dependent* (see Sect. 4).

The solution with the highest $F1$-score is characterized by a first window and a middle window split, with the last window dropped (biologically correct). This solution is revealed almost always with as few as two clusters ($k = 2$), and has a relatively small rank (at most 2) when $k \geq 5$. This is primarily attributed to the use of bounded size windows. However, even with a less stringent filtering of windows ("All" or "Symmetric"), the solution still shows up when the number of clusters k is high enough (see also Footnote 4).

Table 2. Results for CopA-CopT, percentage hit followed by rank and F1-score.

Class	All			Symmetric			Bounded size		
First, middle, last	92.9	2	0.729	99.3	1	0.764	8.5	4.3	0.53
First, middle split, last	7.6	3.5	0.723	88	3.2	0.758	100	1.2	0.701
First, middle, last dropped	87.9	2.1	0.794	97.7	2.3	0.836	73.2	4.7	0.639
First, middle split, last dropped	7.1	2.9	0.838	37	4.8	0.832	100	1.9	0.816
First split, middle, last	2.7	2.8	0.797	61.7	3.6	0.732	0	-	-
First split, middle, last dropped	1.8	4	0.708	20	4.9	0.795	0	-	-

6.3 Circular Interaction

The 4-way junction construct of a hairpin ribozyme complex is shown in Fig. 5b. This structure of four RNAs has a circular interaction Substrate—RzA—FW1—FW2—Substrate (the bipartite interaction graph is a cycle). We distinguish between three types of solutions: Type 1 is the "optimal" where H1 is predicted as a non-symmetric windows, adding a C-G pair and a U-A pair (thus a bulge in H1 on the RzA side). These additional pairs unwind in the actual structure possibly due to high order interactions with other parts of the structure [36]. Type 2 (the second optimal) is when H1 and H2 are predicted as one symmetric window that extends from the 5' end of the Substrate (and the 3' end of RzA) to the center of the 4-way junction (thus with the same additional base pairs reported above). Type 3 is the correct structure as shown in Fig. 5b. We used the same ±3 criteria for window boundaries as described in Sect. 6.2. When symmetric windows are considered, Type 3 will present itself as the second representative (sub-optimal). The results are summarized in Table 3.

---

[4] Since a single non-symmetric window may also represent a split, our percentage hit for window splits is lower than it should be with the no filtering option.

Table 3. Results for the 4-way junction construct in the hairpin ribozyme complex, percentage hit followed by rank and F1-score. T3 is the correct structure.

Class	All			Symmetric			Bounded size		
T1 (H1 as non-symm. window)	100	1	0.974	0	-	-	100	1	0.976
T2 (H1 & H2 as 1 symm. window)	15.6	2.8	0.965	100	1	0.976	0	-	-
T3 (H1 & H2 as 2 symm. windows)	79.7	4.6	0.986	100	2	1	100	4	1

6.4 Star Interactions

We include three more examples of ribozyme complexes from [39] and [22] where one RNA interacts with the rest of the RNAs, thus forming a star bipartite interaction graph (also a path when $m \leq 3$) as in Fig. 7. For the hammerhead ribozyme complex, the correct solution always shows up in the first cluster (see Table 4). For the HP-WT and HP-RJ ribozyme complexes, results are similar to those of the 4-way junction construct, and are shown in Tables 5 and 6, respectively.

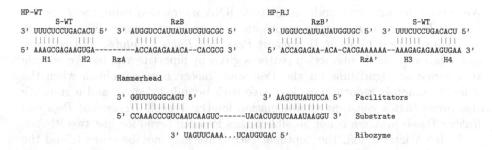

Fig. 7. Three more examples of ribozyme complexes [22,39].

Table 4. Results for the hammerhead ribozyme complex, percentage hit followed by rank and F1-score.

Class	All			Symmetric			Bounded size		
T1 (correct structure)	100	1	0,999	100	1	1	100	1	0.044
T2 (five windows)	63.8	2.3	0.954	98.7	2	0.959	0	-	-
T3 (three windows)	9.4	5	0.892	0.4	5	0.882	0	-	-

Table 5. Results for the hairpin ribozyme HP-WT complex, percentage hit followed by rank and F1-score. T3 is the correct structure.

Class	All			Symmetric			Bounded size		
T1 (H1 as non-symm. window)	100	1	0.955	0	-	-	100	1	0.955
T2 (H1 & H2 as 1 symm. window)	0	-	-	100	1	0.955	0	-	-
T3 (H1 & H2 as 2 symm. windows)	96.4	5	1	100	2	1	100	4	1

Table 6. Results for the hairpin ribozyme HP-RJ complex, percentage hit followed by rank and F1-score. T3 is the correct structure.

Class	All			Symmetric			Bounded size		
T1 (H3 as non-symm. window)	100	2	0.905	0	-	-	100	1	0.905
T2 (H3 & H4 as 1 symm. window)	100	1	0.95	100	1	0.95	0	-	-
T3 (H3 & H4 as 2 symm. windows)	0	-	-	100	3	1	0	-	-

7   Conclusion

We extend our previous work on multiple RNA interaction using the *Pegs and Rubber Bands* formulation to incorporate dependence among different parts of the interaction (*Weight Dependent Pegs and Rubber Bands*), and consider more elaborate RNA interaction patterns given by bipartite graphs. We provide approximation algorithms for the *Pegs and Rubber Bands* problem when the bipartite graph is a path, a cycle, a tree with bounded degree, and a star. We also prove that a constant approximation for the *Weight Dependent Pegs and Rubber Bands* problem is not possible unless P = NP (even for just two RNAs).

In RNA interaction, the "optimal" structure may not be correct, and the correct structure is not necessarily unique. A sampling approach for the above formulations successfully computes optimal and sub-optimal solutions that are truthful representations of the actual biological structures. For instance, it can provide several candidate structures when they exist, e.g. for the U2-U6 complex and its introns in the spliceosome of yeast; and identify structures that are biologically correct, but are not necessarily optimal in the computational sense, e.g. for CopA-CopT in *E. Coli* and several ribozyme complexes.

Appendix

Given a solution S, define $|S|$ as the number of windows in S, and let

$$w(l_1, l'_1, i_1, j_1, u_1, v_1), \ldots, w(l_{|S|}, l'_{|S|}, i_{|S|}, j_{|S|}, u_{|S|}, v_{|S|})$$

be the $|S|$ windows in the order defined by the partial order relation *follow* (from Sect. 2) extended to a total order in a deterministic way.

Each of these windows, say $w(l, l', i, j, u, v)$, defines the two intervals, $[i - u + 1, i]$ in level l and $[j - v + 1, j]$ in level l'. Consider the set of interaction intervals $I(S) = \sum_l I_l(S)$ to be ordered accordingly. Therefore,

$$I(S) = \{I_1, \ldots, I_{2|S|}\} = ([i_1 - u_1 + 1, i_1], [j_1 - v_1 + 1, j_1], \ldots$$

$$\ldots, [i_{|S|} - u_{|S|} + 1, i_{|S|}], [j_{|S|} - v_{|S|} + 1, j_{|S|}])$$

is an ordered set of $2|S|$ intervals. Let $L(S) = \{(l_1, l'_1), \ldots, (l_{|S|}, l'_{|S|})\}$ be an ordered set of $|S|$ pairs, where (l_i, l'_i) is the pair defining the ith window. Therefore, $L(S)$ means that we have the following set of pairwise interactions (not necessarily unique in terms of RNAs): RNA l_1 with RNA l'_1, RNA l_2 with RNA l'_2, ..., RNA $l_{|S|}$ with RNA $l'_{|S|}$. Two solutions that do not agree on this set are considered completely dissimilar; otherwise, their distance is given by the amount of overlap in their interaction intervals (as in the Jaccard metric [21]), hence the following definition of distance:

Given two solutions S_1 with $I(S_1) = \{I_1, I_2, \ldots\}$ and S_2 with $I(S_2) = \{T_1, T_2, \ldots\}$, the distance between S_1 and S_2 is

$$d(S_1, S_2) = \begin{cases} 1 - \frac{\sum_i |I_i \cap T_i|}{\sum_i |I_i \cup T_i|} & L(S_1) = L(S_2) \\ 1 & \text{otherwise} \end{cases}$$

where $\cap$ and $\cup$ represent the standard intersection and union operations on sets respectively, and intervals are treated as sets of integers.

Recall that a symmetric window $w(l_1, l_2, i, j, u, v)$ satisfies $u = v$ (and typically consists of u base pairs). When applying the distance function, a non-symmetric window is first converted to consecutive symmetric windows by maximizing the number of base pairs (but otherwise is still reported as a non-symmetric window in a given solution).

References

1. Ahmed, S.A., Mneimneh, S.: Multiple RNA interaction with sub-optimal solutions. In: Basu, M., Pan, Y., Wang, J. (eds.) ISBRA 2014. LNCS, vol. 8492, pp. 149–162. Springer, Cham (2014). https://doi.org/10.1007/978-3-319-08171-7_14

2. Ahmed, S.A., Mneimneh, S., Greenbaum, N.L.: A combinatorial approach for multiple RNA interaction: formulations, approximations, and heuristics. In: Du, D.-Z., Zhang, G. (eds.) COCOON 2013. LNCS, vol. 7936, pp. 421–433. Springer, Heidelberg (2013). https://doi.org/10.1007/978-3-642-38768-5_38

3. Alkan, C., Karakoc, E., Nadeau, J.H., Sahinalp, S.C., Zhang, K.: RNA-RNA interaction prediction and antisense RNA target search. J. Comput. Biol. 13(2), 267–282 (2006)

4. Alkan, F., et al.: RIsearch2: suffix array-based large-scale prediction of RNA-RNA interactions and siRNA off-targets. Nucleic Acids Res. 45(8), e60 (2017)

5. Andronescu, M., Zhang, Z.C., Condon, A.: Secondary structure prediction of interacting RNA molecules. J. Mol. Biol. 345(5), 987–1001 (2005)

6. Antonov, I., Marakhonov, A., Zamkova, M., Medvedeva, Y.: ASSA: fast identification of statistically significant interactions between long RNAs. J. Bioinform. Comput. Biol. **16**(01), 1840001 (2018)

7. Cao, S., Chen, S.-J.: Free energy landscapes of RNA/RNA complexes: with applications to snRNA complexes in spliceosomes. J. Mol. Biol. **357**(1), 292–312 (2006)

8. Chekuri, C., Kumar, A.: Maximum coverage problem with group budget constraints and applications. In: Jansen, K., Khanna, S., Rolim, J.D.P., Ron, D. (eds.) APPROX/RANDOM - 2004. LNCS, vol. 3122, pp. 72–83. Springer, Heidelberg (2004). https://doi.org/10.1007/978-3-540-27821-4_7

9. Chen, H.-L., Condon, A., Jabbari, H.: An $o(n^5)$ algorithm for MFE prediction of kissing hairpins and 4-chains in nucleic acids. J. Comput. Biol. **16**(6), 803–815 (2009)

10. Chitsaz, H., Backofen, R., Sahinalp, S.C.: biRNA: fast RNA-RNA binding sites prediction. In: Salzberg, S.L., Warnow, T. (eds.) WABI 2009. LNCS, vol. 5724, pp. 25–36. Springer, Heidelberg (2009). https://doi.org/10.1007/978-3-642-04241-6_3

11. Chitsaz, H., Salari, R., Sahinalp, S.C., Backofen, R.: A partition function algorithm for interacting nucleic acid strands. Bioinformatics **25**(12), i365–i373 (2009)

12. Ding, Y., Lawrence, C.E.: A statistical sampling algorithm for RNA secondary structure prediction. Nucleic Acids Res. **31**(24), 7280–7301 (2003)

13. Dirks, R.M., Bois, J.S., Schaeffer, J.M., Winfree, E., Pierce, N.A.: Thermodynamic analysis of interacting nucleic acid strands. SIAM Rev. **49**(1), 65–88 (2007)

14. Durbin, R., Eddy, S.R., Krogh, A., Mitchison, G.: Biological Sequence Analysis: Probabilistic Models of Proteins and Nucleic Acids, Chap. 11. Cambridge University Press, Cambridge (1998)

15. Fukunaga, T., Hamada, M.: RIblast: an ultrafast RNA-RNA interaction prediction system based on a seed-and-extension approach. Bioinformatics **33**(17), 2666–2674 (2017)

16. Gallager, R.G.: Discrete Stochastic Processes, Chap. 4. SECS, vol. 321. Springer, Boston (2012). https://doi.org/10.1007/978-1-4615-2329-1

17. Geman, S., Geman, D.: Stochastic relaxation, Gibbs distributions, and the Bayesian restoration of images. IEEE Trans. Pattern Anal. Mach. Intell. **6**, 721–741 (1984)

18. Hastings, W.K.: Monte Carlo sampling methods using Markov chains and their applications. Biometrika **57**(1), 97–109 (1970)

19. Huang, F.W., Qin, J., Reidys, C.M., Stadler, P.F.: Partition function and base pairing probabilities for RNA-RNA interaction prediction. Bioinformatics **25**(20), 2646–2654 (2009)

20. Huang, F.W., Qin, J., Reidys, C.M., Stadler, P.F.: Target prediction and a statistical sampling algorithm for RNA-RNA interaction. Bioinformatics **26**(2), 175–181 (2010)

21. Jaccard, P.: Etude comparative de la distribution florale dans une portion des Alpes et du Jura. Impr, Corbaz (1901)

22. Jankowsky, E., Schwenzer, B.: Oligonucleotide facilitators may inhibit or activate a hammerhead ribozyme. Nucleic Acids Res. **24**(3), 423–429 (1996)

23. Jiang, T., Li, M.: On the approximation of shortest common supersequences and longest common subsequences. SIAM J. Comput. **24**(5), 1122–1139 (1995)

24. Kolb, F.A., et al.: Progression of a loop-loop complex to a four-way junction is crucial for the activity of a regulatory antisense RNA. EMBO J. **19**(21), 5905–5915 (2000)

25. Kolb, F.A., et al.: An unusual structure formed by antisense-target RNA binding involves an extended kissing complex with a four-way junction and a side-by-side helical alignment. RNA **6**(3), 311–324 (2000)

26. Li, A.X., Marz, M., Qin, J., Reidys, C.M.: RNA-RNA interaction prediction based on multiple sequence alignments. Bioinformatics **27**(4), 456–463 (2011)

27. Metropolis, N., Rosenbluth, A.W., Rosenbluth, M.N., Teller, A.H., Teller, E.: Equation of state calculations by fast computing machines. J. Chem. Phys. **21**(6), 1087–1092 (1953)

28. Metzler, D., Nebel, M.E.: Predicting RNA secondary structures with pseudoknots by MCMC sampling. J. Math. Biol. **56**(1–2), 161–181 (2008)

29. Meyer, I.M.: Predicting novel RNA-RNA interactions. Curr. Opin. Struct. Biol. **18**(3), 387–393 (2008)

30. Mneimneh, S.: On the approximation of optimal structures for RNA-RNA interaction. IEEE/ACM Trans. Comput. Biol. Bioinform. (TCBB) **6**(4), 682–688 (2009)

31. Mneimneh, S., Ahmed, S.A.: Multiple RNA interaction: beyond two. IEEE Trans. Nanobiosci. **14**(2), 210–219 (2015)

32. Mneimneh, S., Ahmed, S.A.: Gibbs/MCMC sampling for multiple RNA interaction with sub-optimal solutions. In: Botón-Fernández, M., Martín-Vide, C., Santander-Jiménez, S., Vega-Rodríguez, M. (eds.) AlCoB 2016. LNCS, vol. 9702, pp. 78–90. Springer, Cham (2016). https://doi.org/10.1007/978-3-319-38827-4_7

33. Mückstein, U., Tafer, H., Hackermüller, J., Bernhart, S.H., Stadler, P.F., Hofacker, I.L.: Thermodynamics of RNA-RNA binding. Bioinformatics **22**(10), 1177–1182 (2006)

34. Newby, M.I., Greenbaum, N.L.: A conserved pseudouridine modification in eukaryotic U2 snRNA induces a change in branch-site architecture. RNA **7**(06), 833–845 (2001)

35. Pervouchine, D.D.: IRIS: intermolecular RNA interaction search. Genome Inform. Ser. **15**(2), 92 (2004)

36. Pinard, R., et al.: Functional involvement of G8 in the hairpin ribozyme cleavage mechanism. EMBO J. **20**(22), 6434–6442 (2001)

37. Salari, R., Backofen, R., Sahinalp, S.C.: Fast prediction of RNA-RNA interaction. Algorithms Mol. Biol. **5**(5), 5 (2010)

38. Sashital, D.G., Cornilescu, G., Butcher, S.E.: U2–U6 RNA folding reveals a group II intron-like domain and a four-helix junction. Nat. Struct. Mol. Biol. **11**(12), 1237–1242 (2004)

39. Schmidt, C., Welz, R., Müller, S.: RNA double cleavage by a hairpin-derived twin ribozyme. Nucleic Acids Res. **28**(4), 886–894 (2000)

40. Sun, J.-S., Manley, J.L.: A novel U2–U6 snRNA structure is necessary for mammalian mRNA splicing. Genes Dev. **9**(7), 843–854 (1995)

41. Tong, W., Goebel, R., Liu, T., Lin, G.: Approximating the maximum multiple RNA interaction problem. Theoret. Comput. Sci. **556**, 63 70 (2014)

42. Wei, D., Alpert, L.V., Lawrence, C.E.: RNAG: a new Gibbs sampler for predicting RNA secondary structure for unaligned sequences. Bioinformatics **27**(18), 2486–2493 (2011)

43. Zhao, C., et al.: Conformational heterogeneity of the protein-free human spliceosomal U2–U6 snRNA complex. RNA **19**(4), 561–573 (2013)

Max-Min Dispersion on a Line

Tetsuya Araki[1] and Shin-ichi Nakano[2(✉)]

[1] Tokyo Metropolitan University, Hachioji, Japan
[2] Gunma University, Kiryu, Japan
nakano@cs.gunma-u.ac.jp

Abstract. Given a set P of n locations on which facilities can be placed and an integer k, we want to place k facilities on some locations so that a designated objective function is maximized. The problem is called the k-dispersion problem.

In this paper we give a simple $O(n)$ time algorithm to solve the max-min version of the k-dispersion problem if P is a set of points on a line. This is the first $O(n)$ time algorithm to solve the max-min k-dispersion problem for the set of "unsorted" points on a line.

If P is a set of sorted points on a line, and the input is given as an array in which the coordinates of the points are stored in the sorted order, then by slightly modifying the algorithm above one can solve the dispersion problem in $O(\log n)$ time. This is the first sublinear time algorithm to solve the max-min k-dispersion problem for the set of sorted points on a line.

Keywords: Dispersion problem · Algorithm

1 Introduction

The facility location problem and many of its variants have been studied [9,10]. Typically, given a set of locations on which facilities can be placed and an integer k, we want to place k facilities on some locations so that a designated objective function is minimized. By contrast in the *dispersion problem*, we want to place facilities so that a designated objective function is maximized.

The intuition of the problem is as follows. Assume that we are planning to open several chain stores in a city. We wish to locate the stores mutually far away from each other to avoid self-competition. So we wish to find k locations so that the minimum distance between them is maximized. See more applications, including *result diversification*, in [7,15,16].

Now we define the *max-min k-dispersion problem*. Given a set P of n possible locations, and a distance function d for each pair of locations, (we assume that d is a symmetric nonnegative function satisfying $d(p,p) = 0$ for all $p \in P$) and an integer k with $k \ll n$, we wish to find a subset $S \subset P$ with $|S| = k$ such that the cost $cost(S) = \min_{\{u,v\} \subset S}\{d(u,v)\}$ is maximized. For convenience if $|S| = 1$ we regard $cost(S) = \infty$. This is the Max-Min version of the k-dispersion problem

© Springer Nature Switzerland AG 2018
D. Kim et al. (Eds.): COCOA 2018, LNCS 11346, pp. 672–678, 2018.
https://doi.org/10.1007/978-3-030-04651-4_45

[15, 18]. For the Max-Sum version see [4–8, 12, 14, 15], and for a variety of related problems see [4, 8]. The max-min k-dispersion problem is NP-hard even when the triangle inequality is satisfied [11, 18]. An exponential time exact algorithm for the max-min k-dispersion problem is known [3]. A geometric version of the problem in d-dimensional space can be solved in $O(kn)$ time for $d = 1$ (if the order of vertices in P on the line is given) and is NP-hard for $d = 2$ [18]. If the order of vertices in P on the line is given the running time for $d = 1$ was improved to $O(n \log \log n)$ [2] by the sorted matrix search method [13] (See a good survey for the sorted matrix search method in [1, Sect. 3.3]), then $O(n)$ [3] by a reduction to the path partitioning problem [13], as explained in Sect. 2. Ravi et al. [15] proved that the max-min k-dispersion problem cannot be approximated within any constant factor in polynomial time, and cannot be approximated within factor of two in polynomial time when the distance satisfies the triangle inequality, unless P = NP. They also gave a polynomial-time algorithm with approximation ratio two when the triangle inequality is satisfied.

In the paper we give a simple $O(n)$ time algorithm to solve the max-min k-dispersion problem if P is a set of unsorted points on a line. This is the first $O(n)$ time algorithm to solve the max-min k-dispersion problem for the set of unsorted points on a line. Then we consider the case if P is a set of sorted points on a line, and the input is given as an array in which the coordinates of the points are stored in the sorted order. We show one can solve the dispersion problem in $O(\log n)$ time, by slightly modifying the algorithm above. This is the first sublinear time algorithm to solve the max-min k-dispersion problem for the set of sorted points on a line.

The remainder of this paper is organized as follows. Section 2 gives an $O(n)$ time algorithm [3] to solve the dispersion problem where P is an ordered set of points on a line, by a reduction to the path partitioning problem. In Sect. 3 we design an $O(n)$ time simple algorithm to solve the dispersion problem when P is a set of "unsorted" points on a line. Section 4 gives an algorithm to solve the dispersion problem when P is a set of sorted points on a line. Finally Sect. 5 is a conclusion.

2 k-Dispersion for Sorted Points on a Line

In this section we show one can solve the k-dispersion problem in $O(n)$ time [3] if P is a set of points on a line and the order of the points on the line is given. The algorithm uses a reduction to the path partitioning problem [13], which can be solved in $O(n)$ time.

Let T be a tree in which each vertex has a nonnegative weight w, and k be an integer. The tree k-partitioning problem is to delete $k - 1$ edges in the tree so as to maximize the lightest weight of the remaining subtree. The tree k-partitioning problem can be solved in $O(n)$ time [13], where n is the number of vertices in the tree. If the input tree is a path then it is *the path k-partitioning problem*, and one can solve the path k-partitioning problem in $O(n)$ time using the algorithm for the tree.

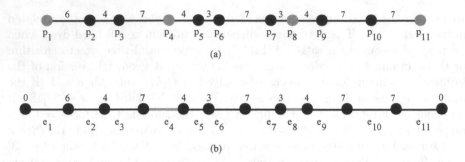

Fig. 1. (a) A 4-dispersion problem on a line, and (b) a path 3-partitioning problem on a line.

Given an instance (P, k) of the max-min k-dispersion problem where P is a set of points on a line, the order of the points in P on the line is given and $k \geq 3$, we can transform it to an instance $(P', k-1)$ of the path $(k-1)$-partitioning problem as follows [3]. First we construct a path $P' = (V', E')$. Assume $P = \{p_1, p_2, \cdots, p_n\}$ and the points appear in this order on the line. Define $V' = \{p_0', p_1', \cdots, p_n'\}$, $E' = \{e_i = (p_{i-1}', p_i') | p_i' \in V'\}$, $w(p_i') = d(p_i, p_{i+1})$ for each $i = 1, 2, \cdots, n-1$, and $w(p_0') = w(p_n') = 0$. See an example in Fig. 1. If the order of the points on the line is given, one can construct the path P' in $O(n)$ time. A solution of the max-min 4-dispersion problem in Fig. 1(a) is $\{p_1, p_4, p_8, p_{11}\}$ and its cost is 17. A solution of the path 3-partitioning problem in Fig. 1(b) is $\{e_4, e_8\}$ and its cost is 17. One can observe that a solution of a max-min k-dispersion problem contains $\{p_1, p_n\}$, and if a solution of a max-min k-dispersion problem is $\{p_1, p_n\} \cup \{p_{i_1}, p_{i_2}, \cdots, p_{i_{k-2}}\}$ then a solution of the path $k-1$-partitioning problem is $\{e_{i_1}, e_{i_2}, \cdots, e_{i_{k-2}}\}$.

One can solve the path k-partitioning problem in $O(n)$ time [13] so one can solve the max-min k-dispersion problem in $O(n)$ time.

Thus one can solve the dispersion problem in $O(n)$ time when all P are on a line and the order of the points in P on the line is given. However the algorithm in [13] is very complicated and hard to implement. In the next section we design a simple $O(n)$ time algorithm to solve the max-min k-dispersion problem even if the points on a line are not sorted.

3 k-Dispersion for Unsorted Points on a Line

In this section we design a simple $O(n)$ time algorithm to solve the k-dispersion problem for a constant k if P is a set of points on a line. Note that we do not assume that the order of the points on the line is given. The idea of our algorithm is a simple divide and conquer algorithm as follows.

Let P be a set of points on a horizontal line and p_ℓ and p_r are the leftmost point and the rightmost point in P. One can find p_ℓ and p_r in $O(n)$ time.

If $k = 1$ then a solution S of the 1-dispersion problem is $\{p_\ell\}$.

If $k = 2$ then the solution S of the 2-dispersion problem is $\{p_\ell, p_r\}$.

If $k = 3$ then the solution S is $\{p_\ell, p_s, p_r\}$ where p_s is the nearest point to the midpoint between p_ℓ and p_r. We can find p_s as follows.

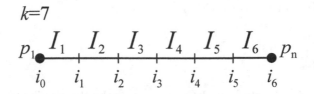

Fig. 2. Illustration of i_j and I_j for $k = 7$.

Let $i_0 = x(p_\ell)$ be the coordinate of p_ℓ on the line, $i_2 = x(p_r)$ the coordinate of p_r, and let i_1 be the coordinate of the midpoint between p_ℓ and p_r. Let I_1 be the interval $(x(i_0), x(i_1)]$, and I_2 be the interval $(x(i_1), x(i_2))$. The solution S consists of p_ℓ and p_r and exactly one more point in either I_1 or I_2. So by pigeonhole principle S has no point in either I_1 or I_2. Thus we have two cases.

Case 1: S has no point in I_1.

In this case, S consists of p_ℓ and the solution of the 2-dispersion problem for the points in $(i_1, i_2]$, which consists of (1) the nearest point to i_1 in I_2 and (2) p_r.

Case 2: S has no point in I_2.

In this case, S consists of p_r and the solution of the 2-dispersion problem for the points in $[i_0, i_1]$, which consists of (1) the nearest point to i_1 in I_1 and (2) p_ℓ.

We can generalize this method for a constant $k > 3$ as follows.

Let $i_0 = x(p_\ell)$, $i_{k-1} = x(p_r)$ and let $i_1, i_2, \cdots, i_{k-2}$ be the coordinates which evenly spaced on the line between p_ℓ and p_r.

Let I_j be the interval $(i_{j-1}, i_j]$ for $j = 1, 2, \cdots, k-2$, and I_{k-1} be the interval (i_{k-2}, i_{k-1}). See an example in Fig. 2. Clearly the cost of the solution is at most $|i_1 - i_0|$.

The solution for the k-dispersion problem consists of p_ℓ and p_r and exactly $k - 2$ points in (i_0, i_{k-1}). So by pigeonhole principle, S has no point in one of $I_1, I_2, \cdots,$ or I_{k-1}. Thus we have $k - 1$ cases as follows.

Case 1: S has no point in I_1.

In this case, S consists of (1) p_ℓ and (2) the solution of $(k - 1)$-dispersion problem for the points in $(l_1, l_{k-1}]$.

Case 2: S has no point in I_2.

In this case, S consists of (1) the solution of s-dispersion problem for the points in $[i_0, i_1]$ and (2) the solution of $(k - s)$-dispersion problem for the points in $(i_2, i_{k-1}]$ for some s with $1 \leq s \leq k - 1$.

Note that the cost of the solution is at most $|I_2|$.

Case 3: S has no point in I_3.

Similar to Case 2.

$\cdots$

Algorithm. Find-dispersion-on-a-line(P, k)

/* p_ℓ and p_r are the leftmost point and the rightmost point in P */
if $k = 1$ then
    $S = \{p_\ell\}$
    return S
end if
if $k = 2$ then
    $S = \{p_\ell, p_r\}$
    return S
end if
/* $i_0 = x(p_\ell)$, $i_{k-1} = x(p_r)$ and let $i_1, i_2, \cdots, i_{k-2}$ be the coordinates which evenly spaced on the line between p_ℓ and p_r */
/* $k \geq 3$ */
/* Case: S has no point in $I_1 = (i_0, i_1]$ */
Let P_R be the set of points of P in $(i_1, i_{k-1}]$.
$S_L = \{p_\ell\}$
$S_R =$ **Find-dispersion-on-a-line**($P_R, k - 1$)
$S = S_L \cup S_R$
/* Case: S has no point in $I_j = (i_{j-1}, i_j]$ for $j = 2, 3, \cdots, k - 2$ */
for $j = 2$ to $k - 2$ do
    Let P_L be the points of P in $[i_0, i_{j-1}]$.
    Let P_R be the points of P in $(i_j, i_{k-1}]$.
    for $s = 1$ to $k - 1$ do
        $S_L =$ **Find-dispersion-on-a-line**(P_L, s)
        $S_R =$ **Find-dispersion-on-a-line**($P_R, k - s$)
        if $cost(S_L \cup S_R) > cost(S)$ then
            $S = S_L \cup S_R$
        end if
    end for
end for
/* Case: S has no point in $I_{k-1} = (i_{k-2}, i_{k-1})$ */
Let P_L be the set of points of P in $[i_0, i_{k-2}]$.
$S_L =$ **Find-dispersion-on-a-line**($P_L, k - 1$)
$S_R = \{p_r\}$
if $cost(S_L \cup S_R) > cost(S)$ then
    $S = S_L \cup S_R$
end if
return S

Case $k - 2$: S has no point in I_{k-2}.
    Similar to Case 2.
Case $k - 1$: S has no point in I_{k-1}.
    In this case, S consists of (1) the solution of $(k - 1)$-dispersion problem for the points in $[i_0, i_{k-2})$ and (2) p_r.
    We (recursively) check all possible cases and choose the best one. See algorithm **Find-dispersion-on-a-line**. (If $|P| < k$ then clearly the subproblem has

no solution so we just discard such cases. For simplicity in the algorithm we omit such cases.)

Thus if we have the solution of at most $2k^2$ smaller child dispersion problems then we can solve the original k-dispersion problem.

We have the following theorem.

Theorem 1. *One can solve the max-min k-dispersion problem in $O(n)$ time when P is a set of unsorted n points on a line.*

Proof. Consider the tree structure of the recursive calls. Each inner node has at most $2k^2$ children and the height of the tree is at most k, so the number of inner node is at most $(2k^2)^k$. Before calling the children one needs to compute p_ℓ, p_r, P_L and P_R by scanning the list of unsorted points with buckets P_L and P_R. So it needs $O(n)$ time, where n is the number of points in current P. Thus each inner node needs $O(n)$ time except for the calls for its children. Therefore the total running time of the algorithm is $O((2k^2)^k n)$. Since k is a constant it is $O(n)$. $\square$

4 k-Dispersion for Sorted Points on a Line

If P is a set of sorted points on a line, and the input is given as an array in which the coordinates of the points are stored in the sorted order, then by slightly modifying the algorithm we can solve the dispersion problem in $O(\log n)$ time.

Before calling the children we need to compute p_ℓ, p_r, P_L and P_R. If the array is given we can compute the index of $x(p_\ell)$ and $x(p_r)$ in the array in $O(\log n)$ time by binary search. Also instead of computing P_L, we can compute the index of the coordinates of the leftmost and the rightmost points in P_L in the array in $O(\log n)$ time by binary search. Similar for P_R. Thus we can call each child with those indices of the leftmost and the rightmost points in P_L and P_R as arguments, instead of P_L and P_R. Now the running time is $O((2k^2)^k \log n)$, which is $O(\log n)$ since k is a constant.

5 Conclusion

In this paper we have designed a simple algorithm to solve the k-dispersion problem when P is a set of unsorted points on a line. This is the first $O(n)$ time algorithm to solve the max-min k-dispersion problem for the set of unsorted points on a line.

Then we show when P is a set of sorted points on a line and their coordinates are given in an array in the sorted order, a slightly modified version of the algorithm above runs in $O(\log n)$ time. This is the first sublinear time algorithm to solve the max-min k-dispersion problem for the set of sorted points on a line.

If P is a set of points on a circle and the order of the points on the circle is given, an $O(n)$ time algorithm to solve the k-dispersion problem is claimed [17]. Can we apply the method in this paper for the circle case?

Can we solve the problem efficiently if P is a set of the corner vertices on a convex polygon?

References

1. Agarwal, P., Sharir, M.: Efficient algorithms for geometric optimization. Comput. Surv. **30**, 412–458 (1998)
2. Akagi, T., Nakano, S.: Dispersion on the line. IPSJ SIG Technical reports, 2016-AL-158-3 (2016)
3. Akagi, T., et al.: Exact algorithms for the max-min dispersion problem. In: Chen, J., Lu, P. (eds.) FAW 2018. LNCS, vol. 10823, pp. 263–272. Springer, Cham (2018). https://doi.org/10.1007/978-3-319-78455-7_20
4. Baur, C., Fekete, S.P.: Approximation of geometric dispersion problems. In: Jansen, K., Rolim, J. (eds.) APPROX 1998. LNCS, vol. 1444, pp. 63–75. Springer, Heidelberg (1998). https://doi.org/10.1007/BFb0053964
5. Birnbaum, B., Goldman, K.J.: An improved analysis for a greedy remote-clique algorithm using factor-revealing LPs. Algorithmica **50**, 42–59 (2009)
6. Cevallos, A., Eisenbrand, F., Zenklusen, R.: Max-sum diversity via convex programming. In: Proceedings of SoCG 2016, pp. 26:1–26:14 (2016)
7. Cevallos, A., Eisenbrand, F., Zenklusen, R.: Local search for max-sum diversification. In: Proceedings of SODA 2017, pp. 130–142 (2017)
8. Chandra, B., Halldorsson, M.M.: Approximation algorithms for dispersion problems. J. Algorithms **38**, 438–465 (2001)
9. Drezner, Z. (ed.): Facility Location: A Survey of Applications and Methods. Springer, New York (1995)
10. Drezner, Z., Hamacher, H.W. (eds.): Facility Location: Applications and Theory. Springer, Heidelberg (2002)
11. Erkut, E.: The discrete p-dispersion problem. Eur. J. Oper. Res. **46**, 48–60 (1990)
12. Fekete, S.P., Meijer, H.: Maximum dispersion and geometric maximum weight cliques. Algorithmica **38**, 501–511 (2004)
13. Frederickson, G.: Optimal algorithms for tree partitioning. In: Proceedings of SODA 1991, pp. 168–177 (1991)
14. Hassin, R., Rubinstein, S., Tamir, A.: Approximation algorithms for maximum dispersion. Oper. Res. Lett. **21**, 133–137 (1997)
15. Ravi, S.S., Rosenkrantz, D.J., Tayi, G.K.: Heuristic and special case algorithms for dispersion problems. Oper. Res. **42**, 299–310 (1994)
16. Sydow, M.: Approximation guarantees for max sum and max min facility dispersion with parameterised triangle inequality and applications in result diversification. Mathematica Applicanda **42**, 241–257 (2014)
17. Tsai, K.-H., Wang, D.-W.: Optimal algorithms for circle partitioning. In: Jiang, T., Lee, D.T. (eds.) COCOON 1997. LNCS, vol. 1276, pp. 304–310. Springer, Heidelberg (1997). https://doi.org/10.1007/BFb0045097
18. Wang, D.W., Kuo, Y.-S.: A study on two geometric location problems. Inf. Process. Lett. **28**, 281–286 (1988)

Miscellaneous

Integer-Programming Bounds on Pebbling Numbers of Cartesian-Product Graphs

Franklin Kenter and Daphne Skipper[✉]

U.S. Naval Academy, Annapolis, MD 21402, USA
{kenter,skipper}@usna.edu

Abstract. Graph pebbling, as introduced by Chung, is a two-player game on a graph G. Player one distributes "pebbles" to vertices and designates a root vertex. Player two attempts to move a pebble to the root vertex via a sequence of pebbling moves, in which two pebbles are removed from one vertex in order to place a single pebble on an adjacent vertex. The pebbling number of a simple graph G is the smallest number π_G such that if player one distributes π_G pebbles in *any* configuration, player two can always win. Computing π_G is provably difficult, and recent methods for bounding π_G have proved computationally intractable, even for moderately sized graphs.

Graham conjectured that the pebbling number of the Cartesian-product of two graphs G and H, denoted $G \,\square\, H$, is no greater than $\pi_G \pi_H$. Graham's conjecture has been verified for specific families of graphs; however, in general, the problem remains open.

This study combines the focus of developing a computationally tractable method for generating good bounds on $\pi_{G \,\square\, H}$, with the goal of providing evidence for (or disproving) Graham's conjecture. In particular, we present a novel integer-programming (IP) approach to bounding $\pi_{G \,\square\, H}$ that results in significantly smaller problem instances compared with existing IP approaches to graph pebbling. Our approach leads to a sizable improvement on the best known bound for $\pi_{L \,\square\, L}$, where L is the Lemke graph. $L \,\square\, L$ is among the smallest known potential counterexamples to Graham's conjecture.

1 Introduction

Graph pebbling, first introduced by Chung [2] in 1989, can be described as a two-person game. Given a connected graph, G, the adversary chooses a *root* vertex r and an allocation of pebbles to vertices. In a *pebbling move*, player two chooses two pebbles at the same vertex, moves one to an adjacent vertex, and removes the other. Player two wins if she finds a sequence of pebbling moves

F. Kenter—Partially funded by the Naval Academy Research Council and by NSF Grant DMS-1719894.

D. Skipper—Partially funded by the Naval Academy Research Council.

D. Kim et al. (Eds.): COCOA 2018, LNCS 11346, pp. 681–695, 2018.
https://doi.org/10.1007/978-3-030-04651-4_46

that results in a pebble at the root vertex r. The *pebbling number* of graph G, denoted π_G, represents the fewest number of pebbles such that, regardless of the initial configuration and root given by the adversary, player two has a winning strategy.

The original motivation for graph pebbling was to solve the following number-theoretic problem posed by Erdős and Lemke [13]: "For any set of n integers, is there always a subset S whose sum is 0 mod n, and for which $\sum_{s \in S} \gcd(s, n) \leq n$?" Kleitman and Lemke [13] answered this question in the affirmative, and Chung [2] translated their technique into graph pebbling. Since then, the study of graph pebbling has proliferated in its own right, inspiring many applications and variations; for an overview see [12]. The translation of the number-theoretic problem to graph pebbling is nontrivial; the reader is referred to [6] for details.

Graham's Conjecture: This study is strongly motivated by famous open questions in pebbling regarding the Cartesian-product (or simply, "product") of two graphs, $G \square H$:

Conjecture 1 (Graham [2]). *Given connected graphs G and H,*

$$\pi_{G \square H} \leq \pi_G \pi_H.$$

Over time, Graham's conjecture has been resolved for specific families of graphs including products of paths [2], products of cycles [9,15], products of trees [15], and products of fan and wheel graphs [7]. It was also proved for specific products in which one of the graphs has the so-called 2-pebbling property [2,15,17].

One of the major hurdles in tackling Graham's conjecture is the lack of tractable computational tools. Milans and Clark [14] showed that the decision problem of determining whether $\pi_G < k$ is Π_2^P-complete. Numerically verifying Graham's conjecture for specific graphs has been extremely difficult; as a result, there does not appear to be a discussion, let alone a consensus, regarding whether or not the conjecture is true.

A more practical intermediate goal is to improve the bounds on the pebbling numbers of product graphs in general, and in special cases. To this end, Auspland, Hurlbert, and Kenter [1] proved that $\pi_{G \square H} \leq \pi_G (\pi_H + |V_H|)$. Since $\pi_H \geq |V_H|$ (the adversary wins by placing a single pebble on each vertex in $V_H \backslash r$), this result gets within a factor of two of Graham's conjecture: $\pi_{G \square H} \leq 2\pi_G \pi_H$.

When seeking a counterexample to Graham's conjecture, it is natural to focus on small graphs that do not possess the 2-pebbling property. The *Lemke graph*, L, shown in Fig. 1, was the first graph of this kind to be discovered [2]. Since then, infinite families of examples have been constructed [16], but the Lemke graph, with $|L| = \pi_L = 8$, is still among the smallest; it was verified in [5] that every graph with seven or fewer vertices has the 2-pebbling property. As suggested in [12], $L \square L$ is a potential counterexample to Graham's conjecture, and would be among the smallest.

Previous Work on Pebbling with IPs: In [11], Hurlbert introduces an integer-programming (IP) technique that uses the weights of spanning trees to

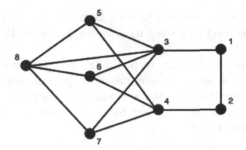

Fig. 1. The Lemke graph, L: a minimum-sized graph without the 2-pebbling property.

bound the pebbling number of a graph. By applying his technique to $L \mathbin{\square} L$, Hurlbert provides strong evidence that $\pi_{L \mathbin{\square} L} \leq 108$, which is an improvement on $\pi_{L \mathbin{\square} L} \leq 2(\pi_L)^2 = 128$, but is still quite far from Graham's conjectured bound of $\pi_{L \mathbin{\square} L} \leq (\pi_L)^2 = 8^2 = 64$. Hurlbert's technique is theoretically sound, but does not scale well to $L \mathbin{\square} L$, which has 64 vertices, 208 edges, and more than 10^{50} spanning trees. In his full model, every subtree corresponds to a unique constraint, so solving the full model is not an option. Still, Hurlbert makes progress by carefully selecting a subset of constraints to include.

Another computational challenge is that, to fully vet a potential upper bound on π_G, the bound must be verified for all possible roots. For example, Hurlbert restricted his search to root candidates that are the most likely to have large root-specific pebbling numbers, but did not verify the bound by exhausting all 64 choices of $r \in V_{L \mathbin{\square} L}$.

Further examples of IP applied to pebbling include [10], in which a targeted IP finds the exact pebbling number of $C_5 \mathbin{\square} C_5$, and [4], which extends Hurlbert's weight method to give bounds on edge counts in Class-0 graphs (in which $\pi_G = |V_G|$).

Our Contribution: We present a novel, computationally scaleable, IP approach to bounding $\pi_{G \mathbin{\square} H}$. Our method improves the best known upper bound on $\pi_{L \mathbin{\square} L}$: from 108 to 91. We leverage the symmetry that is inherent in Cartesian-product graphs via *partial pebbling*, which we introduce as a means of modeling at the level of granularity of H, rather than at the full size of $G \mathbin{\square} H$. In addition to requiring smaller, more computationally viable IP models, our method requires verification by calculating the bound only for each vertex of H, rather for than for all of the vertices of $G \mathbin{\square} H$.

Unlike previous IP approaches to pebbling, we incorporate "2-pebbling" (i.e., pebbling two or more pebbles at a time for a reduced cost). Intuitively, this seems to be a critical ingredient for improving bounds on $\pi_{G \mathbin{\square} H}$, considering the importance of the "2-pebbling property" in previous results. In fact, the inclusion of 2-pebbling greatly improved our bound on $\pi_{L \mathbin{\square} L}$.

The IP solver that we use for our computations, `Gurobi` [8], solves all of our models to integer optimality within a couple of seconds, so there is reason to hope that our approach may be scalable to even larger Cartesian-product graphs. It is also likely that our modestly-sized models can be solved by an exact rational IP solver such as [3].[1]

In Sect. 2, we give an overview of pebbling and introduce partial pebbling. In Sect. 3, we develop a general model based on partial pebbling for bounding $\pi_{G \,\square\, H}$. In Sect. 4, we describe the application of our model to $L \,\square\, L$ and provide computational results. In Sect. 5, we make some closing observations and discuss future directions.

2 Graph Pebbling

In this section, we set the stage by introducing the graph-theoretic notation that we use, as well as concepts and notation from graph pebbling. For a more detailed presentation of graph pebbling, see [12]. We also introduce partial pebbling, which serves as the foundation for our IP model.

Throughout, we assume that our graphs are simple, undirected, and connected. We use the notation $G := (V_G, E_G)$, to indicate a graph with vertex set V_G and edge set E_G. For simplicity, we use $|G| := |V_G|$ to denote the vertex-count of G, and $V_G := \{1, 2, \ldots, |G|\}$ to denote its vertex set. Further, $i \sim_G j$ indicates that $\{i, j\} \in E_G$, and $D_G(i, j)$ represents the graph theoretic distance between vertices i and j in G. If the context is clear, we simply write $i \sim j$ or $D(i, j)$, respectively. The *diameter* of a graph is the maximum distance between any pair of its vertices. We use Δ_G to denote the maximum degree of G.

The *Cartesian-product* (also called the *box-product*, *weak-product*, or *xor product*) graph of G and H, denoted $G \,\square\, H$ has vertex set $V_{G \,\square\, H} := V_G \times V_H = \{(i, j) : i \in \{1, 2, \ldots, |G|\}, j \in \{1, 2, \ldots, |H|\}\}$. For edges of $G \,\square\, H$, we have $(g, h) \sim_{G \,\square\, H} (g', h')$ if $g = g'$ and $h \sim_H h'$, or $h = h'$ and $g \sim_G g'$. For example, $K_2 \,\square\, K_2 = C_4$, the 4-cycle, and $K_2 \,\square\, C_4 = Q_3$, the cube. Although there are other common graph products, in this document "graph product" and "product of graphs" always refers to the Cartesian-product (until Sect. 5 where other product graphs are briefly mentioned).

A natural way to conceptualize $G \,\square\, H$ is to think of it as the graph H (which we call the *frame* graph), with a copy of G at each vertex. For $j \in V_H$, G_j denotes the copy of G at vertex j, so that $V_{G_j} = V_G \times j$. We say that G_j is the *G-slice* of $G \,\square\, H$, or if the context is understood, G_j is simply a *slice*. Similarly, $G \,\square\, H$ has H-slices of the form H_i, for $i \in V_G$. For $i \in V_G$ and $j_1 \neq j_2 \in H$, we have $(i, j_1) \sim_{G \,\square\, H} (i, j_2)$ if and only if $j_1 \sim_H j_2$. In this case, we say that slices G_{j_1} and G_{j_2} are adjacent. Also, the distance between G_{j_1} and G_{j_2} is $D_H(j_1, j_2)$.

2.1 Pebbling G

A *configuration* (or *pebbling configuration*) on G is a vector of nonnegative integers $c = (c_1, c_2, \ldots, c_{|G|})$, where c_i represents the number of pebbles placed on

---

vertex $i \in V_G$. The *support* of c is the set of vertices assigned at least one pebble by c, $\{i \in V_G : c_i > 0\}$. The *size of* c refers to $\|c\|_1$, the number of pebbles allocated by c. The *support-size of* c is $|\{i \in V_G : c_i > 0\}|$.

A *pebbling move* consists of removing two pebbles from one vertex and adding one pebble to an adjacent vertex. More generally, an α-pebbling move consists of removing 2^α pebbles from vertex v and adding one pebble to vertex w, where $D(v, w) = \alpha$. We say that a configuration c is *solvable* if, given any choice of *root* $r \in V_G$, there exists a (possibly empty) sequence of pebbling moves such that the resulting configuration has at least one pebble at r. Otherwise, we say that c is *unsolvable*. The pebbling number of G, denoted π_G, is the lowest positive integer k such that all configurations of size k (i.e., $\|c\|_1 = k$) are solvable.

One variant of the pebbling game is to require the second player to move two pebbles to the root in order to win. A graph G has the *2-pebbling property* if any configuration c of size $2\pi_G - s + 1$ is 2-solvable (i.e., two pebbles can reach the root), provided the support size of c is s. In essence, the 2-pebbling property guarantees that each additional vertex of support provides a discount of one pebble when pebbling twice. We use $\pi_2(G, s)$ to denote the minimum number of pebbles such that if a configuration has support size s, the second player is guaranteed to win.

2.2   Partial Pebbling $G \,\square\, H$

The modest size of our IP formulation results from modeling at the level of *partial pebblings* of box-product graphs. A *partial configuration* with respect to G on $G \,\square\, H$ allocates pebbles to the slices G_j of $G \,\square\, H$, rather than to individual vertices. In a partial configuration $\tilde{c} = (\tilde{c}_1, \tilde{c}_2, \ldots, \tilde{c}_{|H|})$, the nonnegative integer $\tilde{c}_j$ represents the number of pebbles distributed to slice G_j. When a root (r_G, r_H) of $G \,\square\, H$ is chosen, G_{r_H} is known as the G-*root slice*, and H_{r_G} the H-*root slice*.

There is a canonical map from full to partial configurations on $G \,\square\, H$, $\phi \colon \mathbb{Z}_{\geq 0}^{|G||H|} \to \mathbb{Z}_{\geq 0}^{|H|}$. A partial configuration $\tilde{c}$ is unsolvable if there exists some $c \in \phi^{-1}(\tilde{c})$ that is unsolvable. To prove that $\pi_{G \,\square\, H} \leq k$ using partial configurations, one must show that every partial configuration $\tilde{c}$ of size k is solvable.

Pebbling moves cannot always be assumed using the information in $\tilde{c}$ alone. For instance, if $\tilde{c}_j = |G|$, it could be that there is one pebble per vertex of G_j, so that no pebbling move is possible in G_j. On the other hand, if $c_j > |G|$, then at least one vertex in the slice has 2 or more pebbles, and a pebbling move can be made. We say a slice is k-*saturated*, or has a *saturation level of* k, when $\tilde{c}_i \geq k|G|$. If a slice is $(k-1)$-saturated, the pigeonhole principle guarantees that even one "extra" pebble (beyond the first $(k-1)|G|$) implies the existence of a k-*stack*, a collection of k pebbles on a single vertex. This concept is formalized in Lemma 2.

We take this nuance a step further when we capture the support size of each slice (in level II of the model). In this case, we may assume the existence of a k-stack on some vertex without necessarily having $(k-1)$-saturation on the slice.

Finally, we call a collection of π_G pebbles on the vertices of a slice G_j a *set*. The number of sets within a slice is its *set count*; the *total set count* of $G \square H$ is the sum of the set counts over its G-slices.

3   IP Model for Bounding $\pi_{G \square H}$

3.1   Strategy

Let $\mathscr{U}$ be the set of all unsolvable partial configurations on $G \square H$. We describe a relaxation $\mathscr{R}$ of $\mathscr{U}$ ($\mathscr{U} \subseteq \mathscr{R} \subseteq \mathbb{Z}_{\geq 0}^{|H|}$), so that

$$\pi_{G \square H} = 1 + \max\{\|\tilde{c}\|_1 : \tilde{c} \in \mathscr{U}\}$$
$$\leq 1 + \max\{\|\tilde{c}\|_1 : \tilde{c} \in \mathscr{R}\}.$$

So our partial pebbling IP takes the form of $\max\{\|\tilde{c}\|_1 : \tilde{c} \in \mathscr{R}\}$, where $\mathscr{R} \supseteq \mathscr{U}$ is the intersection of $\mathbb{Z}$ with a polytope described by our linear *pebbling constraints*.

Each of our pebbling constraints models a successful pebbling strategy. In other words, any partial configuration that violates a given pebbling constraint may be successfully solved via the strategy modeled by the constraint. In this way, we know that every partial configuration not in $\mathscr{R}$ is solvable, resulting in the relaxation of $\mathscr{U}$ that we require.

We classify each constraint as level I through level III based on how many variables it requires. As we move down the levels, we obtain tighter relaxations of $\mathscr{U}$, and tighter bounds on $\pi_{G \square H}$. At a high level,

level I uses only partial configurations, set counts, and saturation levels;
level II introduces support sizes of slices and k-stacks;
level III introduces 2-pebbling discounts.

We use this classification system with an eye towards applying our model to larger problem instances. The constraint levels are designed so that they build on one another; e.g., the constraints of level II require all of the variables used in level I, but not vice versa. Therefore, if the entire model does not scale well to a particular problem instance, we may choose to exclude level III constraints for a more manageable IP that is still self-consistent.

Our pebbling constraints model three high-level pebbling strategies, which we label A, B, and C,

strategy A: π_H pebbles reach the H-root slice, H_{r_G};
strategy B: π_G pebbles reach the G-root slice, G_{r_H};
strategy C: a targeted attack on the root (r_G, r_H) results from accumulating a 2^k-stack on (r_G, j) in G_j, where $D_H(j, r_H) = k$.

Pebbling constraints rely on counting the number of pebbles *required* at some slice or vertex (to carry out a pebbling strategy), versus the number of pebbles that are *available* there, resulting in the standard form,

$$available + 1 \leq required,$$

where *available* and *required* are restricted to integer values. A partial configuration $\tilde{c}$ violates the constraint (and $\tilde{c}$ is certifiably solvable by the modeled
strategy), only if *available* $\geq$ *required*, i.e., if there are enough pebbles to carry
out the strategy. In order to maintain a relaxation of $\mathcal{U}$ (and thus a valid upper-
bound on $\pi_{G \,\square\, H}$), when exact values are not possible, we use a lower bound
for *available* and an upper bound for *required*.

Pebbling constraints are labeled according to level and strategy. For example,
level I constraint I.A.2(α, β) is the second constraint that models strategy A, and
requires parameters α and β.

3.2 Parameters

The data required for our IP model are as follows.
Parameters:

$\|G\|$	:=	vertex-count of G;
π_G	:=	pebbling number of G;
π_H	:=	pebbling number of H;
r_H	:=	the index (in V_H) of the G-slice that contains the root;
$D(i,j)$	:=	the distance in H between i and j, for all pairs $i, j \in V_H$;
$diam_H$	:=	the diameter of H;
MAX	:=	$2\pi_G\pi_H$, a "big" constant.

Index sets:

V_H	:=	$\{1, 2, \ldots, \|H\|\}$, vertices of H;
$\mathcal{S}$	:=	$\{0, 1, \ldots, \pi_H - 1\}$, possible set counts of a G-slice;
$\mathcal{T}$	:=	$\{0, 1, \ldots, \lfloor \frac{\pi_G\pi_H - 1}{\|G\|} \rfloor\}$, possible saturation levels of a G-slice;
$\mathcal{L}$	:=	$\{1, 2, \ldots, diam_H\}$, possible distances between vertices in H.

We require much more information about H, which we consider the "frame" of
$G \,\square\, H$ in our logic. Since we do not incorporate complete information about
G, we may not expect our model to obtain a tight bound on $\pi_{G \,\square\, H}$. We choose
this course of action intentionally in order to develop a *computationally tractable*
approach to improving the bound on $\pi_{G \,\square\, H}$.

We define MAX as the simple upper bound on $\pi_{G \,\square\, H}$ from [1]. Many
of our constraints are enforced or relaxed based on the value of some binary
variable(s). In these constraints, MAX, or some small multiple of MAX, is
used as the standard "big M" (from integer programming).

The parameter r_H, which is required for strategies B and C, is not introduced
until level II. Thereafter, one must run the model for all values of $r_H \in V_H$, taking
the maximum root-specific bound as the upper bound on $\pi_{G \,\square\, H}$.

3.3 Decision Variables

In this section, we list our decision variables, sorted by level and numeric type,
and discuss a few interesting cases. We manage the behavior of all decision
variables with linear constraints, which are not included in this document. In
the table below, the index j ranges over all $j \in V_H$.

Level I

$\mathbb{Z}_{\geq 0}$	$\tilde{c}_j$	$:=$ number of pebbles assigned to G_j
	set_j	$:=$ set count of G_j: $\left\lfloor \frac{\tilde{c}_j}{\pi_G} \right\rfloor$
	$extra_j$	$:=$ number of extra pebbles on G_j: $\tilde{c}_j \bmod \pi_G$
	sat_j	$:=$ saturation level of G_j: $\left\lfloor \frac{\tilde{c}_j - extra_j}{|G|} \right\rfloor$
	$pair_j$	$:=$ number of pairs in $extra_j$: $\left\lfloor \frac{extra_j}{2} \right\rfloor$
0/1	$x_{t,j}$	$:=$ 1 iff the saturation level of G_j is at least t, for $t \in T$
	y_s	$:=$ 1 iff the total set count is at least s, for $s \in S$

Level II

$\mathbb{Z}_{\geq 0}$	$support_j$	$:=$ support size of $\tilde{c}$ over G_j
	$stack_{\ell,j}$	$:=$ number of 2^ℓ-stacks in G_j, for $\ell \in L$ (lower bound)

Level III

$\mathbb{Z}_{\geq 0}$	$n2peb_j$	$:= \pi_2(G, support_j)$
	nHr_j	$:=$ no. of pebbles that can reach H_{r_G} in G_j (lower bound)
0/1	$can2peb_j$	$:=$ 1 iff G_j can be 2-pebbled (i.e., iff $\tilde{c}_j \geq n2peb_j$)
	***	*(Problem-specific variables to model 2-pebbling in $G \,\square\, H$)*

It is important to note that the pebbles counted by $extra_j$ are "extra" in the sense that they contribute neither to the set count nor the saturation level at G_j.

We calculate the lower bounding $stack_{\ell,j}$ as

$$stack_{\ell,j} = \left\lceil \frac{c_j - (2^\ell - 1)support_j}{2^\ell} \right\rceil,$$

because the maximum possible number of surplus pebbles (that are not part of a 2^ℓ-stack in G_j) is $(2^\ell - 1)support_j$.

Our model is intended for products of graphs that do not have the 2-pebbling property, so there will be problem-specific rules regarding the calculation of $n2peb_j$. For example, the Lemke graph has a special case for 2-pebbling when the support is 5. We handle these special cases with problem-specific binary variables, as described in Sect. 4.

The variable nHr_j takes advantage of the 2-pebbling discount on the first two pebbles. Within G_j, 2 pebbles reach H_{r_G} with the first $n2peb_j$ pebbles assigned to G_j; beyond that, one pebble per set reaches H_{r_G} (as a lower bound).

3.4 Pebbling Constraints

Level 1. Level I constraints model pebbling strategy A, which is based on Lemma 1. The first constraint, I.A.1, is a direct consequence Lemma 1, while the rest of the strategy A constraints model the accumulation of π_H sets via pebbling moves between slices.

Lemma 1. *Any partial configuration that has a total set count of at least π_H is solvable.*

Proof. We can use a set in G_j to move a pebble to any vertex of G_j, and in particular to (r_G, j), the vertex in the intersection of G_j and H_{r_G}. If there are π_H sets across all slices, then we can move π_H pebbles into H_{r_G} in order to pebble the root vertex (r_G, r_H) within H_{r_G} □

Theorem 1. *The inequality*

(I.A.1)
$$\sum_{i \in V_H} set_i + 1 \leq \pi_H$$

is valid for $\mathcal{U}$.

The next lemma relates the number of pebbles on a slice to the distance of between-slice moves that are possible from that slice and follows easily by the pigeonhole principle.

Lemma 2. *If there are at least $(2^\ell - 1)|G| + 1$, pebbles on slice G_j (or equivalently, G_j is $(2^\ell - 1)$-saturated with at least one extra pebble), then it is possible to make an ℓ-pebbling move from G_j.*

Constraints I.A.2$(\eta, \alpha_1, \alpha_2)$ model the accumulation of one or more additional sets by shifting pebbles from a single slice to α_ℓ slices a distance ℓ away, for $\ell = 1, 2$. In particular, for $v \in H$, this pebbling strategy uses extra pebbles in G_v, along with pebbles from $\alpha_1 + \alpha_2 - \eta$ sets in G_v, in order to accumulate $\alpha_1 + \alpha_2$ new sets, increasing the total set count by η. These constraints could be extended to include pebbling moves of greater distances, but in the setting of $G = H = L$, our current focus, a 3-pebbling move from G_v would exhaust an entire set.

Theorem 2. *Fix $\eta, \alpha_1, \alpha_2 \in \mathbb{Z}_{\geq 0}$, with $\alpha_1, \alpha_2 \leq \Delta_H$ and $1 \leq \eta \leq \alpha_1 + \alpha_2 \leq \pi_H - 1$. Let $v \in V_H$. For $\ell = 1, 2$: let $A_\ell \subset V_H$ with $|A_\ell| = \alpha_\ell$, and $D(v, j) = \ell$, for all $j \in A_\ell$. Then, for $\chi = 2^k - 1$, where $k = \max\{\ell \in \{1, 2\} : \alpha_\ell > 0\}$,*

(I.A.2$(\eta, \alpha_1, \alpha_2)$)
$$(\alpha_1 + \alpha_2 - \eta)|G| + extra_v + 1 \leq$$
$$\sum_{\ell \in \{1,2\}} 2^\ell \left(\alpha_\ell \pi_G - \sum_{j \in A_\ell} extra_j\right)$$
$$+ MAX(1 - x_{\chi + (\alpha_1 + \alpha_2 - \eta), v})$$
$$+ MAX(1 - y_{\pi_H - \eta})$$

is valid for $\mathcal{U}$.

Proof. We may assume that the total set count is at least $\pi_H - \eta$ and that G_v is at least $(\chi + \alpha_1 + \alpha_2 - \eta)$-saturated; otherwise, the constraint is relaxed by one of the MAX terms.

Due to the saturation level at G_v, by Lemma 2 there are at least $((\alpha_1 + \alpha_2 - \eta)|G| + extra_v)$ pebbles *available* in G_v to be used in k-pebbling moves. It costs 2^ℓ pebbles to make an ℓ-pebbling move, so the number of pebbles *required* in G_v to complete one set per G_j, for $j \in A_1 \cup A_2$, is $\sum_{\ell \in \{1,2\}} 2^\ell (\alpha_\ell \pi_G - \sum_{j \in A_\ell} extra_j)$.

If the constraint is violated, enough pebbles are available to carry out the strategy: up to $(\alpha_1 + \alpha_2 - \eta)|G|$ pebbles may used from G_v in order to create $\alpha_1 + \alpha_2$ new sets, one in each of the G_j, for $j \in A_1 \cup A_2$. Since $|G| \leq \pi_G$, no more than $(\alpha_1 + \alpha_2 - \eta)$ sets are disassembled at G_v. This strategy increases the total set count by at least η. □

The next set of constraints require a $K_{\alpha,\beta}$ subgraph of H, with disjoint vertex sets A and B. The constraints model the collection of one additional set at each slice G_j, for $j \in B$, using only extra pebbles from the slices G_i, for $i \in A$. This strategy increases the total set count by $|B| = \beta$.

Theorem 3. *Fix positive integers* $\alpha, \beta \leq \Delta_H$, *with* $\alpha + \beta \leq |H|$. *Suppose A and B are disjoint subsets of V_H of sizes α and β, respectively, with $i \sim_H j$ for all $i \in A$, $j \in B$. Then*

$$(\text{I.A.3}(\alpha, \beta)) \qquad \begin{aligned} \sum_{i \in A} pair_i + 1 &\leq \sum_{j \in B}(\pi_G - extra_j) \\ &+ MAX \cdot (\alpha - \sum_{i \in A} x_{1,i}) \\ &+ MAX \cdot (1 - y_{(\pi_H - \beta)}) \end{aligned}$$

is valid for $\mathscr{U}$.

Proof. If the constraint is enforced, the total set count is at least $\pi_H - \beta$, and G_i is at least 1-saturated, for each $i \in A$. The sum $\sum_{i \in A} pair_i$ counts the number of pairs of pebbles in slices indexed by A that can used to 1-pebble to neighboring slices indexed by B (while decreasing neither the saturation level nor the set count at the slices indexed by A). The summation $\sum_{j \in B}(\pi_G - extra_j)$ captures the cumulative number of pebbles required at the slices indexed by B to build a complete set in each. When the constraint is violated, there are enough pebbles to increase the total set count by β. □

Level II. Starting in level II, we use partial-configuration support sizes to get tighter lower bounds on k-stack counts in slices. For the first time, we include constraints that employ strategy B, which require r_H, the index of the G-root slice. The first category B constraint is very straight-forward.

Theorem 4. *The following equation is valid for $\mathscr{U}$:*

$$(\text{II.B.1}) \qquad\qquad set_{r_H} = 0.$$

The next strategy B constraint, II.B.2, models using stacks of pebbles at all non-root slices to build a set in G_{r_H}. Note that in strategy A, it is important to keep track of the number of sets, and to carefully track the increase in total set count that results from a violated A constraint. In strategy B, however, the total set count does not matter, and we use as many pebbles as possible from each slice to build a set in the G-root slice.

Theorem 5. *For $\ell \in \mathcal{L}$, let A_ℓ be the set of vertices that are a distance of ℓ away from r_H in H. Then the inequality*

$$\text{(II.B.2)} \qquad \left(\sum_{\ell \in \mathcal{L}} \sum_{j \in A_\ell} stack_{\ell,j} \right) + 1 \ \leq \ \pi_G - extra_{r_H},$$

is valid for $\mathcal{U}$.

Proof. The variable $stack_{\ell,j}$ provides a lower bound on the number of ℓ-pebbling moves that are possible from G_j. If the constraint is violated, enough pebbles can reach G_{r_H} to complete a π_G set.   □

The previous constraint effectively requires that each stack can pebble to the target slice independently and does not allow for the collection of "loose" pebbles along the way. The next constraint, II.A.4, allows for this possibility along a path (of slices) of length α terminating at G_{r_H}. The proof follows by induction on α and is omitted for brevity.

Theorem 6. *Fix $\alpha \in \mathcal{L}$. Let P be a path of length α in H with $r_H = p_0 \sim_P p_1 \cdots \sim_P p_\alpha$. Then, the inequality*

$$\text{(II.A.4)} \qquad \left(\sum_{i=1}^{\alpha} 2^{\alpha-i}(c_{p_i} - |G|) \right) + 1 \ \leq \ 2^\alpha(\pi_G - c_{r_H}) + 2^\alpha MAX(1 - x_{1,p_\alpha})$$

is valid for $\mathcal{U}$.

Level III. Level III constraints incorporate 2-pebbling discounts. The definition of $n2peb_j$ depends on the graph G. We will discuss the modeling of $n2peb_j$ for the case when $G = L$ in Sect. 4.

The first level III constraint employs strategy A with a 2-pebbling discount "hidden" in nHr_j. The simple proof is omitted for brevity.

Theorem 7. *The inequality*

$$\text{(III.A.4)} \qquad \sum_{j \in V_H} nHr_j + 1 \ \leq \ \pi_H,$$

is valid for $\mathcal{U}$.

Our only strategy C constraint is also the only pebbling constraint that requires modification for $G = L$. The strategy requires pebbling into a slice and then applying a 2-pebbling discount there. Usually 2-pebbling is less expensive with a larger support; however, $\pi_2(L, 5) = \pi_2(L, 4) + 1$. When $G = L$, we include the binary variable $four_v$, which adds a one-pebble penalty for 2-pebbling if $support_v = 4$, to account for the possibility that the support increases to 5 when pebbles move into G_v.

Theorem 8. *Fix* $\alpha \in \mathcal{L}$. *Let* $v \in V_H$ *with* $D(v, r_H) = \alpha$. *Then*

(III.C.1(α))
$$\left(\sum_{\ell \in \mathcal{L}} \sum_{j \in V_H : D_{v,j} = \ell} stack_{\ell,j} \right) + \tilde{c}_v + 1 \leq n2peb_v + [four_v] + (2^\alpha - 2)\pi_G,$$

is valid for $\mathscr{U}$. *("$four_v$" is a binary variable that is required when* $G = L$.)

Proof. If $D(j, v) = \ell$, $stack_{\ell,j}$ counts the number of pebbles that G_j can contribute to G_v. Hence, the total number of pebbles *available* at G_v is the expression on the left side (without the "+1"). If the constraint is violated, enough pebbles are available at G_v to carry out the following strategy: Use $n2peb_v$ (+$four_v$, if $G = L$) of the pebbles to move 2 pebbles to (r_G, v), and use $(2^\alpha - 2)\pi_G$ pebbles to move an additional $2^\alpha - 2$ pebbles to (r_G, v). With 2^α pebbles at (r_G, v), an α-pebbling move places one pebble at (r_G, r_H). □

We return to strategy A for our last set of constraints. The strategy requires an α-star subgraph of H with central vertex v, and comes into play when G_v is highly pebbled. It models pebbling from G_v to build sets in each of α neighboring slices, then finishing out a collection or π_H pebbles in H_{r_G} by pebbling to (r_G, v) within G_v with a 2-pebbling discount.

Theorem 9. *Fix* $\alpha \leq \min\{\Delta_H, \pi_H - 2\}$ *in* $\mathbb{Z}_{\geq 0}$. *For any* $v \in V_H$, *let* $A \subset V_H$, *such that* $|A| = \alpha$, *and* $D(v, j) = 1$, *for all* $j \in A$. *The following is valid for* $\mathscr{U}$:

(III.A.5(α)) $\tilde{c}_v + 1 \leq \left(2 \sum_{j \in A} (\pi_G - extra_j) \right) + n2peb_v + (\pi_H - (2 + \alpha))\pi_G.$

Proof. If the constraint is violated, enough pebbles are *available* in G_v to carry out the following strategy. Keep $n2peb_v + (\pi_H - (2 + \alpha))\pi_G$ pebbles within G_v. Note that $n2peb_v > \pi_G \geq |G|$, so G_v is at least 1-saturated. Use $2(\sum_{j \in A} \pi_G - extra_j)$ pebbles in G_v to transfer $\sum_{j \in A}(\pi_G - extra_j)$ pebbles into the slices indexed by A to complete an additional α sets. With a 2-pebbling discount $n2peb_v + (\pi_H - (2 + \alpha))\pi_G$ is equivalent to $\pi_H - \alpha$ sets in G_v. □

4   Case Study: $L \,\square\, L$

In this section, we describe the specialization of the 2-pebbling constraints to $L \,\square\, L$, and describe the exact collection of constraints that we apply to bound $\pi_{L \,\square\, L}$. Finally, we provide the results of our model when applied to $L \,\square\, L$.

4.1   Model Refinements for $L \,\square\, L$

If G is 2-pebblable, we have the simple formula $n2peb_j = 2\pi_G - support_j + 1$. Otherwise, $n2peb_j$ must be handled on a case-by-case basis, depending on the 2-pebbling table for G. A straight-forward calculation generates the values of $\pi_2(L, s)$, as in Table 1. Note that the 2-pebbling property holds for L, except when $s = 5$ (see [12]). Even when $s = 5$, only 14 pebbles are required to pebble any root twice (as opposed to $2\pi_L = 16$). In order to handle 2-pebbling in the case of $G = L$, we include the following additional 0/1 variables in level III:

Table 1. 2-pebbling table for L

Support-size, s	8	7	6	5	4	3	2	1
$\pi_2(L, s)$	9	10	11	14	13	14	15	16

$$four_j := 1 \text{ iff } support_j = 4; \qquad five_j := 1 \text{ iff } support_j = 5;$$
$$lfour_j := 1 \text{ iff } support_j \le 4; \qquad lfive_j := 1 \text{ iff } support_j \le 5;$$
$$gfour_j := 1 \text{ iff } support_j \ge 4; \qquad gfive_j := 1 \text{ iff } support_j \ge 5.$$

The variables $four_i$ and $five_i$ signal special cases of $\pi_2(L, s)$. The remaining variables are necessary for encoding $four_i$ and $five_i$ using linear constraints. For $G = L$, we have,

$$n2peb_j = 2(\pi_G) - support_j + 1 + 2(five_j),$$

to correct for the special case when the support size is 5. As noted in Sect. 3, $\pi_2(L, 5) > \pi_2(L, 4)$ is a special case that we must handle carefully in constraints III.C.1(α), which is the only place where $four_j$ is used in the model.

4.2   Bounding $\pi_{L \,\square\, L}$

Even for fixed parameter values, many of our constraints actually describe a class of constraints, one for each subset of H with certain characteristics. Adding all constraints for all possible parameter values seems to be an impractical choice. Instead, we construct a much smaller model for $\pi_{L \,\square\, L}$ by iteratively selecting and adding constraints that target the current optimal partial configuration.

In Table 2, we list the constraints added, sorted by level. We also list the results of the model after each level of constraints has been added. After level II constraints have been added, there is a three-way tie for the largest root-specific partial configuration. The results hint that the 2-pebbling constraints of level III are very important in this context.

The level III bound arises as the maximum over all level III root-specific bounds, listed in Table 3. We conclude that for the Lemke graph, L, $\pi_{L \,\square\, L} \le 91$.

Table 2. IP constraints and bounds obtained by level

	Constraints	Maximal configuration(s)	r_H	Bound
Level I	I.A.1	(63, 4, 4, 6, 6, 6, 6, 7)	N/A	104
	I.A.2(1,1,0)			
	I.A.2(1,2,0)			
	I.A.2(1,1,1)			
	I.A.2(2,2,0)			
	I.A.2(2,2,1)			
	I.A.2(1,0,2)			
	I.A.3(2,3)			
	I.A.3(3,2)			
+ Level II	II.B.1	(23, 23, 5, 5, 7, 23, 7, 7)	5	101
	II.B.2.	(15, 15, 5, 5, 23, 7, 23, 7)	6	
	II.B.3(3)	(15, 15, 5, 5, 23, 23, 7, 7)	7	
+ Level III	III.A.4	(40, 13, 13, 5, 5, 5, 5, 4)	8	91
	III.A.5(2)			
	III.C.1(1)			
	III.C.1(2)			

Table 3. Level III root-specific bounds

r_H	1	2	3	4	5	6	7	8
Bound	90	87	84	76	82	82	82	91

5 Conclusion

We have developed an integer-programming model for bounding the pebbling numbers of Cartesian-product graphs. When applied to $L \square L$ our model significantly improves upon the previous best bound on $\pi(L \square L)$. While this work does not directly provide insight on the veracity of Graham's Conjecture, it offers both a new strategy and a new computational tool for addressing it.

We are encouraged by the computational lightness of our model when applied to $L \square L$. There may be computational headroom for incorporating more granularity in our model, resulting in better bounds on $L \square L$ and other Cartesian-product graphs. Another improvement would be to automate constraint selection, so the model may be more easily applied to different box-product graphs.

The concepts introduced in this study may also be altered to provide bounds on the pebbling numbers of different types of graph products, such as cross-product and strong-product graphs [1].

References

1. Asplund, J., Hurlbert, G., Kenter, F.: Pebbling on graph products and other binary graph constructions. Australas. J. Comb. **71**(2), 246–260 (2017)
2. Chung, F.: Pebbling in hypercubes. SIAM J. Discrete Math. **2**(4), 467–472 (1989)
3. Cook, W., Koch, T., Steffy, D.E., Wolter, K.: An exact rational mixed-integer programming solver. In: Günlük, O., Woeginger, G.J. (eds.) IPCO 2011. LNCS, vol. 6655, pp. 104–116. Springer, Heidelberg (2011). https://doi.org/10.1007/978-3-642-20807-2_9
4. Cranston, D.W., Postle, L., Xue, C., Yerger, C.: Modified linear programming and class 0 bounds for graph pebbling. J. Comb. Optim. **34**(1), 114–132 (2017). https://doi.org/10.1007/s10878-016-0060-6
5. Cusack, C.A., Green, A., Bekmetjev, A., Powers, M.: Graph pebbling algorithms and Lemke graphs. Submitted
6. Elledge, S., Hurlbert, G.H.: An application of graph pebbling to zero-sum sequences in Abelian groups. Integers **5**, A17 (2005)
7. Feng, R., Kim, J.Y.: Pebbling numbers of some graphs. Sci. China Ser. A Math. **45**(4), 470–478 (2002). https://doi.org/10.1007/BF02872335
8. Gurobi Optimization Inc.: http://www.gurobi.com
9. Herscovici, D.S.: Graham's pebbling conjecture on products of cycles. J. Graph Theory **42**(2), 141–154 (2003). https://doi.org/10.1002/jgt.10080
10. Herscovici, D.S., Higgins, A.W.: The pebbling number of $C_5 \times C_5$. Discrete Math. **187**(1), 123–135 (1998). https://doi.org/10.1016/S0012-365X(97)00229-X
11. Hurlbert, G.: A linear optimization technique for graph pebbling. arXiv e-prints, January 2011
12. Hurlbert, G.: Graph pebbling. In: Gross, J.L., Yellen, J., Zhang, P. (eds.) Handbook of Graph Theory, pp. 1428–1449. Chapman and Hall/CRC, Kalamazoo (2013)
13. Lemke, P., Kleitman, D.: An addition theorem on the integers modulo n. J. Number Theory **31**(3), 335–345 (1989). https://doi.org/10.1016/0022-314X(89)90077-2
14. Milans, K., Clark, B.: The complexity of graph pebbling. SIAM J. Discrete Math. **20**(3), 769–798 (2006). https://doi.org/10.1137/050636218
15. Snevily, H.S., Foster, J.A.: The 2-pebbling property and a conjecture of Graham's. Graphs Comb. **16**(2), 231–244 (2000). https://doi.org/10.1137/050636218
16. Wang, S.: Pebbling and Graham's conjecture. Discrete Math. **226**(1), 431–438 (2001). https://doi.org/10.1016/S0012-365X(00)00177-1
17. Wang, Z., Zou, Y., Liu, H., Wang, Z.: Graham's pebbling conjecture on product of thorn graphs of complete graphs. Discrete Math. **309**(10), 3431–3435 (2009). https://doi.org/10.1016/j.disc.2008.09.045

On the Complexity of Resilience
for Aggregation Queries

Dongjing Miao[1,2] and Zhipeng Cai[2(✉)]

[1] School of Computer Science and Technology, Harbin Institute of Technology,
Harbin 150001, Heilongjiang, China
[2] Department of Computer Science, Georgia State University,
Atlanta, GA 30303, USA
zcai@gsu.edu

Abstract. Resilience, as an potential explanation of a specified query, plays a fundamental and important role in query explanation, database debugging and error tracing. Resilience decision problem is defined on a database d, given a boolean query q where $q(d)$ is initially *true*, and an integer k, it is to determine if there exists a tuple set Δ such that size of Δ is no more than k and query result $q(d \oplus \Delta)$ becomes *false*, where $\oplus$ can be deletion or insertion operation. Results of this problem on relational algebraic queries have been showed in previous work. However, we revisit this decision problem on aggregation queries in the light of the parametric refinement of complexity theory, provide new results. We show that, this problem is intractable on nested COUNT and SUM query both under data complexity and parametric complexity.

Keywords: Resilience · Aggregation · Database
Parameterized complexity

1 Introduction

Resilience of a given query q with respect to a database d is defined as a set Δ of facts in d, whose deletion will results in a boolean query getting *false* which initially is *true*. Formally, its decision problem can be defined as follow,

RESILIENCE DECISION

INPUT   Given database d, a natural number $k > 0$, a boolean
           query q where $q(d)$ is *true*.

OUTPUT   *yes*, if there exists an subset $\Delta \subseteq d$ of size k such that
           query result $q(d \oplus \Delta)$ is *false*.

Note that, since the conjunctive query is of monotone, operation $\oplus$ can be written as set minus '$-$'. This is a fundamental decision problem in the study of database debugging, cleansing, error tracing, query result explanation and many other applications, since the most important and common task in these applications is to answer the question that given some partial result T of a query

© Springer Nature Switzerland AG 2018
D. Kim et al. (Eds.): COCOA 2018, LNCS 11346, pp. 696–706, 2018.
https://doi.org/10.1007/978-3-030-04651-4_47

q on a database d, why the result T happens here (*why-provenance*). Typically, there are two ways to define the '*why*', as identified in [3], way of source side effect free (ssef) and way of view side effect free (vsef). Intuitively, given a source database d, a query q, its materialized view $q(d)$ and a testing result $t \subseteq q(d)$, the former is to find an r of size k such that $q(d - r) \subseteq q(d) - t$, while the later is to find an r such that $q(d - r) = q(d) - t$. Nevertheless, previous work only focus on the relational algebraic operations. In this paper, we intent to study the complexity of this problem where query q is defined by aggregation operations including COUNT, SUM, MAX, MIN.

Example 1. Suppose an example of resilience in *influential author finding*. Consider an academic database of the research community including two relations, Author(aid) records the basic personal information of each user, and Co-author(aid, cid) records his co-author *cid* for each person *aid* known in this database. There is also a view V defined as an simple aggregation query (SQL-like statement), to show that all the authors who has at least a coauthor.

SELECT *aid*
FROM Author,Co-author,Author
WHERE Author.*aid* = Co-author.*aid* AND Co-author.*cid* = Author.*aid*
GROUP-BY *aid* HAVING COUNT(∗) > 0

	aid cid	
aid	a1 a2	**aid**
a1	a2 a1	a1
Author: a2 Co-author:	a1 a3	*V*: a2
a3	a3 a1	a3
a4	a1 a4	a4
	a4 a1	

First, we want to check if there are some tuples in the database whose absence will result in the query result of V becoming *false*, that are two alternative set of facts in source data d, (a) fact '($a1$)' in Author or (b) all the facts in Co-author. In this case, either of the two facts is the Δ for V.

But if we also want to check if there is a single fact whose absence will make it *false*, then we can simple adjust the requirement of the size of resilience.

As shown by Freire et al. [13], RES can be reduced polynomially to the two above decision problems (ssef and vsef). This is to say that, RES is a more fundamental part of the two problems, the lower bound of RES will also dominate the lower bound of the two problems.

There is still lack of results on the cases with aggregation queries. Therefore, we want to study the complexity of *Aggr*-RES in this paper.

AGGREGATION RESILIENCE DECISION

INPUT   Given database d, a natural number $k > 0$, a boolean
query q defined by aggregate operators COUNT, SUM,
MAX, MIN query q where $q(d)$ is *true*.

OUTPUT   *yes*, if there exists an subset $\Delta \subseteq d$ of size k such that
query result $q(d - \Delta)$ is *false*.

The previous studies provides the pictures of the classical complexity results of these two ways, we summary them in the table above. In total, the previous results is mainly on the classical computational complexity. In this case, the complexity of query languages proposed by Chandra and Merlin has been next to expressibility one of the main preoccupations of database theory ever since two four decades ago. It has been noted rather early that, when considering the complexity of evaluating a query on an instance, one has to distinguish between two kinds of complexity metric: *Data complexity* is the complexity of evaluating a query on a database instance, when the query is fixed, and we express the complexity as a function of the size of the database. The other one is called combined complexity, considers both the query and the database instance as input variables; The *combined complexity* of a query language is typically one exponential higher than data complexity. Of the two, data complexity is somehow regarded as more meaningful and relevant to database if only consider query evaluation.

Table 1. Different cases of view, source side effect free/resilience decision problem

Citations	Complexity: query class
Buneman et al. [3] (vsef)	NP-complete: CQ without *selection*
	PTime: CQ without *projection* and *self-join*
Cong et al. [9] (vsef)	NP-complete: Conjunctive query without *key-preserving*
Freire et al. [13] (ssef/resilience)	NP-complete: Conjunctive query containing *triad*
	PTime: sjf linear CQ
This paper (ssef/resilience)	PTime: non-nested COUNT or SUM query, nested MAX or MIN query
	NP-complete: nested aggregation query without join, projection and union under active domain
	W[1]-hard: the same as queries above

There have been some complexity results on the view side effect free problem [3,8,9,21–23]. On the data complexity of deletion propagation, Kimelfeld et al. [22] showed the dichotomy '*head domination*' for every conjunctive query without self-join, deletion propagation is either APX-hard or solvable (in polynomial time) by the unidimensional algorithm. For functional dependency restricted version, it is radically different from the case without functional dependency (FD), they also showed the dichotomy '*fd-head domination*' [21]. For multiple or group deletion [23], they especially showed the trichotomy for group deletion a more general case including *level-k head domination* and so on; On the combined complexity of deletion propagation, [8,9] showed the variety results for different combination of relational algebraic operators. At the same time, [26] studied the functional dependency restricted version deletion propagation problem and showed the tractable and intractable results on both data and combined complexity aspects.

Besides research on view side effect, there are previous work on source side effect decision problem [3,8,9,13], they show some complexity results on the source side-effect problem on both data and combined complexity. Basically, Freire et al. [13] show that for RES when query is defined by conjunctive query is PTime if q is a conjunctive query without structure of *triad*, NP-complete otherwise. They also extend the dichotomy condition '*triad*' into a more general one '*fd-induced triad*' for case with presence of functional dependencies. All the previous results about view and source side effect free problem showed that, for most cases, the deletion propagation is hard due to the huge searching space.

Additionally, a related topic the view update problem in database has been extensively investigated for more than three decades in the database community, which is stated as follows: given a desired update to a database view, what update should be performed towards the source tables to reflect this update to the view [1,2,10,11,20]. Generally, previous works mainly focus on identifying the condition to make the update unique, and studying under the identified condition how to carry out the update. These works are only effective for very restricted circumstances where there is a unique update Δd to a source database d that will cause a specified update to the view $q(d)$. In practice, an update to d is not always unique. Therefore, an alternative is to find a minimum update to d resulting in the specified update to $q(d)$, which is a more practical task of view propagation. Our results can be applied in some related application, such as complexity analysis in private protecting [5,17,28], reverse detection [14,19,25], error tracing [4,6,7,29] in social network and influence study [15,16,18,24] in wireless sensor networks.

Therefore, in this paper, we want to study the complexity of AGGREGATION RESILIENCE DECISION problem, including data complexity and parametric aspect, since the running time in which n is not raised to a power that depends on q, that is the dependence on n is only permitted as a the n^c where c is a constant independent of the query, and this is the typical paradigm of the parameterized complexity theory.

2 Preparation

We first give a necessary introductions.

Database. A database schema is a finite set $\{R_1, \ldots, R_m\}$ of distinct relations. Each relation R_i has r_i attributes, say $\{A_1, \ldots, A_{r_i}\}$, where r_i is the arity of R_i. Each attribute A_j has a corresponding domain $dom(A_j)$ which is a set of valid values. A domain $dom(R_i)$ of a relation R_i is a set $dom(A_1) \times \cdots \times dom(A_{r_i})$. Any element of $dom(R_i)$ is called a fact. A database d can be written as $\{D; R_1, \ldots, R_m\}$, representing a schema over certain domain D, where D is a set $dom(R_1) \times \cdots \times dom(R_m)$.

Boolean Database Queries. A boolean query q is a function mapping database d to $\{true, false\}$. We limit our study inside the first order query language, so that queries can be written by a certain fragment of the first order query language. We follow the metric using in [27], where the two parameters are, separately, the number of variables x appearing in the query q, and the size of query q which is the number of atoms in the query. The relationship between both parameters is that the query size is no more than the number variables.

Therefore, if the complexity class of the latter case should belong to the class of the former case for our decision problem. However, both are between the data and combined complexity.

W-hierarchy. In parameterized complexity theory, for the problems probably not in f.p.t, W-hierarchy was introduced by Downey and Fellows, which is analogous to the polynomial hierarchy in the classical complexity theory. It contains a series of complexity classes of parametrized problems. They are jointly called the W-hierarchy, which classifies the problems under the parameterized perspective [12]. Concretely, classes in W-hierarchy beyond FPT (in which, every problem can be solved in time of $f(k) \cdot n^c$) are W[i] where $i = 1, 2, \ldots$, and limits to two classes W[P] and W[SAT]. It means that problem in W[i] is at least harder than W[j] if $i \geq j$.

Problems. In the following part, several hard problems necessary to build reduction from and should be introduced here.

VERTEX COVER
INPUT   Given graph $G(V, E)$, an natural number t.
OUTPUT   Yes, if there exists an independent set $C \subseteq V$ of size t such that every vertex of C is not adjacent to any other one of C.

INDEPENDENT SET
INPUT   Given graph $G(V, E)$, an natural number t.
OUTPUT   Yes, if there exists an independent set $I \subseteq V$ of size t such that every pair of vertices $u, v \in I$, $(u, v) \notin E$.

Multicolored Independent Set

INPUT   Given graph $G(V, E)$, an natural number t, and a vertex coloring $\gamma : V \to \{1, \cdots, m\}$.

OUTPUT   Yes, if there exists an independent set $I \subseteq V$ of size t such that every pair of vertices $u, v \in I$, $(u, v) \notin E$ and $\gamma(u) \neq \gamma(v)$.

As we know, Vertex Cover problem is NP-complete, and so does Independent Set problem, while the Multicolored Independent Set problem is W[1]-hard.

3   Results of Aggregation Queries

In this section, we examine the parameterized complexity of different fragments of first-order query on number of variables and query size.

We first show the simple cases, followed by the hard cases even for parameterized complexity.

Theorem 1. RES *is* PTime *for non-nested* COUNT *and* SUM *queries, nested* MIN *and* MAX *queries with selection and projection.*

The proof for nested MAX and MIN queries is based on the observations that (a) nested query can be transformed into a non-nested query with a conjunction condition of MAX and MIN, (b) there are only unary operation in the transformed query, and (c) sorting can be done in polynomial time.

Theorem 2. Aggregation Resilience Decision *is* NP-hard *for non-nested* COUNT, SUM, MAX, MIN *queries with join and union.*

Proof. We build a simple reduction from Vertex Cover to *Aggr*-RES. For given instance $\langle G(V, E), t \rangle$. An corresponding instance $\langle d, q, k \rangle$ can be constructed for *Aggr*-RES as follows.

Database d. We start with two relations $V(x)$ and $E(x, y)$ with respect to V and E of G, denoting each vertex v_i as a unary tuple $(i) \in V$, each edge $(v_i, v_j) \in E$ with two tuples (i, j) and (j, i) inside $E(i, j)$;

Query q. We define a bi-levels nested aggregation query q as

SELECT x
FROM $V(x), E(x, y), V(y)$
GROUP-BY x HAVING COUNT$(y) > 0$

It is easy to verify that query q is *true* initially.

Integers k. At last, we set the integers as $k = t$.

It is easy to see that the correctness follows immediately.

To apply this result into other cases, we can perform a simple transformation. Adding an additional attribute to V, say $V(x, z)$, and set value on z of all tuples as 1, then rewrite the query as

```
SELECT x
FROM V(x, i), E(x, y), V(y, j)
GROUP-BY x HAVING MAX(j) > 0
```

It is easy to verify that query q is *true* initially.

In addition, condition of form "MIN $\leq c$" equals to "MAX $> c$", and vice versa. Our theorem follows immediately.

In the following, to provide the proof, we first define the *"aggregated augment"*, a necessary relation transform which is a query in essentials.

Aggregated Augment. Given a relation $r(\bar{x})$, an aggregated augment $f(r, x)$ where $x \in \bar{x}$, is a query of the form defined as follow,

```
SELECT x̄, COUNT(*) as x'
FROM r(x̄)
GROUP-BY x
```

The result of aggregated augment, say $f(r, x)$, is a new relation which has one more argument than the input.

Theorem 3. AGGREGATION RESILIENCE DECISION *is* NP-hard *for nested* COUNT, SUM *query without projection, join and union, where insertion is only under active domain.*

Proof. We build a reduction from INDEPENDENT SET which is NP-hard to *Aggr*-RES. For given instance $\langle G(V, E), t \rangle$. An corresponding instance $\langle d, q, k \rangle$ can be constructed for *Aggr*-RES as follows.

Database d. We start with a relation $g(x, y)$ with respect to G, (i) Denoting each vertex $v_i \in V$ with tuple (i, i) inside g; (ii) Denoting each edge $(v_i, v_j) \in E$ with two tuples (i, j) and (j, i) inside g; (iii) Let the maximum degree of G is δ, then, for each vertex $v_i \in V$, if it is of degree less than δ, we add $\delta - deg(v)$ pairs of tuples (i, v_j^i) and (v_j^i, i) where $0 < j \leq \delta - deg(v)$, into relation $g(x, y)$; (iv) We add $\delta + t - 1$ pairs of tuples $(0, v_j^0)$ and $(v_j^0, 0)$ where $0 < j < \delta + t$;

Query q. We define a nested aggregation query q over a single relation as

```
SELECT *
FROM
    SELECT x, y, x', y', x'', COUNT(*) as y''
    FROM
        SELECT x, y, x', y', COUNT(*) as x''
        FROM
            SELECT x, y, x', COUNT(*) as y'
            FROM
                SELECT x, y, COUNT(*) as x'
                FROM g
                GROUP-BY x
            GROUP-BY y
        GROUP-BY x'
    GROUP-BY y'
WHERE x = y and x'' < s
```

It is easy to verify that query q is *true* initially.

Integers k and s. At last, we set the integers as follow,

$$k = t(t-1) \text{ and } s = t+1$$

This reduction can be built in polynomial time. Then we prove the correctness of the reduction by showing that it is *yes* for instance $\langle G(V, E), t \rangle$ of INDEPENDENT SET if and only if it is *yes* for its corresponding instance $\langle d, q, k \rangle$ of *Aggr*-RES.

"$\Rightarrow$:" When G has an independent set $I \in V$ of size t, then we can build an tuple set Δ into d such that $q(d \cup \Delta)$ become *false*. We show that add set $\Delta = \{(u, v), (v, u) | u \neq v, u, v \in I\}$ into d. It is easy to see that such tuple set is a solution of size $t(t-1)$. None of tuples in Δ is contained in d. Moreover, $q(d \cup \Delta)$ is *false*: there are three kinds of value in the attribute COUNT (y) of d, respectively, 1, δ, and $\delta + t - 1$. Obviously, there are at least $s = t + 1$ count-one and count-δ tuples. Furthermore, the value x with count $\delta + t - 1$ are all in $I \cup \{v_0\}$. Thus, there are $|I| + 1 = k$ tuples with count $\delta + t - 1$. Therefore,it holds number of tuple pairs is

$$2 \binom{|I|}{2} = t(t-1)$$

"$\Leftarrow$:" When we has a solution Δ to $\langle d, q, k \rangle$ of size $k = t(t-1)$. The following values on COUNT (y) occur in the nested sub-query result: 1, δ, and $\delta + t - 1$. Moreover, it is easy to see that there is only one value v_0 on x with count-$\delta + t - 1$ in d. In $d \cup \Delta$, there must be at least $s - 1 = t$ further count of degree up to at least $\delta + t - 1$ and, hence, each of them has to have at least $t - 1$ new tuples added into δ for each distinct value of x. Thus there are exactly t such values on x, each relates to exactly $t - 1$ tuples in δ. These t vertices form an set of size t in δ and, by construction, the t corresponding vertices form an independent set of size t in G.

This completes the proof of the correctness of the reduction.

For the SUM operator, we can modify the proof above by adding a column of number, and filled by '1' for SUM to mimic COUNT operation. At the same time, rewrite the query by substituting keyword COUNT to SUM. Then the proof accomplished and correctness holds obviously.

Theorem 4. *The parametric complexity of* AGGREGATION RESILIENCE DECISION *for queries stated in the theorem above is* W[1]-*hard, in cases with parameter of the size of resilience.*

Proof. Here, we build a reduction from MULTICOLORED INDEPENDENT SET to *Aggr*-RES.

Given a MULTICOLORED INDEPENDENT SET instance $\langle G(V, E), t, \gamma \rangle$, we build the corresponding instance $\langle d, q, k \rangle$ as follows,

Database d. We start with a relation $g(x, y)$ with respect to G, (i) Denoting each vertex $v_i \in V$ with tuple (i, i) inside g; (ii) Denoting each edge $(v_i, v_j) \in E$

with two tuples (i, j) and (j, i) inside g; (iii) Let the maximum degree of G is δ too, then, for each vertex $v_i \in V$, we add pairs of tuples (i, v_j^i) and (v_j^i, i) into relation $g(x, y)$, where $0 < j \leq t^3 \cdot \gamma(v_i) + \delta$; (iV) At last, for each color $p \in \{1, \cdots, m\}$, we add pairs of tuples (ω_p, u_j^p) and (u_j^p, ω_p) into relation $g(x, y)$, where $0 < j < t^3 \cdot p + \delta + t$;

Query q. We define a nested aggregation query q as the same query above. It is easy to verify that query q is *true* initially.

Integers k and s. At last, we set the integers as follow,

$$k = t(t - 1) \text{ and } s = 2$$

Obviously, the reduction can be built polynomially. Then we prove the correctness by showing that it is *yes* for instance $\langle G(V, E), t, \gamma \rangle$ of MULTICOLORED INDEPENDENT SET if and only if it is *yes* for its corresponding instance $\langle d, q, k \rangle$ of *Aggr*-RES.

"⇒:" When G has an independent set $I \in V$ of size t, then we can build an tuple set Δ into d such that $q(d \cup \Delta)$ become *false*. We can also add tuple set $\Delta = \{(u, v), (v, u) | u \neq v, u, v \in I\}$ into d. Tuple set Δ is a solution of size $t(t - 1)$, there are at least 2 count-1 and count-$(t^3 \cdot p + \delta)$ tuples. Furthermore, the value x with count $t^3 \cdot p + \delta + t - 1$ are all in $I \cup \{\omega_p\}$. Thus, there are 2 tuples with count-$(t^3 \cdot p + \delta + t - 1)$. Number of tuple pairs is $2\binom{t}{2} = t(t - 1)$.

"⇐:" When we has a solution Δ to $\langle d, q, k \rangle$ of size $k = t(t-1)$. The following values on COUNT (y) occur in the nested sub-query result: 1, $t^3 \cdot p + \delta$, and $t^3 \cdot p + \delta + t - 1$, for $p \in \{1, \cdots, m\}$. Observe that only one value ω_p on x has COUNT (y) of $(t^3 \cdot p + \delta + t - 1)$, for each p. Then, size of $|\Delta| \geq t$ due to $s = 2$. Because there is no count of from $(t^3 \cdot p + \delta + t)$ to $(t^3 \cdot p + \delta + t^2 + t - 1)$, for each p, and $k \leq t(t - 1)$, there is no number of tuples up to t^2. Then the tuple set Δ must increase at least a distinct count to at least $h - 1$, so that result can be absent *w.r.t* condition COUNT(∗) $< s$. Then, such value on x must have count up to $t^3 \cdot p + \delta$, there is one pair of new tuples for each p in Δ, and each of them has a count of exactly $t - 1$ new tuples of Δ. Observe that values in Δ is exactly t and already present in d, so that they are also the corresponding vertices of G satisfying the coloring constraints, form a solution of MULTICOLORED INDEPENDENT SET.

Therefore, due to the W[1]-hardness of MULTICOLORED INDEPENDENT SET problem, the W[1]-hardness of *Aggr*-RES follows immediately.

We also have the following two results.

Theorem 5. AGGREGATION RESILIENCE DECISION *is both* NP-hard *and* W[1]-hard *for nested* COUNT, SUM *query without projection, join and union, where insertion is only under finite domain.*

4   Conclusion

We study the complexity of the RES problem by means of parameterized complexity, and provide the results of conjunctive query, positive query and first-order query. The results are summarized in Table 1. In the future work, We plan to investigate the tractable condition for nested COUNT queries and approximation algorithms for intractable cases. Furthermore, we plan to study another objective of this problem which is the side effect on source database. The cases considering other types of dependency constraints on database, such as independent dependencies, also need to be further explored.

Acknowledgement. This work is partly supported by the National Science Foundation (NSF) under grant NOs. 1252292, 1741277, 1704287, and 1829674.

References

1. Bancilhon, F., Spyratos, N.: Update semantics of relational views. ACM Trans. Database Syst. **6**(4), 557–575 (1981). https://doi.org/10.1145/319628.319634
2. Bohannon, A., Pierce, B.C., Vaughan, J.A.: Relational lenses: a language for updatable views. In: Proceedings of the Twenty-Fifth ACM SIGMOD-SIGACT-SIGART Symposium on Principles of Database Systems. PODS 2006, pp. 338–347. ACM, New York (2006). https://doi.org/10.1145/1142351.1142399
3. Buneman, P., Khanna, S., Tan, W.C.: On propagation of deletions and annotations through views. In: Proceedings of the Twenty-First ACM SIGMOD-SIGACT-SIGART Symposium on Principles of Database Systems. PODS 2002, pp. 150–158. ACM, New York (2002). https://doi.org/10.1145/543613.543633
4. Cai, Z., Wang, C., Bourgeois, A.: Preface: special issue on computing and combinatorics conference and wireless algorithms, systems, and applications conference. J. Comb. Optim. **32**(4), 983–984 (2016)
5. Cai, Z., Zheng, X.: A private and efficient mechanism for data uploading in smart cyber-physical systems. IEEE Trans. Netw. Sci. Eng., 1 (2018)
6. Chen, Q., Gao, H., Cheng, S., Li, J., Cai, Z.: Distributed non-structure based data aggregation for duty-cycle wireless sensor networks. In: The 36th Annual IEEE International Conference on Computer Communications (INFOCOM 2017), pp. 1–9, May 2017. https://doi.org/10.1109/INFOCOM.2017.8056960
7. Cheng, S., Cai, Z., Li, J., Gao, H.: Extracting kernel dataset from big sensory data in wireless sensor networks. IEEE Trans. Knowl. Data Eng. **29**(4), 813–827 (2017). https://doi.org/10.1109/TKDE.2016.2645212
8. Cong, G., Fan, W., Geerts, F.: Annotation propagation revisited for key preserving views. In: Proceedings of the 15th ACM International Conference on Information and Knowledge Management. CIKM 2006, pp. 632–641. ACM, New York (2006). https://doi.org/10.1145/1183614.1183705
9. Cong, G., Fan, W., Geerts, F., Li, J., Luo, J.: On the complexity of view update analysis and its application to annotation propagation. IEEE Trans. Knowl. Data Eng. **24**(3), 506–519 (2012). https://doi.org/10.1109/TKDE.2011.27
10. Cosmadakis, S.S., Papadimitriou, C.H.: Updates of relational views. J. ACM **31**(4), 742–760 (1984). https://doi.org/10.1145/1634.1887
11. Dayal, U., Bernstein, P.A.: On the correct translation of update operations on relational views. ACM Trans. Database Syst. **7**(3), 381–416 (1982). https://doi.org/10.1145/319732.319740
12. Downey, R.G., Fellows, M.R.: Parameterized Complexity. Springer, New York (2012). https://doi.org/10.1007/978-1-4612-0515-9
13. Freire, C., Gatterbauer, W., Immerman, N., Meliou, A.: The complexity of resilience and responsibility for self-join-free conjunctive queries. Proc. VLDB Endow. **9**(3), 180–191 (2015). https://doi.org/10.14778/2850583.2850592

14. Han, M., Li, J., Cai, Z., Han, Q.: Privacy reserved influence maximization in GPS-enabled cyber-physical and online social networks. In: 2016 IEEE International Conferences on Big Data and Cloud Computing (BDCloud), Social Computing and Networking (SocialCom), Sustainable Computing and Communications (SustainCom) (BDCloud-SocialCom-SustainCom), pp. 284–292. IEEE (2016)
15. Han, M., Yan, M., Cai, Z., Li, Y.: An exploration of broader influence maximization in timeliness networks with opportunistic selection. J. Netw. Comput. Appl. **63**, 39–49 (2016)
16. Han, M., Yan, M., Cai, Z., Li, Y., Cai, X., Yu, J.: Influence maximization by probing partial communities in dynamic online social networks. Trans. Emerg. Telecommun. Technol. **28**(4), e3054 (2017)
17. He, Z., Cai, Z., Yu, J.: Latent-data privacy preserving with customized data utility for social network data. IEEE Trans. Veh. Technol. **67**(1), 665–673 (2018)
18. He, Z., Cai, Z., Yu, J., Wang, X., Sun, Y., Li, Y.: Cost-efficient strategies for restraining rumor spreading in mobile social networks. IEEE Trans. Veh. Technol. **66**(3), 2789–2800 (2017). https://doi.org/10.1109/TVT.2016.2585591
19. Huang, Y., Cai, Z., Bourgeois, A.G.: Location privacy protection with accurate service. J. Netw. Comput. Appl. **103**, 146–156 (2018)
20. Keller, A.M.: Algorithms for translating view updates to database updates for views involving selections, projections, and joins. In: Proceedings of the Fourth ACM SIGACT-SIGMOD Symposium on Principles of Database Systems. PODS 1985, pp. 154–163. ACM, New York (1985). https://doi.org/10.1145/325405.325423
21. Kimelfeld, B.: A dichotomy in the complexity of deletion propagation with functional dependencies. In: Proceedings of the 31st Symposium on Principles of Database Systems. PODS 2012, pp. 191–202. ACM, New York (2012). https://doi.org/10.1145/2213556.2213584
22. Kimelfeld, B., Vondrák, J., Williams, R.: Maximizing conjunctive views in deletion propagation. ACM Trans. Database Syst. **37**(4), 24:1–24:37 (2012). https://doi.org/10.1145/2389241.2389243
23. Kimelfeld, B., Vondrák, J., Woodruff, D.P.: Multi-tuple deletion propagation: approximations and complexity. Proc. VLDB Endow. **6**(13), 1558–1569 (2013). https://doi.org/10.14778/2536258.2536267
24. Li, J., Cai, Z., Yan, M., Li, Y.: Using crowdsourced data in location-based social networks to explore influence maximization. In: The 35th Annual IEEE International Conference on Computer Communications (INFOCOM 2016), pp. 1–9, 2016. https://doi.org/10.1109/INFOCOM.2016.7524471
25. Liang, Y., Cai, Z., Han, Q., Li, Y.: Location privacy leakage through sensory data. Secur. Commun. Netw. **2017**, 12 (2017)
26. Miao, D., Liu, X., Li, J.: On the complexity of sampling query feedback restricted database repair of functional dependency violations. Theor. Comput. Sci. **609**, 594–605 (2016)
27. Papadimitriou, C.H., Yannakakis, M.: On the complexity of database queries (extended abstract). In: Proceedings of the Sixteenth ACM SIGACT-SIGMOD-SIGART Symposium on Principles of Database Systems. PODS 1997, pp. 12–19. ACM, New York (1997). https://doi.org/10.1145/263661.263664
28. Zhang, K., Han, Q., Cai, Z., Yin, G.: RiPPAS: a ring-based privacy-preserving aggregation scheme in wireless sensor networks. Sensors **17**(2), 300 (2017)
29. Zhang, L., Wang, X., Lu, J., Li, P., Cai, Z.: An efficient privacy preserving data aggregation approach for mobile sensing. Secur. Commun. Netw. **9**(16), 3844–3853 (2016)

Inefficiency of Equilibria in Doodle Polls

Barbara M. Anthony[1]($\boxtimes$)(iD) and Christine Chung[2](iD)

[1] Southwestern University, Georgetown, TX 78626, USA
anthonyb@southwestern.edu
[2] Connecticut College, New London, CT 06320, USA
cchung@conncoll.edu

Abstract. Doodle polls allow people to schedule meetings or events based on time preferences of participants. Each participant indicates on a web-based poll form which time slots they find acceptable and a time slot with the most votes is chosen. This is a social choice mechanism known as *approval voting*, in which a standard assumption is that all voters vote *sincerely*—no one votes "no" on a time slot they prefer to a time slot they have voted "yes" on. We take a game-theoretic approach to understanding what happens in Doodle polls assuming participants vote sincerely. First we characterize Doodle poll instances where sincere pure Nash Equilibria (NE) exist, both under lexicographic tie-breaking and randomized tie-breaking. We then study the quality of such NE voting profiles in Doodle polls, showing the price of anarchy and price of stability are both unbounded, even when a time slot that many participants vote yes for is selected. Finally, we find some reasonable conditions under which the quality of the NE (and strong NE) is good.

Keywords: Doodle polls · Nash equilibria · Approval voting

1 Introduction

Online scheduling apps such as Doodle (www.doodle.com) are an increasingly popular tool for scheduling meetings and other events. In a January 2018 personal communication, Doodle reported more than 27 million polls created per year with total users numbering nearly 30 million. In a Doodle poll, the poll initiator posts a set of possible meeting times, then asks participants to check off the times they are available to meet. The Doodle algorithm simply recommends the time slot(s) with the most checked boxes.

This mechanism employed by Doodle for recommending the best time slot is a *social choice function* equivalent to *approval voting*, where each voter in an election must indicate approval or disapproval of each of the candidates. In a Doodle poll (Fig. 1), the participants are the "voters" and the time slots are the "candidates."

A 2-page extended abstract of an earlier version of this work was published in [2].

© Springer Nature Switzerland AG 2018
D. Kim et al. (Eds.): COCOA 2018, LNCS 11346, pp. 707–721, 2018.
https://doi.org/10.1007/978-3-030-04651-4_48

There has been extensive research done in approval voting dating back to the 1970s. For surveys on approval voting from the voting theory literature see [3, 11, 16].

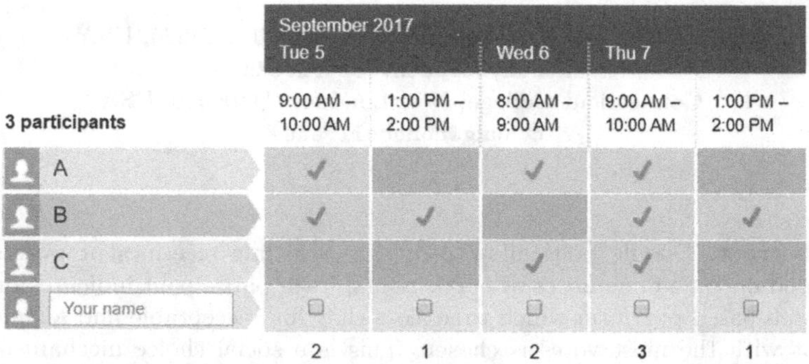

3 participants	September 2017				
	Tue 5		Wed 6	Thu 7	
	9:00 AM – 10:00 AM	1:00 PM – 2:00 PM	8:00 AM – 9:00 AM	9:00 AM – 10:00 AM	1:00 PM – 2:00 PM
A	✓		✓	✓	
B	✓	✓		✓	✓
C			✓	✓	
Your name	☐	☐	☐	☐	☐
	2	1	2	3	1

Fig. 1. An example Doodle poll after three participants have indicated their availability.

However, in contrast to political elections where the voter-to-candidate ratio is very high, Doodle polls are usually conducted on a relatively small scale[1], which allows strategic voting to more easily takes place. Researchers have studied the effect of strategic voting behavior (e.g., [5, 15]), even with respect to approval voting in particular [7, 8, 12, 13]. As in the work of [7], we are interested in how the social welfare of the selected candidate compares with that of the optimal, but we consider Nash Equilibrium (NE) voting outcomes, while they consider randomized embeddings of utility functions into voting rules. The work of [15] also considered voting outcomes that maximize social welfare in comparison with equilibrium outcomes. While their work on plurality voting took a computational approach to finding the NE (and Bayes NE) outcomes, we focus on theoretical worst-case analysis for the approval voting mechanism used in Doodle polls, which are the motivating real-world application of our work.

Similar to the work of [5], which asks 'How bad is selfish voting?', we compare the worst-case NE outcome to the optimal outcome. However, we use social welfare as our metric, while they use a candidate's "honest score" as their metric. They also consider NE that result after a sequence of best response defections from the truthful voting profile, which is a unique voting profile in the three voting systems they consider. In contrast, approval voting does not have a single truthful voting profile, so in this work we study the space of all pure NE. The work of [6] studies team-based coordinated voting in online scheduling polls, giving results regarding computational complexity of finding payoff-improving voter coalitions, and finding NE.

As in [1], we assume each voter has a privately-held, normalized, utility value (or valuation) for each candidate time slot. To measure the quality of a time slot,

---

[1] A sample of over 340,000 polls in a 3-month period in 2011 had a median of about 5 respondents and 12 time slots [17].

we consider the social welfare, or total utility of all voters, for that slot. But while the work of [1] studies the effect of more "protective" voting behavior compared with more "generous" voting behavior on the social welfare of the winning time slot, this work analyzes the *price of anarchy* and *price of stability* in Doodle polls. The *price of anarchy* (POA) (resp., *price of stability* (POS)) is the worst case ratio, over all possible instances of the game, of the social welfare of an optimal slot to the social welfare of the winning slot(s) at the "worst" (resp., "best") pure NE.

The work of [14] analyzes a Doodle game model similar to ours, with the added component of a "social bonus" each voter gets for each time slot they approve of. They also extensively study and compare the two tie-breaking rules: randomized tie-breaking versus lexicographic tie-breaking. They conclude that in the case of "uncapped" social bonus, most Doodle game instances in their model do not admit pure NE under lexicographic tie-breaking. They go on to focus on the case of "capped" social bonus where there are many NE profiles and use a variant of trembling hand perfection to refine them. In our model we do not have a "social bonus" for yes votes at all, and we find under this assumption that most instances do seem to admit a pure NE. We also show randomized tie-breaking also admits pure NE in most instances. As noted, Doodle allows the option for hidden polls, where voters only see their individual responses; such hidden polls negate the presence of a social bonus and thus further motivate the payoff function as defined in our model.

To our knowledge our work is the first to bound the inefficiency of equilibria in Doodle polls. Since Doodle polls are equivalent to approval voting, we note that our results also apply to approval voting in general, but our context for this work is Doodle polls, keeping in mind their idiosyncrasics (like the often low ratio of voters to candidates) that are not commonly found in general approval voting scenarios. First we investigate the space of Doodle poll instances where pure NE exist when assuming voters vote *sincerely* (i.e., they never disapprove of a time slot that they have higher utility for than a time slot they approved of), both under lexicographic and randomized tie-breaking. Then we show that when restricting to the space of instances that admit pure NE, both POA and POS are unbounded. We show this remains the case even when restricted to instances that have at least one time slot with high total valuation. We then give some reasonable conditions under which the POA and POS are good. Finally we present a constant bound on the strong price of anarchy when there are time slots with sufficiently high total valuation.

2   Model and Definitions

We define a Doodle poll instance to be a triple $I = (A, V, U)$, where $A = \{a_1, a_2, \ldots, a_m\}$ is the set of time slots or *alternatives*, $V = \{v_1, v_2, \ldots, v_n\}$ is the set of voters, and U is the $n \times m$ matrix of utility values $0 \le u_{ij} \le 1$ that each voter $i = 1 \ldots n$ privately holds for each alternative $j = 1 \ldots m$. We say voter v_i *prefers* alternative a_j to a_k when $u_{ij} > u_{ik}$.

The utility can be thought of as quantifying how much the voter expects to benefit from attending the meeting at that time (even if derived merely by satisfying some expectation of attendance) minus any inconvenience/cost of attending the meeting at that time. It also may be thought of as representative of how much monetary value a voter would place on attending the event at a given time.

Notice that in Doodle polls, organizers and participants typically expect and believe that participants are not only sincere, but that they are also generally representing their "true" availability, allowing for the selection of a "good" meeting time. Indeed, other studies often assume the most straightforward behavior of a voter is simply to vote "yes" on those time slots for which she is available, and "no" on those she is unavailable. However, we submit that availability is not a binary value; in theory, a participant can make themselves available for *any* time slot, albeit at varying, and potentially quite high, cost. Our model of private cardinal utility values accounts for such a spectrum of "availability," while only requiring individual participants to make a yes/no determination.

Given an instance I, we use an $n \times m$ matrix denoted by $R = [r_1, r_2, \ldots, r_n]$ to represent the *voting profile* (or *strategy profile*), where r_i is a binary vector over the m alternatives in A, representing the *vote* or *strategy* of voter i, with $r_i(j) = 1$ (a *yes vote*) if voter v_i approves alternative a_j, and $r_i(j) = 0$ (a *no vote*) otherwise. When it is clear from the context, we use *vote* to either refer to the full vector r_i, or to the binary value the voter assigns to a specific alternative. We consider only *pure strategies* in this work, so we assume voters will not be randomizing among a set of possible votes.

Let $s(a_j) = \sum_{i=1}^{n} r_i(a_j)$, or the total count of votes of approval for alternative a_j, be the *score* for an alternative a_j. The default Doodle mechanism (approval voting) chooses the set of one or more *winning alternatives*, W, which maximize the total score, that is $W = \arg\max_{a_j \in A} s(a_j)$.

The most commonly-studied tie-breaking rules in the event of multiple alternatives with maximum score ($|W| > 1$) are *lexicographic* tie-breaking, in which the single winning alternative $w \in W$ that comes first in the established tie-breaking order over A is chosen, and *randomized* tie-breaking, which chooses w from the winning alternatives in W uniformly at random. Under lexicographic tie-breaking, we assume without loss of generality that the tie-breaking order on the alternatives proceeds from left to right $a_1 \ldots a_m$ in the poll, and each player i earns a payoff $\pi_i(I, R)$ of u_{ij} if the chosen alternative $w = a_j$. Under randomized tie-breaking, in the case of a tie, the expected payoff is $E[\pi_i(I, R)] = \sum_{a_j \in W} u_{ij}/|W|$, the average utility of the winning alternatives.

Our model most accurately reflects hidden Doodle polls, in which players do not see each others' votes. This removes the motivation for a "social bonus" term in the payoff function as used in [14]. Hidden polls also transform the setting into a simple simultaneous move game.

A *pure Nash equilibrium* (NE) is a strategy profile where no player can unilaterally *defect* to an alternate strategy (i.e. flip some of their voting bits) and strictly increase their payoff. I.e., at a NE no voter i can alter their vote vector $r_i \in R$ to a new vote vector r_i', s.t. $\pi_i(I, R') > \pi_i(I, R)$ (for lexicographic tie-breaking), where R' is R with r_i replaced by r_i'. (For randomized tie-breaking payoffs are taken in expectation.)

We use $OPT(I)$ to denote an *optimal* alternative, which maximizes the social welfare in a given Doodle poll instance I, and $u(a)$ to denote the total utility (social welfare) of alternative $a \in A$. Hence $OPT(I) = \arg\max_{a_j \in A} \sum_{i=1}^{n} u_{ij}$ and $u(OPT(I)) = \max_{a_j \in A} \sum_{i=1}^{n} u_{ij}$.

As justified in many classical and recent works, e.g., [3,4,9,14,17], we assume all voters are *sincere* in their voting, i.e., if $r_i(a_j) = 1$ then $r_i(a_k) = 1$ for all $k \neq j$ where $u_{ik} > u_{ij}$. Let *sincere pure NE* refer to a pure NE where all voters are voting sincerely (and may defect only to sincere strategies) and let $N_s(I)$ denote the set of sincere pure NE for Doodle poll instance I.

Given a Doodle poll instance I, we define *sincere price of anarchy* $POA(I)$ for that instance to be $\frac{u(OPT(I))}{\min_{R \in N_s(I)} u(w(R))}$ and *sincere price of stability* to be $\frac{u(OPT(I))}{\max_{R \in N_s(I)} u(w(R))}$, where $u(w(R))$ is the social welfare of the winning alternative given profile R. Note that in the case of randomized tie-breaking, the *expected* social welfare is used in the denominator. We can then define the sincere price of anarchy (POA) of Doodle polls to be the worst-case $POA(I)$: $\max_{I \in \mathcal{I}} POA(I)$, where $\mathcal{I}$ is the set of all Doodle poll instances. Respectively, we define the sincere price of stability (POS) to be the worst-case $POS(I)$: $\max_{I \in \mathcal{I}} POS(I)$.

3   Existence of Sincere Pure Nash Equilibria

Since we analyze price of anarchy and price of stability only over the space of instances that admit sincere pure Nash equilibria, we investigate in this section what these types of instances look like, under both lexicographic and randomized tie-breaking. To begin with, as noted in [10], "no voter can, by changing her vote only, change the outcome of the game [under] Approval Voting [...when...] one candidate is winning the election with a margin of more than two votes." We state this formally as the following lemma.

Lemma 1. *A voting profile is a sincere pure NE if the two largest scores differ by two or more, under either lexicographic or randomized tie-breaking.*

We refer to an alternative $a_j \in A$ as a *favorite* of voter v_i if $u_{ij} \geq u_{ik}$ for all $k \neq j$. And we say that an alternative $a_j \in A$ is a *k*th favorite of voter v_i if there are exactly $k-1$ alternatives j' for which $u_{ij'} > u_{ij}$.

Corollary 1. *If two or more voters have a favorite alternative in common, then there is a sincere pure NE where the set of winning alternatives W is precisely that favorite alternative, under both lexicograpic and randomized tie-breaking.*

Corollary 2. *If the number of voters exceeds the number of alternatives, that is, $n > m$, then there is a sincere pure NE, under both lexicographic and randomized tie-breaking.*

These corollaries already describe a rather large space of instances where a sincere pure NE always exists. However, sincere pure NE do not always exist, under either lexicographic or randomized tie-breaking. We provide specific instances for each tie-breaking rule, while characterizing further some situations where sincere pure NE do exist.

3.1   Lexicographic Tie-Breaking

Corollaries 1 and 2 collectively ensure the existence of a sincere pure NE under lexicographic tie-breaking whenever two or more voters have the same favorite, and whenever $n > m$. Furthermore, the following lemma ensures the existence of a sincere pure NE when $n = m$. These conditions greatly limit the potential instances without a sincere pure NE, and we provide a specific instance in Table 1 that does not have a sincere pure NE under lexicographic tie-breaking.

Lemma 2. *If the number of voters equals the number of alternatives, that is, $n = m$, then there is a sincere pure NE, under lexicographic tie-breaking.*

Proof. By Corollary 1, assume each voter has a different favorite. Since $n = m$, there must be a voter v whose favorite is alternative a_1. The voting profile in which all votes are no except a single yes vote to a_1 from voter v is sincere. Furthermore, under lexicographic tie-breaking, since the winning alternative is leftmost, no voter has incentive to defect.

Note that Lemma 2 does not hold under randomized tie-breaking, as Table 2 will illustrate. The following lemma will help us to establish the fact that sincere pure NE do not always exist under lexicographic tie-breaking.

Lemma 3. *Given a Doodle poll instance, if there is a sincere pure NE profile with winning alternative w under lexicographic tie-breaking, there is a sincere pure NE where all voters vote yes for alternative w.*

Proof. Suppose we have a NE with winning score $s(w)$ on alternative w. If $s(w) = n$, all voters are already voting yes for alternative w, so we may assume $s(w) < n$. With lexicographic tie-breaking, the scores on all alternatives left of w are strictly less than $s(w)$, and all alternatives to the right of w have score at most $s(w)$. Suppose we take the existing NE profile, and then modify it so that all voters vote yes to alternative w, and anything else required by sincerity. The score for w is now $s(w) + x = n$. Since the yes votes required by sincerity add at most x to the scores for other alternatives, updated scores to the left of w are strictly less than $s(w)+x$, and those to the right are at most $s(w)+x$, with w still the winning alternative. Furthermore, since the only yes votes that are added in alternatives other than alternative w are due to sincerity, if any voter wishes to defect now, they likewise would have prior to the addition, contradicting the assumption that we started with a NE.

Theorem 1. *Sincere pure NE do not always exist in Doodle polls under lexicographic tie-breaking.*

Proof. By Lemma 3, we need only exhibit an instance in which for each alternative, all voters voting yes for that alternative is not a sincere NE profile. Consider the instance in Table 1. For alternatives $j = 2, 3, 4, 5$, observe that if all n votes on alternative j are yes, then by sincerity there are $n - 1$ yes votes on alternative $j - 1$, and each of those $n - 1$ voters who voted yes on $j - 1$ wish to defect by

Table 1. An instance in which no sincere pure NE exists under lexicographic tie-breaking, for $0 < \epsilon \le 1/4$, as described in the proof of Theorem 1.

Voters	1	2	3	4	5
v_1	3ϵ	2ϵ	ϵ	0	4ϵ
v_2	2ϵ	ϵ	0	4ϵ	3ϵ
v_3	ϵ	0	4ϵ	3ϵ	2ϵ
v_4	0	4ϵ	3ϵ	2ϵ	ϵ

saying no to alternative j. Likewise, if all n votes on alternative 1 are yes, then by sincerity, all n votes on alternative 5 are yes; due to lexicographic tie-breaking, alternative 1 would win, but since alternative 5 is preferred by all n voters, they each wish to defect by saying no to alternative 1. Hence, in any sincere voting profile where all voters say yes to a given alternative, at least one voter wishes to defect.

3.2   Randomized Tie-Breaking

We now proceed comparably for randomized tie-breaking, providing a broad categorization of instances which do have sincere pure NE, and again showing that sincere pure NE do not always exist. Recall that we refer to an alternative $a_j \in A$ as a *favorite* of a voter if that voter (weakly) prefers it to all other alternatives. And we say that an alternative $a_j \in A$ is an ith favorite of a voter if there are exactly $i - 1$ alternatives which they strictly prefer to a_j. We say an $n \times n$ instance does not have *distinct ith favorites* if for $i \in 1, 2, \ldots, n - 1$, some alternative is the ith favorite of two or more voters.

Lemma 4. *Consider an $n \times n$ instance. If it does not have distinct ith favorites, then it has a sincere pure NE under randomized tie-breaking.*

Proof. We proceed by using strong induction. Base case: Corollary 1 ensures that any instance where an alternative is the favorite (i.e. $i = 1$) of two or more voters has a sincere pure NE.

Inductive step: Assume that no alternative is the $1, 2, 3, \ldots$, or kth favorite of two or more voters, for $k < n - 1$. We show if some alternative a' is the $k + 1$st favorite of two or more voters, then there is a sincere pure NE.

Since no alternative is the ith favorite of two or more voters for $i = 1, 2, 3, \ldots, k$, and the instance is $n \times n$, each alternative must be the favorite of exactly one voter, the 2nd favorite of exactly one voter, ..., the kth favorite of exactly one voter. Moreover, by assumption, alternative a' is the $k + 1$st favorite of two or more voters. Consider the voting profile consisting of yes votes for all of the 1st favorites, 2nd favorites, ..., kth favorites, as well as any two of the $k + 1$st favorites in alternative a'. Observe that this voting profile is consistent with sincerity. The scores are thus $k + 2$ for alternative a' and k for all of the other $n - 1$ alternatives. Lemma 1 thus guarantees the existence of a sincere pure NE.

Lemma 4 in combination with Corollary 2 indicates that the space of instances that admit sincere pure NE seems quite general and large. For example, in the case of $n = m$, only instances that meet the rather strict structural requirement of distinct ith favorites do not have sincere pure NE. Note that Lemma 4 also applies to lexicographic tie-breaking, but is subsumed by Lemma 2.

Table 2. An instance with no sincere pure NE under randomized tie-breaking, for t and ϵ with $\frac{1}{2} < t < 1$ and $0 < \epsilon < \frac{1}{2}$, as described in the proof of Theorem 2.

Voters	1	2	3
$v_1 = A$	1	t	0
$v_2 = B$	0	1	ϵ
$v_3 = C$	ϵ	0	1

Theorem 2. *Sincere pure NE do not always exist in Doodle polls under randomized tie-breaking.*

Proof. Consider the instance in Table 2, where $\frac{1}{2} < t < 1$ and $0 < \epsilon < \frac{1}{2}$. We consider the voting profiles and corresponding scores, and show by exhaustive cases (of winning score $s = 0, 1, 2,$ or 3) that no pure NE exists under randomized tie-breaking. Observe that there are distinct favorites, which in this instance are on the diagonal, and recall that sincerity must be maintained.

Clearly, there is no NE where the winning score is $s = 0$; that would require all voters to vote no on all alternatives, and they would each have incentive to defect.

Consider what happens when the winning score is $s = 1$. If one voter is voting yes to all three alternatives and $s = 1$, since each voter has a single favorite alternative in this instance, this voter will always have incentive to defect; they prefer to say no to two of the alternatives, voting yes only for their favorite. If one voter is voting yes to exactly two alternatives, and $s = 1$, then there must be a voter who is not voting yes to their favorite alternative who would thus prefer to defect and have their favorite alternative win with two yes votes. Thus, we are left only with the case where each voter is voting yes to at most one alternative, which, due to sincerity, must be their favorite alternative. If only one alternative has a yes vote, then it can easily be verified that either of the other voters would wish to defect and vote yes to their favorite. Likewise, if two of the alternatives have yes votes, in all three such scenarios within this instance, the third voter would prefer to defect to cause a three-way tie. Thus, the only possible NE with $s = 1$ is a three-way tie for the favorites on the diagonal. But this is also not a NE: since $t > \frac{1}{2}$, voter A would rather have t than their payoff in the tie $(1+t)/3$, and thus prefers to defect by voting for alternative 2.

Now, consider the case where the winning score is $s = 2$. Suppose first that there are two or more alternatives with a score of 2. Then some voter is voting yes to two or more of the winning alternatives and such a voter would wish to defect so their favorite alternative alone wins. In the subcase that all three alternatives have a score of 2, if one voter is voting yes for all three alternatives, then that voter would wish to defect by saying no to all but their favorite. Thus, with reported scores on the three alternatives of 2, 2, and 2, each voter must be voting yes for their favorite and second favorite. In this case, voter B would wish to defect since the payoff to voter B when alternatives 1 and 2 are tied, $\frac{1}{2}(1+0) > \frac{1}{3}(1+\epsilon+0)$, its payoff in the three-way tie.

We now consider the case with winning score $s = 2$ and other alternatives' scores are 1 or 0. Due to the structure of the utility values and sincerity, the winning alternative w with score 2 must have a yes vote from the player whose utility is 1 for alternative w (otherwise two alternatives would have a score of 2). If the other yes vote for w is from a player with a utility of 0 or ϵ for w, that voter would always prefer to defect to a vote of no on alternative w, so that their favorite (for which they are already voting yes, due to sincerity) would now be tied for winning; one can verify that with $\epsilon < 1/2$, the (two-way or three-way) tie is better than their current valuation. If the other yes vote for w is from the player with a utility of t, so $w = a_2$, then voter C currently getting 0 payoff must also be voting no on alternative a_1 (otherwise $s(a_1) = s(w) = 2$). So voter C would prefer to defect and vote yes to a_1 to cause the tie between a_1 and a_2.

Finally, consider the case when the winning score is $s = 3$. Note that the structure of the utilities is such that each alternative is the favorite for one voter, the second favorite for another, and the least favorite for another. Thus, unanimous approval of an alternative in conjunction with sincerity implies that one of the other alternatives has at least one yes vote, and the other alternative has at least two yes votes. There cannot be two alternatives with 3 yes votes, since all voters have distinct favorites, and at least one voter would defect to a no vote on their non-favorite so their preferred alternative would win (or be in a tie with fewer winning alternatives). Thus, only the cases where the scores (subject to permutation) are 3, 2, 2 and 3, 2, 1 remain. In both these remaining sub cases, one of the three votes for the winning alternative w must come from a voter for whom w is their least favorite. So that voter would prefer to defect and vote no on w, allowing another alternative to win.

This 3×3 instance can be extended to a $3 \times x$ instance (for $x > 3$) by adding alternatives with 0 valuation. Observe that Table 2 does have a sincere pure NE under lexicographic tie-breaking: namely, the voting profile in which voter A votes yes to alternative 1, and all other votes are no for all voters is sincere and a NE.

4   Price of Stability (and POA) are Unbounded

In this section we exhibit instances that show both price of anarchy and price of stability are unbounded, regardless of which tie-breaking mechanism is used. We first show that POA is unbounded even when the score must be at least $\frac{n}{2}$

and when the utility of OPT must be at least $\frac{n}{2}$. We then provide instances in which POS is likewise unbounded with large score/utility. For the remainder of this work, we use the shortened term "NE" to refer to sincere pure NE. We let $\epsilon > 0$ be an arbitrarily small value.

Table 3. An instance in which POA is unbounded, for an arbitrary odd n, where OPT has utility $\frac{n}{2}$ and score at least $\frac{n}{2}$ for the best NE.

Voters	1	2	3
$v_1 = A$	ϵ	1	0
$v_2 = B$	ϵ	0	1
$v_3 = C$	ϵ	1	0
$v_4 = D$	ϵ	0	1
$\vdots$	$\vdots$	$\vdots$	$\vdots$
v_n	ϵ	1	0

Theorem 3. *The sincere price of anarchy is unbounded in Doodle polls, under both lexicographic and randomized tie-breaking.*

Proof. Consider the instance in Table 3 for an arbitrary odd $n \geq 5$, and $\epsilon > 0$ arbitrarily small. Alternative 2 is optimal with a utility of $\frac{n+1}{2}$. Consider the voting profile where all voters vote yes for all alternatives with non-zero utility, and no to alternatives with zero utility. Clearly this is sincere. It is also a NE, since alternative 1 will have a score of n, and the two other alternatives will have scores of $\frac{n+1}{2}$ and $\frac{n-1}{2}$. Since $n \geq 5$, there is no incentive for any voter to defect. Thus, the utility of this NE is $n\epsilon$. Thus, the price of anarchy is $\frac{n+1}{2}/(n\epsilon) = \frac{1}{2\epsilon} + \frac{1}{2n\epsilon}$ which becomes arbitrarily large for ϵ arbitrarily small.

Notes: For randomized tie breaking, we need $n \geq 5$ since if n was 3, then the scores would be 3, 2, and 1, and a voter who voted yes to an ϵ in the first column could defect. We could easily have an instance with an even number of rows by adding in an additional row with ϵ, 0, 1.

Observe that in the Table 3 instance, POS is 1 since a voting profile where all voters vote yes on alternatives with valuation 1 and no on all other alternatives is sincere, gives the OPT, and can easily be verified to be a NE. Further, not only is the score at least $n/2$ but the total utility of the optimal alternative is also at least $n/2$.

We now provide an instance where OPT has utility of almost $n - 2$, but POS is still unbounded. Note that when the utility of OPT is high, it becomes more likely that OPT itself is a NE, so it becomes harder to find an instance with high POS. The following instance relies on a structure in which for many of the alternatives, all but one voter likes another alternative more, so defections move away from the optimal.

Table 4. An instance with unbounded POS even when OPT has utility close to $n-2$.

Voters	a_1	a_2	a_3	$\cdots$	a_{m-3}	a_{m-2}	a_{m-1}	a_m
v_1	$m\epsilon$	$(m-3)\epsilon$	$(m-4)\epsilon$	$\cdots$	2ϵ	ϵ	0	1
v_2	$m\epsilon$	$(m-3)\epsilon$	$(m-4)\epsilon$	$\vdots$	2ϵ	0	$c+\epsilon$	c
v_3	$m\epsilon$	$(m-3)\epsilon$	$(m-4)\epsilon$	$\vdots$	0	$c+2\epsilon$	$c+\epsilon$	c
v_4	$m\epsilon$	$(m-3)\epsilon$	$(m-4)\epsilon$	$\cdot\cdot^{\cdot}$	$c+3\epsilon$	$c+2\epsilon$	$c+\epsilon$	c
$\vdots$	$\vdots$	$\vdots$	$\vdots$	$\vdots$	$\vdots$	$\vdots$	$\vdots$	$\vdots$
v_{m-3}	$m\epsilon$	$(m-3)\epsilon$	0	$\vdots$	$c+3\epsilon$	$c+2\epsilon$	$c+\epsilon$	c
v_{m-2}	$m\epsilon$	0	$c+(m-3)\epsilon$	$\vdots$	$c+3\epsilon$	$c+2\epsilon$	$c+\epsilon$	c
v_{m-1}	$m\epsilon$	$c+(m-2)\epsilon$	$c+(m-3)\epsilon$	$\vdots$	$c+3\epsilon$	$c+2\epsilon$	$c+\epsilon$	c
v_m	$m\epsilon$	$(m-1)\epsilon$	$(m-2)\epsilon$	$\cdots$	4ϵ	3ϵ	2ϵ	ϵ
$v_n(=v_{m+1})$	$m\epsilon$	$(m-1)\epsilon$	$(m-2)\epsilon$	$\cdots$	4ϵ	3ϵ	2ϵ	ϵ

Theorem 4. *Sincere POS is unbounded in Doodle polls, even when* $|OPT| \approx n - 2$. *Furthermore, the claim holds under both randomized and lexicographic tie-breaking.*

Proof. Consider the instance in Table 4 with $m \geq 2$. Note that $n = m + 1$, and c can be arbitrarily close to 1 as long as $(m - 1)\epsilon < c < 1 - (m - 2)\epsilon$. OPT is the final alternative, a_m, with a utility $1 + (m - 2)c + 2\epsilon$. If we can show that the only NE is one in which the alternative a_1 wins, which has a utility of $m(m + 1)\epsilon$, then we have shown that POS is unbounded.

To verify that the first alternative is the winning alternative in a NE, consider the voting profile that votes yes for all values in a_1, and precisely the other values required by sincerity, namely the valuations of 1 and $c + x\epsilon$ (for any positive x). This yields scores of $m + 1$ for a_1, and $j - 1$ for all other alternatives $j = 2, \ldots, m$. Thus, since all alternatives other than the first have a score of at most $m - 1$, Lemma 1 ensures this is a NE.

We now show that none of the alternatives 2 through m win in a NE. Observe that for $i = 1, 2, \ldots, m - 1$, all but voter i prefers a_{m-i} to a_{m-i+1}. Thus, for $i = m, \ldots, 2$, a_i is not a winner in any NE, because if it were, then by sincerity a_{i-1} must be within one vote of a_i, since anyone other than voter $m - i + 1$ must be voting for a_{i-1} if they are voting for a_i. Hence, one of the voters would prefer to defect and either vote no on a_i or vote yes on a_{i-1}, so that a_{i-1} becomes either tied for winning or the strict winner; with lexicographic tie-breaking from left to right, a_{i-1} would be the strict winner. Hence, an outcome where a_i is winning is not a NE.

The instance can be extended. Additional rows identical to those below the horizontal dashed line can be added: the utility remains arbitrarily close to $m-1$, which is now n minus the number of rows below said dashed line. Additional columns identical to column 1, or with larger multiples of ϵ, can also be added before the vertical dashed line, and POS likewise remains unbounded.

5   Bounds on POS and Strong POA

In this section we describe some situations where price of stability is good.

Since Corollary 1 guarantees that there is a NE which selects an optimal alternative, the following corollaries identify situations in which POS is 1. We then provide a characterization of the set of Doodle polls instances where the expected social welfare in the best NE is optimal.

Corollary 3. *In a Doodle poll instance I, if an optimal alternative is a favorite of multiple voters, then $POS(I) = 1$.*

Corollary 4. *If there are two or more 'indifferent' voters with identical valuations on all alternatives in a Doodle poll instance I, then $POS(I) = 1$.*

Theorem 5. *Given a Doodle poll instance I under randomized tie-breaking, $POS(I) = 1$ if and only if there is no alternative that $n - 1$ voters prefer to an optimal alternative. I.e., for each non-optimal alternative, at least 2 players prefer an optimal alternative to it.*

Proof. If $POS(I) = 1$, and hence an optimal alternative a^* has the most votes at a NE, we assume for the sake of contradiction that there is some other alternative that $n - 1$ voters prefer to a^*. We call this more preferred alternative a', and the one voter who does not prefer it v^*. Then any voter $v \neq v^*$ approving the optimal a^* would also be voting yes on a' by sincerity, which would imply $s(a') \geq s(a^*) - 1$. And in this case, any of those voters $v \neq v'$ would prefer to defect and say no to a^* (if they were voting yes to a^*), or defect and say yes to a' (if they were voting no to it), contradicting the assumption that we are at a NE. Note that if they were all voting no to a^* and yes to a' it would contradict the fact that a^* has the most votes.

We now proceed to show the converse: having no alternative that $n - 1$ voters prefer to an optimal implies $POS(I) = 1$. Let alternative a_s be optimal. Suppose all voters vote yes for a_s, and vote yes for other alternatives as needed to enforce sincerity. We claim this voting strategy is a NE.

Since all voters vote yes for a_s, it has n votes. Suppose voter v_p believes that she can improve her personal valuation by unilaterally changing her vote, while maintaining sincerity. We consider all possible actions she could take, and show that none of them in fact improve her personal valuation, showing that the current solution is indeed a NE.

Action 1: She keeps her yes vote for a_s, but votes yes on another alternative a_x, for which she had previously said no. For this to improve her personal valuation, a_x must now be selected, and $u_{px} > u_{ps}$. But since a_s had n votes, the maximum possible, a_x must also now have n votes and be selected instead of a_s by tie-breaking. For a_x to now have n votes, all voters other than p must have voted yes to it, ensuring that $\forall_{i\neq p, i\in[n]} u_{ix} \geq u_{is}$, since votes for alternatives other than a_s were based solely on sincerity. The social welfare of a_x is $\sum_{i=1}^{n} u_{ix}$. Thus, the social welfare of a_x is also $u_{px} + \sum_{i\neq p, i\in[n]} u_{ix} \geq u_{px} + \sum_{i\neq p, i\in[n]} u_{is} > u_{ps} + \sum_{i\neq p, i\in[n]} u_{is} = \sum_{i=1}^{n} u_{is}$ which is the social welfare of a_s. But this contradicts the fact that a_s was the optimal solution.

Action 2: She keeps her yes vote for a_s, but votes no on another alternative a_x, for which she had previously said yes. This change keeps the vote count at n for a_s, but only lowers the vote count on a_x, so her personal valuation does not improve.

Action 3: She votes no on a_s (and possibly changes her votes on other alternatives as well) and alternative a_x is now selected, for which she is voting no. Since a_s has $n - 1$ yes votes, and a_x can have at most $n - 1$ votes, again, there is at best a tie, so voter p will not switch. Why? For a_x to now be selected as a winning alternative with v_p voting no to it, a_x must have $n-1$ yes votes from all other voters. But since all other voters are voting for a_x based solely on sincerity, then by the same argument as in Action 1, for this to occur, a_s must not in fact be the optimal solution. Hence, the desired contradiction is again reached.

Action 4: She votes no on a_s (and possibly changes her votes on other alternatives as well) and alternative a_x is now selected, for which she is voting yes. For her personal situation to improve, $u_{px} > u_{ps}$. But since she originally said yes to a_s, then by sincerity, she had to originally say yes to a_x as well. Alternative a_s still has $n - 1$ yes votes. If a_x has n yes votes, then in the original voting strategy, it also did, and thus by the same argument as in Action 1, for this to occur, a_s must not in fact be an optimal solution. Thus, we must conclude that a_x has $n - 1$ yes votes, but this contradicts our assumption that no alternative has $n - 1$ voters that prefer it to an optimal slot.

For a given instance I, we note that $POA(I)$ (and hence $POS(I)$) is trivially upper-bounded by $\frac{\max_{a\in A} u(a)}{\min_{a\in A} u(a)}$. However, the alternative $a_{min} = \arg\min_{a\in A} u(a)$ can only be chosen in a NE if there is no other alternative $n - 1$ voters prefer to it. If $n - 1$ voters prefer another alternative, then by sincerity that alternative has a score within one of the chosen alternative, meaning some voter can defect and improve their payoff. This observation gives the following bound on POA.

Proposition 1. *Given a Doodle poll instance I, let a_{low} be the lowest utility alternative s.t. at least two voters prefer a_{low} to a for all other alternatives $a \in A$. Then we have: $POS(I) \leq POA(I) \leq \frac{u(OPT(I))}{u(a_{low})}$.*

5.1   Strong NE

A *strong NE* is a voting profile where no subset (or "coalition") of voters can all simultaneously defect and improve their payoff. All strong NE are NE, and strong NE may not always exist. The *strong POA* (resp, *POS*) is defined as the ratio of the total utility of an optimal alternative to the total utility of the alternative chosen in the worst (resp., best) strong NE, assuming one exists. It has been established (in [4]) that strong NE coincide precisely with those voting profiles that select Condorcet winners. A *weak Condorcet winner* a_c is a candidate such that the number of voters who prefer a_c to a for any other $a \in A$ is at least the number who prefer a to a_c.

Lemma 5 (Adapted from [4]). *Given a Doodle poll instance* $I = (A, V, U)$, *an alternative* $a \in A$ *is a winning alternative in a strong NE if and only if* a *is a weak Condorcet winner.*

We can now show that strong POA is at most 4 when there is an alternative with utility at least $\frac{3n}{4}$, or more generally:

Theorem 6. *Given a Doodle poll instance* $I = (A, V, U)$ *that admits a strong NE, if* $u(a_j) \geq \rho n$ *for some* $a_j \in A$, $1 \geq \rho > 1/2$, *then strong POA, and hence strong POS, satisfy* $sPOS(I) \leq sPOA(I) \leq 1/(\rho - 1/2)$, *which approaches 2 as* ρ *approaches 1.*

Proof. If I admits a strong NE, then there is some weak Condorcet winner $a_c \in A$, where a_c is preferred to any other $a \in A$ by at least half the voters (by Lemma 5). If $u(a_j) \geq \rho n$ for some $a_j \in A$, then more than ρ of the voters have strictly positive utility for a_j. If at least half the voters prefer a_c to a_j, then $u(a_c) \geq \rho n - n/2 = n(\rho - 1/2)$. Since $u(OPT(I)) \leq n$, we thus have $sPOS = u(OPT(I))/u(a_c) \leq 1/(\rho - 1/2)$.

6   Conclusion

Our results have shown that there are many natural Doodle poll instances that admit sincere pure Nash Equilibria. In particular, almost all instances where the number of voters is at least the number of candidates (that is, $n \geq m$) admit sincere pure NE under both randomized and lexicographic tie-breaking. It remains future work to determine when sincere pure NE exist in the case where $m > n$ (the number of candidates exceeds the number of voters), which is not common in standard approval voting settings, but is not so unusual to encounter in a Doodle poll.

Our results have also shown that while the price of anarchy and price of stability are both unbounded, the conditions we found that give rise to these cases seem rather particular and unlucky. We also show that there is also a large set of realistic Doodle poll instances where $POS = 1$; for example, $POS = 1$ when the optimal time slot is the favorite of at least two voters. Finally, we also show that strong POA is reasonable when there is at least one time slot with total utility more than $n/2$. In future work, we hope to expand our understanding of the set of Doodle poll instances where the POA or POS are good, possibly by restricting ourselves to instances where voters cannot vote for all or none of the time slots. We also hope to gather data, and using simulated utility values see if such conditions are likely to be present in most real-life Doodle polls, and also check how commonly the outcomes of the polls are at NE.

References

1. Alrawi, D., Anthony, B.M., Chung, C.: How well do Doodle polls do? In: Spiro, E., Ahn, Y.-Y. (eds.) SocInfo 2016. LNCS, vol. 10046, pp. 3–23. Springer, Cham (2016). https://doi.org/10.1007/978-3-319-47880-7_1
2. Anthony, B.M., Chung, C.: How bad is selfish Doodle voting?: (extended abstract). In: Proceedings of the 2018 International Conference on Autonomous Agents and Multiagent Systems. AAMAS 2018 (2018)
3. Brams, S.J., Fishburn, P.C.: Approval Voting. Birkhauser, Boston (1983)
4. Brams, S.J., Sanver, M.R.: Critical strategies under approval voting: who gets ruled in and ruled out. Elect. Stud. **25**(2), 287–305 (2006)
5. Brânzei, S., Caragiannis, I., Morgenstern, J., Procaccia, A.D.: How bad is selfish voting? In: desJardins, M., Littman, M.L. (eds.) Proceedings of the Twenty-Seventh AAAI Conference on Artificial Intelligence, Bellevue, Washington, USA, pp. 138–144. AAAI Press (2013)
6. Bredereck, R., Chen, J., Niedermeier, R., Obraztsova, S., Talmon, N.: Teams in online scheduling polls: game-theoretic aspects. In: Singh, S.P., Markovitch, S. (eds.) Proceedings of the Thirty-First AAAI Conference on Artificial Intelligence, San Francisco, California, USA, pp. 390–396. AAAI Press (2017)
7. Caragiannis, I., Procaccia, A.D.: Voting almost maximizes social welfare despite limited communication. Artif. Intell. **175**(9–10), 1655–1671 (2011)
8. De Sinopoli, F., Dutta, B., Laslier, J.: Approval voting: three examples. Int. J. Game Theory **35**(1), 27–38 (2006)
9. Endriss, U.: Sincerity and manipulation under approval voting. Theory Decis. **74**(3), 335–355 (2013)
10. Laslier, J.F., Sanver, M.R.: The basic approval voting game. In: Laslier, J.F., Sanver, M.R. (eds.) Handbook on Approval Voting. WELFARE, pp. 153–163. Springer, Heidelberg (2010). https://doi.org/10.1007/978-3-642-02839-7_8
11. Laslier, J.F., Sanver, M.R.: Handbook on Approval Voting. Studies in Choice and Welfare. Springer, Heidelberg (2010). https://doi.org/10.1007/978-3-642-02839-7
12. Laslier, J.F.: The leader rule: a model of strategic approval voting in a large electorate. J. Theor. Polit. **21**(1), 113–136 (2009)
13. Myerson, R.B., Weber, R.J.: A theory of voting equilibria. Am. Polit. Sci. Rev. **87**(1), 102–114 (1993)
14. Obraztsova, S., Polukarov, M., Rabinovich, Z., Elkind, E.: Doodle poll games. In: Proceedings of the 16th Conference on Autonomous Agents and MultiAgent Systems. AAMAS 2017, Sao Paulo, Brazil, pp. 876–884 (2017)
15. Thompson, D.R., Lev, O., Leyton-Brown, K., Rosenschein, J.: Empirical analysis of plurality election equilibria. In: Proceedings of the 2013 International Conference on Autonomous Agents and Multi-agent Systems. AAMAS 2013, St. Paul, MN, pp. 391–398 (2013)
16. Weber, R.J.: Approval voting. J. Econ. Perspect. **9**(1), 39–49 (1995)
17. Zou, J., Meir, R., Parkes, D.: Strategic voting behavior in Doodle polls. In: Proceedings of the 18th ACM Conference on Computer Supported Cooperative Work. CSCW 2015, pp. 464–472. ACM, New York (2015)

Network Cost-Sharing Games: Equilibrium Computation and Applications to Election Modeling

Rahul Swamy[✉], Timothy Murray, and Jugal Garg

University of Illinois at Urbana-Champaign,
104 S Mathews Ave, Urbana, IL 61801, USA
rahulswa@illinois.edu

Abstract. We introduce and study a variant of network cost-sharing games with additional non-shareable costs (NCSG+), which is shown to possess a pure Nash equilibrium (PNE). We extend polynomial-time PNE computation results to a class of graphs that generalizes series-parallel graphs when the non-shareable costs are player-independent. Further, an election game model is presented based on an NCSG+ when voter opinions form natural discrete clusters. This model captures several variants of the classic Hotelling-Downs election model, including ones with limited attraction, ability of candidates to enter, change stance positions and exit any time during the campaign or abstain from the race, the restriction on candidates to access certain stance positions, and the operational costs of running a campaign. Finally, we provide a polynomial-time PNE computation for an election game when stance changes are restricted.

Keywords: Network cost-sharing game · Nash equilibrium Hotelling-Downs

1 Introduction

Network cost-sharing games (NCSGs) are games on a directed graph where each player selects a path from their source to sink, and players sharing an edge divide the utility obtained from that edge. Even though these games are known to possess a pure Nash equilibrium (PNE), computing one is PLS-hard except for simple special cases, e.g., a restricted variant of series-parallel graphs [15]. We study a generalization of these games, NCSG+, where in addition to the shareable utility, each edge incurs a non-shareable player-specific cost (such as a fee or a toll), called the *fixed cost* of traversing that edge. The advantage of studying NCSG+ is that they generalize election games, where a path in the NCSG+ graph corresponds to a campaign strategy in an election. For NCSG+, we show the existence of a PNE using a potential-function argument in any

J. Garg—Supported by NSF CRII Award 1755619.

directed graph. Further, we extend polynomial-time PNE computability for a class of graphs that generalizes series-parallel graphs with multiple source nodes.

In addition to the study of NCSG+, this paper presents a spatio-temporal bi-objective model for an election game with discrete stances and analyzes its PNE computation by utilizing the structural properties of NCSG+. Consider an election where candidates compete to win as many voters as possible. In many real-life elections, a voter has a *stance* on a range of issues that matters to them, and the choice of their candidate is heavily influenced by the candidate's stance on those issues. In the classical Hotelling-Downs model [4], stances on each issue are represented by continuous values in $[0, 1]$, where 0 and 1 are extreme stances on the issue, and a multi-issue hypercube can be constructed containing all the voters' stances. Based on the stance positions in this hypercube, each candidate's objective is to choose their stance to be *close* to the maximum number of voters. When candidates' stances are relatively close to each other, they split their vote share giving rise to a game with spatial competition.

In certain elections, voters' stance positions exhibit natural accumulations of opinions forming clusters. As a candidate deciding what their ideal stance should be, identifying such naturally occurring clusters provides vital information in making a choice that leads to maximal electoral advantage. For example, a 2014 study conducted by Pew Research Center [9] found that 50% of US adults polled believe that climate change is caused by human activity, while 23% believe that it is due to natural patterns, and 25% believe that there is no solid evidence; there are three mutually exclusive clusters of voter opinions. If an election is based only on this one issue, and if there is only one candidate, choosing the stance "caused by human activity" will be their winning strategy. However, if there are 3 other candidates and all of them have picked that as their stance, the winning strategy would then be to pick either of the two smaller clusters. As illustrated, there is a need for election game modeling that extracts the combinatorial structure exhibited by a finite and discrete stance space. Additionally, there is temporal decision-making involved. Since campaigns often cost considerable time, money and resources, the cost of campaigning influences the decisions of *whether* a candidate should even enter the race, and if they do, *when* exactly they should enter the race. Entering early enables them to gain voters from an earlier time, but may incur a higher cost of campaigning given the longer time spent, and vice versa. Hence, there is an inherent trade-off between the accumulation of voters and cost considerations.

The election game presented allows for candidates to (1) decide whether to enter the race or not, (2) decide when to enter and exit the race, (3) choose their stance from a finite set of stances, (4) and also change stances during the race. It also models the trade-off between voter and cost considerations. While some of these modeling aspects have been independently studied in prior work, the flexibility offered by the network-based model provides a unification of these features. Finally, we derive a stronger polynomial-time PNE result for election games with a restriction on stance changes.

1.1 Related Literature

This paper makes contributions in two broad areas of research: PNE analysis in NCSG+, and the spatio-temporal modeling and PNE analysis in election games.

NCSGs naturally model games on a network where the cost of traversing an edge increases with the number of players sharing the edge, and has applications in traffic and communication networks. Introduced by Rosenthal [15], a network congestion game (NCG) is a related game with a general edge latency function which always possesses a PNE. This spurred research in variants of NC[S]Gs and their polynomial-time PNE computability. Syrgkanis [19] showed that PNE computation for NCSGs in general directed graphs is PLS-Complete, while providing polynomial-time algorithms for singleton cost-sharing games (with single-edge paths) and matroid cost-sharing games. Recently, Feldotto et al. [7] considered an extension with two types of costs: latency and bottleneck costs, while players have different preferences for the two. They showed that even though PNE exists for singleton congestion games, deciding on existence is NP-Hard for general matroid congestion games. Along the lines of investigating PNE in various graphs, Fotakis [8] showed that a greedy best-response algorithm computes a PNE for NCGs in series-parallel graphs. However, the question of which broader class of NCG graphs possesses a polynomial-time PNE remains open. This paper provides PNE computation results for a multi-source single-sink graph that generalizes series-parallel graphs.

Modeling election games has early roots in Hotelling's [10] seminal model for spatial competition in which two competing vendors located at two points on a street must decide what prices to charge for their products. He derived closed-form expressions for calculating these price points as a unique PNE. The Hotelling model was brought into the political sphere by Downs [4] as a strategic method for identifying the equilibrium positions which candidates take on an issue. This model has influenced research in modeling electoral politics, including spatial voting models with issue-based stances. Since then, several variations of Hotelling-Downs have been explored [2,17]. However, the difficulty in proving existence and computation of a PNE in general multi-issue elections has led to several specific adaptations of election models. A multidimensional spatial model proposed by Duggan and Fey [5] considered a continuous utility function to obtain equilibrium results under certain special conditions. They show that in two dimensions, when the number of players is odd and when there is symmetry in the utility function, a PNE exists. In elections with proportional representation (where voters submit a preference list of candidates), Ding and Lin [3] formulated a zero-sum game model and show that for two parties (two types of candidates), a PNE exists but computing one is NP-hard. The consequence of choosing stances based on finite clusters is that candidates have influence only within a finite *window* around their chosen stance. A similar idea has previously been modeled by Feldman et al. [6], where voters randomly choose from candidates who are *sufficiently close* to them. This was generalized by Shen and Wang [18] as a model with limited attraction. However, the strategy space in these models is infinite in size due to the continuous nature of the stance space.

Hence, even though a PNE exists in these models, it is unclear how to find one efficiently.

Temporal extensions to Hotelling-Downs have been explored in recent work in modeling election campaigns. Osborne [14] considered the entry of candidates by using the associated campaign cost of doing so. Recently, Kallenbach et al. [11] introduced an optimization problem to compute the optimal cost of campaigning for each candidate, and can be used as a subroutine for equilibrium computations. Sengupta and Sengupta [17] extended Osborne's model to include the option of dropping out from the race. These models possess PNE but only under specific assumptions. Abstention by candidates has been addressed in election modeling from early work by McKelvey and Wendell [12]. In strategic candidacy games (where the choice to enter the election or not is captured by analyzing the incentives), Brill and Conitzer [1] consider a two-stage game: the first stage where candidates decide whether to run or not, and the second where voters decide who to vote for. They show the existence of a PNE when the voter opinions on issues is single-peaked. However, opinion distributions in general are not always single-peaked. This model has been extended by Obraztsova et al. [13] who introduced the concept of *lazy* candidates who will drop out after a certain time period if the campaign costs are too high. A strategy candidacy game proposed by Sabato et al. [16] imposes *restrictions* on each candidate's stance space to within a defined interval and studies its effect using various voting rules. Our model includes the considerations of abstention and dropping out, as well as restricted stance sets for the candidates. While these models individually capture different important aspects of election games, the question remains whether all these can be captured simultaneously. Our model provides a partial answer to this by particularly focusing on elections where voter opinions exhibit natural clusters.

2 Network Cost Sharing Games with Non-sharable Costs

We begin with game theoretic preliminaries. Consider a game with k players, and for each player $j \in [k]$, let $\mathcal{P}_j$ be the set of pure strategies that j can choose from. Further, let $P_j \in \mathcal{P}_j$ denote a strategy that j chooses and let $\mathcal{S} = (P_1, P_2, \ldots, P_k) \in (\mathcal{P}_1 \times \mathcal{P}_2 \times \cdots \times \mathcal{P}_k)$ denote a *strategy profile*, a vector of strategies chosen by all the players. Corresponding to a strategy profile and player j, let $\mathbf{P}_{-j}$ denote the vector of strategies chosen by all the players except j. Further, let $u_j(P_j, \mathbf{P}_{-j}) \in \mathbb{R}$ denote the utility that j receives when j chooses P_j and all the other players choose $\mathbf{P}_{-j}$. Each player tries to choose a strategy that maximizes their utility. A PNE is a strategy profile such that no player can unilaterally increase their utility by deviating from their strategy, i.e., a strategy profile $\mathcal{S} = (P_1, P_2, \ldots, P_k)$ is a PNE if for each player j, there exists no strategy $P_j' \in \mathcal{P}_j$ such that $u_j(P_j', \mathbf{P}_{-j}) > u_j(P_j, \mathbf{P}_{-j})$. The existence of a PNE is not guaranteed in general, e.g., the game Rock-Paper-Scissors does not have a PNE.

An NCSG is a game on a directed graph $G = (V, E)$ with k players, and each player j has a source node s_j and a sink node d_j. Every edge $e \in E$ has a sharable utility u_e, which is equally divided among the players that traverse e. We introduce an NCSG with non-shareable costs (NCSG+), where we also consider a non-sharable player-dependent cost in each edge, called the *fixed cost*. For a player $j \in [k]$ and edge $e \in E$, let f_e^j be the fixed cost for j on edge e. Each player's strategy is a path from their source to their sink. Let P_j be player j's path from s_j to d_j, let $\mathbf{P}_{-j}$ be a vector containing the paths taken by players $[k]\backslash\{j\}$, and let $(P_j, \mathbf{P}_{-j})$ be a vector of paths taken by all the players. Given a strategy profile $(P_j, \mathbf{P}_{-j})$, let n_e be the number of players traversing edge e. The net utility for j is then defined as $u_j(P_j, \mathbf{P}_{-j}) = \sum_{e \in P_j}(\frac{u_e}{n_e} - f_e^j)$. Every player's objective is to choose a path that maximizes their net utility. For network congestion games (with player-independent cost functions), Rosenthal [15] showed that a PNE is guaranteed to exist using a potential function argument. We extend this proof to an NCSG+. A *potential game* is one where there exists a potential function $\phi : (\mathcal{P}_1 \times \ldots \mathcal{P}_k) \to \mathbb{R}$ such that if any player deviates to a better strategy, the change in potential function value is equal to the increase in that player's net utility. We now show that an NCSG+ possesses such a potential function, implying the existence of a PNE.

Lemma 1. *An NCSG+ is a potential game, with its potential function given by* $\phi(\mathcal{S}) = \sum_{e \in E}\left(\sum_{i=1}^{|N_e|} \frac{u_e}{i} - \sum_{i \in N_e} f_e^i\right)$, *where $\mathcal{S}$ is a strategy profile and N_e is the set of players traversing edge e in $\mathcal{S}$.*

Proof. Let $\mathcal{S} = (P_j, \mathbf{P}_{-j})$ with respect to a player j. If j deviates its path from P_j to P_j', let $\mathcal{S}' = (P_j', \mathbf{P}_{-j})$ be the new strategy profile. Then, the set of players on an edge $e \in P_j \cap P_j'$ will remain as N_e and hence considering the difference $\phi(\mathcal{S}') - \phi(\mathcal{S})$ after the deviation, these edges will cancel each other. However, the set of players on an edge $e \in P_j \backslash P_j'$ will be $N_e \backslash \{j\}$, and those on $e \in P_j' \backslash P_j$ will be $N_e \cup \{j\}$. Hence,

$$\phi(\mathcal{S}') - \phi(\mathcal{S}) = \sum_{e \in P_j \backslash P_j'}\left(\sum_{i=1}^{|N_e|-1}\frac{u_e}{i} - \sum_{i \in N_e \backslash \{j\}} f_e^i\right) + \sum_{e \in P_j' \backslash P_j}\left(\sum_{i=1}^{|N_e|+1}\frac{u_e}{i} - \sum_{i \in N_e \cup \{j\}} f_e^i\right)$$

$$- \sum_{e \in P_j \backslash P_j'}\left(\sum_{i=1}^{|N_e|}\frac{u_e}{i} - \sum_{i \in N_e} f_e^i\right) - \sum_{e \in P_j' \backslash P_j}\left(\sum_{i=1}^{|N_e|}\frac{u_e}{i} - \sum_{i \in N_e} f_e^i\right)$$

$$= \sum_{e \in P_j' \backslash P_j}\left(\frac{u_e}{|N_e|+1} - f_e^j\right) - \sum_{e \in P_j \backslash P_j'}\left(\frac{u_e}{|N_e|} - f_e^j\right) = u_j(\mathcal{S}') - u_j(\mathcal{S}).$$

Hence, ϕ is a potential function for an NCSG+. $\qquad\square$

Having shown that a PNE exists in any NCSG+, we focus on polynomial-time PNE computability. A natural question is: what settings of an NCSG+—graphs, utility-cost functions, number of players—permits a polynomial-time PNE computation?

A *series-parallel* (SP) graph is a single-source single-sink directed multi-graph, whose recursive definition is as follows. An *elemental* SP graph consists of a source s, a sink d and the single edge (s, d). Starting from them, any SP graph can be constructed from two other SP graphs G and H, using two composition rules: (a) a *series* composition whose source-sink pair is (s_G, d_H) and d_G is connected to s_H, and (b) a *parallel* composition whose source is s_G and s_H merged into a single node, and whose sink is d_G and d_H merged into a single node. In addition to their extensive applications in electrical networks, SP graphs are of interest to research in computational complexity since many combinatorial problems that are NP-Complete in general graphs are polynomial-time in SP graphs [20].

We now consider a *network congestion game* (NCG) defined as follows. Given a directed graph $G = (V, E)$ with source s and sink d, a cost function l_e for all edges $e \in E$, a NCG is a game where each player $i \in [k]$ sends $w_i \in \mathbb{R}^+$ amount of flow from s to d through G such that their total cost of sending that flow is minimized. An NCSG+ whose fixed-cost values on all the edges are player-independent (i.e. $f_e^i = f_e^j$ for all $i, j \in [k], e \in E$) and the players have a common source and sink (i.e. $s_i = s_j, d_i = d_j \forall i, j \in [k]$) is a special case of an NCG. A natural method to compute PNE in an NCG (and in an NCSG+) is using the *greedy best response* (GBR) algorithm. Starting with an empty set of players, GBR introduces one new player at a time to enter the game where the new player selects their best (highest net utility) strategy that is available to them. This best strategy is also called a *best response* by that player based on the previously introduced players, and the algorithm iteratively finds the best response paths of all the players. An NCG whose edge cost functions are in such a way that the best response is symmetric (player-independent) about all the players is said to possess the *common best response* property. Fotakis et al. [8] showed that for NCGs on SP graphs that possess the common best response property, GBR computes a PNE in $\mathcal{O}(km \log m)$ time, where $m = |E|$ (Theorem 1), even though they produce a simple counter-example where GBR fails for a non-series parallel graph.

Theorem 1 [8]. *Given a series-parallel graph $G = (V, E)$ with source and sink nodes s, d, and a network congestion game that has the common best response property, GBR succeeds and computes a PNE in time $\mathcal{O}(nm \log m)$, where $n = |V|, m = |E|$.*

This result can be extended to a subclass of NCSG+ with player-independent fixed-costs, since these games in SP graphs possess the common best response property.

Corollary 1. *Given a series-parallel graph $G = (V, E)$, and a network cost-sharing game with player-independent fixed-costs, GBR succeeds and computes a pure Nash equilibrium in time $\mathcal{O}(km)$, where $m = |E|$.*

Proof. The success of GBR in computing PNE in a SP graph follows from Theorem 1, since an NCSG+ with player-independent fixed-costs in an SP graph is a special case of an NCG with common best response property in an SP graph. As for the computation time, every new player introduced solves a longest-path problem to find their best response path. This problem takes $\mathcal{O}(m)$ computations on a directed acyclic multi-graph, such as an SP graph. Since there are k

players introduced, it requires $\mathcal{O}(km)$ computations for GBR to find PNE for all the players.   □

In this paper, we consider an NCSG+ with player-independent fixed-costs on a class of graphs where the source nodes are unique for each player even though they share a common sink. This class of graphs generalizes SP graphs and is defined as follows.

Definition 1 (Multi-source Series-Parallel Graph). *A Multi-Source Series-Parallel Graph (MSSP Graph) is given by* $R = (\{G_l\}_{l \in [n]}, \{s_i\}_{i \in [k]}, d, \{H_i\}_{i \in [k]})$, *where* $\mathcal{G} = \{G_l\}_{l \in [n]}$ *is a set of n disjoint series-parallel (SP) graphs, $\{s_i\}_{i \in [k]}$ is a set of k source nodes, and d is a sink node. For each $i \in [k]$, let $H_i \subseteq \mathcal{G}$ be a subset of SP graphs that i has "access" to, i.e., from s_i, let there be edges to the source nodes of all the SP graphs in H_i. Further, from the sink node of each SP graph in $\mathcal{G}$, let there be an edge to d.*

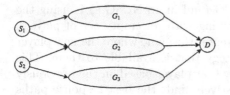

Fig. 1. An MSSP graph with $n = 3$ subgraphs and $k = 2$ sources

Figure 1 depicts an MSSP graph with $n = 3$, $k = 2$, $H_1 = \{G_1, G_2\}$ and $H_2 = \{G_2, G_3\}$. An MSSP graph is a multi-source generalization of SP graphs with a unique source node for each player. Additionally, each player i has access to only a subset of SP subgraphs defined by the collection of sets $H_i \subseteq \mathcal{G}$.

A game on an MSSP graph models the restricted access of players to n resources (SP subgraphs). Even though Corollary 1 states that GBR computes a PNE for an SP graph, it is unclear whether this approach can be extended to an MSSP graph since the common best response property is violated if different players can only access certain graphs (unless $H_i = \mathcal{G}$ for all $i \in [k]$). To do so, we introduce a generalization of GBR, called *greedy best response with reactionary movements* (GBR-RM). In this algorithm, as a reaction to each new player introduced into the game, the players who were introduced earlier may change their previously chosen strategy (termed a reactionary movement) to another strategy that gives them a better net utility, and this may trigger further movements of players, and so on. The success of GBR-RM in computing PNE relies on the eventual *convergence* of players to an equilibrium after every new player introduced (i.e. there are no cycles in reactionary movements).

Theorem 2. *Given a multi-source series-parallel graph $R = (\{G_l\}_{l \in [n]}, \{s_i\}_{i \in [k]}, d, \{H_i\}_{i \in [k]})$, and a network cost-sharing game with player-independent fixed-costs with k players, GBR-RM computes a pure Nash equilibrium in time $\mathcal{O}(km * \min\{n, k\})$, where m is the number of edges in R.*

Proof. The proof proceeds by induction considering the introduction of players by the order of their labels from 1 through n. When player 1 is introduced, it is trivially at equilibrium. Before player $i > 1$ is introduced, let the paths taken by the $i - 1$ players be $(p_1, p_2, \ldots, p_{i-1})$, and let us assume that they are at equilibrium. Let player i's best response path be p_i, and the strategy profile of the system is denoted by $\mathbf{P}^i = (p_1, p_2, \ldots, p_i)$. Further, after i's introduction, for any player $j \leq i$, let the set of paths chosen by all the other players be denoted by $\mathbf{P}^i_{-j}$. Let $u(p_j, \mathbf{P}^i_{-j})$ be the net utility for j to traverse p_j given that the other players chose paths in $\mathbf{P}^i_{-j}$. More generally, let $u(A, \mathbf{P}^i_{-j})$ be the net utility for j to traverse the subset of edges A given that other players chose paths in $\mathbf{P}^i_{-j}$, regardless of whether A is a valid path in the graph. We first show the following claim.

Claim. Consider an NCSG with player-independent fixed-costs with k players. During the GBR-RM, before a player $i \in [k]$ is introduced, let the system be at equilibrium. Let p_i be the best response path chosen by i. Then, the net utility for i will be no more than the net utility for any other player $j < i$ in path p_j, i.e., $u(p_i, \mathbf{P}^i_{-i}) \leq u(p_j, \mathbf{P}^i_{-j})$.

Proof. Since player j was at equilibrium before player i was introduced, the net utility from p_j was better than from p_i. Hence, $u(p_j, \mathbf{P}^{i-1}_{-j}) \geq u(p_i, \mathbf{P}^{i-1}_{-j})$. Let $A = p_j \backslash p_i$ and $B = p_i \backslash p_j$. A and B are disjoint sets, and from the relation above, we have $u(A, \mathbf{P}^{i-1}_{-j}) \geq u(B, \mathbf{P}^{i-1}_{-j})$, where $u(A, \mathbf{P}^{i-1}_{-j})$ denotes the utility derived specifically from the edges of $A \subseteq p_j$. Now consider when player i enters the system and picks a path p_i so as to maximize $u(p_i, \mathbf{P}^i_{-i})$. Clearly $u(p_i, \mathbf{P}^i_{-i}) \geq u(p_j, \mathbf{P}^i_{-i}) \Rightarrow u(B, \mathbf{P}^i_{-i}) \geq u(A, \mathbf{P}^i_{-i})$. However, A and p_i are disjoint, so $u(A, \mathbf{P}^i_{-j}) = u(A, \mathbf{P}^{i-1}_{-j})$, and so are B and p_j so $u(B, \mathbf{P}^i_{-i}) = u(B, \mathbf{P}^{i-1}_{-j})$, which gives us that $u(A, \mathbf{P}^i_{-j}) = u(A, \mathbf{P}^{i-1}_{-j}) \geq u(B, \mathbf{P}^{i-1}_{-j}) = u(B, \mathbf{P}^i_{-i})$. Since p_i and p_j derive the same net utility from the edges of $p_i \cap p_j$, this implies that $u(p_i, \mathbf{P}^i_{-i}) \leq u(p_j, \mathbf{P}^i_{-j})$. $\qquad\Box$

We now show that after player i is introduced, the system reaches equilibrium after at most $\mathcal{O}((i-1) * \min\{n, k\})$ reactionary movements.

In an MSSP graph R, since the series-parallel (SP) subgraphs are disjoint, any path p in R *almost* entirely lies in exactly one of its SP subgraphs, denoted by $G(p)$. We say "almost" since p includes an additional edge from a source node of R to the source node of $G(p)$. For simplicity of notation, if another path q also almost entirely lies in the subgraph $G(p)$, we will denote this as "$q \in G(p)$" instead of $G(p) = G(q)$.

With the introduction of player i to p_i, if a player j_1 wishes to make a reactionary movement, it must be that p_{j_1} is in $G(p_i)$ since j_1 was at equilibrium before i's introduction. Let the new path chosen by j_1 be q_{j_1}. Since we know that GBR computes a PNE (without any reactionary movements) within an SP graph, q_{j_1} must traverse an SP subgraph $G(q_{j_1}) \neq G(p_{j_1})$ such that $G(q_{j_1}) \in H_{j_1} \backslash H_i$. This is because if $G(q_{j_1}) \in H_i$, then it is already a path which player i considered and rejected which is edge disjoint from p_i, i.e., $u(q_{j_1}, \mathbf{P}^i_{-j_1}) = u(q_{j_1}, \mathbf{P}^i_{-i}) \leq u(p_i, \mathbf{P}^i_{-i}) \leq u(p_j, \mathbf{P}^i_{-j})$, where the final inequality comes from Lemma 2. After j_1 moves to q_{j_1}, we set i to traverse p_{j_1}, the path vacated by j_1. By Lemma 2, player i is willing to do so as player j_1 was deriving more utility from it than i currently derives from p_i. The set of strategies chosen by players inside $G(p_i)$ is now the same as it was prior to player i's introduction, and with the exception of i no players have changed strategies. Because this state was a pure equilibrium within $G(p_i)$, all players traversing $G(p_i)$ remain at equilibrium and no new players not currently traversing $G(p_i)$ wish to change their paths to do so. Two reactionary movements have occurred as a result of i's introduction. We continue to iterate this scheme for picking which players make reactionary movements, and our next step is to bound the maximum number of reactionary movements. To do so, we will first show that at each iteration the moving player must switch to a path that none of the previous players who moved had access to using an induction.

Similar to the base case, we next consider the consequences of player j_1 moving to $G(q_{j_1})$. Suppose another player j_2 on $G(q_{j_1})$ wishes to move in reaction to j_1. Then, j_2 must wish to move to $q_{j_2} \in H_{j_2} \setminus (H_{j_1} \cup H_i)$. This comes from the same reasoning as above. So $q_{j_2} \in H_{j_2} \setminus (H_{j_1} \cup H_i)$. We then move j_1 from q_{j_1} to p_{j_2}, restoring the state of $G(q_{j_1})$ to what it was at equilibrium. For the inductive step, assume that players $M = \{i, j_1, j_2, ..., j_m\}$ have all made reactionary movements in that order so far, such that in each case, the new path chosen by player $l \in M$ is $q_l \in G(q_l) \in H_l \setminus H^{-l}$, where $H^{-l} = (H_i \cup H_{j_1} \cup ... \cup H_{j_{l-1}})$. After player j_m switches paths to $q_{j_m} \in G(q_{j_m})$, player j_{m+1} wishes to switch to $q_{j_{m+1}}$. We want to show that $G(q_{j_{m+1}}) \in H_{j_{m+1}} \setminus H^{-(m+1)}$. Assuming the contrary, there exists some player $l \in M$ such that $G(q_{j_{m+1}}) \in H_l$. Then $u(q_{j_{m+1}}, \mathbf{P}^{i_{j_{m+1}}}_{-(j_{m+1})}) = u(q_{j_{m+1}}, \mathbf{P}^{i_l}_{-l})$ where i_l is the set of paths chosen by all players after $l - 1$ has moved to path q_{l-1} and $l - 2$ has moved to p_{l-1}. This is because of the inductive assumption that every player who has moved so far has moved to a path in $H_l \setminus H^{-l}$, and so the state of all players in H^{-l} remains unchanged from when l considered it. We then have $u(q_{j_{m+1}}, \mathbf{P}^{i_l}_{-l}) \leq u(q_l, \mathbf{P}^{i_l}_{-l}) \leq u(p_{l+1}, \mathbf{P}^{i_{l+1}}_{-(l+1)}) \leq \cdots \leq u(p_{j_m}, \mathbf{P}^{i_{j_m}}_{-j_m}) \leq u(p_{j_{m+1}}, \mathbf{P}^{i_{j_{m+1}}}_{-j_{m+1}})$, where the first inequality comes from the fact that l chose q_l instead of $q_{j_{m+1}}$, and all other inequalities come from Lemma 2. Hence, $G(q_{j_{m+1}}) \in H_{j_{m+1}} \setminus H^{-j_{m+1}}$.

Next, we bound the maximum number of reactionary movements. Trivially, no player can move twice since for any player l, $H_l \setminus (H_i \cup H_{j_1} \cup ... \cup H_l \cup ... \cup H_{j_m}) = \emptyset$, and hence there can be at most i moves. Additionally, if $|H_i \cup H_{j_1} \cup ... \cup H_{j_m}| = n$, then for any player l, we have $H_l \setminus (H_i \cup H_{j_1} \cup ... \cup H_{j_m}) = \emptyset$. Since $|H_i \cup H_{j_1} \cup ... \cup H_{j_m}| - |H_i \cup H_{j_1} \cup ... \cup H_{j_{m-1}}| \geq 1$, there are at most $\mathcal{O}(\min\{k, n\})$ movements.

Since k players are introduced, the number of movements that may occur is at most $\mathcal{O}(k * \min\{n, k\})$. Since each subgraph is a series-parallel DAG, we can compute a maximum cost/profit path from s_i to d in $\mathcal{O}(m)$ time. Therefore, we can compute a PNE on an MSSP graph in at most $\mathcal{O}(km * \min\{k, n\})$ time. $\square$

3   Election Game

An election game is between k players (or candidates) competing to appease the maximum number of voters with the least amount of expenditure. A candidate can choose from a finite set of stances $\{1, 2, ..., n\}$, where each stance $s \in [n]$

corresponds to a cluster of voters. Let $p(s) \in [0,1]$ be the fraction of voters *contained* in the cluster corresponding to stance s. Further, for each candidate $j \in [k]$, let $H_j \subseteq [n]$ be the subset of stances that are available to candidate j. H_j models the general condition that j can only choose a stance that is close to their past record or political inclinations.

Single-Period Election Game. First, consider a game where candidates only decide which stance to pick. Let $c_j \in H_j$ denote the stance picked by candidate j, and let $N(c_j)$ be the set of all candidates who picked stance c_j. Assume that there is a certain cost associated with a candidate's expenditure of resources (monetary, personnel, etc.) for choosing a stance. For a candidate j, let $C^j(c)$ denote the cost incurred for j when choosing stance c.

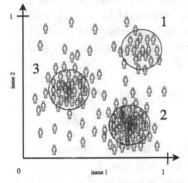

Fig. 2. Stance distribution of voters on 2 issues with 3 clusters.

In order to compare the electoral component of utility (p) and the cost component (C), let $\beta \in \mathbb{R}^+$ be a trade-off parameter between the fraction of voters and the cost (in monetary units). For simplicity, we assume that the cost function C already includes this trade-off.

Then, the net utility obtained by candidate j is given by $u_j(c^j) = (\frac{p(c_j)}{|N(c_j)|} - C^j(c_j))$. This means that candidate j shares their electoral utility with other candidates choosing the same stance, in addition to incurring a non-shareable cost.

Each candidate's goal is to maximize their net utility. Consider an example with the spatial voter distribution in Fig. 2 with 3 clusters. Let there be 3 candidates who will compete to pick 3 stances. Further, let candidate 1 pick stance $2 \in H_1$, and let $p(2) = 0.6$ and $C^1(2) = C^2(2) = C^3(2) = 0.05$. Then, candidate 1 receives a net utility of 0.15 if all three candidates chose stance 2, 0.25 if only one other candidate does so, and 0.55 if no other candidate does so.

Multi-period Election Game. Generalizing the single-period game, we study the election game over T time periods, where each time period is an arbitrary unit of time (a day/week/month or could even be aperiodic like the time between successive state primaries as in US presidential elections). Each candidate must first decide whether they should enter the game or not. If a candidate j decides to enter the game, let the time period at which they enter be $t_1^j \in \{0, 1, \ldots, T-1\}$, the stance that they choose be $c_{j,t_1^j} \in H_j$ and the time period at which they exit the game be $t_2^j \in \{t_1^j + 1, t_1^j + 2, \ldots, T\}$. Further, we assume that candidates are allowed to change stances during the game.

Let $c_{j,t} \in H_j$ denote stance chosen by candidate j at the start of time period $t \in \{t_1^j, t_1^j + 1, \ldots, t_2^j - 1\}$. If a candidate j never exits the race and runs till the end, then $t_2^j = T$. Hence, if candidate j enters the game, their overall strategy is represented by the tuple $(t_1^j, t_2^j, \{c_{j,t}\}_{t=t_1^j}^{t_2^j - 1})$, where $t_2^j > t_1^j$.

Extending the cost function in the single-period game, let $C^j(c,t) \in \mathbb{R}^+$ be the cost associated with candidate j for holding stance c at the start of time period t. As a temporal extension to the voter distribution, let $p(c,t)$ be the fraction of voters with stance c who will affirm their stance at time t. This is a general function that can model any pre-election scenario. For example, if it is an election where most of the voters affirm their stances only just before the election, then the candidates would not gain much utility in entering the race early, as opposed to an election where the opposite trend could occur. Our generic utility function captures either scenario.

Let $J \subseteq [k]$ be the set of candidates who decide to enter the race at some time period, and let $N_t(c) \subseteq J$ be the subset of candidates who enter the race/chose stance c at the start of time period t. On the other hand, if candidate j does not enter the game at all, then let α_j denote the utility obtained by j. This utility is a measure of monetary or political savings when not entering the race. That is, $u_j = \alpha_j$ for every $j \in [k]\backslash J$. Then, the net utility u_j for candidate $j \in J$ is defined as

$$u_j(t_1^j, t_2^j, \{c_{j,t}\}_{t=t_1^j}^{t_2^j-1}) = \begin{cases} \sum_{t=t_1^j}^{t_2^j-1} \left(\dfrac{p(c_{j,t},t)}{|N_t(c_{j,t})|} - C^j(c_{j,t},t) \right), & \text{if } j \in J \\ \alpha_j, & \text{otherwise.} \end{cases} \quad (1)$$

Each candidate's goal is to maximize their net utility. The longer the candidate stays in the race, the more is the electoral utility they will gain from staying in the race. Hence, it is not just sufficient to pick a *good* sequence of stances, but the length of the campaign $(t_2^j - t_1^j)$ also influences the net utility. In other words, even if $t_2^j = T$ (candidate j stayed till the end), the electoral utility is not just defined by the last time period, but accumulates from the time they entered the race. It is also possible that $t_2^j < T$, wherein candidate j drops out before the completion of the race (due to accumulated costs of campaigning dominating electoral gain in utility). The net utility gained by an early drop-out models any amount of political gain resulting from campaigning for the election, even if it may not help them in that particular election.

3.1   Election Game Graph

We now show that an election game reduces to an NCSG+ through the construction of a graph called the election game graph (EGG). This construction transforms a strategy in the election game (to enter the race or not? when to enter? what sequence of stances to choose while in the game? when to quit?) to a path in EGG, constructed as follows.

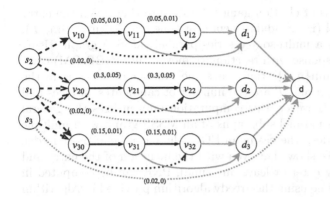

Fig. 3. An EGG with 3 candidates, 3 stances and 2 time periods. Stance-choice, entry, sustain, exit and abstain edges are dashed-black, dotted-black, solid black, solid grey, and thinly dotted-block respectively, along with their (u_e, f_e) values; $(0,0)$ if no label.

Nodes: Each candidate $j \in [k]$ has a source node s_j. A sink node d is common to all the candidates. A terminal node d_c is exists for each stance $c \in [n]$. There are intermediary nodes, called *stance nodes* for every stance and time period. Let v_{ct} be the stance node for stance $c \in [n]$ and time period $t \in \{0, 1, \ldots, T\}$.

Edges: There are six types of edges as outlined below with respect to candidate $j \in [k]$.

1. Stance-choice edge: $e = (s_j, v_{c0})$ for all $c \in H_j$ and represents candidate j's choice of stance $c \in H_j$. The $(u_e, f_e^j) = (0,0)$ of such an edge.
2. Entry edge: $e = (v_{c0}, v_{ct})$ for all $c \in H_j$ and $t \in \{1, \ldots, T-1\}$ represents candidate j having already chosen stance $c \in H_j$, entering the race at time period t. The $(u_e, f_e^j) = (0,0)$ of such an edge for all j.
3. Sustain edge: $e = (v_{c(t-1)}, v_{ct})$ for each stance $c \in [n]$ and time period $t \in \{1, \ldots, T\}$ represents sustaining in the race for time period t. The $(u_e, f_e^j) = (p(c,t), C^j(c,t))$ of such an edge.
4. Stance-change edge: $e = (v_{ct}, v_{c't})$ for each pair of stances $c, c' \in [n], c \neq c'$ and time period $t \in \{1, 2, \ldots, T-1\}$ represents changing stance from c to c' between t and $t+1$. The corresponding $u_e = 0$, and $f_e^j = +\infty$ if $c' \notin H_j$.
5. Exit edge: $e = (v_{ct}, d_c)$ exists for each stance c and time period $t \in \{1, 2, \ldots, T\}$ to represent exiting the race immediately after t. An edge (d_c, d) also exists for each $c \in [n]$ to represent the final exit. The $(u_e, f_e^j) = (0,0)$ of such an edge.
6. Abstain edge: $e = (s_j, d)$ represents abstention, with $(u_e, f_e^j) = (\alpha_j, 0)$.

Figure 3 illustrates an EGG without stance-change edges for simplicity. The construction of the EGG reduces an election game to an instance of NCSG+, thereby implying the existence of PNE in any election game using Lemma 1.

3.2   Computation of PNE

Consider election games under two restrictions: When candidates are not allowed to change their stance (i.e. stance-change edges are removed), and when the non-shareable costs are candidate-independent, denoted by $f_e^j = f_e$ for edge $e \in E, \forall j \in [k]$. We provide a greedy best-response algorithm with reactionary movements for this subclass of election games. First, note that when the entry and exit edges are removed (i.e. candidates enter at time 0 and exit at time T), the corresponding EGG is a multi-source series-parallel (MSSP) graph. Using Theorem 2, greedy best response with reactionary movements (GBR-RM) computes a PNE in $\mathcal{O}(knT^2 \min\{n,k\})$ time since the number of edges is in the order of $\mathcal{O}(nT^2)$, where k, n and T are the number of candidates, stances and time periods respectively. However, this computation only utilizes the general structure of each the series-parallel subgraphs of an MSSP graph with multiple edges between a pair of nodes, whereas in an EGG, there can be at most 1 edge between a pair of nodes. We show that even with the inclusion of the entry and exit edges (candidates may enter or leave any time), PNE can be computed in $\mathcal{O}((k+n)T^2 + (n+T)k^2)$ time using the greedy algorithm provided in Algorithm 1, with a two-order magnitude improvement compared to the same for MSSP graphs.

Corresponding to a path from a source to sink, define a *sustain path* to be a path that consists exclusively of sustain edges. For a stance c, and entering and drop-out time periods t_1 and $t_2(> t_1)$, a sustain path is represented by a sequence of nodes $\{v_{c,t_1}, v_{c,t_1+1}, \ldots, v_{c,t_2}\}$. Let $\mathcal{P}$ be the set of all sustain paths in G. Further, let $A_j = (s_j, d)$ be the abstain path for candidate j. For a path $P \in \mathcal{P} \cup_{j=1}^k A_j$, let $S(P)$ denote the subset of candidates traversing P. For illustration, consider an example with 3 candidates, 3 stances and 2 time periods, and the corresponding EGG in Fig. 3. If candidate 1 enters the race at time period 1 and stays until the end of time period 2 by choosing stance 2, and suppose candidate 2 does the exact same, then $S(\{v_{20}, v_{21}, v_{22}\}) = \{1, 2\}$.

Starting with an arbitrary ordering of candidates, the algorithm assigns a path for each candidate one at a time and ensures that the system settles down to an equilibrium. At each stage, candidates also compare their current utility with the utility in the abstain edge to decide whether they want to abstain or not.

Algorithm 1 finds the best path for each new candidate in $\mathcal{O}(n)$ operations by tracking the best sustain path in each stance. For each stance $c \in [n]$, let $L(c)$ be the current best sustain path and its corresponding net utility. In the example in Fig. 3 before any candidate has entered, it is easy to see that $L(1) = (0.08, \{v_{10}, v_{11}, v_{12}\})$, $L(2) = (0.5, \{v_{20}, v_{21}, v_{22}\})$, and $L(3) = (0.28, \{v_{30}, v_{31}, v_{32}\})$. Once a new candidate has been assigned a path and the system resettles into an equilibrium, we show that for at most one stance $c \in [n]$, $L(c)$ needs to be updated. This can be done in $\mathcal{O}(T^2)$ operations since there are $\frac{T^2+T}{2}$ sustain paths on each stance.

In a general instance, it is possible that a new candidate entering may trigger a chain of candidates to change their paths. We show using two nested loops that the best response does converges to an equilibrium after a finite number of steps. The outer loop is for every new candidate introduced into the game, while the inner loop is for every best response move by an existing candidate in the game.

Theorem 3. *If the system is at equilibrium with $l - 1$ candidates, and the lth candidate is introduced, the sequence of best responses in Algorithm 1 leads to*

Input: An Election Game Graph $G = (V, E)$ and $(u_e, f_e), \forall e \in E$
Output: A pure Nash equilibrium
$\mathcal{P} \leftarrow$ Set of all sustain paths; $A_j \leftarrow$ Abstain path for candidate j;
$S(P) \leftarrow \emptyset, \forall P \in \{\mathcal{P} \cup_{j=1}^k A_j \}$;   // $S(P)$ contains the set of candidates choosing P
$L(c) \leftarrow$ The best sustain path in stance c for new candidate;
$n_e \leftarrow \sum_{P:e\in P} |S(P)|, \forall e \in E$; $U(P) \leftarrow \sum_{e\in P} \dfrac{u_e}{n_e + 1} - f_e$;
for $i = 1, 2, \ldots, k$ **do**
    Assign i to a path in $\arg\max_{c\in H_i} U(L(c))$;
    $l \leftarrow i$; $H \leftarrow H_l$;
    while *system not at equilibrium* **do**
        Choose a candidate j currently not at equilibrium;
        $P_j \leftarrow$ Candidate j's current path;
        Move j to $\arg\max_{c\in H_j \setminus H}\{U(L(c)), U(A_j)\}$;
        Move l to P_j; $l \leftarrow j$; $H \leftarrow H_l \cup H$;   Update $S(P), \forall P \in \mathcal{P}$;
    end
    Update $L(c)$, where c is the stance the last candidate moved to;
end
return S

Algorithm 1. GREEDY PNE COMPUTATION IN AN ELECTION GAME

a PNE in the new game in at most $\mathcal{O}(\min\{l, n\})$ steps, provided that whenever a candidate is indifferent between best response paths, it picks the longest one (most sustain edges).

Define P_i to be the sustain path taken by a candidate i. We first prove Lemma 2.

Lemma 2. *Suppose the system is at equilibrium after $i - 1$ candidates have been introduced via Algorithm 1. When candidate i is introduced, then for every candidate $j \neq i$, either $P_i \subseteq P_j$ or $P_i \cap P_j = \emptyset$.*

Proof. We prove this by contradiction: Assume that $P_i \setminus P_j \neq \emptyset$ and $P_i \cap P_j \neq \emptyset$. First, due to the latter condition, i must have picked the same stance as j did since they have overlapping sustain edges. Second, the former condition implies that there are sustain edges in i's path that are not in j's path. There are two possible cases: i entered the race at an earlier time before j did, or i exited the race after j did. Let t_1^i (t_2^i) and t_1^j (t_2^j) be the entering (exiting) time-periods of candidates i and j, respectively.

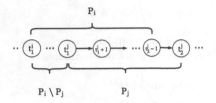

Fig. 4. Candidates i and j overlapping sustain edges, nodes are labeled by corresponding time-periods.

Consider the first case, i.e., $t_1^i < t_1^j$, as depicted in Fig. 4. We claim that the sustain edges in $P_i \setminus P_j$ must together contribute to a net positive utility for candidate i. This is true since otherwise, i would rather not enter the race as early as time period t_1^i, but rather enter at t_1^j for a higher net utility, thereby violating the given condition that P_i is the best stance path i has chosen.

However, this claim implies that before i was introduced into the game by the algorithm, j could have expanded its path to include all the sustain edges in $P_i \setminus P_j$ for a higher net utility, thereby violating the condition that the system was in equilibrium. Hence, the assumption results in a contradiction. We can make a similar argument for the other case where i exits the race at a later time-period than j.   □

Proof **(Theorem 3).** Suppose a new candidate i is introduced when $i - 1$ candidates were previously at equilibrium. Candidate i will join the best (highest net utility) path that it has access to, P_i, which will be a stance path in some stance in H_i. The only candidates who may wish to move are candidates whose path intersects with candidate i's. By Lemma 2, the stance path of such a candidate includes all the edges in P_i. Let a candidate j_1 currently on stance path P_{j_1} wish to move to another stance path P'_{j_1} for better net utility. Any stance path in H_i cannot provide a greater net utility than P_i, since otherwise, candidate i would have picked that path instead. But candidate i found that $u(P'_{j_1}|P_{-i}) \leq u(P_i|P_{-i}) \leq u(P_{j_1}|P_{-j_1}, P_i)$ where $u(P_i|P_{-i})$ is the net utility of path P_i for i given all other candidates. Therefore, it must be that if such a path P'_{j_1} exists, it is on a stance in $H_{j_1} \setminus H_i$. If such a P'_{j_1} exists, the algorithm moves j_1 to it and sets $P_i = P_{j_1}$. Doing so restores the net utility of each of the candidates on the stance path that j_1 just left, to what it was before i joined. Thus, no candidate on i's stance can make a best response move away from it or onto it due to the initial assumption of equilibrium prior to i's introduction. At this point, three movements have occurred (i to P_i, j_1 to P'_{j_1}, i to P_{j_1}).

Suppose that candidates $j_1, ..., j_{l-1}$ have made best response movements and been settled in at most $2i$ movements, and candidate j_l has moved in response $(2l + 1)$. If candidate j_{l+1} wishes to move, then by the same reasoning as in the base case, it will move to a path in a stance set in $H_{j_{l+1}} \setminus H_{j_l}$. However, any path it would consider must also be on a stance set in $H_{j_{l+1}} \setminus H_{j_{l-1}}$, as candidate j_l's current path provides a better net utility or equal utility and greater length than any path in the set $H_{j_{l-1}}$ while being inferior or equal to j_{l+1}'s current path. We extend this line of reasoning back to candidate i and j_{l+1} must choose a path represented by a stance set in $H_{j_{l+1}} \setminus \{H_{j_l} \cup ... \cup H_{j_1} \cup H_i\}$. Candidate j_{l+1} then moves and candidate j_l takes its place, bringing the total number of movements up to $2(l + 1) + 1$. However, this cannot continue indefinitely: once we have a j_l

such that $|H_{j_l} \cup ... \cup H_{j_1} \cup H_i| = n$, there will be no more movements as for any j_{l+1}, $H_{j_{l+1}} \subset (H_{j_l} \cup ... \cup H_{j_1} \cup H_i)$. Similarly, if i is significantly smaller than n such that $|H_1 \cup ... \cup H_{i-1} \cup H_i| < n$, no candidate l moves stance sets twice, as $H_l \subset \{H_i \cup ... \cup H_1\}$ for $l \in \{1, ..., i\}$. This proceeds at most $\min(l, n)$ times, resulting in $\mathcal{O}(\min\{l, n\})$ total movements. $\qquad\square$

Theorem 4. *Algorithm 1 computes a PNE in* $\mathcal{O}((k + n)T^2 + (n + T)k^2)$ *time.*

Proof. Corresponding to each stance, there are $\frac{T^2 + T}{2}$ sustain paths. In total, $n(\frac{T^2 + T}{2})$ operations are needed to compute L. To assign candidate 1, L is checked to find the best path it has access to, which takes $\mathcal{O}(n)$ comparisons. Candidate 1 is assigned to the best path and the net utilities of at most $\frac{T^2 + T}{2}$ paths are updated, bringing the total number of path evaluations to $(n + 1)(\frac{T^2 + T}{2})$. For $2 \leq i \leq k$, suppose the first $i - 1$ candidates have been assigned, the accumulated number of operations till then is $(n+i-1)\frac{T^2 + T}{2} + \sum_{j=1}^{i-1}(j-1)(T + n - 1)$. Candidate i is assigned its best path by checking L, which requires $|H_i| \leq n$ comparisons. We then check if any other candidate on candidate i's chosen path wishes to move. There are at most $i - 1$ such candidates, each taking at most T computations to evaluate the new net utility of their path, and each has to compare the utility of that path to at most $n - 1$ entries of L, implying that at most $(i-1)(T + n - 1)$ operations are necessary to evaluate if some candidate is leaving the stance path that i has joined. Note that if a candidate moves and setting off a chain of movements, there are still at most $(i-1)(T + n - 1)$ operations in total, as there are only $i - 1$ candidates that were previously at equilibrium. As demonstrated in the proof of Theorem 3, if a path was taken by m candidates before introducing a new candidate, it is taken by either m or $m+1$ candidates after all the candidates have settled into an equilibrium. Only for exactly one stance c, the net utilities on the stance-edges of c need to be re-evaluated to update $L(c)$, which takes at most $\frac{T^2 + T}{2}$ computations. Thus, adding candidate i requires at most $|H_i| + \frac{T^2 + T}{2} + (i-1)(T+n-1)$ additional evaluations, bringing the accumulated number of evaluations to $(n + i)\frac{T^2 + T}{2} + \sum_{j=1}^{i}(j-1)(T + n - 1) \Rightarrow O\left((n + i)T^2 + i^2(T + n)\right)$. $\qquad\square$

References

1. Brill, M., Conitzer, V.: Strategic voting and strategic candidacy. In: AAAI (2015)
2. Brusco, S., Dziubiński, M., Roy, J.: The Hotelling-Downs model with runoff voting. Game Econ. Behav. **74**(2), 447–460 (2012)
3. Ding, N., Lin, F.: On computing optimal strategies in open list proportional representation: the two parties case. In: AAAI (2014)
4. Downs, A.: Economic theory of political action in a democracy. J. Pol. Econ. **65**(2), 135–150 (1957)
5. Duggan, J., Fey, M.: Electoral competition with policy-motivated candidates. Games Econ. Behav. **51**, 490–522 (2005)
6. Feldman, M., Fiat, A., Obraztsova, S.: Variations on the Hotelling-Downs model. In: AAAI (2016)

7. Feldotto, M., Leder, L., Skopalik, A.: Congestion games with mixed objectives. In: Chan, T.-H.H., Li, M., Wang, L. (eds.) COCOA 2016. LNCS, vol. 10043, pp. 655–669. Springer, Cham (2016). https://doi.org/10.1007/978-3-319-48749-6_47

8. Fotakis, D., Kontogiannis, S., Spirakis, P.: Symmetry in network congestion games: pure equilibria and anarchy cost. In: Erlebach, T., Persinao, G. (eds.) WAOA 2005. LNCS, vol. 3879, pp. 161–175. Springer, Heidelberg (2006). https://doi.org/10.1007/11671411_13

9. Funk, C., Rainie, L.: Climate change and energy issues. Pew Research Center (2015)

10. Hotelling, H.: Stability in competition. Econ. J. **39**(153), 41–57 (1929)

11. Kallenbach, J., Kleinberg, R., Kominers, S.D.: Orienteering for electioneering. Oper. Res. Lett. **46**, 205–210 (2018)

12. McKelvey, R.D., Wendell, R.E.: Voting equilibria in multidimensional choice spaces. Math. Oper. Res. **1**(2), 144–158 (1976)

13. Obraztsova, S., Elkind, E., Polukarov, M., Rabinovich, Z.: Strategic candidacy games with lazy candidates. In: IJCAI, pp. 610–616 (2015)

14. Osborne, M.J.: Candidate positioning and entry in a political competition. Games Econ. Behav. **5**(1), 133–151 (1993)

15. Rosenthal, R.W.: A class of games possessing pure-strategy Nash equilibria. Int. J. Game Theory **2**(1), 65–67 (1973)

16. Sabato, I., Obraztsova, S., Rabinovich, Z., Rosenschein, J.S.: Real candidacy games: A new model for strategic candidacy. In: AAMAS, pp. 867–875 (2017)

17. Sengupta, A., Sengupta, K.: A Hotelling-Downs model of electoral competition with the option to quit. Games Econ. Behav. **62**(2), 661–674 (2008)

18. Shen, W., Wang, Z.: Hotelling-Downs model with limited attraction. In: AAMAS (2016)

19. Syrgkanis, V.: The complexity of equilibria in cost sharing games. In: Saberi, A. (ed.) WINE 2010. LNCS, vol. 6484, pp. 366–377. Springer, Heidelberg (2010). https://doi.org/10.1007/978-3-642-17572-5_30

20. Takamizawa, K., Nishizeki, T., Saito, N.: Linear-time computability of combinatorial problems on series-parallel graphs. J. ACM **29**(3), 623–641 (1982)

Weak-Barrier Coverage with Adaptive Sensor Rotation

Catalina Aranzazu-Suescun$^{(\boxtimes)}$ and Mihaela Cardei

Department of Computer and Electrical Engineering and Computer Science,
Florida Atlantic University, Boca Raton, FL 33431, USA
{caranzazusue2014,mcardei}@fau.edu

Abstract. Wireless sensor networks have been widely used in environment, climate, animal monitoring, surveillance, and also in the medical field. When deploying sensors to monitor boundaries of battlefields or country borders, sensors are usually dispersed from an aircraft following a predetermined path. In such scenarios sensing gaps are usually unavoidable. We consider a wireless network consisting of randomly deployed sensor nodes and directional border nodes deployed using a random line-based deployment model. In this paper we propose an adaptive distributed algorithm for weak-barrier coverage that allows border nodes to dynamically compute their orientation based on notifications from the sensor nodes, such that to increase the number of intruders detected by the border nodes. We use simulations to analyze the performance of our algorithm and to compare it with a non-adaptive gap mending algorithm.

Keywords: Adaptive algorithm · Directional border nodes
Distributed algorithm · Wireless sensor networks
Weak-barrier coverage

1 Introduction and Related Works

Wireless Sensor Networks (WSNs) have been widely used in event monitoring applications such as environment, climate, animal monitoring, surveillance and also in the medical field. For monitoring boundaries, researchers have used sensor barriers that detect if an intruder trespasses the border.

Deploying a set of sensor nodes in a region of interest where sensors form barriers for intruders is often referred to as the *barrier coverage* [10]. When sensors are deployed to monitor boundaries of large regions such as country borders, sensors are usually dispersed from an aircraft following a predetermined path. Therefore sensing gaps can occur.

Two types of barrier coverage have been addressed in literature: *weak-barrier coverage* and *strong-barrier coverage*. Weak-barrier coverage deals with detecting intruders moving along vertical traversing paths, while strong-barrier coverage detects intruders moving across the field with arbitrary paths. In this paper, we deal with the weak-barrier coverage problem.

© Springer Nature Switzerland AG 2018
D. Kim et al. (Eds.): COCOA 2018, LNCS 11346, pp. 739–754, 2018.
https://doi.org/10.1007/978-3-030-04651-4_50

Paper [9,10] proposes a distributed algorithm which aims to minimize the number of gaps for both weak and strong-barrier coverage. For the weak coverage model, the length of the gaps is minimized as well. One drawback is that the gaps are not monitored at all, and once the orientation angle is computed, the sensor orientation remains unchanged.

For weak and strong-barrier coverage, the sensors can rotate or be static. Works [7] and [8] consider static sensors. In this case, the goal is to minimize the number of mobile sensors needed to cover the gaps in the pre-deployed, stationary sensor barrier. The barrier consists of different types of sensors with different sensing ranges and angles. The barrier sensors cannot rotate after they are deployed in the field. Therefore, additional mobile sensors are deployed to cover the gaps in the barrier.

To determine the type and number of mobile sensor nodes that have to be deployed, clusters are defined by groups of overlapping sensors. Then, a cluster-based directional barrier graph is computed, where vertices are defined by the clusters and edges are weighted. The weight of an edge is calculated by solving the ILP problem of minimizing the cost of mobile sensors needed to cover the gap between two clusters or vertices. Each type of mobile sensor has a cost. The mobile sensors are randomly deployed and they need to move to the barrier gaps. Article [7] uses the Hungarian algorithm to calculate the optimal assignment of mobile sensor to the gaps. However, due to the difficulty of obtaining an optimal solution, in article [8] the authors use a greedy algorithm to move sensors to closer gaps.

Similarly, article [12] uses static sensors for statistical local face-view barrier coverage. In this case, intruder detection is guaranteed if the path has length ℓ, where ℓ is the length of an intruder's path across the barrier, projected parallel to the barrier. Camera sensors are deployed in the field and can rotate to adjust the covered barrier zone, resulting in an adjustment of the length ℓ. Depending on the covered zone, the intruders are detected with a certain probability. This probability also depends on the view angle of the camera, the rotation angle of the camera, the effective angle between the intruder and the camera, and the head rotation angle of the intruder. For a successful detection, there must be a camera detecting the face of the intruder. The authors calculate the probability of intruder detection using the path length, the head rotation angle of the intruder, the effective angle, and the number of cameras.

Article [3] presents an optimization of a full-view barrier coverage using rotating cameras. In this work the cameras do not need to detect the face of the intruder, but rather they rotate to detect the intruder. The authors define the strongly connected Full-view Barrier coverage with Rotable Camera sensors (FBRC) problem, as well as the weakly connected version. For the strongly connected version, they prove that it is NP-hard by reducing the Group Steiner Tree problem to the strongly connected FBRC problem. The problem of minimizing the number of sensors needed to detect an intruder is solved using "subtrees" that are calculated using a modification of the Dijkstra's algorithm, where the edge weights depend on the distance between the nodes. For the weakly connected version, the projection of the arc of the sensing cone of a sensor on the x-axis is used. If the projections of two continuous sensors start in the same point, the one with the larger ending point is selected. If the projections of two continuous sensors start and end in the same point, then the sensor with the larger "y" position is selected.

Similarly, the authors of [2] use rotating sensors for mending gaps in a line-based sensor network. The authors first obtain all sub-barriers in the network using projections of the arcs of the sensing cone of sensors on the x-axis, similar to the work described previously. If the projections of two continuous sensors overlap, then they are part of the same sub-barrier. The authors propose two gap-mending algorithms to fix the barrier gaps. The first algorithm, Simple Rotation Algorithm (SRA), mends the gaps only by rotating the orientation angles of two critical sensors, the sensor on the left of the gap and the sensor on the right of the gap. The second algorithm, Chain reaction-based Rotation Algorithm (CRA), rotates the critical sensors and their neighbors in order to avoid possible new gaps.

Article [11] presents a model that uses both the location and the facing direction of an object in order to detect it using a barrier of sensors. The goal of the algorithm is to minimize sensor overlapping, thus maximizing the coverage of the barrier. At each round of the algorithm, each sensor decides its rotation clockwise or counter-clockwise by a predefined angle. This leads to a reduced overlapping of the view-coverage of a sensor with its neighbors. The algorithm stops when sensors tend to reach a stable state, which means that they oscillate around some specific orientation, which will become the final orientation of the sensor. The authors also implement an improvement of the algorithm using different predefined angles for rotating sensors.

Our work is innovative in the following aspects:

- We use a dynamic distributed algorithm to mend gaps in the line-based barrier. Sensors rotate adaptively in order to mend gaps based on intruders' location. Thus the barrier may exhibit different gaps at different times.
- We use low-cost sensor nodes to detect intruders and notify the border nodes which rotate accordingly.

The rest of the paper is organized as follows. Section 2 presents the network model and describes the problem definition. In Sect. 3 we present our adaptive, distributed algorithm for event detection and gap mending. The performance of our algorithm is illustrated in Sect. 4, where we conduct simulations using MATLAB [5]. The conclusions are stated in Sect. 5.

2 Network Model and Problem Definition

In this paper we propose a solution for the weak-barrier coverage problem, where the objective is to detect events (e.g. intruders) that move north-south. We consider two types of wireless nodes, see Fig. 1: *sensor nodes* and *border nodes*.

Sensor nodes are less expensive, deployed in a larger number and they are used for early detection of intruders. They can be for example small sensors buried in the ground to detect vibrations caused by nearby activity [6]. We assume that n sensor nodes $\{s_1, s_2,...,s_n\}$ are deployed randomly near the border, forming a "band" for early detection of intruders moving north-south. Each sensor node detects a moving object with a certain intensity and has a communication range r_t.

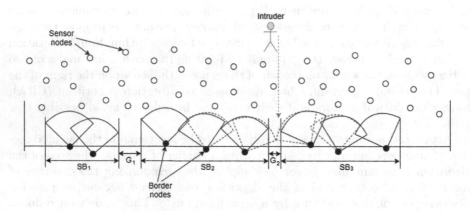

Fig. 1. Network deployment

Border nodes are resource-rich and more expensive nodes, therefore fewer such nodes will be deployed. We assume a model similar to [9], where nodes are equipped with camera, GPS, and are able to run sophisticated detection and classification algorithms. Similar to [9], we assume a directional sensing model represented as a 2D sector-shaped sensing, see Fig. 2b.

When larger borders or perimeters are to be covered, border nodes are remotely deployed, e.g. sprayed from an aircraft. Besides the uniform deployment, the *line-based deployment* model has been used for barrier coverage.

We assume a line-based deployment similar to [9]. A total of N border nodes $\{S_1, S_2, ..., S_N\}$ are deployed in a rectangular field of length L and width H, see Fig. 2a. In the line-based node deployment, nodes are evenly deployed on a horizontal-line (e.g. $y = 0$). Such an example is illustrated in Fig. 2a by the "target" positions, where the coordinate of the ith node is computed as $(2i - 1)L/2N$. This is hard to achieve in practice for remotely deployed sensors, and their "actual" positions have random offsets. We follow the model in [1] where the random offset distances have a Gaussian distribution. A border node S_i's location is denoted by $(S_i.x, S_i.y)$.

Border nodes have the ability to exchange messages with each other and with the sensor nodes. We assume an omnidirectional communication model, where the transmission range of each border node is R_t. In addition, we assume that the border nodes have a directional sensing with a finite view angle. Different from isotropic sensors, they cannot sense the whole circular area.

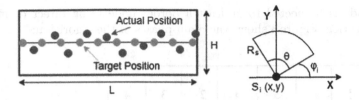

Fig. 2. (a) Line-based random deployment of border nodes; (b) Border node coverage model

Directional sensing can be modeled as 2D sector-shaped sensing [4]. We describe a node S_i using the five tuple $<(S_i.x,\ S_i.y),\ S_i.\theta,\ S_i.\varphi,\ R_s,\ R_t>$, see Fig. 2b, where $(S_i.x, S_i.y)$ are the Cartesian coordinates of S_i, $S_i.\theta$ is the view angle, $S_i.\varphi$ is the orientation angle, R_s is the sensing range, and R_t is the communication range. We assume that all border nodes in the network have the same view angle, sensing range, and communication range.

We assume that intruders move north-south. In the barrier coverage problem, the objective is to construct a barrier such that any north-south path falls into the coverage area of at least a border node. For the *weak-barrier coverage*, the barrier of sensors has to provide coverage when intruders move along vertical traversing paths.

Due to the random deployment of the border nodes, a complete weak-barrier coverage may be impossible to achieve, and in this case the target region is divided into sub-barriers (where coverage is provided) and gaps. Figure 1 shows an initial sensor deployment scenario with sub-barriers (SB) and gaps (G). Intruders crossing through a SB region are detected, while those crossing though a G region are not detected.

Using the distributed algorithm from [9], we can minimize the number and length of the gaps, but the intruders traversing the border through the gaps are undetected. The objective of this paper is to design an *adaptive* mechanism where border nodes rotate dynamically, in order to cover the regions where intruders are expected to trespass the border, based on notification received from sensor nodes.

Problem Definition: *Given a connected WSN with n randomly-deployed sensor nodes $\{s_1, s_2,...,s_n\}$ and N border nodes $\{S_1, S_2,...,S_N\}$ deployed using a random line-based model, design an adaptive, distributed algorithm for weak-barrier coverage that allows border nodes to dynamically compute their orientation angle such to maximize the number of intruders detected by the border nodes.*

3 Adaptive Distributed Algorithm for Weak-Barrier Coverage

In this section, we present our adaptive, distributed algorithm for weak-barrier coverage. We assume that the border nodes form a connected topology and each sensor node is connected to at least a border node using direct or multi-hop communication. Figure 3 shows the main phases of the algorithm:

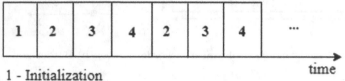

1 - Initialization
2 - Default rotation of border nodes
3 - Sensor nodes detect an event and notify the border nodes
4 - Border nodes rotate for weak coverage of the event

Fig. 3. Network organization

Phase 1 is the initialization phase. Border nodes perform neighbor discovery and exchange location information with neighboring border nodes. Sensor nodes establish a direct or multi-hop path to border nodes. In *phase 2*, border nodes rotate according to the algorithm in [9] that minimizes the number of gaps and the gap distance. We call this the "default" rotation.

In *phase 3*, one or more sensor nodes detect the event and execute the mechanism for reporting the event to border nodes. In *phase 4*, border nodes run the mechanism for computing the orientation angle so that the event is weakly covered by border nodes. After the event crosses the barrier, the border nodes return to the default rotation from phase 2. Phases 2 through 4 repeat in time as new events are detected. Next, we describe each phase in detail.

Phase 1: Initialization
During phase 1 each border node broadcasts a Hello message including its ID and location. For the *weak-barrier coverage* problem, we work with node projections on the x-axis. A border node S_i receiving Hello(S_j, $(S_j.x, S_j.y)$, $hops = 0$), stores S_j in its $BorderNodeNeighbor$ list if $|S_i.x - S_j.x| \leq 2R_s$. Each border node S_i computes its closest right neighbor $S_i.rn$ and its closest left neighbor $S_i.ln$ from the nodes in the $BorderNodeNeighbor$ list.

As Hello messages are initiated by border nodes, sensor nodes form convergecast trees rooted at border nodes. Each sensor node s_k that receives a Hello message for the first time, increments the *hops* field, sets the sending node as its parent, stored in the field $s_k.tp$, and sends a message Hello(s_k, $(s_k.x, s_k.y)$, $hops$). At the end of this step, each sensor node s_k has set-up its parent in a convergecast tree rooted at a border node. If a sensor node receives another Hello messages, then it stores the parent leading to the shortest path, and retransmits the updated Hello message if needed.

Phase 2: Default Rotation of Border Nodes
The border nodes rotate according to the distributed algorithm presented in [9]. The objective is to minimize the number of gaps and the overall gap length.

ALGORITHM 1. Check-Coverage (border nodes S_1, S_2, ..., S_k, event e, BNlist = empty list)

1: $I = [e.x_l, e.x_r]$
2: **for** i = 1 to k **do**
3:　　**if** $S_i.coverage \cap I \neq \Phi$ **then**
4:　　　add S_i to BNlist
5:　　　$I = I - S_i.coverage$
6:　　**end if**
7: **end for**
8: **if** $I == 0$ **then**
9:　　return True
10: **else**
11:　　return False
12: **end if**

Phase 3: Sensor Nodes Detect an Event and Notify the Border Nodes
As specified previously, we assume that border nodes know their location, using GPS. Based on this, we assume that the sensor nodes run a localization mechanism, and thus they are aware of their location too.

A sensor node s_i detecting a moving event e, sends a message EventDetected(e, s_i, ($s_i.x$, $s_i.y$)) to the root border node, following the parent path in the convergecast tree computed in phase 1. More specifically the message is retransmitted by the $s_i.tp$ sensor node. e stores information about the detected event.

Phase 4: Border Nodes Rotate for Weak Coverage of the Event
When a border node S_i receives the EventDetected message from a sensor node s_k, it broadcasts EventCoverageRequest(S_i, e, ($s_k.x$, $s_k.y$)) after a small amount of time. A border node S_j receiving this message will re-broadcast the message if $|s_k.x - S_j.x| \leq R_s$ or if it has received a message reporting the same event from another sensor node.

If S_j has received a report about the same event from another sensor node s_p, then it rebroadcasts EventCoverageRequest(S_i, e, ($s_k.x$, $s_k.y$), ($s_p.x$, $s_p.y$)). When the EventCoverageRequest stops being propagated by border nodes, the left-most and right-most border nodes send back towards S_i an EventCoverageReply message, appending their current orientation angle and a field called *committed*. Committed is 1 if the border node is committed to the current orientation angle for covering a certain event, otherwise it is 0. If committed is 0, then it means that the sensor node can rotate if needed for coverage purpose. These messages are propagated from left and right directions back towards S_i.

After receiving EventCoverageReply messages from the left and right neighbors, S_i knows the orientation angles and the committed attributes for all border nodes that have the potential of covering the event e. S_i executes an algorithm to decide the orientation angle of these border nodes and then sends a message EventRotationCommit to these border nodes. Upon receiving this message,

---

ALGORITHM 2. Compute-Coverage (border nodes S_1, S_2, ..., S_k, event e,
BNlist = empty list)

1: $I = [e.x_l, e.x_r]$
2: **for** j = 1 to k **do**
3:    **if** $S_j.committed == 1$ **then**
4:       **if** $S_j.coverage \cap I \neq \Phi$ **then**
5:          add S_j to BNlist
6:          $I = I - S_j.coverage$
7:       **end if**
8:    **else**
9:       **if** $[S_j.x - R_s, S_j.x + R_s] \cap I \neq \Phi$ **then**
10:          Compute-Orientation-Angle(S_j, I)
11:          add S_j to BNlist
12:          $I = I - S_j.coverage$
13:       **end if**
14:    **end if**
15: **end for**
16: **if** $I == 0$ **then**
17:    return True
18: **else**
19:    return False
20: **end if**

---

the border nodes set their orientation angle and set the committed attribute to 1. The EventRotationCommit message is re-sent periodically by S_i, otherwise after a specific period of time the nodes set the attribute committed to 0 and the orientation angle to the default angle.

Next, we discuss the algorithm used to compute the orientation angles. Let us assume that the border nodes involved in covering the event are S_1, S_2, ..., S_k from left to right, with the orientation angles $S_1.\varphi$, $S_2.\varphi$, ..., and $S_k.\varphi$ respectively. Let us assume that the horizontal projection of the event ranges between $e.x_l$ and $e.x_r$.

The first step is to determine whether $[e.x_l, e.x_r]$ is already covered by the current sensors' orientation, see Algorithm 1. BNlist is used to compute the list of border nodes needed to cover the interval $[e.x_l, e.x_r]$. Lines 2 to 7 iterate through the border nodes, and if new coverage is provided, then the sensor is added to the BNlist. The coverage of a sensor S_j is computed as $S_j.coverage = [\min\{S_j.x, S_j.x + R_s cos(\pi - S_j.\theta - S_j.\varphi)\}, \max\{S_j.x, S_j.x + R_s cos S_j.\varphi\}]$.

If the event is covered by the sensors in the list BNlist, then the algorithm returns True, otherwise it returns False. If Algorithm 1 returns True, then EventRotationCommit message contains the BNlist. The nodes in this list will set-up their committed attribute to 1, in order to guarantee the coverage of the event e. If Algorithm 1 returns False, then S_i runs the Algorithm 2 to compute the sensor rotation angles needed to cover the event e.

ALGORITHM 3. Compute-Orientation-Angle (border node S_j, set of intervals I)

1: $I' = [S_i.x - R_s, S_i.x + R_s] \cap I$
2: **if** $I' == 0$ **then**
3:    return
4: **end if**
5: **if** $I'.x_l \leq S_j.x$ **then**
6:    $S_j.\varphi' = max\{0, \pi - \theta - arccos(\frac{S_j.x - I'.x_l}{R_s})\}$
7: **else**
8:    $S_j.\varphi' = 0$
9: **end if**

Algorithm 2 iterates through the border nodes in order to build the BNlist. BNlist contains the nodes participating in covering the event e and these nodes will set their committed attribute to 1 if it was 0 previously. Lines 3 through 7 address the case when S_j is committed to the current orientation due to the covering of some other event. If the current orientation still covers some parts of I, then S_j is added to the BNlist, and I is updated accordingly. I represents the horizontal interval (or group of intervals) of the event e which are currently uncovered.

Lines 9 through 13 discuss the case when S_j has the committed attribute set-up to 0, that means it can rotate to participate in the covering of the event e if needed. If its rotation can have an impact on I, then its orientation angle is computed in the Algorithm 3. In this case S_j is added to BNlist and I is updated accordingly.

Algorithm 3 starts by computing the horizon interval I' which is the interval (or set of intervals) that can be covered by S_j, based on the sensing range R_s. The orientation angle $S_j.\varphi'$ is computed such that o start covering from the leftmost point of the interval I', denoted by the attribute $I'.x_l$.

Algorithm 2 returns True or False, depending on whether the border nodes in the BNlist are able to completely cover the event e or not. EventRotationCommit message is sent back to the border nodes $S_1, S_2, ..., S_k$, containing the BNlist, and the new orientation angles. The nodes in the BNlist will set-up their committed attribute to 1 (if it was 0 previously), and will set-up their orientation angles accordingly.

Our algorithm does not guarantee that all the events will be completely covered. Some of the border nodes cannot change their orientation angle if they are committed to cover other events. Nevertheless, our adaptive algorithm improves substantially the number of events detected completely or partially.

4   Simulations

We used MATLAB [5] to evaluate the performance of our distributed algorithm, and we compared it with the algorithm proposed in [9].

4.1   Simulation Environment

Following the settings from [9], we deployed the WSN into a square region of length $L = 500$ m and width $H = 100$ m. The initial orientation angle of the border nodes follows a uniform distribution in the range $[0, 2\pi]$. The sensing range is $R_s = 15$ m. The border nodes positions have random offsets following a Gaussian distribution with mean 0 and variance σ^2. We denote δ_i^x and δ_i^y the offset distance of the border node S_i in the horizontal and vertical directions, with $\delta_i^x, \delta_i^y \sim N(0, s\sigma^2)$. In our simulations we set $\sigma = 5$.

We take $n = 100$ sensor nodes $s_1, s_2, ..., s_n$, randomly deployed in the area. In each simulation run we generate events which have a circular area. The center is generated randomly. Five types of events are used in the simulations:

- *small events*, where the event diameter is 10 m
- *small-medium events*, where the event diameter is 30 m
- *medium events*, where the event diameter is 50 m
- *medium-large events*, where the event diameter is 70 m
- *large events*, where the event diameter is 90 m

We assume that the events move north-south, that means that the events move perpendicular to the barrier. The number of events generated follows a Poisson distribution with $\lambda = 5$. The average and maximum speed of the events are indicated in Table 1. The location of an event is selected randomly.

Table 1. Event speed

Average speed (m/s)	Maximum speed (m/s)
1	2
2	3
3	4
4	5
5	6

We generated 100 different sensor deployments. Each data point in our simulation results is an average of 100 experiments. We used the following values when variables are not changed in the experiments:

- number of border nodes = 30
- number of sensor nodes = 100
- view angle = 60° or 120°
- event size = 50 m
- event speed (average) = 2 m/s.

4.2   Simulation Results

In this section we compare the following two algorithms:

- the *Distributed Gap Mending* Algorithm from [9]
- our proposed solution, the Adaptive Distributed Algorithm for Weak-Barrier Coverage, called *Adaptive Distributed W-B Coverage*.

For both algorithms we compare their performance using the following representative metrics:

- the percentage of events which are completely covered by border nodes. Here we count only the events whose horizontal projection is completely covered by border nodes.
- the percentage of event coverage. Some of the events are partially covered, some are completely covered, and some other events are not covered at all. Considering the projection of the events on the horizontal axis, we measure the percentage of event coverage by border nodes. For example, if we have 1 event completely uncovered, 1 event covered only on half of its projection, and 1 event completely covered, then the percentage of event coverage is $1.5/3 = 0.5$, that means 50%.

In our experiments we vary the number of border nodes, the number of sensors, the view angle, the event size, and the event speed.

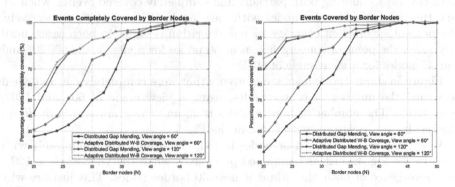

Fig. 4. Results when we vary the number of border nodes. (a) Percentage of events completely covered; (b) Percentage of event coverage.

Figure 4a presents the percentage of events which are completely covered when we vary the number of border nodes. We used two values for the view angle: 60° and 120°. The number of sensor nodes is 100 and the event size is *medium*.

In both cases, our algorithm performs better than the algorithm in [9]. When the view angle is 60°, the algorithms converge when the number of border nodes is close to 50, while for a view angle of 120° the algorithms converge when the

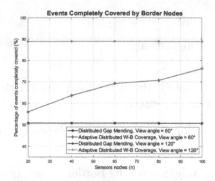 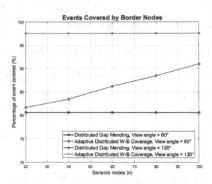

Fig. 5. Results when we vary the number of sensor nodes. (a) Percentage of events completely covered; (b) Percentage of event coverage.

number of border nodes is about 30. When the view angle is bigger, the projection of the coverage sector on the x-axis is larger, thus the node can provide more coverage.

When the number of border nodes is small (near 20), the distance between them is larger, thus more coverage gaps will occur. Contrary, when the number of border nodes is large (near 50), the nodes will become closer, then even with the initial setting of the orientation angle the whole barrier will be covered.

Figure 4b confirms the previous result. This figure shows the percentage of event coverage, counting both partially and completely covered events, when we vary the number of border nodes with view angle 60° and 120°, respectively. The percentage of event coverage by our algorithm is larger in both cases, until they reach the point of convergence: near 30 nodes for a view angle of 120° and near 45 nodes for a view angle of 60°.

Figure 5a shows the percentage of events that were completely covered when we varied the number of sensor nodes, using a view angle of 60° and 120°, respectively. The number of border nodes is 30 and the event size is *medium*. The results are consistent with those in the Fig. 4.

When the number of sensor nodes is 100, the results are comparable with those in the Fig. 4. Just as mentioned previously, when the view angle is 120°, the convergence of both algorithms is near 30 border nodes. This justifies why the percentage of completely covered events is equal even if we vary the number of sensors. When the view angle is 60°, if we decrease the number of sensor nodes, then the two algorithms will converge and have the same performance. This happens because our algorithm will not be able to start *phase 3* and *phase 4*. Figure 5b presents the percentage of event coverage when we vary the number of sensors with a view angle of 60° and 120°, respectively.

Figure 6a shows the percentage of completely covered events when we vary the view angle. We used two measurements: 30 and 40 border nodes. The number of sensors is 100 and the event size is *medium*. As we increase the view angle of the border nodes, the projection of the sensing sector on the x-axis is increasing as well. Therefore, each border node covers a larger portion of the barrier.

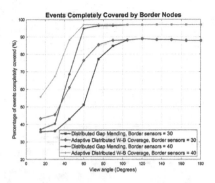
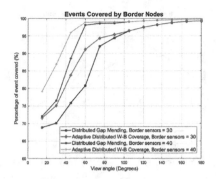

Fig. 6. Results when we vary the view angle of the border nodes. (a) Percentage of events completely covered; (b) Percentage of event coverage.

For 30 border nodes, the converging point of the algorithms is for a view angle of 105°, while for the case of 40 border nodes, both algorithms have a similar performance for a view angle of 97°. Figure 6b presents the percentage of event coverage when we vary the view angle, for 30 and 40 border nodes respectively. Similar to Fig. 6a, our algorithm has better results than the algorithm [9].

When the number of sensor nodes is 100, the results are comparable with those in the Fig. 4. For a 120° view angle, the two algorithms converge near 30 border nodes. This justifies why the percentage of completely covered events is equal even if we vary the number of sensors. When the view angle is 60°, if we decrease the number of sensor nodes, then the two algorithms converge and have the same performance. This happens because our algorithm cannot start *phase 3* and *phase 4*. Figure 5b shows the percentage of event coverage when we vary the number of sensors with a view angle of 60° and 120°, respectively.

Figure 6a illustrates the percentage of events that were completely covered when we vary the view angle, for 30 and 40 border nodes. The number of sensors is 100 and the event size is *medium*. As we increase the view angle of the border nodes, we are also increasing the projection of the sensing sector on the x-axis. Then each border node covers a larger portion of the barrier.

For 30 border nodes, the converging point of the algorithms is for a view angle of 105°, while for the case of 40 border nodes both algorithms have a similar performance for a view angle of 97°. Figure 6b presents the percentage of event coverage when we vary the view angle, for 30 and 40 border nodes respectively. Similar to Fig. 6a, our algorithm performs better than the algorithm in [9].

Figure 7a measures the percentage of completely covered events when we vary the size of the event, for a view angle of 60° and 120°, respectively. We took 30 border nodes and 100 sensor nodes. A small event could fit through a gap, and if the nearby sensors have the *committed* field 1, then the event is completely uncovered. When the size of the event increase to 30 m, the probability to cover a bigger portion of the event increases compared to small events, thus the percentage of detected events increases. As the event size increases, the percentage of events successfully detected decreases. This happens because a larger event

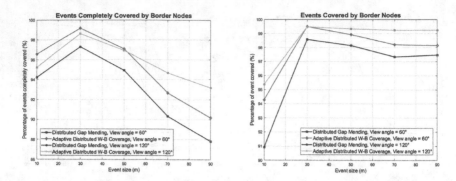

Fig. 7. Results when we vary the event size. (a) Percentage of events completely covered; (b) Percentage of event coverage.

requires coverage from more border nodes. When a new event arrives, some of the senors are committed to another event, and cannot rotate. The number of border nodes play a role too. A smaller number of border nodes results in more gaps with a larger gap distance. Thus it becomes difficult to cover simultaneous events.

When the view angle is 120°, the performance of both algorithms is similar. This is consistent with the results in Fig. 4a. Figure 7b presents the percentage of event coverage when we vary the size of the event. The view angle is 60° and 120°, respectively. The results are consistent with those in the Fig. 7a. When the view angle is 120°, the percentage of events covered is the same when the event size is at least 30 m.

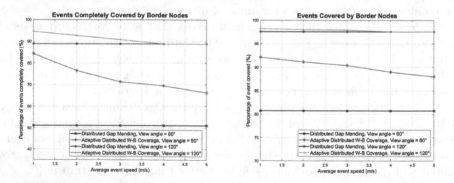

Fig. 8. Results when we vary the event speed. (a) Percentage of events completely covered; (b) Percentage of event coverage.

In Fig. 8a we measure the percentage of completely covered events when we vary the speed of the event. We use two view angles: 60° and 120°. The number of border nodes is 30, the number of sensor nodes is 100, and the event size is *medium*. Varying the speed of the events does not impact the results of the algorithm [9] in any way since the border nodes are static after the first setting of the orientation angles.

For our algorithm, an increase in the event speed results in a decrease in the performance, that means the percentage of events detected is smaller. This is due to the fact that the border nodes are not notified on time to set-up their orientation angle. This notification process is initiated by the sensor nodes which start detecting the event before it crosses the barrier. Figure 8b illustrates the percentage of event coverage. The results are consistent with those in the Fig. 8a. The percentage is larger than the results in Fig. 8a, since this time we measured the percentage of events which are partially or completely covered.

5 Conclusions

In this paper we studied weak-barrier coverage for directional border nodes with random line-based deployment. Since monitoring gaps can occur, we proposed an adaptive, distributed algorithm that allow border nodes to dynamically adjust their orientation angle depending on the events detected by sensor nodes. Simulation results show that our mechanism detects more events compared to the mechanism proposed in [9], where the orientation angles do not change over time.

References

1. Chen, J., Wang, B., Liu, W., Deng, X., Yang, L.T.: Mend barrier gaps via sensor rotation for a line-based deployed directional sensor network. In: High Performance Computing and Communications and 2013 IEEE International Conference on Embedded and Ubiquitous Computing, pp. 2074–2079, November 2013
2. Chen, J., Wang, B., Liu, W., Yang, L.T., Deng, X.: Rotating directional sensors to mend barrier gaps in line-based deployed directional sensor network. IEEE Syst. J. **11**, 1027–1038 (2017)
3. Gao, X., Yang, R., Wu, F., Chen, G., Zhou, J.: Optimization of full-view barrier coverage with rotatable camera sensors. In: IEEE 37th International Conference on Distributed Computing Systems, June 2017
4. Ma, H., Liu, Y.: On coverage problems of directional sensor networks. In: Proceedings of the 1st International Conference on Mobile Ad-Hoc Sensor Networks, December 2005
5. MathWorks. https://www.mathworks.com. Accessed June 2018
6. Quantum Technology. https://www.qtsi.com. Accessed May 2018
7. Wang, Z., Liao, J., Cao, Q., Qi, H., Wang, Z.: Barrier coverage in hybrid directional sensor networks. IEEE 10th International Conference on Mobile Ad-Hoc and Sensor Systems, October 2013
8. Wang, Z., Cao, Q., Qi, H., Chenc, H., Wang, Q.: Cost-effective barrier coverage formation in heterogeneous wireless sensor networks. Ad Hoc Netw. J. **64**, 65–79 (2017)
9. Wu, Y., Cardei, M.: Distributed algorithm for mending barrier gaps via sensor rotation in wireless sensor networks. In: Lu, Z., Kim, D., Wu, W., Li, W., Du, D.-Z. (eds.) COCOA 2015. LNCS, vol. 9486, pp. 293–304. Springer, Cham (2015). https://doi.org/10.1007/978-3-319-26626-8_22

10. Wu, Y., Cardei, M.: Distributed algorithms for barrier coverage via sensor rotation in wireless sensor networks. J. Comb. Optim. (2016). https://doi.org/10.1007/s10878-016-0055-3
11. Yang, C., Zhu, W., Liu, J., Chen, L., Chen, D., Cao, J.: Self-orienting the cameras for maximizing the view-coverage ratio in camera sensor networks. Pervasive Mob. Comput. J. (2015). https://doi.org/10.1016/j.pmcj.2014.04.002
12. Yu, Z., Yang, F., Teng, J., Champion, A.C., Xuan, D.: Local face-view barrier coverage in camera sensor networks. IEEE Conference on Computer Communications INFOCOM, April 2015

Author Index